Fossil Behavior Compendium

Fossil Behavior Compendium

Arthur J. Boucot • George O. Poinar, Jr.

CRC Press
Taylor & Francis Group
Boca Raton London New York

CRC Press is an imprint of the
Taylor & Francis Group, an **informa** business

CRC Press
Taylor & Francis Group
6000 Broken Sound Parkway NW, Suite 300
Boca Raton, FL 33487-2742

First issued in paperback 2019

© 2010 by Taylor & Francis Group, LLC
CRC Press is an imprint of Taylor & Francis Group, an Informa business

No claim to original U.S. Government works

ISBN-13: 978-1-4398-1058-3 (hbk)
ISBN-13: 978-0-367-38428-9 (pbk)

Library of Congress Cataloging-in-Publication Data

Boucot, A. J. (Arthur James), 1924-
 Fossil behavior compendium / Arthur J. Boucot, George O. Poinar, Jr.
  p. cm.
 Includes bibliographical references and index.
 ISBN 978-1-4398-1058-3 (hardcover : alk. paper)
 1. Animals, Fossil. 2. Animal behavior--Evolution. 3. Pathogenic microorganisms. I. Poinar, George O. II. Title.

QE770.B68 2010
591.5--dc22   2009047092

Visit the Taylor & Francis Web site at
http://www.taylorandfrancis.com

and the CRC Press Web site at
http://www.crcpress.com

Contents

Preface

Our purpose here is to document the wealth of information demonstrating that the behaviors of extant organisms extends far back in time and to show that the behavior of extinct organisms closely resembles that of their descendents at the generic and often family level. These issues put to rest any concerns biologists might have about whether behavioral patterns of living organisms occurred in the distant past. The behavioral–evolutionary implications of this documented material are also note-worthy. Based on the fossil record of various organisms from both nonmarine and marine habitats, it is apparent that many genetically determined behavioral characteristics are "hard wired" and very conservative. This implies that behavioral patterns represent an important set of characters that can be applied in taxonomy at the generic and family levels. We have also focused our attention on evidence of pathogens preserved in the fossil record. Such records are of great significance not only for the clues they provide about the antiquity of various disease types but also for their evolutionary implications. Understanding how and when pathogens evolved and became associated with invertebrate vectors is significant to medical researchers, parasitologists, and microbiologists. In the text, we have noted 36 numbered categories. Categories with an asterisk (*1, etc.) are new and were not discussed previously in an earlier work on this subject (Boucot, 1990a).

Acknowledgments

We are greatly indebted to the personnel of the Interlibrary Loan Office of the Oregon State University Library, particularly Deborah Carroll, and to Bonnie Avery, Reference Librarian; without their never-failing friendly help this treatment would not have been possible. We are indebted to the material provided by the Library of the Nanjing Institute of Geology and Palaeontology. Jiří Frýda, in Prague, guided Boucot very kindly to a mass of literature dealing with gastropod protoconchs. J.-C. Gall, in Strasbourg, provided a variety of Grès à Voltzia illustrations dealing with estherian and insect eggs from these unusual deposits. A. Radwański and U. Radwańska, in Warsaw, provided very helpful editorial comment and illustrative materials from their varied studies. Rodney Feldmann, at Kent State University, Kent, Ohio, generously provided his expertise on a number of decapod examples and some other interesting items. Helmut Tischlinger, in Stammham, Germany, provided some elegant colored photos of Solnhofen material, including previously unpublished material. We thank Roberta Poinar for discussions on various entries included herein. We would also like to thank the following persons for providing specimens for this study: Alex Brown, Dan McAuley, Jim Work, Ron Buckley, Ron Cauble, Scott Anderson, Michael Oschin, Ettore Morone, and Andrew Cholewinski/Violetta Grzywacz of Amber Collection, Inc. The authors appreciate the understanding, cooperation, and personal interest from the staff of CRC Press. We are sincerely grateful for their courteous assistance and tolerant guidance.

Authors

Arthur Boucot is an emeritus professor of zoology and geology at Oregon State University, Corvallis, Oregon. He has spent his career working on the taxonomy and morphology of earlier Paleozoic brachiopods, together with their community ecology and global biogeography. He is also involved in considering the contributions that fossils can make to our understanding of behavioral evolution and coevolution. He has carried out field work collecting earlier Paleozoic brachiopods from many locations on five continents, including Antarctica. He has served as president of the Paleontological Society and of the International Palaeontological Association and has published widely on earlier Paleozoic brachiopod taxonomy and morphology, as well as allied biostratigraphic questions, their community evolution, and their global biogeography. More recently, he has produced a work, with colleagues Chen Xu and Christopher Scotese, on global Cambrian through Miocene climate zonation based on the distribution of geologically sensitive guides to climate such as tillites, coal, and evaporites.

George Poinar is currently in the Department of Zoology at Oregon State University, Corvallis, Oregon. He is an emeritus of University of California–Berkeley, where he was in charge of the Insect Disease Diagnostic Laboratory, conducted research on invertebrate parasitic nematodes, and began his search for fossilized remains of parasites and pathogens in amber. His parasitological investigations resulted in research projects in many countries, including England, the Netherlands, France, Italy, Russia, Romania, New Zealand, Australis, Papua New Guinea, New Britain, New Caledonia, Western Samoa, Tahiti, Thailand, the Philippines, Malaysia, Mexico, the Ivory Coast, and Burkina Faso. His studies on the vectors of human diseases in Africa and the South Pacific were conducted while serving as a consultant for the World Health Organization and United Nations. He has served as associate editor for the *Journal of Invertebrate Pathology* for the past 10 years and has published extensively on extant parasites as well as life forms in amber, authoring or coauthoring over 600 publications. He and his wife, Roberta, have written several books based on their amber discoveries, including *The Amber Forest* and *What Bugged the Dinosaurs.*

Introduction

Since 1989, when Boucot's *Evolutionary Paleobiology of Behavior and Coevolution* (1990a) was completed, the evidence supporting the conclusions presented there has more than doubled. In the present work, Boucot has been joined by coauthor George Poinar, who presents a wealth of information from Cretaceous and Tertiary amber deposits, much of it previously unpublished He is an authority on invertebrate diseases (Poinar and Jansson, 1988; Poinar and Thomas, 1978, 1984) and was in charge of the invertebrate disease diagnostic laboratory at Berkeley for a number of years. His findings are of great potential interest to students studying the evolution of modern diseases. Among his numerous contributions are the first fossil records of malarial pathogens (*Plasmodium* and *Haemoproteus*) as well as the first fossil record of the protozoan responsible for leishmaniasis (*Paleoleishmania*).

As mentioned earlier, one of our goals is to show the fossil evidence for the origins and evolution of disease-causing pathogens. Moodie's (1923) treatise on paleopathology emphasized structural abnormalities in vertebrates but also considered basic questions dealing with the history of disease. Additional evidence of skeletal pathology was provided by Tasnádi–Kubacska (1962). Their interpretation of pathological evidence, almost entirely skeletal, may have been superseded, but their emphasis on the potential significance of the data is important. Darrell and Taylor (1993) provided an insightful discussion of symbiosis that includes both modern and fossil examples. Over half a century ago, Haldane (1949) discussed the evolutionary significance of disease, but at that time there was little evidence for pathogens in the fossil record.

In the present work, we emphasize, with a wealth of new material, the overwhelming evidence for the evolutionary conservatism of behavioral and coevolved relationships. We feel strongly that the fossil record has great potential for providing far more insight into certain aspects of the evolutionary process than is provided purely by the morphological evidence of extant species. Much of the material provided here is not accessible on the Internet because most the data are buried within the text of scientific papers and not obvious from titles, abstracts, summaries, or conclusions.

Our Warsaw colleague, Adam Urbanek, pointed out that we possibly should give the credit for originating the concept of coevolution to the late-19th, early 20th-century Nobel Prize winner, Polish novelist Henryk Sienkiewicz, whose Falstaffian character Zaglova, in *With Sword and Fire*, commented in times of dire peril that, "Zginiemy a z nami pchey nasze" ("We will perish and so will our fleas")!

We hope that the information provided here will help specialists in many areas, including ethologists concerned with both nonmarine and marine animals, parasitologists concerned with the evolution and longevity of parasitic relationships, physicians concerned with the history of disease-causing pathogens, ecologists concerned with the evolutionary history of communities, and taxonomists interested in using behavioral characteristics of organisms in addition to the "standard" morphological ones.

RELIABILITY CLASSES

The reliability classes used to rate the evidence are summarized below; they range from very certain (Category 1) to highly speculative (Category 7):

Category 1 is used when a small number of examples exists for which there is no question about the reliability of the interpretation, as in the examples of amber insects preserved *in copula*.

Category 2A is reserved for those examples for which there is little doubt about the reliability of the interpretation, although absolute certainty is absent, such as when two organisms thought to have interacted in one way or another are preserved "close" to each other but not actually in contact.

Category 2B is a widely employed diagnosis where functional morphology is the chief criterion.

Category 3 is used when the behavior is less certain, as in the frass patterns left behind by some leaf miners.

Category 4 is used when there is an even larger degree of uncertainty, as in the attempts to assign certain trace fossils to taxa lacking a body fossil record and where the trace itself might have been convergently made by more than one group of organisms.

Category 5A applies to behaviors for which a high degree of uncertainty exists about the maker, such as with some of the oldest known leaf galls.

Category 5B is used when biogeographic evidence is used to arrive at a behavioral conclusion for an organism.

Category 6 indicates fairly speculative situations, such as those employing the functional morphology of extinct taxa.

Category 7 is used for highly speculative interpretations for which the evidence is clearly controversial at best.

EVIDENCE

COEVOLUTIONARY AND BEHAVIORAL LIMITS OF THE FOSSIL RECORD

Fordyce and Jones (1990) reviewed the fossil record of the penguins; considered the problems involved with trying to determine their ancestry; discussed a possible earliest Eocene–Paleocene New Zealand progenitor; pointed out the convergent nature of the Northern Hemisphere alcids, with a fossil record also extending back to the Eocene; and considered the fossil record of the major penguin feeding types. Fordyce et al. (1986) provided evidence of what may be a penguin, based on Paleocene materials from New Zealand. Tambussi et al. (2005) described what is now the oldest known penguin from the Late Paleocene of Seymour Island on the Antarctic Peninsula. The previous treatment (Boucot, 1990a) emphasized that many important aspects of behavior are not provided by the fossil record, while using the marked distinction shown by Emperor and Adelie penguins in their egg brooding as a typical example.

1 Functional Morphology

(See Figures 1, 2, and 80.) Functional morphology was defined by Lincoln and Boxshall (1987) as "interpretation of the function of an organism or organ system by reference to its shape, form and structure." The examples below follow their definition. As pointed out previously (Boucot, 1990a), whole volumes could be published on vertebrate functional morphology present in the fossil record, based on our extensive knowledge of extant relatives, although for many of the far more poorly known modern groups this is only partly the case, as with most invertebrates.

The most thorough and extensive treatment of marine invertebrate functional morphology using the fossil record is still Dacqué's work (1921), followed by that of Savazzi (1999a), in which a number of very helpful items were collected for various invertebrate groups, most with a good fossil record. Hess et al. (1999) assembled a wealth of functional information regarding crinoids, and Dollo (1910) earlier provided an excellent account of some functional morphological criteria for assessing behavior in certain fishes, trilobites, and eurypterids.

Savazzi (1994) provided an elegant treatment of the functional morphologic evidence provided by various invertebrate skeletons, representing animals that burrow and bore, including bivalves, gastropods, crabs, linguloid brachiopods, and trilobites, plus echinoderms including echinoids and mitrates. Savazzi (1994) made a good case that the "terraces" on bivalve shells may aid in burrowing; however, extending this concept to certain terraced decapods is doubted by Feldmann et al. (1996b), who made the point that the raninid decapod *Lophoranina* possesses a terraced carapace featuring anterior spines (see Vega et al., 2001) that would not have been useful in burrowing but would have helped to conceal the animal in associated algae or debris; however, in some other decapods living in soft sediments such terraces might aid in burrowing. Beschin et al. (1998) illustrated terraces in some Italian Eocene *Lophoranina*.

Bromley and Heinberg (2006) have provided a very helpful account of the attachment mechanisms of many organisms to hard substrates, and Wellnhofer (1991) contains a thorough account of what can be deduced from the functional morphology of pterosaurs.

Poinar and Santiago-Blay (1997; see also Poinar and Poinar, 1999, Figure 22) described a palm bug from Dominican amber whose excessively flattened body and appendages leave little doubt about its habits (Figure 1). A palm beetle was also described from these deposits (Poinar, 1999c) (Figure 80). Although the presence of both insects was indirect evidence that palms existed in the Dominican

amber forest, later discoveries of palm flowers in Dominican amber confirmed their presence (Poinar, 2002b,c). Bechly and Wittmann (2000) described two palm bugs from Baltic amber, and Nel et al. (2004) described an Early Eocene palm bug from the Paris Basin.

The upward pointing spines on the Early Cretaceous Burmese amber tanyderid *Dacochile microsoma* are of interest (Figure 2). The fly has a row of upward-pointing spines on the hind tibiae and first two tarsal segments; they probably supported the insect while it was hanging upside down from vegetation in a manner similar to the habits of extant species of primitive crane flies (Poinar and Brown, 2006a).

Poinar and Doyen (1992) described the flat body of a Mexican amber Miocene termite bug that protects it from attacks by soldiers in the termite colony (Figure 275). Mayr (2004) described a German Early Oligocene bird skeleton of hummingbird type, suggesting that the hummingbird morphology originally was not totally restricted to the New World. Frizzell and Exline (1958) described crustacean

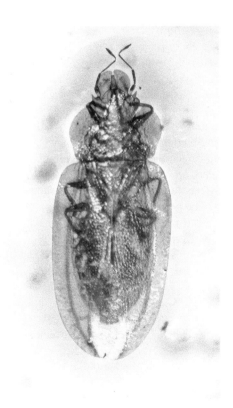

FIGURE 1 A palm bug (*Paleodoris lattini*) in Dominican amber (Poinar amber collection accession no. He-4-49); length of insect = 2.9 mm.

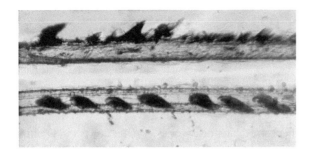

FIGURE 2 Posterior and lateral views of spines on hind leg of *Dacochile microsoma* in Burmese amber (Poinar amber collection accession no. B-D-12); length of spines ranges from 50 to 70 μm.

gastroliths from the Eocene of Texas; such gastroliths today are characteristic of the Nephropsidae (lobsters and crayfish), with an understanding of their function being uncertain. Frizzell and Horton (1961) described some Late Eocene crustacean gastroliths from Louisiana. Bellwood (2003) provided an excellent example of what can be learned about fish feeding behaviors from a functional morphological analysis in his case about the Eocene introduction of algal feeding into the tropical ecosystem.

One of the more amazing examples of the stability of functional morphology through time in some groups is a 100-million-year-old gecko (*Cretaceogekko burmae*) from Burmese amber (see color plate 2B after page 132) that possessed a sophisticated method of toe adhesion that allowed the gecko to climb vertical surfaces (Arnold and Poinar, 2008). The ability of geckos to walk with relative ease on vertical walls and even glass panes is due to the presence of thousands of setae arranged in lamellar arrays on the ventral surfaces of the digits. The tip of each seta is branched into numerous projections with spatulate tips that attach to the substrate by van der Waals force (Autumn et al., 2000). This multidirectional dry adhesion allows geckos to rapidly attach and detach their toes while scaling vertical surfaces. Bioengineers are attempting to duplicate this process with polypropylene microfiber arrays manufactured from stiff polymers. The presence of such pads suggests that the fossil gecko lived in a moist tropical forest with considerable opportunities for climbing.

When geckos acquired their adhesive toe pads that allowed this freedom of movement is unknown. The only other Cretaceous fossil geckos are *Hoburogecko* and *Gobekko* from the Upper Cretaceous of Mongolia, known from fragmentary skeletal remains that lack digits. Although estimates of gecko divergence based on DNA sequence data of living forms suggest that gekkotans originated in the Jurassic, there is no way of knowing whether these forms possessed adhesive toe pads.

The over 1200 species of geckos range in size from a total length of 16 mm in *Sphaerodactylus ariasae* from Hispaniola to around 420 mm in *Rhacodactylus leachianus* from New Caledonia and in *Gecko gecko* of Southeast Asia. Most geckos have relatively big heads and soft scaly skins; usually the eyelids are fused, and the gecko looks out through a large window in the lower one. This makes it look as if the eyelids are lacking (a feature that also turns up sporadically in quite a few other lizards). Many, but not all, geckos have sophisticated adhesive pads on their toes. Geckos naturally occur as far north as the northern shores of the Mediterranean, the Crimea, South and Central Asia and southern Japan. In the United States, they occur naturally in California and Arizona, for example, and introduced *Hemidactylus* geckos range farther north, mainly in towns. The Turkish gecko (*H. turcicus*) reaches at least as far north as Maryland.

An entire volume could easily be devoted to the functional morphology of vertebrates, mammals in particular. Van Valkenburgh (1985) provided a penetrating analysis of locomotor diversity in past and present carnivores that is a model of what can be achieved by analyzing both living and extinct taxa within a guild. A particularly fine example of both functional morphology and convergence is shown by Nevo's (1979) account of subterranean mammals whose many specialized features include such things as major changes in the pectoral girdle, anterior appendages, and vision loss. Included in this group are one marsupial family, two insectivore families, and seven placental families, with a Neogene fossil record for many taxa. Plavcan et al. (2002) assembled a volume on primate behavior that includes both neontological and paleontological evidence.

Thewissen et al. (1996) employed oxygen isotope data to demonstrate that within the primitive toothed whale lineage of the Eocene the change from a freshwater to a saltwater environment took place. Thewissen et al. (1994; see also Berta, 1994) discussed the mid-Eocene archaeocete involved and the problems in placing it both behaviorally and taxonomically. Spoor et al. (2002) described the rapidity within the Eocene with which the semicircular canals of the primitive whales arrived at the "modern" condition, as contrasted with the far slower changes in the postcranial morphological changes (i.e., an excellent example of mosaic evolution).

SUMMARY

Nothing can be added to the 1990 summary, with the possible exception of pointing out that are many examples of convergence indicated in some of the examples in the following materials—for example, the long-fingered mammalian insect seekers (see Section 6CIIk in this volume) and the discussion in Boucot (1990a, Section 19, Social Insects) about anteaters belonging to varied higher taxa, each with similar modifications for getting at ants and termites. Just as with the modern biota, the fossil record is replete with examples of convergence at varied taxonomic levels. (See also Section 3o in this volume, dealing with the independent evolution of herbivory in unrelated tetrapod groups.)

2 Specialized, Potentially Interacting Biological Substrates

2A. MARINE INVERTEBRATE BENTHOS

2Aa. CORAL BARNACLES

Moissette and Saint Martin (1990) discussed Algerian Miocene coral barnacles while reviewing many other occurrences elsewhere. Bałuk and Radwański (1984) provided additional data emphasizing Miocene Korytnica Basin *Creusia* and their coral hosts.

2Ab. WHALE BARNACLES

Seilacher (2005) has provided an elegant review of varied modern whale barnacle morphologies and adaptations. Bianucci et al. (2006a) reviewed the Pliocene and Pleistocene occurrences and suggested the varied migrational routes employed by the whales involved. Bianucci et al. (2006b) described a unique Pliocene–Pleistocene assemblage of *Coronula diadema* from Ecuador and commented on its implications for the range of the presumed humpback host. For anyone interested in "seeing" whale barnacles *in situ* go to the fantastic photographs of Masa Ushioda (SeaPics.com) of modern humpbacks infested by *Coronula diadema*.

2Ac. PLATYCERATIDS

Meek and Worthen (1866, 1868) and Keyes (1888) appear to have been the first to have fully understood the platyceratid–pelmatozoan relationship. They not only emphasized the basic fact that the distal margin of the platyceratid shell molded itself to its pelmatozoan substrate, ruling out the earlier interpretation that the pelmatozoans had perished owing to predation by the platyceratids, but also illustrated the first specimens in which the growth traces of the platyceratid aperture were grooved into the underlying pelmatozoan plates. Baumiller (2003; see also Baumiller, 1990) reviewed the entire question of platyceratid–pelmatozoan relations and made a good case, based on closely associated boreholes in some specimens, for a possibly kleptoparasitic mode of life ("stealing" host food particles, as is done with some capulid gastropods today) or even gametophagy, additional to the still widely accepted coprophagy. He does not favor predation as a possibility.

When Baumiller and Gahn (2002) reviewed the question of the platyceratid–pelmatozoan relationship, they explored the various possibilities and tried to assess the strengths of each. Gahn et al. (2003) provided evidence from a Middle Devonian crinoid, complete with a drillhole through the calyx surrounded by growth lines on the calyx external to the drillhole, with platyceratids present in the same beds that supports a kleptoparasitic life mode for at least some platyceratids. Horný (2000) has unequivocally shown that platyceratids also had the capability of attaching to hard, shelly substrates, best illustrated by his many orthoceroid attached specimens from which they presumably filter fed, as they are tightly molded to the shelly substrate. It is clear that there may be more to the platyceratid story than has been suspected, with the additional problem that there are many localities for platyceratids in which evidence for the presence of pelmatozoans is lacking or for attachment to hard substrates such as Horný has demonstrated; that is, still another feeding mode is possible.

Hengsbach (1990) referred to varied examples from the German literature. Baird et al. (1990) provided some New York Givetian examples, including some with acrothoracid borings. Levin and Fay (1964), in their discussion of a collection of Mississippian blastoids, many of which have attached platyceratids over their anal pyramids, emphasized the aperture of the platyceratid fitting tightly over the host blastoid; that is, this is not some short-term relationship of the kind one might expect from a predator. Gahn and Baumiller (2003), employing a suitably large sample from the Givetian, made a strong case that in their example small crinoids were selectively associated with the platyceratid as contrasted with larger specimens having two host crinoid species being involved. They inferred that nutrient "stealing" from the smaller specimens might have been the controlling factor that allowed larger crinoids to achieve a size refuge. Brett (2003) provided data on host crinoid possible selectivity by some Silurian and Middle Devonian platyceratids.

Bandel (1992) described some early gastropod growth stages from the Late Triassic St. Cassian Formation that he assigned to the platyceratids. In his view, the platyceratids persisted through the Permian extinction event to the Triassic extinction event and were neritaceans distantly related to the post-Paleozoic neritaceans; none has been found to be associated with a pelmatozoan host.

Baumiller (1993) provided evidence consistent with the concept that at least some platyceratids may have made round boreholes in Devonian blastoid hosts, although the evidence is still very circumstantial (only a single platyceratid has been found to be associated closely with a borehole made into a pelmatozoan host; this area deserves more investigation). Baumiller and Macurda (1995; see also Baumiller, 1990, 1993, 1996) described varied circular boreholes in Middle Devonian and Middle Mississippian blastoids, relatively site specific, which they concluded are likely to have been made by platyceratids acting as parasites rather than as predators. Their evidence is circumstantial, as platyceratids caught in the act are still unknown. Using Middle Devonian examples, Baumiller (1996) made a reasonable case that at least some platyceratids drilled holes into blastoid and crinoid hosts before beginning their "parasitic" activities. Baumiller (2002) described several Middle Devonian Silica Shale crinoids with a large number of platyceratids on the crinoid tegmen (medium sized on one specimen and small on another), which does suggest some type of close relationship between the crinoid and the gastropod. Deline (2008) described a presumably predatory or parasitic borehole in an Ordovician stylophoran, the only one recognized to date.

Morris and Felton (1993) discussed a number of Cincinnatian platyceratids in place on crinoids while considering the question about whether or not cornulitids are also symbiotic, in this case on platyceratids. Taylor and Brett (1996) illustrated a Wenlockian platyceratid in place from New York.

In many ways Horný (2000) had the last word on the platyceratid problem when pointing out, using some Czech Silurian and Devonian examples, that some platyceratids molded to small orthocones, with others possibly being coprophagous and still others being neither, among other possibilities out there. In other words, there is no simple answer to the platyceratid problem.

Frýda et al. (2008b) carefully summarized the evidence concerning platyceratid relations. They pointed out that, although many occurrences may be interpreted as coprophagous or kleptoparasitic, others are possibly predatory (as indicated by associated drillholes), whereas others are neither. They also pointed out that the platyceratids are a polyphyletic grouping which further complicates the problem of trying to understand their relations to other, closely associated organisms (Frýda et al., 2009). A particularly well-preserved ontogenetic series from a French Devonian locale made clear the ontogenetic change from merely being attached to a hard substrate to being positioned evenly over a crinoid tegmen. It would be interesting to learn whether individual crinoid families with associated platyceratids involve platyceratid groups distinctive for different crinoid families.

Horný (2005) made a strong case that tergomyan mollusks, in at least one case from the Bohemian Silurian, also lived firmly attached to hard, shelly substrates.

☐ *Reliability*
Category 1, frozen behavior.

2Ad. MANGROVE OYSTERS

Oysters with xenomorphic impressions of a mangrove-like branch from the Late Cretaceous of southern Sweden have been cited previously. One of these unique specimens has finally been discussed a bit and illustrated by Sørensen and Surlyk (2008, Figure 2). The smooth imprint of the mangrove-like branch makes it clear, however, that *Rhizophora* was not involved. Claspers similar to those known from both Eocene and modern mangrove oysters are cited from these specimens.

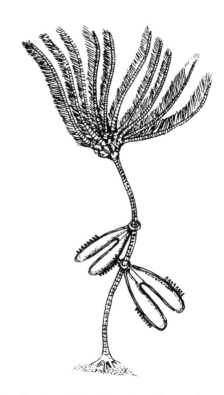

FIGURE 3 Permianellid brachiopod attachment rings on a crinoid stem (about ×0.5) that are presumably homologous with the productid attachment rings cited previously (Boucot, 1990a). (Figure 3 in Shen, S. et al., *Géobios*, 27, 477–485, 1994. Reproduced with permission of Elsevier Masson SAS.)

2Ae. PRODUCTID BRACHIOPOD SPINES

(See Figure 3.) Shen et al. (1994) illustrated and described attachment rings of permianellid productid brachiopods to crinoid stems from the Late Permian of South China. Brunton and Mundy (1993, p. 116) cited a number of Visean genera having clasping spines as juveniles.

☐ *Reliability*
Category 1, frozen behavior.

2Af. *PHOSPHANNULUS* ON PALEOZOIC CRINOID STEMS

Powers and Ausich (1990) illustrated and commented on *Phosphannulus* attached to Mississippian crinoid stems.

☐ *Reliability*
Category 6, due to ignorance concerning the affinities of the organism.

2Ag. PALEOZOIC HOST-SPECIFIC, PIT-FORMING CRINOID EPIZOANS

Feldman and Brett (1998) reviewed the Paleozoic range of the possible crinoid parasite the pit-forming *Tremichnus*, from the Middle Ordovician into the Carboniferous, and described a new occurrence from the Jurassic of Israel that extends the range well into the Mesozoic. Donovan and Pickerill (2002) discussed the nature of *Tremichnus* Brett, 1985, and *Oichnus* Bromley, 1981. They appeared to conclude that examples of both probably represent the same ichnotaxon.

The published examples of *Tremichnus* are from crinoids, whereas those of *Oichnus* involve varied noncrinoidal higher taxa. Donovan and Jagt (2002) described *Oichnus* borings in the outer part of tests of Maastrichtian echinoids, noting a "reactive" blister present internally. These borings resemble those of naticids that failed to penetrate their prey but are clearly nonpredatory. Donovan and Jagt (2004) described additional echinoid tests with pits from the French Upper Cretaceous that do not completely penetrate the test. The nature of these boring organisms is unknown.

2Ah. Hydroid–Serpulid Relationship

Since the publication of Scrutton's (1975) paper suggesting that a number of Jurassic to Pliocene serpulid tubes were infested by a colonial hydroid, a number of papers have appeared that confirm the presumably commensal relation. Prominent among these papers are those by Radwańska (1996, 2004) and Niebuhr and Wilmsen (2005). Citations as far back as the Pliensbachian are presented by various authors. Current evidence suggests that the hydroid *Protulophila* is fairly substrate selective in the sense that it tends to choose certain serpulid species as compared with others present in the same fauna. Evidence for the presence of these tubes within serpulids was noted as early as 1829, although the probable nature of the tube-dwelling organism was first pointed out by Scrutton.

☐ *Reliability*
Category 5, as indicated in Boucot (1990a), but the evidence provided more recently makes Category 2B more reasonable.

2Ai. Crinoid–Tabulate Coral Relationship

Peters and Bork (1998) described some Waldron Shale *Favosites* attached to *Eucalyptocrinites* stems. It is interesting that the crinoid is the same as recorded previously by Brett and Eckert (1982) from the Rochester Shale, although whether this is a favored attachment site or merely reflects the overall abundance of *Eucalyptocrinites* is unknown. On the same specimens Peters and Bork also recorded a number of pedically attached rhynchonellids (*Stegerhynchus indianense*). Peters and Bork concluded that the Waldron Shale represents a soft substrate that many taxa could not accommodate; the use of crinoid stems as a substrate probably indicates that they were among the few available hard substrates for larvae.

☐ *Reliability*
Category 1, frozen behavior.

2Aj. Sipunculid–Coral Towing

In the previous volume, Boucot (1990a) discussed the well-known relationship between certain solitary scleractinians and a sipunculid worm as an example of a specialized, potentially interacting biologic substrate. The most comprehensive account of the sipunculid–coral relation was provided by Yonge (1970), who made it clear that the relation is initiated by the sipunculan first occupying an empty gastropod shell, after which a planula larva colonizes the snail shell, eventually overgrowing it with its contained sipunculid that continues to flourish while the snail shell becomes completely enveloped by the growth of the coral calyx. Additional information (Fisk, 1981; Myriam Preker, pers. comm., 1995) makes it clear that today (and presumably in the past) this is a truly mutualistic relationship in which the sipunculid receives shelter from the solitary coral calyx, whereas the coral is towed about by the sipunculid in a sedimentary regime subject to relatively high sedimentation that would rapidly cover the coral were it not for the activity of the sipunculid. Preker has observed whole areas of sea floor in the Heron Island, Queensland, region covered by the solitary corals and their sipunculid partners. Fisk (1981) also found that the coral *Heteropsammia* is more effective at shedding coarse sediment from its surface and lives in coarser sediment than is the case with *Heterocyathus*, which prefers finer grained sediment.

Stolarski et al. (2001) thoroughly reviewed the sipunculid–coral relation, summarizing a mass of information from the present back to the Albian. They made a strong case for two modes of sipunculid–coral relations. One mode features what they termed a "monoporous" relation with only one pore plus an extension tube, the coral being attached to a gastropod shell. The second involves a coral that overgrows the molluscan base while the sipunculid has one main opening and more then one "pore" opening to the outside. They summarized a number of prior occurrences of the sipunculid–coral relation from the present and the past which indicated the distinctiveness of the tubes generated by hermit crabs involved with gastropod shells and their tubes. Similar to the coral–gastropod shell–sipunculid relation is Zibrowius' (1997) account of modern *Dentalium* shells bearing an attached solitary *Heterocyathus*, one of the corals associated with sipunculids, apically in a similar relation; examples from the fossil record have not yet been recorded.

☐ *Reliability*
Category 2B, functional morphology.

2Ak. Polydorid Mud Blisters in Bivalves

Savazzi (1995) described a number of teratologic features in the shells of an Italian Pliocene bivalve, some possibly being due to serpulid activities, and commented on the possibility that some of the shell features were generated by parasites. Feldmann and Palubniak (1975, Plate II, Figure 4) cited and illustrated Maastrichtian polydorid blisters from North Dakota. These are the oldest known mud blisters. Wisshak

FIGURE 4 Kleemann's photograph of an Eocene Austrian *Favia magnifica* with well-developed, cup-shaped false floors, in a *Lithophaga* borehole; corellita width ≈ 1.3 cm. (Plate 16, Figure 7, in Kleemann, K., *Facies*, 31, 131–140, 1994. Reproduced with permission of Springer Science + Business Media.)

and Neumann (2006) made a good case for polydorid infestation borings on the oral face of Maastrichtian echinoids (*Echinocorys*) from Rügen; their conclusions are very reasonable. Boekschoten (1967) described some *Polydora* shell borings from the Balgian Pliocene. Taddei Ruggiero and Bitner (2008) discussed and illustrated *Polydora* borings in brachiopods from the Italian Pliocene.

□ *Reliability*

Category 2B, functional morphology.

2Al. Boring Bivalves and Corals

(See Figure 4.) Morton (1990) and Kleemann (1990b) have thoroughly reviewed the coral boring bivalves, living and dead, modern and fossil, as well as pointing out the chemical means used by some taxa to facilitate boring, with boring into live corals being a relatively young development. (See Kleemann papers for Mesozoic examples of bivalves boring into live corals, as evidenced by the calcareous "false floors" generated by the bivalve as it kept up with the growth of the coral; also, keep in mind the very restricted surface openings characteristic of many taxa which necessitate modification of said apertures to keep up with growth of the coral.)

Kleemann (1994a) described a boring *Lithophaga* from a Late Triassic Austrian coral (*Pamiroseris*). Kleemann (1980) reviewed and summarized the taxonomy of the coral-boring lithophagids and traced them back to the Liassic. Kleemann (1994b) described additional Cretaceous and Tertiary examples from varied European and a Turkish locality and concluded that this intimate, coevolved relation may have appeared in the Jurassic. Keep in mind that the bivalve larvae needed to evolve a mechanism to avoid being eaten by the polyps. Forteleoni and Eliasova (2000) described symbiotic relationships between *Lithophaga alpina* and two species of *Actinastrea* from the Italian Santonian indicating that the association has been stable for some time at the generic level, Late Cretaceous through Eocene. Žítt and Mikuláš (2006) described Cenomanian *Gastrochaeonolites* from Bohemia.

Wilson and Taylor (2001) described some Maastrichtian bivalves boring into limestone and coral substrates from Oman and the United Arab Emirates, with *Lithophaga* probably being responsible. They reviewed the fossil record of boring *Lithophaga*–coral occurrences, Eocene to present, and used geographic evidence to speculate as to where the coevolved relation first appeared.

Krumm and Jones (1993) described a Late Eocene boring bivalve–coral relationship involving *Lithophaga* and *Actinastrea* from Florida. It is clear, using their evidence, that the living bivalve secreted "false floors" beneath as it elevated its position to keep up with coral growth. Krumm (1999) described some Puerto Rican and Jamaican Oligocene lithophagid bivalves boring into corals.

The intimate *Lithophaga*–coral relation recalls the similarly coevolved coral–barnacle relation; in both cases, the larval capability to avoid the cnidoblasts of the polyps is remarkable. Interest centers on just what compounds or behaviors are involved, as well as the similar problem raised by the trapezoid crab–coral relation (4AIq) and the coralliophilid gastropods (4AIp).

Bałuk and Radwański (1977, p. 81) provided an excellent account of the rock borers present along a coastal rocky shore, including *Gastrochaena*, *Aspidopholas*, *Jouannetia*, and *Lithophaga*, in association with varied non-bivalve rock borers.

Although not in the boring category, the apparently obligate relationship between the pectinid *Pedum* and its scleractinian host is ably summarized by Savazzi (1998); a fossil record is not yet known for this relationship. Kleemann (1990a) also provided extensive information about *Pedum* and its relationship to various tropical colonial coral "hosts."

Related to the recognition of lithophagid borings is the fact that sipunculids of the present have the capability of boring into corals (Por, 1970), and it is possible to confuse their boreholes in the fossil record with those of other boring organisms.

□ *Reliability*

Category 1, frozen behavior.

2Am. Lepadomorph Barnacles and Eurypterids, Other Substrates, and Balanoids

Although not dealing only with barnacle cementation, Bromley and Heinberg's (2006) treatment of marine invertebrate hard substrate attachment mechanisms is well worthy of attention, providing as it does a wealth of information and understanding from the present as well as the past.

Lepadomorphs

Collins and Rudkin (1981) described what may be the oldest lepadomorph from the Middle Cambrian Burgess Shale, an organism with an unmineralized capitulum, and Wills (1963) described one on a eurypterid from the Silurian of Esthonia. Briggs et al. (2005) described metamorphosis in a Silurian Wenlock lepadomorph. Schram (1975) cited the lepadomorph *Illilepas damrowi* attached to a Mazon Creek Pennsylvanian *Myalina* shell.

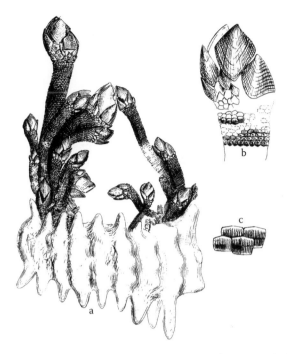

FIGURE 5 Lepadomorph barnacle (*Zeugmatolepas*) attached to an Oxford Clay ammonite (~×0.5). (From Darwin, C., *A Monograph on the Fossil Lepadidae*, Palaeontographical Society, London, 1851, p. 50. Reproduced with permission.)

Although not in the barnacle–eurypterid category, one more example of a lepadomorph that might have been using an organic substrate for attachment is Darwin's (1851, Plate VII, Figure 1, p. 50) *Pollicipes concinnus* (now assigned to *Zeugmatolepas concinnus* [Morris], according to Newman, pers. comm., 1990) attached to a Jurassic, Oxford Clay, ornatus bed, Middle Oolite, Christian Malford, Wiltshire ammonite (Figure 5). The illustrated specimen, with the barnacles attached to one side only of the ammonite, suggests a postmortem hard substrate association rather than a commensal relation. Some other specimens of the species were reported attached to lignite (i.e., a strictly hard substrate relation), although whether the original wood (lignite) or the ammonite was floating or on the bottom is undetermined. Barthel et al. (1990, Figure 7.34) illustrated a Solnhofen ammonite to which are attached a group of lepadomorph barnacles. Drushchits and Zevina (1969) described an Early Cretaceous ammonite from the Caucasus with some attached lepadomorphs.

Jagt and Collins (1999) described a Maastrichtian Belgian log riddled with teredo and closely associated with various lepadid plates belonging to several genera, although barnacle plates are not present in the adjacent sediments—good evidence for lepadomorph attachment to the log. Hattin and Hirt (1986) illustrated some fine specimens of *Stramentum* attached to inoceramid shells from the Kansas Turonian, as well as on other bivalves and an ammonite. Jagt and Collins (1999) discussed varied similar occurrences of Cretaceous and Cenozoic age. Stinnesbeck et al. (2005) briefly cited *Stramentum* attached to a Late Cretaceous Mexican ammonite. Bonde (1987) cited lepadomorphs attached to wood from the Early Eocene of Denmark.

Mizuno and Takeda (1993) illustrated several groups of Japanese Miocene *Lepas* attached to substrates from a probable bathyal depositional environment, although they probably lived at shallow depths.

Withers (1935) illustrated Cenomanian *Stramentum pulchellum* (Plate XLI) attached to an ammonite and Albian *Stramentum syriacum* (Plate XLII) attached to an ammonite.

Grignard and Jangoux (1994) described a unique infestation of the tips of echinoid spines by lepadomorph barnacles which prevented further growth of the spines and left a morphology that could be recognized in the fossil record.

Balanomorphs

Waugh et al. (2004, Table 2) listed a number of well-validated modern occurrences of marine decapods that have epibiontic balanoids attached. Yamaguchi (1980) illustrated some Japanese Pleistocene balanids attached to hard substrates. Donovan (1988) described some Pliocene–Pleistocene Red Crag balanids from East Anglia still attached to their substrate. Seguenza (1876) illustrated some Pliocene balanoids from Messina still attached to varied substrates. Ross (1964) described some Yorktown Formation Virginia balanoids still attached to hard, shelly substrates.

Mayr (1992, Figure 3, p. 141) illustrated a group (14 plus a small oyster according to Mayr, pers. comm., 1997) of *Balanus ornatus* attached to the inner face of a *Glycymeris* shell from the Miocene at Ortenburg bei Passau. (This specimen was in the original collection of Graf zu Münster, is the type of *B. ornatus* Münster, 1840, and survived a World War II bombing; Mayr, pers. comm., 1997.) Münster (1840, Plate VI, Figures 6–12) illustrated additional balanoids, from varied German locales, attached to shelly substrates. Glaessner (1969, p. 429) cited a New Zealand Miocene balanoid on a crab carapace, but without any reference or details. Philippe (1983) discussed and illustrated a balanoid attached to the aboral surface of a Miocene echinoid. Morris (1993) briefly described several Jamaican examples, including *Balanus*, *Eoceratoconcha*, and *Ceratoconcha* from the Pleistocene and Miocene, attached to varied shelly substrates. Doyle et al. (1997) described some Spanish Miocene balanoids attached to the rocks.

Gall and Muller (1975) described a mid-Miocene *Balanus pictus* still attached to a Weissjura–Massenkalk cobble. Yamaguchi (1977) described *Balanus trigonus* still attached to a Japanese Pleistocene *Chlamys nipponensis* and *Solidobalanus* still attached to a late Miocene *Placopecten*. Ziegler (1992) described a *Balanus* "colony" attached to a Miocene rock substrate from the Upper Marine Molasse. Hladilova and Pek (1998, Figure 8) illustrated an Ottnangian, Miocene, Austrian balanid attached to a *Turritella* shell. Feldmann (2003, Figures 6.1 and 6.2, p. 1028) has cited and described some excellent examples of balanoids attached to decapod carapaces from the Miocene of New Zealand and Denmark (Figure 6). Feldmann and Fordyce (1996) described a balanoid attached to Miocene *Lobocarcinus pustulosus* from New Zealand.

Feldmann et al. (2006) cited Miocene New Zealand *Metacarcinus novaezelandiae* with an attached balanoid, and they noted that the attachment sites are ones that the crabs are

2 cm   1 cm

FIGURE 6 Balanoids attached to decapod carapace (*Coeloma*) from the Miocene of Denmark. (Figure 6 in Feldmann, R.M., *Journal of Paleontology*, 77, 1021–1039, 2003. Reproduced with permission.)

unable to clean off. Crawford et al. (2008) briefly described an Argentine Miocene crab carapace from Patagonia with attached balanoids. Waugh et al. (2004) illustrated a balanoid attached to *Trichopeltarion greggi* from the New Zealand Miocene at Glenafric Beach, North Canterbury. Parras and Casadío (2006) briefly discussed an Upper Oligocene–Lower Miocene balanoid attached to *Crassostrea*. Santos and Mayoral (2008) described in detail the settling behavior on an echinoderm test of some Late Miocene *Balanus* from southern Spain. Boekschoten (1967) illustrated a Belgian Pliocene gastropod (*Neptunea*) with some attached balanids. Velcescu (2000) cited an Oligocene balanid attached to oyster shell from Romania.

Among the oldest balanomorphs still attached to a hard substrate are Jim Baichtal's specimens, attached to oyster shells, from the Late Eocene Cowlitz Formation near Longview, Washington (Figure 7). Newman (pers. comm., 2004) concluded that Baichtal's material is "probably *Balanus* cf. *cornwalli* Zullo, 1966," which may belong to *Hesperibalanus* if that genus is properly construed (Southward, 1995).

Ivany et al. (1990) described *Kathpalmeria* attached to a seagrass from the Florida Middle Eocene. Schweitzer (2005a, Figure 7, p. 279) illustrated an African Eocene crab, *Menippe*, with an attached balanid. Withers (1953) illustrated various Tertiary balanomorphs attached to varied substrates. Santos et al. (2005) reviewed the record of Miocene to present balanomorph attachment scars (i.e., trace fossils). Miller and Brown (1979) described some Pleistocene balanid attachment scars, while referring to an earlier description of some Dutch Pliocene examples. Withers (1924) illustrated several balanomorphs attached to New Zealand shells (Pliocene *Megabalanus* on *Coelotrochus* and on a ribbed bivalve; Miocene *Chirona* on *Dentalium*).

Verrucomorphs

Bromley and Martinell (1991) described the trace fossil *Centrichnus concentricus* that is preserved in some cases beneath verrucomorph barnacles and is known from the Miocene to present. Giusberti et al. (2005) described Paleocene *Verruca* attached to echinoid tests from northeastern Italy. Radwański and Wysocka (2004) described Miocene

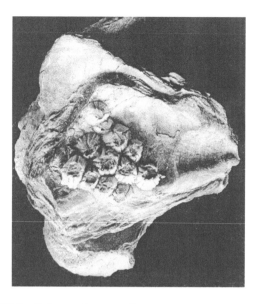

FIGURE 7 *Balanus* cf. *cornwalli* Zullo attached to oyster shell; Late Eocene, Cowlitz Formation, from near Longview, Washington, collected by James Baichtal, Thorne Bay, Alaska (×2). Barnacles identified by William Newman.

Polish specimens of *Verruca* attached to oyster shells. Schram and Newman (1980) described an Albian–Cenomanian *Verruca*, attached to a stomatopod, from Colombia; this is the oldest known verrucomorph. Withers (1935, Plate XLV) illustrated Danian *Verruca rocana* attached to *Ostrea*.

☐ *Reliability*
Category 1, frozen behavior.

*2An.   AULOPORA–LIEOCLEMA ASSOCIATION

McKinney et al. (1990) described *Aulopora* specimens from the Helderberg, Early Devonian Birdsong Shale of Tennessee, from one locality, where specimens of the tabulate coral *Aulopora* were encrusted by the bryozoan *Leioclema* except for the opening necessary for the individual *Aulopora* zooids. They interpreted this as evidence of mutualism and referred to a somewhat similar example from the modern biota; however, the single locality and horizon indicate that additional recognition of this relationship from many more localities and horizons is needed before it can be safely placed in the mutualistic category rather than as an example of a bryozoan employing a convenient hard substrate.

☐ *Reliability*
Category 6, due to the sampling questions.

*2Ao.   SOFT-BODIED MARINE ALGAL SUBSTRATES FOR SHELLY ORGANISMS

(See Figure 8.) Wignall and Simms (1990) provided a thorough account of validated pseudoplanktonic occurrences through time. Soft-bodied marine algae today commonly

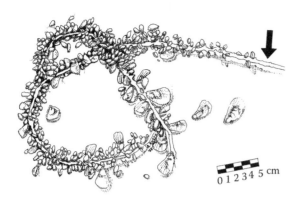

0 1 2 3 4 5 cm

FIGURE 8 McRoberts and Stanley's description is unrivaled in demonstrating an algal frond from the Mississippian Bear Gulch Limestone of Montana, to which are attached a mass of bivalves, with an increase in size indicated from the apparent base of the frond (noted by the arrow) to its end. (Figure 3 in McRoberts, C.A. and Stanley, Jr., G.D., *Journal of Paleontology*, 63, 578–581, 1989. Reproduced with permission.)

serve as substrates for a variety of invertebrates, including well-skeletonized forms. In the Paleozoic such associations are very uncommon. Havlíček et al. (1993) summarized some well-known Middle Ordovician examples from the Prague region (the Barrandian) involving both articulate and inarticulate brachiopods, as well as some potential trilobite examples. The Czech examples are not preserved in such a manner that one can observe the actual alga to which the shells were attached, but the linear invertebrate arrays are very impressive, and the same beds yield isolated, soft-bodied algal fragments that could have served as the substrates. Wright (1968, Plate 2, Figure 1; see also Holland, 1971, Figure 1a) illustrated some specimens of *Dicaelosia* attached to an organic fragment, probably an alga.

McRoberts and Stanley (1989) described an elegant bivalve–algae example from the Late Mississippian Bear Gulch in Montana, where the coiled algae are literally covered with attached bivalves. The sheer mass of the bivalves is more consistent with an epipelagic than with a benthic environment. Barthel et al. (1990, Figure 5.5; see also Frickhinger, 1994, Figure 6, p. 37, for *Phyllothallus* sp. encrusted with small oysters) illustrated an elegant Solnhofen Tithonian specimen of a brown alga encrusted by oysters.

Turek (1990) has described an association of the brachiopod *Valdaria*, an orthotetacid from the Czech Silurian, spatially arranged in such a manner that the original attachment to a soft-bodied "seaweed" or other decomposable substrate is the most rational explanation for the arrangement. The larger shells of the species are arranged in a tight circle, beaks pointing inward, with smaller growth stages present peripheral to the larger in positions suggesting that the smaller were attached to the larger.

Botting and Thomas' (1998, 1999) example of selective colonization of a specific graptolite and of an alga from the British Ordovician by a small inarticulate brachiopod is noteworthy. Pek (1977) described similarly oriented Ordovician

agnostid trilobites which he explained as attachment of some kind to a floating substrate such as an alga.

Sandy (1996) described a Late Ordovician crinoid columnal to which numerous *Zygospira*, an atrypacean brachiopod, are attached, while citing similar material from other horizons and localities; there are other examples of brachiopods attached to pelmatozoan columnals.

Also noteworthy is Seilacher et al.'s (1968) account of crinoids (*Seirocrinus*) attached to floating pieces of German Jurassic, Posidonienschiefer wood and Klug et al.'s (2003) account of a similar Late Devonian, Famennian, Moroccan *Archaeopteris* with many crinoid holdfasts that also indicate a floating crinoid substrate. Wang et al. (2006; see also Wang et al., 2008; Enos et al., 2006, Figure 66), while describing a fine Carnian example of numerous *Traumatocrinus* attached to driftwood from Guizhou in South China, provided an excellent summary of the available evidence of driftwood crinoid substrates. Nye et al. (1975) described a fine examples of bivalves, *Lunulacardium*, attached to a Middle Devonian log from New York.

Although not dealing with a marine organism substrate, Kelber's (1990, Figures 23, 24) comparison of Late Triassic higher land plant leaves from western Germany attached to which are *Spirorbis* shells similar to a modern marine algal example is instructive. *Spirorbis*, of course, is found attached to many marine shells through time, so the Late Triassic leaf example should not be taken to indicate an obligate relation. In connection with the identity of pre-Cretaceous "*Spirorbis*," note should be taken of Taylor and Vinn's (2006) demonstration that they belong to an unrelated, convergent group of possibly lophophorate organisms, *Microconchus*.

☐ *Reliability*

Category 2A, as Havlíček et al.'s (1993) linear arrays of shelly invertebrates preserved in black shale suggestive of an anoxic bottom, associated with soft-bodied algal fragments, are very suggestive of an association but do not actually demonstrate it. One cannot decide whether the soft-bodied alga was attached, benthic, or planktic. Category 1, frozen behavior, for those examples with the algal substrate clearly visible

*2Ap. *Trypanopora* and *Torquaysalpinx*

Weedon (1991) and Bertrand et al. (1993) have described the coiled, tubular organisms *Trypanopora* and *Torquaysalpinx*, which occur within algal bioherms, stromatolites, and chaetetid corals of the Devonian. The biological affinities of these taxa are very uncertain, with opinions varying from tabulate corals to vermetiform gastropods and tentaculitids, but their occurrences within organic substrates are clear. Whether or not they are symbionts or parasites is also unclear.

☐ *Reliability*

Category 6, due to the highly uncertain affinities of the organisms and their relations with the organic substrates.

*2Aq. STROMATOPOROID–CORAL INTERGROWTHS

Kershaw (1987) summarized the literature on stromatoporoid intergrowths with both tabulate and rugose corals and proposed that either symbiosis or commensalism could be a possible explanation. The examples are taken from the Silurian and Devonian; Mistiaen (1984) has compiled the known occurrences. Darrell and Taylor (1993) provided additional, useful materials.

☐ *Reliability*

Category 6, due to uncertainty about the relationship.

*2Ar. FORAMINIFERAL CONSORTIUM

Schmid and Leinfelder (1996) described an intimate association between two foraminiferans (*Lithocodium* and *Troglotella*), one an encruster, with an interpretation that they enjoyed a commensal relation.

☐ *Reliability*

Category 6 insofar as the commensal relation goes, but Category 1 with regard to the Middle Triassic through Cretaceous intimate preferred substrate relation.

*2As. DECAPOD INQUILINISM WITHIN AMMONITE SHELLS

Fraaye and Jäger (1995a,b) documented examples of several decapod types preserved within Jurassic ammonite shells that suggest inquilinism as the best explanation. Fraaije and Pennings (2006) described some Paleocene crab carapaces preserved within nautiloid shells that further support the possibility of inquilinism following molting, while reviewing similar Mesozoic examples preserved in ammonite and nautiloid shells.

☐ *Reliability*

Category 2B, functional morphology.

*2At. MICROPOLYCHAETE–SCLERACTINIAN RELATIONSHIP

Bałuk and Radwański (1997) have carefully described the tubes of a micropolychaete intimately associated with the calyx of a particular species of colonial hexacoral from the Miocene Korytnica Clays, and they provided evidence from recent micropolychaete specimens that strongly support the assignment of the Miocene tubes to *Josephella*. This is a most unusual relationship and appears to be very reliable, as the morphology of the Miocene and modern tubes is strikingly similar, although distinctly different hosts are involved. Roger (1944) described some Moroccan Pliocene solitary corals (*Flabellum*) that show external calyx impressions that he interpreted as the attachment site of an annelid. Fage (1937) described some modern examples involving *Flabellum* with a polychaete (i.e., this is an ancient relationship).

☐ *Reliability*

Category 1, frozen behavior.

*2Au. EPIZOAN–SPONGE RELATIONSHIP

Carrera (2000) has reviewed the literature on sponge encrusters while describing a large amount of Argentine Early Ordovician material. There is evidence of interaction in some cases; however, until additional occurrences from different regions are described it would be too soon to decide whether these relationships are obligate.

☐ *Reliability*

Category 1, frozen behavior.

*2Av. EPIBIONTS

Lescinsky (2001) provided a brief summary of epibionts known to encrust varied shelly substrates during the Phanerozoic; however, this summary does not provide any thoughts about which epibionts might have had obligate relations with the encrusted shell as contrasted with epibionts facultatively selecting any available hard substrate. To determine if a more evolved relation was involved, more information from adequate samples is required.

☐ *Reliability*

Category 7, unreliable.

*2Aw. UMBROPHILIC BRACHIOPODS

Kobluck (1981) summarized much of the Paleozoic record. Spjeldnaes (1976) described some Silurian bryozoans grown in the shade (beneath coral and stromatoporoid heads), mentioned similar occurrences in the Middle Ordovician, and commented that the relationship persists to the present, with spirorbids, presumably *Microconchus*, commonly associated with the bryozoans as well as with a few brachiopods. Prominent among these umbrophils (sciaphils) are certain brachiopods. This mode of life has been adapted by varied, unrelated brachiopod groups through time. What might be the selective advantage involved, and what are the characteristics of this group of shells? Convergent evolution is certainly involved in view of the taxonomically unrelated higher taxa involved during varied time intervals. The shells are commonly cemented to the substrate by the ventral valve, and all tend to be small in the Paleozoic to very small in the post-Paleozoic. Their positions beneath overhangs and in crevices would certainly have shielded them from burial by sediment, but why their small size?

Paleozoic examples include the following: Nield (1986) discussed convincing examples of *Liljevallia* from Gotland for this Silurian form, while Bassett (1984) cited *Leptaenoidea* from the Gotland Silurian. Jones and Hurst (1984) deduced that, in the Silurian, *Dubaria* and a reef-associated species of *Atrypoidea* lived in cryptic habitats. In the latest Silurian and the Devonian, there are at least some of the davidsoniaceans, atrypids with well-developed spiralia. In the earlier Middle Devonian Eifelian there are the possible productid *Auchmerella* and the orthotetacean

Krejcigrafella (for descriptive materials, see Struve, 1964, 1978, 1980). Muir-Wood (1965, Figure 306, p. H451) illustrated an example of *Devonalosia* attached to the underside of a Middle Devonian *Hexagonaria* colony. Brunton and Mundy (1988) described the productid *Sinuatella johnsoni* from beneath *Chaetetes* colonies as attached by "body creeping adherent spines" on the undersides of encrusting *Chaetetes* from the Mississippian of northern Yorkshire. Suchy and West (1988) cited the productids *Cooperina*, *Teguliferina*, and *Heterolosia* occurring beneath laminar Pennsylvanian chaetetid colonies.

Rosen et al. (2000) summarized the abundant information on post-Paleozoic platy coral assemblages, concluding that they normally thrive in relatively deep water where light is a limiting factor. Their conclusions make good sense and may apply to much of the Paleozoic material as well. Many, but not all, of the post-Paleozoic umbrophils are associated with platy corals. Insalaco (1996) provides an excellent set of Late Jurassic examples, while concluding that deeper water, low light, low sedimentation rate, and quiet water conditions are involved.

In the post-Paleozoic there are many genera belonging to the Triassic to present thecidians and to the Late Cretaceous to present megathyrids, including *Argyrotheca*. Asgaard (2008) provided a number of useful insights into the settling behavior of umbrophilic brachiopods, based on modern examples. Gaillard and Pajaud (1971; see also Pajaud, 1974, who also discussed the underside of a Campanian rudistid) provided an excellent description of an Oxfordian example, the thecidian *Rioultina*, while Taylor (1979) illustrated one that may be on the underside of another brachiopod. Jackson et al. (1971) provided evidence and examples from modern low-latitude occurrences while emphasizing their similarity with those from the post-Paleozoic fossil record and provided a body of literature from the present regarding their presence in shaded environments, such as the undersides of platy coral colonies, cracks and crevices of specialized organisms selected for life in semidarkness. Taddei Ruggiero (2001) described a modern example. Palmer and Fürsich (1974) described a typical example of a thecidian (*Moorellina*) occurring underneath an overhang in the British Middle Jurassic. Fürsich (1979) described a number of Jurassic thecidian occurrences in hardground cavities. Schlögl et al. (2008) described some Slovakian Tithonian examples. Manceñido and Damborenea (1990) described some thecideaceans from the Pliensbachian of southern Argentina that lived chiefly on the undersides of platy scleractinians and a few on the undersides of oysters. Wilson (1986) described some Early Cretaceous articulate brachiopods from Britain that lived in boreholes within cobbles that were obviously shaded.

Although this work does not involve brachiopods, the account by Radwańska (2005) of Kimmeridgian Polish comatulid crinoids preserved in a set of alpheid-like shrimp burrows strongly suggests that they used the burrows as shelter during daytime and then emerged at night, as is characteristic today of comatulids "hiding" during the day while roving about at night. Their ultimate preservation was probably due to being overwhelmed by an unexpected flood of suffocating sediment.

☐ *Reliability*

Category 1, frozen behavior.

*2Ax. MEEKOPORELLA–CRINOID RELATIONSHIP

Wyse Jackson et al. (1999, Figure 10) described and illustrated a unique relation between a morphologically very peculiar bryozoan and a crinoid stem to which it was presumably attached.

☐ *Reliability*

Category 2B, functional morphology.

*2Ay. BRACHIOPOD–ORTHOCEROID RELATIONSHIP

Holland (1971, Figure 1b) illustrated a Late Silurian orthocone more or less parallel to which are an aligned group of small strophomenoid brachiopods that presumably were attached to the orthocone, although it is unclear whether this was postmortem or during the active life of the orthocone.

☐ *Reliability*

Category 2B, functional morphology.

*2Az. HELICOSALPINX

Tapanila (2004) described a Late Ordovician occurrence of the curious "coiled" *Helicosalpinx* that occurs in tabulate colonies, between the individual zooids, from the Late Ordovician to the Givetian. The identity of the organism is unknown, but it does appear to have been a commensal.

☐ *Reliability*

Category 7, because the nature of the organism is unknown.

*2Aza. LUMBRINERIS FLABELLICOLA–SCLERACTINIAN RELATIONSHIP

(See Figure 9.) Zibrowius et al. (1975) described the possibly kleptoparasitic relationship between the eunicid annelid *Lumbrineris flabellicola* and three families of scleractinians, as well as a few soft-bodied cnidarians. This relationship is well known from varied Old World, deeper shelf localities. The worms have been shown experimentally to steal food from the polyps. They live in a tube that is incised by unknown means, possibly biochemical, into the coral calyx, and involve both solitary and colonial ahermatypics. They have a well-developed fossil record extending back into the Miocene of the Old World.

☐ *Reliability*

Category 2B, functional morphology.

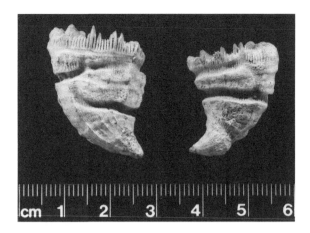

FIGURE 9 A Pliocene Italian solitary coral, *Ceratotrochus*, with grooves in the calyx like those generated today by the kleptoparasitic polychaete *Lumbrineris* (left figure is specimen R45453, and right is R45456). (Figure 3.15 in Darrell, J.G. and Taylor, P.D., *Courier Forschungsinstitut Senckenberg*, 164, 185–198, 1993. Reproduced with permission.)

*2Azb. CARAPUS–HOLOTHURIAN RELATIONSHIP

Bałuk and Radwański (1977) cited the occurrence of otoliths of the fish *Carapus*, which commonly lives in the anus of holothurians. Eeckhaut et al. (2003, Table 2) provided information on some of the many holothurians inhabited by some carapids. Arnold (1956) reviewed the content of the Carapidae.

☐ *Reliability*
Category 2B, functional morphology.

*2Azc. SPHENIA NESTLING

Bałuk and Radwański (1979) described a number of *Sphenia* shape adaptations to nestling in various burrows generated by lithophagids from the Korytnica Miocene. *Sphenia* elsewhere is well known as a nestling bivalve with a record extending back to the Paleocene.

☐ *Reliability*
Category 1, frozen behavior.

SUMMARY

There are many examples down to the generic level of behavioral fixity in this and the 1990 treatment. (See Table 1.)

TABLE 1

Marine Invertebrate Benthos Employing Specialized, Potentially Interacting Substrates (Additions to Table 2 of Boucot, 1990a)

Behavioral Type		Taxonomic Level	Time Duration	Category	Refs.
2Ag.	Paleozoic host-specific, pit-forming crinoid epizoans	Genus–class	Silurian–Jurassic	6	Feldman and Brett (1998)
2Ah.	Hydroid–serpulid relationship	Genus–several genera	Jurassic–Recent	2A	Scrutton (1975); Radwańska (1996, 2004)
2Aj.	Sipunculid–coral towing	Genus–genus	Albian–Recent	2B	Stolarski et al. (2001)
2Ak.	Polydorid mud blisters in bivalves	Order–order	Late Cretaceous–Holocene	2B	Feldmann and Palubniak (1975)
2Al.	Boring bivalves and corals	Varied genera	Late Triassic–Holocene	1	Kleemann (1994a)
2Am.	Lepadomorph barnacles and eurypterids and other substrates				
	Lepadomorphs	Suborder	Silurian–Holocene	1	Darwin (1851); Barthel et al. (1990); Jagt and Collins (1999); Wills (1963)
	Balanomorphs	Suborder	Eocene–Holocene	1	Mayr (1992); Münster, 1840); Gall and Muller (1975); Yamaguchi (1977); Ziegler (1992); Baichtal (unpublished)
2An.	*Aulopora–Lieoclema* association; coral–bryozoan	Genus–genus	Early Devonian	6	McKinney et al. (1990)
2Ao.	Soft-bodied marine algal substrates for shelly organisms; varied taxa	Phyla–phyla	Ordovician–Holocene	2A	Havlicek et al. (1993); McRoberts and Stanley (1989); Barthel et al. (1990); Turek (1990)
2Ap.	*Trypanopora* and *Torquaysalpinx*	Uncertain	Devonian	6	Weedon (1991); Bertrand et al. (1993)
2Aq.	Stromatoporoid–coral intergrowths, stromatoporoids, rugosans	Class–class	Silurian–Devonian	6	Kershaw (1987); Mistiaen (1984)
2Ar.	Foraminiferal consortium; *Lithocodium, Troglotella*	Genus–genus	Middle Triassic–Cretaceous	6	Schmid and Leinfelder (1996)
2As.	Decapod inquilinism within ammonite shells; decapods, ammonoids	Subclass, order	Jurassic	2B	Fraaye and Jäger (1995a,b)
2At.	Micropolychaete–scleractinian relationship; micropolychaete tube, specific hexacoral	Order–species	Miocene–Holocene	1	Baluk and Radwański (1997)
2Au.	Epizoan–sponge relationship	Phylum–class	Early Ordovician	1	Carrera (2000)
2Aw.	Umbrophilic brachiopods	Class	Ordovician–Holocene	1	Kobluck (1981)
2Ax.	*Meekoporella*–crinoid relationship	Genus–class	Early Carboniferous	2B	Wyse Jackson et al. (1999)
2Ay.	Brachiopod–orthoceroid relationship	Class–class	Silurian	2B	Holland (1971)
2Az.	*Helicosalpinx*	Genus–class	Ordovician	7	Tapanila (2004)
2Aza.	*Lumbrineris*–scleractinians	Species–class	Miocene–Holocene	2B	Zibrowius et al. (1975)
2Azb.	*Carapus*–holothurian relationship	Genus–class	Miocene–Holocene	2B	Baluk and Radwański (1977)
2Azc.	*Sphenia* nestling	Genus	Paleocene–Holocene	1	Baluk and Radwański (1979)

3 Mutualism

(See also Sections 2Az and 2Aq.)

3a. MYCORRHIZA

Tester et al. (1987) reviewed the taxonomic distribution of "nonmycorrhizal" plants and considered their significance. Simon et al. (1993) provided ribosomal DNA sequencing data suggesting that endomycorrhizal fungi may have coevolved with vascular plants.

☐ *Reliability*
Category 6, somewhat speculative.

3b. ZOOXANTHELLAE

Stanley and Swart (1995) provided an isotopic method for distinguishing zooxanthellate corals from non-zooxanthellate forms as far back as the Triassic.

☐ *Reliability*
Category 2B, functional morphology.

3c. *VERMIFORICHNUS* AND OTHER EPIBIONTS

(See also Section 9Ao, *Cornulites*.) Brice and Mistiaen (1992) described varied Frasnian brachiopod epizoans from the Boulonnais, providing a good analysis of the possibilities for preferential larval settling. Powers and Ausich (1990) provided some Mississippian epizoan site location data from Indiana. Baird et al. (1990) provided epifaunal information from some Givetian platyceratid gastropods occurring in New York. Bordeaux and Brett (1990) provided an analysis of New York Givetian epibionts that indicates that epibionts did select some taxa over others as substrates; that some epibionts favored one surface type over another, including such things as plications; and that postmortem settling may be an important factor in some cases (i.e., it is not safe to assume that all epibionts were present during the life of the substrate organism).

☐ *Reliability*
Category 2B, functional morphology.

3d. HALECOSTOME–INOCERAMID RELATIONSHIP

Stewart (1990) added considerably to his prior work on the presence of inquilinism among certain halecostome fishes and their inoceramid hosts.

☐ *Reliability*
Category 1, frozen behavior.

*3e. PINNOTHERID CRABS

The commensal–mutualist relations of soft-bodied pinnotherids are well known (Ross, 1983). Feldmann et al. (1996a) reviewed the relationship while describing the only known pinnotherid–brachiopod association. They commonly live as commensals in such things as bivalve mantle margins, other mollusks, annelid burrows, and certain worm-like holothurian burrows. Morris and Collins (1991) described *Pinnixa* from the Pliocene of Sarawak, and Glaessner (1969) cited them as far back as the Eocene. Kato (2005) described a Japanese Miocene pinnotherid and reviewed some of the prior fossil occurrences of the group. Bishop and Palmer (2006) described an Eocene South Carolina pinnotherid. Morris and Collins (1991) also cited *Prepaeduma* from the Sarawak Pliocene, with *Paeduma* being another modern commensal that lives in the tubes of annelids and in hydrozoan cavities. While describing a Colombian Late Cretaceous crab that might be a pinnotherid, Feldmann et al. (1999) pointed out that validated pinnotherids are not known with certainty below the Cenozoic. Blow and Bailey (2003) reported a Pliocene pinnotherid inside of several specimens of *Crepidula fornicata* and commented that, although pinnotherids have been found associated with other crepiduloids, they have not been found in living *C. fornicata*. These (there are many specimens in this occurrence) are the first pinnotherids found intimately associated with such a partner. Zullo and Chivers (1969) described an Oregon Pleistocene example in which *Pinnixa fabo* specimens were found inside articulated valves of *Tresus capax*; these are the only known example of this association, and more careful examination of articulated fossil bivalve shells needs to be done. Bromley (1996, p. 160) discussed the probably commensal relationship between a crab (*Scleroplax granulata*) and the burrow of an echiuran worm in which the crab lives and pointed out a Pliocene example from MacGinitie and MacGinitie (1949).

☐ *Reliability*
Category 2B, functional morphology, as none of these crabs has ever been found in an actual burrow, with only a single Pleistocene occurrence of a bivalve containing a pinnotherid, except for the *Scleroplax*–echiuran burrow example and the *Crepidula* example, which belong in Category 1, frozen behavior.

*3f. FIG WASPS

Poinar (1993, Figure 3) illustrated a Dominican amber fig wasp (Agaonidae), which carries with it the obligately coevolved fig fertilization–fig wasp relationship. Peñalver et al. (2006) described three fig wasps from Dominican amber. Collinson (1989) described the oldest figs and their relatives from the Eocene. Using molecular data, Machado et al. (2001) placed the origination of the fig–fig wasp mutualism back into the Late Cretaceous while emphasizing the very obligate nature of the varied figs with their specific fig wasps. The analogous yucca–yucca moth mutualistic association is, unfortunately, not yet supported by adequate paleontological data, although molecular data (Pellmyr and Leebens-Mack, 1999) suggest a similar story.

☐ *Reliability*
Category 2B, functional morphology.

*3g. TUBE WORMS, BIVALVES, AND RHYNCHONELLID BRACHIOPODS FROM DEEP SEA VENTS, PLUS CRUSTACEANS

Dover (2000) has carefully summarized most aspects of the vent fauna phenomena (and should be carefully read by all concerned with these curious faunas) and dealt with some aspects of their fossil record. Campbell (2006) has provided an excellent overview both geographically and stratigraphically of deep sea vent faunas. Kiel and Goedert (2006c) discussed the problems involved in trying to better understand whether vent animals originated prior to wood-fall animals, which in turn then gave rise to whale-fall animals—very difficult evolutionary questions. Okada and Cadet (1989) provided a wealth of information about seep bivalves and their environments today and some data from the Pliocene. Levin et al. (2000) concluded that methane seep infaunal fauna are not distinct from non-seep infauna. Van Dover et al. (2002) provided an overview of the biogeographic possibilities for the seep and deep sea vent faunas. Little (2001) provided a brief summary of hot-vent and cold-seep faunas through time, while Little et al. (2002; see also Little, 1997) provided an overall account of both cool seeps and hot hydrothermal vents through time. Reitner (2005) provided a useful overview of the bacterial basis for cold-seep faunas based on methane. Jeng et al. (2004) described a unique occurrence of shallow water crabs adjacent to active, shallow vents; the crabs feed actively on plankton, etc. that are killed by the vent products and fall to the seafloor, where they are devoured by the crabs. Vetter (1991) provided an excellent account of the thiotrophic bacteria that form the base of the food chain in these unique environments.

There is an accumulating literature on the occurrence of tube worm remains intimately associated with what are reliably interpreted to have been sulfide-rich, hot, hydrothermal vents situated on ancient spreading centers. Additionally, there is also an accumulating literature on the fauna associated with cold seeps rich in methane and hydrogen sulfide, which supports a biota based on bacteria distinct from the hot seep biotas. Tunnicliffe (1992) and the 1994 issue of *Geo-Marine Letters* involving Aharon (1994; see also Callender et al., 1990) have ably summarized the literature on the geographic, taxonomic, and paleontological distribution of the vent organisms. Tunnicliffe et al. (2003) summarized information on deep sea reducing environments, including vents. Dando et al. (1992) discussed the occurrence of modern vestimentiferan tube worms just about anywhere in the deep seas where sulfide-rich environments prevail (e.g., shipwrecks, dead whales, hot and cold seeps and vents). Ogawa et al. (1996) described some modern abyssal region cold-seep situations from the Japan Trench. McArthur and Tunnicliffe (1998) considered the complex, perplexing questions involved with the antiquity and endemicity of the modern vent faunas as contrasted with the not too well-known ancient taxa. Distel et al. (2000) provided a provocative account of possible evolutionary trends in vent mussels from non-vent ancestors that employed organic substrates within the deep sea. Olu-LeRoy et al. (2004) provided a useful account of some cold-seep, modern mud volcano, fault-associated vent faunas from the eastern Mediterranean.

HOT VENTS

The "hot" type occurrences, associated with ore deposits (see Little et al., 1998, for a compilation that includes those associated with massive sulfide deposits in the Philippines, New Caledonia, Cyprus, Oman, California, and the Urals; see also Little, 1997, 2002), include the following items. Banks (1985) described such worm tubes from the Irish Mississippian, while additional occurrences have been described from beds ranging in age from Late Cretaceous and younger (Haymon and Koski, 1985; Haymon et al., 1984; Oudin and Constantinou, 1984; Oudin et al., 1985). Little et al. (2007) described a very shallow water Late Cretaceous volcanic arc type of vent, with tube-worm-type fossils from Georgia. Scott et al. (1990, Figure 12) illustrate pyritized worm tubes from the Troodos Massif of Late Cretaceous age. Moore et al. (1986) described evidence of tube worms from the Mississippian of Alaska. Kuznetsov et al. (1990, 1991) discussed Middle Devonian tube worms and bivalves from a sulfide deposit in the Urals, as did Zaykov and Maslennikov (1987) for the Uralian Middle Devonian. The oldest described sulfide-type deposit is from the Silurian of the southern Urals, with associated orbiculoid brachiopods, monoplacophorans, and "tubes" of possible vestimentiferan and polychaete types (Little et al., 1997). Little et al. (1999; see also Little et al., 1998a) provided an elegant description and consideration of the oldest known vent faunas, those of Silurian and Devonian age from the Urals. Shpanskaya et al. (1999) described tube casts attributed to vestimentiferan worms from the Uralian Silurian and Devonian occurrences. Boirat and Fouquet (1986) described tube worm tubes from a

Late Eocene massive sulfide deposit in the Philippines. Little et al. (1998b; 2004) discussed a Jurassic Franciscan Complex massive sulfide occurrence from the San Rafael Mountains that includes vestimentiferan worm tubes, *Anarhynchia* cf. *gabbi*, and a trochoidean gastropod. Berkowski (2004) described rugose corals from an Emsian mound at Hamar Laghdad in Morocco which he concluded are the result of hydrothermal vents from underlying volcanics.

Barite Deposit Vents

Poole (1988), Poole and Dutro (1988), and Poole et al. (1991) described occurrences of Late Devonian Famennian barite deposits with associated tube worm cylinders and the rhynchonellid brachiopod *Dzieduszyckia* from deep-water deposits in Nevada and Sonora which may be compared with modern "white-smoker" vent occurrences. It would clearly be profitable to look for similar fossils associated with even older sedimentary barite deposits. Campbell and Bottjer (1995b) considered these barite occurrences to represent cold seeps, but Clark et al. (2004) considered them to be of hydrothermal origin. Naehr et al. (2000) provided evidence that these barite-associated seeps are cold seeps and listed additional occurrences. Greinert et al. (2002) described a modern seafloor, cold-seep-related barite deposit from the Sea of Okhotsk associated with cold-seep organisms, suggesting that ancient sedimentary barite deposits might have a similar origin. Torres et al. (2002) provided additional modern marine barite occurrences of probable cold-seep type. Torres et al. (2003; see also Clark et al., 2004, who favor a hydrothermal origin) reviewed the entire question of cold-seep-generated, stratiform barite deposits from both the Paleozoic and modern data. Peckmann et al. (2007a) described a Moroccan Late Devonian occurrence of *Dzieduszyckia* unassociated with barite that they interpreted as being possibly cold seep related; evidence of petroleum seeping at the locality certainly indicates that the presence of this unique genus in abundance is not correlated 1:1 with barite.

Cold Seeps

The "cold" seep type of occurrences, unassociated with ore deposits of the massive sulfide type associated with ophiolites and mid-ocean ridge spreading centers, are widespread. Martin et al. (2007) outlined an isotopic method for using stable carbon isotopes from Foraminifera to distinguish seep-related forms from non-seep forms. A very comprehensive account of a methane-involved cold seep, using Hokkaido Late Cretaceous material, was provided by Jenkins et al. (2007a,b; see also Kaim et al., 2008) that involves consideration of the associated fossils and of varied isotopic data from the seep rocks, plus a useful discussion of cold-seep nutrition, as well as one limpet whose aperture is modified for life on vestimentiferan tubes.

von Bitter et al. (1990, 1992) discussed a Mississippian cold-seep possibility from Newfoundland. Possible cold-seep-type occurrences of tubes belonging to vestimentiferans have been reported from the Carboniferous of northeastern Siberia (Kuznetsov et al., 1994). Peckmann et al. (2001; see also Gischler et al., 2003) described a Visean, Early Carboniferous cold-seep-type occurrence with rhynchonellid brachiopods, a solemyid bivalve, and microbial carbonate from the Harz.

Peckmann et al. (1999) described French Jurassic and a Miocene Italian cold-seep deposit, with lithological, faunal, isotopic, and organic geochemical supporting data. Gaillard et al. (1992; see also Gaillard et al., 1985, and Rolin et al., 1990, for Late Jurassic from southeastern France) did the same for some French Mesozoic examples, and Gaillard and Rolin (1986) further discussed the southern France Jurassic examples in addition to interpreting the Pierre Shale Tepee Buttes Maastrichtian as a cold-seep example. Kauffman et al. (1996) provided a detailed account of the Campanian Tepee Butte situation. Campbell and Bottjer (1995b) discussed Early Cretaceous occurrences of the terebratuloid brachiopod *Peregrinella*, for which there is good cold-seep evidence, and reviewed many of the potential cold-seep brachiopod occurrences overall. Kiel (2008a) summarized the Cretaceous occurrences of *Peregrinella* while describing an occurrence from the Crimea that had a possibly parasitic worm tube, thought to be a serpulid polychaete, preserved on the interior of the shells. Campbell et al. (2008) described a gastropod from an Early Cretaceous California cold seep. Beauchamp et al. (1989; see also Beauchamp and Savard, 1992) described Aptian–Albian cold-seep-type faunas from the Canadian Arctic. Campbell et al. (1993) described Jurassic and Cretaceous potential cold-seep occurrences from California that involve varied megafossils. Campbell and Bottjer (1994) summarized some of the Mesozoic–Cenozoic brachiopod cold-seep possibilities, while Sandy and Campbell (1994) provided a detailed description of some Tithonian examples from California. Kiel et al. (2008) described gastropods from the California Great Valley Group and Franciscan. Kelly et al. (2000; see also Simon et al., 2000) described some Early Cretaceous, Northeast Greenland, mound-forming limestones with giant bivalve and associated invertebrate fossils that they interpret as cold-seep organisms. Hikida et al. (2002) discussed a Late Cretaceous cold-seep-type bivalve occurrence from Hokkaido, with *Calyptogena*. Gómez-Pérez (2003) discussed an Early Jurassic deep-water stromatolite bioherm interpreted as related to a methane seep. Hikida et al. (2003; see also Kanie and Kuramochi, 1996; Kanie et al., 1996) described in some detail a fossiliferous cold-seep community from Hokkaido, Japan, and also provided carbon isotopic information. Kaim et al. (2008a) described some cold-seep gastropods from a presumed methane-sourced seep of Late Cretaceous age from Hokkaido.

Squires and Goedert (1991; see also Squires and Goedert, 1996) have described the gutless bivalve *Calyptogena* from a methane seep type of deposit in the late Eocene of Washington. Saul et al. (1996) described an Eocene cold seep from western Washington that features a cryptic lucinoid(?) bivalve in an area where other cold-seep-type faunas have been recognized. Goedert and Squires (1990) described an Eocene deep sea occurrence from Washington. Squires and Gring (1996) discussed a possible Eocene cold-seep bivalve fauna from California. Goedert and Kaler (1996) described a species of *Abyssochrysos* from a cold seep in the Middle Eocene of western Washington. Goedert et al. (2003) described a Late Eocene methane seep occurrence from western Washington. Kiel (2006) described Eocene and Oligocene cold-seep gastropods and bivalves from western Washington. Kiel also provided evidence of earlier predation by naticids and possibly by crabs. Kiel (2008a) described a gastropod from the Eocene of a Washington cold seep. Squires and Goedert (1995) described a chiton from Eocene and Oligocene cold-seep limestones on the Olympic Peninsula.

Peckmann et al. (2002) described some Oligocene occurrences from the southwestern Washington Lincoln Creek Formation, with cold-seep-type fossils plus carbon isotope and organic geochemical data confirming a cold-seep methane source for nutrients. Goedert and Campbell (1995) and Squires and Goedert (1996; see also Rigby and Goedert, 1996, for sponges from the same occurrence) described an Early Oligocene cold-seep fauna from the Olympic Peninsula of Washington, with repetition of many of the same taxa encountered elsewhere in extant and fossil communities. Goedert et al. (2000) described a probable cold-seep occurrence of tube worm tubes from the Early Oligocene of western Washington. Goedert and Squires (1993) described *Calyptogena* from the Washington Oligocene, and Goedert and Benham (1999) described *Depressigyra*? from cold-seep carbonates in the Eocene and Oligocene of western Washington.

Kiel and Peckmann (2007) have described four Cenozoic cold-seep occurrences using paleontological and isotopic data from the Late Eocene of California, the Oligocene of Cuba, the Oligocene of Colombia, and the Oligocene and Eocene of Peru. Campbell (1992) discussed cold-seep bivalves from the Miocene–Pliocene of Washington. Peckmann et al. (1999) described a Miocene Italian cold-seep deposit with lithological, faunal, isotopic, and organic geochemical supporting data. The probable cold-seep, methane-based *calcari a Lucina* from the Italian Miocene have been discussed at some length by Taviani (1994), Terzi et al. (1994), and Ricci Lucchi and Vai (1994). Nobuhara (2002) briefly described some Pliocene *Calyptogena* communities from central Japan that are concluded to represent seep-type conditions.

Nobuhara (2003) described a Pliocene Japanese cold-seep carbonate mound with *Calyptogena*. Majima et al. (2005) summarized almost 100 cold-seep occurrences from Japan ranging in age from Albian to Quaternary. Kiel and Little (2006) provided a summary of the stratigraphic ranges of Jurassic and younger cold-seep mollusks. Undescribed crustacean material has previously been cited from cold seeps, but Schweitzer and Feldmann (1008) have described an Eocene galatheiid genus (*Shinkaia*) from a Washington State seep that was previously only known from the present, chiefly from seeps.

WHALE-FALL COMMUNITIES

Smith and Baco (2003) provided a comprehensive account of currently available information about whale falls, all of the taxa involved in recycling whale carcasses, succession, evolutionary questions, and biogeographic questions. Pyenson and Haasl (2007) summarized much of the whale-fall information, including the fact that vesicimyid bivalves are not associated with whale falls until the Miocene, with whale-fall evidence first appearing in the late Eocene, and they described a small Miocene whale fall with vesicomyid bivalves from California. Amano and Little (2005) described a Miocene Japanese "whale-fall" occurrence with some cold-seep-type bivalves that suggests that similar environments were involved, although the nutrients originated in whale bone rather than from a seep. Kiel and Goedert (2006b) summarized information from Washington State, including "Eocene–Oligocene" whale falls. Hachiya (1993) discussed a Japanese Miocene mysticete vertebra closely associated with articulated valves of *Calyptogena* and *Lucinoma* as a similar example. Amano et al. (2007) described another Japanese Miocene whale-fall occurrence involving varied whale bones. Hardly in the whale-fall category, but still of interest, is Marshall's (1994) citation of a New Zealand Middle Eocene occurrence of turtle bones with deep sea gastropods. Hogler (1994) speculated about the many possibilities for reptile and even fish falls in the Mesozoic. Kaim et al. (2008b) documented the occurrence of varied evidence including specialized mollusca and evidence of bacterial activity involving Japanese Cretaceous plesiosaur bones, with the question arising whether this is an example of convergence or of a pre-Cenozoic earlier appearance of the activities. Amano and Kiel (2006) critically reviewed the stratigraphic occurrences of vesicomyid bivalves beginning in the Cretaceous while critically considering those that are and are not associated with whale-fall occurrences.

WOOD-FALL COMMUNITIES

Kiel (2006) provided evidence that whale-wood-substrate-adapted mollusks have evolved separately from the cold-seep taxa, beginning at least in the earlier Tertiary. Kiel and Goedert (2006a) described the fauna from a Late Eocene, Washington wood-fall community that includes several obligate xylophages. Kiel et al. (2009) described a Japanese Late Cretaceous wood-fall, deep-water community while summarizing the overall aspect of the wood-fall associations at present and through time, as well as discussing the taxonomic overlap with other vent and whale-fall communities.

Lindberg and Hedegaard (1996) described a Late Oligocene relation involving the limpet *Pectinodonta*, from Washington State, mentioning that the other known fossil examples involve Miocene occurrences in New Zealand (Marshall, 1985a), and they explored the problems of larval substrate selection involved here. Marshall (1985a) provided excellent evidence from the present that *Pectinodonta* feeds exclusively on wood, based on living specimens attached to blocks of waterlogged wood and of wood fragments in the guts of live specimens. Marshall (1985b) described Miocene and modern specimens of additional marine limpet groups that are presumably wood feeders.

Steineck et al.'s (1990) discussion of deep sea xylophile ostracoda, including a few Tertiary examples, illustrates still again the potential utility of waterlogged wood as a food supply in the seas. Kaim (2006) briefly discussed a Middle Jurassic Polish wood fall closely associated with a variety of mollusks unknown in younger wood-fall communities—very puzzling!

□ *Reliability*
Category 1, frozen behavior, for those limpets found in place on wood, and Category 2B, functional morphology, for those found in a detached manner.

SUMMARY

In all of these examples, there is the strong implication that gutless tube worms and a bivalve have existed in a coevolved condition with symbiotic bacteria of the type discussed by Cary et al. (1993; see also Hentschel and Felbeck, 1993) for a lengthy time interval. The circumstantial evidence provided by the fossil record suggests that the symbiotic relationship has persisted unchanged for lengthy time intervals. This last possibility in turn raises the question about how much time might be involved before a symbiotic relationship would transform into one where the symbiont became an intracellular organelle.

Campbell and Bottjer (1995a) briefly summarized the fossil record of varied hydrothermal and cold-seep communities, including very reliable to unreliable occurrences. Cook and Stokes (1995) described modern vent situations involving tubes belonging to several taxa, while making it clear that when dealing with fossil tubes one may also be dealing with a number of taxa that cannot be distinguished from their tubes alone. In all of the above, from the Silurian to the present, the varied tubes and bivalves occurring in these unique environments represent a very unique type of evidence for life in a most peculiar environment. When coevolution with symbiotic bacteria is a requirement for their existence as a result of the loss of the gut, there is also a good possibility for convergence. Specifically, the very simple morphology of the "worm" tubes, ranging in age from Silurian to present, raises the possibility that there have been repeated adaptive radiations into the unique environments by unrelated, tube-building organisms, rather than this being an example of a Silurian to present phyletic set of organisms. Because of the simple tube morphologies, this is an almost impossible question to investigate with currently available evidence and techniques, although Little et al. (2004) summarized varied evidence of a molecular nature bearing on the problem but considered it not very conclusive. Still, whether taken as evidence of repeated convergence or of phyletic evolution

for almost 400 million years, it is a remarkable example of symbiosis.

Smith et al. (1989) provided evidence suggesting that the dispersal of both warm- and cold-vent faunas may be via "islands" provided by large vertebrate carcasses, using some whale examples from the present in which the decaying environment of the whale has a lot in common with the vent environments. Nesbitt (2005) has reviewed the entire whale-carcass-coevolved faunal question in some detail. Newman (1985) has discussed the questions concerning the ages of the unique vent fauna taxa, concluding that many of them probably date back well into the Mesozoic, and some even into the Paleozoic. This is clearly an area where additional data will be valuable in firming up any conclusions; it is entirely reasonable that some of these curious taxa are very old. Waren and Bouchet (2001), while describing some modern vent and seep gastropods and a monoplacophoran, suggested after reviewing evidence that the modern forms are consistent with a continental-shelf origin, from which they next went "down" to varied seeps before ultimately ending up in the vent fauna. This carries the implication that vents and seeps are not and never have been a source for shelf faunas nor much of a relict refuge over time. Luther et al. (2001) provided data from modern vents suggesting that the distribution about a vent of varied taxa is probably determined by chemical variables, suggesting that in well-preserved vent localities the distribution of the organisms may indicate the distribution of similar chemical variables.

□ *Reliability*
Category 2B, functional morphology, as the symbiotic bacteria have no fossil record.

*3h. CORAL–BRYOZOAN ASSOCIATION

Cadée and McKinney (1994) described a probably mutualistic association involving European Miocene, Pliocene, and modern corals (*Culicia*) and a cheilostome (*Celleporaria*).

□ *Reliability*
Category 2B, functional morphology.

*3i. *BACILLUS*–BEE RELATIONSHIP

Cano et al. (1994) recovered *Bacillus* DNA from Dominican amber stingless bee abdomens having characters similar to those of *Bacillus* fauna known today from the guts of bees that are involved in antibiotic production to preserve food reserves. Cano and Borucki (1995) have actually colonized bacterial spores from similar stingless bees in Dominican amber.

□ *Reliability*
Category 2B, functional morphology, for chemical similarity for the DNA, and Category 1, frozen behavior, for the spores.

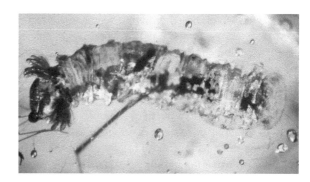

FIGURE 11 *Theope* caterpillar in Dominican amber (Poinar amber collection accession no. L-3-25); length of caterpillar = 5.8 mm.

FIGURE 10 A queen *Brachymyrmex* ant carrying a scale insect in Dominican amber (Poinar amber collection accession no. Sy-10-99); length of ant = 2.3 mm.

*3j.   TEREBRATULOID–BRYOZOAN RELATIONSHIP

Richardson (1987) described a mutualistic relation between the terebratuloid brachiopod *Aulites* and the bryozoan *Selenaria* in which the bryozoan (which has the capability of moving about over the substrate) forms a conical colony beneath which the small terebratuloid is attached by its pedicle. The bryozoan occurs as far back as the late Oligocene, and although the brachiopod is unknown as a fossil its small size might make it easily overlooked.

☐ *Reliability*
Category 6, speculative due to the absence of the brachiopods in the fossil record, associated with the bryozoan.

*3k.   ANT AND SYMBIOTIC SCALE INSECT

Poinar and Poinar (1994, p. 180) illustrated an ant carrying a scale insect (coccid) in its mandibles (Figure 10), in this case a relationship in which the ants farm the scale insects for their sugary exudate; the specimen is from Dominican amber, and the relationship is still extant. Wappler and Ben-Dov (2008) described Messel Eocene angiosperm leaves with attached remnants of female scale insects, Diaspididae, and varied nymphal stages, as well, in their discussion of the fossil occurrences of the group.

☐ *Reliability*
Category 1, frozen behavior.

*3l.   RIODINIDAE BUTTERFLY–ANT SYMBIOSIS

Poinar (1993, Figure 8) illustrated and described a Dominican amber caterpillar similar to those that today have a coevolved relation with certain ants that obtain secretions from paired abdominal glands on the caterpillar's eighth segment and in turn protect the caterpillar from enemies (Figure 11 and

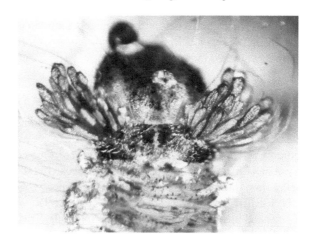

FIGURE 12 Head of Dominican amber *Theope* larva with protruding balloon setae (Poinar amber collection accession no. L-3-25); width of head = 1 mm.

Figure 12). This caterpillar was described later as a member of the genus *Theope* in the family Riodinidae (DeVries and Poinar, 1997). The fossil caterpillar possesses balloon setae and vibratory papillae characteristic of riodinids that possess ant symbionts. This most unusual soft-body preservation, despite the absence of the ants, leaves little doubt about the existence of the coevolved relationship between the butterfly and the ants. Hall et al. (2004), while describing five female riodinids from Dominican amber, remarked that the larvae of this group feed exclusively on the leaves of epiphytic bromeliads and orchids; that is, the presence in the amber of only females makes good sense in terms of ovipositing on leaves of these epiphytes. The orchid possibility underlines the overall rarity of fossil orchids which is puzzling in view of their widespread modern occurrence although undoubtedly related to their herbaceous as contrasted with woody morphology and exceptionally low pollen production (see Stebich, 2006, for an account of fossil occurrences).

Mehl (1986) reviewed the fossil records of the orchids, concluding that the only reliable records begin in the later Miocene. Schmid and Schmid (1977) reviewed the fossil record of the orchids but were skeptical about alleged Eocene and younger occurrences, as was also the case with Arditti (1992), who did not cite Mehl's 1986 evidence, possibly being unaware of its existence. The orchids as a whole are one of the best examples of the rarity of herbaceous plants as fossils, due, in their case, to exceptionally small pollen production in addition to the herbaceous morphology. Probably the best evidence for orchids back in time are provided by such things as the riodinid larvae mentioned here as well as the orchid bees cited later, both of which suggest that at least a good Miocene record for the orchids, if not significantly earlier. Ramírez et al. (2007) described a Dominican amber stingless bee with pollinarium attached to its body, providing very positive evidence for the presence of Miocene or older orchids, depending on the age of the Dominican amber. Molecular evidence also suggested a Late Cretaceous origin of the Orchidaceae.

☐ *Reliability*
Category 2B, functional morphology, which in this example is so well developed that there is little doubt about the conclusions.

*3m. ACARODOMATIA

O'Dowd et al. (1991) described mite domatia from Victorian Australian Eocene angiosperm leaves. Such acarodomatia are small retreats on leaf undersides, some roofed over with spines grown by the leaf, where mites dwell. This is a plant–animal mutualism in which the mites feed on fungi, lichens, or animals that in turn have been using the leaf as a food source. Elaeocarpacean and lauracean types of leaves were recognized at the Australian site. Oribatid mite fragments were also identified at the acarodomatia site. Acarodomatia are widely distributed among the living angiosperms, and one should expect that many more fossil occurrences of this phenomenon will be recognized. Colleague Fred R. Rickson kindly suggested the following selected papers dealing with the varied mite domatia occurring on modern plants: Chevalier and Chesnai (1941), Greensill (1902), Hamilton (1896), Jacobs (1966), Lundstrom (1887), O'Dowd and Willson (1989), Pemberton and Turner (1989), Sampson and McLean (1965), and Shirley and Lambert (1922).

☐ *Reliability*
Category 2A, functional morphology and closely associated animal parts.

*3n. LICHENS

Lichens are commonly thought of as mutualistic associations between varied fungal taxa and an alga, including either cyanobacterial taxa or green algae. However, Ahmadjian (1995) concluded that the relationship may be purely parasitic, with the fungi contributing nothing to the welfare of the autotroph. Taylor et al. (1997) described the allegedly first fossil lichen from the mid-Early Devonian Rhynie chert in cellular detail, with the implication that lichens may well have been present in much older strata due to the very

unusual preservational circumstances of the Rhynie chert, but Poinar et al. (2000b), while describing some lichens in Dominican amber, questioned the nature of the Devonian association. Anzoid lichens occur on the canopy and lower branches as well as the trunks of trees throughout the tropics and subtropics. The Baltic amber *Anzia electra* probably grew in a similar habitat (Rikkinen and Poinar, 2002). The fossil demonstrates that the morphological features of the *Anzia* thallus have remained stable for millions of years. The fossil also provides evidence that the disjunct Laurasian distributions of some modern lichens represent relicts of a formerly wider range.

*3o. HERBIVORY AMONG TETRAPODS

Sues (2000) provided a number of papers reviewing the skeletal evidence favoring the many independent, convergent, adaptive radiations from carnivorous ancestors involving herbivory among tetrapods. These papers carefully reviewed the skeletal evidence, largely dental, but also involving overall size and possession of a more barrel-shaped or elongate form characteristic of herbivores. The latest contribution to this convergent phenomenon is Kirkland et al.'s (2005) account of the Early Cretaceous primitive therizinosauroid from Utah that was well on its way to herbivory, while Fastovsky and Smith (2004; see also Kobayashi et al., 1999) suggest that the edentulous condition of ornithomimids indicates herbivory. Also reviewed are the many possibilities available regarding the coevolution of these herbivore types with microorganisms, protists, and bacteria in their guts to make fermentation of otherwise digestion-resistant plant material possible. There is, unfortunately, little in the way of preserved gut contents available, but a few exceptions are mentioned in this treatment. Rybczinski and Reicz (2001) provided good functional morphological data on a Late Permian herbivore equipped for eating fibrous plant material. A similar work (Mazin and de Buffrenil, 2001) deals extensively and in some detail with the convergent adaptations of tetrapods for life in water. Caldwell (2003), in a similar vein, reviewed those vertebrates in which extreme axial elongation has taken place, another story of convergence.

☐ *Reliability*
Category 2B, functional morphology.

*3p. AZOLLA–ANABAENA SYMBIOSIS

The water fern *Azolla* is first recorded from the Early to mid-Cretaceous, chiefly as spores. It has a symbiotic relation today, and presumably in the past, with the cyanobacterium *Anabaena*. Vajda (2005) cited a Bolivian spore from the Maastrichtian.

☐ *Reliability*
Category 2B, functional morphology.

*3q.   LUMINESCENT FISHES

(See also Section 7, Communication.) Various marine fish belonging to the Monocentridae (Paleocene to Recent) (Carroll, 1988) possess luminescent organs containing symbiotic light-generating bacteria. McFall-Ngai (1991) reviewed the subject, including the varied families of fish that have bacterial symbionts and the internal organs that contain the symbionts. Her summary indicated that, within some families, all or most genera have bacterial symbionts, whereas with others this is not the case. She made it clear that the varied bacterial taxa also occur freeliving in the sea and presumably "inoculate" the larval fish. She also indicated that all of the families with luminous organs today have a fossil record extending only into the earlier Cenozoic, with the exception of one Cretaceous example, and involve higher teleost families only. Not all luminescent fish, however, involve symbiotic bacteria. Haygood (1993) reviewed the topic, with more emphasis placed on the nature of the bacterial taxa involved, while pointing out that many fishes are self-luminescent.

□ Reliability

Category 2B, functional morphology, based on the assumption that the modern examples with symbionts situated in specific internal organs represent familial and generic taxa that presumably had them in the Cenozoic and on one example going back to the Cretaceous.

*3r.   FOSSIL FLATUS: INDIRECT EVIDENCE OF INTESTINAL MICROBES

by George O. Poinar, Jr.

Flatus, commonly known as farting, occurs when gases generated in the stomach or intestine are released from the body through the rectum. It is a common act, but few own up to it (good reason to have a pet in the house—the blame can always be placed on it). While embarrassing to both the perpetrator and the beholder, it is probably one of the most natural acts of not only humans but also all animals with a complete digestive tract. Although not offered as an excuse, rest assured that passing gas has been occurring for millions of years.

Because passed gases are colorless and dissipate fairly quickly, estimating their volume can be a challenge (unless you want to be scientific and hire a flatologist to measure them with a flatometer; see Bolin and Stanton, 1993). With humans and pets, evidence of flatus can be detected by sound and odor. A human produces from 400 to 2400 milliliters of gas per day. Using the average value (1.5 liters), the daily rate of human flatulence the world over would amount to the release of 9.9 billion liters of gas. For cows, each of which releases several hundred liters of methane (the gas that accounts for the majority of flatulence) per day, it is estimated that they are responsible for 60 million tons of this gas annually (Lewin, 1999). Among the insects, termites and locusts are probably the most prolific gas producers, accounting for some 5% of the total annual methane production (Lewin, 1999). Obviously there is no way to document historic evidence of

flatulence, except when it has been released in liquid resin and become fixed in the amber. Insects trapped in amber offer the only opportunity to document gases being expelled from the alimentary tract of ancient animals. Presented here is evidence that not only did insects produce flatus in the past but also, as with humans, these gases originated from the activities of gut microbes.

Rarely do gas bubbles in amber represent flatus. Most trapped gas bubbles are air incorporated into the resin as it flows from the tree and are responsible in large part for the different hues of amber, especially the cloudy type (Poinar, 1992b). Some air is introduced by insects when they land in or make attempts to escape the sticky resin. Such bubbles are frequently trapped under the wings or between the legs of the victims. Care was taken here to be certain that the gas bubbles represent true flatus and originated from the anus of the insect. Evidence that the bubbles are not from general decomposition is twofold: (1) all of the insects depicted here are well preserved and intact, showing no evidence of general decomposition, and (2) if the bubbles originated from general decomposition, they would be associated with other parts of the insect, especially intersegmental membranes and around the oral area.

As with vertebrates (Berg, 1996), gas production in the alimentary tract of insects results from the activity of microorganisms, especially bacteria (prokaryotes) and protozoa (protists). These organisms are in the insect's environment and most enter the digestive tract by way of the mouth. Some bacteria are pathogenic and kill the insect, while others are beneficial in that they serve as food, assist in digestion, supply the host with essential vitamins and nutrients, and even protect the insect from harmful disease-causing microbes (Dillon and Dillon, 2004). Many insects have anatomical modifications such as pouches and folds associated with their alimentary tract, all of which provide homes for various species of microorganisms, many of which are unique symbionts (Brooks, 1963). As with the vertebrates, most of the gases passed by insects probably consist of carbon dioxide, nitrogen, methane, and hydrogen (Berg, 1996). Some scientists claim to be able to identify the gases in amber bubbles. Although there is some debate about whether any gases are really ancient, have been replaced with today's atmospheric gases (Hopfenberg et al., 1988), or are the end products of coalification, analyses of gases in fossil flatulence might assist us in determining the past diet (e.g., fibers, protein) and type of microflora that were present.

All insects with an open alimentary tract certainly carry some microorganisms in their gut, but obtaining evidence of how long these associations have occurred would be impossible without a medium like amber. Based on current knowledge, gut microbes in termites and cockroaches that are vital to their survival would be expected to have been present in the past. By capturing the original bubbles as they emerged from the rectum, amber provides direct evidence of fossil flatus. Yet, even in amber, examples of flatus are rare because most insects die within seconds in the sticky resin, and their body openings are quickly sealed. Also, because gaseous

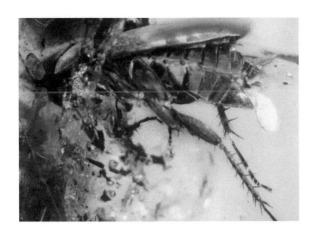

FIGURE 13   Gas bubble emerging from the rectum of a Dominican amber cockroach is evidence of gut microbes, probably bacteria (Poinar amber collection accession no. ICH-3-3A1); length of cockroach = 12 mm.

FIGURE 15   Gas bubble issuing from a Dominican amber worker ant is evidence of gut microbes probably acquired through its diet (Poinar amber collection accession no. ICH-3-3A3); length of ant = 3 mm.

FIGURE 14   Flatus from a Dominican amber midge could be from microbes acquired in its larval stage (Poinar amber collection accession no. ICH-3-3A2); length of midge = 1.2 mm.

FIGURE 16   Flying ants in Dominican amber carry microbes in their alimentary tracts (Poinar amber collection accession no. ICH-3-3A4); length of ant = 4 mm.

emissions are only passed sporadically (when a certain pressure is reached), the insect has to have reached the stage for releasing its gas just when it entered the resin. Finally, the resin has to be thin enough to allow the gas to emerge from the insect's gut.

In some large-bodied insects with enormous numbers of gut microbes, such as termites, the microorganisms, many of which are anaerobic, continue to metabolize after the insect is dead, forming large amounts of flatus. If the resin is thick or the anus is blocked, then the gases expand and stretch the insect's body wall, producing a flatulent condition. Bloated termites are not uncommon in amber.

Cockroaches are another group that probably produces copious amounts of flatus (Figure 13), as most eat decomposing organic matter and host a wide variety of bacteria inside their alimentary tracts. Some suggest that having contact with all these bacteria for so long has bestowed vigorous immune qualities on these cosmopolitan insects (Brooks, 1963). After all, they have survived very well for some 300 million years.

Size is not important in the production of flatulence, as even small midges, which survive only a few days, contain gut microbes that produce gas (Figure 14). Some of these microbes could have been obtained in the larval stage and

carried into the adult gut; however, because the adults normally feed on plant sugars, they could have picked up the microorganisms at that time.

Even ants produce flatus; yet, few scientists have bothered to investigate the microbes in the intestines of these intensively studied, highly social insects, and almost nothing is known about the organisms that could have produced the gases in the worker (Figure 15) and winged (Figure 16) ants shown here. Ants would be expected to ingest a wide range of microbes, especially bacteria, considering the wide range of food items they gather, even those that specialize mostly on plant secretions and honeydew. Some of these microbes apparently colonize the ant gut, which probably provides benefits to both partners. To date, only gut bacteria and fungi have been isolated from the gut of some pollen-feeding ants in South America (Caetano and Cruz-Landim, 1985).

The association between gut microbes and bees is probably extensive, although, again, little is known about the species of microbes involved and how they affect the host. We know that certain gut bacteria are beneficial to some

FIGURE 17   Gas passing out of the anus of a Dominican amber stingless bee is probably from species of *Bacillus* that survived after the bee was entombed (Poinar amber collection accession no. ICH-3-3A5); length of bee = 3 mm.

FIGURE 18   This Dominican amber flat-footed beetle is passing gas that probably was derived from yeast in its alimentary tract (Poinar amber collection accession no. ICH-3-3A6); length of beetle = 4 mm.

FIGURE 19   Blood-sucking insects like this female Baltic amber (40 mya) blackfly could be carrying microbes to assist in the digestion of its blood meal (Poinar amber collection accession no. ICH-3-3A7); length of blackfly = 3.5 mm.

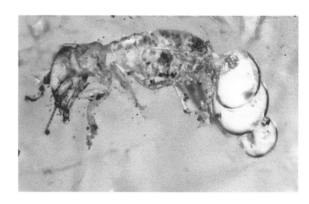

FIGURE 20   This Dominican amber worker termite shows three separate passings of flatus, each possibly originating from different kinds of gut microorganisms (Poinar amber collection accession no. ICH-3-3A8); length of termite = 7 mm.

stingless bees because the workers inoculate their stored supplies and nectar with species of *Bacillus* carried in their guts. The *Bacillus* colonize the food supplies and at the same time produce antibiotics that protect the food from spoilage (Gilliam et al., 1990). This behavior is somewhat similar to our practice of adding bacteria to milk for yogurt production when refrigeration is unavailable. The antiquity of this association was shown when *Bacillus* DNA was recovered from the gut contents of bees in Dominican amber (Cano et al., 1994). Thus, it is very likely that the gas being released from the rectum of the stingless bee shown in Figure 17 came from *Bacillus* cells that continued to survive after the bee became entombed.

Yeasts are another group of microbes that inhabit the gut of insects, especially wood-boring beetles. Flat-footed or platypodid beetles carry around yeasts in their rectum (Baker, 1963), and it is likely that the activity of such yeasts produced the flatus in the representative shown in Figure 18.

Blood-sucking insects often require some assistance in breaking down proteins in their meal (Marshall, 1987).

Also, because blood is deficient in B vitamins, these essential nutrients need to be obtained in some manner. Whereas some insects that feed on blood in both their larval and adult stages have gut microbes to supply these vitamins (Brecher and Wigglesworth, 1944), others such as mosquitoes and biting flies are not known to have such unique organisms (Lehane, 1991). Yet, the female blackfly showing flatus in Figure 19 must have some gut microbes, but the types and the role they perform is unknown today. Some present-day blackfly larvae contain bacteria that could be carried into the adult stage, and others have protozoa in their guts (Laird, 1981; Malone and Nolan, 1978; Wallace, 1966). Possibly, one of that group is the source of the fossil flatus. Adult blackflies also imbibe nectar and plant juices (Lehane, 1991), and microbes certainly could have been acquired from these sources.

As mentioned previously, all termites contain microbes in their hindguts, and these organisms produce copious supplies of gas (Abe et al., 2000) (Figure 20). Often, this active microbial community, with their complicated interactions, makes up over half the weight of the termite hindgut. Most of these microbes are made up of bacteria (prokaryotes) and protozoa (protists), which, as they digest components of the host's diet, supply carbon and nitrogen compounds, as well as other essential nutrients. The principal gases expelled from termites as a result of this mixture of microbes are carbon dioxide, methane, and hydrogen. There are some who claim that termites and locusts together produce about 5% of the total methane production in the world today (some 15% comes from cows) (Lewin, 1999). Occasionally there will be several confluent bubbles at the rear end of amber termites. These could represent flatus from separate groups of microorganisms; each surviving for different lengths of time after the termite became embedded. The last ones to perish would be the most stringent anaerobes.

Flatus in amber provides indirect evidence that microbes have existed in the alimentary tracts of insects for millions of years. It suggests that, as with mammals, these microbes are beneficial to their hosts. Certainly the gut flora represents one of the most basic types of associations between microbes and animals. From these chance encounters, more permanent associations could develop in the insect gut that benefited both partners and which could evolve into obligate relationships, such as the specialized pouches housing microbes within the insect gut. Obviously, more research is needed on microbial symbionts of insects before we can define the exact role of the organisms that produced the fossil flatus depicted here.

*3s. TERMITE AND COCKROACH GUT MUTUALISTS

Lower termites (Mastotermitidae, Kalotermitidae, and Rhinotermitidae) are well known for harboring a variety of gut microbes, including a number of protists. Some of these are flagellates, which have formed unique mutualistic associations with their hosts. Whereas the termite provides food and shelter for the protists, the flagellates digest cellulose in the wood that the termites ingest (Bignell, 2000; Honigberg, 1970). By the Early Cretaceous, when the earliest body fossils of termites appear (Rasnitsyn and Quicke, 2002), mutualistic associations with flagellates had already been established at least with some members of the Kalotermitidae. An Early Cretaceous *Kalotermes* adult preserved in Burmese amber contains a variety of trophic stages and cysts of flagellates

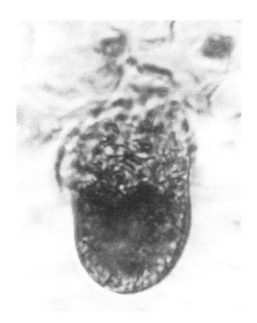

FIGURE 21 A trichonymphid flagellate attached to the gut intima of a kalotermitid termite in Burmese amber (Poinar amber collection accession no. B-I-2); length of protist = 49 µm.

(Figure 21), ciliates, and amoebae in its exposed abdomen. Included are specialized representatives of the Trichomonada, Hypermastigida, and Oxymonada that have mutualistic associations with lower termites today (Brugerolle and Lee, 2000a,b). Two Early Cretaceous Burmese cockroaches contained protists related to mutualistic flagellates occurring in extant *Cryptocercus* cockroaches and lower termites. These protists represent the earliest fossil record of mutualism between microorganisms and animals in the terrestrial environment (Poinar, 2009b,c).

☐ *Reliability*

Category 1, frozen behavior.

SUMMARY

Since 1990, additional examples of mutualism have accumulated, some of which display a level of fixity down to the family and genus levels. (See Table 2A.)

TABLE 2A

Mutualistic Relations with Their Stratigraphic Relations (Additions to Table 3 of Boucot, 1990a)

Behavioral Type	Taxonomic Level	Time Duration	Category	Refs.
3e. Pinnotherid crabs	Genus	Eocene–Holocene	2B	Glaessner (1969)
3f. Fig wasps	Family	Miocene–Holocene	2B	Poinar (1993)
3g. Tube worms, bivalves, and rhynchonellid brachiopods from deep-sea vents	Phyla	Silurian–Holocene	2B	See text.
3h. Coral–bryozoan association	Genus–genus	Miocene–Holocene	2B	Cadée and McKinney (1994)
3i. *Bacillus*–bee relationship	Genus–genus	Miocene–Holocene	1, 2B	Cano et al. (1994)
3j. Terebratuloid–bryozoan	Superfamily–genus	Oligocene–Holocene	1	Richardson (1987)
3k. Ant and symbiotic scale insect	Order–order	Miocene–Holocene	1	Poinar and Poinar (1983)
3l. Riodinidae butterfly–ant symbiosis	Genus–order	Miocene–Holocene	2B	DeVries and Poinar (1997)
3m. Acarodomatia	Order–kingdom	Eocene–Holocene	2A	O'Dowd et al. (1991)
3p. *Azolla–Anabaena* symbiosis	Genus–genus	Cretaceous–Holocene	2B	Vajda (2005)
3q. Luminescent fishes	Genus–family	Cretaceous– and Paleocene–Holocene	2B	McFall-Ngai (1991), Haygood (1993)
3r. Fossil flatus	Insect families	Tertiary	2B	Current work
3s. Termite gut mutualists	Termite families and orders	Early Cretaceous	1	Poinar (2009b)
3s. Cockroach gut mutualists	Cockroach families	Early Cretaceous	1	Poinar (2009c)

4 Host–Parasite and Host–Parasitoid Relationships and Disease

Shields et al. (2006, p. 146), in a treatment of lobster diseases, quoted Stewart (1980) as follows: "The lobster is no exception to the rule that all living things are subject to disease, and that each appears to have its own group of afflictions." The problem, of course, is that the diseases of most organisms are still unknown unless they have medical, veterinary, or economic importance. We have a long way to go in this area of science.

4A. ANIMAL–ANIMAL RELATIONSHIPS

4AI. MARINE

INVERTEBRATES

4AIa. Copepod–Fish

Boucot (1990a) provided a summary of the copepod–fish Brazilian Early Cretaceous relationship. Subsequently, Cressey and Boxshall (1989) provided a very extensive account of the anatomy of the parasitic copepod, and Huys and Boxshall (1991) reviewed the status of copepod taxonomy and phylogeny.

4AIb. Bopyrid Isopod–Decapod

Feldmann et al. (1993, Figure 27-4) illustrated a probable example of Late Cretaceous, Campanian bopyridism in the crab *Torynomma* from the James Ross Basin of the Antarctic Peninsula. Feldmann (1993) illustrated a Late Cretaceous specimen of *Torynomma flemingi* from New Zealand that

suffered bopyridism. Collins and Rasmussen (1992) cited a few bopyridized specimens of *Lyreidus succedanus* from the Campanian–Maastrichtian of West Greenland and *Lyreidus rosenkrantzi* (see their Figure 15A) from the Maastrichtian. Hopkins et al. (1999, Figure 3.6) described *Xanthosia wintoni* from the Late Cretaceous of Texas, with the appropriately inflated carapace margin in one specimen.

Morris and Collins (1991) described a bopyritized crab, *Portunus woodwardi*, from the Pliocene Lower Miri Formation of Sarawak. This example (based on 2 specimens out of 400; see Boucot, 1990a, p. 63) is of more than usual interest because of the paucity of previously described Tertiary, as contrasted with Jurassic and Cretaceous, examples; it has helped to provide more stratigraphic continuity. Tucker et al. (1994) reported bopyridism in a Late Miocene–Early Pliocene crab (*Speocarcinus berglundi*) from southern California, and Jakobsen and Collins (1997) described bopyridized *Protomunida monidoides* from the Danish Middle Danian, both of which add to the Tertiary evidence for continuity. Collins et al. (2003) described a possibly bopyridized crab (*Portunus woodwardi*) from the Sarawak Miocene. Müller (1984, Plate XXIII, Figure 4) illustrated a Miocene bopyridized *Petrolithes* and a *Pisidia* (Plate XXVII, Figure 5) from the Paratethys. (See Table 2B.)

☐ *Reliability*

Category 3.

TABLE 2B
Bopyrid Isopod–Decapod Examples

Decapod	Time	Refs.
Pithonoton, Mesogalathea	Jurassic	Glaessner (1969)
Mesogalathea, Galatheites	Late Jurassic	Housa (1963)
Pithonoton, Goniodromites	Late Jurassic	Housa (1963)
Pithonoton, Notoprosopon	Late Jurassic	Radwański (1972); Bachmayer (1955)
Notopocorystes	Cretaceous	Boucot (1990a, p. 63)
Palaeocorystes	Cretaceous	Bell (1858/1913)
Palaeastacus, Notopocorystes	Early Cretaceous	Glaessner (1969)
Torynomma	Late Cretaceous	Feldmann (1993)
Lyreidus	Late Cretaceous	Collins and Rasmussen (1992)
Xanthosia	Late Cretaceous	Hopkins et al. (1999)
Protomunida	Danian	Jakobsen and Collins (1997)
Branchioplax	Late Tertiary	Rathbun (1916)
Petrolithus, Pisidia	Miocene	Müller (1984)
Speocarcinus	Late Miocene–Early Pliocene	Tucker et al. (1994)
Portunus	Pliocene	Morris and Collins (1991)

4AIc.   Pearl–Bivalve–Ray–Trematode or –Cestode Relationship

Ranson (1959) provided a useful bibliography for many aspects of pearls, including the parasites of living forms. Bachmayer and Binder (1967; see also Hengsbach, 1990) provided entry into the German literature on fossil pearls. Thorne (1973) has commented on many of the previously published fossil pearl occurrences.

Liljedahl (1994) has described what appear to be among the oldest possible bivalve pearls from the Paleozoic, involving some Gotland Silurian *Nuculodonta* (see his Figure 28), while Kříž (1979, Plate XL, Figure 7) has described similarly "pearl-like" structures from a Late Silurian cardiolid bivalve. Their absence elsewhere in the Paleozoic is very puzzling. For example, the unusually rich, very well-preserved, silicified, taxonomically diverse Permian bivalves from the Glass Mountains of Texas have provided no evidence of pearls, nor have the well-preserved, diverse Permo–Carboniferous bivalves from other parts of the world. However, it is notable that post-Paleozoic pearls have been described almost exclusively from Europe and North America; little attention has been paid to their potential presence elsewhere. This peculiar absence of later Paleozoic evidence for pearls may indicate not only that the two Silurian occurrences are unique but also that the process involved (i.e., potential parasites) may be unrelated to the relatively abundant post-Paleozoic examples.

Ozanne and Harries (2002; see also Harries and Ozanne, 1998) described evidence of pearly shell material in Late Cretaceous inoceramids that may be the result of trematode-type parasitism. House (1960) described mound-like growths on the interiors of various Devonian goniatites that may represent pearls generated by some type of parasite.

The stratigraphic distribution of pearls, reliably reported from the Triassic on, with possible very rare Paleozoic occurrences, raises the possibility that the parasites possibly responsible for the relatively abundant post-Paleozoic pearls did not become involved with mollusks until the Triassic. Absence of pearls from modern and fossil brachiopods is notable. Newton (1908; for more recent literature, see Boucot, 1990a, and Ranson, 1959) ably summarized the earlier literature on the molluscan parasites involved with pearl growth. Oakley (1966) has given some detailed descriptions of phosphatic, pearl-like structures intimately associated with Silurian bryozoans from the United Kingdom and Gotland.

□ *Reliability*
Category 1, frozen behavior.

Post-Pliocene, Pre-Recent
   Modiolus modiolus, Scotland (Zilch, 1936)
   Mytilus edulis Sweden (Zilch, 1936)
   Anadara transversa, Maryland (Vokes, 1955)
   Arca transversa, Maryland (Brown, 1946b)
   Volsella modiolus, Scotland (Robertson, 1883)

Pliocene
   Lima scabra, Bahamas (Zilch, 1936)
   Ostrea edulis, England (Zilch, 1936)
   Panope floridana, Florida (Vokes, 1955)
   Melina zealandica, Kansas and New Zealand (Brown, 1940; Marwick, 1922)

Miocene
   Perna oblonga, Germany (Zilch, 1936)
   Perna maxillata, Austria (Zilch, 1936)
   Panopea americana, Maryland (Berry, 1936; Zilch, 1936)
   Isognomon maxillata, Maryland (Vokes, 1955), Austria (Hörnes, 1867)
   Glycymeris subovata, North Carolina (Vokes, 1955)
   Mulinia lateralis, North Carolina (Vokes, 1955)
   Mytilus trigonus?, Argentina (Frenguelli, 1937)
   Arius schafferi, Austria (Kümel, 1935)
   Oncophora socialis, Czech Republic (Ottnangian; Augusta and Remes, 1947, p. 130)
   Probable *Modiolaria sarmatica*, Czech Republic (Sarmatian; Augusta and Remes, 1947, p. 130)
   Mytilus, Austria (Bachmayer and Binder, 1967; Hengsbach, 1990)

Oligocene
   Perna heberti, Germany (Zilch, 1936; Geib, 1952)
   Perna sandbergeri, Germany (Zilch, 1936)
   Ostrea ventilabrum, Belgium (Zilch, 1936)

Eocene
   Pteria sp., Washington (Zilch, 1936)
   Pinna affinis?, England (Zilch, 1936)
   Ostrea tenera, England (Zilch, 1936)
   Ostrea cymbula, Belgium (Zilch, 1936)
   Ostrea sp., England (Zilch, 1936)
   Ostrea cymbula, France (Darteville, 1934)
   Ostrea ventilabrum, Belgium (Rutot, 1879)

Cretaceous
   Perna seeleyi, England (Zilch, 1936; Seeley, 1861)
   Inoceramus sp., England, a number of localities (Zilch, 1936)
   Inoceramus subundatus?, California (Zilch, 1936)
   Inoceramus labiatus, England (Zilch, 1936)
   Inoceramus goldfussianus, Germany (Zilch, 1936)
   Inoceramus sagensis quadrans, New Jersey (Zilch, 1936)
   Inoceramus expansus, Pondoland (Zilch, 1936)
   Inoceramus sp., Japan (Zilch, 1936)
   Exogyra texana, Texas, several localities (Adkins and Winton, 1919; Russell, 1929; Zilch, 1936)
   "*Inoceramus*," Kansas, a number of localities (Brown, 1940; Everhart, 2005b)
   Hippurites sp., no locality given (Klinghart, 1922, in Kirchner, 1927)
   Inoceramus expansus, England (Newton, 1908)

Inoceramus labiatus, England (Newton, 1908)

Inoceramus sagensis, New Jersey (Whitfield, 1885)

Inoceramus subundatus, California (Russell, 1929)

Plicotrigonia, Mongolia (Kolesnikov, 1973)

Platyceramus platinus, Kansas (Kauffman, 1990)

Inoceramus (Inoceramus) cuvieri, Texas (Kauffman, 1990)

Platyceramus or *Cladoceramus*, Texas (Kauffman, 1990).

Volviceramus grandis, Kansas (Kauffman, 1990)

Ostrea vesicularis, Holland (a cobble of Cretaceous aspect in Pleistocene beds; Anderson, 1946)

Jurassic

Gryphaea dilatata, England, several localities (Jackson, 1909; Morris, 1851; Zilch, 1936)

Ostrea eduliformis, Germany (Dorn, 1937)

Ammonite, Germany (Lehmann, 1990, pp. 193–194; see also Keupp, 1986)

Late Triassic

Megalodus amplus, Hungary (Kutassy, 1937)

Pleuronautilus pseudoplanilateralis, Timor (Kieslinger, 1926)

Ammonoid, Germany (Kirchner, 1927)

4Ald.   Sea Urchins and Parasitic Snails

Abdelhamid (1999) provided information on a number of Egyptian Cenomanian–Turonian echinoid tests with boreholes in various test sites. He suggested that many of these specimens are evidence of eulimid gastropod parasitism (see Jangoux, 1990, for discussion of parasitic eulimids), although he also commented on evidence for predation and disease possibilities. Alekseev and Endelman (1989) described some Late Cretaceous Mangyshlak and Usturt echinoid borings that may represent eulimid activity and reviewed other occurrences of echinoid damage. Cross and Rose (1994) discussed and described Late Cretaceous Chalk *Micraster* with evidence of possible gastropod parasitism and predation. Gibson and Watson (1989) described a number of borings into irregular echinoid tests from the Florida Eocene, but with most of these the boring agent is uncertain. Santos et al. (2003) described various Spanish Miocene clypeasteroid borings from the Seville region.

□ *Reliability*

Category 6, uncertain, because the identities of the potential parasites are unknown.

4Ale.   Sea Urchins and Parasitic Crustaceans

(See Figures 22 through 28.) Radwańska and Radwański (2005) thoroughly reviewed, with various new evidence, the different types of copepod parasitism of various echinoids. Good data as far back as the Jurassic further support the presence of copepods as early as the Jurassic based on the presence of bopyrid parasites affecting Jurassic crabs (Boucot, 1990a) in a relationship that today involves a copepod in the parasitic life cycle. Their paper is truly groundbreaking in

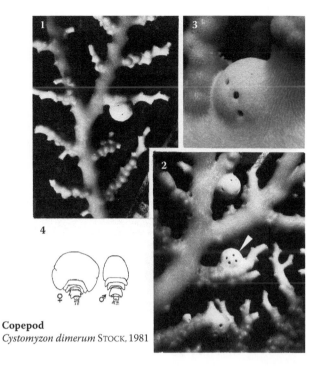

Copepod
Cystomyzon dimerum STOCK, 1981

FIGURE 22   Modern "Halloween pumpkin mask" copepod-generated cysts on the hydrocoral *Stylaster*. (1, 2) Cysts of the copepod *Cystomyzon* on *Stylaster* (×3); (3) same taxa (×8); (4) the copepod *Cystomyzon* associated with the cysts. Potentially copepod-generated cysts associated with modern hydrocorals. All are from Tagula Island, Papua New Guinea. (Figure 8 in Radwańska, U. and Radwański, A., *Acta Geologica Polonica*, 55, 109–130, 2005. Reproduced with permission.)

that they carefully described two distinctive Jurassic structures involving echinoids from France and Poland that are very similar to features of some modern copepods associated with hydrocorals (Figure 22), the copepods being parasitic at least in their earlier growth stages. The first echinoderm-associated exocysts were termed "Halloween pumpkin masks," whereas the second type were termed *Castexia*. Both forms are trace fossils, with the structures being generated by the echinoid in response to the associated organism. The authors also discussed some copepod-type "cysts" found on echinoid spines.

□ *Reliability*

Category 6, uncertain, due to the absence of any modern trace fossils associated with echinoids.

4Alf.   Myzostomid Annelids Parasitic on Crinoids

(See Figure 29.) The modern relations are summarized by Jangoux (1990; see also Eeckhaut and Améziane-Cominardi, 1994). Radwańska and Radwański (2005) thoroughly reviewed the evidence regarding myzostomid trace fossils on echinoids and arrived at the conclusion that the Paleozoic attributions are unproved, whereas the Jurassic to present are reasonable. Franzén (1974) described various Silurian and Devonian examples that differ significantly in

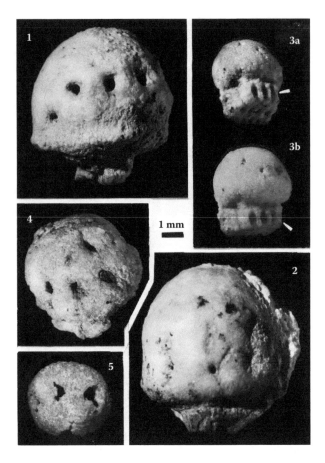

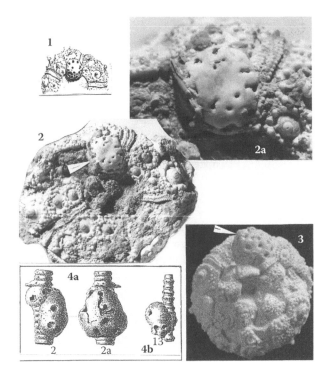

FIGURE 23 "Halloween pumpkin mask" cysts (×10) from the Kimmeridgian of the Holy Cross Mountains. (1, 2) Side views of larger cysts of loaf-like shape. (3a, 3b) Oblique lower and side views of a smaller cyst adhering to the ambulacral column of an echinoid test; note the ambulacral pores. (4, 5) Side views of smaller, spherical cysts. Potentially copepod-generated cysts from Jurassic echinoids. (Figure 7 in Radwańska, U. and Radwański, A., *Acta Geologica Polonica*, 55, 109–130, 2005. Reproduced with permission.)

FIGURE 24 "Halloween pumpkin mask" cysts on echinoids and crinoid columnals, Late Jurassic. (1) Cyst on a cidaroid (×0.5); (2, 2a) cysts on *Plegiocidaris* (×1.0, ×2.5); (3) cyst on *Plegiocidaris* (×1); (4a, 4b) cysts on columnals of *Millericrinus* (×0.5). Potentially copepod-generated cysts associated with modern hydrocorals and Jurassic echinoids. (Figure 5 in Radwańska, U. and Radwański, A., *Acta Geologica Polonica*, 55, 109–130, 2005. Reproduced with permission.)

their morphology from the post-Paleozoic material dealt with by Radwańska and Radwański (2005), who suggested that a trace fossil term is indicated, whereas for the morphologically distinct Paleozoic materials another trace fossil term is indicated.

MacKinnon and Biernat (1970) described an enigmatic tube-constructing organism (*Diorygma atrypophilia*) associated with a Polish Givetian atrypacean.

☐ *Reliability*

Category 6, uncertain.

☐ *Reliability*

Category 2B, functional morphology, for the post-Paleozoic forms, and Category 6, uncertain, for the Paleozoic forms.

4AIg.   Articulate Brachiopod Mantle Dwellers

Although not an articulate brachiopod, the description by Bassett et al. (2004) of a tubular structure in an Early Cambrian inarticulate brachiopod is worth noting as probably the oldest evidence for parasitism. Although an Inarticulate rather than an Articulate, *Linnarssonia constans* (Holmer et al., 2001, Plate 51, Figures 12 to 14) shows the presence of some type of skeletonized tube, presumably laid down by the mantle to wall off an intruding organism of some type or other. The identity of this intruder is, of course, unknown.

4AIh.   Graptolite Tubothecae

Although not involving tubothecae, Teller (1998) cited a number of "abnormalities" observed in some Silurian graptolites. Bates and Loydell (2000) carefully reviewed the scattered but extensive evidence regarding parasitism affecting graptolites, but the identities of the parasites are uncertain.

☐ *Reliability*

Category 6, uncertain.

4AIi.   Echinoid Spines–Gastropod

Boucot (1990a, pp. 74, 76) reproduced Tasnádi-Kubacska's (1962) earlier interpretation of bored cidaroid spines from the Cretaceous as evidence of parasitic gastropod activity; however, Neumann et al. (2007) have shown conclusively that this is not the case. They have shown that the cidaroid and similar borings into Cretaceous and Paleocene sponges are probably the work of sipunculid worms that made them for

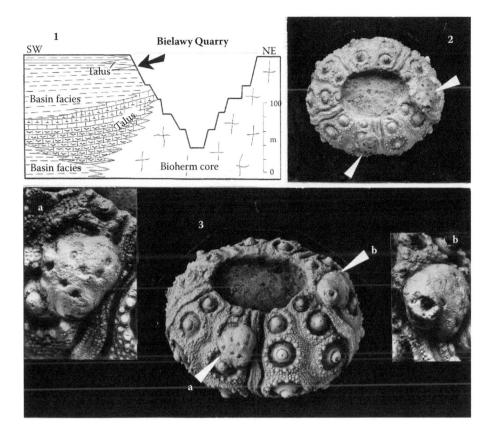

FIGURE 25 Cysts on a northwestern Central Polish *Plagiocidaris* (Bielawy, Upper Oxfordian). (1) Section indicating locality where material was obtained; (2) cyst-bearing *Plegiocidaris* (×0.5), with cyst locations indicated by arrows; (3) same specimen (×0.7); (3a, 3b) cysts (×1.5). Note the adoral location of the cysts. Potentially copepod-generated cysts from Jurassic echinoids. (Figure 6 in Radwańska, U. and Radwańki, A., *Acta Geologica Polonica*, 55, 109–130, 2005. Reproduced with permission.)

shelters from which they could exploit outside resources. The modern parasitic gastropod borings have an entirely different morphology from the sipunculid borings.

□ *Reliability*

Category 2B, functional morphology.

4AIj. Ziegler's Blisters

Brice and Hou Hong-fei (1992) described "blisters" from a Chinese Famennian cyrtospiriferid that they compared with the Ziegler's blisters described previously from Silurian pentameroid brachiopods (Boucot, 1990a, p. 76). The Chinese Late Devonian material may well represent an entirely distinct blister-generating organism, but the overall reaction in terms of blister formation is similar.

□ *Reliability*

Category 2B, functional morphology.

*4AIk. Bivalve–Trematode Pit-Forming Relationship

Ruiz and Lindberg (1989) described the relationship between a parasitic trematode that forms pits inside the shells of certain bivalves. Ruiz (1991) presented additional data indicating that the size of the parasitized host shells becomes progressively smaller through time, as though there is a possibility

that younger and younger individuals were becoming sexually mature to avoid being castrated parasitically. The earliest known examples are Eocene.

□ *Reliability*

Category 6, somewhat speculative.

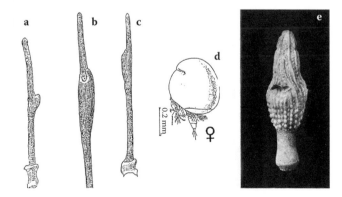

FIGURE 26 Echinoid spines parasitized by copepods. (a–c) Modern echinoid spines parasitized by (d) *Calvocheres*. (e) Oxfordian, Jurassic *Plegiocidaris* spine with trace fossil indicating copepod parasitism (×0.5). (Figure 9 in Radwańska, U. and Radwański, A., *Acta Geologica Polonica*, 55, 109–130, 2005. Reproduced with permission.)

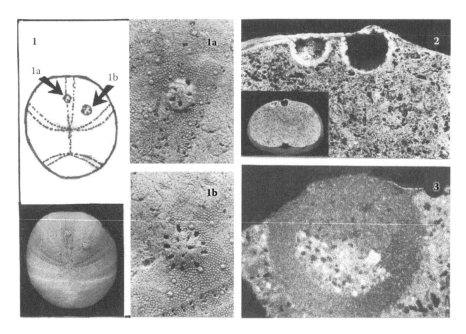

FIGURE 27 Probable Jurassic copepod cysts, *Castexia*, in the cyst-bearing echinoid *Collyrites* from the French Callovian. (1) Generalized view of *Collyrites* (×0.5) showing location of two cysts: (1a) juvenile cyst and (1b) adult cyst. (2) Section through two cysts to show that their walls are made of echinoid calcite (white) (×2.5) (insert is ×0.5). (3) Thin section under crossed nicols showing that the cyst wall (dark) and test of the echinoid are both made of calcite (×10). (Figure 11 in Radwańska, U. and Radwański, A., *Acta Geologica Polonica*, 55, 109–130, 2005. Reproduced with permission.)

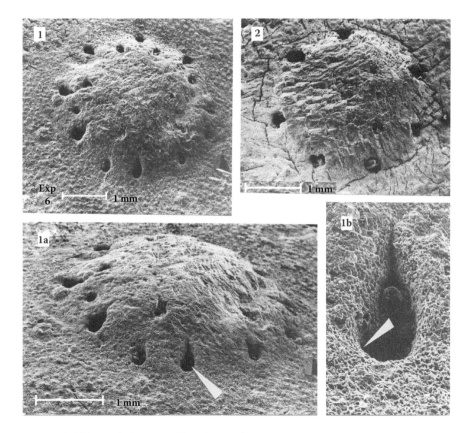

FIGURE 28 *Castexia* cysts with SEM on *Collyrites*. (1) Top view (×35), (1a) oblique view (×35), (1b) close-up of a cyst orifice. (2) Abraded cyst showing the calcite structure to be continuous from the echinoid test through the cyst. Note the continuous cleavage planes of the calcite. Potentially copepod-generated cysts associated with Jurassic echinoids. (Figure 12 in Radwańska, U. and Radwański, A., *Acta Geologica Polonica*, 55, 109–130, 2005. Reproduced with permission.)

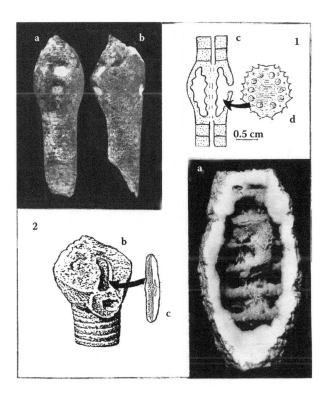

FIGURE 29 Myzostomid parasitism of Mesozoic crinoid stems. (1) Disk-shaped myzostomid infestation producing a gall-like swelling in the stem of *Millericrinus* from the Kimmeridgian of the Holy Cross Mountains; (1a) front view showing two apertures of the myzostomid cyst; (1b) side view showing the extent of the gall-like swelling; (1c) schematic section to show the myzostomid cyst chamber; and (1d) its suggested producer, the modern free-living *Myzostoma*. (2) Worm-shaped myzostomid infestation producing a tunnel and gall-like swelling in the stem of an Oxfordian *Millericrinus* from Central Poland; (2a) section of silicified gall-like swelling to show the preserved wall of the myzostomid tunnel; (2b) schematic section of gall-like swelling to show the myzostomid tunnel and its suggested producer; (2c) modern *Mesomyzostoma*, a potential producer. (Figure 3 in Radwańka, U. and Radwański, A., *Acta Geologica Polonica*, 55, 109–130, 2005. Reproduced with permission.)

*4All. Trilobite Swellings and Borings

Owen (1985), in his comprehensive review of trilobite abnormalities, noted that various swellings and borings may be the result of parasitic activities, with the identity of the potential parasites unknown. He also noted that most trilobite abnormalities, by analogy with living crustaceans, probably were generated during molting. Babcock and Robison (1989a,b) made a good case that there are significantly more predation-type healed scars on the right than on the left side of trilobites, a very enigmatic bit of information that indicates something about predator behavior. Rabano and Arbizu (1999) described additional examples of possible predation and parasitism in some Spanish Ordovician and Devonian trilobites.

☐ *Reliability*

Category 5A, due to maker uncertainty.

*4Alm. Octocoral and Ascothoracican Barnacle

Voigt (1959) described swellings around a cavity in a Maastrichtian octocoral that correspond in form to those made today by parasitic ascothoracican barnacles. Because ascothoracicans are soft bodied, one must rely on the form of the swelling and comparisons with living parasites using similar host substrates. Newman et al. (1969) accepted this ascothoracican example.

☐ *Reliability*

Category 2B, functional morphology.

*4Aln. Ammonoid and Belemnite Paleopathology

Hengsbach (1991a, 1996; see also Keupp, 1985; Kröger, 2002a,b; Monks, 2000; Morard, 2002) provided an excellent summary of the literature on ammonoid teratologies that has accumulated for well over a century (for evidence of damage to aptychi, see Engeser and Keupp, 2002, and Kröger, 2002b). Most of the examples are, as expected, from the Mesozoic, with few from the Paleozoic. Hengsbach pointed out how difficult it is to determine the causes involved with these various teratologies. Our knowledge, or rather ignorance, of disease in modern cephalopods (*Nautilus*, in particular) places us at a distinct disadvantage here. Hengsbach (1991b; see also Hengsbach, 1990) provided an additional, critical summary of many paleopathological problems, again emphasizing ammonoids. In this second paper, he emphasized the importance of recognizing deviance from the symmetrical condition normally present in many animals; ammonoids comprised the bulk of the examples provided, although belemnites, brachiopods, trilobites, and others are also cited. Henderson et al. (2002) discussed skeletal abnormalities in *Baculites* with the possibility of genetic defects being involved.

As always, the causes involved are very difficult to determine. Doguzhaeva (2002) described an Aptian example of skeletal pathology in very young ammonites that might have resulted from an increasingly anoxic environment, a unique example where causation can be better understood.

Merkt (1966) used epibiont oysters on a Sinemurian ammonite to show how the ammonite maintained a vertical position in the water column by twisting later developing parts of its conch to compensate for the tilting caused by the presence of the oysters on another part of the conch. All of this emphasizes the overall difficulty (Boucot, 1981) of trying to determine the causes of paleopathologies in extinct organisms as well as in the majority of living organisms, for which disease causes have been little studied (see also Hengsbach, 1990). Quenstedt (1858, Plate 21, Figure 12) illustrated a teratological Jurassic belemnite with a very distorted guard, but there is no causal information available.

☐ *Reliability*

Category 5A, due to the high level of uncertainty concerning the makers.

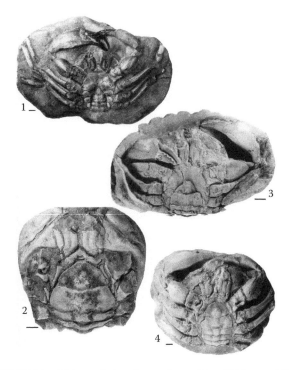

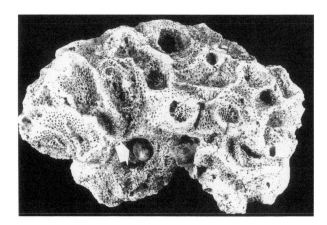

FIGURE 31 The parasitic coralliophilid gastropod *Galeopsis* attached to its hexacoral host from the Miocene of France; length = 112 mm. (From Lozouet, P. and Renard, P., *Géobios*, 31, 171–184, 1998. Reproduced with permission.)

FIGURE 30 Rhizocephalan-decapod parasitism. Feldmann (1998) documented morphological evidence in New Zealand Miocene crabs, *Tumidocarcinus giganteus*, for rhizocephalan parasitism. The evidence consists of the morphology of ventral views of a mature male (1), the abdomen of a mature female (2), the ventral view of an immature female (3), and the ventral view of a feminized male (4) that was parasitized by a rhizocephalan. Scale bar = 1 cm. (From Feldmann, R.M., *Journal of Paleontology*, 73, 493–498, 1998. Reproduced with permission.)

*4AIo. Rhizocephalan–Decapod Parasitism

(See Figure 30.) Feldmann (1998) discovered the first case of rhizocephalan parasitism in the fossil record; the Miocene xanthid crab *Tumidiocarcinus giganteus* from New Zealand was affected by parasitic castration, resulting in feminization of the adult male crab. Feldmann also cited a possible Late Cretaceous example from the Pierre Shale in South Dakota. The presence of feminized male crabs can only be reliably detected in the fossil record when one has a reasonably large sample that includes normal males and females as well as the feminized males.

☐ *Reliability*
Category 2B, functional morphology.

*4AIp. Parasitic Coralliophilidae (Gastropoda)

(See Figure 31 and Figure 32.) Lozouet and Renard (1998) described the fossil record of the parasitic Coralliophilidae with corals and provided an excellent introduction to the abundant literature on parasitic gastropods overall. Included are such novelties as the snails that parasitize certain reef fishes while they are sleeping, including parrotfishes (Bouchet, 1989; Bouchet and Perrine, 1996) and electric rays while they are partially buried in sediment waiting for prey (O'Sullivan et al., 1987). *Cancellaria cooperi* today sucks blood from the ray

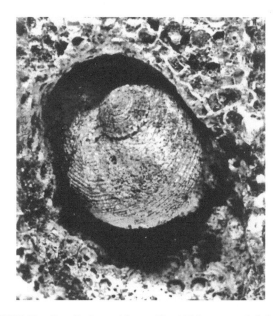

FIGURE 32 Detail of parasitic coralliophilid gastropod *Galeopsis* attached to its hexacoral host from the Miocene of France; enlarged view of area indicated by the arrow in Figure 31. (Figure 8 in Lozouet, P. and Renard, P., *Géobios*, 31, 171–184, 1998. Reproduced with permission.)

Torpedo californica; *Torpedo* has a fossil record as far back as the Paleocene (Cappetta, 1987, 1988). Also addressed are the more well-known pyramidellid parasites of mollusks and annelids (Bandel, 1994, p. 88), which date back as far as the Campanian. Earlier, Boucot (1990a) discussed several examples of eulimid parasitism on echinoids (Eulimidae; see Waren, 1980, 1981, 1983), including on an Albian echinoid from Texas and Chalk *Mucronalia*-type damage to cidaroid spines.

Lozouet and Renard (1998) reviewed the known fossil record of the Coralliophilidae, which extends back into the Oligocene of Aquitaine, and provided examples of the parasitic gastropod still preserved within its burrow within the colonial coral. Massin (1988) described the habit of these coral parasites to have side-by-side burrows within the coral colony, one for the male and the other for the dimorphic female. Fujioka and Yamazato (1983) reviewed the more or less obligate association between certain gastropod genera and certain hermatypic corals. Savazzi (1999d) described some modern coralliophilids, and Petuch (1994, Plate 49, Figure A, p. 288) described a coralliophilid gastropod, *Magilis streami*, from the Florida Pinecrest Beds of Pliocene age.

☐ *Reliability*

Category 1, frozen behavior.

*4Alq. Trapeziidae Crabs and Scleractinians

Knudsen (1967) provided a good account of the habits of the modern Indo-Pacific Trapeziidae crabs, which are symbiotic on acroporid and pocilloporid scleractinians in that they feed on mucus secreted by the corals using specialized "hairs" (a food brush) on some of their distal appendages (feeding setae on the pereiopods). Castro (1976) reviewed in more detail the feeding biology of the trapeziids. Schweitzer (2005b) provided an excellent account of overall trapezoid morphology, stratigraphic ranges, and feeding biology involving nutrient-rich mucus secreted by their coral hosts, as well as cleaning of the coral host, among other things, and their symbiotic relationships with various tropical/subtropical coral groups.

The oldest fossils have been described by Müller and Collins (1991) from Hungarian Late Eocene reef-coral-associated *Tetralia*; see also Müller (1975), who described Hungarian Late Eocene and Badenian Miocene trapeziids, and Schweitzer (2005b). Karasawa (1993) described Miocene Japanese *Trapezia*, and Müller (1984) described a Badenian, later Miocene *Trapezia* from the Paratethys that is coral associated. Garth (1964) described some of the coral-associated modern crabs (for an account of some modern Red Sea forms, see also Galil, 1986/1987). Even though the fossils have not been found on the host coral surfaces where both sexes live with their offspring, the close association (i.e., coral facies) is strong circumstantial evidence.

Glynn (1976) described the activity of *Trapezia* in protecting *Pocillopora* and other corals from predation by *Acanthaster* which provides more evidence of a symbiotic, coevolved relationship. Glynn's (1983) careful account of the protective activities of these crabs finds an apt analogy in the symbiotic ant–swollen thorn acacia relationship.

Here, again, we have another example of the unknown factors involved in some crustacean larvae living on the (Trapeziidae) hermatypic coral surface, where they mature, as do the coral barnacle cyprids that actually develop inside the coral without suffering from the cnidoblasts. Similarly puzzling are the lithophagid bivalves that bore into living colonial corals while suffering no predation, presumably having settled as larvae, as well as the coralliophilid gastropods.

The coral gall-forming crabs, the Hapalocarcinidae, have not yet been found as fossils, although one would predict that the galls inhabited by the crabs would be recognizable in the fossil condition.

☐ *Reliability*

Category 2B, functional morphology.

*4Alr. Hohlenkehle

Harries and Ozanne (1998) and Ozanne and Harries (2002) described the widespread occurrence of an inoceramid teratology, *Hohlenkehle*, that is present in Late Cretaceous shells and increases in frequency into the earlier Maastrichtian prior to the extinction of almost all inoceramid genera. This shell anomaly is thought to represent the activity of an unidentified parasite.

☐ *Reliability*

Category 2B, functional morphology.

*4Als. Abnormal Echinoid Plates

Kier (1957) illustrated and discussed an Eocene echinoid (*Linthias somaliensis*) from former British Somaliland with abnormal plates. Kier suggested that the abnormalities were due to injury of some type, but genetic defects might also be a possibility.

☐ *Reliability*

Category 7, due to ignorance of the cause.

*4Alt. Bald-Sea-Urchin Disease

Radwańska and Radwański (2005) summarized information about bald-sea-urchin disease in modern taxa (regular and irregular, geographically widespread, bacterial in origin) and described a Jurassic specimen whose eroded test is similar to the tests of modern diseased individuals. (For descriptions of the symptoms of the disease and related matters, see Jangoux, 1984, 1990; Maes and Jangoux, 1984; Schwammer, 1989.)

☐ *Reliability*

Category 2B, functional morphology.

*4Alu. Eulima–Holothurian

Bałuk and Radwański (1977) cited the gastropod *Eulima*, a common holothurian parasite (Jangoux, 1990), from the mid-Miocene Korytnica Clays. Gougerot (1969) cited the snail from the Eocene of the Paris Basin, but one needs to recall that eulimids also parasitize non-holothurian echinoderms (Jangoux, 1990).

☐ *Reliability*

Category 2B, functional morphology.

*4Alv. Foraminiferal Teratologies

Bradley (1956) illustrated and discussed a teratological fusuline foraminifer and its occurrences in the fossil record. Nothing is known about causation. Hageman and Kaesler (2002) discussed Permian fusulinid damage that they ascribed to predation rather than disease.

☐ *Reliability*

Category 7, due to ignorance of the cause.

*4AIw.   Shark–Isopod

Bowman (1971) described an Upper Cretaceous isopod from Texas that was associated with a shark skeleton in a relation that might have been either parasitic or scavenging.

☐ *Reliability*

Category 6, due to the level of uncertainty about the potential relation.

*4AIx.   Foraminifer–Echinoid

Neumann and Wisshak (2006) described the relationship between a trace fossil affecting the exterior skeleton of a Maastrichtian echinoid (*Echinocorys*), which they interpreted as being the work of a foraminiferan, with a modern analog.

☐ *Reliability*

Category 6, due to uncertainty about the identity of the trace maker.

Summary

Family-level examples and a few down to the generic level all emphasize overall behavioral fixity. The parasites involved are all probably metazoans, rather than protozoans, bacteria, or viral taxa. (See Table 3.)

VERTEBRATES

*4AIxa.   Avascular Necrosis

Although their examples do not involve organism-caused disease, Martin and Rothschild (1989; Rothschild and Martin, 1987) have suggested that the deep-diving disease known as the bends gave rise to skeletal avascular necrosis in allegedly deep-diving mosasaurs from the Late Cretaceous, as contrasted with shallower water species (Rothschild and Martin, 2005), as well as in some marine turtles of the Cretaceous and Tertiary. However, it must be noted that, in extant vertebrates, the bends, or caisson disease, is restricted to air breathers who continue to ventilate their lungs at depth (e.g., as with SCUBA divers). Rothschild (2006) found that among Cretaceous marine reptiles there was one group that were deep divers and showed evidence of avascular necrosis, while another group lacked such evidence and was presumably restricted to shallow water. Rothschild and Saunders (1997) reported that evidence of avascular necrosis is present in earlier Tertiary cetaceans but not in more recent and modern specimens. Rothschild (1987, 1991) reviewed the evidence of avascular necrosis in turtles and found that the deep divers show it now and back in time as far as the Cretaceous and to the Oligocene in some families and genera, whereas the shallow water forms do not. Rothschild and Storrs (2003) described avascular necrosis in plesiosaurs. Motani et al. (1999a, also Supplemental Information) have found evidence of avascular necrosis in ichthyosaurs, which is in accord with their unusually large eyeballs. Motani et al. (1999b) indicated that ichthyosaurs with the largest eyes also show evidence of avascular necrosis, suggesting that they were deep divers with the bends. Rothschild et al. (1999a) did not find evidence for avascular necrosis in Triassic ichthyosaurs, only in the post-Triassic, with the implication that deep diving was post-Triassic. However, Ruben (pers. comm., 2007) noted that the bends involves breathing at depth (i.e., under greater than atmospheric pressure), which releases nitrogen from body fat into other tissues, which eliminates the bends as a cause of avascular necrosis in deep-diving, air-breathing vertebrates. NIH Publication No. 06-4857 (2006) points out that osteonecrosis (avascular necrosis) can have many distinct causes other than caisson disease. It is clear that although the avascular necrosis of Martin and Rothschild correlates with deep diving there is a cause other than the bends.

TABLE 3
Marine, Animal–Animal Parasitic Relations Known from the Fossil Record (Additions to Table 4 of Boucot, 1990a)

Behavioral Type	Taxonomic Level	Time Duration	Category	Refs.
4AIk. Bivalve-trematode, pit-forming relationship (bivalve–trematode)	Genus–genus	Eocene–Holocene	6	Ruiz and Lindberg (1989)
4AIm. Octocoral and ascothoracican barnacle (octocoral–ascothoracican)	Order–order	Maastrichtian–Holocene	2B	Voigt (1959)
4AIo. Rhizocephalan–decapod parasitism (decapod subclass)	Order–order	Miocene–Holocene	2B	Feldmann (1998)
4AIp. Parasitic Coralliophilidae (Gastropoda) (Coralliophillidae–hexacorals)	Family–class	Oligocene–Holocene	2B	Lozouet and Renard (1998)
4AIq. Trapeziidae crabs and scleractinians	Family–class	Late Eocene–Holocene	2B	Müller (1975)
4AIr. *Hohlenkehle*	Genus–genus	Late Cretaceous	2B	Ozanne and Harries (2002)
4AIs. Abnormal echinoid plates	Class	Eocene	2B	Kier (1957)
4AIt. Bald-sea-urchin disease	Class	Jurassic–Holocene	2B	Radwańska and Radwański (2005)
4AIu. *Eulima*–holothurian	Genus–class	Miocene–Holocene	2B	Bałuk and Radwański (1977)
4AIv. Foraminiferal teratologies	Genera	Permian	7	Bradley (1956)
4AIw. Shark–isopod	Class–order	Late Cretaceous	6	Bowman (1971)

☐ *Reliability*
Category 1, frozen behavior.

*4AIxb. Dipnoan and Chondrichthyan Dentition and Jaw Injuries

Again, although not involving disease organisms, except for those causing caries, Kemp's (2001; see also Kemp, 2005) description of dipnoan tooth plate damage, jaw fractures, and other types of damage and repair is very impressive, with examples taken from the Devonian to present. Kemp (2003a) provided data on caries in Tertiary and Devonian lungfish from various Australian localities as well as abcesses in lungfish jaws and other, less common conditions. Kemp (2003b) reviewed Devonian to Holocene evidence for tooth plate anomalies that may be genetic. Tanke and Rothschild (1997) pointed out the obvious—that caries is not to be expected in most nonmammalian vertebrates where rapid tooth replacement was widespread, although this is not true for broken teeth. However, Duffin (1993, Figure 3, pp. 15–16) found a badly deformed Late Triassic tooth of the chondrichthyan *Rhomphaiodon* from northeastern France and cited another such occurrence from Germany; he ascribed the pathology to damage during feeding.

☐ *Reliability*
Category 1, frozen behavior.

*4AIxc. Schmorl's Nodes

Hopley (2001) described the teratology known as Schmorl's nodes from the vertebrae of an English Lias plesiosaur. This feature has previously been recognized only in humans; its cause is uncertain, although there is some spinal disk involvement with the actual vertebrae. Rothschild (2003b) reported Schmorl's nodes in Late Pleistocene mastodons from New York.

☐ *Reliability*
Category 6, due to a high level of uncertainty about causation.

*4AIxd. Osteoporosis in Pleistocene Deer from Crete

Sondaar (1976) makes a good case for endemic Pleistocene deer populations from Crete having been affected by osteoporosis due to conditions of relative starvation.

☐ *Reliability*
Category 2B, functional morphology.

*4AIxe. Pterodactyloid Pathologies

Bennett (2003) reviewed pathologic evidence occurring in large pterodactyloid pterosaurs, with emphasis on arthritic conditions, fractures, and various healed conditions.

☐ *Reliability*
Category 2B, functional morphology.

4AII. FRESHWATER

4AIIa. Unionids, Actinopterygia, and Glochidia

Bauer and Wächtler (2001) provided reviews of just about all aspects of the glochidial–unionid relationship, including the fossil record, primarily for North America. Araujo and Ramos (2001) discussed the relation between *Margaritifera auricularia* and *Acipenser sturio*, including a Pleistocene occurrence; the markedly decreasing abundance of the bivalve is an excellent example of the potential extinction of an obligate parasite together with its host.

☐ *Reliability*
Category 2B, functional morphology.

*4AIIb. Ilial Frog Tumors

Tyler et al. (1994) described the similarity of some Mio–Oligocene tumors in frog ilia to modern examples of the same phenomenon. The causal agent is suspected to have been an environmental factor.

☐ *Reliability*
Category 2B, functional morphology.

*4AIIc. Late Devonian Fish

Upeniece (1999, 2001) discussed and described circlets of hooks closely associated with Late Devonian Latvian placoderms and acanthodians, as well as some crustaceans. These represent parasitic helminths of some type and are the oldest known vertebrate parasites. (See Table 4.)

☐ *Reliability*
Category 1, frozen behavior.

TABLE 4
Freshwater Animal–Animal Parasitic Relationships Known from the Fossil Record (Additions to Table 5 of Boucot, 1990a)

Behavioral Type	Taxonomic Level	Time Duration	Category	Refs.
4AIIb. Ilial frog tumors	Anura–unknown	Mio–Oligocene–Holocene	2B	Tyler et al. (1994)
4AIIc. Late Devonian fish–helminths (placoderms, acanthodians, helminths)	Class–phylum	Late Devonian–Holocene	1	Upeniece (1999, 2001)

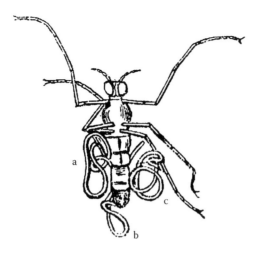

FIGURE 33 Menge's drawing of a midge in Baltic amber harboring three mermithid nematodes. (Figure 7 in Menge, A., *Schriften der Naturforschung Gesellschaft in Danzig*, 1, 1–8, 1866. Reproduced with permission of the LuEsther T. Mertz Library, The New York Botanical Garden.)

FIGURE 34 The extant mermithid nematode *Gastromermis anisotis*, partly emerged from its midge host. Collected at Sagehen Creek, California; length of midge = 6.3 mm.

4AIII. TERRESTRIAL

4AIIIa. Nematode–Planthopper, Nematode–Dipteran, and Nematode–Ant

Poinar (2003a) reviewed the various nematode–insect relationships known at the time and commented on their evolutionary significance. Nematode parasitism of chironomid midges is documented in Cretaceous and Tertiary amber; the first fossil record was reported by Menge (1863) in Baltic amber (Figure 33). His Figure 7 shows three separate mermithid nematodes partially emerged from an adult midge, and they are not that different in appearance from a new specimen in Baltic amber (see the back cover, Figure C). Such associations persist today in many parts of the world (Figure 34).

Poinar (1992b) illustrated a mermithid nematode (family Mermithidae) emerging from a host midge (family Chironomidae) in Dominican amber (his Figure 138) and juvenile allantonematid nematodes (family Allantonematidae) emerging from an infected female drosophilid fly, *Chymomyza primaeva* (family Drosphilidae) (his Figure 139). Poinar (1992b, p. 250) also cited insect parasitic nematodes, Iotonchiidae (Tylenchida), parasitizing mycetophilid flies from Dominican amber. Mermithid nematodes have also been found associated with ants in Dominican amber (Poinar et al., 2006b) (Figure 35). Poinar et al. (1994b; for taxonomic revisions of both host and parasite, see Poinar and Milki, 2001, pp. 32–33) described a mermithid nematode parasitizing an Early Cretaceous, Lebanese amber chironomid (the host was originally mistaken as a ceratopogonid); this is one of the oldest known examples of nonmarine animal–animal parasitism from the fossil record. A second, slightly later example in Burmese amber showed a mermithid emerging from the body cavity of a definite biting midge (Ceratopogonidae) (Poinar and Buckley, 2006) (Figure 36).

FIGURE 35 The nematode *Heydenius myrmecophila*, adjacent to its ant host, *Linepitheme* sp., in Dominican amber (Poinar amber collection accession no. N-3-65); length of nematode = 1.9 mm.

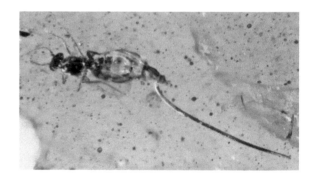

FIGURE 36 *Cretacimermis protus* emerging from an adult biting midge in Burmese amber (Buckley amber collection, Florence, Kentucky); length of exposed portion of nematode = 4.3 mm.

Poinar (1998b, Figure 3A) illustrated a Baltic amber chironomid with two emergent mermithids and cited similar examples from Dominican amber. Weitschat and Wichard (1998, Plate 7, Figures c and e) illustrated a midge and a

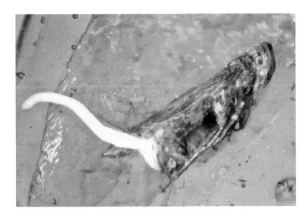

FIGURE 37 *Heydenius brownii* (Mermithidae) parasitizing a planthopper in Baltic amber (Poinar amber collection accession no. N-3-51); length of exposed portion of nematode = 12 mm.

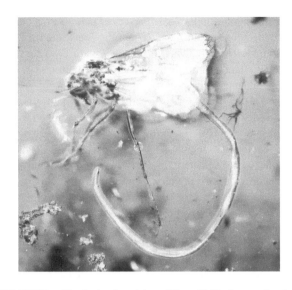

FIGURE 38 *Heydenius formicinus* (Mermithidae) emerging from an ant in Baltic amber (Poinar amber collection accession no. N-3-52); length of exposed portion of nematode = 8 mm.

sciarid fly with emergent nematodes, probably mermithids, from Baltic amber. Janzen (2002, Figure 363) illustrated a Baltic amber chironomid with an emergent nematode. Poinar (1985) described the relationship between a modern tabanid fly and its mermithid parasite, while Grabenhorst (1985) described the same relationship for the Pliocene of Willershausen, Germany.

Poinar (2001c) discovered the first parasitic relationship between a mermithid and a planthopper (Homoptera: Achilidae) from Baltic amber (Figure 37), thus adding still another major insect group to the list of those parasitized by these nematodes in the fossil record. This association still occurs today (Helden, 2008). Poinar (2002a) described an Eocene Baltic amber mermithid nematode parasitizing an ant (Figure 38), the first example of this relationship that is well known in the Recent.

□ *Reliability*
Category 1, frozen behavior.

*4AIIIb. Hairworm–Insect

(See Figure 39.) Poinar (1999a; for erroneous records of hairworms, see also Poinar, 2000a) described a Dominican amber hairworm, the first unequivocal fossil nematomorph, and gave reasons for being sceptical about the reliability of Voigt's earlier example from a cockroach. This is the first validated example of insect parasitism by a nematomorph in the fossil record. The hairworms are closely associated with the host, one being only partly emergent. Poinar and Buckley (2006) described the oldest known hairworm in Burmese Early Cretaceous amber. No host was involved, but the presence of fruiting bodies of a coral mushroom suggested not only a food source but also a moist habitat attractive to roaches, and a member of that group is suggested as a possible host.

□ *Reliability*
Category 2B, functional morphology.

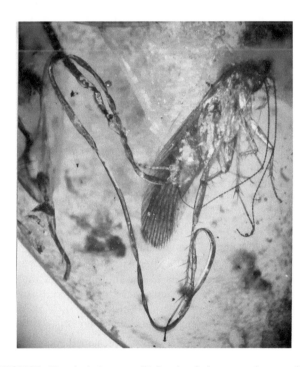

FIGURE 39 A hairworm (*Paleochordodes protus*) emerging from a cockroach in Dominican amber (Poinar amber collection accession no. GOR-3-1); length of portion of hairworm outside of insect = 37 mm.

4AIIIc. Lice and Mammals (Nits) and Birds

(See also Section 4AIIIz, Parasite Eggs on a Bird Feather.) Teerink (1991) provided an excellent summary of what can be done to identify mammals by the scale patterning on their hairs. This is very useful information when one is faced with trying to identify the mammal in which amber nits are preserved associated with mammal hair. Peñalver and Grimaldi (2006) described some details of Miocene

Dominican mammalian hair preserved in amber. In connection with mammalian hair, Repenning (pers. comm., 1983) commented that "fur was a primitive feature possessed by all aquatic land mammals but progressively useless to deep-diving marine carnivores because the insulating value of trapped air in the fur was directly proportional to its thickness, and with increased diving depths the trapped air was compressed to virtually no thickness. Hence the substitution of subcutaneous fat (blubber) as an insulator and hence the random loss of underfur in many marine mammals—all it does is provide a diving chamber for parasites."

Because of the fact that many lice lay eggs in mammalian hair, it is of some interest to trace back the earliest occurrences of fossil hair. Meng and Wyss (1997) described Paleocene mammalian hair from Inner Mongolian coprolites associated with multituberculates. Assuming that the multituberculates are a group related to the modern, hair-bearing mammals, which is reasonable, one can then conclude that hair was probably present as an ancestral character in the common ancestors during the earlier Mesozoic, with implications for the presence of parasitic lice. Franzen (2001) discussed the many examples of Messel, Middle Eocene mammals with well-preserved hair. Mammalian hair is recorded from the Early Cretaceous of northeastern China (Ji et al., 2002). The oldest currently known mammalian hair is from the Middle Jurassic of China (Ji et al., 2006), and Early Cretaceous hair from China has also been documented by Luo et al. (2003) and Ji et al. (2002). Meng et al. (2006) documented mammalian hair from the Jurassic–earlier Cretaceous of Inner Mongolia. Poinar and Poinar (2008) illustrated a mammalian hair in Early Cretaceous Burmese amber. Von Koenigswald et al. (1992b) and Wuttke (1992b) illustrated a number of Messel Eocene mammals with well-preserved hair. Poinar and Poinar (1999, Figure 164, p. 160) illustrated and discussed some Dominican amber mammalian hair, including being able to identify a mammal by the attached ectoparasitic fur mites and fur beetles as well as by the morphology of the hair itself. Krumbiegel and Krumbiegel (2001, p. 68) illustrated some Baltic Amber mammalian hair.

Weitschat and Wichard (1998, Plate 92, Figures a and b) illustrated Baltic amber nits on presumed rodent hairs. Dubinin (1948) described a Pleistocene Siberian sciurid (*Citellus*), a *suslik*, which had been parasitized by an anopluran (*Neohaematopinus*); this was a most unusual find, and preservation in frozen ground was involved.

Wappler et al. (2004) described an Eocene bird louse from Germany associated with some feather fragments. The fossil showed affinities with modern feather lice of aquatic birds, and the authors suggested that it parasitized the ancestors of either water birds or shore birds. This occurrence is slightly older than the previous record-holding feather louse from Baltic amber (Bachofen-Echt, 1949).

☐ *Reliability*

Category 2B, functional morphology, although the association verges on Category 1, frozen behavior.

4AIIId.　Ticks and Mites as Micropredators and Potential Disease Vectors

(See Figures 40 through 50; color plates 3A, 3E, and 4E after page 132.) Poinar and Milki (2001, p. 46) discussed an erythraeid mite taking hemolymph from a biting midge from Early Cretaceous Lebanese amber. Poinar et al. (1991) described two larval erythraeid mites parasitizing (engorged) two different moth taxa (tineid and gracillariid) from Dominican amber. Bachofen-Echt (1949) described a water mite attached to an adult caddisfly from Eocene Baltic amber. Janzen (2002, Figures 115 and 116) illustrated a Baltic amber erythraeid parasitizing a dolichopodid fly. Poinar et al. (1993a) described examples of mite parasitism from Canadian amber (Judith River Group, Late Cretaceous). The parasitized midges belong to the family Ceratopogonidae, and the larval stage mites, attached in feeding position and partly engorged, belong to the family Erythraeidae and the phalanx Trombidia, possibly the family Microthrombidiidae. Koteja

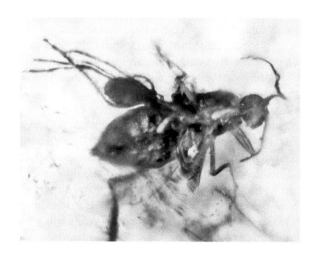

FIGURE 40　An erythraeid mite parasitizing a biting midge in Lebanese amber (Milki amber collection, Beirut, Lebanon); length of biting midge = 1.5 mm.

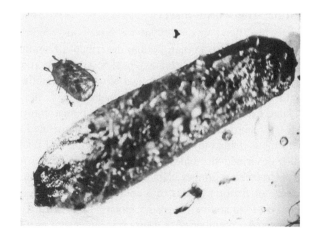

FIGURE 41　Male soft tick, *Ornithodoros antiquus*, adjacent to a coprolite from its probable rodent host in Dominican amber (Poinar amber collection accession no. A-10-73); length of coprolite = 25 mm.

FIGURE 42   Detail of male soft tick, *Ornithodoros antiquus*, in Dominican amber (Poinar amber collection accession no. A-10-73); length of tick = 4.5 mm.

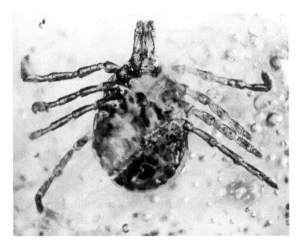

FIGURE 43   A hard tick (Ixodidae) in Dominican amber (Poinar amber collection accession no. A-10-46); length of tick = 4.3 mm.

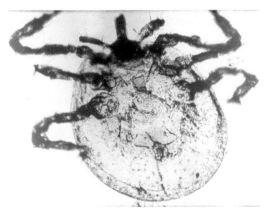

FIGURE 44   One of the oldest described ticks, *Cornupalpatum burmanicum*, from Burmese amber (Poinar amber collection accession no. A-10-260); length of tick = 328 mm.

FIGURE 45   Two parasitic mites attached to the mouthparts of a fly in Baltic amber (Poinar amber collection accession no. SY-1-137); length of fly = 2 mm.

FIGURE 46   Parasitic mites attached to the thorax of a crane fly in Dominican amber (Poinar amber collection accession no. SY-1-71); length of fly = 5 mm.

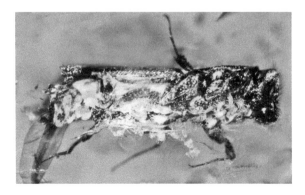

FIGURE 47   Phoretic mites attached to a platypodid beetle in Dominican amber (Poinar amber collection accession no. SY-1-5A); length of beetle = 4.2 mm.

(1998) described Taymyr amber with coccids being parasitized by mites. Poinar (1995b) described elegant evidence from Dominican amber in which is preserved an unfed, male soft tick (Argasidae), together with a probable rodent coprolite and possible rodent hair, all adding up to a picture of a rodent

FIGURE 48 Detail of phoretic mites attached to platypodid beetle (see Figure 47).

FIGURE 49 Thrombitiform mite feeding on the abdomen of a crane fly in Baltic amber (Andrew Cholewinski amber collection); length of fly = 3 mm.

FIGURE 50 Parasitic mite feeding on a springtail in Baltic amber (Andrew Cholewinski amber collection); length of springtail = 1.8 mm.

being micropredated by the tick, while also leaving behind one of its scats and some hair. Another specimen, belonging to the same genus (*Ornithodoros*) is an unfed female. These are the first fully described occurrences of fossil soft

ticks and are truly unique specimens when considered with the evidence for an associated rodent. Grimaldi et al. (2000, Figure 42g) illustrated a larval argasid tick from Turonian New Jersey amber and a cecidomyiid midge with a mite attached to its abdomen (his Figure 42h). The larval argasid tick was subsequently described by Klompen and Grimaldi (2001) as a bird parasite. Poinar and Brown (2003b) and Poinar and Buckley (2008a) reported on the oldest described ticks, *Cornupalpatum burmanicun* and *Compluriscutula vetulum*, respectively, from Early Cretaceous Burmese amber. The specimens were unfed hard tick larvae, which were probably resting on the bark of a tree waiting for a passing host. Morphological characters indicate that they are distantly related to extant ticks that parasitize reptiles. The former specimen has claws on its palpal segments, a character unknown in modern ticks. The role of the claws is unknown, but they could have been used to secure the arachnid to its host.

Poinar et al. (1997) described a Canadian Cretaceous amber specimen that includes an erythaeid mite feeding on a chironomid midge in a manner unknown in any other living or fossil mite. The mite evidently attacked the host in such a way that the host produced "nodes" on its exoskeleton from which the mite then secured host fluids.

Weitschat and Wichard (1998) illustrated a Baltic amber *Ixodes* and various mites (Plate 13, Figures a–h), as well as Baltic amber mites attached parasitically to various flies and a beetle, including Uropodidae on a clerid beetle (with a swarm of mites), *Leptus* on a mycetophilid, *Leptus* on a limoniinid, and *Leptus* on a dolichopodid (Table 14, Figures a–e). Poinar (2003e) illustrated two thrombidid mites attached to the mouthparts of an adult dolichopodid fly in Baltic amber (see Figure 46). Poinar and Poinar (1999, Figure 154, p. 150; see Figure 166 for an illustration of a hard tick) illustrated and discussed a parasitic mite attached to a fungus gnat from Dominican amber. Keirans et al. (2002) described a series of four larval ticks of the genus *Ambylomma* in Dominican amber.

Borkent (1995, p. 121), cited in Grimaldi (1993), discussed a ceratopogonid with an associated parasitic mite (not noted by Grimaldi), but Poinar (pers. comm., 2002) concluded that it is unclear from the figure whether or not the mite is actually attached to the fly. Poinar et al. (1994b) illustrated a Lebanese amber ceratopogonid with the mouthparts of a parasitic mite (erythraeid) attached to them. Krumbiegel and Krumbiegel (2001, p. 53) illustrated an alleged larval, engorged *Ixodes* from Baltic amber, but Poinar suspects that it is a mite. Koteja and Poinar (2004) reviewed the known examples from the fossil record of various coccid life stages subject to parasitism by mites, with the oldest examples extending back to the Early Cretaceous. Selden and Nudds (2004, Figure 239) illustrated a lovely Baltic Amber psocopteran being parasitized on its head by an erythraeid (*Leptus*) mite.

☐ *Reliability*

Category 1, frozen behavior.

4AIIIe. Mosquitoes as Micropredators and Potential Disease Vectors

Poinar (1993, Figure 2) illustrated a Cretaceous Canadian amber mosquito, identified by Ted Pike, that is by far the oldest recognized member of the Culicidae. Poinar et al. (2000c) described this fossil and critically reviewed the fossil mosquito record, thus establishing the first *bona fide* Cretaceous, Campanian culicid from Canadian amber; they only accepted previous Tertiary records as undoubted mosquitoes. The Late Cretaceous Canadian material suggests the possibility of disease vector involvement at least as far back as the Late Cretaceous with various vertebrates. The alleged fossil mosquito (*Burmaculex*) from Burmese amber does not have the diagnostic characters of modern mosquitoes and probably belongs to the family Chaoboridae or a separate group (Poinar, 2006). Poinar and Poinar (1999, p. 91) cited a Dominican amber anopheline mosquito, which was subsequently described by Zavortink and Poinar (2000). This represents the first fossil anopheline from the New World, and amazingly enough two typical anopheline eggs with floats are preserved very close to the female who presumably laid them (Figure 51). All of this is consistent with the presence in the New World of anophelines far back in time, with all that this implies regarding disease transmission. This fossil became even more significant when Poinar (2005a) discovered the bird malarial parasite *Plasmodium dominicana* with several life stages preserved within a Dominican amber female *Culex malariager* mosquito (Poinar, 2005c) (Figure 52 and Figure 53). This amazing discovery provides a minimum age for *Plasmodium* and suggests that some species of primate malaria might have evolved in the Americas, as both culicine and anopheline mosquitoes were present in the mid-Tertiary.

Poinar and Telford (2005) described a Burmese amber Early Cretaceous (Albian) biting midge (Diptera: Ceratopogonidae) with an abdominal cavity that included a number of "*Haemoproteus*-like" oocysts and sporozoites, indicating that a malarial parasite was already about, and structures of the host dipteran suggest a reptilian host.

Various types of malaria infect cold- and warm-blooded vertebrates today (Figure 54). One of the primitive types is *Haemoproteus* in birds and reptiles (Bennett and Peirce, 1988; Telford, 1984). A biting midge of the genus *Protoculicoides* (Diptera: Ceratopogonidae) in Burmese amber contained the

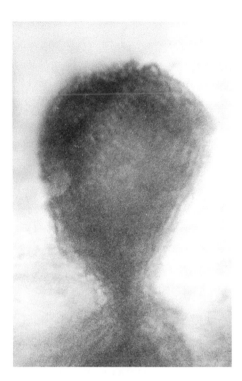

FIGURE 52 Pedunculate oocyst of *Plasmodium dominicanum* in the body cavity of the mosquito *Culex malariager* in Dominican amber (Poinar amber collection accession no. D-7-6b); greatest width of oocyst is 80 μm.

developing stages (see previous paragraph) of an early lineage of malaria (*Paleohaemoproteus burmacis*) (Poinar and Telford, 2005) (see color plate 2D after page 132). The above information makes it possible that human malaria resulted from a "switch" from one phylogenetic host type to another rather than having an original higher primate origin.

Miller et al. (1994) employed biochemical methods to demonstrate *Plasmodium falciparum* in Egyptian mummies with ages back to 5200 years. Baum and Bar-Gal (2003) made the interesting point that the genesis of malaria in humans may

FIGURE 53 Sporozoites emerging from the oocyst of *Plasmodium dominicanum* in the body of the mosquito *Culex malariager* in Dominican amber (Poinar amber collection accession no. D-7-6b); length of sporozoites = 10 to 14 μm.

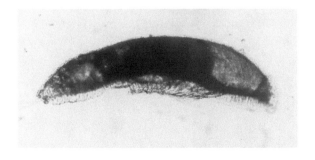

FIGURE 51 An *Anopheles* egg with "floats" in Dominican amber (Poinar amber collection accession no. D-7-6A); length of egg = 430 μm.

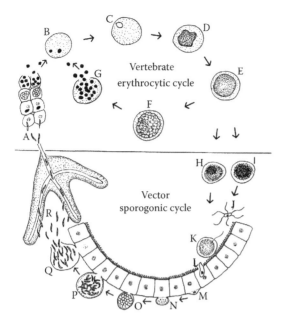

FIGURE 54 Basic life cycle of malarial organisms. (A–G) Erythrocytic cycle in vertebrate blood stream: (A) Sporozoites from vector initiate pre-erythrocytic phase in parenchyma cells of vertebrate and form merozoites. (B) Merozoites released from parenchyma cells enter red blood cells (RBCs) = erythrocytes. (C) Merozoite develops into early trophozoite and forms a ring-shaped stage or signet stage inside RBCs. (D) Trophozoite continues to develop and enlarge inside RBCs. (E) Trophozoite becomes an immature meront within RBCs. (F) Meront subdivides into a mature meront segmenter in RBCs. (G) Mature meront releases merozoites, which are released from the RBCs into the bloodstream. These merozoites then enter new RBCs and continue the cycles in the vertebrate. (H–R) Sporogonic cyle in vector: Immature meronts picked up by the vector during blood feeding develop into (H) macrogametocytes and (I) microgametocytes in the vector's gut lumen. (J) The microgametocyte produces microgametes (six in this case) that escape separately. (K) The macrogamete produced from the macrogametocyte is fertilized by a microgamete, producing a zygote. (L) The zygote becomes a motile ookinete that penetrates the vector's midgut epithelial cells. (M) The ookinete encysts in the basal region of the midgut cells near the hemocoel and forms an oocyst (in some forms of malaria, the oocyst occurs in other areas of the body). (N) The oocyst enlarges and bulges into the hemocoel. (O) The oocyst subdivides and forms spherical immature sporozoites. (P) Elongated immature sporozoites are formed inside the oocyst. (Q) Mature sporozoites are released from the oocyst. (R) The sporozoites migrate from the ruptured oocyst to the salivary gland, enter the salivary ducts, and are introduced by the vector's bite into a vertebrate host (back to A). (Not drawn to scale.)

be related to the origins of human agriculture in West Africa about 10,000 years ago, and they used the prevalence of sickle cell anemia there as part of the evidence. Angel (1966) provided archeological evidence suggesting that malaria has been present as early as 6500 B.C. in marshy areas, but not in dry regions.

□ *Reliability*

Category 2B, based on functional morphology, and Category 1 frozen behavior, for the protozoans with associated vectors.

4AIIIf. Tsetse Flies as Micropredators and Disease Vectors

Meyer (2003, Figure 189) illustrated a Late Eocene Florissant tsetse fly; note that the Florissant has now been redated as Late Eocene rather than Oligocene.

4AIIIg. Parasitoid Wasp–Insect Hosts

(See Figures 55 through 58.) Poinar (1992b, Figure 140; see also Poinar, 2001b) illustrated a homopteran with a parasitoid dryinid wasp larva from Dominican amber. Grimaldi (1996, p. 97) mentioned, but did not illustrate, a Dominican amber leafhopper parasitized by a dryinid wasp larva; the larval sac is attached to the abdomen of the host. Poinar (1992b, p. 253; see also Poinar and Poinar, 1999, Figure 140, p. 136) also cited dryinid-parasitized planthoppers (Fulgoroidea). Olmi and Bechly (2001) reviewed the Baltic amber Dryinidae, and Olmi (2003/2004) revised the world Sclerogibbidae, a family of wasps that are ectoparasitoids of web spinners (Embiidina); included in his revision are several new taxa described from Dominican amber, including *Pterosclerogibba antiqua*, the only known species, extant or extinct, with fully winged females.

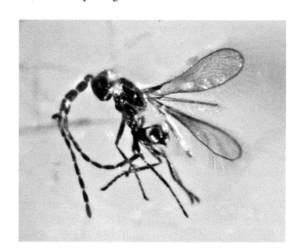

FIGURE 55 A mymarid fairy wasp in Dominican amber. Note the fringe lining the edges of the wings (Poinar amber collection accession no. Hy-10-20); length of wasp = 0.5 mm.

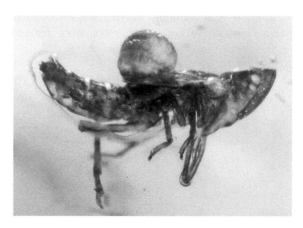

FIGURE 56 "Sac" of a dryinid wasp larva emerging from the body of a planthopper in Dominican amber (approximately ×10).

FIGURE 57   Larva of a neonurine braconid wasp emerging from an ant in Baltic amber (Poinar amber collection accession no. Sy-1-149); length of ant = 3.5 mm.

FIGURE 58   Hymenopteran larva emerging from an adult caddisfly in Baltic amber (Andrew Cholewinski amber collection); length of larva = 640 μm.

Poinar and Miller (2002) described a very definitive Baltic amber specimen with a parasitoid braconid wasp larva emerging from an ant, *Lasius*; this is a most unusual and well-preserved specimen. Poinar later described in detail the behaviors of the parasite and an ant host in a modern example from the same group, the Neoneurinae, and also provided an excellent illustration of the Baltic amber example (Poinar, 2004c, Figure 4B).

Rasnitsyn (1980, Plate VI, Figure 20; see also Rasnitsyn and Quicke, 2002, Figure 351) illustrated an elegant mymarid wasp from the Late Cretaceous of Taymyr. Weitschat and Wichard (1998, Plate 69, Figure b) provided another, as did Janzen (2002, Figure 310) from Baltic amber.

A quite bizarre flea-like creature from the Jurassic was described by Rasnitsyn (1992). The flea-like characters of *Strashila* include the thickened hind legs and strong beak. Pterosaurs were suggested as possible hosts.

Poinar and Anderson (2005) described a Baltic amber adult trichopteran with an emergent larval hymenopteran parasite that may well belong to the Braconidae. This association is currently unknown, as only the larval and pupal stages of trichopterans are parasitized by hymenopterans today.

FIGURE 59   Flea, *Ropalpsyllus* sp., in Dominican amber (Poinar amber collection accession no. Sa-1-1); length = 2.25 mm.

☐ *Reliability*

Category 2B, functional morphology, and Category 1, frozen behavior.

4AIIIh.   Fleas as Micropredators and Disease Vectors

(See Figure 59.) Poinar (1992b, Figure 99) illustrated a Dominican amber flea that Traub thought is a bird flea. Poinar (1995a; see also Poinar and Poinar, 1999, Figure 165) illustrated and discussed another Dominican amber flea and summarized most of what is known of fossil flea occurrences. Weitschat and Wichard (1998, Plate 89) illustrated a Baltic amber flea. Lewis and Grimaldi (1997) described a pulicid associated with mammalian hair from Dominican amber. Panagiotakopulu (2001) documented human fleas, at least 3500 years old, from Pharaonic Egypt in Amarna.

☐ *Reliability*

Category 2B, functional morphology.

4AIIIi.   Parasitoid Wasps and Parasitic Flies–Spiders

Grimaldi (1996, p. 97) illustrated a length of spider web containing cocoons of a wasp parasitic on the spider from Dominican amber. Poinar (1998b, Figure 2B) illustrated a Dominican amber spider fly (Acroceridae), a group with a Jurassic to present range (Carpenter, 1992), that is an excellent example of behavioral fixity. Weitschat and Wichard (1998, Plate 85, Figures g and h) illustrated several Baltic amber acrocerids. Schlinger (1987) described the endoparasitic behavior of the acrocerid larva when finding and parasitizing its spider, and indicated that in at least one instance the relationship is species to species. Poinar (1992b, Figure 141) illustrated an early-stage polysphinctid wasp larva (family Ichneumonidae) adjacent to the abdomen of a Dominican amber spider (family Clubionidae). Poinar (1992b, p. 253) described how the wasp stings the spider into submission, then deposits an egg on its abdomen, following which the hatched larva remains exposed on the spider's abdomen, where it periodically punctures the spider's integument and feeds on the released hemolymph. Poinar (2004a, p. 1878) showed a spider egg case in Baltic amber, with the eggs

FIGURE 60 Female clubionid spider adjacent to an ichneumonid wasp cocoon (arrow) in Baltic amber. (Andrew Cholewinski amber collection); wasp cocoon = 3.2 mm.

depleted and the parent spider adjacent containing a puparium of an ichneumonid wasp, similar to present-day behavior of wasp parasites of spider eggs (see Figure 60).

☐ *Reliability*

Category 1, frozen behavior.

4AIIIj. Human Disease

One needs to keep the proper perspective toward paleopathology when Tanke and Rothschild (1997) pointed out that less than 1% of currently recognized diseases leave any traces on bones! Cockburn (1971; see also Schultz, 1967) provided a thoughtful summary that emphasizes relationships with allied anthropoids, as well as a certain amount of the literature on historical examples derived chiefly from mummies and coprolites. Cockburn and Cockburn (1980) edited a volume on mummies that provides a wealth of information on human disease from the historical past, including Egyptian, Chinese, and Peruvian mummies, with the inference that almost all the diseases detected have been with us for a long time, although this is not "evolutionary time" that is being considered. In the Cockburns' volume, Sandison (1980) pointed out that the interpretation of soft-tissue abnormalities requires coping with the results of embalming (chemical and physical), fungal activity, and postmortem insect attack. Aufderheide (2003) provided another account of various mummies with some emphasis on soft-tissue evidence. Waldron (1994) provided a very readable account emphasizing the problems involved with trying to estimate the prevalence of disease in human skeletal populations, while providing an overall view that is applicable to all organisms. Wilke and Hall (1975; see also Fry, 1985) provided an annotated bibliography of human fecal contents, parasitic and otherwise, chiefly from archaeological sites. Dieppe and Rogers (1993) reviewed the skeletal paleopathology of human rheumatic disorders, and Rothschild and Martin (2006a) provided a wealth of well-documented information regarding paleopathology. Guhl et al. (1997) used molecular techniques to identify evidence of Chagas disease in 4000-year-old human mummies from the Atacama Desert.

Rothschild and Thillaud (1991) discussed the presence of gout in Neanderthals, possibly calcium pyrophosphate deposition disease. Strouhal (1987) described evidence of tuberculosis, including Pott's disease, affecting human vertebrae from an ancient Nubian skeleton and one from Egypt, while reviewing other published occurrences of a few thousand years' antiquity; none appeared to have been recognized in pre-Holocene hominid remains.

Barnes (2005; see also Schultz, 1967) provided an account of human diseases that considered known occurrences of similar diseases in related mammals; for example, among the other higher primates, the following diseases are known: malaria, schistosomiasis, Ebola virus, smallpox, herpes, HIV/AIDS, yellow fever, and dengue. This finding implies that these diseases arise from a common ancestral primate. Her approach has limitations, as we have relatively limited knowledge concerning the diseases present in most wild animals. Also, the problem of disease-causing organisms "switching" from one host organism to another, taxonomically unrelated form is always a possibility.

Miller et al. (1992) provided evidence of schistosomal infections present in Egyptian mummies, as did Deelder et al. (1990). It is also clear that prokaryotic bacterial and viral and some protozoan diseases are not conservative; that is, they are subject to rapid mutation rates and evolve very rapidly (Willis, 1996; Frank, 2002)! This situation contrasts radically with the conservative behavior of metazoan disease-causing organisms.

There is a wealth of information on helminths from mummies and coprolites, and a certain amount on lice obtained from mummies. Ewing (1926) cited head lice from Peruvian Inca mummies. Bresciani et al. (1983) discussed head lice from some 15th-century Greenland Eskimo mummies (see also Bresciani et al., 1991, Figure 163, for an illustration of a hair with an attached egg case of a head louse). Horne (1979) discussed some Aleut mummies of about the same age with head lice. Busvine (1980) reviewed some of the evidence from the present concerning head lice and pubic lice, including the inference that the two species evolved "allopatrically" after humans lost their complete fur coat; he commented that pubic hair is morphologically distinct from that on the head and that the former's density is less. Faulkner (1991) discussed some helminths, including pinworms, roundworms, and a hookworm possibility, plus *Giardia* cysts identified with the aid of monoclonal antibodies specific for the cyst proteins, from some several millenia-old Tennessee cave coprolites. Reinhard et al. (1987) discussed helminth evidence from paleoindian coprolites from the Colorado Plateau. Jones (1986) discussed the presence of whipworms in Lindown Man, dating from about the time of the Roman occupation of Britain. Tankersley et al. (1994) reported on the occurrence of *Giardia* in American Indian paleofeces dated at 2420 B.C. plus or minus 90 years.

Sadler (1990) provided a very interesting account of Viking-age Greenland human fleas and the body louse, as well as of several parasites from associated sheep. Mumcuoglu and Zias (1988) described head lice remains from nit combs derived between the first century B.C. and the 8th century A.D. in Israel, which certainly makes it clear that this nuisance has always been with us. Panagiotakopulu and Buckland (1999) noted 3500-year-old bedbugs (*Cimex lectularius* L.) from Pharaonic Egypt, and Kenward (1999) documented Roman and Medieval pubic lice (*Pthirus pubis* L.) from Britain, making it clear that human parasites have been with us a long time.

Kliks (1990) reviewed occurrences of New World pre-Columbian helminths in feces and cadavers to indicate that many items traveled from the Old World with the dogs of New World people. Harrison et al. (1991) provided an excellent case where skeletal pathology in an American Indian skeleton from Arizona (1000 to 1400 A.D.) could be assigned to coccidioidomycosis, which was confirmed by microscopic preparations that revealed both spherules and endospores of parasitic type belonging to *Coccidioides immitis*. Jouy-Avantin et al. (1999) reviewed a number of occurrences of helminthic parasites associated with sub-Recent and Late Pleistocene human remains.

Clemen (1956) provided evidence of dental caries in various Pleistocene hominids from South Africa (Figure 61). Tillier et al. (1995) described a Middle Palaeolithic example of dental caries that is very convincing and obviously pre-agricultural. Alt et al. (1998; see also Hillson, 1996) provided an elegant summary of caries in hominids (see particularly the Caselitz paper). Swindler et al. (1995) concluded that caries, periodontal disease, and severe dental attrition were present in Egyptian remains from the New Kingdom (1550 to 1070 B.C.); that is, there was little difference from the present. While not involving caries, Hershkovitz et al.'s (1997) description of dental calculus in a Miocene ape, *Sivapithecus*, with scanning electron microscope (SEM) evidence of bacterial presence suggests that higher primate dental problems have been present for a long time. Although not involving caries in humans, the account by Sala Burgos et al. (2007) of a Miocene artiodactyl from Spain with a carious tooth is of interest. Ducrocq et al. (1995), while discussing a Late Eocene dental anomaly in an anthracotherid, also provided a good summary of other mammalian tooth anomalies known from the fossil record, including some primates.

Although not involving disease in the strict sense, the problem of infanticide possibly being a higher anthropoid trait in which stepparents are far, far more heavily involved than genetic parents is significant (Daly and Wilson, 1988a,b). Also relevant here is Wilson and Daly's (1996) finding of a relationship between uxoricide (wife murder) and age of the victim; uxoricide is much, much higher among younger women of reproductive age.

Pérez (1996) summarized and referred to a mass of more recent work on human paleopathology. Evans (1996) described a Late Stone Age coprolite from South Africa that contained eggs of *Trichuris* and *Ascaris*. Pike (1967) provided a useful summary of parasite eggs from human materials, chiefly feces.

Strouhal (1998) reviewed the evidence concerning bony tumors involving the jaw. Rothschild and Rothschild (1995; 1996a,b; 1998a; see also Rothschild, 2005, and Rothschild et

al., 2000) provided a wealth of skeletal data indicating that syphilis evolved from yaws in the New World during pre-Columbian time, and that yaws was formerly cosmopolitan, with bejel, the other treponemal disease, being strictly Old World. The authors employed large populations to determine the skeletal differences between the three diseases, as well as the time when syphilis first appears in the record. Their data strongly indicate that syphilis was brought from the New World to the Old World following the Columbian voyages. Rothschild et al. (1995) provide skeletal evidence for the presence of yaws in the Middle Pleistocene of East Africa in *Homo erectus*. Bourke (1967) discussed evidence of arthritic disease in ancient human remains.

☐ *Reliability*
Category 1, frozen behavior.

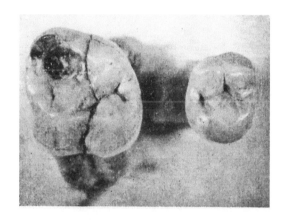

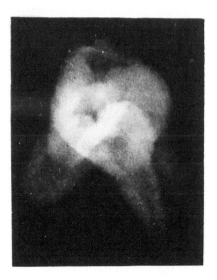

FIGURE 61 (Top) Caries on the occlusal surface of an upper left maxillary second molar of *Paranthropus crassidens* (specimen SK14; ×1.5). The tooth to the right is a modern maxillary molar for comparison (×1.5). Lower photograph is a radiograph of specimen SK14. (Bottom) Caries on the mesial surface of a lower right second molar of *Telanthropus capensis* (specimen SK15; ×1.7). Note proximal attrition on carious surface. (From Clemen, A.J., *British Dental Journal*, 101, 4–7, 1956. Reproduced by permission from Macmillan Publishers Ltd.)

*4AIIIk. Parasitic Insects Other than Wasps and Flies

Poinar and Poinar (1999, Figure 151, p. 147) described and illustrated a Dominican amber parasitic moth larva on a leafhopper. Very unusual are several parasitic beetles whose larvae feed externally on bees or wasps (see Poinar and Poinar, 1999, Figures 152 and 153, pp. 148–149, for Dominican amber examples; see also Poinar, 2009a).

☐ *Reliability*

Category 1, frozen behavior.

*4AIIIl. Trichurids and Caviomorph Rodents

Ferreira et al. (1991) described some coprolites of the caviomorph rodent *Kerodon rupestris* from Piaui, northeastern Brazil, that yielded eggs of the intestinal parasite *Trichuris*. The deposit is alleged to be 30,000 years old.

☐ *Reliability*

Category 2B, based on functional morphology.

*4AIIIm. Protozoan–Deer Relationship

Ferreira et al. (1992) reported oocysts of the parasitic protozoan *Eimeria* in deer (*Mazama* sp.) coprolites from Piaui in northern Brazil; interestingly enough, this parasite had not been previously recorded in living *Mazama*, although the investigators did find similar oocysts in zoo-derived feces of this deer. The initial assignment of the coprolites, obtained from an archeological excavation, was made by comparing their morphology with zoo-derived feces. This example underlines our ignorance of disease in most animals, as the parasite was previously unknown from *Mazama* in the wild.

*4AIIIn. Hyaenid–Nematode Relationship

Ferreira et al. (1993) described larval nematodes preserved in Hyaenidae coprolites from Early and Middle Pleistocene Italian sites. The nematodes were not identified, but the presence of intestinal parasites is reasonable in view of the rich fauna occurring in modern hyaenids.

*4AIIIo. Lizard–Parasitic Nematode Relationship

An equally interesting example is Araujo et al.'s (1982) account of lizard (possibly *Tropidurus*) coprolites, about 9000 years old, from another Brazilian archeological site that yielded fragments of termites and ants, presumably articles of food, together with the eggs of a parasitic nematode (possibly *Parapharyngodon sceleratus*).

*4AIIIp. Vertebrate Pathology Other than Human

Rothschild and Tanke (1992) provided entry into the literature on vertebrate paleopathology and presented a number of new examples. Molnar (2001) surveyed the literature on theropod paleopathology. Rothschild et al. (2003) pointed out, after an extensive fluoroscopic survey, that tumors in dinosaurian vertebrae were only found in hadrosaurs.

*4AIIIq. Tyrannosaur Pathology

Rothschild et al. (1997b,c) described a most interesting case of gout in a tyrannosaur from the Late Cretaceous Hell Creek Formation of Montana and also from Alberta! The bone pathology reported is characteristic of gout and is rarely found in other conditions.

☐ *Reliability*

The reliability of this diagnosis appears high to Rothschild.

*4AIIIr. Tetrapod Osteomyelitis

(See also Section 30c, Bone Fractures.) McWhinney et al. (2001b) described skeletal evidence of osteomyelitis, a bacterially induced skeletal disease, in a number of *Stegosaurus* dermal tail spikes and mentioned the evidence for tetrapod osteomyelitis as far back as the earlier Permian in *Dimetrodon*. Hanna (2002) described allosaur material affected by various pathological items, including healed fractures, osteomyelitis, and various idiopathic items. MacDonald and Sibley (1969) described some Rancho La Brea examples. Dawson and Gottfried (2002) cited osteomyelitis in a Miocene dolphin mandible, and Aufderheide (2003) illustrated a subrecent mammoth example. Although it was not diagnosed as osteomyelitic, Lucas (2000) discussed a fused bony scute situation in a Late Triassic aetosaur that may represent the effects of infection. Hutchinson and Frye (2001) described some possible examples of osteomyelitis from Cenozoic turtles.

☐ *Reliability*

Category 1, frozen behavior.

*4AIIIs. Co-Ossified Vertebrae in Mosasaurs and Whales

Mulder (2002) provided an excellent account of ossified vertebrae caused by infections following predatory attempts in mosasaurs and whales as well as ligamentous calcifications. The examples are very impressive and convincing.

☐ *Reliability*

Category 1, frozen behavior.

*4AIIIt. Lizard–Tick Relationship

Szadziewski (1999, p. 4) cited the first occurrence of a parasite on a vertebrate host, a tick with a lizard, from Baltic amber at Gdansk.

☐ *Reliability*

Category 1, frozen behavior.

*4AIIIu. Hyperdisease in North American Mammoths

Rothschild and Helbling (2001; see also Rothschild, 2003b) provided evidence for hyperdisease in North American Holocene *Mammut americanus* based on a 45% incidence of bone pathology that very likely represents tuberculosis, with such skeletal evidence suggesting that the soft-tissue evidence may have been much higher; that is, the extinction of the North American mammoth may have been caused by disease rather than other factors. Rothschild and Laub (2006) provided evidence of hyperdisease, probably tuberculosis, in Pleistocene mastodons.

Martin et al. (1999; see also Rothschild 2003c) cited DNA evidence for the presence of tuberculosis in Pleistocene North American artiodactyls (bison, musk ox, big horn sheep), 15 to 20,000 years BP. Rothschild et al. (2001) demonstrated the presence of tuberculosis in an extinct North American bison 17,000 years old by using DNA extracted from well-preserved remains. Rothschild and Martin (2006b) discussed skeletal evidence of tuberculosis in Pleistocene to Recent bovids. Martin (2003) pointed out the possibility that human tuberculosis involved "switching" from an ungulate host to humans due to a closer association between humans and such bovids as the bison. Cohen and Crane-Kramer (2003) suggested that in the Old World the presence of tuberculosis in humans correlates with the domestication of cattle, but Rothschild (2003c) raised cogent objections.

☐ *Reliability*

Category 2B, functional morphology.

*4AIIIv.   Osteochondroma

Wang and Rothschild (1992) described skeletal evidence of osteochondroma, which is probably an inherited genetic condition (Figure 62 and Figure 63). They described the condition from various Oligocene and Eocene canids (Canidae, Amphicyonidae), mostly from North American specimens. This is one of those uncommon situations where a distinc-

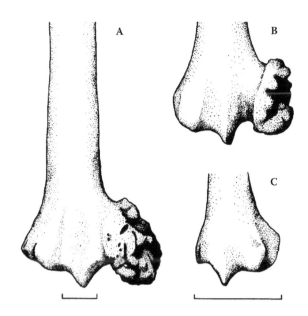

FIGURE 62   Osteochondroma on distal radii of (A) *Daphoenus* sp. and (B) *Hesperocyon gregariou* and (C) normal distal radius of *Hesperocyon*. Scale bars = 10 mm. (Figure 2 in Wang, X. and Rothschild, B.M., *Journal of Vertebrate Paleontology*, 12, 387–394, 1992. Reproduced with permission.)

tive skeletal pathology can be assigned a relatively definitive cause.

☐ *Reliability*

Category 2B, functional morphology.

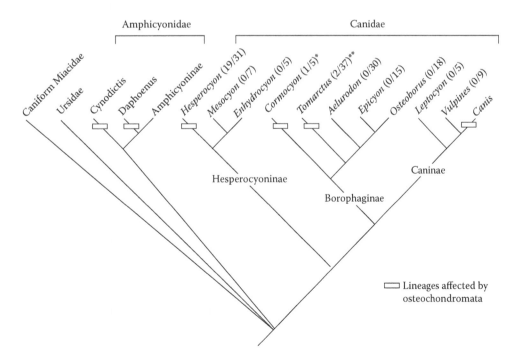

FIGURE 63   Distribution of multiple hereditary osteochondroma (indicated by horizontal rectangles) in Amphicyonidae, Canidae, and other caniform carnivores. Numbers in parentheses are number of affected individuals against total number examined for the taxon. Absence of osteochondroma in some taxa may be a result of small sample size. (Figure 5 in Wang, X. and Rothschild, B.M., *Journal of Vertebrate Paleontology*, 12, 387–394, 1992. Reproduced with permission.)

*4AIIIw. Mycobacteriosis in a Pliocene Kangaroo

Rothschild et al. (1997a) described caudal vertebrae evidence of mycobacterial disease in a Pliocene Queensland kangaroo.

□ *Reliability*

Category 1, frozen behavior.

*4AIIIx. Inflammatory Arthritis (Spondyloarthropathy)

Rothschild and Rothschild (1996d) discussed the presence of post-Paleocene inflammatory arthritis (spondyloarthropathy), while Rothschild et al. (1998a) recognized it in Paleocene and Eocene North American mammals. Rothschild et al. (1994) recognized it in mammoths and cited its occurrence in *Smilodon*. Spondyloarthropathy has also been recognized in Oligocene to Recent equids and rhinoceratids (Rothschild and Rothschild, 1998b), with incidence increasing toward the present.

□ *Reliability*

Category 1, frozen behavior.

*4AIIIy. Vertebrate Intestinal Parasites

Poinar and Boucot (2006) described protozoan cysts, trematode eggs, and ascarid nematode eggs from a Wealden coprolite previously assigned to the predatory dinosaur *Megalosaurus* from the Bernissart *Iguanodon* locality (Figure 64 through Figure 67). These protozoan cysts, trematode eggs, and ascarid nematode eggs are the oldest evidence of terrestrial vertebrate intestinal parasites.

□ *Reliability*

Category 1, frozen behavior.

*4AIIIz. Parasite Eggs on a Bird Feather

Martill and Davis' (1998b) brief citation regarding a Nova Olinda Member, Crato Member, Early Cretaceous Brazilian feather on which are scattered probable acariid eggs is notable.

□ *Reliability*

Category 1, frozen behavior.

*4AIIIza. Trematode Eggs

Jouy-Avantin et al. (1999) described dicrocoelid eggs that may have been derived from bears at a Middle Pleistocene French Pyreneean cave site.

□ *Reliability*

Category 5A, functional morphology.

*4AIIIzb. Spider–Mermithid Relationship

Poinar (2000a) described a unique Baltic amber specimen of a parasitic mermithid associated with a clubionid spider, as is commonly the case today as well (Figure 68).

□ *Reliability*

Category 1, frozen behavior.

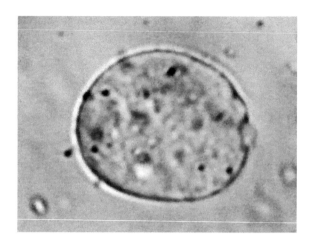

FIGURE 64 Protozoan cyst of *Entamoebites antiquus*. Specimen is from a Wealden-age coprolite from Bernissart, Belgium; length of cyst = 12 μm. (Figure 1A in Poinar, Jr., G.O. and Boucot, A.J., *Parasitology*, 133, 245–249, 2006. Reproduced with permission of Cambridge University Press.)

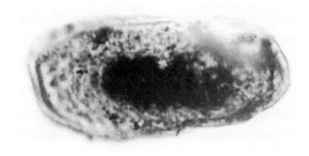

FIGURE 65 Egg of the trematode *Digenites proterus*. Specimen is from a Wealden-age coprolite from Bernissart, Belgium; length of egg = 72 μm. (Figure 1B in Poinar, Jr., G.O. and Boucot, A.J., *Parasitology*, 133, 245–249, 2006. Reproduced with permission of Cambridge University Press.)

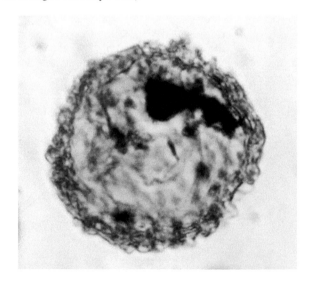

FIGURE 66 Egg of the ascarid nematode *Ascarites priscus*. Specimen is from a Wealden-age coprolite from Bernissart, Belgium; diameter of egg = 45 μm. (Figure 1C in Poinar, Jr., G.O. and Boucot, A.J., *Parasitology*, 133, 245–249, 2006. Reproduced with permission of Cambridge University Press.)

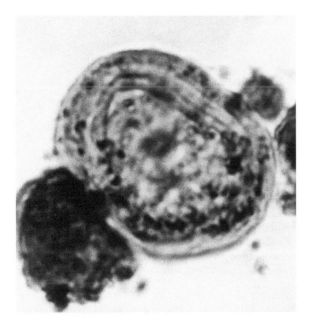

FIGURE 67 Egg of the ascarid nematode *Ascarites gerus*. Specimen is from a Wealden-age coprolite from Bernissart, Belgium; diameter of egg = 52 μm. (Figure 1D in Poinar, Jr., G.O. and Boucot, A.J., *Parasitology*, 133, 245–249, 2006. Reproduced with permission of Cambridge University Press.)

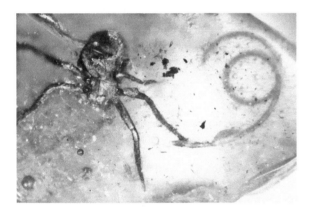

FIGURE 68 The mermithid *Heydenius araneus*, adjacent to its spider host in Baltic amber (Andrew Cholewinski amber collection); length of mermithid = 24 mm.

*4AIIIzc.   Pentastomida

The Pentastomida are commonly considered to be either a phylum close to the Arthropoda or a subclass of the Crustacea (Schmidt and Roberts, 1989). The adults are commonly parasitic in the respiratory tract of vertebrates (reptiles, birds, amphibians, and mammals), whereas the intermediate stage may be parasitic in fishes, amphibians, reptiles, insects, or mammals. The earliest fossil records of vertebrate parasites are isolated pentastomid larvae from the Cambrian (Waloszek and Müller, 1994). These specimens could be similar to the phosphatized material from the Tremadocian, Early Ordovician, of Öland, that were recovered by Reimer (1989) and considered to be "Vermutliche Pentastomida."

□ *Reliability*

Category 6, rather speculative, although Reimer's morphological evidence is attractive.

*4AIIIzd.   Strepsipteran Parasitism

The Strepsiptera are an order of parasitic insects (for a brief, introductory account of their habits, see Pierce, 1964) that are known from the Cretaceous onwards (Kinzelbach and Lutz, 1985; Grimaldi et al., 2005). They parasitize various groups of insects internally, with the female being restricted to the host during life. They are known in the larval and adult condition as fossils (Carpenter et al., 1994). Lutz (1992, p. 64) reported them within a host ant from the Messel Eocene as puparia. Poinar and Poinar (1999, Figure 150, pp. 145–146) illustrated and described from Dominican amber parasitic strepsipteran larvae emerging from their mother who had parasitized a planthopper (see Figure 76). Kathrithamby and Grimaldi (1993) reviewed the strepsipteran fossil record. Pohl and Kinzelbach (2001) described a Baltic amber female stylopterid parasitizing an ant, which is unusual as modern males only are known to parasitize ants whereas the modern females favor mantodes and ensiferans. Poinar (2004d) illustrated a Dominican amber specimen parasitizing a halictid bee, and two families of plant hoppers. A strepsipteran puparium in a stingless bee in Dominican amber is shown in Figure 69.

□ *Reliability*

Category 2B, functional morphology, except for Lutz's material, which belongs in Category 1, frozen behavior.

*4AIIIze.   Phorid Dipteran–Allantonematid Nematode

Poinar (2003d) described a parasitic relation between allantonematid nematodes from the body cavity of a phorid fly from Baltic amber (Figure 70).

□ *Reliability*

Category 1, frozen behavior.

FIGURE 69 A stingless bee, *Proplebeia dominicana*, in Dominican amber with a puparium of a strepsipteran parasite protruding from its abdomen (Poinar amber collection accession no. Sy-1-16); length of bee = 3.5 mm.

FIGURE 70 *Howardula helenoschini* (Allantonematidae) emerging from a phorid fly in Baltic amber (Poinar amber collection accession no. N-3-64); nematode on right = 780 μm.

*4AIIIzf. Insect Vector–Trypanosome Relationship

Poinar and Poinar (2004a,b) described a trypanosomatid leishmanial parasite (*Paleoleishmania proterus*) from an Early Cretaceous Burmese sand fly (*Palaeomyia burmitis*) (Poinar, 2004f) and provided evidence of the parasite developing inside reptilian blood cells ingested by the sand fly (see Figure 71 through Figure 73. Some rare sand fly larvae (Diptera: Phlebotomidae) in Burmese amber were trapped while feeding on the fruiting bodies of a coral fungus (*Palaeoclavaria burmitis*) (Poinar et al., 2006a) (see color plate 3D after page 132). Extant sand fly larvae are very difficult to locate in nature but are known to feed on rotting plants material containing fungi (Wermelinger and Zanuncia, 2001).

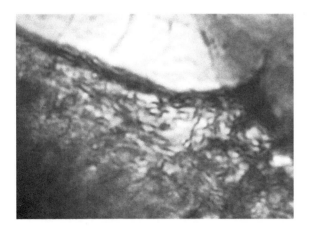

FIGURE 71 Developing promastigotes of *Paleoleishmania proterus* in the midgut of the Burmese amber sand fly *Palaeomyia burmitis* (Poinar amber collection accession no. B-D-16); length of promastigotes ranges from 5 to 10 μm.

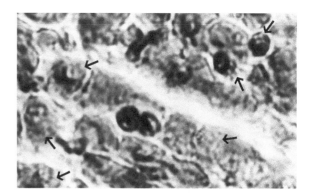

FIGURE 72 Developing amastigotes (arrows) of *Paleoleishmania proterus* in parasitophagous vacuoles within reptilian blood cells in the foregut of the Burmese sand fly, *Palaeomyia burmitis* (Poinar amber collection accession no. B-D-16); diameter of parasitophagous vacuoles ranges from 4 to 8 μm.

Poinar (2005b) described a reduviid vector (*Triatoma dominicana*) of a trypanosome (*Trypanosoma antiquus*) in Dominican amber. A bat was determined to be the likely vertebrate host based on mammalian hairs with bat-like features adjacent to the hemipteran. This is the first fossil record of a hematophagous hemipteran (see Figure 74 and Figure 75).

Poinar and Telford (2005) detected malarial parasites in the body cavity and midgut lumen of a Burmese amber biting midge (Ceratopogonidae), and Poinar (2008a,b) described a second specimen of *Paleoleishmania* from a Dominican sand fly and a *Paleotrypanosoma burmanicus* from a biting midge (*Leptoconops* sp.) in Burmese amber.

□ *Reliability*

Category 1, frozen behavior, as this is strong evidence for the conservatism of protozoan-generated diseases.

*4AIIIzg. Viruses in Biting Midges and Sand Flies

Poinar and Poinar (2005) described an Early Cretaceous (Albian) Burmese amber biting midge (Ceratopogonidae) infected with a cytoplasmic polyhedrosis virus (see color plate 3C after page 132) and a nuclear polyhedrosis virus in the midgut epithelial cells of a Burmese amber sand fly.

□ *Reliability*

Category 1, frozen behavior.

*4AIIIzh. Hypermetamorphosis

Hypermetamorphosis is a condition where the first-stage insect larvae are radically different from subsequent stages. They occur in the life cycle of some parasitic groups where successful development is dependent upon the first-stage larvae locating a suitable host. It occurs in all members of the Strepsiptera (Figure 76), Mantispidae, Meloidae, Rhipiphoridae, Acroceridae, Bombyliidae, Nemestrinidae, Perilampidae, Eucharidae, Cyclotornidae, and Epipyropidae and in some representatives of the families Gracillariidae and Staphylinidae. In hypermetamorphosis, the first larval stage is free living and modified for movement and the

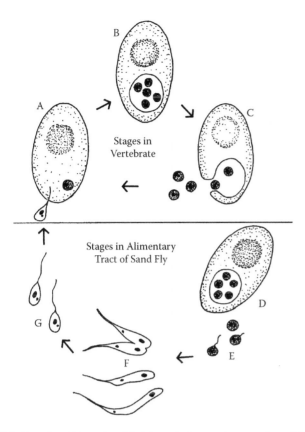

FIGURE 73 Life cycle of *Leishmania* in vertebrate and sand fly vector. (A) Paramastigote stages of *Leishmania* introduced into the vertebrate from the bite of an infected sand fly invade a vertebrate macrophage (a form of nucleated white blood cell that functions as a phagocyte to ingest and destroy foreign bodies); a flagellum is located at the anterior end of these protozoans. (B) Within the infected macrophage, the paramastigotes pass through a developmental cycle, leading to the production of round, nonflagellated amastigotes inside localized areas known as parasitophorous vacuoles. (C) When released from the macrophage, mature, spherical amastigotes are phagocytized by healthy macrophages and continue the cycle in the vertebrate (go back to A) or (D) infected macrophages can be taken up by a feeding sand fly. (E) Inside the alimentary tract of the sand fly, amastigotes released from the infected macrophages acquire flagellae. (F) These amastigotes become slender promastigotes that multiply by simple asexual division (fission). (G) Eventually, flagellated short paramastigotes (also called metacyclic promastigotes) appear and migrate to the anterior portion of the sand fly's alimentary tract. They are introduced into the vertebrate when the sand fly takes its next blood meal. The cycle continues at A.

detection and selection of hosts, while the succeeding stages are parasitic, only slightly mobile, and often grub-like in appearance. Because of their unique structure these first-stage larvae are often referred to as planidia (a term most frequently used with hypermetamorphic Diptera, Lepidoptera, and Hymenoptera) or triungulins (used for hypermetamorphic larvae of Strepsiptera, Meloidae, and Mantispidae). In the Meloididae, the triungulins are so specific that they have been used in classifying the family (MacSwain, 1956).

Although some adult members of extant hypermetamorphic groups have fossil records dating back to the Cretaceous, actual evidence of this condition is dependent on discovering

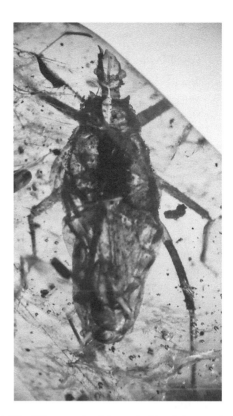

FIGURE 74 *Triatoma dominicana* in Dominican amber (Poinar amber collection accession no. HE-4-73); length of insect = 19.5 mm.

FIGURE 75 Metatrypanosomes of *Trypanosoma antiquus* in a fecal droplet of *Triatoma dominicana* in Dominican amber (Poinar amber collection accession no. P-3-3); length of metatrypanosome flagellates ranges from 11 to 20 μm.

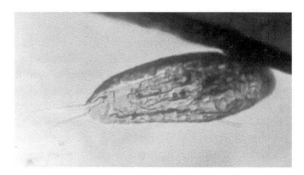

FIGURE 76 Strepsipteran triungulin that emerged from the body of a delphacid planthopper in Dominican amber (Poinar amber collection accession no. HO-4-33a); length of triungulin = 156 μm.

TABLE 5
Fossil Evidence of Triungulins and Planidia

Parasite	Host	Location	Refs.
Meloidid	Unkown	Baltic amber	Larsson (1978)
Meloidid	Bee (Megachilidae)	Baltic amber	Engel (2005)
Meloidid	Bee (Apidae)	Dominican ember	Poinar (1992b, 2009); Poinar and Poinar (1999a)
Strepsiptera	Planthopper (Delphacidae)	Dominican amber	Poinar (2004d)
Strepsiptera (?)	Unknown	Eocene brown coal	Kinzelbach and Lutz (1985)
Strepsiptera (?)	Unknown	Burmese amber	Grimaldi et al. (2005)
Rhipiphoridae (?)	Unknown	Burmese amber	Grimaldi et al. (2005)

the planidia or triungulins. This is not an easy task because most of these larvae are under 2 mm in length, and unless they are attached to specific insects they are often overlooked or misidentified. Only members of the Strepsiptera and Meloidae (Coleoptera) have been reported as fossils (see Table 5).

*4AIIIzi. Gregarine Infections in Insects

by George O. Poinar, Jr.

Primigregarina burmanica n. gen., n. sp., an Early Cretaceous gregarine (Apicomplexa: Eugregarinorida) parasite of a cockroach (Insecta: Blattodea).

Introduction—Cockroaches (Blattaria) are one of the most successful insect groups today, and their presence has dominated subtropical and tropical habitats since their appearance in the Carboniferous (Bell et al., 2007; Rasnitsyn and Quicke, 2002). Their success can be attributed in large part to an omnivorous diet consisting of a variety of living and dead organic matter, a body architecture built for concealment and escape from predators, and a natural immunity against a host of potential parasites and pathogens.

Gregarines occur in a wide variety of marine and terrestrial invertebrates and are probably one of the most ancient groups of protozoan parasites. Cockroaches may well have been the first insects to host gregarines, which in turn were one of the few protozoan lineages that developed successful parasitic associations with blattids. These relationships persist today, with at least five families (Diplocystidae, Stenophoridae, Gregarinidae, Hirmocystidae, and Actinocephalidae) of Eugregarinorida parasitizing extant cockroaches (Roth and Willis, 1960; Clopton, 2000). The present study describes an eugregarine parasite of an Early Cretaceous cockroach.

The amber piece (accession no. B-OR-1C) is rectangular in shape, measuring 100 mm long, 8 mm wide, and 2 mm deep. Observations, drawings, and photographs were made with a Nikon SMZ-10 R stereoscopic microscope and Nikon Optiphot compound microscope with magnifications up to 1060×. Photographs of the protists were processed in Adobe Photoshop under various degrees of contrast and brightness to reveal details depicted in the accompanying figures.

The amber was obtained from a mine first excavated in 2001 in the Hukawng Valley, southwest of Maingkhwan, in Kachin State (26°20′N, 96°36′E), Myanmar (Burma). On the basis of paleontological evidence, the Noije Bum 2001

Summit Amber Site was assigned to the Early Cretaceous Upper Albian (Cruikshank and Ko, 2003), placing the age at 97 to 110 mya. Nuclear magnetic resonance (NMR) spectra and the presence of araucaroid wood fibers in amber samples from the same site indicate an araucarian (possibly *Agathis*) tree source for the amber (Poinar et al., 2007a). Terminology and systematics follow those of Clopton (2000).

Results—A trophozoite and three gametocysts of an eugregarine protozoan were immediately adjacent to a half-eaten cockroach in Burmese amber. Because of its immature state and the damage it sustained, the cockroach host (4.2 mm in length) could not be assigned to a specific family (Figure 77a). The dorsum of the abdomen and pronotum had been removed, and almost all internal tissues (gut and fat body) were gone. While the tissues were being pulled from the body cavity and devoured, some of the gut contents spilled out and remained associated with the corpse. Among the gut contents were a trophozoite and three gametocysts of an eugregarine parasite. These stages are described below as a new genus of eugregarine parasites. The classification system followed is based on that of Clopton (2000):

Apicomplexa Levine, 1970
Eugregarinorida Léger, 1900
Stenophaoricae Levine, 1984
Monoductidae Ray and Chakravarty, 1933
Primigregarina n. gen.

Type species—*Primigregarina burmanica* n. sp.

Diagnosis—Trophozoite consisting of epimerite with short neck expanding into button-like disc surrounded by epicytic collarette; protomerite transverse; deutomite semihemispherical; gametocysts spherical, with single sporoduct; spores ellipsoidal.

Primigregarina burmanica Poinar

Type host—An Early Cretaceous cockroach of undetermined family status.

Type locality—Hukawng Valley, southwest of Maingkhwan, in Kachin State (26°20′N, 96°36′E), Myanmar (Burma).

Type material—Holotype (accession no. B-OR-1C) deposited in the Poinar amber collection maintained at Oregon State University, Corvallis, OR 97331.

Description—See Figure 77 and Figure 78.

Characters as listed under generic diagnosis:

Trophozoite (*n* = 1): Total length, 187 µm. Epimerite with short neck expanding into a button-like disc surrounded by an epicytic collarette; epimerite length, 67 µm; width at top of collar, 72 µm; epimerite–protomerite septum, 29 µm. Protomerite transverse length, 24 µm; greatest width, 93 µm; protomerite–deutomerite septum faint, 93 µm. Deutomerite hemispherical; deutomerite length, 94 µm; greatest width, 112 µm. Deutomerite filled with packets varying from 8 to 12 µm in diameter containing elongate bodies ranging from 1.7 to 2.5 µm in length; nucleus adjacent to protomerite–deutomerite septum, 24 µm in diameter.

Gametocyst (*n* = 3): Spherical, dark-brown gametocysts, specimen one (113 µm in diameter) lacking basal discs and sporoducts; specimen two (190 µm in diameter) with basal disc and initial sporoduct; specimen three (215 µm in diameter) with single extruded sporoduct. Length of bulbous enlargement at base of extruded sporoduct on specimen three, 23 µm; length of sporoduct, 143 µm. All gametocysts surrounded by gelatinous wall varying in thickness from 20 to 38 µm. Spores ellipsoidal, hyaline; length, 3 to 5 µm; width, 1 to 2 µm. Sporozoites not observed.

Remarks—*Primigregarina* is tentatively placed in the family Monoductidae based on the single sporoduct of the gametocysts and ovoid, hyaline oocysts (Clopton, 2000). Unique characters are the epimerite with a short neck expanding into a button-like disc surrounded by an epicytic collarette. The epimerites of extant members of this family have prongs, and the hosts are insects and millipedes; however, at least one species, *Phleobum gigantium* (Haldar and Chakraborty, 1974), lacks an epimerite and parasitizes Orthoptera (grasshoppers). The indistinct septum between the protomerite and deutomerite is similar to that of some grasshopper eugregarines illustrated by Hyman (1940). Some eugregarines, such as species of *Garnhamia* Crusz and the monoductid *Schneideria* Léger, lack a protomerite–deutomerite septum completely. In species of *Gamocystis* Schneider, the protomerite is transitory and absent in mature trophozoites (Clopton, 2000). The rod-shaped bodies in the deutomerite are interpreted as paraglycogen bodies and are similar in size and shape to paraglycogen bodies reported in the gamont of other eugregarinids (Landers, 2002).

The gametocyst of *Primigregarina burmanica* shares many characters with the extant cockroach parasite *Gregarina blattarum* von Siebold and the extant grasshopper parasite *Gregarina garnhami* (Canning, 1956). A bulbous

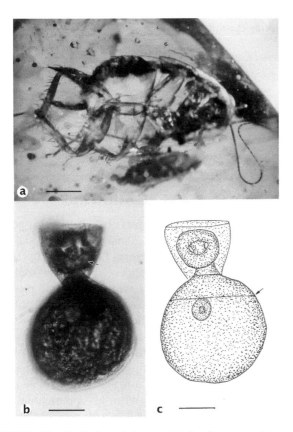

FIGURE 77 (a) Cockroach host in Early Cretaceous Burmese amber. (b) Trophozoite of *Primigregarina burmanica* n. sp.; arrow points to the button-like disc portion of the epimerite. (c) Drawing of trophozoite of *P. burmanica* n. sp. showing faint protomerite–deutomerite septum (arrow) and nucleus (Poinar amber collection accession no. B-OR-1C). Scale bars: a, 930 µm; b, 40 µm; c, 40 µm.

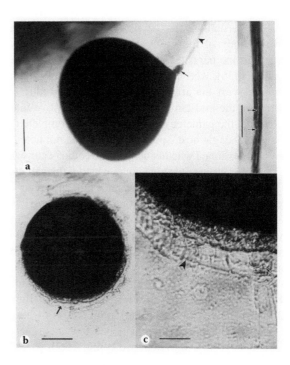

FIGURE 78 *Primigregarina burmanica* n. sp. (a) Dehiscing gametocyst. Arrow shows bulbous enlargement at the base of the sporoduct, and the arrowhead indicates the everted sporoduct. Insert on right shows portion of sporoduct containing a row of spores (arrows). (b) Gametocyst of specimen two; arrow shows gelatinous deposit on outer wall. (c) Detail of portion of specimen two gametocyst (Poinar amber collection accession no. B-OR-1C). Arrowhead indicates the outer gelatinous deposit surrounding the wall of the gametocyst. Note the additional spores adjacent to the gametocyst wall. Scale bars: a, 55 µm; a insert, 20 µm; b, 58 µm; c, 35 µm.

enlargement at the base of the sporoduct in *P. burmanica* is almost identical to that in the gametocysts of *G. blattarum* (Sprague, 1941). Gametocysts of *G. garnhami* are spherical and range in diameter from 114 to 470 μm, which compares with the range of 113 to 215 μm for *P. burmanica*. The width of the gelatinous layer surrounding the gametocysts ranges from 11 to 50 μm in *G. garnami*, which compares with the range of 20 to 35 μm in *P. burmanica*. Canning (1956) noted that the gelatinous material was deposited in layers, a pattern also noted on the gelatinous layer of *P. burmanica* (Figure 78).

While *Gregarina garnhami* appears to have a consistent number of eight sporoducts, evidence of only one sporoduct was noted on two gametocysts of *Primigregarina burmanica*. The diameter of the basal discs of the spore ducts of *G. garnhami* varied from 20 to 35 μm depending on the size of the gametocysts. The diameter of the spore disc on the gametocyst of specimen two of *P. burmanicus* is 30 μm. A difference between the two species was noted in the spore size, with those of *G. garnhami* ranging from 6.5 to 7.0 μm in length, while those of *P. burmanica* were only 3.0 to 5.0 μm.

Discussion—The structure of the epimerite is unique for *Primigregarina burmanica*. It is likely that the button-like disc was horizontal in position with the short neck protruding as a short style that made contact with the host cell. The presence of a dehiscing gametocyst of *P. burmanica* is interesting, as, normally, gametocysts dehisce after being passed from the body of the host. The question of autoinfection by eugregarine parasites of dictyopterids was discussed by Canning (1956), who noted that gametocysts of *Gregarina garnhami* with mature spores and basal sporoduct discs occurred in the mid-gut of the grasshopper host. While this showed that the gametocysts could undergo complete development in a single host, spore expulsion was not noted, and Canning questioned whether dehiscence could occur without an external stimulus. It is not known whether the gametocyst of *P. burmanica* that was in dehiscence indicates precocious development and autoinfection or if the resin provided an external stimulus that triggered dehiscence. Because gametocysts are thick walled and structured to survive in the immediate environment, they could have persisted in the resin long enough to dehisce.

Although there is no doubt that the fossils described here are gregarines, consideration was given to the possibility that they represented the larval stages of cestodes and acanthocephalans, both of which utilize cockroaches as intermediate hosts today (Roth and Willis, 1960). Regarding acanthocephalans, three larval stages may occur in the body cavity of cockroaches: the acanthor, acanthella, and cystacanth. The latter two stages are 1 mm or greater in size, which eliminates them from further consideration. Even without the size limitation, the acanthella possesses a rudimentary proboscis, proboscis and ligament sheaths, and gonad anlage, which are not present in the fossils. The acanthor stage (which exits from the egg) varies from 117 to 428 μm, a range that overlaps in size with the gametocysts

described here. However, this stage is characterized by rostellar hooks, which are not present in the fossils (Cheng, 1964; Chandler and Read, 1961).

Cystode larvae or cystercoids are over 1 mm in length and possess an inverted or withdrawn scolex with a definite rostellum containing partially developed hooks and clearly defined suckers (Cheng, 1964; Chandler and Read, 1961). None of these characters is present on the fossils.

Primigregarina burmanica represents the first fossil gregarine and shows that cockroaches harbored these parasites in the Early Cretaceous.

☐ *Reliability*

Category 1, frozen behavior.

*4AIIIzj.   Association between Fly Planidium and Mites

Various stages of insect parasites actively search for their hosts. Modified first-stage larva (planidia) of acrocerid flies actively search for spider hosts, which they enter to complete their development. One such planidium was found associated with an amystidid mite in Baltic amber (Kerr and Winterton, 2008). Although the authors suggest possible mite parasitism by this fly group, the body size of the mite would restrict parasite development, and no extant mites are known to be successfully parasitized by acrocerid flies. It is likely that this association was accidental, possibly the acrocerid planidium mistaking the mite for a spider.

☐ *Reliability*

Category 6, considerable uncertainty about the relation.

SUMMARY

Some of the examples are at the family level, with only a few getting down to the generic level. (See Table 6.)

4B.   ANIMAL–PLANT RELATIONSHIP

Scott et al. (1992) and Labandeira (1998d) provided general accounts, with some specific examples, of arthropod–plant interactions, Paleozoic to Cenozoic. Labandeira (2007) provided a number of specific examples (especially for some leaf mine and leaf gall types) and reviewed the overall problems involved. Krassilov (Krassilov and Rasnitsyn, 2008) provided a ground-breaking account of leaf mine and gall form morphology, as well as predation scar morphologies made on the mine- and gall-forming organisms, using specific examples from the Cretaceous of Israel; this account places the description of these traces (his "phyllostigmas") on a usable ichnogenus basis for the first time. When his approach has been applied to the Cenozoic there will be a far better understanding of mining and galling relations, as well as of the predators involved in attacking the mine- and gall-forming animals.

Krassilov and Bacchia (2000) make the interesting suggestion that early angiosperms suffered higher levels of galling, mining, and leaf cutting than later forms.

TABLE 6

Terrestrial Animal–Animal Parasite Relationships Known from the Fossil Record (Additions to Table 6 of Boucot, 1990a)

Behavioral Type	Taxonomic Level	Time Duration	Category	Refs.
4AIIIa. Nematode–insect				
Nematode–drosophilid	Phylum–phylum	Eocene–Holocene	1	Poinar (1992a); Poinar and Poinar (1999a)
Nematode–biting midge	Family–family	Early Cretaceous	1	Poinar et al. (1994b); Poinar and Milki (2001)
Nematode–planthopper	Family–family	Eocene–Holocene	1	Poinar (2001a, 2001c)
Nematode–ant	Family–family	Eocene–Holocene	1	Poinar (2002a)
4AIIIb. Hairworm–insect	Family–family	Eocene–Holocene	1	Poinar (2000a)
4AIIId. Ticks and mites as micropredators				
Erythraeid–biting midge	Family–family	Early Cretaceous–Holocene	1	Poinar and Milki (2001)
Erythraeid–moths	Family–family	Eocene–Holocene	1	Poinar et al. (1991)
Water mite–caddisfly	Family–family	Eocene–Holocene	1	Bachofen-Echt (1949)
Ceratopogonidae–Erythraeidae	Family–family	Late Cretaceous–Holocene	1	Poinar et al. (1993a)
Mite–coccid	Family–family	Cretaceous–Holocene	1B	Koteja (1998); Koteja and Poinar (2004)
Argasidae–rodent	Family–order	Oligocene–Holocene		Poinar (1995b)
4AIIIe. Mosquitoes as micropredators	Family–class	Late Cretaceous–Holocene	2B	Poinar (1993); Poinar et al. (2000a)
4AIIIg. Parasitoid wasp–insect hosts				
Dryinid–homopteran	Family–suborder	Oligocene–Holocene	1	Poinar (1992, 2001b)
Dryinid–planthopper	Family–family	Oligocene–Holocene	1	Poinar (1992b); Poinar and Poinar (1999)
Adult trichopteran–hymenopteran parasite	Family–family(?)	Eocene	1	Poinar and Anderson (2005)
Braconid–ant	Family–genus	Eocene–Holocene	1	Poinar and Miller (2002)
4AIIIi. Parasitoid wasps and parasitic flies–spiders				
Acroceridae–spiders	Family–order	Jurassic–Holocene	2B	Carpenter (1992)
Polysphinctinid–Ichneumonidae	Subfamily–family	Oligocene–Holocene	1B	Poinar (1992b)
4AIIIt. Lizard–tick	Suborder–family	Eocene–Holocene	1	Szadziewski (1999)
4AIIIx. Inflammatory arthritis	Class	Paleocene–Holocene	1	Rothschild et al. (1998a)
4AIIIy. Vertebrate intestinal parasites	Order	Early Cretaceous	1	Poinar and Boucot (2006)
4AIIIz. Parasite eggs on a bird feather	Class–order	Early Cretaceous	1	Martill and Davis (1998a)
4AIIIza. Trematode eggs	Phylum	Pleistocene	5A	Jouy-Avantin et al. (1999)
4AIIIzb. Spider–mermithid relation (clubionid–mermithid)	Family–family	Eocene–Holocene	1	Poinar (2000a)
4AIIIzd. Strepsipteran parasitism	Order–order	Cretaceous–Holocene	1, 2B	Grimaldi et al. (2005); Lutz (1992)
4AIIIze. Phorid fly–Allanonematid nematode	Family–family	Eocene–Holocene	1	Poinar (2003d)
4AIIIzf. Insect vector–trypanosome relationships	Family–family–class	Mid-Cretaceous	1	Poinar and Poinar (2004a,b)
4AIIIzg. Biting midge–cytoplasmic polyhedra virus	Family–family	Early Cretaceous	1	Poinar and Poinar (2005)
4AIIIzh. Hypermetamorphosis	Order–order	Cretaceous–Holocene	1	MacSwain (1956); Rasnitsyn and Quicke (2002)
4AIIIzi. Eugregarines of an Early Cretaceous cockroach	Order–order	Early Cretaceous	1	See text
4AIIIzj. Association between fly planidium and mites	Class–class	Eocene–Holocene	6	Kerr and Winterton (2008)

4Ba. HICKORY APHID–LEAVES AND APHID–PLANT RELATIONSHIPS

While describing a Dominican amber aphid, Heie and Poinar (1988) pointed out that the modern aphids really begin in the Cretaceous with the angiosperms, whereas the older aphids occurred first on gymnosperms. Poinar and Brown (2005, 2006b) described three new genera of Early Cretaceous aphids in Burmese amber that had the hind wings reduced to stubs or hamulohalters. Such hamulohalters are characters of scale insects and are unknown in extant aphids. Moran (1989) used the modern distribution of the aphid *Melaphis*, which generated galls on *Rhus* in China and North America, to deduce the Eocene presence of this coevolved relation because of the presence of the plant in the Alaskan Eocene; the post-Eocene disjunct distribution pattern reminds one of the *Nothofagus*–aphid relationship (Boucot, 1990a) that may date from the Cretaceous.

☐ *Reliability*

Category 5B, for the aphid–plant relation, functional morphology.

4Bb. ARTHROPOD LEAF MINERS

(See color plate 4A after page 132.) Krassilov (Krassilov and Rasnitsyn, 2008) described various Israeli Cretaceous leaf mines while making an important advance by using ichnogenera and species (his "phyllostigmas") to describe the material. Stephenson and Scott (1992) reviewed evidence of leaf mining. Castro (1997) described leaf and rhachis mine type of damage from the Spanish Stephanian. Srivasta (1996) described some Lower Gondwana structures that may be leaf mines.

Scott et al. (2004) described various South African Late Triassic leaf mines from beds with associated insects, but they were not able to attribute the mines to any specific insect types, although the relative abundance of the mines is in contrast to that from most of the younger Mesozoic. Labandeira and Allen (2007) cited what they concluded to be acceptable evidence of Triassic leaf mining.

Labandeira et al. (1994) described what they considered to be reasonable leaf-mining evidence from mid-Cretaceous Dakota Formation localities that can be assigned to several lepidopteran families, including two generic identifications. They concluded that the adaptive radiation of the lepidoptera occurred in the Late Jurassic and Early Cretaceous, significantly earlier than the appearance of the angiosperms; that is, switching to angiosperm leaves occurred in the earlier to mid-Cretaceous.

Labandeira (1998a) cited Dakota Sandstone leaf mines. Rasnitsyn and Quicke (2002, Figure 475) illustrated two Turonian leaves from Kazakhstan with alleged nepticulid mines and discussed other insect-generated plant-feeding traces. Labandeira (2002a, Figure 210) provided various data on leaf mining with examples extending back to the Jurassic.

Jarzembowski (1989) described several late Paleocene microlepidopteran leaf mines from southern England. Wilf and Labandeira (1999) discussed levels of leaf mining in the late Paleocene and Early Eocene of Wyoming. Wilf et al. (2006) provided illustations of a number of Paleocene leaf mines.

Johnston (1993) illustrated a middle Eocene leaf mine from northern Mississippi. Labandeira (2002b, Figure 4) discussed various examples of leaf mining from the Middle Eocene of Republic, Washington, with lepidopterans implicated in some of the examples, and discussed nepticulids in particular. Lang et al. (1995) described some English Eocene leaf mines, and Meyer (2003, Figure 44) illustrated a Late Eocene Florissant leaf mine.

Givulescu (1984) described various leaf mines from the late Miocene (Pannonian) of Chiuzbaia. Poinar and Brown (2002, Figure 12) illustrated the skeletonization of a *Hymenaea* leaflet from Tertiary Mexican amber, with the moth family Gracillariidae being a good possibility.

☐ *Reliability*

Category 2B, based on functional morphology.

4Bc. LEAF GALLS

Dreger-Jauffret and Shorthouse (1992) provided an excellent account of gall-forming diversity among the insects, while Roskam (1992) reviewed the high level of morphological convergence shown by many unrelated groups of gall formers (Figure 79). Bernays and Chapman (1994) provided excellent background on just how many arthropods actually manage to find their way to the galling sites present on various plants and their organs.

As noted earlier, Krassilov (Krassilov and Rasnitsyn, 2008) described various Israeli Cretaceous leaf galls while making an important advance by using ichnogenera and species (his "phyllostigmas") to describe the material. Castro (1997) described gall-type evidence on Stephanian, Pennsylvanian leaves from Spain that are among the older evidence of this type of activity. Labandeira and Allen (2007) described some Texas Early Permian leaf galls. Vasilenko (2007) described some Russian Permian leaf galls, using the Vialov trace fossil nomenclature for insect leaf damage and galls.

Srivasta (1996) discussed some possible leaf galls from the Lower Gondwanas, Permo–Carboniferous. Vishnu-Mittre (1957) described an Indian Jurassic leaf gall, with mites possibly being involved. Ash (1997; for more Chinle Formation leaf margin insect damage, see also Ash, 1999) summarized various animal–plant interactions involving Late Triassic, Chinle Formation leaves, and showed that galls and various feeding traces on leaves are present. He also commented about the relatively low frequency of affected leaves and discussed several potential explanations.

Scott et al. (1994) provided an excellent overview of fossil galls, with emphasis on leaf galls, including new data on Late Cretaceous and Tertiary materials; they suggested that the pre-Cretaceous gall examples are somewhat questionable,

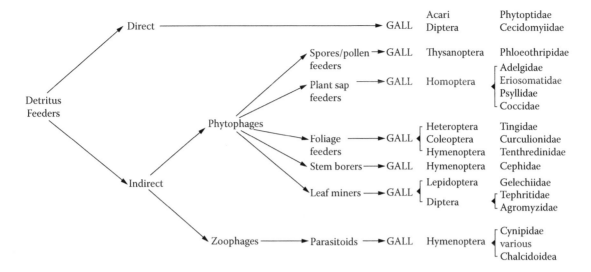

FIGURE 79 Ancestral trophic types of gall-inducing animals. (Figure 3.1 in Roskam, J.C., in *Biology of Insect-Induced Galls*, Shorthouse, J.D. and Rohfritsch, O., Eds., Oxford University Press, Oxford, 1992, pp. 34–49. Reproduced with permission.)

and their Table 1 provides insight by emphasizing that the modern gall-forming arthropods appear in the Jurassic and younger which further supports this conclusion. Larew (1992) has provided an authoritative review of leaf galls in the fossil record, including some very helpful comments about Early (not Late) Cretaceous material from Maryland. Vasilenko (2005) described some leaf gall material from the Jurassic–Cretaceous of Transbaikalia.

While describing an early Miocene Spanish phylloxerid aphid, Heie and Penalver (1999) summarized known occurrences of leaf galls from the Arkansas Eocene that can be attributed to this group (Lewis, 1989; Lewis and Heikes, 1991; Wittlake, 1969, 1981) plus additional less certain occurrences from the Paleocene and Late Cretaceous (Brown, 1962; Wittlake, 1969). Wilf and Labandeira (1999) discussed levels of leaf galling in the late Paleocene and Early Eocene of Wyoming. Erwin and Schick (2007) described cynipid wasp oak leaf galls from the Miocene of western North America.

Labandeira (2002b, Figure 3) discussed various Middle Eocene leaf galls from Republic, Washington. Stephenson and Scott (1992) illustrated Eocene leaf galls from England and from the Claiborne Formation in the Gulf Coast region. Archibald and Mathewes (2000) cited some Early Eocene leaf galls from British Columbia. Jarzembowski (1992) described an unidentified "larva" from inside a gall-like structure from the early Eocene London Clay. Meyer (2003, Figure 43) illustrated a Late Eocene Florissant leaf gall.

Givulescu (1984) described various leaf galls from the late Miocene (Pannonian) of Chiuzbaia. Waggoner and Poteet (1996) described an elongate set of galls on a mid-Miocene oak leaf from Nevada that resemble those made by one group of cynipid wasps; their evidence appears to be very convincing, and this is a truly unique specimen. Waggoner (1999) described a similar oak leaf gall from the Oregon Miocene. Diéguez et al. (1996a,b) described and discussed various late Miocene leaf galls from Spain; a number of plant leaf genera were involved, and, based on a number of very well-preserved

specimens, mites, flies, and hymenopteran formers were also suggested. Marty (1894) briefly described a French Pliocene leaf gall ascribed to *Cecidomyia* on a leaf assigned to *Fagus*.

☐ *Reliability*

Category 2B, functional morphology.

4Bd. STEM AND PETIOLE GALLS

Krassilov (Krassilov and Rasnitsyn, 2008) described various Israeli Cretaceous petiole mines while making an important advance by using ichnogenera and species (his "phyllostigmas") to describe the material. Larew (1992) provided an excellent summary of fossil stem and petiole galls. Labandeira (1991, 1993; Labandeira and Phillips, 1996a) described a Pennsylvanian example of stem mining that antedates evidence for holometabolous development (holometaboly is unknown from body fossil evidence in the Pennsylvanian) and features petioles with frass-containing tunnels that show evidence for subsequent development of parenchymatic tissue growth. Labandeira and Phillips (1996a; see also Labandeira and Phillips, 2002, Table 1, for excellent coverage of various Paleozoic to Cretaceous examples) further documented this Late Pennsylvanian material from *Psaronius* petioles contained in Illinois Basin coal balls and provided illustrations of coprolites and frass from within the galls, as well as other critical features. This is the best documented evidence of a plant gall in the Paleozoic. Labandeira (2002a, Figure 2.11) reviewed Pennsylvanian and younger evidence of various gall types, with an emphasis on leaf galls. Grauvogel-Stamm and Kelber (1996) described stem galls from the Keuper of western Germany and eastern France. Labandeira (2002b, Figure 2) discussed some stem galls from the Middle Eocene of Republic, Washington.

☐ *Reliability*

Category 2B, functional morphology.

4Be. CONE GALLS

Larew (1992) provided an excellent summary concerning fossil cone galls.

□ *Reliability*

Category 1, frozen behavior, and Category 2B, functional morphology.

4Bf. SCALE INSECTS

Scale insects of adult female and nymphal armored scale insects (Hemiptera: Coccoidea: Diaspididae) were found in dicotyledonous and monocotyledonous fossil leaves in Middle Eocene deposits from Germany. The structure of the fossil scale covers closely resembles that of extant members of the subfamily Aspidiotinae.

□ *Reliability*

Category 1, frozen behavior.

*4Bg. ACORN GALLS

Larew (1987, 1992) discussed and illustrated galled acorns from the Rancho La Brea deposits that were first noted by Templeton in an unpublished dissertation (see Boucot, 1990a, p. 231). Larew ascribed the damage to the cynipid *Callirhytis milleri*, which generates similar damage today.

□ *Reliability*

Category 2B, functional morphology.

*4Bh. SEED AND SPORE BORING

Scott (1992; for seeds, see also Collinson, 1999a) summarized much of what has been reported about borings attributed to insects into seeds and spores. Collinson (1990, Figure 46a) illustrated an insect exit hole from the Late Eocene of southern England and an insect exit hole from the Messel Eocene. Taylor and Scott (1983, Figure 7) illustrated

British Pennsylvanian seeds and spores with evidence for insect boring. Poinar (1999c) described a seed predating Dominican amber palm beetle (Figure 80), as well as a Cretaceous member of the group (Poinar, 2005d), which indicates that seed predation of palms has a long history. Labandeira (2002a, Figure 2.12) provided various evidence, beginning in the Pennsylvanian, of seed boring. Mikuláš et al. (1998) described bored Miocene and Pleistocene *Celtis* seeds from Bohemia.

□ *Reliability*

Category 2B, functional morphology, in the absence of seeds preserved with their "predators."

*4Bi. BARK BEETLE MYCANGIA

Batra (1963) reviewed the ecology and morphology of the ambrosia beetles and their fungal-containing mycangia that are used for fungal inoculation (Figure 81). Grimaldi (1996, pp. 122, 124, 125) described Dominican amber bark beetles with preserved mycangia that still contained spores and hyphae of the fungus used to innoculate the trees used by the beetles as a source of their fungal ambrosia.

□ *Reliability*

Category 1, frozen behavior.

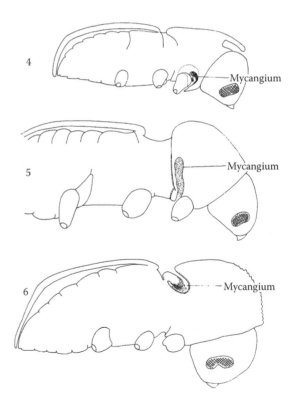

FIGURE 81 Mycangia present in three beetle types: (4) *Monarthrum fasciatum*, (5) *Trypodendron lineatum*, and (6) *Anisandrus dispar* (all ×8). (Figures 4 to 6 in Batra, L.R., *Transactions of the Kansas Academy of Science*, 66, 213–236, 1963. Reproduced with permission.)

FIGURE 80 A palm beetle, *Caryobruchus dominicanus*, in Dominican amber (Poinar amber collection accession no. C-7-104A); length of beetle = 8 mm.

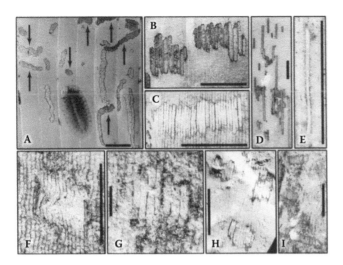

FIGURE 82   Recent and fossil *Cephaloleichnites strongi* hispine damage on Zingeriberales leaves. (A) Live and (B) dried leaves; (C, E, H) early Eocene leaves; (F, G) latest Cretaceous. (A) *Chelobasis perplexa* larva feeding on a *Heliconia curtispatha* leaf. (D) Hispine damage on a modern *Renealmia cernua* leaf. (E), (H), and (I) are early Eocene. Scale bars in all panels equal 5 mm, except in G, where the scale bar is 1 mm. (Figure 1 in Wilf, P. et al., *Science*, 289, 291–294, 2000. Reproduced with permission from AAAS.)

*4Bj.   Hispine Beetle–Ginger Grazing

Wilf et al. (2000) provided an excellent example from the Maastrichtian Hell Creek Formation in North Dakota of hispine beetle feeding on ginger leaves (*Zingiberopsis*), as evidenced by their unique trace fossil *Cephaloleichnites strongi* (Figure 82). Similar examples are also known from the Early Eocene Wasatch and Golden Valley formations in Wyoming. In the modern biota there are many examples of relatively obligate relationships between hispine beetle herbivory, producing the identical type of trace and specific ginger species. This relationship is strictly Neotropical today, and at the subfamily level has persisted intact since the later Cretaceous. The earliest hispine body fossils are Middle Eocene; however, García-Robleda and Staines (2008) demonstrated, employing modern material, that the same trace fossil can be produced by other, totally unrelated insect groups (several lepidopterans and an unrelated beetle).

□ *Reliability*
Category 2B, functional morphology.

*4Bk.   Araucaria–Beetle Relationship and Araucaria Seed Cone Damage

Labandeira (2000) outlined the biogeographic evidence suggesting a long-term coevolutionary relationship between a Chilean–Argentinian *Araucaria* and the beetle *Palophagopsis*, as well as a closely related Australian *Araucaria* with the beetle *Palophagus*, which is closely related to the South American *Palophagoides* (Kuschel and May, 1996). They also discussed a similar beetle–cycad relation between

a South African and Australian pair. Sequeira and Farrell (2001) proposed that members of the Araucariaceae were the first trees that bark beetles (Scolytidae) attacked.

While fossil evidence of insect damage to leaves and other external plant parts is fairly obvious, past evidence of insect activity inside plant organs is rare (see Section *4Bl, Nematode–Plant Relationship, and Section 9Bc, Termite Borings in Wood). Insect frass filling a seed cavity of a permineralized seed cone of *Araucaria mirabilis* from Patagonia shows that this food resource was already utilized in the Jurassic (see color plate 4C after page 132). Additional evidence of insect activity on Arucariacaea in the Cerro Cuadrado of Argentina includes larval insect borings in permineralized boughs (Jung et al., 1992).

□ *Reliability*
Category 2B, functional morphology.

*4Bl.   Nematode–Plant Relationship

(See color plate 1D after page 132.) The earliest evidence of an association between terrestrial plants and animals includes eggs, juveniles, and adults of the nematode *Palaeonema phyticum* (Poinar et al., 2008) developing in the stomatal chambers of the Early Devonian land plant *Aglaophyton major*. The presence of family clusters within the plant tissues represents an early stage in the evolution of plant parasitism by nematodes.

□ *Reliability*
Category 1, frozen behavior.

Summary

There are a number of family-level examples and some at the generic level. Of particular interest is the commonly unsolved problem of trying to figure out just which organisms were responsible for the pre-Cretaceous leaf mines and galls. (See Table 7.)

4C.   MYCOTA–PLANT RELATIONSHIPS

Stubblefield and Taylor (1988) provided a review of advances in paleomycology. Taylor et al. (1992a) reported the first instance of fungal parasitism in algae, an Early Devonian Rhynie Chert *Palaeonitella* with an endobiontic fungus, a member of the Plasmodiophoromycetes, and commented about the relation: "What is perhaps most surprising is that the host symptoms in at least one of these interactions has remained unchanged for at least 400 million years." Remy et al. (1994) described an Early Devonian Rhynie Chert blastocladalean fungus associated with a tracheophyte. Hass et al. (1994) described mycoparasitism from the Rhynie Chert. Taylor et al. (1992b) described the relations of Chytridiomyces with the tracheophyte *Aglaophyton major* from the Rhynie Chert, and Taylor et al. (1992c) described

TABLE 7

Animal–Plant Parasitic Relations Known from the Fossil Record (Additions to Table 7 of Boucot, 1990a)

Behavioral Type	Taxonomic Level	Time Duration	Category	Refs.
4Ba. Hickory aphid–leaves	Genus–genus	Eocene–Holocene	2B	Moran (1989)
4Bb. Leaf miners	Orders	Jurassic–Holocene	2B	Labandeira (2002a)
4Bc. Leaf galls	Various families	Early Cretaceous–Holocene	2B	Larew (1992)
4Bd. Stem and petiole galls	Uncertain	Late Pennsylvanian	2B	Labandeira (1991); Labandeira and Phillips (1996a)
4Bh. Seed and spore boring	Insecta–seeds and spores	Pennsylvanian	2B	Taylor and Scott (1983); Labandeira (2002a)
4Bi. Bark beetle mycangia	Bark beetles–fungal spores and hyphae	Oligocene–Holocene	1	Grimaldi (1996)
4Bj. Hispine beetle–ginger grazing	Subfamily–genus	Maastrichtian–Holocene	2B	Wilf et al. (2000)
4Bk. *Araucaria*–beetle relationship	Genus-genus	Probably Cretaceous–Holocene	2B	Labandeira (2000)
4Bl. Nematode–plant relationship	Genus–genus	Early Devonian–Holocene	1	Poinar et al. (2008)

the relations of three Rhynie Chert fungi with the charophyte *Palaeonitella cranii*. Taylor et al. (1994) described the oldest known *Allomyces*, a member of the Blastocladiales, from the mid-Early Devonian Rhynie Chert. The host, on which the fungus is a saprophyte, is the tracheophyte *Aglaophyton major*. They commented: "Finally, the presence of this fungus by early Devonian time not only underscores the antiquity of the order but, perhaps more important, demonstrates that thallus morphology and reproductive systems in some terrestrial fungi were established very early and apparently have changed little since then."

Poinar and Brown (2003a) described some unique Burmese Cretaceous fungi. Krassilov and Makulbekov (2003) described a Late Cretaceous gasteromycete from Mongolia. Currah and Stockey (1991) described a British Columbian Eocene angiosperm parasitized by a smut fungus. They assigned the fungus to the Ustilaginales. Lindqvist and Isaac (1991, Figure 9) illustrated and briefly discussed fungal hyphae, including a possible clamp connection, from New Zealand Miocene *Podocarpus* wood.

Wang (1997, pp. 272–273) discussed the fossil record of leaf spot fungi, dating back to the Late Carboniferous, while considering some Chinese Permian materials. Johnston (1993) discussed and illustrated a middle Eocene leaf spot fungus from northern Mississippi. Lange (1978) described some early Tertiary Australian spot fungi. Phipps and Rember (2004) described epiphyllous fungi from the Miocene of Clarkia, Idaho.

Poinar (2003b) described a Mexican amber Miocene leaf spot fungus, *Asteromites mexicanus*, with pycnidia, and some Dominican epiphyllous Coelomycetes (Figure 83). Currah et al. (1998) described an Eocene tar spot on a fossil palm and its fungal hyperparasite from the British Columbia Princeton Chert.

Phipps and Taylor (1996) described arbuscular mycorrhizae from the Triassic of Antarctica. Rikkinen and Poinar (2000) described an ascomycete (*Chaenothecopsis*) from Bitterfeld amber that thrives on resin; a number of growth stages are present, and it differs little from the modern examples of this

FIGURE 83 Pycnidia (arrow) of the leaf spot fungus *Asteromites mexicanus* on a petal in Mexican amber. (Poinar amber collection accession no. SY-1-116A); diameter of pycnidia = 200 μm.

taxon. Krassilov (1982, Plate 1, Figure 15) illustrated what may be one of the oldest known agarics, a mushroom-like form from the Early Cretaceous of Mongolia. Hibbett et al. (1997) described Miocene and Cretaceous (Turonian, Raritan Formation) amber mushrooms, Homobasidiomycetes, from New Jersey.

Magallon-Puebla and Cevallos-Ferriz (1993) described a Neogene, possibly earlier Neogene, basidiocarp and spores of an earthstar (Geasteraceae: Gasteromycetes) from Puebla, Mexico. Cantrill and Poole (2005) described an Antarctic Eocene araucarian wood with remains of a saprophyte, possibly an ascomycete, present. Taylor and Osborn (1992) reviewed the record of fungal saprophytism in wood, with a record extending back to the Devonian. Lutz and Herbst (1992) described a Permian example from Paraguay. Dörfelt and Schmidt (2006) described a myxomycete slime mold from Baltic amber.

☐ *Reliability*

Category 1, frozen behavior.

TABLE 8

A Summary of Fossil Evidence for Fungus–Plant Interactions (Additions to Table 9 of Boucot, 1990a)

Type of Parasite	Taxa Known	Nature of Evidence	Time
Plasmodiophoromycetes	*Palaeonitella*	Fungal endobiontic on alga	Early Devonian–Holocene
Blastocladalean	Tracheophyte	Fungus–tracheophyte	Early Devonian–Holocene
Chytridiomycetes	Tracheophyte	Fungus–*Aglaophyton major*	Early Devonian–Holocene
Fungi	Charophytes	Fungus–charophyte	Early Devonian–Holocene
Smut fungus, Ustilaginales	Angiosperm	Fungus–angiosperm	Eocene–Holocene
Tar spot and fungal hyperparasite	Monocot	Fungus–fungus–monocot	Eocene–Holocene
Saprophytes	Tracheophytes	Fungus–tracheophyte	Early Devonian–Holocene

FUNGUS–PLANT, FUNGUS–ALGA, FUNGUS–FUNGUS, AND FUNGUS–ANIMAL FOSSIL ASSOCIATIONS

(See Table 8.) Since the review of Sherwood-Pike (1990) on the fossil evidence of plant–fungus interactions, a number of new discoveries have been made not only in the fungus–plant category but also with regard to fungus–fungus, fungus–alga (lichen), and fungus–animal associations. Some of the major recent discoveries are briefly noted here.

SYMBIOTIC ASSOCIATIONS

Fungus–Plant

Many of the recent discoveries pertain to the fascinating Devonian Rhynie Chert deposits in Aberdeenshire, Scotland, where all major groups of modern fungi occur in association with various land plants, especially *Aglaophyton* (Taylor and White, 1989; Taylor et al., 1992a,b). These and other discoveries strongly suggest that higher beneficial plant–fungal associations may have been essential for the successful establishment and colonization of land by terrestrial plants (Blackwell, 2000). Many of these early associations appear to resemble extant mycorrhizal associations, with the oldest dating from the Ordovician of Wisconsin (about 460 mya) (Redecker et al., 2000). Evidence of a lichen-like symbiosis from Ordovician deposits (Yuan et al., 2005) represents the oldest association between a fungus and an alga. Fossilized fungal hyphae and spores from this era resemble modern arbuscular mycorrhizal fungi (Glomales, Zygomycetes) and indicate that glomales-like fungi had formed associations with land flora that probably only consisted of bryophyte-like plants.

These findings support molecular estimates of fungal phylogeny that place the origin of the major groups of terrestrial fungi (Ascomycota, Basidiomycota, and Glomales) around 600 million years ago (Redecker et al., 2000). Evidence of ectomycorrhizae on rootlets in Dominican amber is shown in Figure 84. The first fossil lichen belonging to an extant genus (*Parmelia*) was discovered in Dominican amber (Poinar et al., 2000b) (Figure 85).

FIGURE 84 Ectomycorrhizae in Dominican amber (Poinar amber collection accession no. AF-9-24); length = 5 mm.

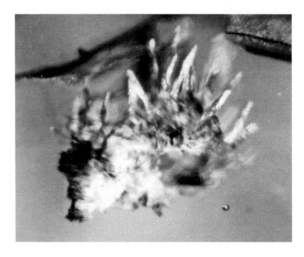

FIGURE 85 The lichen *Parmelia* in Dominican amber (Poinar amber collection accession no. AF-9-17); length = 3 mm.

PARASITIC ASSOCIATIONS

(See Table 9.)

Fungus–Plant

Recent records of fossil leaf-inhabiting parasitic fungi include representatives of the genera *Colletotrichum*, *Uncinula*, and *Phragmothyrites* recovered from sauropod dinosaur dung in

TABLE 9

Fossil Evidence of Fungus Interactions with Plants, Animals, and Other Fungi

Type	Taxa	Nature of Evidence	Time
Symbiotes			
Mycorrhizae	Various	Hyphae, resting spores; arbuscules in rhizomes	Ordovician–Holocene
Lichens	Various	Fungus and alga	Ordovician–Holocene
Parasites			
Oomycota	Unknown	Fungal hyphae and resting spores in plants	Pennsylvanian–Holocene
Leaf spot	Various	Stages on leaves, coprolites, petals	Eocene–Holocene
Mycoparasites	Hyphomycetes	Stages in fungal tissue	Cretaceous–Holocene
Hyper-mycoparasites	Phycomycetes	Hyphae	Cretaceous–Holocene
Insect parasites	Various	Fruiting bodies with spores	Cretaceous–Holocene
Saprophytes			
"Lower fungi" on wood and rhizomes	Oomycota	Hyphae and resting spores in decaying wood	Devonian–Holocene
"Higher fungi" on wood	Polyporales, Clavariaceae, Nidulariales, Agaricales	Hyphae, fruit bodies, resting spores in wood	Devonian–Holocene
Aquatic	Dermatiaceae	Isolated spores in wood	Cretaceous–Holocene
Hyphomycetes	Dermatiaceae	Isolated spores	Silurian–Holocene

Note: See also Table 9 in Boucot (1990a).

the Indian Late Cretaceous. These are all leaf-inhabiting species that were thought to develop on the leaves of the food plants of the sauropods (Kar et al., 2004). Other leaf spot fungi (see Figure 83) developing on plants parts preserved in Dominican and Mexican amber were assigned to fossil genera (Poinar, 2003b, Figure 4c).

Fungus–Fungus

Discovery of the oldest known mushroom (*Palaeoagaracites antiquus*) (Figure 86) in Early Cretaceous Burmese amber was made even more significant when a mycoparasite (*Mycetophagites atrebora*) was discovered within its tissues and a hypermycoparasite (*Entropezites patricii*) occurred within the hyphae of the mycoparasite (Figure 87) (Poinar and Buckley, 2007). This discovery shows that sophisticated patterns of fungal parasitism had been established some 100 mya. This is the first find of hyperparasitism in the fossil record.

Fungus–Plant Saprophytic Associations

There are numerous reports of saprophytic fungi growing on the remains of fossil plants. One of the earliest is the presence of wood-decaying fungi in *Callixylon newberryi* from the Upper Devonian (Stubblefield et al., 1985). The xylem of Triassic *Araucarioxylon* and Permian remains of *Vertebraria* in the Antarctic were degraded by wood-decaying fungi that showed patterns of decay similar to those of extant forms (Stubblefield and Taylor, 1986). Shelf-like, woody fruiting bodies of Polyporaceae were reported from Turonian–Maastrichtian deposits in India (Kar et al., 2003), and immature fruiting bodies of *Ganoderma*, *Xylaria*, and *Favolaschia* in Dominican amber show the presence of extant genera of saprophytic fungi in the mid-Tertiary (Figures 88 through 90). Sporocarps of the club mushroom *Palaeoclavaria burmitis* in Burmese amber (Poinar and Brown, 2003a) represent

FIGURE 87 The mycoparasite *Mycetophagites atrebora* parasitizing the agaric *Palaeoagaracites antiquus* in Burmese amber (Buckley amber collection accession no. AB 368, Florence, Kentucky). Arrow indicates a hypha of the hyperparasite *Entropezites patricii*. This discovery shows that sophisticated patterns of fungal parasitism were well developed some 100 mya.

FIGURE 86 Cap of *Palaeoagaracites antiquus* from Early Cretaceous Burmese amber. (Buckley amber collection accession no. AB 368, Florence, Kentucky); diameter of cap = 2.2 mm.

FIGURE 88 *Ganoderma* in Dominican amber, determined by E. Both (Poinar amber collection accession no. AF-9-13); width of cluster = 2 mm.

FIGURE 89 *Xylaria* in Dominican amber, determined by E. Both (Poinar amber collection accession no. AF-9-12); length of dark portion = 10 mm.

FIGURE 90 Bracket fungus (*Favolaschia* sp.) in Dominican amber, determined by E. Both (Poinar amber collection accession no. AF-9-22); diameter of specimen = 2.3 mm.

the oldest fossil record of the Clavariaceae (Figure 91). As with many of these associations, it is difficult to determine whether *Palaeoclavaria* was saprophytic or a weak parasite. Diverse fungi, most of which appear to be saprophytic, occur in Carboniferous peat deposits and coal balls (Taylor and

FIGURE 91 Two sporocarps of *Palaeoclavaria burmitis* in Burmese amber (Poinar amber collection accession no. B-P-1); length of tallest sporocarp = 3.5 mm. (From Poinar, Jr., G.O. and Brown, A., *Mycological Research*, 107, 763–768, 2003. Reproduced with permission from Elsevier.)

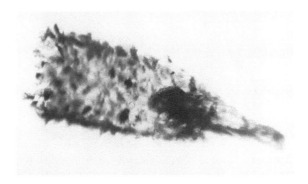

FIGURE 92 Bird's nest fungus in Dominican amber (Poinar amber collection accession no. AF-9- 20A); length of specimen = 2 mm.

White, 1989). Some first examples of fossil saprophytic fungal groups include representatives of the Nidulariales (bird's nest fungi) in Dominican (Figure 92) and Baltic (Figure 93) amber and a morel in Dominican amber (Figure 94).

Fungus–Animal Saprophytic and Parasitic Associations

Fungal mycelium can often be found on animal remains, especially arthropods in amber. Examples in Dominican amber include fungi on insects (Figure 95) and arachnids (Figure 96). Most of these associations are presumed to result from saprophytic fungi whose spores germinated after the insect died but before it was completely immersed in the resin. However, in some cases the growth patterns of the fungi are

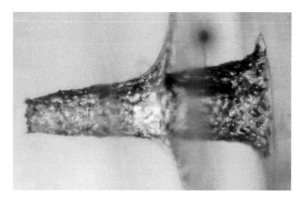

FIGURE 93 Bird's nest fungus in Baltic amber (Poinar amber collection accession no. AF-9-20B); length of specimen = 2.7 mm. (A) Lateral view; (B) ventral view.

FIGURE 94 Morel in Dominican amber, determined by E. Both (Poinar amber collection accession no. AF-9-10); length of specimen = 2 mm.

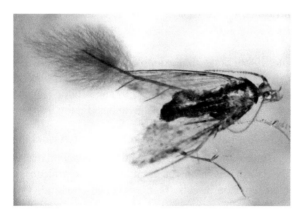

FIGURE 95 Fungus growing on moth leg in Dominican amber (Poinar amber collection accession no. AF-9-19); length of moth body = 3 mm.

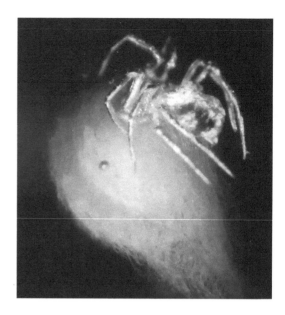

FIGURE 96 Fungus growing on a spider in Dominican amber (Poinar amber collection accession no. AF-9-14); length of spider = 1 mm.

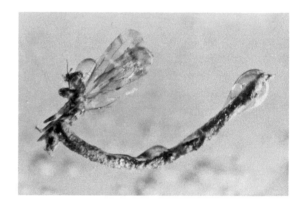

FIGURE 97 Synnematia on a psocid (Troctoposocidae: *Troctoposocpsis* sp.) in Dominican amber (psocid determined by E.L. Mockford) (Poinar amber collection accession no. Sy-1-13); length of synnematia = 3.5 mm.

so regular that it suggests some continued development in the resin after the food source was completely immersed.

Fossil evidence of animal parasitism by fungi is rare. Synnematia emerging from a psocid in Dominican amber (Figure 97) and a scale insect in Burmese amber (Figure 98) are a few examples. Other recent ones are an entomophoran infection of a fly and thalli on a culicine mosquito resembling those of present-day Trichomycetes, both of which have been reported from Dominican amber (Poinar and Poinar, 2005) (Figure 99). Poinar (2001a, Figure 4d) described a series of puffballs in Mexican amber, including what is the oldest reliable puffball (Figure 100), a gasteromycete, Early Miocene–Late Oligocene, and reviewed the fossil record of the puffballs; the Mexican puffball was actually growing on wood, thus providing evidence of a decay organism on rotting wood. A fossil ascomycete, described as *Stigmatomyces succini*, that is attached to the thorax of a stalk-eyed fly

FIGURE 98 Synnematia of *Paleoophiocordyceps coccophagus* emerging from the body of a scale insect (Hemiptera: Coccoidea) in Burmese amber (Poinar amber collection accession no. B-Sy-18); length of synnematia = 2 mm. (From Sung, G.-H. et al., *Molecular Phylogenetics and Evolution*, 49, 495–502, 2008.

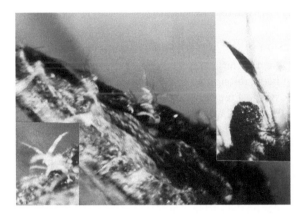

FIGURE 99 Thalli of a possible Trichomycete protruding from the abdomen of a culicine mosquito in Dominican amber. Lower left insert shows detail of several thalli, and right insert shows different types of thalli (Poinar amber collection accession no. AF-9-31); length of thalli = 85 to 105 μm. (From Poinar, Jr., G.O. and Poinar, R., *Journal of Invertebrate Pathology*, 89, 243–250, 2005.)

(Diopsidae: *Prosphyracephala succini*) in Baltic amber is the first fossil record of the Order Laboulbeniales (Rossi et al., 2005).

A fungal parasite, *Paleoophiocordyceps coccophagus*, of a male scale insect (Hemiptera: Albicoccidae) in Early Cretaceous Burmese amber provides the oldest fossil evidence of animal parasitism by fungi (Figure 98). The fossil was used to calibrate the multigene phylogeny of insect pathogens in three families of the Hypocreales. It appears that insect parasitism was the ancestral character of the Clavicipitaceae, Cordycipitaceae, and Ophiocordycipitaceae, which originated in the Jurassic (Sung et al., 2008).

RESINICOLOUS FUNGI

Some fungi are resinicolous and grow on resin deposits on the bark of trees. Spores of resicolous fungi germinate in fresh resin, although it is not clear how they utilize resin

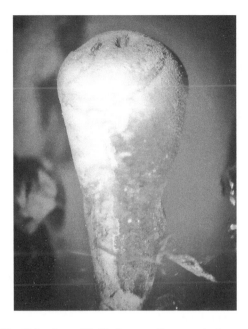

FIGURE 100 A puffball *Lycoperdites tertiarius*, growing on wood in Mexican amber (specimen deposited in the Amber Museum of Tales of the Earth, London); length of puffball = 14 mm. (From Poinar, Jr., G.O., *Historical Biology*, 15, 219–221, 2001. Reproduced with permission.)

compounds as a carbon source. A fossil resinicolous fungus in Bitterfeld amber shows that this association occurred some 20 to 30 million years ago (Rikkinen and Poinar, 2000). The fossil *Chaenothecopsis bitterfeldensis* has extracellular material on the surface of the hyphae that was interpreted to represent calcium oxalate crystals, possibly used to assist in penetrating the semisolid resin, a character found in extant forms.

4D. PLANT–PLANT RELATIONSHIPS

4Da. *ARCEUTHOBIUM* (DWARF MISTLETOE)

Wesley Wehr (pers. comm., 1990) informed Boucot that the late Jack Wolfe recognized a mistletoe, aff. *Dendrophythora*, in the Middle Eocene Republic flora of northeastern Washington. Lindqvist and Isaac (1991, Figure 10) illustrated a mass of rootlets in a coniferous wood that they suggest are those of epiphytic mistletoe from the New Zealand Miocene.

☐ *Reliability*

Category 5A, due to uncertainty about the taxonomic identity of the leaf in question.

4E. MYCOTA–ANIMAL RELATIONSHIPS

*4Ea. BARK LOUSE–FUNGUS

Poinar (1993, Figure 12; see also Poinar and Poinar, 1999, Figure 157A, p. 153) illustrated and discussed a bark louse (*Troctopsocopsis*, Troctopsocidae: Psocoptera) parasitized by a fungus that has produced a synnema similar to those found today in *Hirsutella*. Poinar and Poinar (2005) described an ectoparasitic fungus on a culicine mosquito from Dominican amber and another ectoparasitic fungus on a fungus gnat from Dominican amber.

□ *Reliability*
Category 1, frozen behavior.

*4Eb. CADDISFLY–MOLD

Kosmowska-Ceranowicz (2001, p. 65) illustrated a Baltic amber caddisfly with mold on one leg.

□ *Reliability*
Category 1, frozen behavior.

NOTES ON THE ORIGINS AND EVOLUTION OF *BACILLUS* IN RELATION TO INSECT PARASITISM

by George O. Poinar, Jr.

INTRODUCTION

Bacteria are the oldest organisms known, dating back to about 3.6 billion years. The fossil record of *Bacillus* extends back to the Carboniferous, some 320 million years ago. The original host of *Bacillus* species might have been annelids and crustaceans, although insect parasitism could have arisen as early as the Triassic, some 235 million years ago, when the beetle pathogenic strains could have evolved. The main stem of adaptive radiation of *Bacillus thuringiensis* probably occurred in the Cretaceous when the Lepidoptera were evolving in conjunction with the diversification of the Angiosperms.

ANTIQUITY OF BACTERIA

Origins in a Marine Environment

Bacteria are the oldest known organisms in the fossil record, with the earliest records (pertaining to the filamentous bacteria) reported from the 3.5-billion-year-old carbonaceous cherts of the early Archaean Warrawoona Group of northwestern Australia (Awramik et al., 1983). Thus, back in the Precambrian, after the major episode of meteorite impacts ended, four types of filamentous bacteria appeared: narrow filamentous forms (*Archaeotrichion*), small septate filamentous forms (*Eoleptonema*), tubular sheathed forms (*Siphonophycus*), and large septate filamentous forms (*Premaevifilum*) (Taylor and Taylor, 1993). All of these forms closely resemble extant species of cyanobacteria.

Probably the first associations between bacteria and eukaryotes occurred with marine invertebrates. This was some time after the period of eubacterial fusion when purple photosynthetic bacteria invaded prokaryotic cells to form mitochondria and cyanobacteria invaded red algae to form chloroplasts (Darnell et al., 1990). Various bacteria must have occurred in the intestinal tract and intersegmental folds of marine creatures. In fact, in the soft-bodied Burgess Shale Cambrian fossils there is ample evidence of decay, which almost certainly

was due to bacteria (Whittington, 1985). In the Burgess Shale fossil onychophoran *Aysheaia pedunculata*, many of the specimens show a dark stain, which has been attributed to decay (Whittington, 1978). This decay most likely was caused by bacteria, and these bacteria could have been facultative parasites with the ability to invade the body cavity of *Aysheaia*. The present-day onychophoran *Peripatopsis moseleyi* from Natal is susceptible to bacterial infections, and two specimens kept in captivity died as a result of what appeared to be bacterial infections (Holliday, 1942). Perhaps this is one of the oldest continuous symbiotic associations at the phylum level. Many, if not all, of the Burgess Shale fossils had some type of association with bacteria. Even at that time bacteria had probably evolved into three basic associations with marine invertebrates: the generalist saprophytic forms that occurred on the integument of a range of species, specialists that could be found inside the alimentary or reproductive tracts of select genera or species, and pathogenic forms that could invade the body cavity and cause disease.

Invasion of the Terrestrial Environments

Just when bacteria invaded the land is unknown but they must have been among the first, if not the first, terrestrial organisms in the Precambrian. In fact, they undoubtedly inhabited the intertidal zone and then adapted as dry land became available. Many taxa were probably brought ashore with their invertebrate hosts and then coevolved with the latter as they adapted to various terrestrial niches. Both autotrophic and heterotrophic bacteria probably lived in moist habitats along the borders of water courses, then in decaying vegetation, and later on the surfaces of plants and animals that had taken up terrestrial residence.

Because of their small size, the fossil record of terrestrial bacteria is sparse and often acquired from indirect evidence. Thus, the presence of scars on pollen and spores has been used to determine the presence of terrestrial bacteria on plant surfaces in the Carboniferous (Taylor and Taylor, 1993). Direct evidence of bacteria comes mainly from amber deposits (Blunck, 1929; Katinas, 1983; Poinar, 1992b; Poinar et al., 1993b; Schmidt and Schäfer, 2005; Waggoner, 1993, 1994, 1996).

It is unknown when bacteria became associated with terrestrial invertebrates; however, it is logical to assume that all invertebrates that occurred in the Early Paleozoic had some type of symbiotic relationship with bacteria. Most of these were probably cases of inquilinism, with the bacteria occurring in the same habitat as the invertebrate. Later, as bacteria became established in the alimentary and reproductive tracts of invertebrates various commensal, mutualistic, and parasitic relationships developed.

Origin of the Genus *Bacillus* and Early Associations with Invertebrates

Terrestrial fossil records of *Bacillus* extend back to the Carboniferous (see Table 10). The precursors of *Bacillus* are unknown, although certain characters found in members of

TABLE 10
Reports of *Bacillus* in the Fossil Record

Bacillus Species	Deposit	Millions of Years	Refs.
B. circulans	Salt deposit	250	Dombrowski (1963)[a]
B. sphaericus	Amber	25–40	Cano and Borucki (1995)[a]
B. subtilis	Amber	25–40	Cano et al. (1994)
	Amber	40	Hamamoto and Horikoski (1994)[a]
	Sedimentary	40	Bradley (1931)
Bacillus spp.	Sedimentary	29	Renault (1896)
	Sedimentary	100	Ellis (1914)
	Sedimentary	320	Renault (1896)
	Amber	40	Galippe (1920)[a]

[a] Claimed to have been revived and cultured.

the genus *Lactobacillus* indicate that they might be the stem group of primitive *Bacillus* species. Many lactobacilli occur on the surfaces of plants and in decaying plant tissues, which would bring them into contact with various invertebrates; however, lactic acid bacteria do not form endospores and are normally immobile. Endospore formation was a significant development in *Bacillus* in both motile and nonmotile forms. Infections of amphipods by *B. subtilis* (Duncan, 1981) and earthworms by *B. thuringiensis* (Heimpel, 1966) suggest that annelids or crustaceans might have been early hosts of *Bacillus* species, possibly in a marine environment. A possible *Bacillus* sporangium associated with a mycetophagous nematode in Mexican amber is shown in Figure 101.

Associations of *Bacillus* with Insects

The genus *Bacillus* today contains species that are both facultative and obligate parasites of insects as well as a single species that parasitizes mammals. The insect parasites occur predominantly in holometabolous insects and obviously coevolved with the latter.

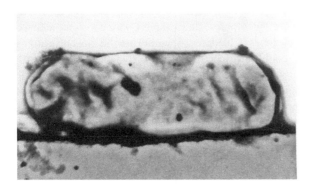

FIGURE 101 A possible sporangium of a *Bacillus* sp. in Mexican amber from inside the body cavity of the mycetophagous nematode *Oligaphelehchoides atrebora*, photographed by Roberta Poinar (Poinar electron micrograph collection accession no. 011993); length of sporangium = 2 μm.

Symbiotic Associations with *Bacillus*

The only known symbiotic associations between *Bacillus* species and insects occur in honey bees (*Apis mellifera*) and in stingless bees. In these associations, the bacteria occur in the alimentary tract of both the adult and larval stages. Adult worker bees release the bacteria from their guts onto pollen storage. Enzymes, fatty acids, and antibiotics produced by the bacteria are responsible for the predigestion, metabolic conversion, partial fermentation, and preservation of the pollen (Gilliam, 1992). The *Bacillus* species involved are all non-crystal-producing strains that have no detrimental effect on the bees, as far as is known. In fact, because the bacteria benefit by being carried to a food source and the bees profit by having their pollen supplies protected and predigested for larvae consumption, the association is considered mutualism. The *Bacillus* species involved in this relationship are *B. cereus*, *B. subtilis*, *B. megaterium*, *B. pumilus*, *B. sphaericus*, and *B. circulans* (Gilliam, 1992).

This association has been in existence for millions of years. Evidence for this comes from studies with extinct stingless bees preserved in amber. Amber from the Dominican Republic contains numerous specimens of the stingless bee *Proplebeia dominicana*. The amber preserves both the actual tissue and the DNA in select bees (Poinar, 1994).

Cano et al. (1994) recovered *Bacillus* DNA from the abdomen of *Proplebeia dominicana* preserved in 25- to 40-million-year-old amber from the Dominican Republic. Cano and Borucki (1995) claim to have cultured a strain of *B. schadericus* from the same species of extinct bee in Dominican amber. These studies corroborate earlier work by Hamamoto and Horikoski (1994), who characterized a bacterium they reportedly isolated and grew from Baltic amber and which had a 99.6% homology to *B. subtilis*. Even back in 1920, Galippe claimed to have revived bacteria (some of which appear to be *Bacillus*) from Baltic amber, although none was fully characterized or named.

On the subject of reviving ancient *Bacillus*, it is necessary to point out that in 1963 Dombrowski reported reviving

Paleozoic, Permian *Bacillus circulans* from the Kali and Zechstein salt deposits in Germany. If these reports (summarized in Table 10) on reviving ancient spores are accurate, then it demonstrates the amazing resilience of *Bacillus* spores over time. It is interesting that *Bacillus* species have never been reported as intracellular symbiotes of animals or plants.

Parasitic Associations with *Bacillus*

The earliest fossil record of *Bacillus* is from the Carboniferous some 350 million years ago (see Table 10). It is interesting to note that the earliest known insects (not including the Lower Devonian Collembola, which are no longer considered insects, although they are hexapods) are also from the Carboniferous (Carpenter and Burnham, 1985; Rasnitsyn and Quicke, 2002). Although these early *Bacillus* species had endospores, the acquisition of parasporal bodies apparently came later, when associations of insects became more specialized. The association between strains of *B. cereus* with insects today probably was similar to the relationships that occurred in the Carboniferous, namely facultative parasitism. These early *Bacillus* species existed saprophytically in the environment, yet had the ability to enter and develop within the body cavity of insects, especially when the latter were under stress or molting (Poinar and Thomas, 1984). Such facultative parasites are generalists in their selection of potential hosts and probably occurred in a wide range of insects, just like *B. cereus* today, which has been recovered from the body cavity of representatives in the insect orders Coleoptera, Homoptera, Diptera, Lepidoptera, and Hymenoptera (Thomas and Poinar, 1973).

Associations of *Bacillus* with Vertebrates

It is interesting that *Bacillus cereus* also causes opportunistic infections in vertebrates, including humans. This species, as well as *B. subtilis*, can cause infections of the eye, soft tissue, and lungs, sometimes resulting in bacteremia and endocardites. In humans, the infection is often associated with immunosuppression, trauma, an indwelling catheter, or contamination of medical equipment such as an artificial kidney. Strains of *B. cereus* isolated from human abscesses are capable of producing destructive pyogenic toxins. This *Bacillus* species can also cause food poisoning by means of enterotoxins (Ryan, 1990). The only species of *Bacillus* which causes a primary disease in vertebrates is *B. anthracis*, which is responsible for the zoonosis anthrax in mammals, including humans.

Just when *Bacillus anthracis* became pathogenic is unknown; however, the disease is most common in herbivores such as horses, sheep, and cattle and may have evolved with these groups. Horses (Perissodactyla: Equidae) appeared in the Eocene with *Hyracotherium* and evolved into various post-Eocene taxa, including the modern horse (*Equus caballus*), which only appeared in the Pleistocene. Cows and sheep, both in the order Artiodactyla, arose in the Miocene, and the modern forms appeared in the Pleistocene (Colbert, 1969). It would appear that these ungulates probably encountered spores of a pre-*B. anthracis* that had been carried up onto the leaves of grasses and other herbs. Infection leading to a fatal septicemia could have resulted from injuries to membranes of the mouth, which were probably frequent in grass-eating ungulates. The pathogen could have been spread by spores occuring in the feces of infected animals or during the decomposition of dead hosts. Spores of *B. anthracis* are known to remain viable in the soil for decades (Ryan, 1990). It is possible that at least some of the extinctions of North American ungulates in the Late Pleistocene were caused by *B. anthracis* infections.

Specializations of *Bacillus* in Relation to Toxin Production

As members of the genus *Bacillus* diversified, some formed symbiotic associations with insects (bees), others became facultative parasites of insects, and several became insect pathogens through the production of various toxins. Toxin production can be associated with various processes in *Bacillus* species.

In the case of *Bacillus sphaericus*, the vegetative cells of some strains produce toxin unrelated to sporulation, while in other strains toxin production is a sporulation-associated event and can be attributed to the formation of crystal-like polyhedral bodies (Berry et al., 1991). In both cases, toxicity was mainly effective against nematoceran Diptera in the families Culicidae and Simulidae.

In the case of *Bacillus thuringiensis*, all of the known naturally occurring strains obtain their pathogenicity from insecticidal crystals (parasporal bodies) produced during spore formation. When insects ingest these crystals, the combination of a high pH and proteases in their gut dissolves the crystals and activates their subunits into insect-toxic proteins. These activated toxins then bind to sites in the midgut, causing gut paralysis, cell lysis, and mortality (Adang, 1991). Products of the Cry genes determine the potency and toxic spectra of each *B. thuringiensis* strain against specific insects. There are currently four classes of Cry genes: (1) Lepidoptera-specific genes, (2) Lepidoptera- and Diptera-specific genes, (3) Coleoptera-specific genes, and (4) Diptera-specific genes (Adang, 1991). These specific genes were certainly acquired over a long period of association with specific insect groups.

On the basis of the known first appearance of these host groups in geological time (Table 11) we can assume that class 3 Cry genes (Coleoptera-specific) are the most primitive type and could have occurred in chrysomelid beetles during the Triassic, some 235 million years ago. With the rise of the Angiosperms in the Early Cretaceous or Late Jurassic, there appeared diversification of the Lepidoptera. Their feeding habits, coupled with the physiological conditions of their midgut, were ideal for the development of the class 1 Cry genes, and certainly the adaptive radiation of these bacterial strains would have occurred widely in the Cretaceous and Early Tertiary as they coevolved with various groups of emerging Lepidoptera.

Class 4 strains against nematoceran Diptera could well have appeared by the Early Tertiary, as this suborder is considered to contain the most primitive group of flies. Strains of *Bacillus thuringiensis* would have been expected to evolve

TABLE 11

Geological Origin of the Major Orders and Families of Insects Susceptible to *Bacillus* Species[a]

Order and Family	First Undisputed Fossil Record	Millions of Years
Coleoptera	Permian	256
Chrysomelidae	Triassic	235
Scarabaeidae	Jurassic	160
Diptera	Triassic	235
Culicidae	Cretaceous	90
Simuliidae	Jurassic	160
Lepidoptera	Cretaceous	130
Noctuidae	Eocene	40

[a] Geological data taken from Carpenter and Burnham (1985) and Rasnitsyn and Quicke (2002).

relations with those orders or families of insects with one of the following attributes: (1) feeding on exposed surfaces of plants (stems, branches, leaves) that could retain the spores and crystals of *Bacillus*, (2) feeding externally on roots closely associated or in contact with *Bacillus* spores and crystals, (3) feeding on particulate matter that contains associated spores and crystals, or (4) possessing receptive gut conditions.

Exhibiting the above types of feeding behavior does not ensure infection by *Bacillus*; however, there are many insect groups that have these feeding characteristics that must have ingested *Bacillus* spores and crystals for countless eons and never become infected. An example would be the members of the order Orthoptera which have feeding habits similar to those of the leaf-feeding Lepidoptera and Coleoptera. It is obvious that *Bacillus* species could never evolve mechanisms to adjust to the physiological conditions found in the Orthopteran gut and possibly hemocoel; thus, all families of this relatively primitive group of insects have remained essentially immune to all species and strains of *Bacillus*. When the host ranges of the hundreds of *Bacillus thuringiensis* strains that have now been isolated by the various commercial companies are eventually tested, perhaps some will show activity against insects other than those in the holometabolous groups.

Very little is known about the coevolution of bacterial pathogens with their hosts or evolution of bacteria in general. The fossil record of bacteria is very sparse, and when forms are found it is difficult to associate them with any definite group. The rare exceptions to this are the claims that the bacteria can actually be revived as in the cases with the few *Bacillus* species discussed above. Thus far, this phenomenon seems to be restricted to specialized microenvironments such as amber and salt deposits.

5 Density and Spacing

(See also Section 2As, Decapod Inquilinism within Ammonite Shells.) As discussed previously (Boucot, 1990a), one type of spacing is clumping. In some brachiopods, clumping behavior involves the penetration of "threads" from the pedicle into a carbonate substrate, resulting in a trace fossil *Podichnus*. Cited below are examples of when this type of clumping involves brachiopods.

Robinson (2005; see also Robinson and Lee, 2008, who describe the correlation between *Podichnus* types and their relations to varied brachiopod higher taxa) reports that there are at least three distinct *Podichnus* types. With regard to "making" *Podichnus*, Bromley and Surlyk (1973) made it clear that just about any carbonate substrate would do, including such things as belemnite rostra and oyster shells—not just other brachiopods belonging to the same species. Cooper (1975, Plate 4, Figures 15–21) illustrated some modern *Podichnus*, including "blisters" on shell interiors covering the spots where the pedicle threads penetrated the shell interior in *Macandrevia africana*; these are the best evidence for the mantle reaction to the penetration of pedicle threads into the shell interior. Bromley (2008) provided additional understanding of the complexities involved in *Podichnus* morphology and construction while describing examples from a Pleistocene coral.

Some of the oldest reported examples of *Podichnus*, the trace produced by articulate brachiopod pedicle threads boring into each other for attachment that results in clumps, are from the Silurian (Bromley, 2004), Middle Devonian of New York (Vogel et al., 1987), Mississippian of Utah, and the Permian of Idaho (Alexander, 1994), followed by the Rhaetian, latest Triassic of the Carpathians, in *Zeilleria* (Michalik, 1977). Bromley (2005) cited a Silurian occurrence, while reviewing the nature of the relationship. Nekvasilová (1976) described Early Cretaceous *Podichnus* from Štramberk, Czechoslovakia.

Nekvasilová (1975) described Late Cretaceous from Bohemia, as well as Late Cretaceous Bohemian *Podichnus* surrounded on three sides by a horseshoe-shaped groove presumably produced by the rocking back and forth of the closely attached brachiopod that made the attachment scar. Bromley (1994, Figure 5.3D) provided an elegant illustration of *Podichnus* borings into a Campanian belemnite from Norwich, England. Moosleitner (2000, Plate 8, Figure 1) illustrated an Early Cretaceous French *Podichnus*. Hanger (1992) described *Podichnus* in some Texas Cretaceous brachiopods. Jagt et al. (2007) described a Belgian Maastrichtian *Podichnus* on an echinoid test. Bromley and Heinberg (2006, Figure 12) provided an elegant Cretaceous example together with a resin cast of the structure that shows the radiating nature of the individual borings (see Figure 102). Sørensen and Surlyk (2008), while describing a Late Cretaceous Swedish *Podichnus* example, cited much of the literature dealing with this trace fossil. Taddei Ruggiero and Annunziata (2002) described some Early Pleistocene *Podichnus* from southern Italy, as did Taddei Ruggiero and Bitner (2008).

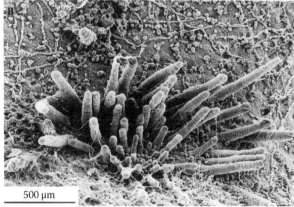

FIGURE 102 The root-like attachment scars, *Podichnus*, of a brachiopod on a Cretaceous belemnite; the lower figure is an epoxy cast of a brachiopod *Podichnus* on a Pleistocene coral. (Figure 12 in Bromley, R.G. and Heinberg, C., *Palaeogeography, Palaeoclimatology, Palaeoecology*, 232, 429–453, 2006. Reproduced with permission from Elsevier.)

5a. BENTHIC SHELL EXAMPLES

OYSTERS

Feldmann and Palubniak (1975) described some Late Cretaceous oyster banks from North Dakota. Watkins (1990) provided an excellent example of brackish water oyster banks from the Pliocene of the Salton Trough region in southern California. Aguirre (1998) provided an excellent account of a Spanish Pliocene brackish water oyster bank. Pufahl and James (2006) described a Pliocene Australian brackish water example.

BRACHIOPODS

Boucot (1990a) provided a number of examples of clumping by brachiopods. Laurin (1984) provided an additional example from the French Jurassic involving a rhynchonellid.

OPHIUROIDS, "BRITTLE STARS"

(See Figure 103.) Fujita (1992) collected a mass of information on ophiuroid aggregations indicating that they are most dense in the modern bathyal regions, although they also occur in the neritic; he also described a Pleistocene occurrence. Mikuláš et al. (1995b; see also Petr, 1989) described what may be the oldest example of an ophiuroid aggregation from the Middle Ordovician of Bohemia. Liddell (1975) described a Middle Ordovician, Trenton-age ophiuroid occurrence from Ontario. Goldring and Stephenson (1972) described a Ludlow, Late Silurian ophiuroid bed occurrence in Wales. Strauch and Pockrandt (1985) described an Early Devonian example from the Eifel in Germany. Lehmann

FIGURE 103 Radwański described a Polish Anisian ophiuroid occurrence with typically aggregated individuals from the lower Muschelkalk of the Holy Cross Mountains. (Plate 1 in Radwański, A., *Acta Geologica*, 42, 395–410, 2002. Reproduced with permission.)

(1951) described a number of Hunsrückschiefer examples, but the individuals are so scattered that they do not qualify as aggregations. Sass and Condrate (1985) described a Famennian ophiuroid assemblage from the Catskill delta. Kesling and Le Vasseur (1971) described an Early Mississippian ophiuroid bed. Radwański (2002) carefully described an Anisian ophiuroid bed from the Holy Cross Mountains, including extensive comments about similar occurrences elsewhere.

Zaton et al. (2008) described several Polish Middle Triassic ophiuroid aggregations. Twitchett et al. (2005) described some aggregated, probable ophiuroid, Early Triassic trace fossils (*Asteriacites*). Feng (1985, Plate I, Figures 3a,b) illustrated some Early Triassic aggregations from Guizhou while illustrating other materials of Late Permian and Middle Triassic age. Goldring and Stephenson (1972) described an Early Jurassic, Lias, British ophiuroid bed occurrence. Hess (1960) described a mid-Jurassic, French ophiuroid aggregation (see also Bachmayer, 1960, Plate 1). Meyer (1988b) described a French Kimmeridgian ophiuroid aggregation. Blake and Aronson (1998, Figure 7-2; see also Aronson and Blake, 1997) illustrated what appears to be a normal ophiuroid aggregation from the Eocene of Seymour Island, Antarctic Peninsula. Rasmussen (1972, Plates 13 and 14) illustrated without comment a Lower Eocene occurrence that may reflect ophiuoid aggregations from the Danish Mo-Clay Formation and another from the British Bartonian Late Eocene. Merriam (1921) described what appears to be an ophiuroid aggregation in a California limestone from the Miocene Santa Margarita Formation.

Prakfalvi (1992) described an Early Miocene (Eggenburgian) Hungarian ophiuroid aggregation consisting of casts. Ishida et al. (1998) described an *Ophiura* aggregation from the late Miocene of central Japan while also referring to additional occurrences from the Pleistocene. Ishida et al. (1999) described Late Miocene to Early Pliocene Japanese ophiuroid aggregations from varied localities, and Ishida and Inoue (1994) described a Pleistocene ophiuroid aggregation. Allman (1863) described a Pleistocene Scottish ophiuroid bed. Goldring and Stephenson (1972) discussed and described varied British Lias, Ordovician, Silurian, and Devonian "starfish" beds (including ophiuroids). Aronson and Sues (1987) collected a number of ophiuroid aggregation occurrences, Middle Ordovician through Late Jurassic, and inferred that post-Jurassic predation accounted for Cretaceous and younger absence, except in unusual, predator-free environments, but enough post-Jurassic occurrences exist that appear to falsify this conclusion. The overall rarity of ophiuroid occurrences makes their absence here and there through time easy to understand.

Ophiuroid aggregations are still one more evidence of behavioral fixity. Lefebvre (2007) considered the Cambrian and younger occurrences of stylophoran echinoderms which sometimes occur with ophiuroid aggregations. He pointed out that ophiuroid occurrences may be in large part related to oligotrophic waters and low predation pressures, which are more commonly found in deep-water, bathyal, cold,

FIGURE 104 Part of a herd of Pennsylvanian echinoids (*Archaeocidaris brownswoodensis*) from Texas (×0.11). (Figure 5 in Schneider, C.L. et al., *Journal of Paleontology*, 79, 745–762, 2005. Reproduced with permission.)

FIGURE 105 Group of edrioasteroids colonizing Pennsylvanian pebbles from Oklahoma. (Figures 3-11 and 3-12 in Sumrall, C.D. et al., *Journal of Paleontology*, 80, 229–244, 2006. Reproduced with permission.)

oligotrophic conditions. Although not in the ophiuroid category, Lefebvre (2007) provided a summary of many Middle Cambrian–Silurian stylophoran echinoderm occurrences that mimic the high-density ophiuroid occurrences and are sometimes associated with ophiuroids. These, however, are overall relatively rare items.

Echinoids "Sea Urchins" and Edrioasteroids

(See Figure 104 and Figure 105.) Nebelsick and Kroh (2002) reviewed the evidence of sand dollar aggregations and observed that most represent transported assemblages involving originally *in situ* aggregations. Radwański and Wysocka (2004) described Miocene aggregations of the regular echinoid *Psammechinus* from the Polish Holy Cross Mountains that are best interpreted as storm-disturbed herds. Schneider et al. (2005) described a truly elegant Pennsylvanian example from Texas dominated by *Archaeocidaris*, and Sumrall et al.

(2006) described some Oklahoma Pennsylvanian edriosteroids densely attached to pebbles. Meyer (1990) reported on a Late Ordovician Kentucky edrioasteroid occurrence in which the numerous echinoderms were restricted to the relatively flat shells of certain brachiopods, avoiding the soft substrate mud between the brachiopod shells.

Balanomorph and Lepadomorph Barnacles

In Section 2Am, a number of examples of attached barnacle aggregations are considered; this is clearly an ancient behavior. Presumably the aggregation of larval, planktonic cyprids is involved followed by further increase in population size from brooded offspring of the initial planktonic settlers. Knight-Jones (1953) explored some of the factors involved in the settling of balanoid cyprids from the plankton in a manner that results in gregariousness.

Decapods

Crawford et al. (2008) described a unique Argentinian Miocene mass mortality occurrence of crabs from a Patagonian locality. Combining data from matrix petrography (volcanic ash), isotope paleothermometry, taphonomic evidence, and basic knowledge of decapod respiratory anatomy and behavior, they were able to conclude that numerous, fully articulated individuals of one crab species died due to the ash clogging their respiratory apparatus, which suffocated them. This is the first example of a shallow-water, marine-environment mass mortality caused by volcanic ash. Other examples of marine mass mortalities, such as pelmatozoan thickets overwhelmed by a sudden sediment influx and the possibilities of anoxia, have been documented, as well as the occurrences of fish schools. In general, however, it has been difficult to understand the causes of death for nonpredation-generated marine benthic fossils.

☐ *Reliability*

Category 1, frozen behavior.

5b. BELEMNITE SHOALS

Doyle and MacDonald (1993) thoroughly reviewed the evidence on belemnite "battlefields" and observed that there are at least five possible modes of formation, only two of which involve the true preservation of shoals. Some of the other possibilities include such things as stomach content aggregates preserved within a predator, condensation accumulates, and resedimented accumulates. Their paper provided useful criteria for discriminating among the varied accumulation types. Although not in the belemnite "battlefield" category, the description by Soja et al. (1996) of a massive accumulation of very small nautiloids from the Eifelian of southeast Alaska strongly suggests a very near-shore nursery. This possibility is supported by the presence of abundant calcareous algae and a shallow-water molluscan facies featuring both gastropods and bivalves.

FIGURE 106   Fungus gnat swarm in Baltic amber; the piece of amber is 37 mm long. (Figure 1 in Ross, A., *Amber*, The Natural History Museum, London, 1998. Reproduced with permission.)

☐ *Reliability*

Category 6, fairly speculative at this stage. Until a careful paleoecological study is performed, one cannot be sure that the Alaskan example represents a nursery, although the possibility sounds reasonable.

5c.   CRANE FLY AND FUNGUS GNAT SWARMS: INSECT SWARMS

Ross (1998) illustrated a swarm of Baltic amber fungus gnats (Figure 106); see Boucot (1990a) for a crane fly reference. A swarm of biting midges in Dominican amber is shown in Figure 107.

☐ *Reliability*

Category 1, frozen behavior.

5d.   SHRIMP SCHOOLS

Chlupač (1995), while describing a new Early Cambrian Czech phyllocarid (*Vladicaris*), cited a number of occurrences, additional to the data for *Vladicaris*, of gregarious phyllocarids from the Cambrian and Ordovician. The *Vladicaris* occurrence may be in a nonmarine environment, but the others cited are from the marine environment. Dzik

FIGURE 107   Swarm of biting midges in Dominican amber (Poinar amber collection accession no. SY-1-90); midge length = 1 mm.

et al. (2004) described evidence for Early Devonian marine shrimp schools using well cores from northern Russia.

☐ *Reliability*

Category 1, frozen behavior.

5e.   FISH SCHOOLS

Trewin (1986) described Scottish Middle Devonian mass mortality occurrences of *Osteolepis*, *Palaeospondylus*, *Mesacanthus*, and *Dipterus* that may represent schools. Evans (1998) briefly cited a mass mortality involving a school of palaeoniscoids from the Mississippian Waaiport Formation, Witteberg Group from South Africa. Tintori (1990) described schooling in the Italian Triassic actinopterygian *Prohalectites*. Chang et al. (2003, Figure 101) illustrated a basal Cretaceous slab covered with similar sized specimens of *Lycoptera* that presumably represent schooling in a freshwater environment.

Espinosa-Arrubarrena and Applegate (1996, p. 545) referred to the presence of a schooling clupeomorph at the Albian Tlaya Quarries in Puebla, southern Mexico. Siber (1982, p. 9) illustrated a fine Green River Eocene bedding surface covered with a large group of *Knightia*, presumably the mass death of a school. Bonde (1987) briefly discussed schools of argentinoids from the Danish Lower Eocene, and Christiansen and Brock (1981, pp. 6–7) illustrated a Danish Eocene example. Nury and Schreiber (1997, Figure 6.23) illustrated a core with numerous examples of *Gobias aries* from the Oligocene of southern Provence that appear to represent schooling.

☐ *Reliability*

Category 2B, based on functional morphology.

5f.   DINOSAUR HERDS

Cotton et al. (1996) provided more data on dinosaur herd possibilities, including the presence of older and younger individuals. Zhao et al. (2006, 2007) provided a brief description of an Early Cretaceous Chinese ceratopsian occurrence involving older and younger individuals. In a paper that provides a good set of references dealing with the gregariousness topic, Lockley et al. (1994) described some Portugese Late Jurassic trackways that suggest gregarious behavior shown by seven small sauropods. Coria (1994) described data for an Argentine Middle Jurassic sauropod indicating more of the same. Currie (1980) discussed evidence concerning the presence of both ceratopsian and hadrosaurian herds, including young individuals, from the Alberta Cretaceous.

Varricchio et al. (2008) described a death assemblage of Mongolian Cretaceous juvenile ornithomimids that apparently died in a drying up pond or lake environment. This find tells us something about the behavior of a group of juveniles separate from any adults. Lockley and Matsukawa (1999) summarized what is known about trackway evidence of gregarious behavior in dinosaurs, with adequate references to the literature concerning the groups involved. Currie and Dodson

(1984) commented about a herd of ceratopsians involved in a mass death situation. Day et al. (2002) discussed some Middle Jurassic sauropod trackways from Britain that suggest herding and multispecies groups as well. Dollo (1923) illustrated and discussed evidence regarding *Iguanodon* footprints and a tailprint from the Wealden of southern England. Lockley and Meyer (2000) summarized a great deal of data concerning dinosaur herds. Lockley et al. (2002) discussed a Bolivian, Late Cretaceous dinosaur track example of gregariousness. Day et al. (2004) described a British Jurassic set of dinosaur trackways that involve gregarious behavior for the sauropods.

☐ *Reliability*
Category 1, frozen behavior.

5g. MAMMALIAN HERDS

Haynes (1991) reviewed information on Quaternary proboscidian aggregations and from modern aggregations, providing evidence that herd life was probably characteristic of both the mammoth and the mastodon, with the population structure of the Quaternary aggregations providing the evidence from the past. Voorhies (1981) described Miocene *Teleoceras* herd evidence, with an ashfall potentially being the cause of death for the herd.

☐ *Reliability*
Category 1, frozen behavior.

*5h. DICYNODONT HERDS: MAMMAL-LIKE REPTILES

Bandyopadhyay (1988) described circumstantial evidence suggesting the presence of a herd of kannemeyeriid dicynodonts from the Middle Triassic Yerrapalli Formation in India.

☐ *Reliability*
Category 2b.

*5i. DIAPSID AGGREGATION: REPTILE

Smith and Evans (1996) described an aggregation of young *Youngina* from the uppermost Permian Beaufort Group in South Africa (see Figure 279).

☐ *Reliability*
Category 1, frozen behavior.

*5j. PTEROSAUR COLONY

Bell and Padian (1995) provided evidence from the Chilean Neocomian for the presence of an inland pterosaur colony.

☐ *Reliability*
Category 2B, functional morphology.

*5k. ACRIDID AGGREGATION: GRASSHOPPERS

Arillo and Ortuno (1997) discussed the relative abundance of acridids at the Oligocene Izarro locality in Spain and concluded that it probably represents evidence of aggregation, as it is so characteristic of the grasshoppers today.

☐ *Reliability*
Probably Category 3 or 4, although the relative rarity of fossil acridids elsewhere is consistent with the authors' explanation.

*5l. MASS MOTH MIGRATION

Rust (2000) documented several mass moth migrations from offshore deposits of the Paleogene Fur Formation in the lower Tertiary of Denmark. This is a unique occurrence of what must have been a common phenomenon, normally denied to us by taphonomic factors.

☐ *Reliability*
Category 1, frozen behavior.

*5m. ANT IMAGO SWARMS

Martínez-Delclòs et al. (2004, Figure 3C) illustrated a trapped Dominican amber ant imago swarm.

☐ *Reliability*
Category 1, frozen behavior.

*5n. TERMITE SWARMS

Martínez-Delclòs et al. (2004, Figure 3D) illustrated detached termite wings preserved in Dominican amber following a nuptial flight.

☐ *Reliability*
Category 1, frozen behavior.

*5o. PLATYPODID SWARMS

Martínez-Delclòs et al. (2004, Figure 3E) illustrated a platypodid swarm trapped in Dominican amber.

☐ *Reliability*
Category 1, frozen behavior.

*5p. CRYPTIC TRILOBITE BEHAVIOR

Chatterton et al. (2003) described a number of examples in which trilobites, sometimes a large number, sometimes isolated individuals, occur within burrows probably generated by other organisms. The significance of this behavior is not easy to determine but may include such things as susceptibility to predation after molting (for a discussion, see Paterson et al., 2008). Cherns et al. (2006) described

TABLE 12
Spacing Data from the Fossil Record (Additions to Table 10 from Boucot, 1990a)

Behavioral Types	Taxonomic Level	Time Duration	Category	Refs.
5e. Fish schools	Genus	Triassic–Holocene	2B	Tintori (1990)
5f. Dinosaur herds	Orders	Late Triassic–Cretaceous	6	See text
5h. Dicynodont herds	Family	Middle Triassic	2B	Bandyopadhyay (1988)
5i. Diapsid aggregation	Genus	Late Permian	1	Smith and Evans (1996)
5j. Pterosaur colony	Order	Early Cretaceous	2B	Bell and Padian (1995)
5k. Acridid aggregation	Family	Oligocene–Holocene	3 or 4	Arillo and Ortuno (1997)
5l. Mass moth migration	Order	Paleogene–Holocene	1	Rust (2000)
5m. Ant imago swarms	Order	Miocene–Holocene	1	Martínez-Delclòs et al. (2004)
5n. Termite swarms	Order	Miocene–Holocene	1	Martínez-Delclòs et al. (2004)
5o. Platypodid swarms	Family	Miocene–Holocene	1	Martínez-Delclòs et al. (2004)
5p. Cryptic trilobite behavior	Class	Silurian	6	Chatterton et al. (2003)
5q. Juvenile millipede behavior	Class	Pennsylvanian–Holocene	6	Wilson (2006)

a Llanvirn occurrence of the trace fossil *Thalassinoides* from Jemtland in which there were occurrences of *Asaphus*; the burrows were interpreted as the work of the trilobite. Although not involving trilobites, the Portuguese Lower Cretaceous occurrence of *Thalassinoides* burrows associated with abundant *Mecochirus* crustaceans is notable (Neto de Carvalho et al., 2007), as it also is indicative of the higher preservation potential of burrow-inhabiting crustaceans as contrasted with epibenthic taxa, calling to mind the similar relation of regular and irregular echinoids in biasing the fossil record.

□ *Reliability*
Category 6, speculative.

*5q. JUVENILE MILLIPEDE AGGREGATION

Wilson (2006) described an aggregation of juvenile millipedes and discussed the possible aggregation causes while reviewing the various factors involved in modern millipede aggregations.

□ *Reliability*
Category 6, speculative.

SUMMARY

Post-1990 data further support the behavioral constancy through time of many groups, with a few extending down to the generic level. (See Table 12.)

6 Predation and Feeding Behaviors

Keep in mind that any and all obligate predator–prey relationships remain at equilibrium during specific ecological–evolutionary units and subunits and that the same applies to obligate host–parasite relationships.

6A. MARINE

6AI. INVERTEBRATE

Skovsted et al. (2007) reviewed the record of the Ediacaran and Cambrian, with an emphasis on the Early Cambrian evidence of predation affecting well-skeletonized marine invertebrates; their examples make it clear that predators, including both shell borers and shell damagers, have been with us since the beginnings of any evidence of well-skeletonized marine invertebrates. Harper (2006) provided an excellent account of evidence of post-Paleozoic marine invertebrate predation that highlights the many problems involved in evaluating the reliability of the scrappy evidence. Zaton and Salamon (2008) described some good evidence for durophagous molluscan shell predation from the Polish Jurassic, emphasizing the presence of angular shell fragments with sharp, unabraded margins; their class of evidence should be used more widely when evaluating shelled invertebrate predation possibilities up and down the column.

6AIa. Naticid–Muricid–Cassid Borehole Position and Boring: Gastropods

Bromley (1981) reviewed the "small round holes in shells" known from the fossil record while considering the identities of the potential boring organisms and erected *Oichnus* for them. A note of caution for interpreting all of the Paleozoic boreholes as the work of predators was provided by Richards and Shabica (1969), who showed in at least their Ordovician dalmanellid brachiopod borehole sample that a nonpredatory explanation is reasonable.

Taylor et al. (1980) carefully reviewed the entire spectrum of gastropod predation by prosobranchs beginning, according to them, in the Aptian. Hoffmeister et al. (2004) provided borehole frequency data from the silicified Glass Mountains Permian brachiopods and bivalves indicating a lower frequency than in the post-Paleozoic and noted that there is more evidence of bivalve than of brachiopod boring.

Kase and Ishikawa (2003) made a good case for naticid predation only beginning seriously by the Campanian–Maastrichtian and pointed out, using modern material, that true naticids only date from that time. Taylor et al. (1983a) provided detailed samples of Albian, mid-Cretaceous bivalves bored by naticids and muricids. Allmon et al. (1990) used a large sample of turritelline gastropods, some bored and some peeled, to make the case that there has not been an "arms race" type of change in drilling during the Cenozoic. Harries and Schopf (2007) provided extensive data from the Western Interior Maastrichtian to indicate that there was considerable variation in naticid-type drilling intensity affecting bivalve shells, so the utility of the "escalation" concept is still in doubt. Daley et al. (2007) used Florida Pleistocene examples to make it clear that the interpretation of drilling frequencies is very complex and is not usable in a simplistic manner.

Jonkers (2000) made the very good point, using pectinid examples, that muricid and naticid predation on bivalves can be heavily influenced by the behaviors of the bivalves, with both different bivalve species and different growth stages of the same species being involved, as evidenced by data taken from living and fossil material. In an insightful paper, Amano (2006) used muricid boreholes into Late Cenozoic *Glycymeris yessoensis* to make the important point that changes from one predatory muricid to another through time can only be inferred by considering co-occurring muricids and that the switching from one prey type, as well as borehole position, to another is also involved; thus, it is difficult to interpret the niceties of muricid predation.

Zardini (1985, Plate 2) illustrated various Late Triassic boreholes, successful and unsuccessful, into *Cassianella* and also into a bivalve (*Palaeocardita*). Zardini's data supplement the earlier report of Fürsich and Jablonski (1984; see also Boucot, 1990a) on Late Triassic naticid-type boreholes. Arua (1989) provided excellent examples of selective predation of various bivalve and gastropod prey examples from the Nigerian Eocene. Fortunato (2007) employed some Gatun Formation Miocene data from Panama to point out that prey selection is a very important factor in naticid predation. Cosma and Baumiller (2005) described a borehole in a Silurian bivalve while considering the question as to why there are so few boreholes reported in Paleozoic bivalves and gastropods (for an Ordovician example, see Rohr, 1991).

The danger of estimating gastropod predation in terms of borehole numbers is illustrated by Frey et al.'s (1986) account of some modern naticids preying on razor clams without any evidence of boring; they merely begin feeding at the more exposed portions of the open ends of the bivalve. Additional to this problem is that predation by shell crushing is not easily recognized in the fossil record.

Walker (2007) provided the most extensive review to date of gastropod predation on molluscan prey, adding the necessary caveats about how best to apply data from the present to the fossil record; her account makes it clear that overly simplistic interpretations of the record have all too often been employed in the past. This is a very complex topic potentially involving such things as poisons and muscle relaxants as well as the radular boring of shells sometimes aided by acidic materials. Kabat (1990) reviewed naticid predatory shell boring, including boring locales, mechanisms, and prey types. Naticids are burrowers largely restricted to soft-substrate bivalves and gastropods. The countersunk boreholes are easily recognized, with fossils known from the Cretaceous to present and a possible Triassic "experiment." Kabat (1990) provided a summary of fossil occurrences of naticid boreholes and reviewed the question of naticid site specificity; an excellent account of borehole character in the different gastropod borers is included, and the parallel-sided borings of muricids are contrasted with the chamfered borings of naticids. He rejects nassarids as shell borers. Reyment (1967) described evidence for boring naticids and muricids involving Nigerian modern and Paleogene ostracodes and mollusks, with some evidence for prey selectivity being involved; juvenile gastropods are implicated as the predators due to the small size of the boreholes and of the prey bored. Reyment et al. (1987) described and illustrated in some detail bored Late Cretaceous Israeli and Early Paleocene Nigerian ostracods (see Figure 108, Figure 109, and Figure 110).

Kabat and Kohn (1986) made a good case for naticid predators being size correlated with their prey and also emphasized that naticids have a much higher level of success in predation than do crustaceans, as judged by healed fractures due to unsuccessful crustacean predation. Dietl and Kelley (2006) made it very clear, based on experiments with living naticids and their bivalve prey, that it is very difficult to use the boreholes to identify the predator.

Kelley and Hansen (1993, 1996, 2003, 2006) provided the most comprehensive studies yet made of naticid predation borehole evidence through time (later Cretaceous and Paleogene of the Coastal Plain in the southeastern United States). They do not find evidence for the continuing escalation that some had hypothesized but rather a level of complexity that will require still more work to completely disentangle; they noted the marked changes in drilling frequencies associated with extinction events. Hansen et al. (1996) provided additional data concerning the sporadic changes in drilling frequency and intensity and relations to extinctions, all of which add up to a very complex story through post-Paleozoic time (i.e., there is not a smooth increase in drilling intensity). Kelley (2007) made a good case that gastropod boreholes in bivalve shells do not make the shells more susceptible to breakage.

Dietl and Alexander (2000) used a change in preferred borehole drilling position for Miocene to Recent naticids as evidence for an evolutionary, behavioral shift. Bellomo (1996) assembled a very large sample of *Polinices lacteus* from the Calabrian Pleistocene that provides an excellent view of changing predation intensity on the gastropod with growth, although the identity of the different predators was not determined. The smaller size classes are those most

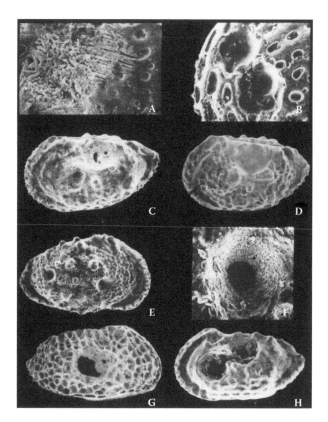

FIGURE 108 (A) *Buntonia* sp. in early phase of drilling, displaying parallel radula scratches and the effects of dissolution of the shell by the accessory boring organ (ABO) (Paleocene, Nigeria, ×50). (B) *Leguminocythereis lagaghiroboensis* Apostolescu with partially completed hole; the surface texture of the pits shows evidence of solution (Paleocene, Nigeria, ×55). (C) *Veenia fawwarensis fawwarensis* Honigstein with abandoned hole; the broken-through parts of the hole, particularly the crescent-shaped perforation, point to a naticid predator (Santonian, Israel, ×22). (D) *Veenia fawwarensis dividua* Honigstein with incompletely bored muricid hole (Santonian, Israel, ×22). (E) *Trachyleberis teiskotensis* (Apostolescu) specimen showing two muricid(?) boreholes, the anterior of which is complete (Paleocene, Nigeria, ×30). (F) *Ovocytheridea* sp. cf. *O. pulchra* Reyment with muricid borehole (Paleocene, Nigeria, ×100). (G) *Anticythereis*(?) *bopaensis* Apostolescu with almost complete naticid borehole (Paleocene, Nigeria, ×25). (H) *Veenia fawwarensis fawwarensis* Honigstein with twin, completed naticid boreholes (Santonian, Israel, ×22). (Figure 2 in Reyment, R.A. et al., *Cretaceous Research*, 8, 189–209, 1987. Reproduced with permission of Elsevier.)

subject to predation. Harper et al. (1998) discussed varied Jurassic, United Kingdom bivalve boreholes of muricid type that make the point about borehole convergence, as muricid shells are unknown anywhere before the earlier Albian. They also reviewed the known pre-Jurassic examples of muricid-type boreholes, including Paleozoic materials.

Arua and Hoque (1989) described a number of Nigerian Eocene examples and concluded that muricids are more active below the sediment–water interface and naticids more active above it. Zlotnik (2001) provided a wealth of information from the Korytnica Clay Miocene concerning prey selectivity sites in terms of specific molluscan species. Pickerill

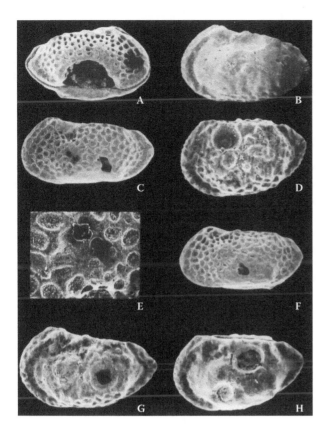

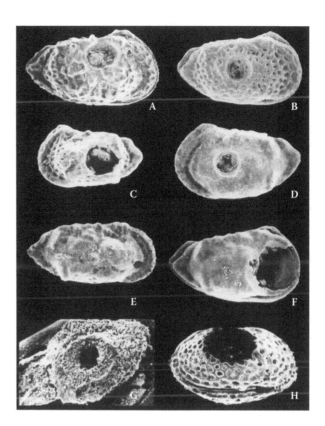

FIGURE 109   (A) *Anticythereis judaensis* Honigstein with ventrally located borehole, presumably initiated at the ventral margin
contact and drilled obliquely; this unusual siting could have been
made on a moribund individual (Santonian, Israel, ×22). (B) *Veenia
fawwarensis fawwarensis* Honigstein with incomplete muricid borehole (Santonian, Israel, ×22). (C) *Cythereis cretaria dorsocaudata*
Honigstein with two partially completed boreholes; the one with
two small perforations was made by a naticid (see Part E, this figure) (Santonian, Israel, ×25). (D) *Veenia fawwarensis fawwarensis*
Honigstein with incomplete muricid(?) borehole (Santonian, Israel,
×20). (E) *Cythereis cretaria dorsocaudata* Honigstein, for which
one of the incomplete borings of Part C, above, is shown at higher
magnification; the broken-through parts of the excavation present
one of the criteria of naticid drilling as understood by Ziegelmeier
(1954) (Santonian, Israel, ×123). (F) *Cythereis cretaria dorsocaudata* Honigstein with incomplete borehole showing a broad annular
zone over which the reticular ornamentation has been scraped away
(Santonian, Israel, ×22). (G) *Veenia fawwarensis* Honigstein with
almost completed borehole with a broad zone of frosting (Santonian,
Israel, ×22). (H) *Veenia fawwarensis fawwarensis* Honigstein with
two holes, one complete and one incomplete, both probably from
muricids (Santonian, Israel, ×25). (Figure 3 in Reyment, R.A. et
al., *Cretaceous Research*, 8, 189–209, 1987. Reproduced with permission of Elsevier.)

FIGURE 110   (A) *Cythereis cretaria* van den Bold with a naticid
hole drilled in a reticulated surface; the hole is breached dorsally
(Santonian, Israel, ×22). (B) *Cythereis cretaria* van den Bold with
naticid hole drilled in a reticulated surface (Santonian, Israel, ×22).
(C) *Cythereis diversereticulata* Honigstein with borehole possibly
made by a naticid (Santonian, Israel, ×25). (D) *Veenia fawwarensis
fawwarensis* Honigstein with borehole probably made by a naticid
(Santonian, Israel, ×22). (E) *Cythereis diversdereticulata* Honigstein
with muricid hole in a rather anterior location (Santonian, Israel,
×20). (F) *Veenia fawwarensis* Honigstein with entire anterior
zone excavated by a naticid borehole (Santonian, Israel, ×25). (G)
Leguminocythereis lagaghiroboensis Apostolescu with muricid
hole drilled through secondary calcite, presumably deposited postmortem (Paleocene, Nigeria, ×109). (H) *Anticythereis*(?) *bopaensis*
Apostolescu with large borehole drilled in a strongly reticulated
surface (Paleocene, Nigeria, ×20). (Figure 4 in Reyment, R.A. et
al., *Cretaceous Research*, 8, 189–209, 1987. Reproduced with permission of Elsevier.)

and Donovan (1998) described and discussed a number of
Miocene boreholes into molluscan shells, with the implication that most are gastropod generated.

Kowalewski et al. (1998) surveyed the Phanerozoic borehole record in shelled marine invertebrates, pointing out that
there is a very low level of such evidence for the Permian
through Jurassic, with no good explanation. Harper et al.
(1999) discussed the difference in opinion concerning whether

or not there is a considerable drop in drill-hole frequency in
the later Paleozoic and earlier Mesozoic; this is clearly a sampling problem that will eventually be solved with far larger
samples. Kowalewski et al. (2000) reviewed occurrences of
Permian drilled bivalve and brachiopod shells, noting that
the frequency is considerably below that of the post-Jurassic.
Hoffmeister et al. (2001) commented on the low level of predatory boreholes in the vast Cooper collection of West Texas
Permian brachiopods and bivalves, the far lower level overall
of such boreholes in Paleozoic benthos, and its contrast with
the post-Paleozoic situation; presumably, the Paleozoic borehole generators may have had far less effect on the ecosystem
than has been the case in the post-Paleozoic. Hary (1987)
described some Sinemurian boreholes drilled into an oyster

from the Paris Basin. Stone (1998) made a good case that attached bivalves with strong spines use them as an antimuricid predation device. Walker (2001) reported on widespread deep water boring in gastropod shells from the Late Pliocene of Ecuador.

In a very insightful paper, Delance and Emig (2004) described some modern boreholes into *Gryphus* shells made by predatory gastropods from the Mediterranean and suggested that brachiopod predation by gastropods or gastropod-like drillers may well be badly overemphasized due to ignorance about the activities of soft-bodied predators, as well as the activities of predators that fragmented prey brachiopod shells.

Harper (2003) reviewed the entire question of reliability in Paleozoic and Mesozoic drilling predation examples, with attention being given to sample size as a measure of reliability, and Reyment (1999) provided some very useful information about gastropod drilling. Kabat (1990) reviewed other borehole generators, including nudibranchs boring into spirorbid polychaetes, some pulmonates boring into other gastropods, some polyclad turbellarian flatworms boring into oysters, and even a cottid fish that uses vomeral teeth to punch boreholes into gastropod prey. Kabat (1990) also briefly discussed the varied Paleozoic boreholes of uncertain origin into varied shell types, with the inference that many are of predatory origins. Kowalewski (1993) provided data illustrating the problems involved in trying to assign molluscan-generated predatory boreholes to suprafamilial level taxa, including the use of borehole form and borehole size; his data should be taken into account by anyone concerned with these problems. Norton (1988) provided excellent illustrations of the ragged "boreholes" made by certain predatory fishes into gastropod shells; such holes should be easily recognizable in the fossil record. Harper (1994) described some Miocene incomplete boreholes into bivalves and considered the functional significance of conchiolin sheets in discouraging predation.

Simões et al. (2007) found that drilling predation affecting rhynchonellid brachiopods and bivalves in a southern Brazilian shelf environment resulted in far more bivalve drilling than brachiopod drilling, although the overall levels are relatively low. When combined with the known taxonomic selectivity in many drilled brachiopod samples this emphasizes the need for caution when trying to interpret predation intensity from drilled shell frequencies.

Although not in the gastropod category, the account by Wetzel (1960, Figure 4) of a Maastrichtian scaphopod from Chile shows boreholes attributed to gastropod predation. Müller (1969b) ascribed boreholes in some Belgian Maastrichtian serpulid polychaete tubes (*Ditrupa*) to naticid activities. The oldest recognized predatory-type borehole is one from the Late Precambrian present in the tubular organism *Cloudina* (Hua et al., 2003).

In general, one may conclude that gastropod drillers seem to prefer thin-shelled prey; they may utilize many prey species, if they are available, and more than one size class; predation rates are very variable; muricids tend to feed on epifauna whereas naticids prefer infauna; and overall one needs to be aware that predation by gastropods may be relatively minor in some instances as compared to other predaceous groups. Kelley and Hansen (2006) have largely falsified the Vermeij concept of "escalation" in which it was hypothesized that predators, including boring gastropods, became more effective from the Cretaceous to present.

6AIb.   Crabs–Mollusks and Gastropod–Bivalve

Alexander and Dietl (2003; see also Preston et al., 1996) reviewed the question of shell-breaking predation on marine bivalves and gastropods and concluded that it is too soon to arrive at firm conclusions about the overall history of shell breaking. It is clear that the many complications affecting sampling reliability within the Phanerozoic have not been evaluated reliably enough to provide much guidance. Additionally, compilations of potential predator-defeating forms over time at the species or genus level provide no data to tell us anything about such morphologies as a function of abundance of the varied morphologies over time, a critical failure in the currently available data. The attempts to assess differing levels of predation through time are badly flawed in this regard. Ebbestad and Stott (2008) discussed the evidence provided by percentages of repaired shell injuries using Late Ordovician Canadian examples, numbers of shell injuries per shell, and overall incidence of such injuries, with the caveat that fragmented shells might also reflect predation evidence; clearly, the predation intensity through time affecting gastropods is still a work in progress. Cadée (2008, p. 47) described shell breakage in some modern *Hydrobia* (Gastropoda) that had passed through shelduck guts without actually killing the snail!

Savazzi (1991) provided an excellent account of predation affecting strombids (see his Figure 11) with a discussion of their morphological features involved with predation defense. Dietl and Hendricks (2006) found that sinistrally coiled gastropods suffered far less from crab predation (peeling) than was the case with dextrally coiled gastropods. They used both modern and Plio–Pleistocene materials in their work. This being the case, it is puzzling that so few gastropods are sinistrally coiled. Walker and Yamada (1993) provided experimental evidence from the present indicating that predacious crabs attack empty gastropod shells in search of prey while ignoring disarticulated bivalve shells, which represents a potential source of error in interpreting the fossil record of predation.

Cadée (1999) pointed out, using an Antarctic limpet example, that not all cases of shell repair need involve predation. In the Antarctic example, the shell damage that was repaired was due to intertidal shell breakage caused by ice and stones battering the shells. Cadée (2000) went on to cite evidence from predatory, modern herring gulls on razor clams that involves a high level of shell breakage; thus, shell breakage as contrasted with gastropod boring is a real possibility when reviewing evidence of predation. Cadée's Table 1 listed the following from over 700 razor clams: no damage, 26.4%; two valves slightly damaged, 20.8%; one valve broken in the middle, 26.1%; and both valves broken in the middle, 26.7%.

Skovsted et al. (2007) described evidence of predatory attempts, as shown by interrupted growth lines, on a probable mollusk (*Marocella*) from the South Australian Early Cambrian, together with a review of earlier Paleozoic predation. Lindström (2003) described shell damage ascribed to predatory activities in two Texas Pennsylvanian pleurotomarians. Lindström and Peel (2005) described various Ordovician to Carboniferous pleurotomarian shell damage featuring healed fractures and suggested that some shell morphologies are more predator resistant than others.

Ozanne and Harries (2002) described later Campanian–Maastrichtian inoceramids with marginal evidence of shell crushing; they demonstrated a marked increase in the later Maastrichtian before the ultimate extinction of the shells, which may well be the result of crab activities. Crampton (1996) discussed a Late Cretaceous *Inoceramus* from which a predator had taken a wedge-shaped "bite," following which repair occurred. Kauffman (1972) described a Cretaceous inoceramid with marginal shell indentations that might have been made by a ptychodid shark. Darragh and Kendrick (1994, pp. 14–15) described evidence of peeling in a Maastrichtian gastropod from the Carnarvon Basin.

Papp et al. (1947) discussed and illustrated numerous examples of predation by the peeling activities of hermit crabs involved chiefly with Tortonian and Eocene gastropod prey plus a few scaphopod examples. Walker (2001) cited widespread shell repair from deep-water Late Pliocene gastropods in Ecuador. Boekschoten (1967) provided some Pliocene examples of crab damage to gastropod shells. Rust (1997, Plate 1) discussed and illustrated various crab-damaged, repaired gastropods shells from the Late Miocene of Greece, where the damage may have been due to a single brackish water crab. Boekschoten (1967) illustrated some Belgian Pliocene repaired gastropod shells that presumably represented failed predation attempts by crabs.

Dietl (2003) carefully discussed the predator–prey relation between a modern conch as predator and a modern, thick-shelled bivalve as prey, with a record for both going back into the Miocene. He made it clear how difficult it is to be sure that morphological changes in one or both of the pair can be viewed as strictly coevolved predator–prey relationships due to the need to consider other predators of both members of the pair as well as a variety of other factors; this is a very mature consideration of the complexities inherent in natural situations.

Although this account is not in the predator–prey category, Horný's (1998b) vivid description of a Czech Late Silurian tremanotid with a fractured and repaired area on its final whorl does make sense as the possible result of an impact from a volcanic bomb, as the Barrandian area had active volcanoes present at this time.

6Alc. Echinoid Lantern Scratches: Aristotle's Lantern Grazing Traces

Bromley (1975) described a variety of post-Paleozoic echinoid grazing traces and established their taxonomy. Smith (1984, Figure 3.31) illustrated an early Bajocian (Middle Jurassic) example of a grazing trace, *Gnathichnus pentax*, from the Cotswolds.

Michalik (1977) described Rhaetic examples, which he named *Roderosignus*, from Carpathian *Zeilleria*. Radwańska (1999, Figures 11, 12) reviewed the grazing traces made by Rhaetic to present echinoids (see Figure 111 and Figure 112). She emphasized that the Aristotle's lantern produces a five-rayed grazing trace (see her Figure 11.1a) when an isolated trace is preserved but that normally the grazed substrates are so thoroughly grazed that their five-rayed character is not readily obvious. De Gibert et al. (2007) described some Pliocene *Gnathichnus* and provided a summary of their stratigraphic record extending back to the Middle–Upper Triassic.

☐ *Reliability*
Category 3.

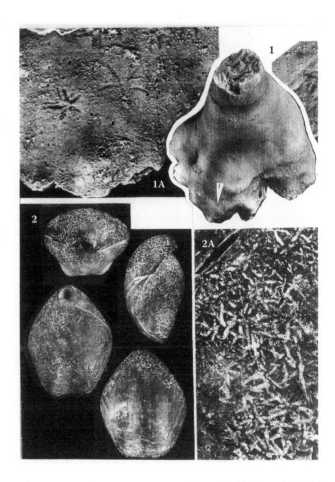

FIGURE 111 Grazing traces of regular echinoids from the Polish Kimmeridgian. (1) *Apiocrinites* holdfast (×1) partly scratched by echinoid grazing traces (arrow indicates area of 1A), partly overgrown by successive growth layers of the crinoid. (1A) *Apiocrinus* holdfast with pentagonal grazing traces (×5). (2) Terebratulid brachiopod, *Sellithyris*, showing areas with pentagonal grazing traces (×1.5). (2A) Close-up showing overlapping grazing traces in the umbonal region of the dorsal valve (×5). (Figure 11 in Radwańska, U., *Acta Geologica Polonica*, 49, 287–364, 1999. Reproduced with permission.)

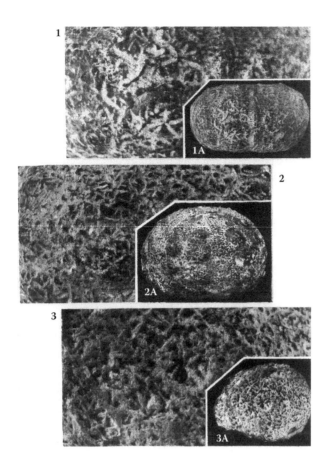

FIGURE 112 Grazing traces of regular echinoids on other regular echinoid tests. (1) Overlapping traces on test of *Rhabdocidaris*. (1A) View of entire test (×1.5). (2) Overlapping traces on *Gymnocidaris* (×5). (2A) View of entire test (×1.5). (3) Overlapping traces on *Gymnocidaris* (×5). (3A) Heavily, echinoid, scratch-eroded test that is hardly recognizable (×1.5). (Figure 12 in Radwańska, U., *Acta Geologica Polonica*, 49, 287–364, 1999. Reproduced with permission.)

6Aid. Chiton and Gastropod Radular Grazing Traces

De Gibert et al. (2007) described some Pliocene *Radulichnus* generated by polyplacophorans and provided a summary of their stratigraphic record extending back to the Mesozoic.

☐ *Reliability*

Category 3.

6Ale. Nematode Predation on Foraminifera

Kabat (1990) briefly commented on nematode predation on Foraminifera (see also Culver and Lipps, 2003).

6Alf. Starfish Feeding on Mollusks

Blake and Guensburg (1994) described a Late Ordovician asteroid from Ohio that is wrapped about a bivalve in the characteristic feeding position, and they also provided an excellent account of and introduction to the literature on modern asteroid predation and feeding. Although not involving bivalves, Jell's (1989) example of a starfish in close association with a trilobite is concluded by him to be evidence of predation.

☐ *Reliability*

Category 1, frozen behavior.

6Alg. Position of Boreholes in Ostracodes

Reyment and Elewa (2003; see also Reyment et al., 1987) reviewed the record of boreholes in fossil ostracodes that can be attributed to predators, presumably immature naticids and muricids, Santonian to present. In the prior treatment (Boucot, 1990a), Silurian material was discussed for which the nature of the predator is uncertain. (See Figures 108 through 110.)

☐ *Reliability*

Category 2, functional morphology.

6Alh. *Cruziana–Teichichnus–Halopoa* Community and *Cruziana–Teichichnus* Nutritional Relationship

Brandt et al. (1995) described a fine *Rusophycus–Palaeophycus* specimen from the Late Ordovician of Ohio that further supports the conclusion that some trilobites fed on infaunal "worms." Their trilobite was clearly *Isotelus*, and they summarized the evidence provided by previous Cambrian and Silurian finds of the same sort. Pratt (1996) provided a similar *Rusophycus* story, associated with vertical burrows that might have been made by "worms," in which the *Rusophycus* was definitely a nontrilobite arthropod. Rydell et al. (2001) provided new data from the Swedish Early Cambrian indicating that at least in some cases the trilobite relationship to the underlying trace fossil organism was not predatory.

6Ali. Octopus Boreholes

Kabat (1990) briefly commented on octopus boreholes and predation. Guerra and Nixon (1987) provided additional information on the nature of octopus boreholes into mollusks. Bromley (1993) carefully reviewed the whole matter of octopod boreholes, including their morphology, while illustrating some Pliocene examples from Rhodes; his suggestion that these holes be looked for more assiduously should be heeded. Harper (2002) reported on museum collections of octopod boreholes in Pliocene scallop shells from Florida, emphasizing the fairly stereotypic site for boring over the adductor muscle area and noting that modern octopods generally pry bivalve shells apart to get at the flesh, using boreholes plus venom only as a last resort with larger, recalcitrant specimens. It is clear from the fairly widespread Pliocene evidence that there should be a concerted effort to examine pre-Pliocene materials for evidence of the characteristic boreholes. Taddei Ruggiero and Bitner (2008) described some octopus-generated *Oichnus* from the Polish Miocene in brachiopods.

☐ *Reliability*

As pointed out earlier (Boucot, 1990a), the form and size of octopus boreholes provide a very reliable basis, so Category 1, for characterization.

6AIj. Capulid Gastropods as Commensals on Bivalves

Bongrain (1995) made a good case that there are Miocene and younger specimens of varied scallop species bearing the traces of *Capulus* species that were commensals around the rim of the scallop. However, the capulids presumably responsible for generating the traces have not been found with the fossils, although such associations are well known in the recent. Matsukuma (1978) made a strong case for some capulids having made boreholes in the posterior portion of certain scallops, living and Pleistocene, and that they acted as commensals rather than as parasites. Kabat (1990) commented on capulid boreholes in echinoids and mollusks, where they are ectoparasitic.

☐ *Reliability*
Category 2B, functional morphology, and Category 1, frozen behavior, for those with the capulid still attached.

6Aik. Squid–Fish

Fuchs (2006, p. 7) cited a Late Cretaceous *Dorateuthis* from Lebanon with "fish remains" in the gut region. This supplements the previously cited (Boucot, 1990a) Late Jurassic *Belemnoteuthis* from the Callovian of England.

☐ *Reliability*
Category 1, frozen behavior.

6AIl. Juliidae–*Caulerpa* Relation

Bałuk and Radwański (1977; see also Bałuk, 1968) documented the presence of the juliid bivalve gastropod *Berthelinia* in the mid-Miocene of the Korytnica Basin which is good evidence for the presence of the Indo-Pacific alga *Caulerpa* on which it feeds. Keen and Smith (1961) reviewed *Berthelinia*, including its fossil record that extends back into the Eocene (see Figure 113). Le Renard et al. (1996) compiled the stratigraphic record of the taxa included in the Juliidae, mid-Eocene to present (see also Schneider et al., 2008).

☐ *Reliability*
Category 2B, functional morphology.

6AIm. Scaphopod Feeding on Foraminifera

Culver and Lipps (2003) carefully reviewed the available evidence concerning foraminiferal feeding and predation, modern and fossil, while making it clear how difficult it is in the fossil record to be sure about the nature of potential predators.

☐ *Reliability*
Category 6, very uncertain.

*6AIn. Paleozoic Predation on Gastropods

(See also Section 6AIb, Crabs–Mollusks and Gastropod–Bivalve, and Section 6AIa, Naticid–Muricid–Cassid Borehole Position and Boring: Gastropods.) Aside from boreholes generated by presumed predators, there is some information about other predatory modes for the Paleozoic. Horný (1997b)

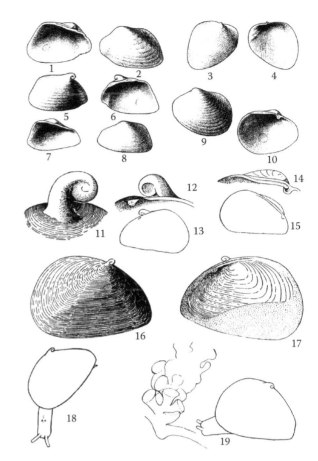

FIGURE 113 Eocene and modern Juliidae. Figures 1 to 10 are Eocene, and Figures 11 to 19 are Recent. (1, 2, 5, 6) *Berthelinia elegans elata* Cossmann; (3, 4) *Ludovicia squamula*; (7, 8) *Berthelinia elegans*; (9, 10) *Anomalomya corrugata*; (11–15) *Berthelinia (Anomalomya)* sp.; (16, 17) *Berthelinia (Edentiellina) corallensis*; (18, 19) *Berthelinia chloris belvederica*. (From Keen, A.M. and Smith, A.G., *Proceedings of the California Academy of Sciences*, Series 4, 30(2), 47–66, 1961. Reproduced with permission of the California Academy of Sciences.)

summarized records extending back to the Ordovician of predated and also repaired gastropod shells (see also Ebbestad and Peel, 1997). Ebbestad (1998) described some Middle Ordovician Swedish bellerophontids that probably underwent predator attacks, which they survived. Horný (1997a, Plate 14, Figures 4, 5) illustrated an excellent specimen of *Lophospira*(?) *debganensis* showing good evidence of shell repair from the Moroccan Ordovician. Isakar and Ebbestad (2000) described *Bucania* from the Estonian Ordovician that shows evidence of repaired peeling-type attacks.

Horný (2001) described failed predation attempts on Silurian *Novakopteron* from Bohemia and illustrated the healed "cracks" in the shell. Although obviously not involving crabs or mollusks, Peel et al. (1996) discussed shell breakage and repair in the Silurian gastropod *Oriostoma* that is reminiscent. Lindström and Peel (1997) elaborated on the *Poleumita* (probably = *Oriostoma*) theme with some excellently illustrated Gotland examples of shell repair. Peel (1984) offered spectacular evidence of predation in Silurian *Euomphalopterus* from Gotland. Lindström and Peel (2003) described a number of Early Devonian, Czech platyceratids showing evidence of repaired predation attempts in a small percentage of the specimens. Tapered boreholes in presumed Paleozoic shelly prey are very rare, but Heidelberger and Amler (2002) described a tapered borehole in a specimen of *Macrochilina drozdzewski* from the Pragian Massenkalk of the Dornap complex in Bergisches Land; the identity of the presumed predator is unknown, although it simulates the post-Paleozoic muricid type. Horný (2002) described a good example of failed predation affecting a Konéprusy, Early Devonian Barrandian specimen of *Anarconcha pulchra*. Brett and Cottrell (1982) described some Middle Devonian New York evidence of peeling type damage in the gastropod *Palaeozygopleura*.

Schindel et al. (1982) provided evidence of failed predation attempts affecting some Pennsylvanian snails from Texas in terms of failed crushing attempts recorded by healed shells, as well as information about the frequency of attacks by shell-crushing predators and the differences shown for various taxa and morphologies. The evidence is just as convincing as with the post-Paleozoic, despite ignorance about the identities of the predators. For a useful discussion of potential Paleozoic predators, vertebrate and invertebrate, see Signor and Brett (1984). In all of these Paleozoic examples, the identity of the predators is unknown, although Lindström and Peel (1997) provided a useful review of the potential actors.

☐ *Reliability*

Overall, Category 6, due to uncertainty about the identity of the predators.

*6AIo. Stomatopod Predation on Gastropods

Geary et al. (1991) provided excellent evidence of stomatopod crustacean predation in Florida Plio–Pleistocene gastropod shells, as evidenced by the similarities of holes found in varied fossil snails and modern examples of the same morphology where the identity of the predator was known (see Figure 114). Pether (1995) provided considerable insight into the appearance of modern stomatopod predation holes, including those formed by stabbing and clubbing. Bałuk and Radwański (1996) reviewed the entire scope of stomatopod predation on gastropods and documented various European Miocene occurrences. They carefully described the predatory damage and discussed and illustrated "repaired" gastropod shells that survived attacks. One presumes that considerably older evidence of stomatopod predation will eventually be found, since the stomatopods have a fossil record extending back at least to the Late Jurassic; they even report evidence of a "kitchen midden" (Plate 8, Figure 3) suggesting that some Miocene stomatopods brought prey back to their holes as is sometimes the case today. Although not involving gastropods, Schwarzhans' (2007) account of stomatopod burrows from the Danish Eocene that contain fish remains accords well with their modern habits of preying on small fish.

☐ *Reliability*

Category 2B, based on functional morphology.

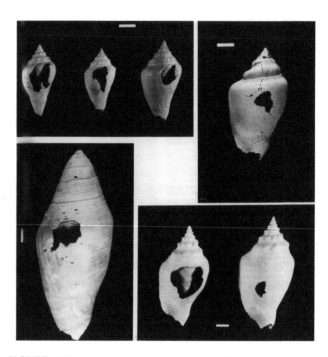

FIGURE 114　Stomatopod-made holes in *Strombus floridanus* from the Pliocene Florida Pinecrest beds (scale bar = 1 cm). (Figure 2 in Geary, D.H. et al., *Journal of Paleontology*, 65, 355–360, 1991. Reproduced with permission.)

*6AIp. Boreholes in Brachiopods and Predation on Brachiopods in General

Leighton (2003b) reviewed most of the literature dealing with predation evidence on brachiopods, while making it clear that little is known about the nature of the predators. While describing varied boreholes from Canadian Ordovician, Silurian, and Devonian brachiopods, Daley (2008) considered the problem of discriminating among shell borings made by predators, possible parasites, and postmortem boring. Taddei Ruggiero et al. (2006) provided evidence from the modern Mediterranean of brachiopods being predated by gastropods and octopuses that involved boreholes as well as probable predation by fishes and crabs based on broken shells, in addition to a reasonable level of prey selection.

Boreholes

Baumiller and Macurda (1995) collected a number of references dealing with circular boreholes in Paleozoic brachiopods, although their suggestion that these are "gastropod-like boreholes" does not deal with the problem that the boreholes lack the morphologies characteristically made by post-Paleozoic predatory gastropods. Baumiller et al. (1999) made a strong case that two spiriferid genera from the Mississippian Fort Payne Formation were overwhelmingly bored in the ventral valve, with borehole size and form not varying too much; whether or not this borehole activity may be blamed on platyceratids is still a question. They illustrated one Middle Devonian *Atrypa* with a borehole in one area of the shell and a platyceratid pressed against the same valve in another spot; the evidence is very circumstantial as yet, although by no means impossible.

Harper and Wharton (2000) provided data, chiefly British, indicating that predation on Jurassic and Cretaceous brachiopods indicated by boreholes is more common than has been previously suggested. Leighton (2001) provided borehole information from Frasnian, Late Devonian brachiopods indicating that the presence of spines lowered the incidence of boring. Leighton's is the most thorough review of Paleozoic brachiopod boreholes and their implications. She makes it clear that there was site stereotypy (over muscle fields and on the valve lying against the substrate); the nature of the predator is uncertain. Leighton also pointed out that not all genera in the fauna were bored (indicating predator preference). Baumiller and Bitner (2004) described boreholes present in Middle Miocene Polish brachiopods, some samples intensely bored and others not. Kowalewski et al. (2005) summarized a body of evidence suggesting that boreholes in Paleozoic brachiopods increased from very, very low frequency in the Silurian to a bit higher frequency in the Late Paleozoic, while assuming that such boreholes were made by a predatory organism, but provided no taxonomic evidence other than some for Late Paleozoic *Composita*, despite the fact that such boreholes, where they do occur, tend to be restricted to only certain taxa (i.e., a level of "prey" selection is involved).

There is a certain body of literature about presumably predator-generated boreholes into Paleozoic brachiopods, although not much for the post-Paleozoic. These boreholes tend to be simple, cylindrical structures that lack the distinctive morphological features associated with many post-Paleozoic gastropod-generated predatory boreholes. Conway Morris and Bengtson (1994) collected much of the literature on the topic. The common restriction of these Paleozoic brachiopod boreholes to one valve makes it unlikely that they were merely made by a sediment-burrowing organism rather than a predator. These brachiopod boreholes are commonly valve specific but not tightly site specific in many instances. It is notable in some of these Paleozoic brachiopod examples that by no means all the brachiopod genera and species from a specific sample are bored, suggesting that the predator was being reasonably prey specific. Hoffmeister et al. (2003) described site-specific boreholes from the very small Pennsylvanian brachiopod *Cardiarina*, with high site and valve specificity (see also Brett, 2003). Leighton (2003b), using a Middle Devonian strophomenoid example (see Figure 115), provided evidence that lamellose specimens of a species are less subject to drilling than are relatively smooth specimens. Deline et al. (2003) made a good case for drilling near the anterior shell margins of the Mississippian brachiopod *Perditocardinia*, with the possibility that this might reflect a kleptoparasitic life mode akin to that of some pinnotherid crabs, additional to evidence for predation (see Figure 116).

Examples are provided from the Cambrian by Conway Morris and Bengtson (1994), Miller and Sundberg (1984), Ushatinskaya (2003), and Robson and Pratt (2007); from the Ordovician by Bucher (1938), Cameron (1967), Carriker and Yochelson (1968), Holmer (1989), and Fenton and Fenton (1930–1931); from the Silurian by Wright (1968), Rohr (1976), Liljedahl (1985), and Chatterton and Whitehead

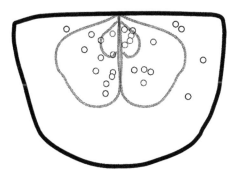

FIGURE 115 Plot of the boreholes into the muscle field of a Middle Devonian strophomenoid brachiopod indicating the site specificity of the drilling organism. (Figure 5 in Leighton, L.R., *Palaeogeography, Palaeoclimatology, Palaeoecology*, 201, 221–234, 2003. Reproduced with permission of Elsevier.)

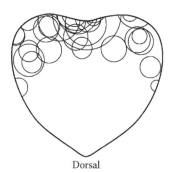

Dorsal

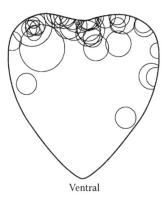

Ventral

FIGURE 116 Site-specific drilling behavior of a predator for the anterior margin of a Mississippian *Perditocardinia* brachiopod. (Figure 4 in Deline, B. et al., *Palaeogeography, Palaeoclimatology, Palaeoecology*, 201, 211–219, 2003. Reproduced with permission of Elsevier.)

(1987); from the Devonian by Sheehan and Lespérance (1978), Lespérance and Sheehan (1975), Buehler (1969), Smith et al. (1985), Brett and Bordeaux (1990), Fenton and Fenton (1930–1931), and Racheboeuf and Lespérance (1995); and from the Carboniferous by Brunton (1966, with borehole frequencies in different taxa), Cooper (1956, Plate 61, Figure 1), and Watkins (1974a).

Bassett and Bryant (1993) cited boreholes in micromorphic Belgian Tournaisian *Lambdarina*. Mottequin and Sevastopoulo (2007) briefly described boreholes in some

Belgian Tournaisian brachiopods, mostly *Crurithyris*, with a high level of site specificity and also prey specificity. Grant (1988) discussed similar boreholes from Late Paleozoic materials belonging to the Cardiarinidae. Ausich and Gurrola (1979) and Legrand-Blain and Poncet (1991) described some boreholes from Carboniferous brachiopods. Surlyk (1972, Figure 24) noted muricid(?) boreholes on some Maastrichtian brachiopods, while noting that Steinich (1965, p. 196) had noted that bored brachiopods occurred in the Rügen Chalk. Taddei Ruggiero (1990) described some Pliocene examples. Baumiller et al. (2006) provide a fine study of boreholes in some Algerian Pliocene *Megerlia*, accompanied by an excellent discussion of borehole occurrence in the Cenozoic; their information makes it clear that there was some evidence of intense borehole presence on Cenozoic brachiopods. Also see Harper (2005), who similarly addressed a Pliocene occurrence in *Apletosia* with some specimens showing other types of presumed predatory damage. Taddei Ruggiero and Annunziata (2002) described some Italian Early Pleistocene examples, and Taddei Ruggieero and Bitner (2008) described some repaired shell damage found in Spanish and Italian Pliocene brachiopods. Donovan and Harper (2006) described some Pleistocene borings from Jamaica and Barbados, including evidence about selectivity of taxa depending on their environment. Wiedman et al. (1988, p. 453, Figure 2.18–20) briefly noted an Eocene Antarctic lingulid with boreholes.

Although not dealing with positive evidence of boring by gastropods, the Conway Morris and Bengtson (1994) paper on known boreholes made by unknown predators into varied Cambrian shells raises a number of interesting questions about the times when the borehole-drilling type of predation first appeared.

Non-Borehole Damage

Małkowski (1976) described a number of repaired shell injuries in the Albian brachiopod *Cancinithyris biplicata* from Poland and concluded that the actual injuries were caused by physical factors, unspecified, rather than predation or other biological factors, although the evidence is inconclusive in our opinion. Additional descriptions of repaired shell injuries in brachiopods include Williams and Rowell (1965, pp. 73–74); Sarycheva (1949), who interpreted some of the malformations as being the result of damage caused by shark teeth in failed attacks; Fenton and Fenton (1932); Pajaud (1977); Schmidt (1951, Figure 6) and Ivanova (1949), for Moscow Basin Carboniferous; Clarke (1921, Figure 105), for the Late Ordovician; and Moodie (1923, Figure 3e), for an Ordovician strophomenoid. In none of these, however, is there good evidence of the causes. Sulser and Meyer (1998) reported on evidence of predation on a Late Jurassic *Sellithyris subsella* group sample from northwestern Switzerland. Alexander (1981) described a number of predation scars in some Mississippian brachiopods that he concluded are due to shell-crushing shark predation. Tasnádi-Kubacska (1962) described a number of Liassic specimens of *Pygope* with shell deformations that he ascribed to marine

predator attempts. Elliott and Brew (1988) described shell damage to a Pennsylvanian productid from Arizona that they attributed to attempted predation.

Donovan and Gale (1990) reviewed the varied potential predators of fossil brachiopods, as well as examples of brachiopod predation for the present, while suggesting that asteroids may have been important in the post-Paleozoic. Blake and Hagdorn (2003) commented on the Donovan and Gale conclusion and suggested that post-Paleozoic asteroid predation may not be quite as important as they suggested, but they do not dismiss it. Alexander (1986) reviewed various types of potential predation on Late Ordovician brachiopods and illustrated a strophomenoid with a possible nautiloid rhyncholite fragment still embedded. Except for Alexander's nautiloid possibility, the identities of the potential predators remain speculative.

Bałinski (1993) reviewed the available evidence regarding potential predation on brachiopods, including a review of individuals known to have had their shells repaired following damage that might have been caused by a potential predator. The amazing thing about the articulate brachiopod predation question is how few specimens showing potentially predator-generated evidence are known; either brachiopods were unsuitable food for predators or they were eaten and then their shells somehow show no evidence of predation or were fragmented beyond recognition. Jones (1982) described one Late Silurian Canadian Arctic occurrence of *Atrypoidea* with a significant number of badly deformed shells for which predation seems unlikely, although an unknown, soft-bodied epibiont might have been the cause of the shell deformation—a most unusual occurrence for a widely distributed genus. Malzahn (1968) and Schaumberg (1979) described two petalodont, skate-like fish, obviously bottom feeders, with fragmented pieces of the Late Permian productid *Horridonia* in their visceral content, together with other prey items; this is obviously an example where a fragmented brachiopod would provide no evidence that it had been predated, and there must be many others.

Ebbestad and Hogstrom (2000) described two elegant examples of Late Ordovician predator-type, repaired injuries on Swedish strophomenoid brachiopods, while considering the overall predation situation among shelled marine invertebrates. Godefroid (1999, Figure 6) illustrated a Moroccan Givetian atrypid (*Invertrypa*) with shell damage that could have been caused by a fish biting the shell and causing near puncture-type damage to the mantle. Taddei Ruggiero and Annunziata (2002) described some Early Pleistocene Italian shell damage that may represent predator activity. Harper (2005) described some Pliocene damage made by a presumed predatory attempt.

In much of the above there is good evidence for mantle damage leading to very irregular shell growth, but in none of this is there solid evidence of the causes of the damage. Elliott and Bounds (1987) provided a note of caution in the interpretation of "crushed" brachiopods as caused by predators rather than by nonorganic compaction processes, using specimens of *Composita* from the Arizona Pennsylvanian.

Simões et al. (2007) provided evidence from the modern Brazilian fauna that the frequency of drilled brachiopods is significantly lower than for drilled bivalves, which accords well with somewhat similar evidence from the fossil record. Mahon et al. (2003) conducted experiments where the Antarctic terebratulid *Liothyrella* was fed as tissue extracts and whole organisms to starfish and fish; their findings suggest a high level of unpalatability, which provides a potential explanation for the overall low level of boreholes present in most brachiopod-rich samples.

☐ *Reliability*
Category 6, because the nature of the predator is commonly unknown.

*6Alq. Possible Ophiuroid, Brittle Star Predation
Petr (1989) described and illustrated Middle Ordovician ophiuroids from Bohemia that may have been preserved either scavenging or predating carpoid, conularid, and sphenothallid materials.

☐ *Reliability*
Category 6, speculative.

*6Alr. Ammonite Feeding
Jäger and Fraaye (1997; see also Jäger, 1991) reviewed the available information on possible ammonite "stomach" contents and concluded that there is convincing evidence from European Jurassic specimens of their having fed on thinly skeletonized decapod crustaceans and also on smaller ammonites.

☐ *Reliability*
Probably Category 2A.

*6Als. Graptolite Predation
Loydell et al. (1998) provided evidence of graptolites, now regarded as hemichordates, having been predated in the Silurian by an unknown predator. The evidence consists of ball-like graptolite fragments that probably originated as a result of either regurgitation by a predator or as coprolitic masses.

☐ *Reliability*
Category 2B, functional morphology.

*6Alt. Predation on Echinoids
Good evidence of predated sea urchins, regular or irregular, is uncommon. Kroh and Nebelsick (2006) provided a good review of the varied predators, vertebrate and invertebrate, that affect echinoids using both modern and fossil materials. Smith (1984, pp. 11–14) reviewed evidence from the past and present, as have Kowalewski and Nebelsick (2003). McNamara (1991) discussed some Australian Miocene echinoids concluded to have been bored by cassids. The paper that best documents predated irregular urchins is Zinsmeister's (1980) treatment of some Miocene clypeasteroids from Patagonia, many of them with massive, marginal damage of

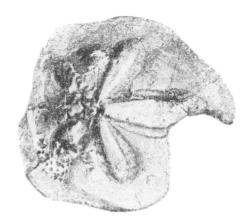

FIGURE 117 Middle Miocene *Clypeaster* with a healed, marginal "bite" (Felsö-Orbó, Hungary); the bite measures 30 × 25 mm. (Figure 1 in Vadasz, M.E., *Centralblatt für Mineralogie, Geologie und Paläontologie*, 9, 283–288, 1914. http://www.schweizerbart.de. Reproduced with permission.)

the type discussed by Smith. Fish or crabs could have been responsible, as well as other potential predators. Vadasz (1914) described several Hungarian Miocene echinoids with a massive bite or bites taken out of their periphery that healed over subsequently (see Figure 117); see also Santos et al. (2003) for some Spanish Miocene examples.

Tasnádi-Kubacska (1962) discussed a number of damaged echinoids. Saint-Seine (1950, 1959) documented varied test damage that might have involved predators and/or parasites. Ernst (1971, p. 97) suggested that some Late Cretaceous *Echinocorys* from the North German chalk have outer test grooves made by predatory crabs. Schormann (1987) discussed and illustrated some Maastrichtian and Danian echinoids with grooves that he suspects were made by rays. Ali (1982) illustrated several Egyptian Neogene clypeasteroids with healed "bites" taken out of their margins, but as usual the identity of the predators is unknown; he also noted boreholes of uncertain origin and aboral damage to the test. Neumann (2000) summarized much of the literature on echinoderm predation from the fossil record.

Sohl (1969, Figures 7, 8) illustrated Late Eocene cassid boreholes in Cuban echinoids, and Beu et al. (1972) documented Oligocene and Miocene New Zealand cassid boreholes in echinoids. McNamara (1994b) described some Australian Oligocene and younger examples of cassid predation on echinoids, while reviewing many aspects of cassid predation overall. Cross and Rose (1994) cited possible gastropod predation holes on Late Cretaceous *Micraster* from Britain, and Santos et al. (2003) described some clypeasteroid evidence from the Spanish Miocene of the Seville region that may be due to fish predation. Mitrović-Petrović (1964) described varied Miocene evidences of echinoid test damage interpreted as being due to various predators and parasites.

Ceranka and Złotnik (2003) described some predatory boreholes present in Polish Miocene clypeasteroids and considered that they were probably the result of cassid gastropod activities. Donovan and Pickerill (2004) reviewed this paper and concluded that either cassid or eulimid activities might have been responsible for the boreholes, which they assigned to a species of the ichnogenus *Oichnus*.

☐ *Reliability*

Category 2B, functional morphology.

*6AIu. *Ostenocaris* Predation or Scavenging

Pinna et al. (1985) described fish remains, both bony and cartilaginous; coleoid hooklets; and crustacean fragments in the gut area of the strange Jurassic arthropod *Ostenocaris*. They interpreted the gut area material as being the result of bottom scavenging, whereas Rolfe (1985) preferred mid-water, pelagic predation. Regardless, there certainly is reasonable evidence for ingestion by the arthropod.

☐ *Reliability*

Category 1, frozen behavior.

*6AIv. Invertebrate Predation on Ammonoids and Nautiloids

(See also Section 6AIIb, Vertebrate Tooth Puncture Marks and Potential Invertebrate Correlated Shell Injuries and Gut Contents; Section 6AIId, Shark Feeding; and Section 6AIIe, Vertebrate Predation on Cephalopods.) Not too much information has been published about evidence of predation on ammonoids and nautiloids; however, Kröger and Keupp (2004; see also Kröger, 2004) discussed Ordovician nautiloid examples of repaired evidence of predatory activity. Bond and Saunders (1989) did the same for some Mississippian ammonoids, and Mapes et al. (1995) for some Late Paleozoic examples. The nature of the predators is uncertain. Radwański (1996) reviewed evidence of both repaired ammonoid shells, where predation attempts might have been involved, and ammonoid shell fragment accumulations that may have been the result of successful invertebrate predation, while describing some Polish Maastrichtian examples of punctures in ammonoid shells that might have been caused by decapod crustaceans.

Kröger (2002a,b) discussed similar evidence from some Mesozoic ammonoids showing repair from predatory activity (see also Radwański, 1996). Bukowski and Bond (1989) described a very well-documented ammonite from the Maastrichtian Severn Formation of Maryland with bite marks clearly of marine reptilian origins. Ward and Hollingworth (1990) described a Callovian, Middle Jurassic ammonite with predation-generated holes that they could not assign to any specific predator after reviewing the available predator types, including various plesiosaurs and marine crocodiles, although a marine reptile is concluded to have been most likely (ichthyosaurs and fish being unlikely). They commented on the rarity of such predated ammonites overall, although there is always the possibility that thoroughly fragmented ammonoid shell might have been scattered and remain unrecognized as evidence for predatory activity.

☐ *Reliability*

Category 6, somewhat speculative.

*6AIw. Predation on Bryozoans

McKinney et al. (2003) discussed the very limited evidence concerning predation on bryozoans.

*6AIx. Invertebrate Predation on Trilobites

Babcock (2003, 2007) provided a thorough review of the evidence for predated trilobites, including repaired damage to varied trilobites and gut contents within predators, plus coprolites containing trilobite fragments. He pointed out that there is an overall tendency for the evidence of healed attacks being preferentially located on the right posterior portions, particularly in the Cambrian. Predators containing trilobite debris within their guts are very rare, as are coprolites. Babcock also makes a good case for the spiny proximal portions present on some trilobites reflecting a predacious mode of life involving the capture and trituration of soft-bodied prey.

While describing evidence of Middle Cambrian predation on eodiscid trilobites by an unknown arthropod, Zhu et al. (2004) summarized the evidence for predation in the Cambrian, limited as it is. Also pertinent is Vannier and Chen's (2005) unique paper on Maotianshan coprolite types, some of which contain trilobite fragments, together with an account of Cambrian soft-bodied predators that included trilobite material in their visceral regions.

☐ *Reliability*

Category 6, due to ignorance about the identity of the predators.

*6AIy. Predation on Crinoids

Baumiller and Gahn (2003) considered predation in crinoids through time. The only positive evidence consists of varied levels of arm regeneration here and there through time and the later Cretaceous to present presence of stalked crinoids almost entirely in deep waters.

☐ *Reliability*

Category 6, due to ignorance about the identity of the predators.

*6AIz. Potential Cephalopod Predation of Lobsters

Tshudy et al. (1989) made a good case, using circumstantial evidence, that lobsters have commonly been predated by cephalopods. The evidence (see their Figure 2) consists of the dominance of cephalothorax, or large parts of it, over the abdomens of fossil lobsters, combined with observations of modern *Nautilus* eating lobster abdomens and largely ignoring the cephalothorax.

☐ *Reliability*

Very reasonable, although in the absence of actual bite marks a Category 2B or Category 6 position could be used.

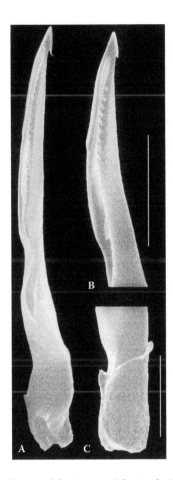

FIGURE 118 *Conus californicus*, radular teeth. (A) Entire tooth; length 0.96 mm. (B) Detail of apex of tooth showing adapical opening and first three barbs. (C) Detail proximal to B, showing fourth and fifth barbs. Scale bars in B and C = 0.1 mm; specimen is 19 × 12 mm (Dike Rocks, La Jolla, California). (Figure 4 in Kohn, A.J. et al., *Journal of Molluscan Studies*, 65, 461–481, 1999. Reproduced by permission of Oxford University Press.)

*6Alza. Large Abalones (*Haliotis*) and Coldwater Kelps

Estes et al. (2005) described and documented the co-occurrence today of relatively cool-/coldwater kelp forests in the Northern and Southern Hemispheres with kelp forests that involve different taxa in both hemispheres. They pointed out that following the Maastrichtian appearance of *Haliotis* there are only relatively small species until the Miocene–Pliocene boundary, when large species appeared abruptly in both hemispheres. They employed molecular data to make a good case for the uniqueness of the Northern and Southern Hemisphere species of both large and small shells. They interpreted the relatively sudden appearance of the large shells with the appearance of cold water and the use of the kelps as food—an entirely reasonable interpretation, unique in that it employs both paleontological and molecular information.

☐ *Reliability*

Difficult to categorize using the numbering system employed elsewhere in this treatment, but very reasonable as an explanation.

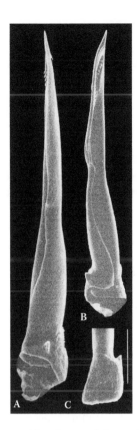

FIGURE 119 *Conus abbreviatus*, radular teeth. (A, B) Entire set of teeth of the same individual shown at different magnifications; length = 0.65 mm. The blade is oriented toward the viewer in A and toward the left in B. (C) Details of the base, showing spur. Scale bar in C = 0.1 mm; specimen is 31 × 20 mm (Makapuu Point, Oahu, Hawaii). (Figure 6 in Kohn, A.J. et al., *Journal of Molluscan Studies*, 65, 461–481, 1999. Reproduced by permission of Oxford University Press.)

*6Alzb. *Conus*

The predaceous gastropod *Conus* has an Early Eocene to present (Kohn, 1990), tropical–subtropical record. It subdues its prey infaunally, epibenthic or demersal, by means of a harpoon-like, modified radular tooth equipped with potent venom (Kohn et al., 1999) (see Figure 118 and Figure 119). It is nocturnal (Kohn, 1990). Different species tend to specialize on different prey, including varied annelid families, fish, and other gastropods, plus a few other taxa such as hemichordates (Duda et al., 2001; Kohn 2001). Marshall et al. (2002) described the venom producing and injecting apparatus in some detail (see also Hinegardner, 1958).

☐ *Reliability*

Category 2B, functional morphology, as none of the many fossils has been found in the act of actually attacking its prey.

*6Alzc. Boreholes in Hederellid Bryozoans

Wilson and Taylor (2006) described some predatory boreholes in the putative bryozoan *Hederella* from the Givetian.

☐ *Reliability*

Category 7, as the identity of the predator is unknown.

*6AIzd. Predation on *Mobergella*

Bengtson (1968) described evidence of predation on the enigmatic Early Cambrian genus *Mobergella* involving both boreholes and growth line anomalies.

□ *Reliability*

Category 7, as the identity of the predator is unknown.

*6AIze. Ostracode Scavenging

Willey et al. (2006) made a good case for ostracode scavenging on a Late Carboniferous Scottish shark.

□ *Reliability*

Category 1, frozen behavior.

*6AIzf. Predation on Dacryoconarids

Berkyová et al. (2007) described evidence of failed predation on some Early Devonian *Nowakia* as shown by their growth lines.

□ *Reliability*

Category 2B, functional morphology.

*6AIzg. Loosely Attached, Limpet-Like Foraminifer

Cossey and Mundy (1990) described a unique set of Mississippian rugose corals from England closely associated with what appears to be a loosely attached foraminifer, *Tetrataxis*, in what appears to have been a feeding relation possibly involving grazing on algae present on the surfaces of the corals.

□ *Reliability*

Category 1, frozen behavior.

Summary

There is enough family-level information to make the point about behavioral constancy but little at the generic level (see Table 13).

6AII. Vertebrate

6AIIa. Ray Holes

De Gibert et al. (2001) discussed probable ray holes, from the Spanish Cretaceous, that were generated by skates or their relatives while obtaining prey infaunal invertebrates.

□ *Reliability*

Category 2B, functional morphology.

6AIIb. Vertebrate Tooth Puncture Marks and Potential Invertebrate-Correlated Shell Injuries and Gut Contents

Massare (1987) presented an exceptionally good analysis of the functional morphology of the teeth of Mesozoic marine reptiles and the evidence provided by known stomach contents that bears on the problem.

Mosasaurs

The discussion by Martin and Bjork (1987) about some Campanian Pierre Shale mosasaur gastric residues from South Dakota provides a good summary of what is currently known about mosasaur feeding; they emphasized the fact that teleosts, possibly sharks, and smaller mosasaurs plus the maritime bird *Hesperornis* were utilized. They noted the opportunistic aspect of mosasaur feeding, as is also the case with so many of the larger marine carnivores. Lingham-Soliar (1998) discussed a Late Cretaceous mosasaur specimen

---

TABLE 13

Marine Invertebrate, Predator–Prey Examples from the Fossil Record (Additions to Table 12 from Boucot, 1990a)

Behavioral Type		Taxonomic Level	Time Duration	Category	Refs.
6AIf.	Starfish–mollusks	Class–class	Late Ordovician	1	Blake and Guensburg (1994)
6AIh.	*Cruziana–Teichichnus–Halopoa* community	Genera	Cambrian–Silurian	5	See text
6AIo.	Stomatopod predation on gastropods	Order–class	Miocene–Holocene	2B	Bałuk and Radwański (1996)
6AIp.	Boreholes in brachiopods and predation on brachiopods in general	Class	Cambrian–Holocene	2B	Leighton (2003b)
6AIt.	Predation on echinoids	Class	Cretaceous–Holocene	2B	Smith (1984); Kowalewski and Nebelsick (2003)
6AIv.	Predation on ammonoids and nautiloids	Class	Ordovician to present	2B	Kröger and Keupp (2004)
6AIx.	Predation on trilobites	Class	Cambrian–Permian	6	Babcock (2003, 2007)
6AIy.	Predation on crinoids	Class	Ordovician–Holocene	6	Baumiller and Gahn (2003)
6AIzb.	*Conus*	Family	Eocene–Holocene	2B	Kohn (1990)
6AIzc.	*Hederella*	Genus	Givetian	7	Wilson and Taylor (2006)
6AIzd.	*Mobergella*	Genus	Early Cambrian	7	Bengtson (1968)
6AIze.	Ostracode scavenging	Genus–genus	Late Carboniferous	1	Willey et al. (2006)
6AIzf.	Predation on dacryoconarids	Genus	Early Devonian	2B	Berkyová et al. (2007)
6AIzg.	Loosely attached foraminifer	Genus–genus	Mississippian	1	Cossey and Mundy (1990)

in which ramming behavior by a conspecific might account for the evidence of brain damage. Russell (1967, p. 68) briefly referred to some aspects of mosasaur feeding. Everhart (2004b) described a Late Cretaceous mosasaur from Kansas that contains plesiosaur remains in its visceral region and reviewed similar evidence from the literature concerning mosasaur visceral contents. Lingham-Soliar (1992) cited an earlier work of Dollo (1887, p. 520) in which turtle remains were found in the visceral region. Williston (1899) cited fish bones in a mosasaur visceral region.

Kauffman (1990) provided an analysis of mosasaur predation on ammonoid cephalopods, and Kauffman (2004) described a Late Cretaceous California nautiloid conch bearing what he interprets as mosasaur bite marks; in the same paper, he cast doubt on many of the alleged "limpet scars" on some ammonite conchs as contrasted with mosasaur bite marks, while reviewing the entire problem critically.

Martin and Fox's (2004) abstract cited fragmented inoceramids and *Anomia*(?) in the visceral region of a Late Cretaceous crushing toothed mosasaur, *Globidens*. Tsujita and Westermann (2001) carefully described numerous specimens of *Placenticeras* from the Late Cretaceous of southern Alberta, making a very strong case that the perforations in a significant percentage of the conchs are due to mosasaur rather than limpet activity; they summarized the many papers on this topic. Ludvigsen and Beard (1994, Figures 64, 111; 1997) illustrated bite marks on a Maastrichtian ammonite from the Lambert Formation and one on a nautiloid from the Haslam Formation, Late Cretaceous, both from Vancouver Island. Hewitt and Westermann (1990) described some western Canadian Late Cretaceous mosasaur bite marks on ammonites.

In view of the above information, interpreted as evidence for mosasaur predation on ammonites, one needs to pay careful attention to Kase et al.'s (1998) work indicating that at least some of the circular pits on ammonite conchs represent the home sites of limpets that foraged on the ammonite for algae, always returning to their home site. The authors diagram one such depression, which shows scratch marks similar to those made by modern limpets. This being the case, it is clear that the pits on Cretaceous ammonite conchs, previously all interpreted as evidence of mosasaur predation, may be in part evidence for limpet activity.

☐ *Reliability*
Category 6 for Russell's item, due to its speculative nature, and Category 1 for the other information.

Ichthyosaurs
Wang et al. (2008) cited Chinese Carnian ichthyosaurs with visceral contents including shelled invertebrates, fish remains, and lithic gastroliths. A large body of information is available on numerous Jurassic ichthyosaur gastric residues (Keller, 1976; Pollard, 1968), with emphasis on teleost and dibranchiate cephalopod hooklet content. Pollard (1968) discussed the problem of the identity of the cephalopod materials and concluded that the ichthyosaurs were mid-

water feeders analogous to sperm whales. Böttcher (1989) described a Posidonienschiefer ichthyosaur with abundant cephalopod hooklets in its visceral region plus some small ichthyosaur centra, interpreted to be the young of another genus. Pollard (1968) also described a lower Lias ichthyosaur replete with different types of dibranchiate hooklets that probably belonged to belemnoids and reviewed the literature on ichthyosaur stomach contents; he found that dibranchiate hooklets are most common but that fish remains also occur, primarily in coprolites. Perkins (2002) reported on a Queensland Cretaceous ichthyosaur said by investigators to have fish and hatchling turtles in its visceral cavity. Kear et al. (2006) described the gut contents of this Early Cretaceous Queensland specimen as also containing the remains of a bird and fish.

☐ *Reliability*
Category 1, frozen behavior.

Plesiosaurs
When considering the problems of plesiosaur diet in an objective manner, including the difficulty of reliably reconstructing same, Martill (1992) described a pliosaur's stomach contents from the Oxford Clay, with dominant cephalopod hooklets plus some additional vertebrate debris and a few pebbles (gastroliths), which adds still more to the story. Storrs (1995) cited bony fish debris and some small-size gravel in the visceral region of a small English Liassic plesiosaur. Tarlo (1959) cited cephalopod hooklets from a Kimmeridge Clay *Pliosaurus* (i.e., prey). Patterson (1975) mentioned in passing the presence of *Pholidophorus* and leptolepid remains with a pliosaur from the English Lower Lias. Wahl (1998) reported several thousand coleoid hooklets in the visceral region of an Upper Redwater Shale, Sundance Formation plesiosaur from Wyoming. Zhuravlev (1943) reported a Late Jurassic pliosaur from southern Russia with fish remains, cephalopod hooklets, and gastroliths in the visceral region.

Sato and Tanabe (1998) documented a Late Cretaceous Japanese plesiosaur with gastroliths, ammonite jaws, mollusk shells, and a shark's tooth, adding more to the view of plesiosaur diet as being an opportunistic one (see also Matsumoto et al., 1982). Martin and Kennedy (1988) briefly described stomach contents of a Late Cretaceous plesiosaur from South Dakota with gastroliths and fish remains. Sato et al. (2006) cited gastroliths associated with a Japanese Late Cretaceous plesiosaur.

Thulborn and Turner (1993, 1995) provided good evidence for an Australian, Early Cretaceous elasmosaur having been predated, with appropriate tooth marks on the prey skull, by a very large pliosaur. Cicimurri and Everhart (2001, Table 1) described a Kansas Late Cretaceous elasmosaur with identifiable teleost and shark debris in its visceral region, as well as gastroliths, and provided a table listing all of the plesiosaur and pliosaur remains known to them that are associated with visceral contents and gastroliths. Everhart (2005b, pp. 135–139) provided further discussion concerning elasmosaur gastroliths. McHenry et al. (2005) described coprolitic-type evidence from Australian Aptian and Albian plesiosaurs for bottom feeding involving varied shelly marine invertebrates. Druckenmiller and Russel (2008), while describing an Alberta Albian plesiosaur (*Nichollsia*), briefly discussed its gastroliths accompanied by a few fish vertebrae and also mentioned some in the South Australian *Umoonasaurus*.

☐ *Reliability*
Category 1, frozen behavior.

Crocodilians

(See also Section 6AIIf, Crocodilian Turtle Feeding, and Section 6De, Crocodilian Mammal Feeding.) Martill (1985, 1986a,b) cited a Lower Oxford Clay *Metriorhynchus* with associated cephalopod hooklets and an Oxford Clay *Leedsichthys* that had been attacked by a crocodile, as indicated by an embedded tooth over which bone had grown subsequently (i.e., the fish escaped). Phipps (2008) described some Yorkshire Jurassic *Gryphaea* with indentations in some specimens that possibly represent crocodilian predation.

Chelonians

(See also Section 6CIIz, Turtle–*Celtis* Feeding.) Kear (2006) described the visceral region contents of an Australian Albian sea turtle (cf. *Notochelone*) that include many inoceramid fragments; associated coprolites also contain similar inoceramid fragments. *Celtis* seeds are known (see Section 6CIIz) from two turtle specimens from the Oligocene.

Fish

(See also Section 6AIId, Shark Feeding.) Janvier (1996, pp. 297–300) provided a general discussion of feeding evidence in Early Paleozoic fish. Denison (1956, pp. 425–426) cited a British Early Devonian acanthodian with the head shield of a small *Cephalaspis* in its visceral area that probably represents predation. Zakharenko (2008) described a Late Devonian Russian occurrence of one placoderm's visceral region yielding bony debris belonging to another type (i.e., probable evidence for predation). Miles and Westoll (1968, p. 462) cited evidence of bone and acanthodian scales in the visceral area of a Middle Old Red Sandstone Scottish *Coccosteus cuspidatus*, and Davidson and Trewin (2005) reported acanthodian scales and possible dipnoan remains plus coarse sand grains in the visceral area of some Scottish Middle Devonian specimens of this species.

Martill (1990) described Middle Jurassic ammonite fragments from Britain that bear tooth marks similar to those of certain contemporary fish that had crushing plate type teeth. Doyle and MacDonald (1993, Figure 2) illustrated a fine Posidonienschiefer *Hybodus* shark with belemnite gut contents. Neumann (2000) reviewed the evidence of predation by varied fish types on starfish, emphasizing some German Cretaceous examples, including bite marks and coprolite occurrence. In at least one case, a specific chondrichthyan predator is implicated. Gripp (1929) described bite marks on some Late Cretaceous echinoids from northern Germany

FIGURE 120 Tuna centra (*Thunnus*) punctured by billfish predators (*Tetrapterus albidus* and *Makaira nigricans*) (×1). (Figure 1 in Schneider, V.P. and Fierstine, A.L., *Journal of Vertebrate Paleontology*, 24, 253–255, 2004. Reproduced with permission.)

that may well have been made by fish; the marks consist of relatively deep, subparallel grooves, suggesting the anterior dentition of a fish. Thies (1985) documented varied "bite marks" on *Echinocorys* tests from the German Maastrichtian (for some Spanish Miocene evidence, see also Santos et al., 2003).

Williams (1990) described conodont assemblages in the visceral region of Upper Devonian Cleveland Shale fish, and Lund (1990) cited conodont animal remains within Late Mississippian fish from Montana. These are, of course, very unusual preservation situations. Maisey (1994) described Lower Cretaceous Brazilian fish with crustacean remains in their visceral area (see also Wilby and Martill, 1992), and Lund (1990) did the same for some Late Mississippian fish from Montana—again, very unusual preservation situations due to the low preservation potential of poorly mineralized crustacean exoskeletons.

Kriwet (2001, Table 1) summarized what is known about Mesozoic and Paleogene pycnodont visceral region contents, mostly shelly, while considering the dentitions and jaw functioning of the group. He made the point that many fishes reject the shelly prey debris from their mouths which obviously affects the nature of the sample.

Schneider and Fierstine (2004) described some Early Pliocene Yorktown Formation tuna caudal vertebrae with evidence of puncture wounds inflicted by billfish (marlins and their relatives) and provided examples of billfish feeding from the present that include "spearing" tunas (see Figure 120). The holes in the tuna vertebrae fit the bills of the istrophorids very nicely; these holes cannot be blamed on other predators, and fossil billfish remains are present in the Yorktown Formation. This is a very positive example of specific predation, with the implication that the predation type has remained unchanged since at least the Early Pliocene.

Marine Mammals

In view of the relative wealth of visceral region "gut" contents known from piscivorous fish and from marine reptiles, it is surprising that almost nothing is known from marine mammals. The only exceptions known to us are Uhen's (2004, p. 163) report of fish remains in the Eocene archaeocete *Dorudon atrox* and the abstract by Swift and Barnes (1996) reporting fish debris in *Basilosaurus*. The overall rarity of visceral remains in terrestrial carnivores is also noteworthy and requires explanation. Are mammalian carnivores while decaying more apt to somehow eliminate skeletal remains of their vertebrate prey? Equally puzzling is why the uncommon occurrences of mammalian fetal remains are more common than are visceral prey remains—is there different preservational potential?

☐ *Reliability*
Category 3.

6AIIc. Arthrodire–Ctenacanth Shark

Although not concerning relations between an arthrodire and a ctenacanth, the description by Capasso et al. (1996) of a Middle Devonian median dorsal *Dunkleosteus* plate with a "broad perforation" is of interest and was interpreted by them as an intraspecific interaction.

☐ *Reliability*
Category 6, somewhat speculative.

6AIId. Shark Feeding

(See also 6AIIb, Vertebrate Tooth Puncture Marks and Potential Invertebrate Correlated Shell Injuries and Gut Contents, and Section 6Dd, Mosasaur–Shark.) Mapes et al. (1995) documented predator damage (suitably spaced holes of appropriate morphology) in a Pennsylvanian ammonite from Kansas and Nebraska that can reasonably be interpreted as being due to a chondrichthyan, and a symmoriid shark, in particular. Bond and Saunders (1989) previously had documented repaired injuries, probably of predatory origins, from Mississippian ammonoids from Arkansas. Mapes et al. (1989) described some predated Pennsylvanian conularids from Oklahoma that may have been predated by sharks. Brunton (1966) documented evidence of shell damage to some Viséan productids that might have been caused by sharks.

Cione and Medina (1987) discussed the presence of a shark tooth with a Late Cretaceous Antarctic plesiosaur, suggesting that scavenging might have been involved, and reviewed a number of other occurrences where associated skeletons and teeth might be similarly interpreted, as well as examples of scratch marks on bone (including varied cetaceans, penguins, and a turtle). Dortangs et al. (2002), while describing a Late Cretaceous mosasaur from the Netherlands, cited numerous associated shark teeth as evidence for postmortem scavenging. Forrest (2003) described a *Metriorhynchus* tooth mark on a plesiosaur vertebra that he interpreted as scavenging evidence. Corral et al. (2004) described some puncture marks on Campanian, Late Cretaceous mosasaur bone from Alava in the Vasco–Cantabrian region, but as usual whether scavenging or predation is involved is unclear.

Bowman (1971) cited an occurrence of shark teeth belonging to one genus associated with the skeleton of another genus in a relationship that suggested scavenging. Schwimmer et al. (1997) provided convincing evidence from bite marks on Late Cretaceous vertebrate bone, as well as embedded shark teeth and marks on dinosaur bone preserved in a marine environment, for scavenging activity within *Squalicorax* in at least some instances, including sharks, bony fish, turtles, mosasaurs, plesiosaurs, and a dinosaur. Shimada (1997) made a good case that the 5- to 6-meter-long shark *Cretoxyrhina mantelli* was an omnivorous feeder (the large teleost *Xiphactinus* and possibly plesiosaurs), as Everhart (1995, 1999, 2004) provided evidence that this shark, as well as *Squalicorax falcatus*, was either predating or scavenging a mosasaur. Shimada et al. (2002) cited evidence that *Cretoxyrhina* predated or scavenged ichthyodectid fish and protostegid turtles, also from the Western Interior Late Cretaceous. Shimada and Everhart (2004) provided further evidence about the probability of *Cretoxyrhina* predation on *Xiphactinus* in terms of a broken tooth of the latter embedded in the third vertebra of the former. Shimada and Hooks (2004) described *Cretoxyrhina* teeth imbedded in protostegid, marine turtles from the Late Cretaceous Mooreville Chalk in Alabama, with predation or scavenging being indicated; they reviewed a variety of evidence for shark feeding from the fossil record.

Everhart and Hamm (2005) described a nodosaur limb bone with bite marks suggesting that *C. mantelli* probably scavenged the item. Everhart (2005) described a Late Cretaceous plesiosaur, elasmosaurid paddle with evidence for *C. mantelli* predation on the humerus. Everhart and Hamm (2005) described some nodosaur remains from the Kansas Santonian that bear evidence of *Cretoxyrhina* scavenging in terms of tooth scratches and etching. Martin and Rothschild (1989; see also Rothschild et al., 2005) provided good evidence that a mosasaur was bitten by a shark but survived with a shark's tooth still imbedded; the mosasaur revealed evidence of wound healing and possible infection. Druckenmiller et al. (1993) found a *Squalicorax* closely associated with presumed prey bones of *Ichthyodectes*, a fish, turtle, and mosasaur remains. Schumacher and Everhart (2005) described more evidence of Late Cretaceous feeding on plesiosaur remains by *Squalicorax*, Everhart (2005c) discussed Late Cretaceous evidence of probable scavenging of a mosasaur by *Squalicorax*, Kaddumi (2006) provided evidence of shark scavenging from the Maastrichtian of Jordan, and Sato et al. (2006) provided evidence of predation or scavenging affecting a Japanese Late Cretaceous plesiosaur.

Bigelow (1994) described tooth marks on a pinniped bone (*Allodesmus*) from the Miocene of Washington and associated squaloid shark teeth that suggest either predation or scavenging. Nomura et al. (1991) described a Japanese Miocene mysticete whale skeleton occurrence associated with over 600 shark teeth! Dawson and Gottfried (2002) cited a shark tooth imbedded in a Miocene dolphin lumbar vertebrae in such a manner as to suggest scavenging.

Morgan (1994) discussed widespread Miocene–Pliocene evidence of predation and/or scavenging by *Carcharodon megalodon* from Maryland to Florida, with baleen whales providing the bulk of the evidence, chiefly tooth marks and tooth fragments. Walsh and Hume (2001) illustrated a spheniscid right coracoid with cut marks from a shark (predation or possibly scavenging) from the Neogene of north-central Chile. Cigala-Fulgosi (1990) described some very convincing "grooved" scars on Italian Pliocene bottlenosed dolphin bones that might have been made by serrated teeth, like those of the great white shark, that may represent either scavenging or predating.

McKee (1987) described some Pliocene New Zealand penguin bones with vertebrate tooth marks indicative of either predation or scavenging. Aguilera et al. (2008) described a Venezuelan Pliocene occurrence of a large shark tooth imbedded in cetacean bone, as well as both cetacean and sirenian bone with shark bite marks, with the usual problem regarding whether predation or scavenging was involved. Esperante and Brand (2002) mentioned Chilean Mio–Pliocene mysticete whale remains associated with abundant shark teeth. Some shark teeth are actually imbedded in some of the bones. It is unclear whether scavenging or predation was involved. Barnes and McLeod (1984) described evidence of a Late Pleistocene gray whale that was either predated or scavenged by sharks. Finally, Repenning and Packard (1990) described a unique California Miocene desmostylian, a marine mammal similar in some regards to the manatees, that was probably predated by sharks, leaving the skeleton associated with numerous shark teeth; shark teeth are absent elsewhere in the vicinity within the same beds. It is clear that the Cretaceous and younger sharks are about as omnivorous in their treatment of vertebrate prey and carcasses as they are today!

☐ *Reliability*

Category 2B, functional morphology, based on proximity of the shark teeth to a potential carcass and of scratch marks similar to those made by sharks. The "absence" of Jurassic citations of shark predation is presumably a sampling artifact.

6AIIe. Vertebrate Predation on Cephalopods

(See also Section 6AIIb, Vertebrate Tooth Puncture Marks and Potential Invertebrate Correlated Shell Injuries and Gut Contents.) Mehl (1978) described a Late Jurassic German Solenhofen coprolite containing broken ammonoid aptychii but no other prey remains. He ascribed the material to the activity of holostean fish. The absence of any fragments of the ammonite conch might be owing to its aragonitic composition as contrasted with calcite for the aptychii. Martill (1986b, p. 182) cited a later Jurassic Oxford Clay *Caturus* with ink sacs in its visceral area, an indication of teuthoid feeding. Martill (1986a) discussed a Jurassic *Metriorhynchus* with cephalopod hooklets, a belemnite rostrum, and a questionable pterosaur bone in its visceral region. Mapes and Chaffin (2003) thoroughly reviewed the evidence on predated cephalopods from the fossil record and considered the

insight that modern *Nautilus* provides. Their data make it clear that we are far from understanding the overall nature of this predation, including the identity of the predators and the levels of predation on shelled cephalopods through time. Escalation hypotheses have a long way to go before being accepted.

*6AIIf. Crocodilian Turtle Feeding

Boucot (1990a, p. 10) expressed skepticism about the presence of any good evidence for crocodilians feeding on turtles. Carpenter and Lindsey (1980) did suggest that the short, blunt teeth of a Late Cretaceous crocodilian suggested chelonivory, but in the absence of visceral-region turtle remains this possibility remains unproved. More recently, J.-P. Leeming (pers. comm.) called attention to Erickson's (1984) work on evidence for chelonivory in a late Paleocene crocodile (*Leidyosuchus formidabilis*) from North Dakota. Erickson described traumatic injuries to the turtles, including evidence of healed, thickened bone around injuries, puncture wounds for which the holes correspond in size to the largest teeth of the associated crocodilian, punctures spaced in a manner similar to the spacing of the crocodilian teeth, and wounds on the posterior and posterodorsal surfaces suggesting that the turtle was traveling away from the crocodilian. Escape presumably occurred while the crocodilians were trying to orient the turtle for feeding. It is notable that the dentition of *Leidyosuchus* is not specialized for crushing; thus, chelonivory need not involve specially flattened teeth. However, Erickson (1984) provided some circumstantial evidence from the North Dakota Paleocene that a crocodilian with a normal crocodilian dentition might well have eaten turtles, based on presumed tooth marks preserved on turtle carapace specimens. Note, too, that the posterior teeth of crocodilians tend to be relatively flat and suitable for crushing. Sawyer and Erickson (1987) provided more information about crocodilian predation on turtles with both modern and fossil examples.

Joyce (2000) provided evidence of probable predation by a Late Jurassic German crocodilian on a turtle, as evidenced by healed bite marks, while citing similar evidence from the Swiss Jurassic. Hutchinson and Frye (2001) described some possible turtle injuries made by crocodilians from the Cenozoic. Jiménez-Fuentes et al. (1987) discussed an Eocene Spanish turtle that may have been attacked by a crocodilian.

Although not necessarily involved with crocodilian activity, Wieland (1909) described a huge, Late Cretaceous specimen of *Archelon* with excellent evidence provided by the missing distal portion of the right flipper (only a healed tibia and fibula remain) of the individual having been predated when much younger and survived.

☐ *Reliability*

Category 3.

*6AIIg. Plankton Feeding

Planktivores are one of the important elements in both the marine and freshwater ecosystems today, yet our knowledge of just what plankton taxa were present during the past, since so few plankton modern and past are well skeletonized, is very defective. One can think of such things as radiolarians since the Cambrian, globigerinid Foraminifera beginning in about the Jurassic, tintinnids beginning in the earlier Paleozoic, coccoliths since about the mid-Mesozoic, diatoms beginning in about the Jurassic, heteropods and pteropods beginning in the Jurassic, acritarchs, chitinozoans, and so forth, but many currently important groups such as copepods and euphausids are missing from the fossil record. While discussing the bopyrid isopod parasites of varied crustaceans, Boucot (1990a) made the case for marine copepods being present at least since the earlier Jurassic by using as evidence bopyrid isopods parasitic on crustaceans and involving an intermediate parasitic lifecycle stage in a normal marine copepod today.

Vannier et al. (2007) described arrow worms from the Chinese Lower Cambrian that provide evidence for their having fed on plankton at that time. Fryer and Stanley (2004) made the very valid comment that the presence of floating jellyfish in the column, from the Late Proterozoic on, is witness to the presence of available planktonic food of one unknown type or another. Röper et al. (1999) illustrated various Late Jurassic jellyfish from southern Germany that make the point about the presence of plankton. Moore (1956) provided a review of the older literature on fossil jellyfish. Nützel and Frýda (2003) commented that the change in the mid-Paleozoic from openly coiled, early gastropod growth stages to tightly coiled suggests a major change in planktonic predators. The presence in the Paleozoic of skeletonized forms such as radiolaria, acritarchs, and chitinozoans does not really solve the problem. (Anderson, 1996, Figure 3, reviewed the foods utilized by modern radiolaria, but we obviously cannot extrapolate this directly to extinct taxa.)

It is worthwhile to explore the presence and stratigraphic ranges of various vertebrate plankton feeders as evidence for the past presence of soft-bodied plankton lacking a fossil record, as is also the case with the normal marine copepods. Sanderson and Wassersug (1993) provided a useful account of vertebrate suspension feeding. The plankton-feeding Clupeomorpha have a fossil record extending back at least to the Late Jurassic. Martill (1988) discussed the Middle and Late Jurassic *Leedsichthys* as a plankton feeder based on its gill raker morphology, with the nature of the plankton being uncertain but probably in the shrimp-size class. Martill (pers. comm., 2001) now has information about *Leedsichthys* in the Oxfordian of Chile.

The plankton-feeding basking shark *Cetorhinus*, of the Cetorhinidae, is known as far back as the Eocene (Cappetta, 1987; for a diagram of a gill raker, see Daniel, 1934, Figure 44). Possibly the oldest plankton-feeding shark evidence is based on teeth and was discussed by Duffin (1998) with regard to Rhaetic, latest Triassic cetorhinid-type teeth. The time gap between the Rhaetic and the Eocene later appearance of cetorhinid-type teeth may indicate that the Triassic material represents an independent, much earlier radiation into the plankton-feeding niche that did not persist for very long, as there is no positive evidence between the late Triassic and the Eocene.

Megachasma, of the Megachasmidae, is known from the North Carolina Pliocene (Purdy et al., 2001, p. 106, Figures 21a–i). Cappetta (pers. comm., 2006) mentioned undescribed specimens from the Miocene of Chile and California, and Purdy et al. (2001) cited undescribed Miocene material from California also cited by Taylor et al. (1983). The centimeter-sized teeth arrayed around the outer rim of the huge mouth appear to be vestigial in this filter feeder, as suggested by Taylor et al. (1983), who mentioned euphausids in the stomach contents of the first one described.

Stewart (1991) discussed the "reduced" lamnoid morphology of the relatively small megachasmid teeth and also provided information on the California Miocene megachasmid teeth. The currently oldest known megachasmid evidence consists of teeth from the Colorado Cenomanian (Shimada, 2007). Molecular evidence (Martin and Naylor, 1994) suggests that the cetorhinid and megachasmid lamnid sharks evolved independently. Cappetta (1987) cited the Rhincodontidae sharks, Early Eocene to present (*Rhincodon, Palaeorhincodon*), and the batoids of the Mobulidae, Late Paleocene to recent. It is puzzling why so few records of the plankton-feeding sharks have been provided; their fossil record has probably been underestimated.

Whalebone whales (mysticetes), known from the Oligocene to present, are capable of using larger plankton such as euphausids (Carroll, 1988), but gray whales (Eschrichtiidae), known from the Miocene, feed by sieving out ampeliscid amphipods from muddy sand using their whalebone as a sieve rather than feeding on plankton (Bisconti and Varola, 2006; Nelson and Johnson, 1987). Oliver et al. (1984) concluded that gray whales probably secure their amphipods by suctioning sediment from the substrate following which the amphipods are filtered out by the whalebone rather than actually scooping sediment from the bottom. Nerini (1984) reviewed stomach content information for gray whales and an overall view of their feeding ecology; they are not plankton feeders.

Bonaparte (1971) described *Pterodaustro guinazui* from the Argentine Early Cretaceous and concluded that it was a probable plankton feeder, based on its dentition. Chiappe et al. (2000) provided excellent functional morphological dental evidence that the Early Cretaceous Argentinian pterosaur *Pterodaustro guinazui* was a freshwater plankton feeder, although the nature of the plankton is unknown.

Richter and Baszio (2001a) carefully described the gut contents found in the Eocene Messel teleost *Thaumaturus intermedius* as consisting largely of fragments of cladocerans, culicid and chaoborid larvae, and some additional insect fragments; clearly, this fish was a planktivore/insectivore. Richter and Baszio (2001b) provided a very detailed food web for the Messel Lake, beginning with the phytoplankton and ending with a crocodilian.

Mark-Kurik (1992; see also Mark-Kurik and Carls, 2004) concluded that the homostian arthrodires were probably Early and Middle Devonian planktivores, presumably

TABLE 14

Marine Vertebrate, Predator–Prey Examples from the Fossil Record (Additions to Table 13 from Boucot, 1990a)

Behavioral Type		Taxonomic Level	Time Duration	Category	Refs.
6AIIb.	Vertebrate tooth marks on invertebrates	Order–class	Jurassic	6	Martill (1990); Ward and Hollingworth (1990)
	Shark-ammonoid	Order–order	Pennsylvanian	6	Mapes et al. (1995)
6AIId.	Shark feeding	Order–order	Cretaceous	2B	Cione and Medina (1987); Schwimmer et al. (1997)
6AIIe.	Vertebrate predation on cephalopods	Order–order	Jurassic	1	Wahl (1998)
6AIIf.	Crocodilian turtle feeding	Order–order	Paleocene	3	Erickson (1984)
			Jurassic	3	Joyce (2000)
6AIIh.	*Allosaurus–Stegosaurus* relationship	Genus–genus	Jurassic	2B	Carpenter et al. (2005)
6AIIi.	Branchiosaur feeding	Genera-classes	Lower Permian	1	Werneburg et al. (2007)

using unknown plankton. The independent evolution of varied planktivores over time is reminiscent of the appearance and independent evolution of varied herbivorous tetrapods over time (Sues, 2000). Also of interest here is Bandel and Hemleben's (1987) account of heteropod and pteropod planktonic gastropods first appearing in the Early Jurassic and early Tertiary, respectively.

□ *Reliability*

Category 2B, functional morphology.

*6AIIh. *Allosaurus–Stegosaurus* Relationship

Carpenter et al. (2005) described an injured *Allosaurus* caudal rib with a morphology consistent with damage caused by a *Stegosaurus* tail spike, with some evidence of post-trauma bone infection and bone reaction. The presence of stegosaur tail spikes with tips broken off is further evidence.

□ *Reliability*

Category 2B, functional morphology.

*6AIIi. Branchiosaur Feeding

Werneburg et al. (2007) summarized visceral region contents in Lower Permian branchiosaurs showing distinct changes in food type with growth. Small forms might have fed on freshwater plankton; mid-sized forms with remains of conchostracans and ostracods are known; and large-sized forms with small vertebrates ingested palaeoniscoid fish and small branchiosaurids. Werneburg (1986, Plate II, Figure 1) illustrated a specimen with four conchostrans in the visceral region, and Werneburg (1988, p. 20) reviewed further information citing conchostracans and triopsids as well as specimens with small fish (*Rhabdolepis*) and branchiosaur larvae (Figure 10; see also Werneburg, 1989, Figure 5). Boy and Sues (2000) discussed feeding in branchiosaurs and provided food-web interpretations.

□ *Reliability*

Category 1, based on actual gut contents.

Summary

Most of the information is at the ordinal and class level, with little below, which is consistent with the omnivorous habits of many of the marine vertebrate predators; that is, a level of omnivorous behavioral constancy is manifest (see Table 14).

6B.   FRESHWATER

6Ba.   Carboniferous Scorpion Decomposition

The "nematodes" described by Størmer (1963) as *Scorpiophagus baculiformis* Størmer and *S. latus* Størmer from the decomposed remains of the Lower Carboniferous scorpion *Gigantoscorpio willsi* Størmer were further examined by Poinar. These objects are very peculiar because the posterior end is always decayed or missing. The "anterior" end contains a short, narrow, darkened invagination thought to represent a buccal cavity, but there is no evidence of an associated alimentary tract or any other structure that indicates these objects are nematodes. Careful examination of the photographs and drawings show that the measurements of *S. baculiformis* and *S. latus* overlap with those of setae and spines associated with the scorpion remains. For these reasons, both *S. baculiformis* and *S. latus* are considered as types of setae and/or spines dislodged from the scorpion during burial.

□ *Reliability*

Category 2B, functional morphology.

6Bb.   Predation Marks on Estheriids

Bi (1986) described growth line abnormalities that are interpreted as due to bite marks made by fish. The specimens are from the Early Cretaceous Yantang Formation of Anhui and belong to *Yanjiestheria*. Tasch (1961) described valve injury and repair in living and fossil conchostracans that may have to do with predation.

□ *Reliability*

Category 6, somewhat speculative although reasonable.

*6Bc. Ciliate Feeding

Poinar et al. (1993b, Figure D) illustrated a freshwater ciliate feeding on a cyanobacterial filament, from Carnian amber occurring in the Bavarian Raibler Sandstone, which has subsequently been shown to be of Cretaceous age (Schmidt et al., 2001).

□ *Reliability*

Category 1, frozen behavior.

*6Bd. Crayfish Predation

Feldmann and May (1991) described some predated Quaternary crayfish. The identity of the predator is uncertain but they considered mammals to have been most likely.

□ *Reliability*

Category 6, due to uncertainty about the identity of the predator.

*6Be. Unionid Predation

Kear and Godthelp (2008) described some Early Cretaceous, New South Wales unionids with good evidence of predator-generated bite marks. The identity of the predator is considered, but no definite conclusion was possible.

□ *Reliability*

Category 6, due to uncertainty about the identity of the predator.

6C. TERRESTRIAL

6CI. Invertebrates

Hotton et al. (1996) provided evidence suggesting Early Devonian plant feeding by arthropods and possible mycophagy, based on trace fossils, while Labandeira (1996) and Labandeira and Phillips (1996b) provided evidence from the Pennsylvanian. Banks and Colthart (1993) described good evidence of later Early Devonian feeding, possibly by insects or other invertebrates, on trimerophytes.

6CIa. Spider Webs, Spinnerets, and Bundled Prey

See also Section 4AIIIi, Parasitoid Wasps and Parasitic Flies–Spiders). Nentwig and Heimer (1987; see additional papers in the same volume for more spider silk information) provided additional information about spider web functions and design. Zschokke (2003) discussed a piece of Early Cretaceous, Lebanese amber spider silk replete with araneoid-type sticky droplets (see Figure 121); this is cited as the earliest example of this silk type and all that it implies behaviorally. Peñalver et al. (2006) describe some Spanish Early Cretaceous spider web with sticky droplets and prey. Brasier et al. (2008, p. 18) cited Early Cretaceous spider web from amber in Sussex. Selden et al. (1991) described the earliest evidence for spider silk with

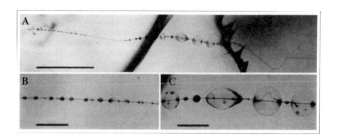

FIGURE 121 Fossil araneoid spider thread with glue droplets (Lebanese Early Cretaceous amber). Scale bars: A, 500 μm; B, 100 μm; C, 100 μm. (Figure 1 in Zschokke, S., *Nature*, 424, 636–637, 2003. Reproduced by permission from Macmillan Publishers Ltd.)

their *Attercopus*, which has a spinneret, from the Givetian, later Middle Devonian of New York. Selden (1990) described several orb-web-weaving spiders from the Early Cretaceous of Montsech, northeastern Spain. Selden and Nudds (2004) illustrated a Baltic Amber spider spinneret with silk streaming out (their Figure 247), as well as more spider web from Baltic Amber (their Figure 248).

Weitschat (1980, Figure 9; see also Weitschat and Wichard, 1998, Plate 20, Figures a, b) illustrated a very fine Baltic amber spider ventrally, showing silk streaming from its spinnerets. Codington (1992) described a unique Eocene, Parachute Creek Member, Green River Formation leaf associated with which are an assemblage of miscellaneous insects (90 plus) and two spiders (Anon., 1993), all of which resemble a spider web dumped into the water with its load and preserved nearly in place. The *Geotimes* article (Anon., 1993) also mentioned that flies, beetles, ants, and wasps are involved, including some insect fragments that might have been fed on, and that the leaf belonged to a sycamore on which the web was woven. Weitschat and Wichard (1998, Figures 36, 37) illustrated a Baltic amber dysderid with an ant in its grasp and an ant bundled up in silk, discussed some of the web types reported from Baltic amber, and illustrated a well-preserved web (Plate 20, Figure d). Koteja (1998) cited coccids entrapped in spider web. Janzen (2002, Figures 104–106) illustrated an ant "wrapped" in spider silk from Baltic amber, a spider web with a tunnel from which the spider can emerge on the attack, and a spider cocoon showing both its outer, loosely woven layer and its inner, more tightly woven layer.

Many examples of spider behavior, including camouflage (see front cover, Figure A), webs, courtship, mimicry, moulting, cannibalism, parasitism, phoresy, and predation in amber and copal, were supplied in a two-volume work by Wunderlich (2004). Poinar (1995c, p. 41) illustrated a spider in Dominican amber close enough to an ant to conclude that it represents a predator–prey relationship. Poinar and Poinar (1999, Figure 70) illustrated a moth fly caught in a Dominican amber araneid spider web. Poinar (1998b, Figure 1C) illustrated a spider web preserved in Dominican amber. A biting midge trapped in a spider web (see Figure 122) in Early Cretaceous Burmese amber was illustrated in Poinar

FIGURE 122 A biting midge caught in a spider web in Early Cretaceous Burmese amber (Poinar amber collection accession no. B-Sy-15); length of midge = 1.2 mm.

and Poinar (2008). Vander Wall (1990) summarized information about spider prey storage in prey bundles.

☐ *Reliability*

Category 2B, based on the structures present in two of the above examples, two modern orb-web-weaving higher taxa are present.

Although apparently not involving spider web, Grimaldi (1996, p. 95) illustrated a jumping spider with a prey millipede in amber from an unspecified locality and horizon.

☐ *Reliability*

Category 1, frozen behavior.

6Clb.   Dung Beetles

Dung beetles are included here under "Predation and Feeding Behaviors" because they involve larval feeding on the nutrients remaining in the dung. Hanski and Cambefort (1991) summarized a wide variety of material on modern dung beetles, and Cambefort (1991) provided a solid background on the evolutionary relations of the group, including their historical biogeography. Genise and Laza (1998) redescribed probable dung beetle brood chamber traces from the Late Cretaceous–Early Tertiary of Uruguay, where they are associated with other traces that they interpret as the work of dung beetle egg parasites. Three species of dung beetles have been described from the Miocene of Kenya (Paulian, 1976); it is surprising that fossil dung beetle burrows and balls have

almost never been recognized in Africa in view of its rich fauna of modern dung beetles. Kitching (1980) described some South African Plio–Pleistocene dung beetle balls.

Lister (1993c) cited abundant *Aphodius prodromus* dung beetle remains from a Late Pleistocene English mammoth site. Duringer et al. (2000a,b) described an amazing set of burrows and presumed dung beetle balls from the Pliocene of Chad. Their arguments for scarabid, dung beetle origins are convincing, although they require a beetle whose activities are unknown today. The presence of a grub chamber and internal laminae is notable. Vignaud et al. (2002, Figure 2) cited Late Miocene examples.

Bruet (1950) cited later Quaternary dung beetle balls and provided a brief morphological description from Ecuador with some possible evidence of a parasitic hymenopteran accompanying. The individual balls are about 70 mm in diameter (i.e., very large). Chin and Gill (1996) described what they concluded are pads of dinosaur dung from the Late Cretaceous of Montana that are associated with boreholes of various sizes, which they interpret as being the work of dung beetles. Despite the desirability of recognizing pre-Cenozoic dung beetle activity, however, their evidence needs support before being accepted. Using a modern model, they inferred that dung beetle brood chambers might have been present beneath the alleged dung pats, although such structures were not present.

There are several dung beetle types reported from the Jurassic and Cretaceous that probably represent different types of dung beetle behavior than we have today. *Geotrupoides* and *Cretogeotrupes* were probably tunnelers, whereas *Proterocarabeus* and *Cretaegialia* could have been dwellers and *Holocorobius* possibly a roller (Krell, 2000; Nikolayev, 1993; Rasnitsyn and Quicke, 2002).

☐ *Reliability*

Category 2B, functional morphology.

6Clc.   Flesh-Eating Insects

Zumpt (1965) provided an excellent overview of "flesh"-eating flies. Roberts et al. (2007) described some presumably insect-generated boreholes in dinosaur bones from the Late Cretaceous of Madagascar and Utah and reviewed the entire question of the stratigraphic ranges, occurrences, and morphology of these trace fossils, Triassic to present. Britt et al. (2008) presented a comprehensive account of dermestid beetle bone-boring activity and consideration of nondermestid possibilities, while describing some Upper Jurassic dinosaur borings from Wyoming. Schwanke and Kellner (1999) described some probably insect-generated boreholes in Triassic synapsid bones from the Brazilian Santa Maria Formation.

Hasiotis and Fiorillo (1997) and Laws et al. (1996) cited boreholes in Late Jurassic dinosaur bones that they concluded are due to dermestid activity. Hasiotis et al. (1999) provided more detail about alleged dermestid evidence on these Late Jurassic dinosaur bones, as well as a good review of the potential that dermestids have for boring into bone. Rogers (1992) interpreted some borings in Late Cretaceous dinosaur bones from the Two Medicine Formation of northwestern Montana as being due to dermestid or other beetles with similar feeding types and the ability to bore into hard substrates. He also cited an unpublished account of possible carrion beetle "egg cases." The evidence is reasonable. Paik (2000) discussed bored dinosaur bone from the Early Cretaceous of Korea that he ascribed to dermestid activity. McAlpine (1970) described some Alberta Late Cretaceous puparia (?Calliphoridae) from a shoreline region where they might have been working over some marine animal materials.

Laudet and Antoine (2004) described a possibly dermestid-bored rhinoceratid bone from the Oligocene Quercy phosphorite. Tobien (1965) described some Pliocene materials while summarizing previous work on purported holes in mammal bone ascribed to dermestid activities. Martin and West (1995) described dermestid pupation chambers in some detail and provided various North American Neogene examples.

Germonpré and Leclercq (1994) beautifully illustrated empty puparia associated with various Pleistocene mammals, as well as some of the larvae. Kitching (1980) suggested that apparent blowfly puparia, possibly calliphorids, trogid pupae, and possibly dermestid borings are present in the South African Plio–Pleistocene. Gautier and Schumann (1973) described late Eemian, Late Pleistocene blowfly larvae from a Belgian bison skull. Gautier (1974) described a flesh-eating fly puparia (*Protophormia terraenovae*) from a Late Pleistocene Belgian wooly rhinoceros. Lister (1993c) cited puparia of *Phormia terraenovae* from a Late Pleistocene English site. Heinrich (1988) described some Pleistocene blowfly puparia associated with mammoth skull fragments from Bottrop in the Rhineland. The oldest known dermestid body fossils are from Early Cretaceous Burmese amber (Cockerell, 1917) and also are known from the Eocene (Zhantiev, 2006).

☐ *Reliability*
Category 2B, functional morphology.

6CId.   Reduviid Bug–Ants

Ambrose (1999) provided an introduction to the extensive literature on the biology and behavior of assassin bugs, as well as references to their fossil record. Poinar (1991b; see also Poinar, 1993, Figure 6) described a second fossil reduviid belonging to the Holoptilinae, which largely specialize on ants, from Dominican amber of Miocene–Oligocene age (see Figure 123). Poinar (1993) pointed out that the long stiff hairs on the legs and antennae are a defense against being bitten by attacked ants. Weitschat and Wichard (1998, Plate 43, Figure a) illustrated a closely related *Proptiloceras* with an ant carcass in Baltic amber.

☐ *Reliability*
Category 2B, functional morphology.

FIGURE 123 A reduviid ant bug, *Praecoris dominicana*, in Dominican amber (Poinar amber collection accession no. He-4-26); length of bug is 4.7 mm. (Figure 1 in Poinar, Jr., G.O., *Entomologica Scandinavica*, 22, 193–199, 1991.)

6CIe.   Gardening Ants: Leafcutter Ants and Bees

Sarzetti et al. (2008) described circular leaf margin damage from the Patagonian Eocene that they consider to be due to megachilid bees, although similar damage has previously been considered the work of leafcutter ants as well as of megachilid bees (Boucot, 1990a). The ichnogenus *Phagophytichnus*, to which this type of damage belongs, has been recognized as far back as the Late Paleozoic (i.e., it is not characteristic of any one group of insects). Adami-Rodrigues et al. (2004) provided examples of Permian leaf margin feeding, presumably by insects.

☐ *Reliability*
Category 6, due to uncertainty about the makers.

6CIf.   Xyelidae Feeding

Krassilov (1987) provided additional information on xyelid feeding in the Mesozoic. Krassilov et al. (1997b; see also Krassilov, 2003) summarized the available data on pollen feeding in fossil insects, including Permian and Jurassic and citing Cretaceous; they suggested that coevolutionary specialization on pollen is a real possibility for the insects involved. Although their work does not involve xyelid pollen feeding, Rasnitsyn and Krassilov (1996) described a number of Permian, Kungurian specimens featuring various insect taxa with pollen in their guts, making a very positive case for the pollen-feeding habit this far back in time; they also illustrated a Middle Jurassic, Karatau example. Krassilov and Rasnitsyn (1997) described Early Permian insects (Hypoperlidae, Grylloblattida) with pollen in their guts and summarized what is currently known about evidence for pollen feeding from the fossil record. Krassilov et al. (1997b) described grains of *Classopolis* pollen from the guts of two Kazakhstan, Jurassic orthopterans and reviewed the possibility that the primitive orthopterans were pollenivorous.

Labandeira (1996, 1998c) has summarized much of the data concerning pollen and spore feeding from the fossil record. Labandeira (1997b) provided additional thoughts on Late Paleozoic pollenivorous insect possibilities, particularly aimed at discounting the presence of pollen in insect guts as accidental ingestion as contrasted with purposeful. Krassilov et al. (1999) described Kungurian pollen feeding in a Uralian booklouse, with an analysis of the various pollen grains in the gut of the insect. Afonin (2000) described another Kungurian example involving a Grylloblattid. Labandeira (2002a, Figure 2.15) provided evidence of pollen and spore feeding from the Late Silurian onwards, and Labandeira (1998c) summarized much of the literature on latest Silurian and younger isolated coprolites containing abundant spores and pollen, with the caveat that the identity of the feeders is uncertain, although insects and mites are undoubtedly involved. Caldas et al. (1989) described an Early Cretaceous, Brazilian xyelid wasp with angiosperm pollen (*Afropollis*) in its gut, from an Araripe Basin locality. Krassilov et al. (2003) presented detailed evidence of xyelid species from the Cretaceous with several pollen species in their intestinal tracts, providing further support for the antiquity of this relation.

☐ *Reliability*

Category 1, frozen behavior. These are very important examples that provide truly reliable evidence of feeding habits.

*6CIg.   Petioles with Cavities Containing Coprolites

Lesnikowska (1990) described some Stephanian, Late Pennsylvanian, marattialean tree-fern *Psaronius chasei* petioles from Illinois that have cavities containing herbivore coprolites. The cavities are bordered by "extensive proliferation of the cortical parenchyma," which suggests wound tissue. She concluded that insects of some uncertain group are the most likely cause of the cavities and the coprolites. This is one of the earliest positive examples of herbivory, as contrasted with possible detritivory, because the coprolites composed of plant debris are found in the possible source cavity with accompanying wound tissue. Rex and Galtier (1986) described varied Mississippian, Visean coprolites involving occurrences with higher land plant materials from the Massif Central; these are probably the oldest coprolites intimately associated with higher land plant materials. The sources of the coprolites are unknown but presumably involve arthropods of some type.

☐ *Reliability*

Category 2A.

FIGURE 124   Resin is still attached to the front legs of this apiomarine resin bug in Dominican amber (Poinar amber collection accession no. He-4-21); length of bug = 2.5 mm.

*6CIh.   Reduviid Bug Using Resin and Stingless Bee with Resin and Pollen

Poinar (1992a; see also Poinar and Poinar, 1999, pp. 123–124) described a very innovative discovery of a Dominican amber reduviid bug belonging to the Apiomerinae, a group that attaches resin to their fore tibia, which have long, erect hairs that serve to hold the resin; these are so-called "resin bugs" (see Figure 124). The resin helps to hold prey. Abundant in the Dominican amber are workers of the stingless bee *Proplebeia dominicana*, which today are preyed upon by resin bugs (see Figure 125). The older workers specialize in collecting resin for the nest (for good discussion and illustration, see Poinar

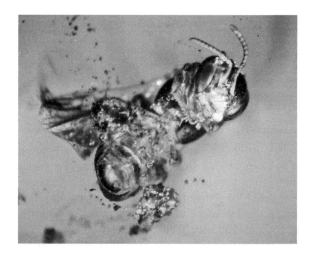

FIGURE 125   Stingless bee in Dominican amber with both pollen grains and resin attached to its hind legs. It is very unusual to have a specimen that has this combination, as the great majority of stingless bees in Dominican amber are resin collectors (Poinar amber collection accession no. Sy-1-106); length of bee = 3 mm.

and Poinar, 1999, Figure 123, p. 123, and jacket illustration). Only 21 bees out of 750 trapped in the amber had "resin balls" attached to their corbiculae (i.e., got trapped after collecting). This stingless trigonid bee–resin bug relationship has apparently persisted for at least 15 to 40 my. Poinar and Poinar (1999, Figure 124, p. 124) illustrated a resin bug in Dominican amber that had its right front leg over a stingless bee. The bee, obviously the intended victim, has a puncture hole in one eye, quite possibly made by the resin bug in its initial attack.

□ *Reliability*

Category 2A, because none of the resin bugs has been observed with its prey caught on the resin.

*6CIi. Protorthopteran Spore Feeding

Scott and Taylor (1983; Scott et al., 1985; Taylor and Scott, 1983) discussed the protorthopteran *Eucaenus* from the Mazon Creek Pennsylvanian with *Lycospora*(?) spores in its gut, as well as coprolites with spores.

□ *Reliability*

Category 1, based on frozen behavior.

*6CIj. Palm Flowers with Microlepidopteran Coprolites Containing Palm Pollen

Schaarschmidt and Wilde (1986) described Messel Eocene palm flowers with feeding trails on which there are microlepidopteran coprolites containing palm pollen.

□ *Reliability*

Category 1, frozen behavior.

*6CIk. Beetle Containing Pollen

Schaarschmidt (1988) described a Messel Eocene beetle with pollen in its gut.

□ *Reliability*

Category 1, frozen behavior.

*6CIl. Praying Mantis Attacked by Ants

Grimaldi (1996, p. 93) illustrated a praying mantis attacked by ants in amber from an unspecified locality and horizon.

□ *Reliability*

Category 1, frozen behavior.

*6CIm. Whip Scorpion and Insect Prey

Grimaldi (1996, p. 95) illustrated a whip scorpion with insect prey in its jaws from an unspecified locality and horizon.

□ *Reliability*

Category 1, frozen behavior.

*6CIn. Plant-Feeding Snail

Poinar (1998a, Figure 2E) illustrated a Dominican amber leaf with "bites" removed from its margin that might have

been made by a snail (*Strobilops*) also preserved nearby in the same specimen.

□ *Reliability*

Category 2B, functional morphology.

*6CIo. Dolichopodid Fly with an Enchytraeid Worm Fragment

Ulrich and Schmelz (2001) carefully described a Baltic amber specimen of a closely associated female Dolichopodidae fly and some Enchytraeidae worm fragments with the conclusion that the fly had been predating the worm prior to their joint engulfment by resin. They provided modern examples of similar predator prey examples that strongly support their interpretation of the Baltic amber specimen. Weitschat and Wichard (1998, Plate 7, Figures g, h) illustrated and noted that the same dolichopodid fly closely associated with the enchytraeid worm fragment had probably been eating when overwhelmed.

□ *Reliability*

Category 1A, almost frozen behavior, with the presumed prey fragments situated very close to the presumed predator.

*6CIp. Coccid Salivary Sheaths

Koteja (1998) illustrated and discussed the unique salivary sheaths generated by the mandibulary and maxillary stylets of Homoptera, certain Baltic amber, and Bitterfeld amber coccids in particular. These structures are generated by the drying out of the saliva produced by a series of attempts at feeding by the stylets.

□ *Reliability*

Category 1, frozen behavior.

*6CIq. Elaterid Feeding

Kinzelbach and Lutz (1985) cited a stylopid larva within the gut of a Geiseltal Eocene elaterid beetle.

□ *Reliability*

Category 1, frozen behavior.

*6CIr. Insect Mouthparts

Labandeira (1997a) defined and reviewed the many distinct insect mouthpart types involved in different feeding types and provided information about their first known appearances within the fossil record. Labandeira and Phillips (1996b) provided evidence for insect fluid feeding from the Late Pennsylvanian.

*6CIs. Oribatid Mite Feeding

Labandeira et al. (1997) reviewed the fossil record of oribatid mites, both body and trace fossils, and observed that there is an extensive trace fossil record within the Carboniferous. Their evidence suggests that these "soil" mites largely feed on plant debris within the soil profile or possibly on the associated fungi rather than directly on plant tissues. Ash (2000) described some Late Triassic plant damage involving hollowing out of stems and leaving behind a mass of minute coprolites that may belong to oribatid mites; the evidence indicates that the causal organism was eating the plant tissue while the plant was alive (i.e., was an herbivore rather than a detritivore). A possible feeding relationship was indicated between an oribatid mite belonging to the genus *Teleioliodes* (Liodidae) and the mushroom *Coprinites dominicana* (Coprinaceae) in Dominican amber. Many extant oribatids are mycophagous, and *Teleioliodes* was probably feeding on the mushroom when both were covered with resin (Poinar and Poinar, 1999, Figures 29, 30, p. 40).

☐ *Reliability*

Category 2B, functional morphology, as the trace fossil burrows and frass are not totally direct evidence.

*6Clt.   Insect Herbivory

Labandeira (2002, Figure 2.7; 2002b, Figure 6) illustrated a number of leaf-feeding types including leaf buds from varied parts of the fossil record. Elsewhere in this and the earlier treatment various examples through time of folivory types ascribed to insect activities have been discussed. Stephenson and Scott (1992) and Chaloner et al. (1991) reviewed some of the evidence of leaf margin feeding. Labandeira (1998c) reviewed much of the evidence regarding folivory in the Paleozoic and also with some post-Paleozoic information. It is obviously very difficult to identify the leafcutting insect with the pre-Cenozoic examples, and few have really been identified with certainty.

Labandeira (1998a) cited Pennsylvanian evidence of leaf herbivory that might be due to insects. Castro (1997) described leaf margin damage from the Spanish Stephanian which represents yet more evidence for Carboniferous activity. Scott and Taylor (1983) mentioned Mazon Creek Pennsylvanian leaf damage. Beck and Labandeira's (1998) account of varied Early Permian leaf damage from north-central Texas indicated that varied taxa, probably insects, were active this early; the dominance of leaf feeding in this example involving gigantopterids is of interest. The authors also considered the overall question of insect folivory through time. Labandeira and Allen (2007) provided additional Early Permian leaf herbivory evidence from Texas, and Srivasta (1996) discussed some leaf damage from the Lower Gondwanas.

Ash's (1999) Late Triassic, Chinle Formation leafcutting type of damage to *Sphenopteris* is typical of insect-generated leafcutting damage. Ash pointed out that repair to the edges of the wounds is good evidence that this is evidence of herbivory rather than detritivory. Grauvogel-Stamm and Kelber (1996) described leaf margin damage that appears to have been done by insects of one kind or another from the lower Keuper, Triassic of western Germany and adjacent France. Geyer and Kelber (1987) discussed and illustrated a fine example of Late Triassic leaf margin damage from Germany. Scott et al. (2004) described varied leaf margin damage from the South African Late Triassic. Kelber's (1990, Figures 99,

FIGURE 126   Chewing damage to a sepal of *Hymenaea mexicana* in Mexican amber (Poinar amber collection accession no. SD-9-98B); width of sepal = 3.5 mm. (From Poinar, Jr., G.O. and Brown, A.E., *Botanical Journal of the Linnean Society*, 139, 125–132, 2002. With permission.)

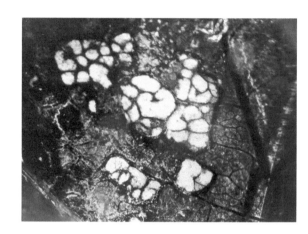

FIGURE 127   Skeletonized leaflet of *Hymenaea mexicana* in Mexican amber (Poinar amber collection accession no. SD-9-125A); lower blotch area = 1.8 mm. (From Poinar, Jr., G.O. and Brown, A.E., *Botanical Journal of the Linnean Society*, 139, 125–132, 2002. With permission.)

100) figures of leaf damage that may well be due to insect activity of one kind or another, from the Late Triassic of western Germany, are notable.

Rasnitsyn and Krassilov (2000) found leaf tissue in the gut of Late Jurassic insects (*Brachyphyllum* in *Brachyphyllophagus phasma*) from Kazakhstan. Rasnitsyn and Krassilov (2000) provided positive evidence for phyllophagy in insects, with identified plant parts in a Late Cretaceous Karatau insect. Vasilenko (2006) provides evidence of Late Jurassic–Early Cretaceous leaf margin feeding. Wilf and Labandeira (1999) discussed levels of folivory in the Late Paleocene and Early Eocene of Wyoming. Johnston (1993) described some Middle Eocene leaf damage of possible insect origin. Labandeira (2002b, Figure 5) illustrated varied leaf feeding examples from the Middle Eocene of Republic, Washington. Evidence of insect damage to the legume *Hymenaea mexicana* in Mexican amber is shown by chewing damage to sepals, skeletonization of leaves, and chewing damage to petals (Poinar

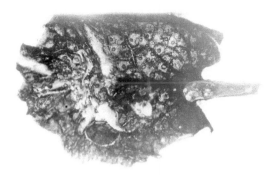

FIGURE 128 Feeding damage to a petal of *Hymenaea mexicana* in Mexican amber (Poinar amber collection accession no. SD-9-67A); length of petal = 12 mm.

and Brown, 2002) (see Figure 126, Figure 127, and Figure 128). Berger (1953) discussed some Pliocene Vienna Basin leaves (*Castanea* and *Salix*) with holes presumably due to insect folivory.

□ *Reliability*

Category 5A, as the identity of the maker is all too commonly uncertain, although there are some exceptions.

*6Clu. Ant–Pseudoscorpion Relationship

Poinar (2001b) discussed and illustrated an ant (*Azteca*) and a pseudoscorpion locked in combat from Dominican amber (see Figure 129), and Poinar and Poinar (1999, Figure 115, p. 111) illustrated an encounter between an *Azteca* ant and a pseudoscorpion that occurred on the bark of the resin-producing algarrobo tree.

□ *Reliability*

Category 1, frozen behavior.

*6Clv. Blood-Feeding Dipterans

(See also Section 4AIIId, Ticks and Mites as Micropredators and Potential Disease Vectors, and Section 4AIIIe, Mosquitoes as Micropredators and Potential Disease Vectors.) Lehane (1991) reviewed the blood-sucking insects: nematocerans, including Psychodidae (Triassic to present) (Rasnitsyn and Quicke, 2002); some Miocene Dominican Phlebotominae (Peñalver and Grimaldi, 2006); Culicidae (Cretaceous to present); Ceratopogonidae (Cretaceous to present); Simuliidae (Jurassic to present); brachycerans, including Rhagionidae (Jurassic to present); Tabanidae (Jurassic to present) (Rasnitsyn and Quicke, 2002); Muscidae (Cretaceous to present); Hippoboscidae (Oligocene to present), Calliphoridae (Cretaceous to present, with larval blood feeding); and Glossinidae (Late Eocene to present). They also cited the groups that feed on bats (Streblidae and Nycteribiidae). Lehane gave some idea of the various vertebrate groups on which the blood-sucking members of the above families feed. Dukashevich and Mostovski (2003) reviewed the varied groups of blood-feeding insects, including the dipterans, with their known stratigraphic ranges.

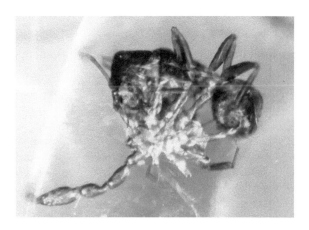

FIGURE 129 A pseudoscorpion and ant locked in combat in Dominican amber (Poinar amber collection, accession no. Sy-1-38); length of ant = 4 mm.

Black (2003) provided a succinct account of the many variables involved in blood feeding by different terrestrial arthropod vectors and the disease parasites they spread.

Borkent (2000) reviewed the ceratopogonid fossil record, beginning with Early Cretaceous Lebanese Amber examples, and pointed out the groups that may well have fed on vertebrate blood early on. Grogan and Szadziewski (1988) described a New Jersey Late Cretaceous ceratopogonid, reviewed the question of blood feeding among the ceratopogonids, and provided evidence that the behavior arose during the Jurassic by citing a British Purbeck example; it is likely that warm-blooded animals were fed upon, with birds and mammals being the available choices. Borkent (1995, p. 29) mentioned that modern lizards are fed on by some ceratopogonids. They also observed that various modern taxa feed on vertebrates (fish to mammals) and other insects, and some are even nectar and pollen feeding (p. 129), in addition to noting that the mandibles of insect blood feeders differ from those using vertebrates (p. 131).

Currie and Grimaldi (2000) reviewed the fossil record of the black flies (Simuliidae) which extends back to about the Jurassic–Cretaceous boundary; not all female simuliids are blood feeders. Kalugina (1991) described Simulidae and other blood-sucking types from Siberian examples and suggested that blood sucking dates from the Jurassic. Labandeira (1998c) discussed the record of potentially blood-feeding fossil insects. Comments elsewhere in this treatment deal with mosquitoes; for other blood feeders, see Section 4AIIIf, Tsetse Flies as Micropredators and Disease Vectors, Section 4AIIIc, Lice and Mammals (Nits) and Birds and Fleas, and Section 4AIIIi, Parasitoid Wasps and Parasitic Flies–Spiders.

□ *Reliability*

Category 2B, functional morphology.

*6Clw. Piercing and Sucking

Labandeira (2002a, Figure 2.8) provided evidence about the presence of piercing and sucking of various plant tissues, presumably by a variety of insects as well as mites, with

TABLE 15

Terrestrial Invertebrate Predator–Prey and Feeding Examples from the Fossil Record (Additions to Table 15 from Boucot, 1990a)

Behavioral Type	Taxonomic Level	Time Duration	Category	Refs.
6CIa. Spider webs, spinnerets, and bundled prey	Order	Middle Devonian–Holocene	2B	Selden et al. (1991)
6CId. Reduviid bug using resin	Subfamily–class	Late Eocene	2A	Poinar (1992a)
6CIj. Palm flowers with microlepidopteran coprolites	Family–class	Eocene	1	Schaarschmidt and Wilde (1986)
6CIo. Dolichopodidae–Enchytraeidae	Family–family	Eocene–Holocene	1A	Ulrich and Schmelz (2001)
6CIp. Coccid salivary sheaths	Order–order	Eocene	1	Koteja (1998)
6CIq. Elaterid feeding	Order–order	Eocene	1	Kinzelbach and Lutz (1985)
6CIt. Insect herbivory	Order–phylum	Jurassic	1	Rasnitsyn and Krassilov (2000)
6CIx. Empidid fly and a chironomid	Family–family	Eocene	1	Janzen (2002)
6CIy. Phorid fly attacked by insect larva	Family–class	Eocene	1	Janzen (2002)
6CIz. Ground sloth dung–sciarid larvae	Genus–family	Subrecent	1	Waage (1976)

the requisite mouthparts, as far back as the Rhynie Chert. Labandeira and Phillips (1996b) provided a useful discussion of piercing and sucking of plant fluids by insects, with an emphasis on a Pennsylvanian marattialian example. Wilf and Labandeira (1999) discussed levels of piercing and sucking on Late Paleocene and Early Eocene leaves from Wyoming.

☐ *Reliability*

Category 2B, functional morphology, as the affected plants with appropriate features are known as well as insects with the necessary mouthparts, but the two have not yet been found together, as in the aphid–plant association discussed earlier (Boucot, 1990a).

*6CIx. Empidid Fly and a Chironomid

Janzen (2002, Figure 385) illustrated a Baltic amber empidid fly with a chironomid prey fly—a very unique specimen.

☐ *Reliability*

Category 1, frozen behavior.

*6CIy. Phorid Fly Attacked by an Insect Larva

Janzen (2002, Figure 399) illustrated a Baltic amber phorid fly being attacked by an unidentified insect larva.

☐ *Reliability*

Category 1, frozen behavior.

*6CIz. Ground Sloth Dung–Sciarid Larvae

Waage (1976) described some Subrecent ground sloth dung from Gypsum Cave, Nevada, containing sciarid fly larvae that were presumably feeding on nutrients preserved in the dung.

☐ *Reliability*

Category 1, frozen behavior.

Summary

Some family level, and higher, taxa are involved (see Table 15).

6CII. Vertebrates

Munk and Sues (1993) employed some Late Permian Kupferschiefer examples from Hesse to point out how difficult it is with small samples to be certain of the dietary significance of vertebrate visceral-region stomach content possibilities deduced from the fossil record. Their particular examples involved several reptiles, with gravel in one case that might represent either gastroliths or incidentally swallowed material, contrasting with a second example in which woody seeds are present that could be interpreted either as woody gastroliths or as evidence of an herbivorous diet, which in the latter case would not fit very well with the carnivore-type teeth of the animal. They cited a Tertiary example where several specimens provided equally conflicting results. *Caveat emptor.* Their *Parasaurus* specimen represents plant debris in the visceral region combined with teeth similar to those of modern herbivorous iguanodons, whereas their *Protorosaur* combines carnivore-type dentition with ovules of the conifer *Ullmania* plus small quartz pebbles in the visceral area (i.e., possible gastroliths plus ovules combined with a carnivore-type dentition).

6CIIa. "Fighting" Dinosaurs

While not having to do directly with fighting dinosaurs, Erickson et al.'s (1996) conclusions regarding bite marks on a ceratopsian pelvis make a good case that tyrannosaurs were capable of bite forces as great as those belonging to any modern carnivore and had strong, impact-resistant teeth. Erickson and Olson (1996) provided good evidence of *Tyrannosaurus rex* feeding activity (bite marks on ceratopsian bones). Varricchio (2001) provided evidence that tyrannosaurids in the Montana Campanian fed on hadrosaurs, based on alleged stomach contents that have been etched by stomach acids. Fiorillo (1991b) provided evidence suggesting that bone utilization by dinosaurs was not very prevalent as contrasted with mammals and that bone crushing was absent. Chen et al. (1998) provided good evidence from Liaoning of an Early Cretaceous theropod, *Sinosauropteryx*, with a prey lizard in its visceral region. The highly predaceous *Baryonyx walkeri* described by Charig and Milner (1997) retains fish scales of *Lepidotes* and some *Iguanodon* dinosaur bones in its visceral region, within the rib cage; when combined with its dentition and other features, this suggests that catching and eating fish may have been a major feature of its behavior. Their evidence sounds convincing for predation rather than scavenging.

Jacobsen's (1998, p. 24) paper about rehealed tooth marks on some tyrannosaurid skull bones is consistent with fighting between tyrannosaurids. Jacobsen discussed the various evidence for carnivorous dinosaur predation shown by marks on dinosaur bones, using Alberta Late Cretaceous samples, and found that hadrosaur distal limb bones were far more predated than ceratopsians, and other carnosaurs were far less scarred. Fiorillo (1991a; see also Jacobsen, 1998) reviewed information on tooth marks left on dinosaur bones of herbivore type and concluded that bone crushing and bone "feeding" were not involved. Buffetaut and Suteethorn (1989) reviewed occurrences of sauropod skeletons associated with theropod teeth that suggest either predation or scavenging. Buffetaut et al. (2004) described some pterosaur cervical vertebrae from the Santana Early Cretaceous that have an embedded spinosaur tooth, but as usual whether this represents scavenging or predation is unclear.

The Late Triassic theropod *Coelophysis* has been described by Colbert (1989) as having possibly been cannibalistic or, less likely, a scavenger based on two specimens that contain the bones of smaller individuals within the visceral area, but Gay (2002), after carefully inspecting the specimens, concluded that they are actually "beneath" the rib cage rather than within it; that is, this is *not* a case of cannibalism or anything related. Nesbitt et al. (2006) reviewed the specimen in question and demonstrated that the visceral region of the dinosaur contained crocodilomorph bones, thus it represents a good case of predation. Rogers et al. (2003), however, made a strong case for cannibalism in Late Cretaceous Madagascar *Majungatholus atopus* based on tooth marks on bone of young specimens, although scavenging is another possibility. Chin et al. (1998) described a very large Late Cretaceous Saskatchewan coprolite filled with bone fragments and inferred that it was deposited by *Tyrannosaurus*; the attribution is based on very circumstantial evidence, but the bone content does suggest a carnivore, while the size is compatible with a large dinosaur.

Rothschild and Tanke (1997) described a pathologic ceratopsian skull with a "wound" that suggests an assault by an intraspecific, possibly suggesting combat of some sort. Sues et al. (2003) described a Late Triassic crocodylomorph archosaur from North Carolina that may have been involved in a lethal fight with an associated sphenosuchian.

□ *Reliability*
Category 1, frozen behavior.

6CIIb. Hadrosaurian Dinosaur Diet

Coe et al. (1987, p. 226) summarized what was known about herbivorous dinosaur gut contents at the time. Currie et al. (1995) have considered the matter of hadrosaurian diet, based mostly on two "mummified" specimens from Wyoming,

including a previously unstudied specimen. The results of the study of the plant material in the visceral cavity are equivocal. Although not involving hadrosaurs, Molnar and Clifford's (2001) description of an ankylosaurian gut content with varied plant debris suggesting a diet of soft vegetation is noteworthy and reliable, although based on only a single example. Prasad et al. (2005) described some Late Cretaceous Indian nonmarine coprolites containing varied plant debris and grass-type phytoliths, which they suspected were from titanosaurs, although this possibility is speculative, with hypsodont sudamericid gondwanathere mammals also being a possibility.

□ *Reliability*
Category 1, frozen behavior.

6CIIc. Owl Pellets

Tyler (1992) ascribed a Holocene cave accumulation of frog bones as resulting from owl predation and an accumulation of owl pellets. Tyler (1977) described modern owl predation on frogs. Andrews' (1991) elegant account of small mammal taphonomy includes a mass of information relating to modern and Quaternary owl pellet occurrences. Tobien (1977) described a pellet-like structure, from the Late Miocene of Ohningen in Baden, that yielded small vertebrate bones as well as bituminous material that may represent degraded nonskeletal materials. This occurrence very likely represents a raptor pellet, with owls being the prime candidates. Mourer-Chauviré (1994) reviewed the oldest known occurrences of fossil owls, with the conclusion that the first appearances of the group occurred in the later Paleocene of Europe and North America. Dauphine et al. (1997) provided information on the levels of solution and destruction found in various owl pellets as a means of ascertaining the nature of the predator. Worthy and Holdaway (1994) described a New Zealand Late Quaternary bone accumulation ascribed to owl activity.

Sanz et al. (2001) described what appears to be a regurgitated "pellet" from the Spanish Early Cretaceous that contains the bones of three distinct bird species belonging to four juvenile birds. The identity of the predator is very uncertain, although a theropod or a pterosaur is seriously considered. Alexander and Burger (2001) provided evidence from the Middle Eocene for possible raptor predation on mammals, aside from owls, and Thornton and Rasmussen (2001) reviewed the types of damage incurred on varied mammalian remains by carnivores. Storch et al. (1996) cited some German Oligocene rodent teeth preserved in owl pellets.

Although not in the owl pellet category, Berger and Clarke's (1995) analysis of the faunal remains from a Pleistocene South African cave occurrence provided good evidence for predation by a large raptor, possibly an eagle. Trapani et al. (2006) described the types of remains left behind by a Uganda crowned hawk-eagle. Also of interest is Laudet and Selva's (2005) account of raven pellets from modern samples with applications possible for the Neogene.

□ *Reliability*
Category 2B, functional morphology.

6CIId. Felid Activities

Akersten (1985) provided an excellent discussion of *Smilodon* functional morphology and skeletal pathology, all leading to the conclusion that it may have been a social cat and that it attacked the abdominal region of its prey, rather than the neck region, with the normal carnivorous killing bite. Van Valkenburgh et al. (1990) described microwear molar details indicating that *Smilodon* used little bone in its diet and provided an excellent illustration of the complexities involved in interpreting tooth microwear patterns. Van Valkenburgh and Ruff (1987) provided an analysis of canine tooth strength in carnivores and suggested that sabrecats are closer in this regard to dogs than to felids; they also considered other carnivores, both living and fossil. This is still one more way of extracting data of functional significance from the mammalian record. Van Valkenburgh (1988; see also Van Valkenburgh and Hertel, 1993) provided an analysis of tooth breakage among carnivores that will help us to better understand the fossil record. Biknevicius and Van Valkenburgh (1996) critically reviewed the craniodental adaptations of predators. Although his work did not address felids specifically, Courville (1953) discussed a number of cranial injuries, some of which involved Pleistocene mammals, including the dire wolf, with the implication that the Pleistocene material was not involved with predation as contrasted with combat.

☐ *Reliability*

Category 2B, functional morphology.

6CIIe. Sloth Diets

(See Table 16.) Van Devender et al. (1980) provided a wealth of information, based on the analysis of some dung balls, about the diet of the extinct ground sloth *Nothrotheriops shastense*, indicating that browsing on xerophytes, including *Ephedra*, was important. The useful data came from cuticle analysis, whereas pollen data provided some information about regional pollen rain but not about diet. Mead and Agenbroad (1989) provided a useful summary of extinct herbivore dung research from the late Pleistocene of the Colorado Plateau (also see Section 6CIIu, Early Cretaceous Seed-Eating Bird from China). McDonald (2003, p. 10) pointed out the obvious in that the large accumulations of sloth dung in some North American caves indicates that the caves were used as latrines. In this connection, the well-known Holocene packrat middens

of the Southwest should be mentioned (Betancourt et al., 1990). Poinar et al. (1998b) identified DNA of the following plant groups from Gypsum Cave, Nevada, coprolites: Poaceae, Liliaceae, Chenopodiaceae, Malvaceae, Vitaceae, Epheraceae, and Hydrophylaceae. Also of interest concerning this dung is Schmidt et al.'s (1992) report of the parasites recognized in it, including nematodes, helminth eggs, and coccidian oocysts.

☐ *Reliability*

Category 1, frozen behavior.

6CIIf. Bite Marks on Fossil Nuts and Mammal Bones

(See also Section 4Bk, *Araucaria*–Beetle Relationship and *Araucaria* Seed Cone Damage.) Gregor (1982) described mid-Miocene pine cones of the Ries Crater in southern Germany that were stripped of their seeds by small rodents and gnawed down to their central spindle. Probably the oldest example to date is Collinson's (1990, Figure 46c) find of a presumably rodent-gnawed seed from the Late Eocene of southern England. In a very thorough paper on the topic, Collinson and Hooker (2000) extensively considered the nature of the gnaw marks on Late Eocene *Stratiotes* seeds from the Solent Group in the Hampshire Basin and provided an account of the probable feeder as well as general considerations. Yoshikawa (2000) provided excellent evidence of stereotypic holes gnawed in Middle Pleistocene *Juglans ailanthifolia* nuts by field mice (*Apodemus speciosus*) from Japan (see Figure 130 and Figure 131). He showed that over 80% of the nuts bear stereotypic, gnawed holes; his data and their analysis showed what can be done with a large sample. His SEM illustrations of the marks left by the gnawing mice on both modern and fossil nut shells are very convincing. Yokoyama and Kodaira (1921) cited gnawed *Juglans* nuts. Scott and Jepsen (1936, Figure 2, p. 2) illustrated a rodent-gnawed *Titianotherium* bone from the Oligocene of the White River Formation. Although his example is hardly in the gnawed mammalian bone category, Fiorillo (1991b) made a good case for nonutilization of vertebrate bone as a nutritional resource by predatory dinosaurs and provided an excellent account of bone utilization by mammalian carnivores.

☐ *Reliability*

Category 2B, functional morphology, with the gnawed spindles looking like modern examples, and Category 1, frozen behavior, for Yoshikawa's material.

TABLE 16
Sloth Diets

Site	Main Dietary Components	Material	Refs.
Aden Crater, New Mexico	*Xanthocephalum, Atriplex*	Megafossils	Eames (1930)
Gypsum Cave, Nevada	*Yucca brevifolia, Agave utahensis*	Megafossils	Laudermilk and Munz (1934)
Muav Caves, Arizona	*Phragmites communis, Fraxinus*	Cuticles	Hansen (1978)
Rampart Cave, Arizona	*Sphaeralcea ambigua, Ephedra*	Cuticles	Hansen (1978)
Shelter Cave, New Mexico	*Ephedra, Agave*	Cuticles	Thompson et al. (1980)

Source: Thompson, R.S. et al., *Quaternary Research*, 14, 360–376, 1980. Reproduced with permission of Elsevier.

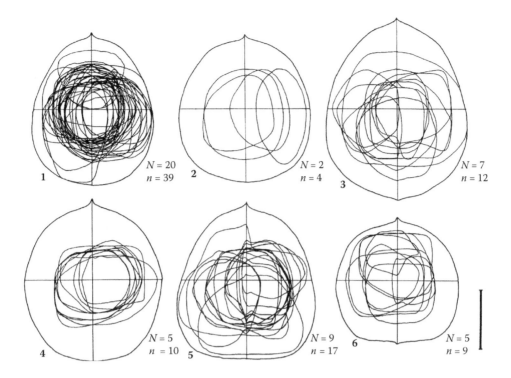

FIGURE 130 Positions of gnawed holes on Pleistocene and modern Japanese walnuts. (1) Fossil walnuts from the Sahama mud; (2–6) modern walnuts gnawed by living *Apodemus speciosus* from five Japanese localities. The outlines of the walnuts are based on average size at each locality. *N*, number of walnuts; *n*, number of gnawed holes. Scale bar = 1 cm. (From Yoshikawa, H., *Science Report of the Toyohashi Museum of Natural History*, 10, 23–30, 2000. Reproduced with permission.)

6CIIg. Insectivorous Bats

Sigé (1991) reviewed the microchiropteran story, including paleoecology and biogeography, while describing a Tunisian Early Eocene bat. Rayner (1991) thoroughly reviewed many aspects of microchiropteran behavior and noted their distinctions from the megachiropterans. Simmons et al. (2008) reviewed bat phylogeny while describing a primitive Green River bat that suggests that flight preceded the development of sonar in the microchiropterans.

Edinger (1926) described several Tertiary microchiropteran endocasts and emphasized that their form does not differ from the modern condition, particularly in the enlarged inferior colliculi involved with auditory function, just as parallel enlargement of optic areas is so prominent

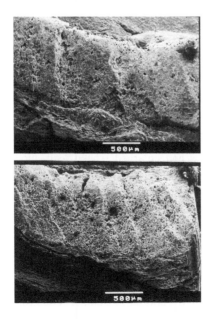

FIGURE 131   SEM photographs of (left) tooth marks on walnuts from the Pleistocene and (right) modern marks made by *Apodemus speciosus*. (From Yoshikawa, H., *Science Report of the Toyohashi Museum of Natural History*, 10, 23–30, 2000. Reproduced with permission.)

in birds and pterodactyls. When one adds this information to what was summarized earlier (Boucot, 1990a) about the evidence for preening, insect predation, and cochlea, taken together with what can be said about the functional morphology of the skeleton, all as early as the Middle Eocene, it is clear that these aspects of microchiropteran behavior have been unchanging.

In one of the most original, insightful papers ever published on behavioral evidence from the fossil record, Richter (1993) provided strong evidence that the Messel, Eocene bats *Palaeochiropteryx tupaiodon* and *P. spiegeli* fed largely on caddisflies and moths belonging to particular taxa and that they were specialized as water bats. Habersetzer and Storch (1989) provided an elegant analysis of Messel bat ecology based on morphology, including that of their cochleas and stomach contents, leading to conclusions about varied, specialized behaviors. Richter and Wedmann (2005, Figure 11) described a Messel coprolite filled with lepidopteran scales that might have been produced by *Palaeochiropteryx* (for associated midge larvae, see Wedmann and Richter, 2007). Richter provided additional evidence about the stomach contents of several other taxa of Messel, Eocene bats. Habersetzer et al. (1994, 1992) provided convincing evidence that Messel bats belonging to two families had distinctive feeding strategies, one hunting high in the canopy and the other much nearer ground level, as judged from both wing characteristics and stomach contents (with a good sample for stomach contents—it is rare in the fossil record to uncover such a statistically useful sample for deducing feeding strategies). Storch (2001) provided an overview of the complex Messel bat story involving a number of taxa and feeding habits.

☐ *Reliability*

Category 2B, functional morphology.

6CIIh.   Pangolin Feeding on Ants and Termites

Storch and Richter (1992a) provided additional evidence on pangolin feeding on ants and termites from the Messel Eocene, derived chiefly from stomach content data.

6CIIi.   Beaver Wood Cutting and Beaver-Gnawed Mastodon Molars

Harington (1996; see also Hutchinson and Harington, 2002; Tedford and Harington, 2003) cited beaver-gnawed sticks associated with a beaver (*Dipoides*) from Pliocene beds, dated by means of associated mammalian fossils, in the Strathcona Fiord area of Ellesmere Island. This is the oldest known occurrence of beaver-gnawed wood, with the inference that many of the other, older beavers lacked this habit. Whether or not this *Dipoides*-gnawed wood is closely related to the work of Holocene *Castor* depends on the phylogenetic relations of these two genera. It should be noted in passing that most beavers from the fossil record do not appear to have had aquatic habits. Rybczynski (2008) provided a penetrating analysis of wood-cutting behavior in fossil and modern beavers, emphasizing the possibility that the behavior may have arisen independently in at least two lineages as well as

the possibility that it was present as far back as the Oligocene in several other, related lineages.

☐ *Reliability*

Category 2B, functional morphology.

*6CIIj.   Venom-Conducting Reptilian Teeth

Sues (1991) described some Late Triassic reptilian teeth from Virginia that are grooved in the same manner as those of modern venomous reptile teeth. This is an example of convergence. The generic term *Uatchitodon*, after the ancient cobra goddess, *Uatchit*, of Lower Egypt, is most appropriate. Sues (1996) described an additional venom-conducting type of reptile tooth from the Late Triassic Chinle Formation of Arizona and summarized other known occurrences, including one from a Late Cretaceous Mongolian varanid lizard (Norell et al., 1992). Reynoso (2005) described a Middle Jurassic Mexican sphenodontid with venom-injecting type of teeth. Sues remarks that the relatively high incidence of modern venomous reptiles is not echoed in the evidence from the fossil record, with the implication that we are missing something here. Schierning (2005) described possible venom-conducting teeth from a Tamaulipas Jurassic sphenodontian, and Fox and Scott (2005) described a venom-conducting type of Paleocene, Canadian mammalian canine teeth. There are only a few living mammals with venom-conducting, grooved canines; they are unrelated to the new Paleocene specimens (i.e., convergence would be involved). Folinsbee et al. (2007) and Orr et al. (2007) documented abundant evidence for being skeptical of the Fox and Scott conclusion.

☐ *Reliability*

Category 2B, functional morphology, with a vengeance! In connection with the antiquity of venomous lizards and snakes, Fry et al. (2006) presented convincing molecular evidence of their close relationships; these two groups have a fossil record only extending back into the Late Cretaceous, suggesting that the pre-Late Cretaceous venomous tooth examples are due to convergent evolution.

*6CIIk.   Long-Fingered, Mammalian Insect Seekers

Von Koenigswald (1990; von Koenigswald and Schierning, 1987) described the Middle Eocene *Heterohyus nanus*, an Apatemyiden insectivore, from Messel that has long fingers and enlarged frontal teeth modified for grubbing out wood-boring insects (see Figure 132). He compared the Messel material with the convergently similar modern phalangeroid marsupial *Dactylopsila* and the lemuroid primate *Daubentonia*. These mammals occupy a niche very similar to that of the woodpeckers. Von Koenigswald et al. (2005) described an Early Eocene apatemyid from Wyoming that has the same elongated fingers as the Messel form, presumably indicating similar feeding habits. Senter (2005) suggested a similar anteater function for the enlarged manus of a Late Cretaceous dinosaur, but without more confirmatory information about its dentition (it is not edentulous) this is a very speculative item.

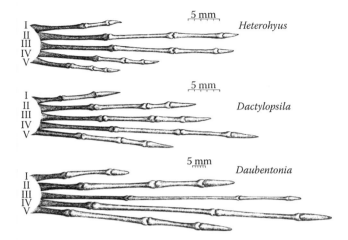

FIGURE 132 Long-fingered convergence for grubbing out insect larvae from tree trunks, woodpecker style, with the Middle Eocene apatomyid *Heterohyus* from Messel, compared with the modern marsupial *Dactylopsila* and the lemuroid *Daubentonia*; note the differences in scale. (Figure 21 in von Koenigswald, W., *Palaeontographica Abteilung A*, 210, 41–77, 1990. http://www. schweizerbart.de. Reproduced with permission.)

☐ *Reliability*

Category 2B, based on functional morphology for the mammalian examples.

*6CIIl.   Frozen Pleistocene Mammals

Guthrie (1990) provided an excellent summary of the late Pleistocene and subrecent frozen "mummies" from the Beringian regions of northern Siberia and Alaska. Included in his summary are information and references to the appropriate Soviet literature, as well as details on an Alaskan bison that he concluded suffered lion predation. Many of the specimens have had their stomach contents, when available, studied. All of the specimens are herbivores. Guthrie's work on "Blue Babe," his Alaskan bison, is probably the most ambitious of its type ever undertaken by a paleontologist and provides a careful analysis of bison behavior and overall ecology as it applies to his Alaskan specimen.

☐ *Reliability*

Category 2B, functional morphology.

*6CIIm.   Mammoth Diet

Davis et al. (1985) studied mammoth dung from a Subrecent (13,500 to 11,700 BP) Utah cave and found that the bulk of the material indicated a grazing rather than a browsing diet, in accord with the mammoth tooth type (for a discussion of mammoth and mastodon diets, see Haynes, 1991; for mastodon diet, see King and Saunders, 1984).

☐ *Reliability*

Category 1, frozen behavior.

*6CIIn.   *Propalaeotherium* Stomach Contents

Collinson (1990, 1999b) discussed the soft diet, evidenced by gut contents, of the horse relative *Propalaeotherium* from

the Messel Eocen; Vitaceae seeds and leaves were present in the visceral cavity.

☐ *Reliability*

Category 1, frozen behavior.

*6CIIo.   Insectivore Diets

Storch and Richter (1994) reviewed the visceral area contents of three Messel insectivores and found them to have been relatively omnivorous, including one species that ate fish as a regular part of its diet (for additional data, see Habersetzer et al., 1992). Von Koenigswald et al. (1992a) cited varied insectivore gut content evidence from the Messel Eocene, including fish debris from the otter-like pantolestid *Buxolestes*. Storch (2001) mentioned that *Pholidocercus* had insect cuticles, pulpy fruit tissue, and leaves in its visceral area (i.e., it was omnivorous), whereas *Macrocranion* had fish remains in its visceral region, as well as seed shells, tissue from fruits, insect cuticle plus leaves, stalks and fungus (i.e., it was another omnivore).

☐ *Reliability*

Category 1, frozen behavior.

*6CIIp.   Piciform Bird with Stomach Contents

Peters (1992, Figure 217) cited a piciform bird from the Messel Eocene with seeds in its visceral area.

☐ *Reliability*

Category 1, frozen behavior.

*6CIIq.   *Eurotamandua* Feeding

Storch and Richter (1992b) provided evidence for termite feeding in the Messel Eocene anteater, including both termite fragments and bits of termite carton.

☐ *Reliability*

Category 1, frozen behavior.

*6CIIr.   Rodent, Horse, and Even-Toed Ungulate Feeding

The Messel Eocene provides a variety of evidence from these groups (Collinson, 1999b; Franzen, 1992; Franzen and Richter, 1992; von Koenigswald et al., 1992a). Laub et al. (1994) described various late Pleistocene–Subrecent mastodon occurrences closely associated with masses of plant debris, with an emphasis on conifer twigs, suggesting intestinal and fecal materials; they also discusssed the feeding possibilities and mammoth and ground sloth data of a similar sort.

☐ *Reliability*

Category 1, frozen behavior.

*6CIIs.   *Diprotodon* and *Thylacoleo*

Runnegar (1983) described a Pleistocene *Diprotodon* ulna with marks indicating that it had been gnawed by *Thylacoleo*.

☐ *Reliability*

Category 2B, functional morphology.

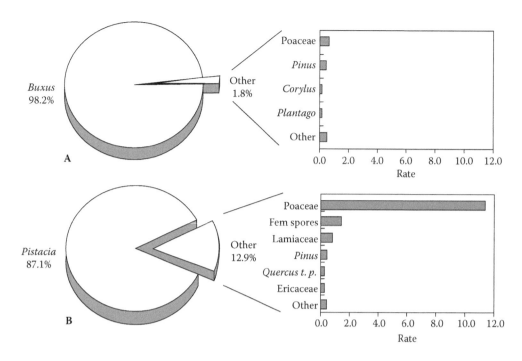

FIGURE 133 (A) Pie diagram of pollen content of *Myotragus balearicus* from coprolites (Cova Estrela, Balearic Islands). (B) Pie diagram of pollen content of recent *Capra hircus* dung pellets (Cova Estrela, Balearic Islands). (From Alcover, J.A. et al., *Biological Journal of the Linnean Society*, 66, 57–74, 1999. With permission from Wiley-Blackwell.)

*6CIIt. Velociraptorine Feeding on a Pterosaur

Currie and Jacobsen (1995) described pterosaur bones from the Alberta Late Cretaceous, one of which has some tooth marks of theropod type and also the imbedded tip of a dromaeosaurid velociraptorine tooth, suggesting the involvement of scavenging. They also mentioned a Late Cretaceous example from Montana involving a theropod tooth tip imbedded in a hadrosaur bone.

□ *Reliability*

Category 1, frozen behavior.

*6CIIu. Early Cretaceous Seed-Eating Bird from China

Zhou and Zhang (2002) described an Early Cretaceous bird from Liaoning that has "dozens" of well-preserved ovules in its visceral area—a unique discovery. The bird is almost edentulous.

□ *Reliability*

Category 1, frozen behavior.

*6CIIv. Late Pleistocene–Holocene Caprinid Diet

Alcover et al. (1999) discussed the abundant box (*Buxus*) pollen found in the dung of the Pleistocene–Holocene hypsodont, endemic caprinid *Myotragus* from the Balearic Islands (see Figure 133 and Figure 134). Of interest here is that when box is fed to modern artiodactyls it makes them sick and can even cause death; thus, there must have been an evolutionarily very rapid dietary change involved with the Balearic caprinid. This an unusual example in that it not only provides information about an ungulate's feeding habits but also represents an evolutionarily very rapid physiological change

from not using an obnoxious plant to using it under insular conditions of very small population size.

□ *Reliability*

Category 1, frozen behavior.

FIGURE 134 Photograph of cave with *Myotragus* dung pellets (Cova Estrela, Balearic Islands); see vertical bar for scale. (Figure 3 in Alcover, J.A. et al., *Biological Journal of the Linnean Society*, 66, 57–74, 1999. Reproduced with permission from Wiley-Blackwell.)

*6CIIw. Jurassic Salamander Diet

Gao and Shubin (2003) cited conchostracan remains in the visceral region of some Chinese Bathonian salamanders.

☐ *Reliability*

Category 1, frozen behavior.

*6CIIx. Cope's Rule and Hypercarnivory

Van Valkenburgh et al. (2004) presented a reasonable case that among North American canids the evolution of large taxa led to extinction due to their hypercarnivorous habits essentially putting them out of business.

☐ *Reliability*

Category 1, frozen behavior; a statistical type of evidence that is reasonable.

*6CIIy. Ursid Activities

Capasso (1998) described a traumatic injury to a cave bear skull that was followed by healing. He also reviewed some of the extensive literature on cave bear trauma and suggested in his Spanish example that the trauma was either intraspecific or might have involved a cave lion.

☐ *Reliability*

Category 2B, functional morphology.

*6CIIz. Turtle–*Celtis* Feeding

Marron and Moore (2006) described two specimens of *Stylemys* from the Oligocene Brule Formation of South Dakota with hackberry seeds situated in their visceral regions (i.e., evidence of fruit feeding).

☐ *Reliability*

Category 1, frozen behavior.

*6CIIza. Eocene Mammalian Predator–Prey Example

In a very innovative paper, Vasileiadou et al. (2007) used evidence of both stomach acid bone etching and predator tooth puncture marks on prey bone to make a strong case for Eocene carnivore predation on rodents from England.

Summary

There are both family- and genus-level examples here (see Table 17).

6D. MARINE, FRESHWATER, AND TERRESTRIAL

6Da. VERTEBRATES SWALLOWING OTHER VERTEBRATES

(See Table 18 and Table 19.) Concerning birds, Storrs L. Olson (pers. comm., 1981) wrote, "There is abundant reference to fish turning behavior in birds. Many species, such as herons and anhingas, spear their prey clean through (rather than catching it by the tail) and then flip the impaled fish into the air and catch it head first (I have witnessed this). I don't know about learned vs. instinct." Dr. Ralph Schreiber, who works extensively with fish-eating birds, particularly pelicans, has shown, for example, that young pelicans can't catch fish as well as more experienced adult pelicans.

Hart (1997) provided information and access to the literature on piscivore capture and handling of their prey, including their turning them into a head-first position for swallowing. Gay and Rickards (1989, p. 28) provided a cautionary example against interpreting the presence in the fossil record of a "predaceous" fish with a smaller fish only half swallowed in the larger one's mouth as an example of death following an attempted swallowing of a too large of a prey fish:

TABLE 17
Terrestrial Vertebrate Feeding Examples from the Fossil Record (Additions to Table 20 from Boucot, 1990a)

Behavioral Type	Taxonomic Level	Time Duration	Category	Refs.
6CIIa. "Fighting" dinosaurs	Order–order	Cretaceous	1	Charig and Milner (1997)
6CIIf. Bite marks on fossil nuts and mammal bones	Family–family	Eocene–Holocene	2B	Collinson (1990)
6CIIi. Beaver wood cutting and beaver-gnawed mastodon molars	Genus–class	Plio–Miocene	2B	Harington (1996)
6CIIj. Venom-conducting reptilian teeth	Class	Late Triassic–Holocene	2B	Sues (1991, 1996)
6CIIk. Long-fingered, mammalian insect seekers	Genus	Middle Eocene	2B	von Koenigswald (1990)
6CIIm. Mammoth diet	Genus	Subrecent	1	Davis et al. (1985)
6CIIo. Insectivore diets	Genera	Middle Eocene	1	Storch and Richter (1994); Habersetzer et al. (1992)
6CIIp. Piciform bird with stomach contents	Genus	Middle Eocene	1	Peters (1992)
6CIIq. *Eurotamandua* feeding	Genus–order	Middle Eocene–Holocene	1	Storch and Richter (1992b)
6CIIs. *Diprotodon* and *Thylacoleo*	Genus–genus	Pleistocene	2B	Runnegar (1983)
6CIIt. Velociraptorine feeding on a pterosaur	Family–order	Late Cretaceous	1	Currie and Jacobsen (1995)
6CIIv. Late Pleistocene–Holocene caprinid diet	Family–genus	Pleistocene–Recent	1	Alcover et al. (1999)
6CIIw. Jurassic salamander diet	Family–order	Jurassic–Recent	1	Gao and Shubin (2003)
6CIIy. Ursid activities	Family	Pleistocene–Recent	2B	Capasso (1998)
6CIIz. Turtle–*Celtis* feeding	Genus–genus	Oligocene–Recent	1	Marron and Moore (2006)
6CIIza. Eocene mammalian predator–prey example	Genus–genus	Eocene	2B	Vasileiadou et al. (2007)

TABLE 18

Piscivore Examples from the Fossil Record (Additions to Table 23 from Boucot, 1990a)

Predator	Prey	Age	Prey Orientation	Reference, Catalog Number, and/or Collection
Glyptolepis	Smaller *Glyptolepis*	Middle Devonian	Prey headed posteriorly	Ahlberg (1992)
	Cheiracanthus	Middle Devonian	Prey headed posteriorly	Ahlberg (1992)
Onychodus sp.	Arthrodire	Late Devonian	Prey headed anteriorly	Long (1991; 1995, p. 189)
Euthenopteron foordi	*Homalacanthus concinnus*	Late Devonian	Prey headed posteriorly	Arsenault (1982)
Nematoptychius	*Acanthodes*	Mississippian	Prey headed posteriorly	Traquair (1879)
Belonostomys tenuirostris	*Notagogus*	Late Jurassic	Prey headed posteriorly	Saint-Seine (1949)
Rhacolepis sp.	*Vinctifer* sp.	Early Cretaceous	Prey headed posteriorly	Wilby and Martill (1992)
Calamopleurus cylindricus	*Vinctifer*	Early Cretaceous	Prey headed posteriorly	Maisey (1994); AMNH 1189
	Vinctifer	Early Cretaceous	Prey headed posteriorly	Maisey (1994); DNPM 777LE
	Vinctifer	Early Cretaceous	Prey headed posteriorly	Maisey (1994); see also Viohl (1990, Table 23, no. 19)
	Vinctifer	Early Cretaceous	Prey headed posteriorly	Maisey (1994); Saurier Museum, Aathal, Switzerland
Dastilbe elongatus	*Dastilbe*	Early Cretaceous	Prey headed posteriorly	Maisey (1994); AMNH 12736
Notelops brama	*Rhacolepis*	Early Cretaceous	Prey headed posteriorly	Maisey (1994, 1996); AMNH 11919
Xiphactinus audax	*Gillicus arcuatus*	Late Cretaceous	Prey headed posteriorly	Beamon (2001)
Xiphactinus	*Gillicus*	Late Cretaceous	Prey headed posteriorly	Andrew Neuman (pers. comm., 1997); TMP 83.19.1, Royal Tyrell Museum, Drumheller, Alberta
Goulmimichthys	*Enchodus*	Late Cretaceous	Prey headed anteriorly	Cavin (1999)
Cimolichthys	*Enchodus*	Late Cretaceous	Prey headed posteriorly	Everhart (2005b, p. 87)
Paranguilla tigrina	Unidentified	Eocene	Prey headed posteriorly	Sorbini (1972, Plate XV, Figure 1)
"Smelt"	Argentinoid	Lower Eocene	Prey headed posteriorly	Bonde (1987, p. 11)

B.R. was pike fishing in the Fens with Bob Church for his ITV (Midlands) film series. As insurance, though it proved unnecessary, several good pike, averaging 7 to 9 lb, had been caught the day before by B.R. One of these inexplicably died (only the third time this has happened to B.R. in some forty years of piking). The fish was around 8 lb, and it assumed the normal arched-back death profile, mouth agape. We decided to leave it in a large (soft-mesh) keepnet with the other fish as we could use it to demonstrate to the cameras the morphological features of a pike without it leaping about all over the place. On removing the net from the water later on, we were staggered to find that a fish of 7 lb or so had thrust its head inside the gaping jaws of the 8-lb fish and couldn't get out. It may, of course, have gone in there for partly disgorged food items in the throat of the dead fish—in any event, it could not get out and a few hours later there would have been two dead pike, and any observer would have recorded "death by combat struggle." Yet, *we* know it was not the case. Why *should* a pike choke on a fish it is perfectly capable of swallowing? It seems to us more likely that it was dead in the first place.

Maisey (1994; see also Wilby and Martill, 1992) summarized predator–prey relationships among Santana Formation, Early Cretaceous fishes and cited examples of crustacean prey (see Figure 135, Figure 136) (Maisey and Carvalho, 1995, described the sergestid shrimp). They also worked out trophic web relations for the Santana fishes plus those from the Solnhofen based on information provided earlier (Viohl, 1990). Maisey's paper is an elegant example of what can be done in working out some aspects of trophic relations from the fossil record when unusually reliable information is available. Lund (1990) cited an example of a Late Mississippian elasmobranch that had crustaceans in its visceral region. Richter and Baszio (2001a; see also Richter and Wedmann, 2005) discussed Mid-Eocene Messel evidence of planctivory/insectivory in a teleost, the first example known for this type of feeding behavior. Richter and Baszio (2001b) also provided a very detailed food web for the Messel Lake from phytoplankton to a crocodilian. Although not involving the head-first swallowing of a prey vertebrate by a vertebrate predator, the account by Lund and Lund (1985) of a Montana Mississippian coelacanth (*Caridosuctor populosum*) with a large shrimp (*Tyrannophontes theridion*) swallowed head-first in its intestine is impressive.

TABLE 19

Reptilian Predator Examples from the Fossil Record (Additions to Table 24 from Boucot, 1990a)

Predator	Prey	Age	Prey Orientation	Ref.
Stenopterygius	*Euthynotus* cf. *incognitus*	Liassic	Prey headed posteriorly	Bürgin (2000)

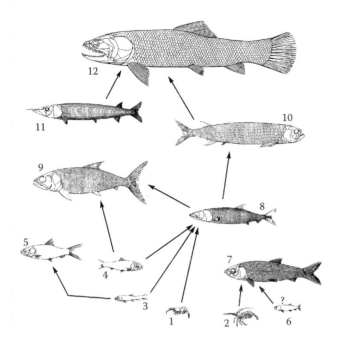

FIGURE 135 Trophic relations among Lower Cretaceous Santana Formation fishes and crustacean prey from Brazil. (Figure 12 in Maisey, J.G., *Environmental Biology of Fishes*, 40, 1–22, 1994. Reproduced with kind permission of Springer Science + Business Media.)

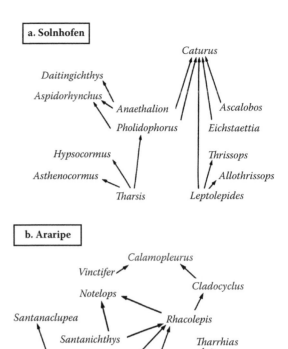

FIGURE 136 Trophic relations between Solnhofen Limestone, Late Jurassic fishes from Germany and Santana Formation and Lower Cretaceous fishes from Brazil (Figure 15 in Maisey, J.G., *Environmental Biology of Fishes*, 40, 1–22, 1994. Reproduced with kind permission of Springer Science + Business Media.)

Though hardly in the same league as the specimens discussed previously in this unit, Dalla Vecchia et al.'s (1988) account of a presumed gastric pellet, from the earlier Norian of northern Italy, has been interpreted as the remains of the oldest known pterosaur that was ejected from the stomach of an unknown fish. Such vertebrate-containing pellets are very rare anywhere, except for owl pellet remains. Additionally, Werneburg (1989, Figure 5) briefly discussed a Lower Permian example of the branchiosaur *Melanerpeton sembachense* that he interpreted as having cannibalized a smaller specimen of the same species. The smaller specimen is shown with its skull and more anterior vertebrae protruding from the area of the predator's skull; presumably, it was in the process of being swallowed tail-first or it had been caught by the posterior before being turned into a head-first swallowing position.

6Db. Gastrolith-Mediated Digestion?

(See also Section 6AIId, Shark Feeding.) Whittle and Everhart (2000) provided a very comprehensive review of which living and fossil vertebrates, fishes to mammals, are associated with gastroliths and presented evidence that in some groups the habit appears to be ubiquitous; their compiled evidence for plesiosaurs and crocodilians is overwhelming and that for sauropods and prosauropods is convincing. Taylor (1993) provided a thoughtful review of gastrolith occurrences, past and present, and arrived at the conclusion that buoyancy control is heavily implicated for those marine tetrapods that "fly" underwater (plesiosaurs, penguins, otariids) and that crocodilians use the gastroliths for buoyancy control as

well. However, Platt et al. (2006) made a good case that in at least one modern crocodilian the presence of gastroliths in all growth stages suggests that, at least in the earlier growth stages, buoyancy control is unlikely. Wings and Martin Sander (2007) provided evidence that discounts the use of gastroliths in sauropod dinosaurs.

Zhuravlev (1943) reported a Late Jurassic pliosaur from southern Russia with gastroliths in its visceral region. Druckenmiller and Russel (2008) described an Alberta Albian plesiosaur (*Nichollsia*) and briefly discussed its gastroliths accompanied by a few fish vertebrae; they also mentioned some in the South Australian plesiosaur *Umoonasaurus*. Ernst et al. (1996) provided an interesting account of a Cenomanian pebble bed from the Harz region where the pebbles are interpreted as scattered gastroliths and reviewed some of the gastrolith literature; their accumulation of circumstantial evidence sounds reasonable.

Yang and Yang (1994) cited a rallid, *Youngornis gracilis*, with "well-rounded gizzard stones" from the Miocene beds in Linqu, Shandong Province. Yeh (1981) cited the same bird, with sand in the position of the gizzard. Zhou and Zhang (2003) documented gastroliths in an Early Cretaceous bird from Liaoning, northeastern China. Zhang (2001, Figures 102–104) illustrated a mass of gastroliths in the visceral area of *Caudipteryx* from the Early Cretaceous of Jehol.

Ludvigsen and Beard (1994, 1997) cited a Campanian plesiosaur from Vancouver Island with associated gastroliths. Cicimurri and Everhart (2001) reviewed many gastrolith occurrences and vertebrate and cephalopod (ammonite conchs) food items (Table 1) associated with plesiosaurs. They discussed a Kansas Late Cretaceous plesiosaur with gastroliths, and Everhart (2000) described the gastroliths associated with this plesiosaur. Welles and Bump (1949) detailed the nature of the gastroliths associated with a Late Cretaceous elasmosaur from South Dakota (e.g., the evidence that their 253 stones, weighing 8249 grams, are gastroliths is excellent). Matsumoto et al. (1982) cited a Japanese Late Cretaceous plesiosaur with gastroliths. Wings (2007) thoroughly reviewed the functions of gastroliths.

Denton et al. (1997), while describing some Late Cretaceous–Early Tertiary *Hyposaurus* crocodilians, mentioned the occurrence of gastroliths associated with one occurrence. Long et al. (2006) provided a penetrating review of the gastrolith problems and described some from a Late Triassic Chinese ichthyosaur specimen while emphasizing that gastroliths are rarely associated with ichthyosaurs. Keller (1976) recorded some gastroliths in a Posidonienschiefer ichthyosaur. Keller and Schaal (1992) cited gastroliths from Messel Eocene crocodilians (*Diplocynodon*; see also von Koenigswald, 1987, Figure F29, p. 89). Munk and Sues (1993) cited a Kupferschiefer reptile, a Late Permian pareiasaur from Hesse with gastroliths. De Klerk et al. (2000) cited gastroliths in an Early Cretaceous theropod from South Africa and gastroliths known from other theropods elsewhere. Sanders et al. (2001) discussed gastroliths from the Early Cretaceous sauropod *Cedarosaurus*. Stokes (1987) reviewed some of the gastrolith literature attributed to dinosaurs and was skeptical about some of the occurrences. Galton (1973) cited a Late Triassic German prosauropod dinosaur with gastroliths. Kobayashi et al. (1999) discussed in some detail the gastroliths (grit mass) preserved in a Late Cretaceous ornithomimid from Mongolia and suggested that it must have been an herbivore. Calvo (1994) reviewed the occurrences of gastroliths in sauropods and pointed out that very few are valid; however, Christiansen (1996) thoroughly reviewed the gastrolith problem and concluded that Calvo's skepticism about most of the sauropod-associated gastroliths is unwarranted.

☐ *Reliability*
Category 1, frozen behavior.

*6Dc. PETALODONTID GUT CONTENTS

Malzahn (1968) documented the gut contents (brachiopods, crinoids, Foraminifera, crustacean fragments) of a Late Permian, Zechstein *Janassa* from western Germany, while Schaumberg (1979) documented the brachiopod (*Horridonia*) and bryozoan fragments from another locality.

☐ *Reliability*
Category 1. Frozen behavior.

*6Dd. MOSASAUR–SHARK

Wright (1991) discussed a Late Cretaceous mosasaur from Kansas in which Rothschild and Martin (see Martin and Rothschild, 1989, for a description of the evidence for infectious spondylitis caused by the wound) found abnormal tail vertebrae that were produced by reparative bone deposition around a shark-bitten area, as evidenced by the presence of a completely enclosed shark tooth. This specimen provides unequivocal evidence that shark predation on mosasaurs occurred, rather than mere scavenging; without this specimen, one could not determine whether shark damage to mosasaur bone had been merely evidence for scavenging. The two workers also found evidence of the bends in both mosasaur and marine turtle bone from the same collections (but see Section 4AIxa, Avascular Necrosis).

☐ *Reliability*
Category 1, frozen behavior.

*6De. CROCODILIAN MAMMAL FEEDING

Pickford (1996) discussed and illustrated a Miocene climacoceratid long bone with puncture marks of crocodilian form in a Miocene deposit replete with crocodilian remains. Alexander and Burger (2001) cited a crocodilian with "a nearly complete left hind-limb of the perissodactyl *Helaletes* sp. lodged entirely within its rib cage" from the Eocene Bridger Formation with some parallel grooves and scratches on the bones that match those of the teeth of the predator *Pristiochampsus vorax*; they suggested that some associated, articulated body portions may represent similar crocodilian behavior in the tearing apart of prey bodies. Franzen (2001) referred to a Messel, Middle Eocene primate bone with a still imbedded crocodilian tooth. Franzen and Frey (1993) described the posterior part of a Messel, mid-Eocene *Europolemur* skeleton (see Figure 137). They presented evidence, including an imbedded small crocodilian tooth, that the primate was seized by the reptile, presumably after the mammal had fallen into the water, and was bitten and shaken so severely as to separate a large portion of the posterior of the carcass from the anterior portion—a very impressive piece of paleobiology. They summarized various crocodilian techniques for prey disposal, including prey maceration by letting it decay enough to enable easy fragmentation for eating; torsion, involving rapid rotation that "screws out" a prey fragment; and shaking, where the heavily bitten prey is dismembered by shaking.

☐ *Reliability*
Category 1, frozen behavior.

SUMMARY

Family and class levels are present (see Table 20).

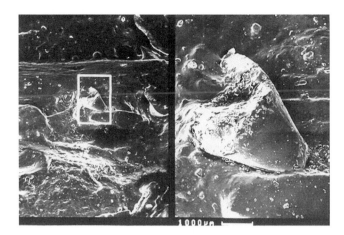

FIGURE 137   Caudal vertebra 8 of *Europolemur* from Grube Messel displaying the tip of a conical crocodile tooth stuck in a small fissure of the bone. Magnification is noted at the bottom of the right-hand SEM photograph. (Figure 16 in Franzen, J.L. and Frey, E., *Kaupia*, 3, 113–130, 1993. Reproduced with permission.)

6E. FUNGAL

*6Ea. Nematophagous Fungi

Nematophagous fungi are those that are capable of parasitizing nematodes (Barron, 1977; Poinar, 1983). Most extant nematophagous fungi belong to the deuteromycetes (Deuteromycotina) or basidiomycetes (Basidiomycotina). There are two main types: predators and parasites. The predatory forms are mostly facultative, construct trap devices in the environment, and invade the nematode's body after it is secured in the trap. The parasitic forms are mostly obligate and have spores that attach to the nematode's cuticle and then directly penetrate through the host's body cavity. One type of predatory trap consists of a ring that constricts the nematode when it attempts to crawl through it. The second type consists of sticky clubs or short strands of fungal mycelium that catch the nematodes as soon as they make contact.

Evidence of these fungi in the Tertiary is shown with fossil nematodes in Mexican amber (see Figure 138). Not only are mycelia present inside the bodies of the nematodes but spores and mycelium of the fungi also occur in the surrounding amber (Jansson and Poinar, 1986). More recently, evidence of Early Cretaceous nematode-trapping fungi has been

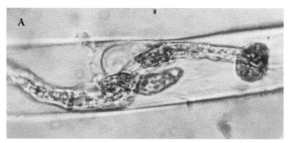

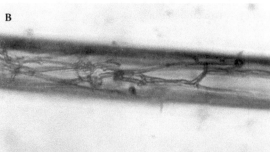

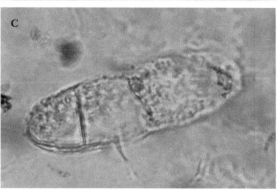

FIGURE 138   (A) Fungi developing in the body cavity of the nematode *Oligaphelenchoides atrebora* Poinar in Mexican amber; width of nematode = 15 µm. (B) Fungal mycelium in the body cavity of the nematode *Oligaphelenchoides atrebora* Poinar in Mexican amber; width of nematode = 30 µm. (C) Conidium of a nematophagous fungus adjacent to the nematode *Oligaphelenchoides atrebora* Poinar in Mexican amber; spore length = 50 µm. (All specimens are in block no. B-7053-4, Museum of Paleontology, University of California, Berkeley.)

reported in Late Albian amber from southwestern France. This fungus, described as *Palaeoanellus dimorphus* gen. et sp. nov. (Deuteromycotina), possessed unicellular hyphal rings as trapping devices and formed blastospores from

TABLE 20
Marine, Freshwater, and Terrestrial Predatory Gut Contents and Gastroliths

Behavioral Type	Taxonomic Level	Time Duration	Category	Refs.
6Da. Vertebrates swallowing other vertebrates	Class	Devonian–Holocene	1	Boucot (1990a); present text
6Db. Gastrolith-mediated digestion	Class	Permian–Holocene	1	Boucot(1990a); present text
6Dc. Petalodontid gut contents	Family–misc.	Permian	1	Malzahn (1968)
6Dd. Mosasaur–shark	Family–class	Cretaceous	1	Wright (1991)
6De. Crocodilian mammal feeding	Order–class	Eocene	1	Alexander and Burger (2001); Franzen and Frey (1993)

which a yeast stage developed. The fossil is thought to represent an anamorph of an ascomycete belonging to an extinct
lineage of carnivorous fungi with regular yeast stages. The
specimens of *P. dimorphus* represent the only fossil record of
a fungus that developed trapping devices. The microhabitat
was considered to be part of a highly diverse coastal forest
floor biocoenosis (Schmidt et al., 2008).

□ *Reliability*
Category 1, frozen behavior.

7 Communication

AUDITORY

The most elegant example of a stridulating organ known to us from the fossil record is McKeown's (1937, Plate IV; see also Carpenter, 1992, Figure 114) Middle Triassic specimen of the orthopteroid *Mesotitan* from New South Wales (Figure 139). It is clearly fine evidence for a noisy environment, and its color pattern is well preserved. Guinot and Breton (2006, Figure 13) described the stridulating structures present on a Cenomanian crab. Krzeminska and Krzeminski (1992, Figure 161) illustrated the stridulating organ on the wings of two Baltic amber ceratopogonids.

Rust et al. (1999) described in some detail the stridulatory apparatus, sending and receiving parts, of a Danish Paleogene bushcricket (*Pseudotettigonia amoena*) (Figure 140) and compared evidence regarding the frequency of its sound-producing capabilities with that of modern members of the group. They also noted the use of the sounds in courtship. This is a truly exemplary case of functional morphology.

FIGURE 139 Wing of Middle Triassic *Mesotitan*, McKeown's (1937) *Clatrotitan*, showing color pattern and stridulatory apparatus; length of wing = 138 mm. (Plate IV in McKeown, K.C., *Records of the Australian Museum*, 20, 31–37, 1937.)

Feldmann and Bearlin (1988) described and discussed the stridulatory structure of a palinurid lobster (*Linuparus*) from the Eocene of New Zealand, and Schweitzer et al. (2003) described another example of that genus and its stridulatory

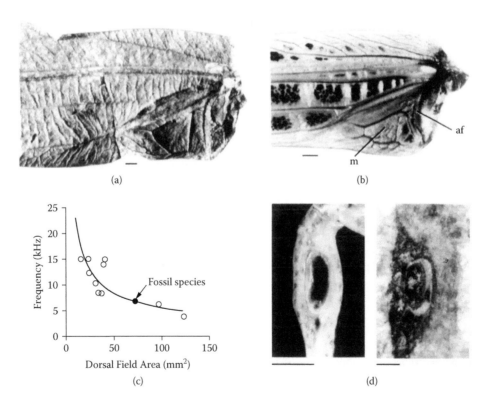

FIGURE 140 Data of Rust et al. (1999, Figure 1c) showing the probable frequency of the sound made by the male stridulatory apparatus of *Pseudotettigonia amoena*, based on a comparison of its stridulatory apparatus with similar living members of the same group. The fossils are from the Danish Paleogene. Females today are silent and lack the stridulatory structures of the males.

structure from the Late Cretaceous of Vancouver Island, the oldest known at this time.

Fritzsch et al. (1988) considered the many problems involved in the auditory systems of amphibians, living and extinct, and the implication that an auditory function has been present since their first appearance. Robinson et al. (2005) described the stapes from a Westphalian A temnospondyl amphibian (*Dendrerpeton acadianum*) from Nova Scotia as being of the type used for sound reception and having a presumed tympanic membrane in a system similar to that used by anurans. Kalmijn (1988; see also Rogers and Cox, 1988) reviewed the functioning of acoustic features in the aquatic environment.

Thewissen and Hussain (1993) provided skeletal evidence for the intermediate nature of hearing (normal cetacean vs. ancestral artiodactyl) in the Eocene whale *Pakicetus*. Features of the dentary and incus provide excellent evidence for the intermediate nature (cetacean-type underwater sound reception vs. airborne artiodactyl sound reception) of hearing in this key form. The rate information provided by the cetaceans is truly impressive, with evidence for a rapid, probably later Eocene transition from the *Pakicetus* auditory situation to the modern type in the Late Eocene, followed by essentially no subsequent change. Edinger (1955) surveyed the cetacean endocast data, Eocene to present, and concluded that a number of changes took place involving the auditory function between the Paleogene and Neogene cetacean condition, although Jerison (1973) questioned some of her conclusions. In any event, of interest is Jerison's comment about the differing rates of evolution of the expanded neocortex, such that the modern condition was attained by porpoises in the Miocene whereas that in hominids is relatively recent. Nummela et al. (2004) provided more evidence about the rate of change affecting cetacean hearing capabilities and noted the fact that there has been mosaic evolution involved in the varied skeletal changes affecting cetaceans since their Eocene movement into the sea.

Ducrocq et al. (1993) described a megachiropteran tooth from the late Eocene of Thailand; this is the oldest megachiropteran known, and they suggested that it derived from an earlier Eocene eochiropteran ancestor. Edinger (1926) described Late Eocene microchiropteran endocasts with enlarged inferior colliculi that indicate greatly increased auditory capabilities. The widespread presence since at least the Devonian of fish otoliths testifies to the presence of sound receptors in the sea (Nolfe, 1985); the waters have presumably been noisy for a long time.

☐ *Reliability*

Category 2B, based on the functional morphology of both endocasts and auditory structures.

The previous section was devoted to communication by means of sound transmission and reception; however, sense organs other than auditory are involved in communication and are discussed in the following text.

VISION

Eyes for both light/dark and image perception are well known in many groups of invertebrates, as well as in almost all vertebrates. Cronin (1988) addressed vision in aquatic animals, with an emphasis on invertebrates. Fernald (1988) dealt with fish, and Sivak (1988) with amphibians, while emphasizing the changes needed between the aquatic and terrestrial environments. Poinar (2008d) described elongate ozophores in a mite harvestman (Opiliones: Cyphophthalmi: Sironidae) from Burmese Early Cretaceous amber. The ozophores contain microvilli and are capped by a transparent cuticular dome, raising the possibility that the ozophores are photoreceptors in addition to being involved with the production of noxious scent from exocrine glands.

LUMINOUS ORGANS

(See also Section *3q, Luminescent Fishes.) Luminous organs are well known in subphotic zone fish (for typical examples, see Arambourg, 1921, 1927; Danilchenko, 1946, 1947, 1960; Gregorová, 2004; Kalabis, 1948; Kalabis and Schultz, 1974; Prokofiev, 2005). See Table 38 in Chapter 37 for myctophid fish occurrences (myctophids have luminescent organs).

CHEMORECEPTORS AND TACTILE ORGANS

Chemoreceptors in terms of nasal and other organs are also well known in various vertebrate and invertebrate groups, as well as different tactile sense receptors. Finger (1988) discussed the functioning of the olfactory bulb in fish. The presence of this structure in various terrestrial animals is well known, including evidence from the fossil record for its presence in brain endocasts as well as its diminution in both volant and fully aquatic forms. Edinger (1926) observed that the reduction in olfactory portions of the cetacean endocasts correlates with a complex story that differs between mysticetes and odontocetes, as well as the more primitive condition present in the archaeocetes.

ELECTRICAL ORGANS

Not so well known are the electric sense organs. Bullock et al. (1982) provided a summary of electroreception among the major groups of fishes, including the conclusion that it is a very ancient feature. Kalmijn (1988b) dealt with the nature of electric fields in the aquatic environment. Klembara (1994) provided skeletal evidence for the presence of electroreceptors in an Early Permian seymouriamorph and reviewed the literature concerning the fossil record of other lower vertebrates that have been considered to also have had electrical receptors. Cappetta (1988; see also Cappetta, 1987) provided a description of torpedo teeth from the Moroccan Paleocene which indicate the presence of electric organs used to subdue prey. *Torpedo* is also known from the Monte Bolca Eocene as the whole fish.

LATERAL LINE ORGANS

Janvier (1996) commented on the distribution of sensory lines within various groups of fish and provided some information about their stratigraphic distribution, and Coombs et al. (1989) provided extensive information about lateral line organs and their functions. Werneburg (2004) described post-cranial evidence of lateral line organs in Rotliegend branchiosaurs. These are organs sensitive to movement in water. Witzmann (2006, p. 158) described similar evidence for the Permo–Carboniferous Saar-Nahe Basin temnospondyl *Archegosaurus*, and similar evidence is known from some fossil fish. Boy and Sues (2000) discussed the presence of lateral lines in branchiosaurs, and Bleckmann (1986) discussed the varied functions of lateral lines in fish. Lund (1977, 1985) described lateral line canals in several Late Mississippian elasmobranchs from Montana. Kalmijn (1988a) reviewed the nature and functioning of the lateral line system, and Coombs et al. (1988; see also Denton and Gray, 1988) reviewed their evolutionary relations with an emphasis on fish.

8 Trace Fossils and Their Formers

The difficulty of assigning truly unique trace fossils to the activities of specific organisms is epitomized by Yochelson and Fedonkin's (1993) detailed account of the Late Cambrian form *Climactichnites* (Figure 141 and Figure 142). This form has been considered to reflect the activities of varied, unrelated major animal groups but still remains difficult to assign. Knox and Miller (1985) provided illustrations and a description of the trace produced by the modern carnivorous gastropod *Polinices* in loose, upper intertidal sand. Although it is one-third to one-quarter narrower than typical *Climactichnites*, it is remarkably similar in most respects; one wonders whether an unshelled gastropod is responsible for the Late Cambrian enigmatic trace. The trace fossil *Plagiogmus* is also similar (see MacRae, 1999, p. 3), and similar uncertainty obtains as to its maker. Getty and Hagadorn (2008) provided excellent illustrations of *Climactichnites* from New York, Wisconsin, and Missouri while describing a presumably shallow burrowing "relative"; they suggested that a molluscan or "mollusk-like" animal was probably responsible for these unique trace fossils.

Seilacher's (2007) volume on trace fossils is very useful for the many insights and for developing a better understanding of the significance of many trace fossil types. He included many specific examples of the trace fossils dealt with in our treatment. Buffetaut and Mazin (2003) edited a volume on pterosaurs that includes several innovative papers dealing with the trace fossils left by the limbs of pterosaurs; it serves as an excellent example of what can be done with a wholly extinct group.

Also worth noting is the fact that some trace fossils are elegant examples of evolutionary convergence, including such things as *Daemonelix*, discussed elsewhere in this treatment, which was produced by both mammals and reptiles and *Cruziana*, a "characteristic" trilobite trace fossil present in the Triassic nonmarine environment, where it obviously had nothing to do with trilobites (Bromley and Asgaard, 1979, 1991; MacNaughton and Pickerill, 1995; Shone, 1978).

8A. MARINE

8Aa. Limuloid Trails

Barthel et al. (1990, Figure 5.5) illustrated a Solnhofen *Mesolimulus* preserved together with a circular track that may be concluded to be a "death track" due to the animal apparently being very confused immediately prior to death. Babcock et al. (1995) described some Late Devonian limuloid trace fossils from western Pennsylvania and made a good estimate about the nature of their formers. Hasiotis and Demko (1996) described a nonmarine Jurassic limuloid trail.

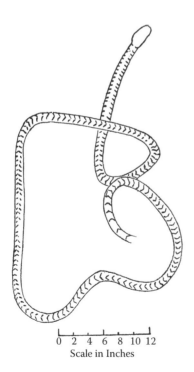

0 2 4 6 8 10 12
Scale in Inches

FIGURE 141 *Climactichnites wilsoni* showing the landing spot (above) and the takeoff (below); Late Cambrian, Quebec. (Figure 1 in Clark, T.H. and Usher, J.L., *American Journal of Science*, 246, 251–253, 1948. Reproduced by permission.)

FIGURE 142 *Climactichnites* from Wisconsin showing the probable landing trace followed by the movement trace on the substrate; centimeter scale below. (Figure 44 in Yochelson, E.L. and Fedonkin, M.A., *Smithsonian Contributions to Paleobiology*, 74, 1–74, 1993. Courtesy of the Smithsonian Institution.)

☐ *Reliability*
Category 1, frozen behavior.

8Ab.   *OPHIOMORPHA* AND *CALLIANASSA*: CRUSTACEA

Herbig (1993) extended the lower range of the anomuran coprolite *Favreina* into the Frasnian–Famennian, based on a datable clast from a Moroccan Carboniferous conglomerate, thus extending the generic range down from the Early Permian. Häntzschel (1975) extended the range of *Ophiomorpha* down from the Jurassic to the Early Permian. Masse and Vachard (1996) described the anomuran coprolite *Palaxius salataensis* from the Moscovian, Late Carboniferous of the southern Urals.

Becker and Chamberlain (2006) described anomuran coprolites associated with *Ophiomorpha* burrows and *Protocallianassa* from the Maastrichtian Navesink Formation of New Jersey—a useful association. Stilwell et al. (1997; see also Schweitzer and Feldmann, 2000) described a rare occurrence of callianassids within their *Ophiomorpha*-type burrow from Eocene-age beds in the McMurdo Sound region of Antarctica, further confirming the relationship back in time. Peckmann et al. (2007b) described *Palaxius* coprolites associated with callianassid claws from the Eocene Humptulips Formation of Washington.

☐ *Reliability*
Category 2B, based on functional morphology, and Category 1, frozen behavior, for the callianassid–*Ophiomorpha* example.

8Ac.   DECAPOD TRAILS

Barthel et al. (1990, Figure 5.12) illustrated a fine Solnhofen *Mecochirus* track with the animal in place at the end of the track. Pirrie et al. (2004) described a Late Cretaceous Antarctic decapod trackway and commented on some others.

☐ *Reliability*
Category 1, frozen behavior.

8Ad.   CRAB BURROWS

Anderson and Feldmann (1995, p. 929) cited *Lobocarcinus* burrows from the Eocene in the Fayum, including one with part of the crab still in place. From the Jurassic, Sellwood (1971) and Bromley and Asgaard (1972) described *Thalassinoides suevicus* with remains of the palinuran shrimp *Glyphea* still in place (for the similar relation of *Thalassinoides* and *Mecochirus*, see also Neto de Carvalho et al., 2007). Förster et al. (1987) described some Early Miocene homolodromid crabs from the Cape Melville Formation, King George Island, West Antarctica, that are still in their burrows. Schweitzer and Feldmann (2000) described callianassid burrows from erratics of later Eocene age in the McMurdo Sound region of Antarctica. Meyer (1988a, p. 110) cited the trace

fossil *Rhizocorallium* containing "more or less complete" *Protaxius*, a callianassid worth noting. Lewy and Goldring (2006) provided an excellent account of some Campanian crustacean burrow systems and a review of crustacean burrows in general.

☐ *Reliability*
Category 1, frozen behavior.

8Ae.   ECHINOID BURROWS AND TRACES

Asgaard et al. (1996) experimentally unraveled the feeding style that explains heart urchin behavioral traces. Bromley and Asgaard (1975) described the trace fossil *Scolicia* with the maker, *Echinocardium cordatum*, still at the end of the burrow, from the Pleistocene. Very important behaviorally in regard to *Echinocardium* is Radwański and Wysocka's (2001) account of Middle Miocene Ukrainian *Echinocardium* with spines not very different from those of modern *E. cordatum* and associated with burrows similar to those of the modern species; this is an excellent example of behavioral fixity within a genus. Bown (pers. comm., 1993) discovered typical heart urchin burrow traces in the Early Miocene of the Qattara Depression in northwestern Egypt (Hasiotis and Bown, 1993). One of the associated beds, below the bed with the echinoid traces, yielded a fossil urchin with "a trace of the trace" still attached and is a most unusual occurrence.

☐ *Reliability*
Category 1, frozen behavior, for urchin occurrences within their burrows.

8Af.   OPHIUROID RESTING TRACES

Mikuláš (1990) described a close association of the brittle star trace *Asteriacites* with body fossils of *Taeniaster bohemicus* from the Late Ordovician of Bohemia, on the same slab.

☐ *Reliability*
Category 2A, close association.

8Ag.   ORTHOCEROID TRACES

Osgood (1970) rejected Flower's (1955) interpretation of certain nautiloid cephalopod traces as being those of orthoceroid tentacles in favor of a feeding burrow interpretation.

☐ *Reliability*
Category 3, due to the uncertainty of the evidence.

*8Ah.   FISH SCRAPING

Chumakov (1998, p. 320) described several dropstones from the Cambridge Greensand that bear scratches similar to those made in the Cenozoic by parrotfish (scarids). Although somewhat speculative, this is a possibility worth considering, although it is unclear what group of Cretaceous fish might have been involved. There are other scratch marks on some other stones in his collection that may reflect the activities of scraping invertebrates, but the evidence is still speculative; we have too little basic data about the morphologies of organically generated scratch marks.

□ *Reliability*

Category 6.

*8Ai.   MACROBORING INTO HARD SUBSTRATES

Ekdale and Bromley (2001a), while describing a boring into an Early Ordovician hardground, reviewed the record of macroboring in marine hardgrounds. The borings involve unknown borers.

□ *Reliability*

Category 6, due to uncertainty about the identity of the borers.

*8Aj.   BIVALVE TRACE FORMER

Ekdale and Bromley (2001b) described a remarkable specimen involving a Pennsylvanian clam that was burrowing and feeding while also moving through the sediment. The clam thus produced three trace fossil types present on a single slab. The taxonomic identity of the bivalve is unknown.

□ *Reliability*

Category 1, frozen behavior.

*8Ak.   ISOPOD TRACES

Gaillard et al. (2005) described some Jurassic isopod traces with the isopods dead in their tracks, a very reliable interpretation.

□ *Reliability*

Category 1, frozen behavior.

*8Al.   TELLINOIDEAN BIVALVE TRACE

Bromley et al. (2003) analyzed a very complex California Paleocene trace and made a strong case for its having been produced by a tellinoidean bivalve.

□ *Reliability*

Category 2B, functional morphology.

*8Am.   SABIA PITS

Vermeij (1998) made a very good case that the unique pits developed on Miocene to present gastropod shells have been excavated by the hipponicid gastropod *Sabia*.

□ *Reliability*

Category 2B, functional morphology.

5 mm

FIGURE 143 Anomiid bivalve attachment scars from the Spanish Pleistocene. (Figure 10 in Bromley, R.G. and Heinberg, C., *Palaeogeography, Palaeoclimatology, Palaeoecology*, 232, 429–453, 2006. Reproduced with permission of Elsevier.)

*8An.   LIMPET TRACES

Seilacher (1998, Figure 3) made a good case for the preservation of limpet rasping traces since the Late Cretaceous.

□ *Reliability*

Category 2B, functional morphology, reliable comparison with modern examples.

*8Ao.   ANOMID BIVALVE TRACES

Bromley and Martinell (1991; Bromley, 1999; Bromley and Heinberg, 2006) described the unique trace fossil from the Spanish Pleistocene *Centrichnus eccentricus*, preserved beneath the anomiid bivalves attached to carbonate substrates (Figure 143). The bivalve byssus dissolves the substrate to form the characteristic trace fossil. Taddei Ruggiero and Bitner (2008) described similar *Centrichnus* and *Anomia* generated from the Italian Pleistocene.

□ *Reliability*

Category 2B, functional morphology.

SUMMARY

The data are essentially at the family and higher level, except for the generic level with the echinoid burrows (see Table 21).

8B.   FRESHWATER

8Ba.   CADDISFLY CASES

Weaver and Morse (1986) provided an account of caddisfly evolution, including the development of varied case types. Wiggins (2004) provided a comprehensive overview of caddisflies, including a classification of their larval cases. Greenwood et al. (2006) pointed out that caddisflies that

TABLE 21
Marine Invertebrate Trace Fossils with Well-Established Makers (Additions to Table 25 from Boucot, 1990a)

Behavioral Type	Taxonomic Level	Time Duration	Category	Refs.
8Aa. Limuloid trails	Order	Devonian–Holocene	2B	Babcock et al. (1995)
8Ab. *Ophiomorpha* and *Callianassa*	Superfamily	Late Devonian–Holocene	2B	Herbig (1993)
8Af. Ophiuroid resting trace	Asteriacites	Late Ordovician	1	Mikuláš (1990)
8Ag. Orthoceroid traces	Order	Ordovician	3	Osgood (1970)
8Ah. Fish scraping	Class	Cretaceous	6	Chumakov (1998)
8Ak. Isopod traces	Class	Jurassic–Recent	1	Gaillard et al. (2005)
8Al. Tellinoidean bivalve trace	Family	Paleocene–Recent	2B	Bromley et al. (2003)
8Am. *Sabia* pits	Genus	Miocene–Recent	2B	Vermeij (1998)
8An. Limpet traces	Family	Late Cretaceous–Recent	2B	Seilacher (1998)
8Ao. Anomiid bivalve traces	Family	Late Cretaceous–Recent	2B	Bromley and Martinell (1991)

construct cases are more likely to represent slower to quiet water bodies rather than fast-moving waters.

Sukacheva (1994) described some Late Jurassic caddisfly cases from Mongolia. Hasiotis and Demko (1996) and Hasiotis et al. (1998b) cited Late Jurassic caddisfly cases from the Morrison Formation in Colorado. Vinogradov (1996) illustrated and discussed some Russian and Kazakh caddisfly cases from the Late Jurassic–Early Cretaceous which employed phylactolaemate bryozoan statoblasts in their construction. (It would be interesting to learn what cues caddisfly larvae use in selecting their case materials!) Coram and Jarzembowski (2002) discussed a Late Jurassic caddisfly case made of ostracod shells from southern England. Rasnitsyn and Quicke (2002) illustrated an Early Jurassic Siberian case made of small sand grains (Figure 287) and showing evidence of attachment bands (Figure 291), as well as Late Jurassic or Early Cretaceous cases made of ostracode shells. Sukacheva (1990) described a number of Mongolian Neocomian caddisflies and caddisfly cases. Ivanov (2006) described a number of Neocomian caddisfly cases from Siberia that still have the larvae preserved within their cases. Jarzembowski (1995) described some Early Cretaceous caddisfly cases from southern England. Heads (2006) described a Wealden caddisfly case from the Isle of Wight composed of fish scales and fish bones, a most unusual and presumably freshwater fish remains.

Rasnitsyn and Quicke (2002) cited an Early Cretaceous case made of miscellaneous mineral grains (Figure 289), Late Jurassic or Early Cretaceous Siberian *Folindusia undae* made of plant fragments, Transbaikalian Early Cretaceous *Pelindusia conspecta* made of bivalve shell fragments (Figure 292), and Early Cretaceous Mongolian *Frugindusia karkenia* made of gingoalean seeds from *Karkenia* (Figure 293). They also illustrated an Early Cretaceous caddisfly pupa from Siberia (Figure 201) and Paleocene Maritime Province *Secrindusia pacifica* silk cases (Figure 294). Sukacheva (2005) described some Late Cretaceous caddisfly cases from the Amur region.

Lewis (1992, Plate 3, Figures A, B) illustrated more trichopteran larval cases from the Eocene of the Republic, Washington area, assigning them to the Phryganeidae. Lutz (1992, Figure 108) discussed and illustrated some spun silk cases from the Messel Eocene. Archibald and Matthewes (2000) cited some Early Eocene caddisfly cases from British Columbia. Lutz (1990, Plate 6, Figure 4) illustrated one composed of sand grains from the Messel Eocene. Johnston (1993) discussed and illustrated a middle Eocene caddisfly case from northern Mississippi in which the anterior portion of the larva is still preserved. Leggitt and Cushman (1999, 2001) and Biaggi et al. (1999) briefly cited caddisfly-case Eocene bioherms, with cases made from ostracode shells. Loewen (1999) further considered the four case types present. Leggitt et al. (2007) described caddisfly cases in some detail within Eocene, Green River Formation carbonate mounds. Leggitt and Loewen (2002) described Eocene, Green River Formation masses of caddisfly cases in almost reef-like masses that were previously misidentified by Wilmot Bradley as algae; these Green River Formation examples have no modern reef-like counterparts, but their construction, form, and size leave little doubt about their identity. Meyer (2003, Figure 192) illustrated a Late Eocene Florissant caddisfly case made from ostracod shells and biotite grains.

Rasnitsyn and Quicke (2002, Figure 295) cited Oligocene Russian Far East *Folindusia* (*Spirindusia*) *kemaensis* made of spirally arranged plant fragments. Lewis and Heikes (1990) described Miocene, Latah Formation caddisfly cases from northern Idaho. Hugueney et al. (1990) described caddisfly cases from the French Miocene that are encrusted with snails and some cases that contain preserved larvae as well. Vinogradov (1996) illustrated and discussed some Russian and Kazakh caddisfly cases from the Miocene that employed phylactolaemate bryozoan statoblasts in their construction. Sukacheva (1989) described a number of Cenozoic caddisfly cases from Primorye.

Some extant caddisflies leave the aquatic habitat to forage on adjacent vegetation, and that is probably how some caddisfly larvae within their cases occur in amber deposits. This behavior is well illustrated by a larval caddisfly still within its case from Baltic amber (Figure 144) as well as another from the Lower Cretaceous of Siberia (Figure 145).

☐ *Reliability*

Category 1, frozen behavior, for the French Miocene material, and Category 2B, functional morphology, for that from Idaho.

8Bb. LUNGFISH BURROWS

Hasiotis and Mitchell (1993) and Hasiotis et al. (1993b) provided the evidence for discounting earlier accounts of the Chinle Formation, Late Triassic lungfish burrows from the Colorado Plateau since they are actually those of freshwater

FIGURE 144 Baltic amber caddisfly case with larva (Poinar amber collection accession no. TR-2-17); length of case = 9 mm.

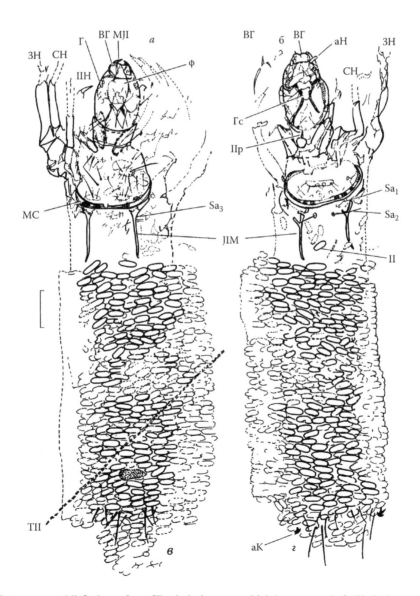

FIGURE 145 Lower Cretaceous caddisfly larva from Siberia in its case, which is composed of elliptical coprolites; scale line = 1 mm. (From Ivanov, V.D., *Paleontologicheskii Zhurnal*, 2, 62–71, 2006. Reproduced with permission of Springer Science + Business Media.)

crayfish. Hasiotis and Hannigan (1991), however, with the use of a biochemical test, were able to reliably recognize some lungfish burrows from the same region with morphologies distinct from those of associated crayfish burrows. The morphological comparison of the fossil crayfish and lungfish burrows with those made by modern lungfish and crayfish was instrumental in sorting out this trace fossil dilemma. Carlson (1968) described some Permian lungfish burrows with included occupants, while Berman (1976) described additional examples from other localities with references to other occurrences, and Boucot (1990a) discussed a Triassic example.

☐ *Reliability*

Category 1, frozen behavior, as examples with the maker *in situ* are known.

*8Bc. CRAYFISH BURROWS

Hasiotis and Mitchell (1993) and Hasiotis et al. (1993a) provided excellent documentation for three types of freshwater crayfish burrows from the Late Triassic Chinle Formation of the Colorado Plateau. Hasiotis (pers. comm., 1993) has recovered crayfish burrows from the Late Cretaceous and Paleocene of the Rocky Mountain region and the Late Cretaceous of the Colorado Plateau.

☐ *Reliability*

Category 1, frozen behavior, due to additional comparisons with modern burrow counterparts, and Triassic and Cretaceous examples with the maker in place have also been found.

*8Bd. FISH TRACES

Wisshak et al. (2004), while describing probably Early Devonian acanthodian traces from Spitsbergen, summarized much of the literature on the traces left on soft bottom sediments by fish swimming near the bottom.

☐ *Reliability*

Category 6, as the identity of the makers is not too definite.

SUMMARY

The data summarized above are mostly at about the superfamily to ordinal levels (see Table 22).

8C. TERRESTRIAL

8CI. INVERTEBRATE

8CIa. Mud Wasp Nests

Freeman and Donovan (1991) qualified the taxonomic assignment of the wasp made by Genise and Bown (1990), who described mud dauber wasp nests (*Chubutolithes*) from the Eocene of Patagonia.

☐ *Reliability*

Category 2B, functional morphology.

8CIb. Leafcutting Bees

Berry (1921) discussed a Kentucky Eocene example that may reflect bee leafcutting activity. Lewis (1994) described a mid-Eocene, Klondike Mountain Formation leaf from Washington with margin damage similar to that generated today by leafcutting bees. Meyer (2003, Figure 45) illustrated Late Eocene Florissant leafcutting bee activity, and Wappler and Engel (2003) illustrated an excellent mid-Eocene example of megachilid-type leaf cutting from Eckfeld Maar, Germany. Givulescu (1984) described varied leaf margin damage of uncertain origins from the late Miocene (Pannonian) of Chiuzbaia. Sarzetti et al. (2008) described Middle Eocene Patagonian leaf margin trace fossils ascribed to leafcutter bees, although one wonders about the alternative possibility of attid ants.

☐ *Reliability*

Category 6, due to the difficulty of being certain about the agent.

8CIc. Mining Hymenopterans

Mader (1999) provided an extensive account of holes driven into present-day rock exposures—heavily weathered, of course—by varied solitary bees that use them for laying their eggs; presumably similar trace fossils will turn up in the fossil record. Elliott and Nations (1998) described a Cenomanian set of possible bee burrows, possibly halictines, from northeastern Arizona and summarized the literature on similar trace fossils from elsewhere. Genise et al. (2002) provided an extensive description and discussion of Late Cretaceous Patagonian halictine bee burrows and associated coleopteran burrows; their halictine bee material is very convincing (Figure 146). Rasnitsyn and Quicke (2002, Figure 467) illustrated an alleged aculeate hymenopteran nest fragment

---

TABLE 22
Freshwater Trace Fossil with Well-Established Makers (Additions to Table 26 from Boucot, 1990a)

Behavioral Type	Taxonomic Level	Time Duration	Category	Refs.
8Bc. Crayfish burrows	Infraorder	Late Triassic–Holocene	1	Hasiotis and Mitchell (1993); Hasiotis et al. (1993a)
8Bd. Fish traces	Class	Early Devonian–Holocene	6	Wisshak et al. (2004)

FIGURE 146 Late Cretaceous, Laguna Palacios Formation solitary bee burrows from central Patagonia assigned to *Cellicalichnus chubutensis*, which are interpreted to represent an extinct bee taxon. (From Genise, J.F. et al., *Palaeogeography, Palaeoclimatology, Palaeoecology*, 177, 215–235, 2002. Reproduced with permission from Elsevier.)

(*Desertiana mira*) tentatively attributed to Vespidae, from the Coniacian of Uzbekistan.

Genise and Bown (1996) described some Late Cretaceous or Early Tertiary plus Miocene trace fossils that they interpreted as the burrows of solitary bees, but unlike the burrows of modern solitary bees. Melchor et al. (2002) described

Celliforma from the Argentine Eocene. Meehan (1994) discussed and illustrated various hymenopteran-type burrows from the Late Oligocene of northwestern Nebraska. Alexander and Burger (2001) discussed a number of Middle Eocene insect burrows similar to those made by solitary bees. Fejfar and Kaiser (2005, Figure 6) described some Czech Oligocene hymenopteran-type solitary brood cells.

Blair and Armstrong (1979) cited solitary bee burrows from the late Miocene Muddy Creek Formation of southeastern Nevada and northwestern Arizona. Thackray (1994) described an extensive set of burrows ascribed to sweat bees from the Miocene of Kenya; the burrows strongly resemble those of modern sweat bees (Halictinae). Genise and Bown (1994a) described various Patagonian Miocene examples, and Uchman and Alvaro (2000) described some Spanish Miocene *Celliforma* and *Rosellichnus*. Dominguez Alonso and Coca Abia (1998) described what they concluded are Miocene digger wasp nests from Honduras and assigned them to *Celliforma*.

Houston (1987) described some unique Pleistocene stenotritid bee burrows from the Pleistocene Upper Bridgewater Formation of South Australia that were misinterpreted earlier by Retallack, who had confused them with anthophorid bee cells. The unique form of these burrows with their "concave septa" are notable (Figure 147). Michaux et al. (1991), while describing some endemic shrews

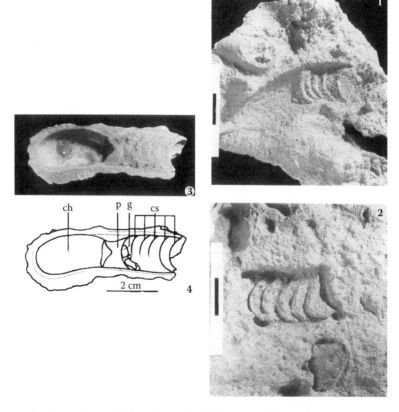

FIGURE 147 Brood burrows of solitary Stenotritid bees from the Pleistocene of South Australia. Figures 1 and 2 are cross-sections of burrows embedded in calcrete (scale = 1 cm), Figure 3 is a cross-section of another burrow, and Figure 4 is a diagram of the same. Note the concave septae. (Figures 3 to 5 in Houston, T.F., *Transactions of the Royal Society of South Australia*, 111, 93–97, 1987. Reproduced with permission.)

and rodents from the Pleistocene on Fuerteventura, Canary Islands, mentioned anthophorid burrows accompanied by terrestrial gastropods from several nearby localities in what must be a fossil soil. The anthophorid material was not illustrated or described. Ellis and Ellis-Adam (1993) described what they concluded are abundant Subrecent brood cells of solitary bees (anthophorids) from the Canary Islands, as well as evidence of predation on some of the brood cell larvae; they also critically reviewed relevant literature dealing with solitary hymenopteran cells.

Edwards and Meco (2000) described varied Quaternary, ground-dwelling solitary bee brood cells from the Canary Islands. Edwards et al. (1998) described some nonmarine, cocoon-like trace fossils from the Late Eocene on the Isle of Wight that they interpreted as possibly being the brood chambers of hymenopterans. The material is still uncertain but may well be due to insect activity of one kind or another, and their careful description will probably eventually permit a reliable identification.

☐ *Reliability*
Category 2B, functional morphology.

8CId. Aleyrodidae Pupal Case

Jarzembowski and Ross (1993) described a whitefly pupal case attached to a leaf from the Eocene of the Isle of Wight. This is the first published account, with illustration, of a fossil occurrence of a whitefly pupal case (for a 30,000-year-old olive whitefly occurrence from Santorini, see Boucot, 1990a, 4Bj, Olive Whitefly). Allied to these pupal cases is Poinar's (1999b, Figure 11, p. 11) account of a case-bearing Dominican amber chrysomelid larva.

☐ *Reliability*
Category 1, frozen behavior.

*8CIe. Coleopteran Pupal Chambers and Possible Scarabid Beetle Burrows

Johnston et al. (1996) made a very convincing case for the presence of varied beetle pupal chambers in the Late Cretaceous of Mongolia and adjacent Inner Mongolia. Hasiotis et al. (1993a) described the trace fossil *Scaphichnium hamatum* from the early Eocene of northwestern Wyoming and provided evidence that it was made by scarabid beetles, as modern scarabids make similar burrows in which they deposit eggs. Genise and Bown (1994a) described varied Patagonian Miocene scarabeid pellets.

☐ *Reliability*
Category 2B, functional morphology.

*8CIf. Caterpillar Coprolites Misidentified as Araliaceae Fruits

Lancucka-Srodoniowa (1964) described caterpillar coprolites from various Tertiary sites extending back to the Eocene that had been misidentified as the fruits of Araliaceae. The plant

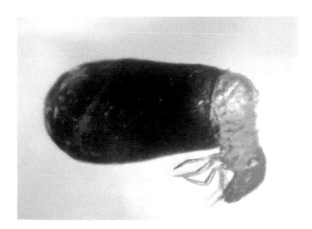

FIGURE 148   A leaf beetle larva (Coleoptera: Chrysomelidae) protruding from its case in Dominican amber (Poinar amber collection accession no. C-7-52); length of case = 1 mm.

debris present inside some of these coprolites, plus their morphology and lack of seeds internally, leaves little doubt as to their true nature. These coprolites provided good evidence for the presence of lepidopterans.

☐ *Reliability*
Category 1, frozen behavior.

*8CIg. Neuropteroid Cocoon

Poinar (1998b, Figure 1B) illustrated a neuropteroid cocoon preserved in Dominican amber.

☐ *Reliability*
Category 2B, functional morphology.

*8CIh. Chrysomelid Larval Case

Poinar (1998b, Figure 1E) illustrated a larval chrysomelid case, with the actual larva still inside, from Dominican amber (Figure 148).

☐ *Reliability*
Category 1, frozen behavior.

*8CIi. Earthworm Burrows

Meehan (1994) discussed evidence for the presence of Late Oligocene earthworm burrows from northwestern Nebraska and used the Risk and Szczuczko (1977) staining technique to prove the presence of some type of saccharides in the burrow walls, thus eliminating the possibility of arthropods as the burrow formers.

☐ *Reliability*
Category 2A.

Summary

Most of the examples are at the family or higher levels, with a few at the generic level (see Table 23).

TABLE 23
Terrestrial Invertebrate Trace Fossils with Well-Established Makers
(Additions to Table 27 from Boucot, 1990a)

Behavioral Type	Taxonomic Level	Time Duration	Category	Ref.
8Cli. Earthworm burrows	Superfamily	Late Oligocene	2	Meehan (1994)

8CII. VERTEBRATE

8CIIa. Vertebrate Tracks

The Lockley and Meyer volume (2000) on European fossil tracks is an elegant example of how difficult it is to ally tracks and other vertebrate traces with their makers at a satisfyingly low taxonomic level. They commented on the lovely exception provided by a German Rotliegende diplocaulid trace (Walter and Werneberg, 1988) (Figure 149); Beerbower (1963) illustrated the curious "winged" skull of a diplocaulid like that responsible for the Rotliegende trace (Figure 150). Lockley (2001) summarized the evidence provided by dinosaurian tracks and dinosaur locomotion. Manning et al. (2006; see also Manning, 2004) made it very clear that the ichnotaxonomy of vertebrate tracks undoubtedly is burdened with synonyms due to the basic fact that a foot pressing down on soft sediment will leave varied track shapes at different bedding planes. Only tracks showing skin imprint should be referred to as "footprints." Although involving crustaceans rather than vertebrates, Uchman and Pervesler's (2006) experiments with living amphipods and isopods walking on varied substrate types demonstrates that trace types of different morphologies can be produced by the same organism. Wellnhofer (1978) reillustrated (from Stokes, 1957) an alleged pterosaur trackway (Figure 17, *Pteraichnus*) from the Late Jurassic Morrison

Formation in Arizona that makes very good sense as evidence for quadrupedal locomotion. Unwin (1997) made a good case for pterosaurs being quadrupedal when on the ground; he used their unique trackways to provide the evidence while summarizing the currently available evidence. Buffetaut and Mazin (2003) contains several papers that make a strong case for pterodactyls having been quadrupedal while on land; their volume also includes the unique paper by Frey et al. (2003), who made a strong case that a stout leaf caused the death of a Brazilian Cretaceous pterosaur; it is very, very rare for the cause of death to be reasonably deduced for a fossil vertebrate.

Cotton et al. (1996) provided further data regarding both herding and locomotion in dinosaurs. Lockley et al. (1994b) provided an excellent account of some Late Jurassic Portugese sauropod tracks, together with an excellent set of references to the topic (for some Argentine, Middle Jurassic data, see also Coria, 1994). Gatesey et al. (1999) analyzed and described Late Triassic Greenland theropod tracks in a very innovative manner that tells a lot about the nature of the theropod foot. Considerable literature has attempted to use dinosaur trackways and skeletons for estimating the speed at which they were capable of walking and running (e.g., Alexander, 1976; Farlow, 1981; Thulborn, 1982), but there does not appear to be much agreement about the different approaches. Day et al. (2002b) provided additional evidence concerning dinosaur locomotion styles from the British Jurassic with some possibility of estimating theropod running speed. Lockley and Meyer (2000) provided a wealth of dinosaur track data plus material from many other groups. Platt and Hasiotis (2006) described some Upper Jurassic Morrison Formation sauropod tracks that even preserve impressions of scales on the distal portions of the limb impressions.

FIGURE 149 Illustration of a Rotliegende, earlier Permian German resting trace of a morphologically very curious, specialized, freshwater diplocaulid amphibian, one of the few vertebrate trace fossils where one can be certain of the maker's identity (approximately ×2). (Figure 2.13 in Lockley, M. and Meyer, C., *Dinosaur Tracks and Other Fossil Footprints of Europe*, Columbia University Press, New York, 2000. This image was first published in Walter, H. and Werneburg, R., *Freiberger Forschungshefte (Leipzig)*, C419, 96–106, 1988. Reproduced with permission.)

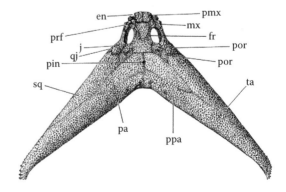

FIGURE 150 Dorsal view of the skull of *Diploceraspis*, Early Permian Dunkard Group (×0.8). (Figure 1 in Beerbower, J.R., *Bulletin of the Museum of Comparative Zoology*, 130, 31–108, 1963. Reproduced with permission from Harvard University.)

Yang et al. (1995) described Late Cretaceous and early Tertiary, flamingo-like and duck-like bird tracks with webs from Korea. Renders (1984) provided data on Pliocene *Hipparion* tracks from Laetoli, Tanzania (two adults, one juvenile) that agree with modern horse tracks and their gaits—a "running walk," in this instance. Motani (2002a) provided evidence that ichthyosaurs propelled themselves through the water in a manner similar to modern tunas, and like them may well have had elevated temperatures. Motani (2002b) provided the best currently available analysis of marine reptile swimming speeds.

☐ *Reliability*

Category 1, frozen behavior.

8CIIb.   *Daemonelix*

Meyer (1999) provided a wealth of additional data on the corkscrew burrow of the mid-Tertiary beaver *Palaeocastor fossor*. Smith (1987) described the helical burrow, *Daemonelix*, of the South African Permian dicynodont reptile *Diictodon*; this occurrence provides an excellent example of behavioral convergence, as the Miocene beaver (Boucot, 1990a) and the Permian dicynodont have nothing to do with each other.

8CIIc.   Dabble Marks and Accompanying Tracks, Plus Associated Skeletal Material

You et al. (2006) described Aptian–Albian bird remains from northwestern China that feature evidence of webbed feet.

☐ *Reliability*

Category 2B, functional morphology.

8CIId.   Pocket Mouse and Kangaroo Rat Burrows

Szafer (1957) described a system of hamster burrows from the Subrecent of Poland, many of which included seed stores. Collinson (1999b), summarizing Manchester (1987) and Wehr (1995), cited a cache of *Carya* nuts from a Miocene *Platanus* stump found in Washington and noted that it might represent sciurid work. Vander Wall (1990) provided an excellent summary account of food hoarding behaviors, including those of rodent types that have provided paleontologically preserved materials, including those cited earlier (Boucot, 1990a). Vander Wall (1990) considered the varied evolutionary advantages of food hoarding and observed that rodents in highly seasonal environments, as contrasted with tropical nonseasonal environments, are highly involved.

☐ *Reliability*

Category 2B, functional morphology.

*8CIIe.   Cicioniiformes-Like Tracks

Mustoe (1993) described some Eocene bird tracks resembling those of modern Cicioniiformes-type birds. Varied walking activity is indicated.

☐ *Reliability*

Category 2B, based on functional morphology.

*8CIIf.   Artiodactyl Tracks

Fornós et al. (2002) described some Pleistocene *Myotragus* tracks in aeolinite from the Balearic Islands and discussed how their modified legs have given rise to the trackway design. Bromley (2001) described some modern musk ox trackways in soft sediment that eventually trapped the animal.

☐ *Reliability*

Category 2B, functional morphology.

*8CIIg.   Human Footprints

Bennett et al. (2009; see also Crompton and Pataky, 2009) described undoubted human, hominin footprints from the earlier Quaternary (1.5 million years old) of Kenya. These footprints may have been made by *Homo ergaster* or *Homo erectus*. They are the oldest known hominid footprints. Avanzini et al. (2008) described some much younger Middle Pleistocene human footprints from southern Italy. In both examples, the stride lengths are similar to those of present-day humans, as is the orientation of the big toe. Kim et al. (2009) described some Late Quaternary Korean human footprints and discussed a variety of Neogene, chiefly Pleistocene and Holocene, human footprints from elsewhere.

☐ *Reliability*

Category 1, frozen behavior, as the footprints in all regards are similar to those of modern humans, although they presumably were made by different hominid species.

Summary

Most of the information is at a high taxonomic level, with little at the generic level (see Table 24).

TABLE 24
Terrestrial Vertebrate Trace Fossils with Well-Established Makers (Additions to Table 28 from Boucot, 1990a)

Behavioral Type	Taxonomic Level	Time Duration	Category	Refs.
8CIIa. Vertebrate tracks				
Diplocaulid	Genus	Permian	1	Walter and Werneburg (1988)
Dinosaur tracks	Family–superfamily	Triassic–Cretaceous	1	Lockley and Meyer (1999)
8CIIc. Dabble marks and tracks	Class	Cretaceous–Recent	2B	You et al. (2006)
8CIIe. Cicioniiformes-like tracks	Superfamily	Eocene–Recent	2B	Mustoe (1993)
8CIIf. Artiodactyl tracks	Genus	Pleistocene–Recent	2B	Fornós et al. (2002)
8CIIg. Human footprints	Genus	Pleistocene–Recent	1	Bennett et al. (2009)

COLOR PLATE 1 (A) *Thrissops formosus* from the Solnhofen Plattenkalke (length = 349 mm); note the bands of color spots. (From Tischlinger, H., *Archaeopteryx*, 16, 1–18, 1998. With permission.) (B) Same specimen as Part A enlarged to emphasize the color pattern. (C) Head of a *Sphaerodactylus* gecko in Dominican amber; note the small scales and staring eyes lacking lids, which are characteristic of this genus (Jim Work amber collection, Ashland, Oregon). (D) An adult female of the Early Devonian nematode *Palaeonema phyticum*, in a stomatal chamber of the primitive land plant *Aglaophyton major*; length of nematode = 890 μm.

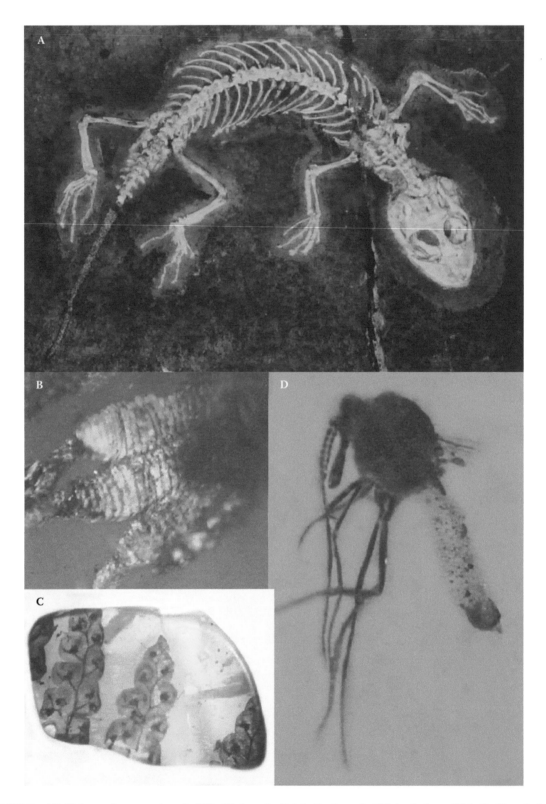

COLOR PLATE 2 (A) *Eichstaettisaurus schroederi* (Broili) from the Solnhofen Jurassic, Wintershof near Eichstät, Bavaria (collection of the Bayerische Staatssamlung für Paläontologie, München, No. 1937 I 1). The length of the specimen without regenerated tail = 120 mm, and the length of the regenerated tail = 50 mm. Note the absence of vertebrae in the regenerated tail, as is also the case with modern lizards. (Photograph kindly provided by Helmut Tischlinger, Stammham.) (B) Toes of the Burmese amber gecko, *Cretaceogekko burmae*, with sophisticated pads that would have allowed it to climb vertical surfaces (Poinar amber collection accession no. B-V-4); length of middle toe = 1.8 mm. (C) Four partial pinnules of the Burmese amber fern *Cretacifilix fungiformis* (Deniz Eren amber collection, Istanbul, Turkey); length of amber piece = 12 mm. (D) Burmese amber biting midge (*Protoculicoides* sp) containing the developing stages of the malarial parasite *Paleohaemoproteus burmacis* in its body cavity (Poinar amber collection, accession no. B-D-32); length of biting midge = 1.5 mm.

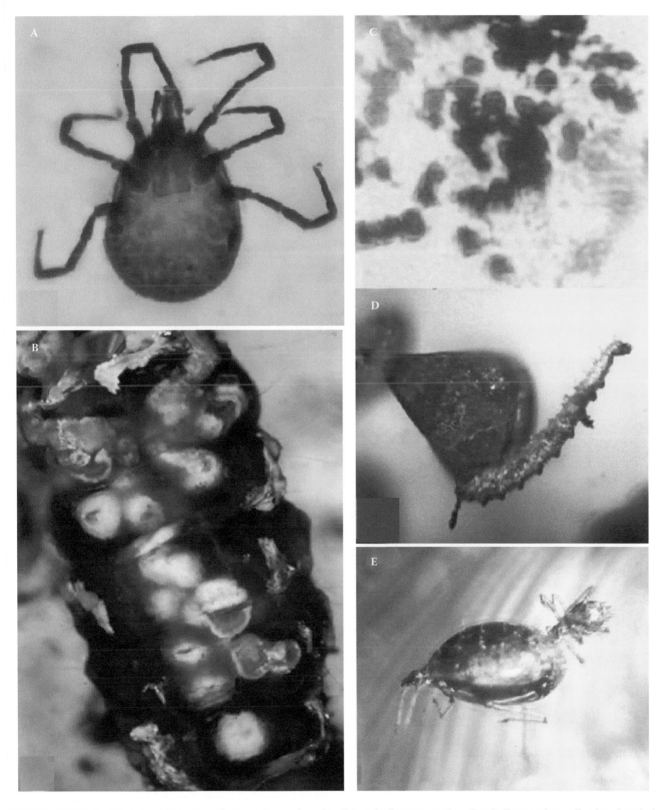

COLOR PLATE 3 (A) A hard tick, *Compluriscutula vetulum* (Ixodidae), in Burmese amber (Deniz Eren amber collection, Istanbul, Turkey); length of tick = 0.53 mm. (B) Tumors in a caterpillar (Lepidoptera) in Mexican amber (Poinar amber collection accession no. Sy-1-103); diameter of tumors ranges from 380 to 615 μm. (C) Polyhedra of a cytoplasmic polyhedrosis virus (CPV) in the midgut of an Early Cretaceous biting midge (*Protoculicoides* sp.) in Burmese amber (Poinar amber collection accession number B-D-36); diameters of polyhedra range from 2.3 to 5.2 μm. (D) Sand fly larva adjacent to its food source, a fruiting body of a coral fungus in Burmese amber (Poinar amber collection accession no. B-D-24); length of larva = 2.8 mm. (E) A parasitic mite feeding on a larger mite in Baltic amber (Andrew Cholewinski amber collection, Redondo Beach, California).

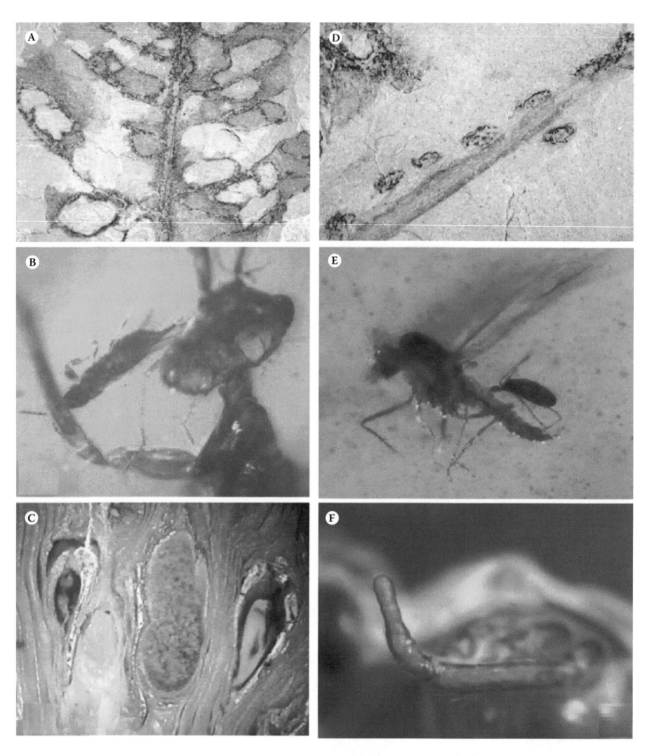

COLOR PLATE 4 (A) Cretaceous insect blotch leaf mine (*Distigmonoma variegata*). (Plate XXXIX, Figure 4, in Krassilov, V.A. and Rasnitsyn, A.P., Eds., *Plant–Arthropod Interactions in the Early Angiosperm History: Evidence from the Cretaceous of Israel*, Pensoft Publishers/Brill, Boston, MA, 2008, p. 173. Reproduced with permission.) (B) Ant holding a thrips in its mandibles in Burmese amber (Scott Anderson amber collection, accession no. AB0016); length of thrip ≈ 2 mm. (C) Three seed cavities in the Jurassic permineralized cone of *Araucaria mirabilis* with the middle cavity filled with frass from an herbivorous insect (Poinar collection); length of damaged seed cavity = 7.0 mm. (D) Cretaceous insect (*Costoveon admatum*) egg scars associated with leaf midvein. (Plate VII, Figure 2, in Krassilov, V.A. and Rasnitsyn, A.P., Eds., *Plant–Arthropod Interactions in the Early Angiosperm History: Evidence from the Cretaceous of Israel*, Pensoft Publishers/Brill, Boston, MA, 2008, p. 141. Reproduced with permission.) (E) Parasitic mite on a Burmese amber midge (Diptera: Chironomidae) (Poinar amber collection accession no. B-Sy-1); length of midge = 1 mm. (F) Regenerated patella of the spider *Caudasinus regeneratus* (Theridiidae) in Baltic amber; length of regenerated patella = 0.35 mm. (From Wunderlich, J., *Beiträge zur Araneologie*, 5(1), 855, 2008. With permission.)

9 Specialized Substrates

9A. MARINE

9Aa. SHELL-BORING FUNGI AND ALGAE

Wisshak and Tapanila (2007) gathered together papers that provide a wealth of information on fungal and algal borers, including microborers, together with extensive biostratigraphic information (Benner et al., 2007; Golubic and Radtke, 2007; Pawlowska et al., 2007; Tapanila, 2007; Tribollet, 2007; Wisshak et al., 2007). Vogel (1993) summarized the Precambrian to present record of shell borers and other endoliths, and Vogel (1987) described various New York Givetian microborings attributed mostly to algae and fungi. Glaub (1994) provided a wealth of information on European Jurassic and Early Cretaceous microborers, Glaub and Konigshof (1997) discussed various Devonian and Early Carboniferous microborings in conodonts, and Glaub and Bundschuh (1997) provided more information on Silurian and Jurassic–Early Cretaceous microborings, with careful attention being paid to both light penetration and depth correlations of the microborings. Glaub et al. (2007) more recently have provided a comprehensive summary of microborings and microbial endoliths with a good account of their Phanerozoic stratigraphic ranges. Underwood et al. (1999a) described some mid-Cretaceous microborings into marine fish teeth. Bromley et al. (2007) described dendriniform microborings that may be the work of certain Foraminifera with Recent and fossil examples. Vénec-Peyré (1996) described the work of bioeroding Foraminifera on varied substrates, with a Jurassic to present range.

☐ *Reliability*

Category 2B, functional morphology.

9Ab. CLIONID BORING SPONGES

Vogel (1993) described Early Cambrian boring sponges for which "even typical clionid chips are preserved" and emphasized that boring sponges appear to have played a very minor role during the Paleozoic, but they do begin to be important in the later Triassic. Perry and Bertling (2000) considered the record of Mesozoic and Cenozoic coral reef macroborers and concluded that there is a dominance in the Neogene of sponge borers. Wisshak and Tapanila (2007) provided papers with a variety of data concerning boring sponges in space and time (Bromley and Schönberg, 2007; Calcinai et al., 2007; Schönberg, 2007; Schönberg and Shields, 2007; Wisshak, 2007).

☐ *Reliability*

Category 1, frozen behavior.

9Ac. BRYOZOAN–SNAIL–HERMIT CRAB COMPLEX AND HYDROZOAN–GASTROPOD COMPLEX

Darrell and Taylor (1989; see also Allmon, 1993, Figure 10, and text) described scleractinian symbionts of some Florida Pliocene hermit crabs, as well as associated bryozoan *Hippoporidra* and the spionid annelid-generated trace fossil *Helicotaphrichnus*. The corals extend as a tube-like extension of the hermit-crab-inhabited snail shell in a manner similar to the "tubes" formed by some bryozoans. Taylor and Schindler (2004) described a Florida Late Eocene *Hippoporidra* involved with a gastropod shell as a hermit crab symbiont; this is the oldest known bryozoan–hermit crab symbiosis known (Figure 151). Taylor et al. (2004a, Figure H) illustrated some New Zealand Neogene hermit crab symbiont examples. Taylor (1991) described varied bryozoan–hermit crab symbioses, as well as various New Zealand modern occurrences in which the bryozoan extended the hermitted gastropod aperture significantly in a tubular manner. He also refers to Miocene New Zealand examples of the

FIGURE 151 Taylor and Schindler (2004) described the oldest known occurrence of hermit crab symbionts, the bryozoan *Hippoporidra*, from the Late Eocene, Florida Ocala Limestone. Figures 1 through 3 show the opening in the bryozoan colony for the hermit crab, and Figure 4 is a section through the long axis of the bryozoan tube (scale bar is in millimeters). (From Taylor, P.D. and Schindler, K.S., *Journal of Paleontology*, 78, 790–794, 2004. Reproduced with permission.)

same relationship that remain to be described. Olivero and Aguirre-Urreta (1994) described a number of coiled hydractinian "tubes," each having an internal gastropod shell as the initial beginning, associated in the same beds with claws of *Paguristes*, which are best interpreted as the coevolved relation between the hermit crab and the hydractinian—all of this from Maastrichtian-age beds.

Walker (1992, Table 7) reviewed, among many other phyla, the published occurrences of bryozoan-covered gastropod shells and added a few new Tertiary occurrences from North America. It is obvious that additional observations will add a wealth of new material from various localities and horizons, Jurassic to Pleistocene, for the bryozoan–gastropod situations. Feldmann et al. (1993) described some Late Cretaceous *Paguristes* chelipeds from the Antarctic Peninsula and referred to Aguirre-Urreta and Olivero's (1992) work regarding the presence of some pagurized gastropod shells encrusted by bryozoans in a symbiotic manner, additional to the closely associated chelipeds.

Taylor (1994) provided a thorough review of bryozoan–hermit crab symbioses, both modern and fossil. Evidently the bryozoans are not always obligate with the hermitted snails today. Occurrences of both tubular outgrowths from the gastropod aperture and branched colonies are present today. The bryozoan taxonomy of the symbionts is critically reviewed, and Taylor suggested that species-level specificity between the hermit crabs and the bryozoans is low. Seldom does an obligate relationship seem to be involved. It is unclear whether only shells inhabited by crabs are colonized by the bryozoans in most cases. Tubular colony growth is the easiest way to be certain of the relationship among fossils. Worn colony bases are also reviewed, as are some bryozoan–gastropod relations from the Ordovician and Silurian (p. 176) that may represent similar, although not hermit crab, relations with an unknown organism. Morris et al. (1991) described a Middle Devonian gastropod–bryozoan, possibly symbiotic, relationship with an unknown inhabitant of an empty gastropod shell, while reviewing the literature on similar occurrences as far back as the Ordovician; they favor a sipunculid occupant, although there is no positive evidence of such. The potential selective advantages to both the hermit crabs and the bryozoans are reviewed. Taylor (1994) found no support for coevolution in the examples known at present.

El-Hawat and Abdel Gawad (1996) described some mid-Miocene Libyan hermit crab symbioses with the gastropod shells being heavily encrusted by *Membranipora* in a somewhat ellipsoidal manner. Buge and Fischer (1970) described bryozoan encrustacean of a gastropod shell that is apparently involved with a hermit crab occupant of the snail shell from the Bathonian.

Although unassociated with bryozoans, it is worth noting that Walker (2001) found abundant evidence of outer shelf to bathyal gastropod shells occupied by hermits, as well as Pliocene shells occupied by modern hermits, a mixing of different age materials. Barnes (2001) provided further examples of hermit crabs using fossil snail shells. Fraaije (2003) described an Early Cretaceous hermit crab from England,

10 μm

FIGURE 152 Jagt et al. (2006) reillustrated the specimen of *Palaeopagurus*, shown here, from the English Early Cretaceous and housed within an ammonite conch. This is the same specimen described by Fraaije (2003). (Figure 6 in Jagt, J.W.M. et al., *Revista Mexicana de Ciencias Geológicas*, 23, 364–369, 2006. Reproduced with permission.)

occupying an ammonite conch (Figure 152). Walker et al. (2003) described the trace fossil trackway of a Holocene land hermit crab. Karasawa (2002) described a Late Oligocene Japanese *Pagurus* within a *Turritella infralirata* shell. Hu and Tao (1996, Plate 6, Figures 2 and 7, pp. 61–62) briefly described *Pagurus* sp. B within a high-spired Miocene gastropod from Taiwan. Collins and Jakobsen (2003) described a Danish Eocene hermit crab claw within a gastropod shell and reviewed many of the earlier descriptions of hermit crabs present within gastropod shells.

The oldest hermit crab preserved within a molluscan shell is within an ammonite from the Pliensbachian, Early Jurassic, of southern Germany (Jagt et al., 2006), described in a paper that reviews the known occurrences of hermit crab remains preserved within gastropod and ammonite shells. Jagt et al. (2006) made it very clear that the characteristic hermit crab behavior of occupying a vacant, coiled shell, gastropod or ammonite, appeared at just about the same time as isolated hermit crab body fossils first appear. It would be interesting to learn just what selective forces were involved that ended up coupling empty, coiled molluscan shells with the heavily skeletonized anterior claw of the hermits with the poorly skeletonized body and posterior appendages.

While describing some Central European Jurassic hermit crabs, van Bakel et al. (2008) suggested that the initial shell inhabiting by hermit crabs in the Mesozoic, probably beginning in the Jurassic, involved ammonite shells rather than gastropod shells. Although not involving hermit crabs, Unal and Zinsmeister's (2006) brief account of Lower Cambrian hyolithids with encrusting oncolitic materials that left the hyolithic opening clear is strongly suggestive of an unknown occupant.

☐ *Reliability*

Category 2B, functional morphology, and Category 1, frozen behavior, for those with the crab claw still located within the gastropod shell.

9Ad. *HELICOTAPHRICHNUS*: TRACE FOSSIL

Darrell and Taylor (1989) documented hermit-crab-associated, spionid-polychaete-generated trace fossil *Helicotaphrichnus* from the Florida Pliocene. Walker (1992) illustrated an Eocene occurrence from Mississippi. One can expect the *Helicotaphrichnus* record to go much further back in time after more attention has been paid to it. Walker (2001) reported on Late Pliocene gastropods from Ecuador, from outer shelf to bathyal depths, with *Helicotaphrichnus*. Bromley (2004) summarized the previously published accounts of this genus. Ishikawa and Kase (2007) cited some Pliocene Japanese *Helicotaphrichnus* while describing another spionid polychaete trace fossil (their *Polydorichnus subapicalis*) associated with hermitted gastropod shells from the modern environment, with fossils extending back to the Late Miocene.

☐ *Reliability*

Category 2B, functional morphology.

9Ae. *THYLACUS*–LARGER GASTROPOD RELATIONSHIP

Although not pertaining to the gastropod *Thylacus*, Vermeij's (1998) description of the relation between the hipponicid gastropod *Sabia* and various tropical gastropods and hermitized gastropod shells is similar in that the commensal presumably used debris from the host gastropod or hermit crab as food. The scars left on the gastropod shells are characteristic, indicating long-term occupancy. Shells as old as the Miocene with *Sabia*-type scars have been recognized.

☐ *Reliability*

Both Category 1, frozen behavior, and Category 2B, functional morphology, have been observed.

9Af. ROCK- AND WOOD-BORING BIVALVES

Bromley and Asgaard (1993) discussed and illustrated rock-boring bivalve traces as well as those produced by rock-boring sponges. Kennedy (1974) provided an excellent account of the pholads while describing those from the West American Cenozoic. Savazzi (1999) summarized a wealth of information regarding bivalves boring into varied substrates. Kelly and Bromley (1984) carefully defined the trace fossils *Gastrochaeonolites* and *Teredolites*. Donovan and Hensley (2006) described Cenozoic *Gastrochaeonolites* from the Antillean region, including some with both siphon impressions preserved.

Rock Borings

The oldest bivalve rock boring is documented in a Late Ordovician mytilacean by Pojeta and Palmer (1976). Kleemann (1983) provided a fairly comprehensive catalog of modern and fossil lithophagid occurrences and taxa, extending back through the Carboniferous. Wilson and Palmer (1998) described some Pennsylvanian rock-boring traces. Carter and Stanley (2004) described some gastrochaenid and lithophaginid borings from the Late Triassic of Nevada and Austria and reviewed several aspects of the boring bivalve record.

Fürsich et al. (1994) described some reef-boring *Lithophaga* and *Gastrochaena* from the English Jurassic. Hölder (1972) described some borings into a belemnite (*Megateuthis*) from the German Bajocian that were made by *Hiatella* (*Pseudosaxicava*), as well as some *Lithophaga* boreholes. In his important paper on Jurassic hardgrounds, and hardgrounds in general, Fürsich (1979) described a number of *Lithophaga* and *Gastrochaena* occurrences.

Zakharov (1966, p. 103) cited an Early Cretaceous northern Siberian example of *Gastrochaena* borings into a *Liostraca* valve (an excellent example is on exhibit in the Tomsk Geological Museum). Akpan (1991) discussed Albian stromatolite borings by *Lithophaga*. Bandel et al. (1997) described some Albian, mid-Cretaceous, pholad-bored amber from Jordan. Bien et al. (1999) described lithophagid borings from Late Cretaceous oysters from Delaware. Wilson and Taylor (2001) described lithophagid borings from the Cretaceous of Oman.

Domenech et al. (2001) discussed a Spanish Miocene bored rocky substrate involving traces produced by bivalves, sponges, annelids, and sipunculids. Masuda (1968) described some Miocene pholads boring into andesite, and Masuda and Noda (1969) described others (*Zirfaea*) boring into Pliocene basaltic–andesitic substrates. Itoigawa (1963) described rock boring by varied Japanese Miocene lithophagid taxa. Jones and Pemberton (1988) described some Grand Cayman *Lithophaga* borings from the Pleistocene and provided references to lithophagid activity back to the Jurassic. Boekschoten (1967) described some rock boring by *Hiatella* from the Belgian Pliocene.

Teredinid and Pholad Bored Wood

Clapp and Kenk (1967) provided an extensive, annotated bibliography on modern wood borers. MacEachern et al. (2007, Figure 4.9) illustrated some elegant examples of wood infested with teredo. Vahldiek and Schweigert (2007) described the oldest example of *Teredolites* from the German Pliensbachian, which is one more example of the Jurassic incoming at about the superfamily level of the modern marine ecosystem.

Evans (1999) provided an excellent summary of the hardpart morphologies of the wood borers, living and extinct, plus their known stratigraphic ranges from the latest Jurassic to present. Hoagland and Turner (1981) also provided a useful account of the wood borers, with many examples from the fossil record. Lopes et al. (2000) provided an excellent discussion of the digestive organs and activity of teredoes.

Savrda and King (1993) provided information on Campanian *Teredolites*, and Savrda and Smith (1996) provided an overview of the ichnogenus. Rogers (1998) cited a *Teredolites* occurrence from the Campanian Two Medicine Formation of Montana, and Ferrer and Gibert (2005) described a Spanish Early Cretaceous occurrence of *Teredolites*. Hatai (1951) described a Japanese Early Cretaceous "*Teredo*" with a few of the tubes containing "shells," and Hart et al. (1996)

mentioned "log-grounds" from the Late Cretaceous (Turonian–Coniacian) of the Cauvery Basin, southeastern India, replete with teredo-bored wood. Ludvigsen and Beard (1997) illustrated Late Cretaceous, Lambert Formation, teredo-bored conifer wood from Vancouver Island, *Teredina* in a piece of driftwood from the same formation, *Martesia clausa* in coalified wood from the Trent River Formation, and *Lithophaga* in Late Cretaceous, Haslam Formation *Odontogryphaea*. Kelly et al. (2000; for some Antarctic Cretaceous examples and a careful review of examples known from the fossil record elsewhere, see also Kelly, 1988) discussed some Early Cretaceous teredo-bored wood (with *Turnus in situ*). Jagt and Collins (1999) described a Maastrichtian Belgian log riddled with teredo and discussed similar examples of Cretaceous and Cenozoic age. MacRae (1999, p. 224) briefly cited teredo-bored Cretaceous wood from KwaZulu-Natal. Bromley et al. (1984) described some pholad-bored wood from the Late Cretaceous of Alberta, with *Martesia* still in place. Crampton (1990) described a New Zealand Late Cretaceous species of *Martesia* still *in situ* within araucarian wood. Wilson and Taylor (2001) described teredinid borings from the Cretaceous of Oman. Mikuláš et al. (1995a) described some Cenomanian wood with *Teredolites* from Moravia, and Mikuláš (1993) described some Cenomanian examples from Bohemia. Kříž and Mikuláš (2006) described *Teredolites* from the Upper Cretaceous of the Bohemian Cretaceous Basin.

Cvancara (1970) described pallets from some North Dakota, Paleocene, teredo-bored wood, and Elliott (1963) described some Iraqi Paleocene wood bored by *Banksia* with pallets, tubes, and valves—a very unique occurrence. Savrda (1990) discussed wood-boring traces (*Teredolites*) from the Alabama Paleocene, and Savrda et al. (1993) discussed them further, pointing out that they may occur *in situ* within wood, in peats, as bundled remains after the wood has decayed, or as isolated tubes following wood decay.

Huggett and Gale (1995) illustrated fine specimens of teredo burrows, some with pallets and calcareous linings, from the English Eocene. Francis (2000) illustrated *Teredolites* borings in Eocene wood from the McMurdo Sound region of Antarctica, and Savrda et al. (2005) described *Teredolites* from the Eocene of Alabama. Bonde (1987) cited some teredo-bored wood from the Danish Lower Eocene. Noda and Lee (1989) described *Martesia*-bored Miocene wood from Korea, Itoigawa (1963) described varied Japanese Miocene wood-boring bivalves, and Campbell (1993) briefly described varied wood- and rock-boring bivalves from the Virginia Pliocene, including *Cyrtopleura*, *Pholas*, *Martesia*, and *Teredina*.

□ *Reliability*

Category 1, frozen behavior.

9Ag. ACROTHORACICAN BARNACLES

Seilacher (1969) reviewed acrothoracican characteristics and many occurrences through time (Pennsylvanian, Permian, Triassic, Jurassic, Cretaceous, Miocene). Häntzschel (1962,

pp. W228–W232) described some of the acrothoracican boring trace fossil genera. Baird et al. (1990) provided a well-documented case of Givetian acrothoracican borings into *Naticonema* associated with *Taxocrinus* in the usual coprophagous condition whereas the associated *Platyceras* are unbored, suggesting that the barnacle was only associated with the coprophagous platyceratids. Some of the barnacle borings produced a shell-secreting response in the host snail. The stratigraphic range of this platyceratid–acrothoracid relationship is Early Devonian through Pennsylvanian. The Givetian specimens provide good evidence that the barnacles were boring into live gastropods, which reacted to their presence by secreting inner shell layers. Rodriguez and Gutschick (1977) described Famennian acrothoracican borings in varied hosts.

Legrand-Blain and Poncet (1991) described some Carboniferous acrothoracican borings from Algeria, Ettensohn (1978) described some Mississippian acrothoracican borings from Kentucky and Alabama, and McKinney (1968) described some from Tennessee that he misidentified as bryozoans. Gundrum (1979) described some acrothoracican borings into sponges from the Missouri Pennsylvanian, and Hoare and Steller (1967) discussed some Pennsylvanian acrothoracican borings into a brachiopod (*Schizophoria*). Merrill (1979) described some Pennsylvanian acrothoracican borings into algal oncolites. Sumrall et al. (2006) described some Early Pennsylvanian acrothoracican borings from Oklahoma.

Cooper and Grant (1974, Plate 82, Figures 1 and 2; Plate 87, Figures 1 and 2; Plate 99, Figures 40 and 41) illustrated acrothoracican borings into three genera of Permian brachiopods from the Glass Mountains and showed the "blisters" on the interiors of the shells that were presumably mantle reactions to the intruding barnacles activities. Schlaudt and Young (1960) illustrated some acrothoracican borings into Texas Permian brachiopods and bryozoans and also into mid-Cretaceous Texas gastropods. Rodda and Fisher (1962) described some Texas Permian acrothoracican borings into brachiopods, Simonsen and Cuffey (1980) described some Midcontinent Permian acrothoracican trace fossils, and Tomlinson (1963) described some elegant Pennsylvanian and Permian acrothoracicans in bivalves from Kansas, Texas, and Oklahoma.

Baird et al. (1990) provided a good summary of other Paleozoic acrothoracican borings. Jahnke (1966) described some German Oberkants Terebratelbank acrothoracican borings. Hollingworth and Wignall (1992) described some Callovian–Oxfordian acrothoracican borings into *Gryphaea*. Pugaczewska (1970) described some Polish Jurassic acrothoracid borings into bivalves. Hary (1987) described Sinemurian borings into *Liogryphaea* from the Paris Basin. Hallam (1963) described some Early Jurassic British acrothoracican borings into bivalves.

Codez and Sainte-Seine (1958) described Jurassic, Cretaceous, Miocene, and Pliocene acrothoracids from France, and Saint-Seine (1951, 1954, 1955) described still more. Cross and Rose (1994) described some Late Cretaceous acrothoracican-type borings into *Micraster* tests. Moosleitner (2000) illustrated some from French Early Cretaceous acrothoracican borings. Seilacher (1968) discussed acrothoracican boreholes from some Spanish Early Cretaceous belemnites. Turner (1973) briefly cited a Late Cretaceous, New Jersey acrothoracican boring into a bivalve shell that contained actual acrothoracican teeth—a truly unusual occurrence. Wilson and Taylor (2001) cited acrothoracican borings from the Cretaceous of Oman, and Fürsich and Pandey (1999) described some Late Cretaceous acrothoracicans from the Indian Cauvery Basin. Hattin (1986) described some Late Cretaceous acrothoracicans on *Platyceramus* from the Western Interior Seaway. Lambers and Boekschoten (1986) described some acrothoracid borings from the German Early Cretaceous, and Pugaczewska (1985) decribed some acrothoracican borings into some Late Cretaceous Polish belemnites. Joysey (1959) described some possible acrothoracican borings from the English Chalk. Voigt (1967) described some German Cenomanian acrothoracican borings, and Taylor (1965) described some Aptian acrothoracican borings into Antarctic belemnites. Seibertz and Spaeth (2005, Figure 4) illustrated an elegant Albian belemnite from Mexico, riddled by acrothoracican borings. Abletz (1993) described two Eocene acrothoracican borings, one into a solitary coral, and the second into an oyster. Bałuk and Radwański (1991) described some Miocene acrothoracican borings in gastropod shells from Korytnica (see also Section 9An, Savazzi's "Leaning Tower of Pisa" Morphology, below). Zapfe (1936) described some fine Miocene acrothoracican borings into Miocene gastropods from Hungary and Austria.

□ *Reliability*

Category 2B, based on functional morphology; Turner's (1973) example is Category 1, frozen behavior.

*9Ah. Acrothoracican Barnacle–Hermit Crab Shell

Bałuk and Radwański (1991) made a good case for commensalism, similar to that existing today, between a Miocene Polish gastropod shell inhabited by a hermit crab and the position of acrothoracican borings. Walker (1992, Table 5) reviewed the diversity of boring barnacles associated with pagurid-inhabited gastropods.

□ *Reliability*

Category 2B, functional morphology.

*9Ai. Arachnostega

Bertling (1992) described a Late Jurassic trace fossil from northern Germany that occurs in marine bivalves as a ramifying burrow in the outer shell layer. In the modern environment, a similar trace is formed by some errant polychaetes. The similarity of the modern and Jurassic traces is striking.

□ *Reliability*

Category 3.

*9Aj. Algal and Fungal Paleozoic Microborings in Corals

Elias and Lee (1993) described and summarized currently available information on microborings into tabulate and rugose corals interpreted to be due to boring algae and fungi. It is unsure whether or not these are anything more than commensals. In any event, it appears that they probably are far more than rock borers that merely selected a calcium carbonate substrate; that is, they probably favored a coralline substrate. Occurrences are recorded from the mid-Ordovician (Caradoc equivalents) to the Permian.

□ *Reliability*

Category 6, because of the varied uncertainties involved.

*9Ak. Limpets and Bone Substrates

Marshall (1994; see also Marshall, 1987, and Haszprunar, 1988) described some modern limpets associated with whale bones on the deep seafloor and discussed an Eocene New Zealand limpet example associated with turtle bone.

□ *Reliability*

Category 2B, functional morphology, for the Eocene example.

*9Al. Limpet Depressions in Ammonites

Kase et al. (1994) described depressions in Japanese and Sakhalin Late Cretaceous ammonites formed by limpets that employed the ammonite as a substrate in life. Although not involving an ammonite, Noda's (1991) description of a limpet attached to a Pliocene gastropod with a homing scar is notable.

□ *Reliability*

Category 1, frozen behavior.

*9Am. Asteriastoma cretaceum Breton, 1992

Breton (1992) described a unique Late Cretaceous trace fossil from the French chalks, *Asteriastoma cretaceum*, occurring chiefly in asteroids but sometimes in echinoids; he ascribed it to the activities of phoronids making unique boreholes into hard substrate echinoderm skeletons (Figure 153). No explanation is known concerning why this trace is restricted to skeletonized echinoderms, but future work may be helpful.

□ *Reliability*

Category 6, due to the uncertainty of the generating organism.

*9An. Savazzi's "Leaning Tower of Pisa" Morphology

Savazzi (1999c) briefly reviewed some of the benthic invertebrates living on soft substrates that have a life form leading them to adopt an ever-changing, somewhat curved to circular/helical morphology, which permits their aperture

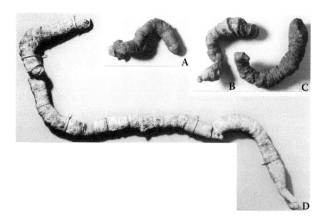

FIGURE 153 The trace fossil *Asteriostoma* on echinoderm remains from France is a unique trace interpreted to be phoronid activity on asteroid and echinoid skeletal materials (Figure 1, ×14; 2, ×15; 3, ×30; 4, ×18; 5, ×12; 6, ×18; 7, ×22; 8, ×7; 9, ×7). (Figure 12 in Breton, G., *Bulletin Trimestriel de la Société Géologique Normandie et Amis Muséum du Havre, fascicule hors série*, 78(4), 1–592, 1992. Reproduced with permission of Bibliothèque nationale de France.)

FIGURE 154 Scrutton (1998) illustrated a set of rugose corals that have twisted and turned in response to changing positions on a soft substrate: (A) *Amplexus coralloides* (×0.1) from the Lower Carboniferous of Ireland; (B, C) ?*Spongophylloides* (×.0.1) from the Canadian Silurian; (D) *Caninia* (×0.8) from the Early Carboniferous of Somerset, England. (Figure 11 in Scrutton, C.T., *Proceedings of the Yorkshire Geological Society*, 52, 1–57, 1998. Reproduced by permission of the Yorkshire Geological Society.)

to always be within the water column rather than being buried in mud. Savazzi mentioned some serpulids, tube-dwelling bivalves, and barnacles. In the Paleozoic, a number of rugose corals adopted this type of curved morphology. Neuman (1988) discussed and illustrated some fine Silurian rugose coral examples. Scrutton (1998, Figure 11) illustrated some fine Silurian and Early Carboniferous examples (see Figure 154), and Sorauf (2001, Figure 3) illustrated some elegant Devonian examples (see Figure 155). Sando's work (1984) on Mississippian rugosan epibionts further supports the "Leaning Tower of Pisa" conclusion, as the epibionts are preferentially located on the concave side of the corals.

Some organisms actually specialize on certain hard substrates, whereas the majority employ just about any available hard substrate, biologic or nonbiologic, for larval attachment and metamorphism. Zakharov (1966) illustrated various terebratuloid brachiopods that used *Chlamys* as a hard substrate as well as serpulid tubes adherant to a *Chlamys* valve (Plates XII and XV) and a *Liostraca* valve with adherent terebratuloids (Plate XLII) (see Figures 156 through 158). A puzzling example is the presence of Pennsylvanian *Crurithyris* attached to the proximal ends of *Archaeocidaris* spines (Schneider, 2003; see also Schneider et al. 2005). Alvarez and Taylor (1987) illustrated and discussed *Aulopora* relations to *Anathyris* that suggest that the coral has a preferred growth position on the anterior regions of the brachiopod.

□ *Reliability*
Category 2B, functional morphology.

*9Ao. "Hard" Substrates

Any number of organisms use scattered "hard" substrates as contrasted with "soft" level-bottom sediment substrates.

FIGURE 155 *Heliophyllum halli* from the New York Middle Devonian that beautifully illustrate Tower of Pisa morphologies (all, ×0.25). (Figure 3 in Sorauf, J.E., *Journal of Paleontology*, 75, 24–33, 2001. Reproduced with permission.)

FIGURE 156 Terebratuloids (Early Cretaceous of Siberia) that used *Chlamys* valves as a hard substrate (×0.5); there is no evidence that this use of the hard substrate was obligate (Zakharov, 1966, Plate XII, Figure 1).

Cornulites

Vinn and Mutvei (2005; see also Zhan and Vinn, 2007) reviewed the biological affinities of the cornulitids and arrived at the conclusion that the biology and systematic position of this group are still very enigmatic. Richards (1974) provided an extensive discussion, with examples of *Cornulites* and its varied invertebrate substrates, Middle Ordovician through Mississippian and Pennsylvanian. Most of the examples suggest that just about any hard substrate, living or dead shell, will do (Watkins, 1981; Lebold, 2000), except for some late Middle Devonian Silica Shale examples oriented in such a manner on the brachiopod *Paraspirifer*'s valves, with the open end of the *Cornulites* at the commissure suggesting that a living association is involved. Kesling et al. (1980) were convinced that the *Cornulites–Paraspirifer* association was parasitic because of deformation of the host's shell margin consistent with the *Cornulites* having fed on mantle tissue. Kesling et al. (1980, p. 1142) also pointed out that at the northern Ohio locality, where most of their material came from, the *Paraspirifer* occur in clumps that are not always infested with *Cornulites*.

Spjeldnaes (1984, Plate 1, Figure 5) discussed a *Meristina* with two *Cornulites* oriented in an anteriorward manner. Sparks et al. (1980) provided a wealth of information concerning the epifauna of *Paraspirifer* (see Figure 159), while Kesling and Chilman (1975; see also Hoare and Steller, 1967) provided more documentation (see Figure 160). Schumann (1967) provided excellent illustrations of a Late Devonian Canadian example of *Cornulites* similarly oriented on *Mucrospirifer*, which makes it clear that this not a purely *Paraspirifer* relationship (see Figure 161). Gekker (1935) illustrated and discussed some Late Devonian and Main Devonian Field *Cornulites* (identified as *Serpula devonica* Pacht) attached to a spiriferid and a rhynchonellid in an oriented manner as well as on a bivalve, aperture anteriorward. Hurst (1974) described Late Silurian *Cornulites* from

FIGURE 157 Serpulid tubes (Early Cretaceous of Siberia) that used *Chlamys* valves as a hard substrate (×0.5); there is no evidence that this use of the hard substrate was obligate (Zakharov, 1966, Plate XV, Figure 2).

FIGURE 158 Oysters and serpulid tubes (Early Cretaceous of Siberia) that used *Chlamys* valves as a hard substrate (×0.5); there is no evidence that this use of the hard substrate was obligate (Zakharov, 1966, Plate XLII, Figure 4).

FIGURE 159 *Cornulites* (×1) oriented toward the anterior margin of an Ohio, Middle Devonian *Paraspirifer*, demonstrating shell deformation of the brachiopod. (Plate 7, Figure 8, in Sparks, D.K. et al., *Epizoans on the Brachiopod* Paraspirifer bownockeri *(Stewart) from the Middle Devonian of Ohio*, Papers on Paleontology No. 23, University of Michigan Museum of Paleontology, Ann Arbor, 1980. Reproduced with permission.)

FIGURE 160   *Cornulites* (×1) oriented toward the anterior margin of an Ohio, Middle Devonian *Paraspirifer*, demonstrating shell deformation of the brachiopod. Note that in none of these examples is there proof for parasitism rather than the use of a hard substrate, with the preferred orientation possibly indicating the use of suspended food particles brought in by currents generated by the brachiopods for their own use—possibly a bit of stealing but not real parasitism. (Plate 31, Figure 4, in Kesling, R.V. and Chilman, R.B., *Strata and Megafossils of the Middle Devonian Silica Formation*, Papers on Paleontology No. 8, University of Michigan Museum of Paleontology, Ann Arbor, 1975. Reproduced with permission.)

Gotland brachiopods that are oriented in a manner similar to the *Mucrospirifer* and *Paraspirifer* described above. Holland (1971, Plate 2, Figure a) illustrated a Gotland Silurian *Eospirifer* with a *Cornulites* in the middle of the ventral sulcus, aperture oriented anteriorly. Frey (1987) discussed some Late Ordovician *Cornulites* occurring on the bivalve *Byssonychia* in a manner oriented toward the "anterior" part of the bivalve. Lescinsky (1997) described orientations on Carboniferous brachiopods that indicate a degree of orientation on some plicate shells but far less on smooth shells.

Morris and Felton (2003) provided more examples from the Late Ordovician of Ohio; some are attached to crinoid stems, suggesting a life mode for filter/suspension feeding well above the sediment–water interface. Hoare (2003, Plate 1, Figure 2) illustrated some Ohio Mississippian *Cornulites* oriented in a parallel manner on a *Composita*, with their open ends in one direction and possibly two generations being represented, but the epizoans are not at the commissure. Powers and Ausich (1990) also reported a relatively obligate relation between *Cornulitella* and *Composita* from the Mississippian of Indiana, with many of the cornulitids having their apertures oriented anteriorly on the brachiopods. Richards (1972) illustrated random orientation of *Cornulites* on Late Ordovician *Rafinequina*. Morris and Rollins (1971) described some Late Ordovician *Cornulites* from Ohio that make a good case for orientation in life on a living bivalve and on a living bellerophontid, as well as on some stony bryozoan colonies; this is clearly a complex business. Prantl (1948) described a later Ordovician *Conchicolites* from the Barrandian attached to an orthoceroid; the apertures of the cornulitid open anteriorly in a manner oriented toward the orthoceroid aperture (see Figure 162).

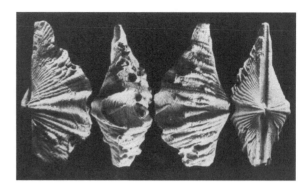

FIGURE 161   *Cornulites* (×1) oriented toward the anterior margin of a Late Devonian, Canadian *Mucrospirifer*; note that the anterior margin of the brachiopod has been deformed by some, but not all, of the *Cornulites*. (From Schumann, D., *Palaeogeography, Palaeoclimatology, Palaeoecology*, 3, 381–392, 1967. With permission by Elsevier.)

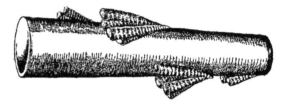

FIGURE 162   Reconstruction of *Conchicolites* specimens attached to a later Ordovician orthoceroid (×0.5). The cornulitid apertures are pointing anteriorly on the orthoceroid. (Figures 2 and 3 in Prantl, F., *Vestnik Kralovske česke Společnosti Nauk*, 9, 1–7, 1948.)

Bordeaux and Brett (1990) provided an account of epibiont frequencies on varied Givetian brachiopods from New York and discussed the various possibilities involving the relationships, including a rugophilic relationship for *Cornulites*, which was also advocated by Hurst (1974). Alexander and Scharpf (1990), however, discussed cornulitids occurring in an unoriented manner on varied Late Ordovician, Richmond brachiopods from southeastern Indiana. Hall (1888) illustrated various Ordovician cornulitids attached to brachiopod valves in an unoriented manner, making it clear that the oriented attitude documented for the Devonian examples is only one possibility. Hall (1888) also cited New York Trenton Limestone cornulitids, presumably of Caradoc age, which may be the earliest members of the group known. The *Cornulites* data might be interpreted as suggesting that more than one taxon was involved; that is, some taxa behaved like those on *Paraspirifer* and *Mucrospirifer*, whereas others favored an essentially more random, unoriented manner. The lack of attention given to *Cornulites* taxonomy–substrate relations prevents any conclusion, however. Taylor and Brett (1996) described several *Cornulites* attached near the margin of a Silurian *Arctinurus* pygidium. Gabbott (1999) described some Late Ordovician South African cornulitids attached somewhat randomly to an orthocone. Vinn (2006) discussed cornulitids from the Caradoc of Baltica and the possibility of their presence in the Chinese Arenig. The *Cornulites* described by Morris and Felton (1993) from the Late Ordovician of the Cincinnati region are attached to the older whorls of gastropods which themselves are attached to the tegmens of crinoids. While describing some Ashgill-age Chinese cornulitids attached to brachiopods, Zhan and Vinn (2007) reviewed many aspects of cornulitid ecology. Manceñido and Gourvennec (2008, Figure 3) illustrated specimens of Middle Devonian *Spinocyrtia* with the tabulate coral *Aulocystis* oriented parallel to the plications of the brachiopod in the same manner as the cornulitids cited above (i.e., anteriorly open).

☐ *Reliability*

Category 1, frozen behavior.

*9Ap. DENDROID GRAPTOLITE SUBSTRATES

Bull (1987, Figure 7) made a good case that at least some dendroid graptolites used a hard substrate, even including a Silurian *Leptaena* "*rhomboidalis*" for attachment.

☐ *Reliability*

Category 1, frozen behavior.

*9Aq. PYGMAEOCONUS–HYOLITHID

In a very innovative paper, Horný (2006) carefully described the morphology of the tergomyan gastropod *Pygmaeoconus* attached to the dorsum of a hyolithid, as are some corals and bryozoans. This mollusk–hyolith relationship is apparently obligatory. In the same vein, Galle and Parsley (2005; see also Marek and Galle, 1976) detailed the obligate relationship between the hyolith dorsum and the attached tabulate coral *Hyostragulum* in the Devonian, as well as some bryozoan hyolith epibionts; the relationships have also been recognized in the Spanish and Moroccan Devonian (see Figure 163) and strongly support hyolithids as epibenthos resting on their ventral sides and oriented into currents with their opening. Galle and Plusquellec (2002) summarized what is currently known about the hyostragulids.

☐ *Reliability*

Category 1, frozen behavior.

*9Ar. MEIOFAUNA

In Section 6AIIg, Plankton Feeding, it is made clear that the fossil record of the plankton biota is very inadequate, even when considering such things as plankton feeders and their specialized morphologies. The aquatic meiofauna, in many ways analogous to the terrestrial soil biota, is also very, very poorly represented in the fossil record (for an account of the modern meiofauna, see Higgins and Thiel, 1988). One of the few records is commented on by Danielopol and Wouters (1992), who discussed Pokorny's (1989) discovery of some morphologically distinctive Cenomanian ostracod genera known today only from the meiofauna. In the Paleozoic, there is the possibility that the very minute arthropods present in

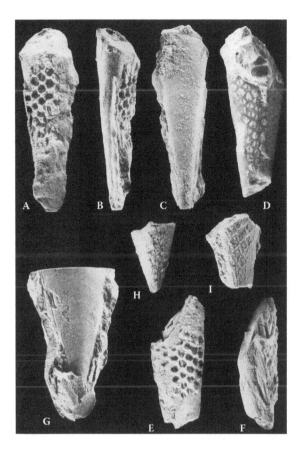

FIGURE 163 (A–F) *Hyostragulum* attached to the dorsum of "*Pterygotheca*" and (G) ventral side of hyolith (*Ottomarites*), all showing the relation of the coral to the dorsum of hyolithids from the Devonian of the Barrandian (all, ×1.5). (Figure 3 in Marek, L. and Galle, A., *Lethaia*, 9, 51–64, 1976. Reproduced with permission from Wiley-Blackwell.)

the Late Cambrian "Orsten" fauna of southern Sweden may be meiofaunal, but there is the other possibility that they were planktonic (Walossek and Müller, 1990).

SUMMARY

Mostly at the family level with a few genus level examples (Table 25).

9B. TERRESTRIAL

9Ba. BEETLE BORING IN WOOD

Elias (1994, p. 71) commented on the British elm decline coinciding with the presence of the elm bark beetle *Scolytus scolytus* in the mid-Holocene, while wondering if the cause might not have involved Neolithic activities. Although not pertaining to beetle borings, it is worth noting Labandeira and Phillip's (2002) account of various borings in Pennsylvanian *Psaronius* woods with frass-filled tunnels and gallery networks that are very impressive and presumably reflect insect activity. Although their example cannot be ascribed to beetle activities, Taylor and Scott's (1983) account and illustration

TABLE 25

Marine Organisms Employing Special Organic Substrates but Not Always with Biologic Interaction (Additions to Table 29 from Boucot, 1990a)

Behavioral Type	Taxonomic Level	Time Duration	Category	Refs.
9Ab. Clionid boring sponges	Family	Cambrian–Holocene	1	Vogel (1993)
9Ad. *Helicotaphrichnus*	Family	Eocene–Holocene	2B	Walker (1992)
9Af. Rock- and wood-boring bivalves				
Rock	Family	Carboniferous–Holocene	1	Kleemann (1983)
	Superfamily	Late Ordovician–Holocene	1	Pojeta and Palmer (1976)
Wood	Family	Early Cretaceous–Holocene	1	Kelly et al. (2000)
9Ag. Acrothoracican barnacles	Family	Middle Devonian–Holocene	2B	Baird et al. (1990)
9Ah. Acrothoracican barnacle–hermit crab shell	Family–superfamily	Miocene–Holocene	2B	Bałuk and Radwański (1991)
9Ak. Limpets and bone substrates	Family–class	Eocene–Holocene	2B	Marshall (1994)
9Al. Limpet depressions in ammonites	Family–subclass	Late Cretaceous	2B	Kase et al. (1994)
9An. Savazzi's "Leaning Tower of Pisa" morphology	Varied genera	Silurian–Holocene	1	Scrutton (1998)
9Aq. *Pygmaeoconus*–hyolithid and *Hyostragulum*	Genus–order	Ordovician, Devonian	1	Galle and Parsley (2005)
9Ar. Meiofauna	Class	Cretaceous–Holocene	2B	Danielopol and Wouters (1992)

of obvious frass in Pennsylvanian wood from Ohio suggest the presence of either insect or mite wood boring. Weaver et al. (1997) described some wood borings with frass from the Late Permian of Antarctica; the makers are unknown.

Labandeira (2002a, Figure 2.9) provided varying evidence of oribatid and insect borings from the Triassic and younger. Zhou and Zhang (1989) described Chinese Jurassic borings and frass suspected to be of coleopteran origin. Kierst and Wiesner (1975) briefly described a German Dogger (mid-Jurassic) complex boring in conifer wood that they ascribed to the Scolytidae (this requires confirmation). Krzeminska and Krzeminski (1992, Figure 112) illustrated a Baltic amber block having the impression of what appear to be scolytid borings. Labandeira et al. (2001) described a bark engraving by the scolytid *Dendroctonus* on a Middle Eocene *Larix*, a conifer which even today has a similar relation at the family level.

Rajchel and Uchman (1998) provided an excellent summary and account of wood boring by beetles, anobiids in particular, while describing a Polish Oligocene example, and they provided an extensive review of previously published occurrences. Freess (1991) discussed a Middle Oligocene example (assigned to *Anobium*) from near Leipzig. Poinar and Poinar (1999, Figure 58, p. 61) illustrated and discussed a platypodid borer preserved with a "plug" of frass, together ejected from its boring, presumably by resin flow. Grimaldi (1996, pp. 82, 105) illustrated a Dominican amber platypodid beetle with a sawdust plug that was presumably pushed out of one of its tunnels. Guo (1991) described an elegant scolytid boring from the Miocene Shanwan Formation of Shandong Province. Selmeier (1984) described examples from German Miocene wood of *Anobium* with the borings still filled with frass. Nel (1994) described wood borings from the Late Miocene of Uganda that he ascribed to beetle activities, as well as the alleged larva of a wood-boring beetle in wood. Claus (1958) described a late Pleistocene scolytid-type boring from Thuringia.

☐ *Reliability*

Category 2B, functional morphology.

9Bb.　TERMITE BORINGS IN WOOD

Noirot and Noirot-Timothée (1969) outlined the anatomical features in the termite anal region that give rise to the hexagonal cross-section of their frass pellets. Rozefelds (1990; see also Rozefelds and de Baar, 1991) provided an excellent example of Kalotermitidae borings and tunnels, with accompanying frass, from the mid-Tertiary rainforest of central Queensland. Although not involving any actual borings Collinson's (1990, Figure 46b) illustration of a Late Eocene termite bit of frass from southern England is worth noting. Francis and Harland (2006) described an excellent example of termite borings from the Early Cretaceous of the Isle of Wight.

One of the two oldest known evidences of termite social behavior known is the Maastrichtian, Lambert Formation drywood termite nest fragment from Vancouver Island, complete with frass (Ludvigsen, 1993; Ludvigsen and Beard, 1994, Figure 92; Ludvigsen and Beard, 1997, Figure 104). Poinar (1998a, Figure 2B) illustrated a drywood termite (Kalotermitidae) actually excreting several drywood fecal pellets, presumably under the stress of being caught in the tree resin.

Genise (1995) described what he interpreted as termite nests, with frass, in cycads from the Patagonian Late Cretaceous. Solórzano Kraemer (2007, Plate 2, Figure A) illustrated a Mexican Miocene amber *Calcaritermes* associated with pellets resembling termite frass.

☐ *Reliability*

Category 2B, based on functional morphology.

TABLE 26
Terrestrial Organisms Employing Special Organic Substrates, but Not Always with Biologic Interaction (Additions to Table 30 from Boucot, 1990a)

Behavioral Type	Taxonomic Level	Time Duration	Category	Refs.
9Ba. Beetle boring in wood	Family–genus	Jurassic–Holocene	2B	Freess (1991); Zhou and Zhang (1989)
9Bb. Termite borings in wood	Family	Maastrichtian–Holocene	2B	Ludvigsen and Beard (1997)
9Bd. Wood-boring bees	Order	Oligocene	2B	Rajchel and Uchman (1998)
9Bf. Wood-boring mites	Class	Pennsylvanian–Recent	2B	Labandeira et al. (1997a)

9Bd.   WOOD-BORING BEES

Rajchel and Uchman (1998) described Polish Oligocene wood borings similar to those of the larvae of the wasp *Sirex*.

☐ *Reliability*
Category 2B, functional morphology.

*9Bf.   WOOD-BORING MITES

Labandeira et al. (1997) discussed and illustrated mite borings with frass in Pennsylvanian wood. Kellogg and Taylor (2004) illustrated alleged Antarctic Late Paleozoic and Mesozoic mite-bored wood and reviewed the entire question about whether the borings and the contained coprolites can reliably be ascribed to mite activity. Goth and Wilde (1992) described some Early Permian German wood borings that suggest mite activities.

☐ *Reliability*
Category 2B, functional morphology.

SUMMARY

Some family and genus level evidence, but most is at a higher taxonomic level (Table 26).

10 Sexual Behavior

Sexual dimorphism correlates with a number of behavioral attributes, some of which can be recognized in the fossil record, as noted below for placental mammals (Kurten, 1969), and even for some fossil reptiles (Olson, 1969). Sullivan et al. (2003) invoked sexual dimorphism to explain some aspects of the dentition of a South African Permian cynodont (*Diictodon*). Størmer (1969) described dimorphism, presumably sexual, in eurypterids and pointed out that it is unclear which dimorphs are male as contrasted with female. The potential for sexual dimorphism in ammonoid cephalopods has been recognized since the 19th century (see Davis et al., 1969; other papers in Westermann, 1969; Lehmann, 1966; for another review with good illustrations of varied forms, including putative females with lappets, see Makowski, 1962). Etter (2004) briefly reviewed sexual dimorphism in tanaidacean crustaceans. Sexual dimorphism is widespread among varied decapods (Feldmann, 1998, 2003; Feldmann and de Saint Laurent, 2002; Schweitzer, 2003; Schweitzer Hopkins and Feldmann, 1997). Feldmann (1998, Figure 3) provided truly elegant views of male, female, and castrated male ventral abdominal development in some New Zealand Miocene crabs (see also Section 4AIo, Rhizocephalan–Decapod Parasitism). Sexual dimorphism is, of course, widespread among varied spider and insect taxa preserved in amber. Stubblefield and Seger (1994) commented extensively on varied evidence for sexual dimorphism in different hymenopteran groups.

With placental mammals, there is a strong tendency for the males of polygenous taxa to have larger canines and larger body size than do the monogamous taxa, as emphasized in the following citations. Lincoln (1994) reviewed the sexual dimorphism present in many ungulate groups that involves various antler types that are shed annually, as well as the probability of ancestrally primitive large canines in the males; most of the antler bearing involves males only and is used to obtain access to females, with increase in antler size being correlated with polygyny. Coombs (1975) discussed the evidence for sexual dimorphism in chalicotheres, a group of perissodactyls. Martin et al. (1994) provided an excellent account of sexual dimorphism, concluding that it is prominent in Old World monkeys, apes, and humans; body size and canine size are prominently involved. There is little reliable evidence from the fossil record, however, because of the problem of distinguishing specific differences from sexual differences, but Fleagle et al. (1980) made a good point about canine-based Fayum Oligocene dimorphism in early anthropoids representing sexual dimorphism which also correlates today with polygyny. Poole (1994) discussed body size and tusk size differences between male and female African elephants. Gittleman and Van Valkenburgh (1997) reviewed information on dimorphism in modern carnivores and noted the many differences in canine size, with those of males being significantly larger than females, and Van Valkenburgh and Sacco (2002) used Rancho La Brea canine data to make a good case for neither the dire wolf nor *Smilodon* having been polygynous. O'Leary et al. (2000) indicated the presence of sexual dimorphism in two specimens of the mesonychid mammal *Ankalagon* from the Paleocene of New Mexico. McHenry (1991) noted the presence of sexual dimorphism in *Australopithecus afarensis*, and Krishtalka et al. (1990) reported evidence from canine size indicating the same thing for Eocene *Notharctus* (i.e., polygyny as a good possibility). Kurten (1969) discussed size differences in Pleistocene bears (see Figure 164 and Figure 165), and Grandal-d'Anglade and López-González (2005) provided morphometric data on their skulls.

10A. TERRESTRIAL

10AI. PLANTS

10AIa. Reproduction in Lower Plants

A eupolypod fern, *Cretacifilix fungiformis* Poinar and Buckley, 2008, in Burmese amber, shows that "modern" ferns already existed in the Early Cretaceous (see color plate 2C after page 132). Assignment to the eupolypodes is by the presence of a vertical annulus that partially encircles the spore case (Figure 166). The thick indusial, with cells filled with bacilliform bodies, that covers and protects the

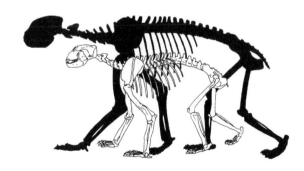

FIGURE 164 Kurten (1969) noted size differences in male and female Pleistocene bears. Figure compares the North American *Tremarctos floridanus*. (Figure 1 in Kurten, B., in *Sexual Dimorphism in Fossil Metazoa and Taxonomic Implications*, G.E.G. Westermann, Ed., Schweizerbart, Stuttgart, 1969, pp. 226–227. http://www.schweizerbart.de. Reproduced with permission.)

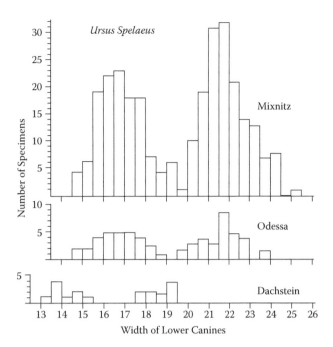

FIGURE 165   Kurten (1969) noted size differences in male and female Pleistocene bears. Figure indicates the differences between samples of Pleistocene cave bears (*Ursus spelaeus*) from Mixnitz and Odessa, together with a dwarf sample from Dachstein. (Figure 4 in Kurten, B., in *Sexual Dimorphism in Fossil Metazoa and Taxonomic Implications*, G.E.G. Westermann, Ed., Schweizerbart, Stuttgart, 1969, pp. 226–227. http://www.schweizerbart.de. Reproduced with permission.)

sporangia (Figure 167) is also unusual. The deeply impressed sori may indicate periods of aridity, as this character is also possessed by the extant "resurrection fern," which becomes dormant during dry periods. There are many types of vegetative reproduction. Gemmae are bud-like cells or cell clusters, often resembling a miniature plant, borne on the parent plant (the gametophyte in the case of ferns, mosses, and liverworts). Gemmae are capable of reproducing vegetatively after disengaging from the parent plants; they indicate the presence of a dry climatic cycle. Fossil examples of gemmae

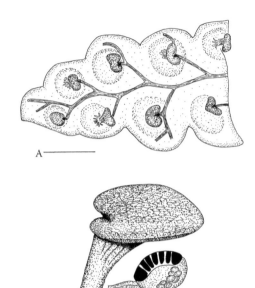

FIGURE 167   (A) A pinnule of *Cretacifilix fungiformis* with veins and indusia; scale bar = 0.6 mm. (B) Partial reconstruction of a stalked indusium and two sporangia, one of which contains eight spores, of *Cretacifilix fungiformis*; length of sporangium = 160 μm.

on a moss and liverwort in Dominican amber are shown in Figure 168 and Figure 169.

☐ *Reliability*

Category 2B, functional morphology.

10AII.   INVERTEBRATES

10AIIa.   Spider Sperm Pumps and Copulation

Menge (1856, p. 7) mentioned, but did not illustrate, a Baltic amber theridian spider pair: "*wahrscheinlich wahrend der begattung begraben wurden*." Wunderlich (2004, pp. 694–695) further described copulatory materials for oonopids.

☐ *Reliability*

Category 1, frozen behavior.

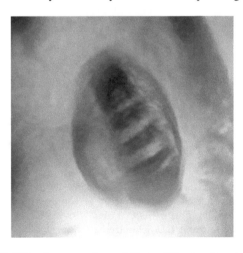

FIGURE 166   A sporangium of *Cretacifilix fungiformis* with a short vertical annulus (Deniz Eren amber collection); length of sporangium = 160 μm.

FIGURE 168   Gemmae on the leaves of the liverwort *Bryopteris* (Lejeuneaceae) indicating that a dry cycle occurred in the original Dominican amber forest (Poinar amber collection accession no. B-1-19); length of leaf = 2 mm.

FIGURE 169　Gemmae on moss leaves (Poinar amber collection accession no. B-1-20); length of leaf = 1 mm.

10AIIb.　Mating Insects

Fossilized mating insects are not common but do occur from time to time, especially in amber when the pair was rapidly preserved before separation could occur. Although it is likely that mating behavior remained fairly constant in select groups over the eons, comparisons between fossil and extant mating patterns can be difficult to make because a considerable amount of flexibility in mating positions can occur within insect species (Gwynne, 2003).

Mating behavior of insects includes stages during which the pair can demonstrate three separate positions. The first involves pair formation, where the male grasps the female; the second is assumed during copulation when sperm is transferred to the female; and the third occurs during the process of disengaging when the union between the pair is broken. Whether the fossil insects were preserved in the grasping, copulation, or disengaging positions can only be ascertained after these positions have been determined in extant members of the same genus or family. Many possible positions exist during the copulation period, when the posterior ends of the insects are conjoined end to end. The male can be on top or under the female (Figure 170), both sexes can be facing in the same direction or in opposite directions (see Figure 171 and Figure 172), one sex can have its venter oriented to the substrate while the other has its dorsum facing the substrate (upside down), or the male can mate while remaining at the side of the female.

The most common fossilized mating insects are those that tend to have a long copulation period and initiate mating in flight, such as midges (Chironomidae and Ceratopogonidae), scavenger flies (Scatopsidae), and fungus gnats (Sciaridae). Most scavenger flies mate end to end, face in opposite directions, and have the same body orientation (ventral surfaces in the same direction; see Figure 173). This position is commonly seen in mating specimens in amber, such as gall gnats (Figure 174), biting midges (Figure 175 and Figure 176), chironomid midges (Figures 177 through 179), long-legged

FIGURE 170　Mating marine water striders (Hemiptera: Veliidae) in Dominican amber (Poinar amber collection accession no. HE-4-28); length of female = 1.5 mm. Arrow shows male mounted on female (head and part of thorax of male are missing).

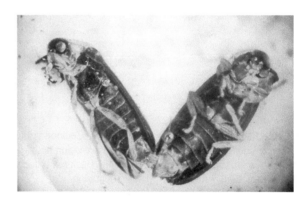

FIGURE 171　Mating lightning bugs (Coleoptera: Lampyridae) in Dominican amber (Morone amber collection, Torino); length of each individual is about 8 mm. Each sex is facing in opposite directions with the same body orientation.

FIGURE 172　Mating platypodid beetles (Coleoptera: Platypodidae) in Dominican amber (Poinar amber collection accession no. Sy-1-100); length of each beetle is 4 mm. Each sex is facing in opposite directions with the same body orientation.

FIGURE 173   Mating scavenger flies (Diptera: Scatopsidae) in Dominican amber (Poinar amber collection accession no. D-7-36); length of larger specimen = 1 mm. Specimens are facing in opposite directions with same body orientation.

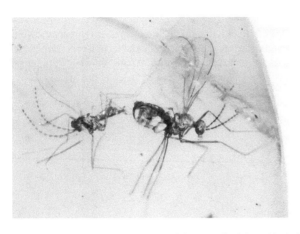

FIGURE 174   Mating gall gnats (Diptera: Cecidomyidae) in Dominican amber (Poinar amber collection accession no. D-7-72); length of female (larger specimen) = 1.2 mm. Each sex is facing in opposite directions with the same body orientation.

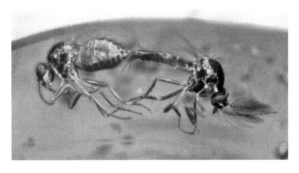

FIGURE 175   Mating biting midges (Diptera: Ceratopogonidae) in Dominican amber (Poinar amber collection accession no. D-7-146A); length of larger specimen = 1.2 mm. Each sex is facing in opposite directions with the same body orientation.

flies (Figure 180), bethylid wasps (Figure 181), and fungus gnats (Figure 182 and Figure 183). However, one sex may be oriented differently from the other with the ventral surface twisted 45 degrees from the other, as in some chironomid midges (Figure 184) and scavenger flies (Figure 185) (Wu, 1996, Figure 492). Or the body orientation of one sex may be

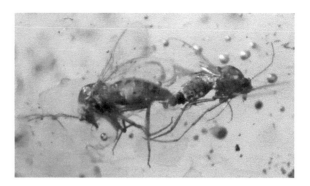

FIGURE 176   Intersex male mating with a normal female biting midge in Baltic amber (Andrew Cholewinski amber collection, accession no. COP-022); length of larger specimen = 1 mm. Each sex is facing in opposite directions with the same body orientation.

FIGURE 177   Mating midges (Diptera: Chironomidae) in Baltic amber (Poinar amber collection accession no. D-7-137A); length of specimens is about 1 mm. Each sex is facing in opposite directions with the same body orientation.

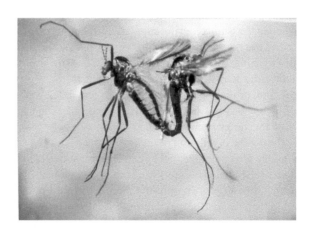

FIGURE 178   Mating midges (Diptera: Chironomidae) in Dominican amber (Poinar amber collection accession no. D-7-137B); length of larger specimen = 2 mm. Each sex is facing in opposite directions with the same body orientation.

completely reversed from that of the opposite sex as in some pairs of biting midges (Figure 186) and fungus gnats (Figure 187). It is difficult to say whether these differences are the result of entombment artifact, variability of mating patterns within genera, or distinct mating differences.

An unusual case of two male midges attempting to mate is shown in Figure 188. When a female midge enters

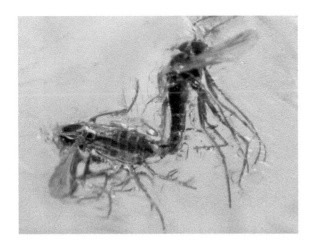

FIGURE 179 Mating midges (Diptera: Chironomidae) in Baltic amber (Andrew Cholewinski amber collection, accession no. COP-001); length of specimens = 1 mm. Each sex is facing in opposite directions with the same body orientation.

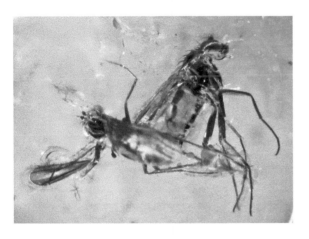

FIGURE 180 Mating long-legged flies (Diptera: Dolichopodidae) in Baltic amber (Andrew Cholewinski amber collection, accession no. COP-014); length of larger specimen = 3.2 mm.

FIGURE 181 Mating bethylid wasps (Hymenoptera: Bethylidae) in Dominican amber (Poinar amber collection accession no. H-10-151); length of larger specimen = 2 mm.

FIGURE 182 Mating fungus gnats (Diptera: Sciaridae) in Baltic amber (Andrew Cholewinski amber collection, accession no. COP-002); length of specimens = 2 mm. Specimens are facing in opposite directions with same body orientation.

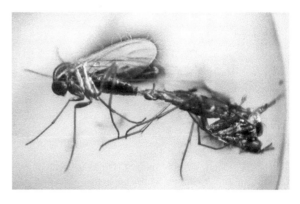

FIGURE 183 Mating fungus gnats (Diptera: Sciaridae) in Baltic amber (Andrew Cholewinski amber collection, accession no. COP-016); length of specimens = 2 mm. Specimens are facing in opposite directions and with different body orientations.

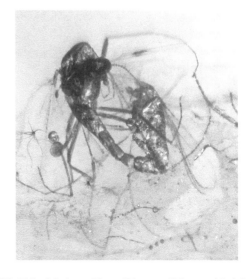

FIGURE 184 Mating midges (Diptera: Chironomidae) in Baltic amber (Andrew Cholewinski amber collection, accession no. COP-018); length of larger specimen is 1.6 mm. The specimens are facing each other.

a male mating swarm, males sometimes become confused and attempt to mate with each other. An intersex male biting midge mating with a normal female is shown in Figure

FIGURE 185 Mating scavenger flies (Diptera: Scatopsidae) in Baltic amber (Andrew Cholewinski amber collection, accession no. COP-004). Specimens are facing in opposite directions with reverse body orientation.

FIGURE 186 Mating biting midges (Diptera: Ceratopogonidae) in Dominican amber (Poinar amber collection accession no. D-7-146B); length of specimens = 1.5 mm. Each sex is facing in opposite directions but with their body orientation reversed.

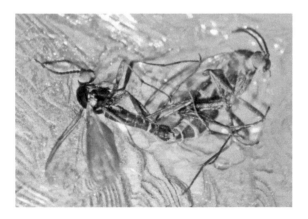

FIGURE 187 Mating fungus gnats (Diptera: Sciaridae) in Baltic amber (Andrew Cholewinski amber collection, accession no. COP-026); length of specimens = 2 mm. Specimens are facing in opposite directions with reverse body orientation.

176. Such intersexes today are often the result of parasitism, especially by mermithid nematodes. These male intersexes have female-like antennae and mouthparts but male genitalia (Poinar and Monteys, 2008). Males of most strepsipterans (Figure 189) mate with the female extruding from her host.

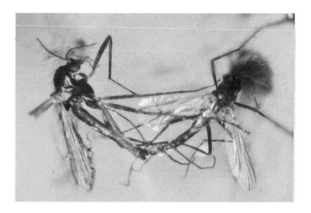

FIGURE 188 Two male midges (Diptera: Chironomidae) attempting to mate in Baltic amber (Andrew Cholewinski amber collection, accession no. COP-023); length of specimens = 1.2 mm.

FIGURE 189 Sexual dimorphism in the parasitic twisted-wing insects (Strepsiptera) (Poinar amber collection accession no. St-1-4); length of specimen = 1.3 mm. The males are winged, as shown here in this Dominican amber specimen, but the females are wingless and most never leave their host insect. Mating thus occurs when the male alights on the host harboring the female.

Side-by-Side Mating

One type of behavior that would appear to be impossible to discover from fossil evidence is known as "side-by-side mating." This type of behavior is practiced by some insects, especially members of the Order Hemiptera, and it probably evolved in insect groups confined to small, flattened habitats where normal top–bottom mating would be impractical (Schuh and Slater, 1995). The palm bug, *Paleodoris lattini* (shown in Figure 1), belongs to a family (Thaumastocoridae) whose extant members practice side-by-side mating. Males have an asymmetrical genital capsule that attaches to the symmetrical genital area of the female. The fossil shown in Figure 1 is a female with symmetrical terminal abdominal segments; the flattened body and legs indicate a concealed life style (presumably it lived between unopened palm leaves as does its closest extant descendant), and it probably practiced side-by-side mating.

In the Hemipteran infraorder Leptopodomorpha side-by-side mating occurs in all included families, and there is often a modification on the wing of the female that is grasped by

TABLE 27

Instances of Fossilized Mating Insects (Baltic, Dominican, Bitterfeld, and Lebanese Citations Refer to Amber Deposits)

Order/Family	Source	Refs.
Coleoptera		
Cantharidae	Baltic	Wichard and Weitschat (2004)
Lampyridae	Dominican	Poinar and Poinar (1999, Figure 135); same mating pair is depicted in Figure 171
Platypodidae	Dominican	Present work, Figure 172
Undetermined	Limestone deposits	Andrée (1937, p. 78)
Diptera		
Bibionidae	Lacustrine deposits	Fujiyama and Iwai (1974); see also Boucot (1990a, Figure 319, p. 389)
Cecidomyiidae	Dominican	Poinar (1993, Figure 4); same mating pair is depicted in Figure 174
Ceratopogonidae	Baltic	Ross (1998, Figure 92); Wichard and Weitschat (2004, Figure 3)
	Dominican	Present work, Figures 175 and 186
Chironomidae	Baltic	Dahlström and Brost (1996, p. 29); Kobbert (2005, p. 142); Kosmowska-Ceranowicz and Konart (2005, p. 208); Ross (1998, Figure 138, p. 58); Wichard and Weitschat (2004, p. 24); present work, Figures 177–179 and 184
	Limestone deposits	Heer (1850, Plate XIV, Figures 13, 13a, p. 190); Müller (1957, Figure 141; 1979, Figure 7)
	Florissant	Abel (1935, Figure 3)
	Lebanon	Whalley (pers. comm., in Boucot, 1990a, p. 385)
	Dominican	Present work, Figure 178
Dolichopodidae	Baltic	Arillo (2007, Figure 1B, p. 163); Kobbert (2005, p. 149); Krumbiegel and Krumbiegel, 2005, p. 92); Weitschat and Wichard (1998, Figure 86e); Wichard and Weitschat (2004, p. 24); present work, Figure 180
Limoniidae	Baltic	Janzen (2002, Figure 346, p. 141); Kosmowska-Ceranowisz (2001, Figure 85)
	Dominican	Grimaldi (1996, p. 86)
Lygistorrhinidae	Baltic	Scheven (2004, p. 45)
Mycetophilidae	Baltic	Arillo (2007, Figure 1A, p. 163); Janzen (2002, Figure 349, p. 141)
	Dominican	Montgomery de Merette (1984, p. 38)
Phoridae	Baltic	Weitschat and Wichard (1998, Plates 87g,h, p. 213)
Scatopsidae	Dominican	Cook (1990a, Figure 321, p. 390); Grimaldi and Engel (2005, Figure 12.32); Montgomery de Merette (1984, p. 38); Ross (1998, Figure 63, p. 23); Wu (1996, Figures 4 and 493, p. 196); present work, Figure 173
	Baltic	Present work, Figure 185
Sciaridae	Baltic	Present work, Figures 182, 183, and 187
Hemiptera		
Cicadellidae	Dominican	Grimaldi and Engel (2005, Figure 2.27, p. 57)
Cicadoidea	Baltic	Andrée (1937, p. 78)
Pityococcidae	Bitterfeld	Koteja (1998, Figure 13, p. 209)
Fulgoroidea	Limestone deposits	Heer (1850, Plate 13, Figure 10)
Gerridae	Dominican	Andersen and Poinar (1992, Figures 1 to 4, p. 257); present work, Figure 222
Velidae	Dominican	Andersen and Poinar (1998, Figure 1, p. 3); present work, Figure 170
Hymenoptera		
Bethylidae	Dominican	Present work, Figure 181.
Formicidae	Baltic	Heer (1865, p. 386); Hölldobler (1976, Figure 119, p. 125); Hölldobler and Haskins (1977)
Platygasteridae	Dominican	Wu (1996, Figure 565, p. 209)
Undetermined	Lake deposits	CoBabe et al. (2002)

a special apparatus of the male during copulation (Schuh and Slater, 1995). The front cover, Figure D, shows a female *Palaeoleptus burmanicus* (Poinar and Buckley, 2009) in Burmese amber that has a disc-like modification at the base of the wing margin, as well as an asymmetrical genital segment, indicating that it probably practiced side-by-side mating. While *Paleodoris* probably lived in a confined habitat that necessitated side-by-side mating, the leptopodomorph is not especially flattened, and it is unlikely that its lifestyle was restricted to narrow spaces. This is also the case for many of the extant Leptopodomorpha and even some extant members of the Thaumastocoridae. Although they practice side-by-side mating, they live in fairly open spaces. Apparently, side-by-side mating, along with its morphological modifications, was established early in the basal lineages and then continued after the habitat changed and this type of behavior was no longer essential for survival. Reports of fossilized mating insects, most of which occur in amber, are presented in the Table 27.

□ *Reliability*

Category 1, frozen behavior.

Eggs, Oviposition, and Maternal Care in Amber

by George O. Poinar, Jr.

See below the many arthropod egg categories into which some of Poinar's examples fit comfortably.

Introduction

Fossil evidence of eggs, oviposition, and maternal care are rare, and two examples in amber were presented previously (Boucot, 1990a). Here are presented additional examples in amber, along with notes on similar behaviors found in extant members of the same taxonomic groups. The items covered are eggs, egg cases, oviposition, hatching, and maternal care.

Eggs and Egg Cases

This category includes eggs and egg cases that are still within their parent or are solitary in amber. Examples include:

1. Figure 190 shows the Baltic amber terrestrial turbellarian *Micropalaeosoma balticus* (Platyhelminthes: Typhloplanoida) containing numerous eggs within its body cavity. The eggs represent thin-shelled, subitaneous types that are self-fertilized and develop into juveniles that rupture the parent's body to exit (Poinar, 2003c).

2. Populations of the mycetophagous nematode *Oligaphelenchoides atrebora* (Aphelenchidae: Aphelenchoididae) in Mexican amber consisted of adults, juveniles, and eggs (Figure 191), as well as an abundant supply of its probable food source, fungal mycelium (Poinar, 1977). Extant members of the genus *Aphelenchoides* deposit their eggs in the feeding environment, and hatching is dependent on temperature. The eggs are usually deposited individually and take only a few days to hatch.

3. A spider's (Arachnida: Araneae) triangular egg sac in Baltic amber was attached to a web by silk strands (Figure 192). The type of spider that constructed this interesting egg case has not been identified but is probably a member of the Theridiidae. From the thick, leathery walls and solid construction of the sac, the eggs would probably be protected for a long

period. Such durable egg cases can be left in exposed areas in contrast to the more friable ones that have to be hidden or camouflaged with various materials. The hole in the center of the sac may be an exit hole; if so, the spiderlings would have already left the sac. Other spider egg sacs in amber are illustrated by Wunderlich (2004).

4. The large solitary egg in Dominican amber shown in Figure 193 has characteristics of present-day walking stick or phasmid eggs (Phasmidia). Walking sticks drop their eggs as they feed in the canopy. Their sturdy construction protects them when they strike limbs and other objects as they descend to the forest floor. Obviously, this one fell into a resin pool. These eggs are built to last for long periods; it may be months before the small cap opens and the young stick insect makes its way back up the host plant.

5. The eggs shown in Figure 194 are attached to the edge of a petal of *Hymenaea protera* in Dominican amber; they belonged to a lepidopteran, probably a moth, whose larvae developed on the floral and vegetative parts of this extinct resin producer. It is impossible to say whether this insect was polyphagous (i.e., the larvae could develop on a number of different plants in the amber forest) or monophagous and thus restricted to this particular plant species.

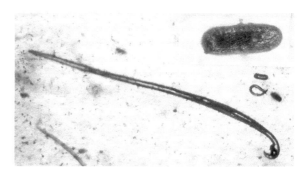

FIGURE 191 Eggs of the nematode *Oligaphelenchoides atrebora* in Mexican amber (University of California–Berkeley Museum of Paleontology, block no. B-7053-4); length of egg = 62 µm.

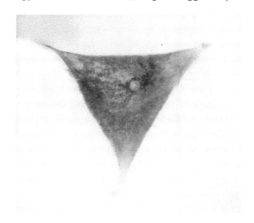

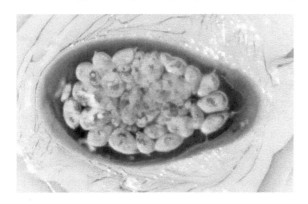

FIGURE 190 The turbellarian *Micropalaeosoma balticus* in Baltic amber (Poinar amber collection accession no. PT-3-1); length of body = 1.54 mm.

FIGURE 192 Triangular egg case of spider (Theridiidae?) in Baltic amber (Poinar amber collection accession no. Sy-1-140); each side of case = 3 mm.

FIGURE 193 Phasmid egg in Dominican amber (Poinar amber collection accession no. O-2-13); length of egg = 5 mm.

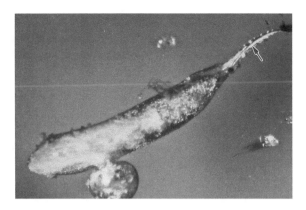

FIGURE 195 Possible muscid egg with respiratory tube (arrow) in Dominican amber (Poinar amber collection accession no. D-7-177); length of egg = 3 mm.

FIGURE 194 Moth(?) eggs (arrows) on *Hymenaea* petal in Dominican amber (Poinar amber collection accession no. Sy-1-179); length of eggs = 0.8 mm.

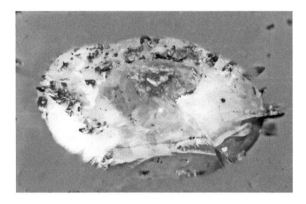

FIGURE 196 Reptilian egg in Mexican amber (Poinar amber collection accession no. R-3-13); length of egg = 16 mm.

6. Figure 195 shows an egg with a respiratory horn, which is typical of various flies, in Dominican amber. Such eggs are deposited in rotting vegetation, mature fruit, animal dung, or other soft, moist materials. The respiratory horn always protrudes out of the substrate to obtain an unrestricted flow of air.

7. A reptilian egg in Mexican amber (Figure 196) shows a hatch hole on one side and probably belonged to a lizard. The size and shape of this egg are similar to those of present-day anoles, and these iguanids have been captured in both Mexican and Dominican amber. Anoles will deposit their leathery shelled eggs on tree forks, and the young emerge through a hole similar to the one on the side of this egg. The eggshell could have been dislodged during the hatching process and fallen into a pool of resin.

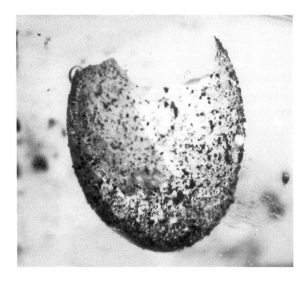

FIGURE 197 Bird eggshell in Dominican amber (Jim Work amber collection, Ashland, Oregon); length of eggshell = 7 mm.

8. A diminutive bird eggshell in Dominican amber probably came from a hummingbird (Trochilidae) (Figure 197). The calcareous nature of the shell and the type of breakage identify it as a bird egg and that a chick hatched from it. The parent bird could have

ejected the shell from the nest, a predator could have knocked the shell out as it attacked the hatchling, or some type of disturbance could have shaken the shell out of the nest (Poinar et al., 2007c).

Oviposition and Hatching

This category includes eggs that are in the process of being deposited, have just been laid, or are in the process of hatching. Although it is not a common occurrence, the question arises why some females oviposit immediately after becoming entrapped in resin. Some possible explanations for this behavior include the following: (1) the desire to reproduce is so strong that when females are dying they attempt to release their remaining eggs; (2) the trauma of falling into the resin produces a type of involuntary oviposition, possibly resulting from relaxation of the sphincter muscles that normally close the opening of the ovipositor; (3) the pressure from the resin physically forces out any eggs that were in the oviduct; or (4) the insect was in the process of laying eggs when it was engulfed by a resin flow or was blown into an adjacent pool of resin. This category includes the following examples:

1. Figure 198 shows three nymphiparous uropodid or tortoise mites (Arachnida: Acari) hatching from eggs in Mexican amber. Mite hatchlings generally are six-legged larvae, but occasionally a molt occurs within the egg and eight-legged nymphs emerge, as in this case. Tortoise mites live in habitats rich in organic matter, feed on microscopic plants and animals, and eventually molt to the adult stage.

2. A springtail or collembolan, *Sminthurus longicornus* (Collembola), deposited an egg in Baltic amber, and a second egg behind the female had already been laid (Figure 199). The female is covered by a cuticle, showing that it was in the process of molting at the time it was captured. Extant springtails continue to molt after reaching sexual maturity. The rough debris on the fossil eggs is probably fecal material

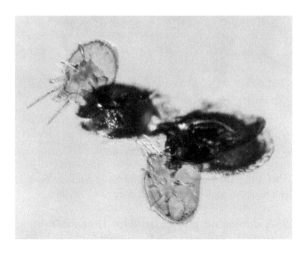

FIGURE 198   Nymphiparous uropodid mites hatching in Mexican amber (Poinar amber collection accession no. Ac-10-20); length of mites = 350 µm.

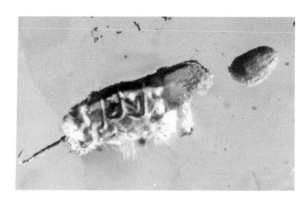

FIGURE 199   A springtail, *Sminthurus longicornis*, depositing an egg in Baltic amber (Poinar amber collection accession no. CL-1-4); length of springtail = 1.8 mm. A second egg behind the springtail was laid previously.

FIGURE 200   A cockroach ootheca in Dominican amber (Poinar amber collection accession no. O-2-18A); length of ootheca = 6 mm.

voided through the anus at the time of oviposition, a behavior employed by extant species of *Sminthurus*. Apparently, this deposit protects the eggs from desiccation and fungal parasites (Poinar, 2000b).

3. Cockroaches produce their eggs in an egg capsule called an *ootheca*. The ootheca is formed when secretions produced by the cells of the oviduct glue together a group of eggs. The eggs are arranged in symmetrical double rows and protected by the hardened outer surface of the capsule. The roach can deposit the ootheca just after formation (Figure 200), and the eggs will develop on their own, or, as in this case, the adult carries the ootheca until the eggs are ready to hatch (Figure 201), thus demonstrating maternal behavior. The ootheca has a thick wall that protects the developing embryos. When the eggs are ready to hatch, the two halves of the ootheca partially separate, allowing the young to escape. An Early Cretaceous ootheca in Burmese amber is shown in Figure 202 (see also Figure 200).

4. A biting midge or ceratopogonid (Diptera: Ceratopogonidae) that deposited a series of eggs in Dominican amber is shown in Figure 203. Extant midges deposit their eggs in numerous locations depending on their life history. Often the preferred

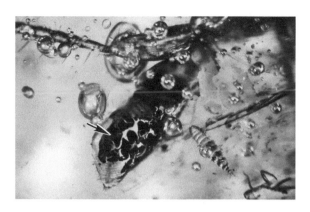

FIGURE 201 Cockroach with ootheca (arrow) still attached to end of abdomen (Poinar amber collection accession no. O-2-18B); length of ootheca = 4 mm.

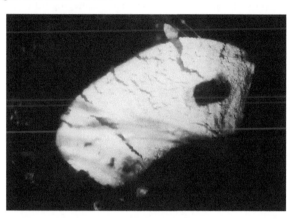

FIGURE 202 An Early Cretaceous cockroach ootheca in Burmese amber (Poinar amber collection accession no. B-OR-8); length of ootheca = 9 mm.

FIGURE 203 Biting midge (Diptera: Ceratopogonidae) laying eggs in Dominican amber (Poinar amber collection accession no. S-1-61); length of biting midge = 1.3 mm.

FIGURE 204 A coccid (Hemiptera) with attached egg sac (Poinar amber collection accession no. HO-4-17); length of adult = 3 mm. Arrow points to young emergent from brood sac.

FIGURE 205 Eggs laid by a moth captured in Dominican amber (Poinar amber collection accession no. S-1-95); length of moth = 3.5 mm.

habitat is moist sand or soil; however, some of the mammalian feeders are known to deposit eggs in the dung of their victims. These eggs have little protection and would have hatched shortly after deposition.

5. In Figure 204, a female coccid (Hemipera) in Dominican amber has its egg sac attached to the posterior portion of its body. The adults carry the eggs along with them while they feed. Within the sac are eggs in various stages of development.

Hatching often occurs sequentially within the sac, and the young are released one by one. A recently liberated larva can be seen in the amber behind the egg sac (arrow).

6. A moth (Lepidoptera) that laid a group of eggs in Dominican amber is shown in Figure 205. These eggs are similar to those shown on the petal of the Dominican amber tree (see Figure 194), although there is no way of knowing whether they belong to the same species. Normally, the adult deposits her eggs on a plant that supports the development of the caterpillars, but occasionally a mistake is made, and then it is a test for the young to develop on an unnatural host or die.

7. A crane fly (Diptera: Tipulidae) that deposited a clump of eggs in Dominican amber is shown in Figure 206. Crane flies deposit their eggs in various habitats—soil, rotting wood, mud along water sources, and even semi-aquatic habitats. The eggs are unprotected and usually hatch soon after being deposited. The larvae are long lived and can take up to a year to complete their development while feeding on living or decomposing vegetation.

8. Figure 207 shows a sand fly or phlebotomine fly (Diptera: Psychodidae) that just deposited a group of eggs in Dominican amber. According to Dr. Jake

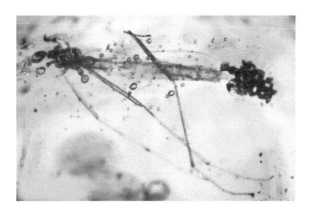

FIGURE 206   A cranefly (Diptera: Tipulidae) with deposited eggs in Dominican amber (McAuley collection, Windsor, VT); length of egg = 0.5 mm.

FIGURE 207   A sandfly (phlebotomine) with deposited eggs in Dominican amber (Poinar amber collection accession no. S-1-95); length of sandfly = 1.3 mm.

Jacobson (Department of Parasitology, The Hebrew University–Hadassah Medical School, Jerusalem), the fossil eggs are similar in shape and size to those of both Old and New World sand flies. The white glistening and gently curved appearance of the freshly laid eggs and the darkening that comes later are typical of phlebotomine eggs. Female sand flies die after laying their last batch of eggs, and, according to Dr. Jacobson, the fossil scene is identical to what occurs in breeding pots. All female sand flies feed on vertebrate blood, and the ovipositional behavior and larval habitats of most species are unknown. Apparently the larvae develop on fungi and other organisms in decomposing organic matter (Poinar et al., 2006a). The eggs are unprotected and probably hatch soon after being deposited.

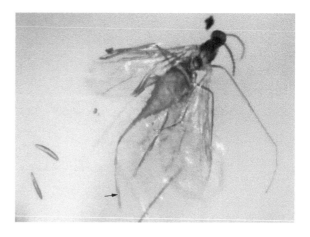

FIGURE 208   A gall midge (Diptera: Cecidomyiidae) with several deposited eggs and one (arrow) that is emerging from her ovipositor, in Dominican amber (Poinar amber collection accession no. S-1-95); length of fly = 2 mm.

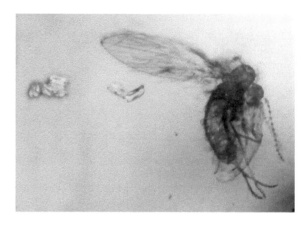

FIGURE 209   A moth fly (Diptera: Psychodidae) with deposited eggs in Dominican amber (Poinar amber collection accession no. S-1-95); length of fly = 2 mm.

9. A gall midge (Diptera: Cecidomyiidae) with several eggs already laid and another one in the process of coming out of the elongate ovipositor is preserved in Dominican amber (Figure 208). Note how much shorter the deposited eggs are in contrast to the elongate one that is emerging from the ovipositor (arrow). Insects with long, narrow ovipositors as seen here (and especially in the parasitic Hymenoptera with their narrow sclerotized ovipositors) can only pass eggs that can be compressed enough to squeeze through the narrow lumen of the ovipositor. The eggs resume their original shape after deposition, which explains why the two free eggs are shorter and thicker than the egg emerging from the ovipositor in the fossil.

10. Figure 209 shows a moth fly (Diptera: Psychodidae) that laid a mass of eggs in Dominican amber. Moth flies are scavengers and normally deposit their eggs on decomposing organic matter, often under the bark of damaged trees, which explains how their

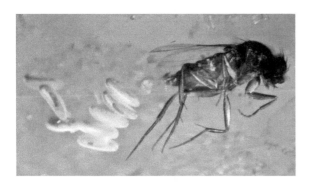

FIGURE 210   A scuttle fly (Diptera: Phoridae) with eight deposited eggs in Dominican amber (Poinar amber collection accession no. S-1-95); length of fly = 2 mm.

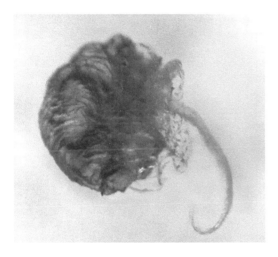

FIGURE 211   A tadpole emerging from an egg in Dominican amber (Poinar amber collection accession no. Am-3-7); length of egg = 1.3 mm.

larvae sometimes become entrapped in amber. This female might have been close to its breeding site under the bark of an amber tree.

11. A phorid fly (Diptera: Phoridae) that laid eight eggs was preserved in Dominican amber (Figure 210). Phorid or scuttle flies are fairly common in amber. Extant larvae feed on decomposing organic matter in a wide range of habitats, such as under bark, in various animal nests, in soil high in organic matter, on corpses, etc. The eggs hatch soon after oviposition, and most larvae develop rapidly. Their biodiversity and wide range of habitat preferences contribute to their success in the world today.

12. Figure 211 shows a tadpole found emerging from an egg in Dominican amber. This fertile egg probably came from a phytotelmata, a water reservoir formed by the leaves and stems of various plants—in this case, probably an epiphytic bromeliad growing on the branch of an amber tree. Some disturbance probably dislodged the plant and caused the water and its contents to splash out. Obviously, some of the contents fell into a pool of resin. Whether the

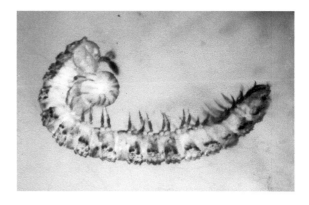

FIGURE 212   A millipede with her young in Dominican amber (Poinar amber collection accession no. DC-9-21); length of adult = 3 mm.

FIGURE 213   An isopod with young in her brood pouch in Dominican amber (Poinar amber collection accession no. Cr-2-13); length of isopod = 4 mm.

tadpole had already begun to emerge while still in the phytotelmata or did so only after the egg fell into the resin is unknown.

Maternal Care

In this category is actual evidence of maternal care in amber. It includes the following examples:

1. Figure 212 shows a millipede with its newborn young that was preserved in Dominican amber. Some millipedes remain with their eggs after laying them in various types of nests. This amounts to guarding the eggs until they hatch and, in the fossil case, even tending the newly hatched. With few exceptions, millipedes are vegetarians and feed on both living and decaying plant matter. Some occur under bark or in rotting wood, and these would be more prone to encountering resin and ending up in amber.

2. An isopod was found in Dominican amber with young clustered in a brood chamber on the underside of her body (Figure 213). Most isopods carry their eggs in a brood chamber formed by the flat projections of the thoracic legs. This holds the eggs closely adpressed to the underside of their bodies. After the

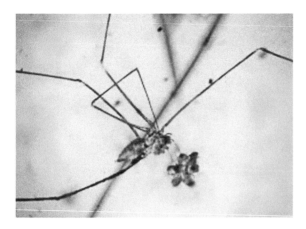

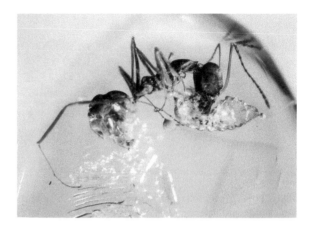

FIGURE 214   A pholcid spider (Arachnidae: Araneae) holding her egg sac in Dominican amber (Poinar amber collection accession no. SY-1-44); diameter of egg sac = 1 mm.

FIGURE 216   An ant carrying its larva in Dominican amber (Poinar amber collection accession no. S-1-48); length of ant = 2 mm.

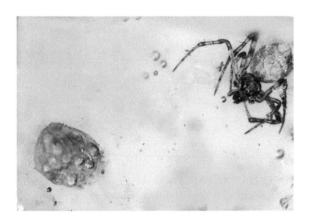

FIGURE 215   A spider (Arachnida: Araneae) tending its egg sac in Dominican amber (Poinar amber collection accession no. Sy-1-86); diameter of egg sac = 0.5 mm.

FIGURE 217   Baltic amber ant that was carrying an egg (arrow) and newly hatched larvae in Baltic amber (Poinar amber collection accession no. Sy-1-139); length of ant = 3 mm.

eggs hatch, the mother may continue to carry the young in the brood chamber (as in the fossil) until they can fend for themselves. The adult's armored bodies extend almost to the ground and provide a protective covering for both eggs and young. Food consists of various organisms in decaying organic matter.

3. Figure 214 shows a pholcid spider holding its egg sac in Dominican amber. These spiders spin a very loose web and, when disturbed, will shake the web by swinging around. The egg sac, which is suspended by the chelicerae, is made with thin strands of silk that are almost transparent, thus giving the impression that they are glued together.

4. A mother spider tending its egg sac is shown in Dominican amber in Figure 215. The spider may have been transporting the egg sac when it was entombed. Such protective behavior is typical of many spiders.

5. A worker ant carrying an ant larva was fossilized in Dominican amber with a larva in its jaws (Figure 216). By the careful way the worker is carrying the

uninjured larva we can surmise that it was transporting it to a new nest and not bringing home prey. The urge to protect their brood is so strong with ants that even as they perish in the resin some will not release their hold.

6. An ant in Baltic amber was discovered with an egg and several newly hatched larvae in the close vicinity of its head (Figure 217). This suggests that the ant was carrying several developmental stages when it encountered the resin.

7. The tadpole egg shown in Figure 218 was found in Dominican amber. The shape and nature of this egg suggest that it was a trophic egg deposited by the female for nourishment of the tadpole, which was in the same piece of amber.

10AIIc.   Other Evidence of Insect Egg Laying

Grimaldi (1996, p. 86) illustrated a midge extruding a string of eggs, from Dominican amber. Koteja (1998) illustrated a Baltic amber coccid larva in the process of hatching from its eggshell and a larva in the process of molting. Koteja also illustrated varied Baltic amber coccids with everted male

FIGURE 218   A trophic egg of a frog in Dominican amber (Poinar amber collection accession no. Am-3-4); length of egg = 1.2 mm.

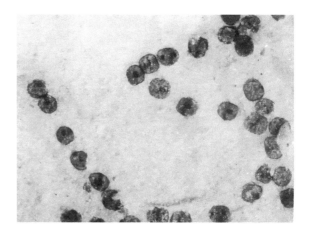

FIGURE 219   Gall and Grauvogel (1966) interpreted several Late Triassic, French Grès à Voltzia occurrences as insect egg masses; shown is *Monilipartus tenuis* (×14). (Plate A, Figure 1, in Gall, J.C. and Grauvogel, L., *Annales de Paléontologie, Invertébrès*, 52, 155–161, 1966.)

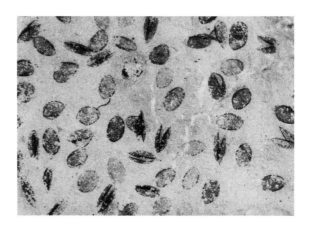

FIGURE 220   *Clavapartus latus* (×28). (Plate B, Figure 2, in Gall, J.C. and Grauvogel, L., *Annales de Paléontologie, Invertébrès*, 52, 155–161, 1966.)

sexual organs from which spermatophores have been, or are in the process of being, ejected, presumably due to the trauma of being caught in the gum; examples include Baltic amber, New Jersey amber, and Bitterfeld amber. Koteja (2000) illustrated several New Jersey Turonian coccids with spermatophores and other reproductive structures, as well as examples of coccid larvae. Weitschat and Wichard (1998) illustrated a Baltic amber plecopteran associated with her eggs (Plate 30, Figure a), "normal" adult male morphologies and "larval" adult female morphologies of Baltic amber coccids (Plate 48, Figures a–h), and an Anisopodid fly with a string of eggs (Plate 82, Figure a).

Rasnitsyn and Quicke (2002) illustrated an Early Permian Order Caloneurida Paleuthygrammatidae (*Paleuthygramma tenuicornis*) female with "large sclerotised eggs" in the abdomen (their Figure 111) and an unidentified Middle Miocene mastotermitid winged female with an egg-filled abdomen (their Figure 378). Gall and Grauvogel (1966) illustrated and discussed several egg mass types from the Late Triassic Grès à Voltzia from northeastern France that they interpret as insect eggs (see Figure 219 and Figure 220). Martínez-Delclòs et al. (2004, Figure 2G) illustrated a Dominican amber Keroplatidae female extruding eggs. Johnston et al. (1996) described pupal cases, from the Late Cretaceous of the Gobi Desert and earlier mistaken for dinosaur eggs, that were constructed from siliciclastic materials and surrounded the actual developing pupa, similar to those in at least three families of modern beetles.

□ *Reliability*

Category 1, frozen behavior; Category 2B, functional morphology.

10AIId.   Spider Cocoons, Eggs, and Spiderlings

(See also Section 6CIa, Spider Webs, Spinnerets, and Bundled Prey.) Weitschat and Wichard (1998, Plate 20, Figures e–h) illustrated Baltic amber spider cocoons with very visible eggs and spiderlings as well. Krzeminska and Krzeminski

(1992, Figure 144) illustrated a spider cocoon in amber sectioned to show the eggs very clearly (see Figure 221). Poinar and Poinar (1999, Figure 72) illustrated a pholcid female, in Dominican amber, who was carrying her eggs. Wunderlich (2004) illustrated a number of spider cocoons in Baltic and Dominican amber.

10AIIe.   Caterpillars and Dipteran Larvae and Nymphs: Immature Insect Stages

Grimaldi (1996, p. 36) illustrated a caterpillar from Neocomian Lebanese amber, the oldest recognized to date. Kuhbandner and Schleich (1994) described some Miocene stratiomyid larvae from Germany. Duncan et al. (1998) described a trichopteran larva from the Late Oligocene/ Early Miocene of Riversleigh, Queensland. MacKay (1969) described several microlepidopterous larvae from Late Cretaceous Canadian amber, and Kinzelbach and Lutz (1985) described a Geiseltal Eocene stylopid larva. Leakey (1952) illustrated a Miocene lepidopteran caterpillar from Kenya,

FIGURE 221 A Baltic amber spider cocoon showing the eggs inside; length of amber piece = 32 mm. (Figure 144 in Krzeminska, E. and Krzeminski, W., *Les Fantômes de l'Ambre*, Musée d'Histoire Naturelle de Neuchâtel, Switzerland, 1992. Photograph by D. Rapin. Reproduced with permission.)

and Meyer (2003, Figure 198) illustrated a Late Eocene Florissant lepidopteran larva. Rasnitsyn and Quicke (2002, Figures 340, 341) illustrated a Jurassic or Early Cretaceous Siberian caterpillar that may belong to the Xyelidae and an Early Cretaceous Xyelidae caterpillar. Mitchell and Wighton (1979) illustrated and discussed a number of Paleocene larvae from Alberta, including numerous beetles, as well as some insect nymphs and pupae. Grimaldi (2000, Figures 2, 3) illustrated a raphidiopteran larva from the New Jersey Cretaceous. Rundle and Cooper (1970) described an Early Eocene beetle pupa from the London Clay. Meyer (2003, Figures 138, 140) illustrated Late Eocene Florissant odonata nymphs and chironomid larvae.

Rasnitsyn and Quicke (2002) illustrated and cited a number of larvae: Early Permian mayfly *Kukalova americana* (Figure 88); Late Jurassic Siphlonuridae *Stackelbergisca sibirica* (Figure 91); Late Jurassic or Early Cretaceous Siphlonuridae *Proameletus caudatus* (Figure 92); Late Cretaceous mayfly *Ephemeropsis melanurus* (Figure 93); Early Jurassic Epeoromimidae *Epeoromimus kazlauskasi* (Figure 94); Early Jurassic Mesonetidae *Mesoneta antiqua* (Figure 95); Early Cretaceous Hemeroscopidae *Hemeroscopus baissicus* (Figure 102); Early Jurassic Campterophlebiidae *Samarura gigantea* (Figure 106); Late Miocene libullulilids, Middle Carboniferous Mischopteridae *Mischoptera douglassi* (Figure 107); Early Cretaceous Coptoclavidae *Coptoclava longipoda* (Figure 202); Late Jurassic Parahygrobiidae *Parahygrobia natans* (Figure 223); Early Jurassic Liadytidae *Angaragabus jurassicus* (Figure 225); Early Jurassic Hydrophilidae *Angarolarva aquatica* (Figure 227); Early Cretaceous Chauliiodidae *Cretochaulus lacustris* (Figure 244); Late Permian dobsonfly-like *Permosialis* (Figure 246); Early Cretaceous Chauliodidae *Cretochaulus lacustris* (Figure 247b); Early Permian Atactophlebiidae *Gurianovaella silphidoides* (good ontogenetic set) (Figure 387); Early Permian intermediate between grylloblattidan and stonefly *Sylvonympha*

tshekardensis (Figure 390); Early Permian Atactophlebiidae *Gurianovaella silphidoides* (Figures 392, 401); Late Permian Euryptilonidae *Euryptilodes cascus* (Figure 404); Early Permian Tshekardoperlidae *Sylvoperlodes zhiltzova* (Figure 410); Late Permian Nemouromorpha *Barathronympha victima* (Figure 411); Early or Middle Jurassic Mesoleuctridae *Mesoleuctra tibialis* (Figure 413); Early Jurassic Platyperlidae *Platyperla platypoda* (Figure 415); Early or Middle Jurassic Perlariopseidae *Spinoperla spinosa* (Figure 418); and Late Eocene unidentified Forticuloidea (Figure 426).

From Baltic amber, Janzen (2002) illustrated an elateroid larva (Figure 223), beetle larva (Figure 252), tenthredinid (Figure 269), lepidopteran (Figures 335, 337), and pupae (Figures 336, 338). Johnson and Borkent's (1998) description of Chaoboridae pupae from the middle Eocene of Mississippi is worthy of note. Weitschat and Wichard (1998) illustrated varied Baltic amber insect growth stages: Ephemeropteran larva (Figure 42 and Plate 27, Figure f); imago emerging from its subimagine exuvium (Plate 27, Figure e); larval plecopteran (Plate 29, Figures a–d; Plate 30, Figure b); larva (Figure 47); split larval exuvium Dermapteran larva (Plate 32, Figure b); Blattodean wingless larvae (Plate 34); Isopteran larvae (Plate 35, Figure g); Phasmatodean larva (Figure 51 and Plate 36); larval Orthopterans (Figure 52 and Plate 37); larval Coxidian (Plate 40, Figure f); a previously published larval gerrid (p. 120); larval cicadas, Fulgoromorpha and Cicadomorpha (Plate 46, Figures a–h); larval aphids (Plate 47, Figures e, f); larval Aphalarids (Plate 49, Figure h); Planipennid pupa (Plate 56, Figure g); Formicid pupae (Plate 71, Figures d, f); Tipulid pupa (Plate 80, Figures g, h); Anisopodid pupa (Plate 82); larval Sialid (Plate 51, Figure c); larval raphidiopteran (Plate 52, Figure c); larval Coniopterygid (Plate 53, Figure d); larval Neurorthidae (Plate 54, Figures d, e); larval Psychopsids (Plate 55, Figures e, g); Planipennid pupa (Plate 56, Figure g); larval Ascalaphid (Plate 56, Figure h, and Figure 69); larval dytiscids (Plate 57, Figures c–f); larval Gyrinid (Figure h); larval Scirtidae and Cleridae (Plate 62, Figures a, h); larval Scraptids (Plate 63, Figure e); ant larvae and pupae (Plate 71, Figures c–f); larval Lepidoptera (Plate 78, Figures a–d); larval lepidopteran "cases" (Plate 79, Figures a–h); larval Tipulinids (Plate 80, Figures g, h); and pupal exuvium and imago of an Anisopodid fly (Plate 82, Figure d); pupa and larva (Figures b, c).

Weidner (1958) described well-preserved odonate, coleopteran, megalopteran, and plannipennid larvae from Baltic amber. Lukashevich (1995) described Jurassic and Cretaceous dipteran pupae of Eoptychopteridae and Ptychopteridae from Siberia. Selden and Nudds (2004, Figure 240, p. 137) illustrated a Baltic amber wood gnat, *Mycetobia* (Diptera: Anisopodidae), emerging from its pupal case (according to Seldon, pers. comm., 2006).

□ *Reliability*

Category 2B, functional morphology.

10AIIf.   Beetle Eggs Deposited on a Leaf

Lewis and Carroll (1991; see also Lewis, 1992, Plate 3, Figures D and C [mislabeled as E]) described a Republic Formation, Eocene alder leaf from northeastern Washington on which a flea beetle (Chrysomelidae) had laid eggs in a distributional pattern similar to that of extant *Altica*. Lewis and Carroll (1992) described similar material from the Oligocene of the John Day Formation in Oregon. Johnson (1993) discussed and illustrated some possible beetle eggs on a middle Eocene leaf from northern Mississippi.

□ *Reliability*

Category 1, frozen behavior.

*10AIIg.   Cockroach Ootheca

(See Figure 200, Figure 201, and 202.) Poinar (1992b, Figure 50) illustrated the egg packet (ootheca) of a Dominican amber cockroach (see also Boucot, 1990a, p. 660, for reference to an Eocene Geiseltal ootheca). Vishnyakova (1980a) discussed ootheca specimens from the Commonwealth of Independent States (CIS) Late Cretaceous. An Early Cretaceous cockroach ootheca in Burmese amber is shown in Figure 202. Labandeira (1998c) suggested that oothecate cockroaches first appeared in the Cretaceous, as the pre-Cretaceous members of the group have prominent ovipositors. Basibuyuk et al. (2002) described a hymenopteran from Early Cretaceous Lebanese amber of the same group that parasitizes modern cockroach oothecae, further supporting an Early Cretaceous appearance of oothecate cockroaches. Anisyutkin et al. (2008) described a Cretaceous Israeli ootheca of somewhat uncertain taxonomic affinity, possibly cockroach or praying mantis.

□ *Reliability*

Category 2B, based on functional morphology.

*10AIIh.   Mate Guarding in Gerrids

Poinar (1992b, color plate 7, bottom figure) illustrated a pair of Dominican amber Gerridae entombed in the mate-guarding posture (Figure 222); see Andersen and Poinar (1992) for a description of the specimen and consideration of

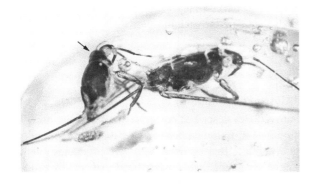

FIGURE 222   Mate guarding by the male (arrow) of a pair of water striders in Dominican amber (Poinar amber collection accession no. He- 4-20); length of female (in front) = 3 mm.

the behavior. Thornhill and Alcock (1983) provided discussions of mate guarding among insects from an evolutionary perspective.

□ *Reliability*

Category 1, frozen behavior.

*10AIIi.   Phasmida Eggs

(See Figure 193.) Clark Sellick (1994) described mid-Eocene Clarno Formation phasmida eggs from Oregon. The morphology of the eggs is similar to that of modern members of the group, including their convergence to a grass seed form that induces some modern ants to carry them into their nests. Poinar and Poinar (1999, p. XIV) illustrated a phasmid egg in Dominican amber.

□ *Reliability*

Category 2B, functional morphology.

*10AIIj.   Odonata Eggs Laid on Leaves

Krassilov and Rasnitsyn (2008) provided ichnofossil definitions and descriptions of varied Israeli Cretaceous odonate leaf scars (see color plate 4D after page 132); their system, when applied in the Cenozoic, will advance the understanding of these ichnofossils considerably. Vasilenko (2008) described various odonate-type leaf scars from the Late Cretaceous of the pre-Amur region. Hellmund and Hellmund (1996a,b; see also 1991, 1993, 2002) described various Cretaceous to modern odonate (damselflies, in particular) egg masses laid on leaves. Schaarschmidt (1992, Figure 78) illustrated a particularly fine specimen of a leaf with insect eggs arrayed in a highly patterned manner, although the identity of the insect was not discussed. Hellmund and Hellmund (1996a,b) described additional odonate eggs, laid out in distinctive patterns, from the Late Miocene of Baden–Wurttemberg and the Middle Oligocene of Saxony. Hellmund and Hellmund (1998) described more damselfly eggs laid on leaves from the Early Oligocene of Saxony.

Grauvogel-Stamm and Kelber (1996) described various Keuper, Late Triassic eggs of insect type that may well represent Protodonata (very large eggs, much larger than those of dragonflies), laid in patterns on plant parts similar to those of living dragonflies, as well as egg masses, presumably water laid, that might also represent Odonata or even chironomids (see also Gall and Grauvogel, 1966; Geyer and Kelber, 1987). Van Konijnenburg-van Cittert and Schmeissner (1999) described some Early Jurassic odonate eggs on leaves and provided excellent illustrations of the egg-laying patterns. Although their account probably has nothing to do with Odonata, Kelber and Geyer (1989) described probable insect eggs laid on *Equisetites* from the lower Keuper of Franconia. Labandeira (2002a, Figure 2.19; Labandeira, 2002b, Figure 6) provided several examples of oviposition, mostly into leaves. Vasilenko (2005) described some unidentified Jurassic–Cretaceous insect egg material from Transbaikalia that was laid on leaves. Sarzetti et al. (2008) provided extensive

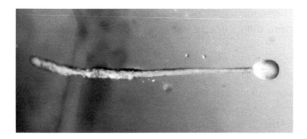

FIGURE 223 A stalked spermatophore of the springtail *Sminthurus longicornis* in Baltic amber (Poinar amber collection accession no. CL-1-4); length of spermatophore = 2.20 mm. (Adapted from Poinar, Jr., G.O., *Historical Biology*, 14, 229–234, 2000.)

discussions, illustrations, and a summary of previous work on odonate-type trace fossils, while commenting on the "stasis" of this trace fossil type from the Patagonian Eocene.

☐ *Reliability*

Category 2B, functional morphology.

*10AIIk. Coleopteran Pupal Cases

Johnston et al. (1996) described pupal cases, from the Late Cretaceous of the Gobi Desert and earlier mistaken for dinosaur eggs, that were constructed from siliciclastic materials and surrounded the actual developing pupa, similar to those in at least three families of modern beetles.

☐ *Reliability*

Category 2B, functional morphology.

*10AIIl. Collembolan Sperm and Insect Spermatophores

Poinar (2000b) very convincingly described a Baltic amber occurrence of a collembolan spermatophore closely associated with a collembolan and with collembolan sperm cells—the last being what one would normally assume to be an impossible item for preservation (see Figure 223 and Figure 224). This find establishes the existence of indirect-type sperm transfer in the fossil record (40 my). Although not a collembolan, the illustration in Martínez-Delclòs et al. (2004, Figure 8C) of a Mymarommatidae wasp spermatophore from Early Cretaceous Spanish amber is notable. Koteja (2000) illustrated several New Jersey Turonian coccids with spermatophores and other reproductive structures and provided examples of coccid larvae.

☐ *Reliability*

Almost Category 1, frozen behavior, because of the close association.

*10AIIm. Mosquito and Biting Midge Mating Swarms (Leks)

Poinar et al. (2000c) suggested that the male antennae of a Campanian, Late Cretaceous mosquito from Canadian amber are consistent with the antennae having been used to receive the "flight tone" of the females which is consistent with mating having taken place in swarms. Mating swarms or lekking

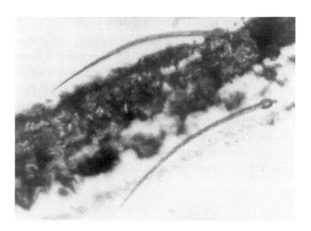

FIGURE 224 Two spermatozoa associated with the tip of a collembolan spermatophore in Baltic amber (same specimen as shown in Figure 223); length of spermatozoa = 0.085 and 0.095 mm.

of biting insects can be found in amber, as is shown with a cluster of flies in Dominican amber (see Figure 106 and Figure 107).

☐ *Reliability*

Category 2B, functional morphology.

*10AIIn. Isopod with Young

Poinar and Poinar (1999, p. 80) cited an isopod with a cluster of young in Dominican amber (see Figure 213).

☐ *Reliability*

Category 1, Frozen behavior.

*10AIIo. Oviposition Notches

Rasnitsyn and Quicke (2002, Figure 468) illustrated and discussed alleged odonatan oviposition notches on an Early Cretaceous Mongolian plant remain.

☐ *Reliability*

Category 2B, functional morphology.

*10AIIp. Opilione Sexual Organs

Dunlop et al. (2003) described a penis and an ovipositor from Rhynie Chert harvestmen—a most unusual preservation that extends the taxon far back in time.

☐ *Reliability*

Category 2B, functional morphology.

*10AIIq. Platyhelminth Eggs

Poinar (2003c) described a rhabdocoel turbellarian (Platyhelminthes) in Baltic amber showing that these animals had established a terrestrial habit and were producing subitaneous eggs (i.e., eggs that self-fertilized and rapidly produced under favorable conditions) by the Eocene (see Figure 190). Platyhelminthes living in varied environments are present in the record from the Devonian on. Table 28 provides the currently available data.

TABLE 28
Fossil and Subfossil (Less than 10,000 Years Old) Records of Platyhelminthes

Fossil	Location	Age	Refs.
Bothriocephalus, eggs	Bog cadaver, Germany	1500 years	Szidat (1944)
Dicrocoelium, eggs	Roman remains, England	~2000 years	Taylor (1955)
Diphyllobothrium, eggs	Coprolite, Indian, Peru	~4000 years	Callen and Cameron (1960)
Turbellarian eggs	Lakes, United States	Pleistocene–Holocene	Frey (1964)
Turbellarian eggs	Lakes, England	Pleistocene–Holocene	Harmsworth (1968)
Schistosoma, eggs	Mummy, Egypt	5000 years	Ruffer (1910)
Schistosoma, eggs	Mummy, Egypt	5000 years	Deelder et al. (1990)
Dicrocoelid eggs	Bear coprolites, France	5500 years	Jouy-Avantin et al. (1999)
Metacercarial pits	Marine bivalves, global	Eocene–Holocene	Ruiz and Lindberg (1989)
Turbellarians	Hypersaline lake beds, California	Miocene	Pierce (1960)
Micropalaeosoma balticus	Baltic amber, Germany	Eocene	Poinar (2003c)
Terricolichnus permicus	Trail, Italy	Permian	Alessandrello et al. (1988)
Cestode eggs	Shark, United States	Pennsylvanian	Zangerl and Case (1976)
Platyhelminthes(?)	Hooks in fish, Latvia	Devonian	Upeniece (2001)

□ *Reliability*

Category 1, frozen behavior.

SUMMARY

Mostly at the ordinal level (see Table 29).

10B. AQUATIC INVERTEBRATES

Westermann (1969) reviewed the sexual dimorphism recognized in many aquatic fossil groups, as well as in some tetrapods. The obvious dimorphism of decapods, however, was not covered in his discussion. Godfrey (1995) described some strings of eggs from the Mazon Creek Pennsylvanian for which the maker is unknown but is presumed to be aquatic.

10Ba. DIMORPHISM AND BROOD CARE IN OSTRACODES

Lethiers et al. (1997) provided extensive evidence regarding brooding in Permian nonmarine ostracodes, including a brood preserved within a female, and summarized some additional occurrences of broods (Adamczak, 1968, 1991). It is clear that the brooding capability has developed independently in several ostracode lineages. Sylvester-Bradley (1969) provided an excellent discussion of ostracod sexual

TABLE 29
Terrestrial Invertebrate Sexual and Ovipositional Behavior Evidence and Protective Larval Cases from the Fossil Record (Additions to Table 31 from Boucot, 1990a)

Behavioral Type	Taxonomic Level	Time Duration	Category	Refs.
10AIIe. Caterpillars and dipteran larvae	Orders	Early Cretaceous	1	Grimaldi (1996)
10AIIf. Beetle eggs deposited on a leaf	Order	Eocene	1	Lewis and Carroll (1991)
10AIIg. Cockroach ootheca	Order	Late Cretaceous	2B	Vishnyakova (1980a)
10AIIh. Mate guarding in gerrids	Order	Miocene–Holocene	1	Anderson and Poinar (1992)
10AIIi. Phasmida eggs	Family	Eocene–Holocene	2B	Clark Sellick (1994)
10AIIj. Odonata eggs laid on leaves	Order	Cretaceous–Holocene	2B	Hellmund and Hellmund (1996a,b); Krassilov and Rasnitsyn (2008)
		Jurassic–Holocene	2B	van Konijnenburg-van Cittert and Schmeissner (1999)
10AIIk. Coleopteran pupal cases	Order	Late Cretaceous–Holocene	2B	Johnston et al. (1996)
10AIIl. Collembolan sperm and spermatophores	Class	Eocene–Holocene	1	Poinar (2000b)
10AIIm. Mosquito mating swarms	Family	Late Cretaceous–Holocene	2B	Poinar et al. (2000c)
10AIIn. Isopod with young	Order	Miocene–Holocene	1	Poinar and Poinar (1999)
10AIIo. Oviposition notches	Order	Early Cretaceous	2B	Rasnitsyn and Quicke (2002, Figure 468)
10AIIp. Opilione sexual organs	Order	Early Devonian	2B	Dunlop et al. (2003)
10AIIq. Platyhelminth eggs	Class	Eocene	1	Poinar (2003c)

dimorphism and brooding of larvae. Siveter et al. (2006) provided an elegant example of brooding in myodocopids from the Silurian.

☐ *Reliability*
Category 1, frozen behavior.

10Bb. CREPIDULID GASTROPOD SEX CHANGES

Lozouet (1997) discussed examples of French Early Miocene crepidulids (*Bicatillus* and *Crepidula*) still attached to hard-substrate gastropods, although not in stacks. Lozouet interpreted the relationship as commensal because the examples of *Crepidula* occur in the aperture of pagurized gastropod shells, whereas *Bicatillus* lived on both *Aporrhais* and turritellid exteriors. Bałuk and Radwański (1985) described a large sample of gastropods from the Eocene of the Bavarian Alps and the Polish Miocene for which the evidence is excellent that the *Crepidula* settled on the interiors, near or almost at the apertures of empty gastropod shells, with some described individuals featuring presumably small males attached to larger females as is customary for the genus. It is interesting that the stratigraphic range extends back to the Eocene as is also the case in North America.

10Bc. AMMONOID EGG SACS

Lewy (1996) collected the references to potential ammonoid eggs (Dreyfuss, 1933; Lehmann, 1966, 1981; Müller 1969a, 1978b) and discussed various reproductive possibilities in ammonoids. Lehmann (1990, p. 60) illustrated and briefly described a possible ammonoid egg sac from the German Jurassic that contains what may be developing young.

☐ *Reliability*
Category 3.

10Bd. TRILOBITE AND CRAB CLUSTERS

Powell and Nickerson (1965) described "piles" of juvenile king crabs from Alaska, and Stevcic (1971) discussed laboratory observations on aggregations of the spiny spider crab. These examples may indicate similar significance for trilobite clusters. Karim and Westrop (2002) described some Oklahoma Ordovician Bromide Formation trilobite clusters interpreted as breeding aggregations similar to those dealt with earlier by Speyer (1990). Chatterton et al. (2003) reviewed the question of cryptic behavior in trilobites, including the possible use of various types of cavities for molting and reproduction as well as to escape predation. They provided a wealth of examples, including some truly spectacular ones from the Burgess Shale that occur in worm tubes (Plate 1). Although not strictly "clusters," the "queues" (Radwański et al., 2009) of blind phacopid trilobites (*Trimerocephalus*) from the very shallow water Polish Late Devonian suggest a possible arthropod behavior not previously noted. The single-file trilobite queues possibly represent migratory behavior

under the environmental stress of a sudden sea-level drop leading to emersion. The trilobites were suffocated and "frozen" in place by overlying soft sediment. The specimens are not entombed in a burrow as are some other linear trilobite arrays (see Section 5p, Cryptic Trilobite Behavior).

10Be. GASTROPOD EGGS AND BIVALVE BROOD

Russell et al. (1992) described fossilized embryonic stages of the venerid *Transennella* from the Miocene–Pliocene Etchegoin Formation of the Kettleman Hills in California. They found the presence in most specimens of only one size class to be a taphonomic artifact, based on experiments with a modern species of the genus where they demonstrated that earlier growth stages that show the presence of asynchronous brooding are preferentially destroyed, although a single fossil specimen was discovered with a range of developmental stages present. Their evidence suggests that the brooding style has not changed since the Miocene–Pliocene.

Hoagland (1977) reviewed the taxonomy of the American and European crepidulids and pointed out many examples of both planktotrophic and nonplanktotrophic lifestyles. Palmer (1958) described viviparity in a Yorktown Formation (Pliocene) *Turritella* from Virginia, cited another from Maryland, and commented on the viviparous habit. She also drew attention (Palmer, 1961) to an earlier account of a viviparous *Turritella* from the Choctawhatchee Formation in Florida (Sutton, 1935). Burns (1899) described several *Turritella* with brood from the Maryland Miocene at Plum Point. In all of these examples, numerous brood were found within the adult shell. Herbert and Portell (2004) described brooded *Crepidula* from the Florida Pleistocene (Figure 225 and Figure 226), a first for this group, and summarized many of the previously published occurrences of gastropod brooding known in the fossil record. Marwick (1971, Figure 6) described an excellent example of brooding in a New Zealand Miocene turritellid (Figure 227), and Tampa (1976) described brooding in a Kansas Pleistocene pulmonate. Dockery (1996) described brooding in a Mississippi Late Cretaceous *Gyrodes*(?).

☐ *Reliability*
Category 1, frozen behavior.

*10Bf. PATAGONIAN OYSTER REPRODUCTION

Iribarne et al. (1990) described an unusual late Pliocene or late Miocene situation in which small male oysters attached themselves in the umbonal region of the female until they reached a moderate size, following which they became detached. As they grew, they eventually underwent a sex change to the female condition, just as is the case with a living species from the same region. This situation is reminiscent of that in *Crepidula* (Boucot, 1990a).

☐ *Reliability*
Category 1, frozen behavior.

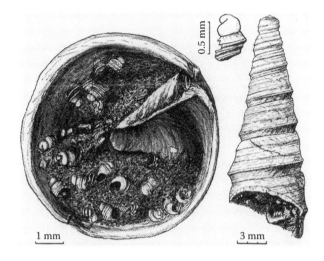

FIGURE 227 Brooded Miocene turritellids from the New Zealand Miocene. (Figure 6 in Marwick, J., *New Zealand Journal of Geology and Geophysics*, 14, 66–70, 1971. Reproduced with permission.)

FIGURE 225 Brooded *Crepidula* from the Florida Pleistocene; scale bar = 500 μm. (Figures 2.1 through 2.5 in Herbert, G.S. and Portell, R.W., *Journal of Paleontology*, 78, 424–429, 2004. Reproduced with permission.)

associated with a nematode. Hooker et al. (1995) mentioned similar clitellate cocoons from the late Eocene of southern England. Manum et al. (1991, 1994; Manum, 1996) recognized the cocoons of clitellate hirudineans and oligochetes in palynological preparations with stratigraphic ranges extending from the Late Triassic to the Pliocene in freshwater deposits. Hooker et al. (1995, Figure 8) illustrated and briefly discussed some probable Late Eocene examples from southern England. McLoughlin et al. (2002, p. 87) described some Victorian Early Cretaceous material of this type. Jansson et al. (2008) described clitellate cocoons from the Australian Early Jurassic.

□ *Reliability*

Category 1, frozen behavior, for some of the specimens in which the actual worm is still preserved, and Category 2B, functional morphology, for the mostly empty cocoons.

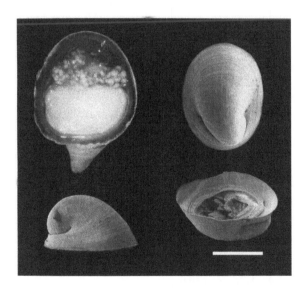

FIGURE 226 Brooded *Crepidula*, modern examples; scale bar = 500 μm. (Figures 3.1 through 3.4 in Herbert, G.S. and Portell, R.W., *Journal of Paleontology*, 78, 424–429, 2004. Reproduced with permission.)

*10Bg. PROBABLE HIRUDINEAN AND EARTHWORM COCOONS

In a very well-documented paper, Manum et al. (1991) made an excellent case for the presence in certain freshwater deposits of cocoons made by clitellates, probably hirudineans, with the oldest examples being from the Triassic. Manum et al. (1992) described more clitellate cocoons and provided some very detailed structural information; they also contributed more to the clitellate cocoon data, including Paleocene specimens. Manum et al. (1994) provided still more information regarding a cocoon from the Early Cretaceous of Spitsbergen

*10Bh. CTENOPHORE GONADS

Stanley and Stürmer (1987) described a most unusual specimen from the Early Devonian Hunrückschiefer of the Rhineland that displays, in x-ray evidence, typical ctenophore gonads associated with the comb rows, as is the case with the living animals.

□ *Reliability*

Category 1, frozen behavior.

*10Bi. GRAPTOLITE "REGENERATION"

Han and Chen (1994) described regeneration in a specimen of Ordovician *Cardiograptus* and provided other examples previously described from the Silurian. Regeneration does not, naturally, involve any sexual behaviors but is involved with an asexual form of reproduction in colonial organisms.

□ *Reliability*

Category 1, frozen behavior.

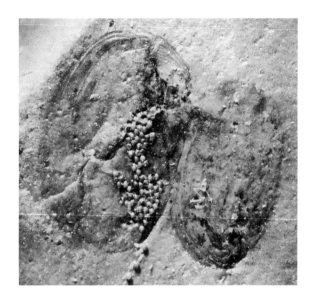

FIGURE 228 French, Late Triassic Grès à Voltzia estheriid with brooded eggs (×17). (Plate E in Gall, J.C. and Grauvogel, L., *Annales de Paléontologie, Invertébrès*, 52, 155–161, 1966.)

*10Bj.   CRAB LARVAE

Maisey and de Carvalho (1995) described the oldest examples of brachyuran crab protozoea (from the Early Cretaceous Brazilian Santana Formation), obtained from the gut contents of a teleost (*Tharsis araripis*). The antiquity of this larval stage, only one period younger than the initial appearance of brachyurans, is a notable example of developmental conservatism.

☐ *Reliability*

Category 2B, based on functional morphology.

*10Bk.   CLADOCERAN EGGS

Lutz (1989, Figure 4.12) illustrated a group of *Daphnia* ephippia from the Rott Oligocene. Richter and Wedmann (2005) described some cladoceran resting eggs from the Messel Eocene. Although not in the cladoceran category, British Silurian ostracod with eggs described by Siveter et al. (2008) are notable.

☐ *Reliability*

Category 2B, functional morphology.

*10Bl.   ESTHERIAN CRUSTACEAN EGG BROODING

Gall and Grauvogel (1966, Plate E) illustrated an Early Triassic French estherid with numerous eggs present within the valves (see Figure 228). Dechaseaux (1951) illustrated some Stephanian estherids with eggs and reviewed similar Pennsylvanian occurrences as well as others in the fossil record. Vannier et al. (2003) described French Stephanian branchiopods with eggs still retained and visible within the organisms.

☐ *Reliability*

Category 1, frozen behavior.

*10Bm.   FORAMINIFERAL PLASTOGAMY

Mukhopadhyay (2003) described an example of plastogamy (union of two cells without nuclear fusion) in the protozoan *Nummulites boninensis* Hanzawa from the Middle Eocene of the Indian Cambay Basin; this is not an invertebrate but does represent a sexual item.

☐ *Reliability*

Category 1, frozen behavior.

*10Bn.   ARGONAUT PAPER NAUTILUS EGG CASES

Saul and Stadum (2005 described some Southern California Miocene argonaut eggcases, the "paper nautilus" in which the female Argonaut deposits her eggs (see also Gust, 2000; Morgan and Brown, 1998; Raschke, 1989; Stadum, 1997; Wagner and Marks, 1990). Their fossil record goes back to the Oligocene (Engeser, 1990). Holland (1988) thoroughly reviewed the body of information dealing with argonauts, fossil and modern. While describing a Miocene *Argonauta* from Cyprus, Martill and Barker (2006) reviewed the occurrences known at this time, including various European, North African, Japanese, Sumatran, and New Zealand occurrences, and also commented on some aspects of their biology.

☐ *Reliability*

Category 2B, functional morphology.

*10Bo.   CUPULADRID BRYOZOAN REPRODUCTION

O'Dea et al. (2008) illustrated and discussed in some detail sexual (aclonal) and asexual (clonal) reproduction in cupuladrid bryozoans of Cenozoic age.

☐ *Reliability*

Category 1, frozen behavior.

*10Bp.   COPULATING GASTROPODS

Dieni (2008) illustrated and described a number of gastropods *in copula* (*gasteropodi e amore*) from the Eocene and Oligocene of northern Italy (see Figure 229). Reproduction in gastropods varies from depositing their eggs and sperm in water, where fertilization takes place (see Section 10Bc, Ammonoid Egg Sacs, in Boucot, 1990a, for sex change in *Crepidula*), to forms that employ copulation.

☐ *Reliability*

Category 1, frozen behavior.

SUMMARY

Family and higher levels (see Table 30).

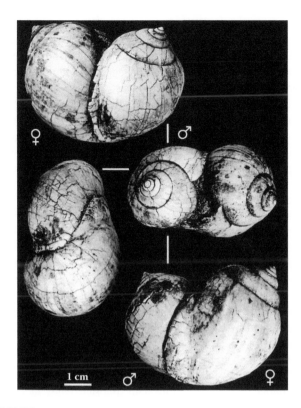

FIGURE 229 Gastropods *in copula* (*gasteropodi e amore*) from the Eocene and Oligocene of northern Italy. (Figure 2 in Dieni, I., *Rivista Italiana di Paleontologia e Stratigrafia*, 114, 505–514, 2008. Reproduced with permission.)

10C. VERTEBRATES

10Ca. VERTEBRATE EGGS AND EGG CASES

Kohring and Hirsch (1996) provided an excellent review of avian and reptilian shelled egg microstructure and described some Messel Eocene crocodilian and avian eggshells. Zelenitsky and Hirsch (1997) provided a penetrating review of the characteristics and classification of the shelled eggs. Hirsch (2001) discussed pathological fossil eggshells of varied types. Some of the papers in Currie et al. (2004)

deal extensively with dinosaur and bird egg characteristics. Hirsch (1994) provided a careful review of fossil vertebrate eggs, emphasizing the shell structure distinctions between the many groups known in a fossil condition.

Vianey-Liaud et al. (1997) provided extensive data regarding Peruvian Late Cretaceous eggs, as well as additional understanding of the similarities and differences between eggshells of different types. Deeming (2006) made a good case for at least some dinosaurian eggs having been buried shallowly, as contrasted with most bird eggs, in a manner similar to that of crocodilians and many other reptiles. Zelenitsky and Thierrien (2008) provided a penetrating account of maniraptoran theropod eggs and their classification that considers those forms associated with skeletal remains of various types, embryos to adults. They avoided the use of previously defined parataxa. Although their work has nothing to do with eggs or egg cases, the description by Thomson et al. (2003) of the enigmatic Middle Devonian, Scottish *Palaeospondylus* as a larval lungfish is well worth noting with regard to what it implies about the conservatism of lungfish larvae.

Chondrichthyan "Cartilaginous Fish"

Müller (1978a) provided a useful review of both morphologic and stratigraphic range data for chondrichthyan egg cases, including the form genera *Fayolia*, *Palaeoxyris*, *Spirangium*, and *Vetacapsula*. Schneider and Reichel (1989) provided an excellent summary of certain Paleozoic chondrichthyan egg cases, as well as documentation about the stratigraphic ranges of the form genera. Crookall (1928) reviewed data chiefly on European and eastern North American *Fayolia* from the Carboniferous and considered egg case information from the Paleozoic and Mesozoic worldwide involving varied genera (Crookall, 1929).

Rössler and Schneider (1997) described some German Early Carboniferous chondrichthyan egg cases of the *Fayolia* type. McGhee and Richardson (1982) described the first North American occurrence at Mazon Creek of a Pennsylvanian example of the chondrichthyan egg case *Vetacapsula*. Sroka and Richardson (1997) described and

TABLE 30

Aquatic Invertebrate Sexual Behavior Evidence from the Fossil Record (Additions to Table 33 from Boucot, 1990a)

Behavioral Type	Taxonomic Level	Time Duration	Category	Refs.
10Ba. Dimorphism and brood care in ostracodes	Order	Ordovician–Holocene	1	Adamczak (1991); Boucot (1990a, p. 397); Lethiers et al. (1997)
10Bf. Patagonian oyster reproduction	Family	Pliocene–Holocene	1	Iribarne et al. (1990)
10Bg. Probable hirudinean and earthworm cocoons	Superfamily	Triassic–Holocene	1	Manum et al. (1992)
10Bj. Crab larvae	Infraorder	Early Cretaceous–Holocene	2B	Maisey and Carvalho (1995)
10Bk. Cladoceran eggs	Order	Oligocene–Holocene	2B	Lutz (1989)
10Bl. Estherian egg brooding	Suborder	Pennsylvanian–Holocene	1	Dechaseaux (1951); Gall and Grauvogel (1966)
10Bn. Argonaut egg cases	Superfamily	Oligocene–Holocene	2B	Holland (1988)
10Bo. Cupuladrid bryozoan reproduction	Family	Cenozoic	1	O'Dea et al. (2008)
10Bp. Copulating gastropods	Class	Cenozoic	1	Dieni (2008)

illustrated varied Mazon Creek chondrichthyan egg cases. Bock (1949) described a Late Triassic chinaeroid egg case. Sze (1954) described some Early Jurassic Chinese chondrichthyan egg cases but raised questions about their animal vs. plant identity; these specimens appear to be similar in all regards to those identified by others as chondrichthyan egg cases. Frickhinger (1994, Figure 400, p. 185) illustrated a Solnhofen Tithonian chimaeroid eggcase. Jarzembowski (1991) described the spiral eggcase *Spirangium jugleri* from the English Wealden. Dean (1909) described a Late Cretaceous North American chimaeroid egg case. Brown (1946, 1950) described some Jurassic and Cretaceous chimaeroid egg cases, as well as an Oligocene example. It is of interest that some of the descriptions of these egg cases mistook them for higher land plant organs of one kind or another.

While describing an egg case similar morphologically to those of chimaeroids from the probable mid-Devonian (probably Givetian) of the Bokkeveld Group, South Africa, Chaloner et al. (1980) reviewed the known occurrences of chimaeroid egg cases in the fossil record (see also Bessels, 1869; Brown, 1946a; Gill, 1905; Jaekel, 1901; Obruchev, 1967; Vakrameev and Pushcarovskii; 1954; Voronets, 1952; Warren, 1947). Chimaeroids are not known from the fossil record in beds as old as the Devonian, and Chaloner et al. (1980) suggested that a placoderm origin is conceivable because of the relations between placoderms and chimaeroids. It is puzzling that the relatively tough skate egg cases ("mermaid's purses") have not yet been recognized in the Jurassic to present record.

Although not dealing with eggs, the figure by Hagadorn (2001, Figure 9.3) of two Late Mississippian stethacanthid sharks from the Montana Bear Gulch Formation "likely preserved in a pre- or post-copulatory position" is not only unique but is also probably the only known occurrence of this sort of thing.

Amphibian

A trophic frog egg in amber and a tadpole emerging from its egg, also in amber, have been recognized by Poinar (see Figure 211 and Figure 218).

Reptilian

Although not strictly an egg, a nothosaur embryo from the Middle Triassic of the Swiss Alps, illustrated in Sanders (1988) in its curled attitude, probably reflects a nonmineralized egg with contained late embryo.

Dinosaur

Hirsch et al. (1989) discussed a unique Late Jurassic dinosaur egg from Utah that contains features indicating an embryo-like structure and teratologies compatible with retention in the female's oviduct that are similar to those present in modern reptiles as contrasted with birds. MacRae (1999, pp. 204–205) cited a clutch of Jurassic dinosaur eggs with

embryos from the Elliot Formation in Free State Province. Park et al. (2004) described a Late Cretaceous Korean egg-bearing deposit involving egg clutches. Mueller-Töwe et al. (2002) described varied dinosaur eggs, providing information on possible hatching techniques.

Mikhailov (1994) described dinosaurian eggs from Mongolia and Kazakhstan, and Mikhailov (2000) reviewed the record of Cretaceous dinosaur and bird eggs from Mongolia. Grigorescu et al. (1990) described Late Cretaceous dinosaur eggs from Romania, associated with calcrete. Dunham et al. (1989, Table 1) provided information on dinosaur egg clutch sizes, and Coombs (1989) summarized data on dinosaur eggs in a behavioral context. Varricchio et al. (1999) described a very well-preserved Late Cretaceous nest with well-disposed eggs from Montana that was attributed to *Troodon*. Norell et al. (1994) described Late Cretaceous theropod eggs from Mongolia that include skeletal remains of actual embryos. The eggshell is similar in structure to that of ratites. The problem of assigning associated skeletal remains to the taxon represented by the eggs is considered. Chiappe et al. (1998) described some Patagonian Late Cretaceous sauropod eggs complete with embryos.

The volume edited by Carpenter et al. (1994) contains a number of papers dealing with dinosaur eggs, including their occurrences geographically (just about every continent except Antarctica) and stratigraphically (Late Triassic to Late Cretaceous, with most occurrences being in the Cretaceous and the possibility that this represents a correlation with arid climate sites of preservation rather than a biological phenomenon), as well as the nature of egg nests and nest relations to each other at a number of sites, plus detailed information on eggshell morphology using varied analytical techniques. Dodson (1995) provided a realistic view of what may be currently said about dinosaur nests yielding eggs and their behavioral implications. Norell et al. (1995) described a Late Cretaceous Mongolian *Oviraptor* nest with eggs arranged in a radial manner around the rim (see also Section 10Cd, Nesting and Parental Care Among Dinosaurs and Crocodilians). Chiappe et al. (2004b) described what they interpret to be a Late Cretaceous sauropod nest with eggs from Patagonia. Sander et al. (2008) described, compared, and contrasted Late Cretaceous egg clutches (*Megaoolithus*) identified with titanosaurs, from France, Spain, India, and Argentina, while considering their environmental and behavioral implications.

Sahni and Khosla (1994) described some Indian Maastrichtian dinosaurian eggshell features, as well as nesting characteristics. Vianey-Liaud and Lopez-Martinez (1997) provided a detailed description of some Late Cretaceous, Spanish dinosaur eggshells. Grellet-Tinner et al. (2006) provided a variety of additional data about dinosaur eggs and their occurrences. Buffetaut et al. (2005) described some very small theropod-type eggs from the Early Cretaceous of Thailand that have a shell structure similar to that of birds rather than dinosaurs but other characters that are more dinosaurian.

Pterosaur

Wang and Zhou (2004) described an actual Early Cretaceous pterosaur egg with an included, well-skeletonized embryo within from Liaoning. Ji et al. (2004) described a leathery pterosaur egg, and Chiappe et al. (2004a) described one with a calcite shell. Distinctly different taxa are involved here, and both are Early Cretaceous.

Lizards

Kohring (1991) described an occurrence of lizard eggshells from the Early Cretaceous of Cuenca Province, Spain, and reviewed the fossil record of lizard eggs. A probable *Anolis* egg in Mexican amber is illustrated in Figure 196.

Gecko

Hirsch et al. (1987) described a gecko-type egg from the early Eocene of Wyoming. Hirsch and Harris (1989) described an *Achatina* shell from the Early Miocene of Kenya that is filled with calcified eggs. Study of the eggshell structure indicated that the eggs are those of a gecko rather than those of a gastropod. Modern geckos are reported to use cavities as egg-laying sites, although none has been reported from snail shells. Packard and Hirsch (1989) provided a thorough treatment of geckonid shell structure.

Turtles

Rothe and Klemmer (1991) described an occurrence of terrestrial turtle eggs (Testudinidae) from the Pliocene of Fuenteventura on the Canary Islands. Kohring (1990) described Late Jurassic chelonian eggshell from central Portugal.

Crocodilian

Rogers (2000) described an Early Cretaceous, Albian crocodilian egg from the Glen Rose Formation, Texas. Oliveira et al. (2008, p. 72) cited Upper Cretaceous crocodilian eggs from Brazil.

Bird

(See Figure 197.) Sauer (1972) discussed struthid and aepyornithid eggshell characteristics from the Canary Islands, as well as the possibility that the latter on Madagascar and the Canaries are the biogeographic relics of earlier, far more widespread populations. MacRae (1999, pp. 201–249) illustrated and briefly discussed some Miocene to Recent giant eggshells from Namibia. Mikhailov (1996) described some bird eggs from the Santonian–Campanian of Mongolia, with several form genera and species being erected. Schweitzer et al. (2002) described some Argentine Late Cretaceous bird eggs with embryos still preserved within, as well as detailing something about the eggshell structure in thin section. Zhou and Zhang (2004) described an unusually well-preserved precocial bird embryo from the Early Cretaceous of Liaoning. The eggshell of a hummingbird was recovered from Dominican amber; this is the first New World record of this bird lineage. Poinar et al. (2007a) described the material and discussed how it could have "arrived" in the amber.

Uncertain

Although the example is of uncertain affinities, Godfrey (1992) illustrated and discussed an elegant Mazon Creek, Illinois, Pennsylvanian specimen indicating the presence of eggs included in a gelatinous strand of the same type deposited by many amphibians and some fish. What may be the oldest known vertebrate eggs, possibly fish eggs laid on leaves, were reported from the Pennsylvanian of New Mexico (Mamay, 1994). Mamay et al. (1998) described what may be Early Permian amphibian eggs associated with transported leaves from north-central Texas.

☐ *Reliability*

Category 2B, functional morphology.

10Cb. CLASPERS AND PREGNANT CHONDRICHTHYANS

Janvier (1996; see also Long, 1995) provided extensive information about the presence of claspers in varied taxa. Bright (2002, pp. 23, 26) discussed reproduction involving the male grabbing the female by a pectoral fin and also illustrated actual insertion of a clasper into a cloaca. Pratt and Carrier (2005) provided a very extensive account of elasmobranch courtship and mating behavior. Grande and Buchheim (1994, Figures 9A,B) illustrated two specimens of an unnamed stingray, presumably *Heliobatis*, from the Eocene Green River Formation—one with a well-developed embryo in place, anterior pointing anteriorly, and the other with several similarly sized small specimens outside the female with either a newborn or postmortem delivery being indicated; see Grande (2001, Figure 3B) and varied figures in de Carvalho et al. (2004) for several taxa with embryos and one male with claspers (see Figure 267 in Chapter 14). Saint-Seine (1949, Plate IV, Figure A) illustrated a specimen of *Belemnobatis sismondae* with a well-preserved pair of claspers. Klug and Kriwet (2006) described an Early Jurassic shark with well-developed claspers. Lund (1982) described a Bear Gulch Mississippian Chondrenchelyiformes with claspers in the males but none in the females. The frontal tentaculum of chimaeras is presumed to be used by the male to grasp a pelvic fin of the female prior to copulation involving a clasper. Chimaeroid frontal tentaculae are known beginning with the Devonian (for a good example, see Lund, 1990).

☐ *Reliability*

Category 1, frozen behavior.

10Cc. ANURAN POLLUX, EGG-LAYING, AND LARVAL AMPHIBIANS

The oldest described anuran tadpole is from the Middle Jurassic of Inner Mongolia (Yuan et al., 2004). Báez (1981) described a Cretaceous frog. Chipman and Tschernov (2002) described an Early Cretaceous larval pipid frog growth sequence from Israel, and Roček and van Dijk (2006) further described this material while comparing it to other Cretaceous anuran material. Báez and Pugener (2003) discussed an Argentine Paleogene pipid frog growth series from tadpole to adult. Wuttke (1992a) illustrated Messel Eocene frogs with associated spawn (Figures 153, 154, and 365), as well as a tadpole (Figure 156). The Messel frog spawn is the oldest yet recognized. Although not involved with sexual evidence, Franzen's (2001, Figure 3) illustration of a Messel, Middle Eocene frog with skin outline, eyes, liver, and veins on the medial side of the thighs is notable as an example of soft-tissue preservation.

Wuttke (1989, Figure 6.5) illustrated an *Eopelobates* tadpole from the Rott Oligocene with the skin outline of the body still preserved. Roček (2003) described a developmental Oligocene larval sequence from the Czech Republic, and Roček and Rage (2000) summarized various Tertiary occurrences, except for those from South America.

Sun (1995, p. 1) cited a developmental series of frogs from tadpoles to adults from the Shanwang Miocene lake beds of northeastern China. Wassersug and Wake (1995) described several Miocene tadpoles from Turkey (see also Paicheler et al., 1978). Roček et al. (2006) described giant Miocene tadpoles from Germany. McNamara et al. (2006) discussed a Spanish Miocene set of tadpoles in which the lentic environment is conserved. See Eggs, Oviposition, and Maternal Care in Amber section, above (Figure 211 and Figure 218) for an account of several Dominican amber tadpole eggs, with one beginning to hatch, that probably came from bromeliad phytotelmata.

Olson (1985) described a larval Early Permian temnospondylous amphibian from Texas. Boy (1971, 1974) described a number of European Late Paleozoic larval amphibians (branchiosaurs). Milner (1982) described some larval temnospondyl amphibians from the Mazon Creek Pennsylvanian, with external gills, tails, and other larval features.

Gao and Shubin (2001) discussed and described several Chinese Late Jurassic salamander larval and adult forms that have undergone metamorphosis as in extant taxa, although the forms involved are primitive (i.e., the larval and neotenous phenomena involved are very ancient). Gao and Shubin (2003) described some relatively modern-type Bathonian, Jurassic salamander larvae from China associated with adult forms, external gills and other features being present.

☐ *Reliability*
Category 1, frozen behavior.

10Cd. NESTING AND PARENTAL CARE AMONG DINOSAURS AND CROCODILIANS

Coombs (1989; see also Dodson, 1995) provided the most thorough review to date of nesting and parental care in modern crocodilians, ratites, and megapode birds as it might apply to the dinosaurian record as well as to the crocodilian record. The overall point is that the modern archosaurian descendants have varied behaviors that can provide significant guidance in evaluating the tantalizing clues coming out

of dinosaur studies. Winkler and Murry (1989) provided clues to dinosaurian social behavior from a Texas Early Cretaceous site. The authoritative work here is certainly the volume edited by Carpenter et al. (1994). Norell et al.'s (1995) paper on a nesting *Oviraptor* sitting on a clutch of eggs provided strong evidence for the primitive archosaurian character of nesting behavior, whether it be birds, dinosaurs, or crocodilians. Meng et al.'s (2004) description of a unique Liaoning specimen featuring a number of young *Psittacosaur* associated with the remains of a single adult is suggestive of adult care. Norell et al. (1994), while describing some Mongolian Cretaceous theropod dinosaur egg nests, provided evidence favoring parental care. Sander et al. (2008) provided an extensive account of dinosaur egg clutches and their behavioral implications.

☐ *Reliability*
Category 1, frozen behavior, for the example in Norell et al. (1995).

10Ce. ICHTHYOSAUR AND MOSASAUR BIRTH DELIVERY ATTITUDE

Böttcher (1990) has eloquently summarized the overwhelming evidence from 46 skeletons for live birth, tail first, in Posidonienschiefer ichthyosaurs. Brinkmann (1996) described a Middle Triassic mixosaur from Switzerland in which at least three embryos are present and a tail-first delivery position is indicated. Massare and Callaway (1988) briefly noted tail-first delivery for Nevada Late Triassic ichthyosaur material. Dal Sasso and Pinna (1996) described a Middle Triassic Italian shastosaurid (*Besanosaurus leptorhynchus*) with contained embryos. Deeming et al. (1995) provided a thorough analysis, with new data, regarding ichthyosaur delivery position, coming to the same (tail-first) conclusion as earlier workers. Wang et al. (2008) briefly cited and illustrated a Chinese Carnian ichthyosaur with associated embryos. Kear et al. (2003) cited an Early Cretaceous Australian ichthyosaur with an enclosed fetus preserved with its skull in an "anterior" position relative to its posterior. Maxwell and Caldwell (2003) described an Albian ichthyosaur from the Northwest Territories, Canada, with an associated fetus.

Although not concerned with ichthyosaurs, Caldwell and Lee's (2001) account of a Cretaceous mosasaur (*Carsosaurus*) with four embryos having a tail-first orientation is notable (see Figure 230). Bell et al. (1996) also briefly discussed evidence supporting live birth in mosasaurs, with evidence of dogfish scavenging being provided by numerous, scattered teeth. Shine (1985) reviewed the many living reptilian groups in which viviparity has independently developed; he concluded that the taxa he considered, all terrestrial, are more likely cold-climate items, but this correlation does not do for completely aquatic taxa that are viviparous, as are all of the known occurrences from the fossil record. Purely aquatic mammals, such as whales and manatees, also deliver posterior first to prevent drowning of the offspring. Cheng et al.

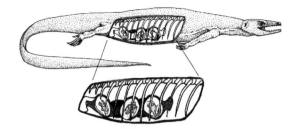

FIGURE 230 Evidence for live-bearing in a Cretaceous mosasaur with embryos. (Figures 1c,d in Caldwell, M.W. and Lee, M.S.Y., *Proceedings of the Royal Society London B*, 268, 2397–2401, 2001. Reproduced with permission.)

(2004) described several Chinese Middle Triassic pachypleurosaurs (*Keichousaurus hui*) containing several embryos to further document the situation.

□ *Reliability*
Category 1, frozen behavior.

10Cf. POSSIBLE NURSING

Voorhies (1981) discussed the possibility that some of the Miocene rhinoceratid *Teleoceras* from Nebraska were preserved in such a position that nursing was involved. Sigé et al. (1998) illustrated mid-Eocene, Messel microchiropterans with twin embryos; it is probable that the earlier example from the same locality (Maugh, 1975) for which nursing was considered as a possibility actually belongs here. These specimens are also of great behavioral interest with the anterior part of the deciduous dentition in a "highly derived condition found in living bats for clinging to the mother's fur."

□ *Reliability*
Category 1, frozen behavior.

10Cg. OTARIOID SEAL ROOKERY

Renouf (1991) provided an excellent account of pinniped behavior, including breeding, which helps to better understand the fossil record of same (see Boucot, 1990a).

10Ch. INTERNAL FERTILIZATION IN PLACENTAL MAMMALS (BACULA AND PREGNANT FEMALES)

Dixson (1995) considered the correlations between baculum length and intromission time in varied placentals and reviewed the difficult question of potential baculum function. Wang and Rothschild (1992) cited baculae from *Hesperocyon* and another canid genus from specimens in the Oligocene lower Brule Formation of Brule County, South Dakota. Harington (1996) described a Pliocene locality on Ellesmere Island where the baculum of *Plesiogula* has been found. Franzen (2000) cited a mid-Eocene, Messel primate baculum, and Storch et al. (2002) cited a baculum present in a Messel, mid-Eocene bat.

Franzen (1992) provided additional information on Messel Eocene pregnant horses (for pregnant *Propalaeotherium*, see von Koenigswald et al., 2004, Figures 3, 4). Storch (2001) noted that the Messel horses had only a single fetus, as is the case with modern horses, and MacFadden (1992, Figure 4.7) illustrated a Miocene *Protohippus supremus* from Nebraska with a single fetus. This is also the case for a pregnant pangolin, whereas the bats have two fetuses; Storch (2001) mentioned the curiously hooked deciduous teeth of the younger bats that presumably aid in clinging to the female's fur (see also Sigé et al., 1998, for details). Franzen (1997) briefly described an unusual Wyoming Oligocene oreodont *Miniochoerus gracilis* (Leidy 1851) with a single fetus. Mead (2000) illustrated fetal bones within the pelvis of a *Teleoceras* specimen from the Nebraska Miocene and commented on prior discussion of some skull wounds in males as having been made during butting contests.

Gingerich et al. (2009) described and analyzed several unique Pakistan Middle Eocene protocetid whale specimens, one a female with a near-term fetus and the other a slightly larger male. The female specimen contains the remains of a nearly full-term fetus. The fetus is positioned for head-first delivery, as in all terrestrial mammals and in presumed contrast to more modern whales, including slightly younger archaeocete whales with limb morphologies that would have confined them to the water, as contrasted with the protocetids with their fully developed hind limbs. The tooth development stage of the fetus suggests that it was precocial.

□ *Reliability*
Category 2B, functional morphology, and Category 1, frozen behavior, for the pregnant females.

10Ci. FIGHTING PHYTOSAURS

Ruben (1990) discussed the possibility that fighting capable of inflicting puncture wounds in both phytosaur and crocodilian skeletons was a primitive archosaurian character. While describing similar evidence from an Eocene crocodilian, Buffetaut (1983) reviewed the possibility and came to the same conclusion. Williamson (1996) has described a Late Cretaceous, Campanian example of such crocodilian skeletal trauma that had healed following the injury (to the dentary). Sawyer and Erickson (1998) described varied Paleocene examples of injuries possibly associated with intraspecific fighting and provided additional pathologic skeletal information. While describing evidence of possible Pliocene intraspecific fighting from Queensland, Mackness and Sutton (2000) reviewed a number of examples of possible aggressive behavior recorded from fossil crocodilians of earlier Cenozoic and Late Cretaceous age. Katsura (2004) described evidence (e.g., anterior third of the mandible amputated with signs of healing, right tibia and fibula fractured and later fused, plus puncture marks on several scutes) of probable intraspecific fighting in a Middle Pleistocene crocodilian from Japan and provided an excellent review of the

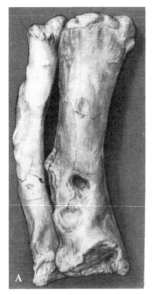
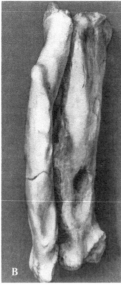

FIGURE 231 Bitten right epipodials of *Gavialosuchus* showing penetration and distortion of tibia/fibula. (A) Distorted doral view; (B) distorted more lateral view (about 0.5 natural size). (Figure 25 in Erickson, B.R. and Sawyer, G.T., *Monographs of the Science Museum of Minnesota*, 3, 1–47, 1996. Courtesy of the Science Museum of Minnesota.)

paleontological evidence known elsewhere for intraspecific crocodilian fighting. Erickson and Sawyer (1996) described fractured limb bones and deep bite marks in *Gavialosuchus* from the South Carolina Oligocene (see Figure 231 and Figure 232). Sawyer and Erickson (1987) provided more evidence of crocodilian behaviors resulting in varied wounds to limbs (including amputation) and jaws. This is clearly an area where the accumulated evidence is beginning to be very convincing. One should now more carefully examine other archosaurian groups for similar evidence.

☐ *Reliability*
Category 2B, functional morphology.

10Cj. Fish Nests

There is an extensive literature on fish nests and associated parental care (Blumer, 1979; Gebhardt, 1987; Gross and Sargent, 1985; Perrone and Zaret, 1979). Gross and Sargent (1985) pointed out that parental care in the fishes is far more varied than in the other vertebrate classes. This same point could be made about reproductive behaviors from merely skimming Breder and Rosen (1966). What this means is that the class Pisces is far more diverse than is the case for the other vertebrate classes. Feibel (1987) described a Kenyan

FIGURE 232 Alignment of skull teeth with punctures in tibia showing relationship to other bones of the right hind limb and lower jaw of *Gavialosuchus*. (Figure 26 in Erickson, B.R. and Sawyer, G.T., *Monographs of the Science Museum of Minnesota*, 3, 1–47, 1996. Courtesy of the Science Museum of Minnesota.)

Plio–Pleistocene trace fossil that is reasonably interpreted as a fish nest.

*10Ck. Viviparity or Ovoviviparity in Fishes

The well-known ovoviviparous condition in the modern coelacanth *Latimeria* (Levett Smith et al., 1975) and in the Solnhofen Late Jurassic coelacanth *Holophragmus* (=*Birgeria*) (Watson, 1927) is worth mentioning. Beltan (1997) has described a Madagascar Early Triassic actinopterygian in which viviparity or ovoviviparity is also indicated. Breder and Rosen (1966) provided an extensive account of reproductive styles known in fishes (although preceding the 1975 coelacanth discovery). Rieppel (1985) reported viviparity in the Middle Triassic from Switzerland in *Saurichthys*. Long et al. (2008) described a remarkable Late Devonian occurrence of a ptychodont placoderm from Western Australia with a well-preserved fetal specimen still connected to its umbilical cord and egg sac, in a chondrichthyan-type relation, plus a second specimen of a second taxon with a similarly contained fetus (i.e., viviparity and evidence of internal fertilization).

☐ *Reliability*
Category 1, frozen behavior.

Summary

Information at class and ordinal levels (see Table 31).

TABLE 31

Vertebrate Sexual Behavior Evidence from the Fossil Record (Additions to Table 34 from Boucot, 1990a)

Behavioral Type		Taxonomic Level	Time Duration	Category	Refs.
10Ca.	Vertebrate eggs and egg cases				
	Chondrichthyan	Class	Middle Devonian–Holocene	2B	Chaloner et al. (1980)
	Reptilian				
	Lizard	Suborder	Early Cretaceous–Holocene	2B	Kohring (1991)
	Gecko	Family	Eocene–Holocene	2B	Hirsch et al. (1987)
	Dinosaur	Orders	Late Triassic–Cretaceous	1, 2B	Carpenter et al. (1994)
	Crocodilia	Order	Early Cretaceous–Holocene	2B	Rogers (2000)
	Chelonia	Order	Late Jurassic–Holocene	2B	Kohring (1990)
	Aves	Class	Late Cretaceous–Holocene	1	Schweitzer et al. (2002)
10Cb.	Claspers and pregnant chondrichthyans	Class	Late Cretaceous–Holocene	1	Boucot (1990a, p. 415)
10Cc.	Anuran pollux, egg-laying, and larval amphibians	Class	Pennsylvanian–Holocene	1	Boucot (1990a, p. 417)
10Ce.	Ichthyosaur birth delivery attitude	Class	Middle Triassic–Jurassic	1	Brinkmann (1996)
10Ck.	Viviparity or ovoviviparity in fishes	Class	Triassic–Holocene	1	Levett Smith et al. (1975)

11 Parental Care

*INVERTEBRATES

McNamara (McNamara and Barrie, 1992) described a new South Australian Miocene marsupiate echinoid and redated without comment Lambert's (1933) marsupiate echinoid from Madagascar as Paleocene rather than Cretaceous. The oldest known marsupiate regular echinoid is Blake and Zinsmeister's (1991; see also Smith and Jeffery, 2000, p. 187) *Amucidaris durhami* from the Maastrichtian of Seymour Island. Smith and Jeffery (2000) pointed out that the oldest irregular is from the Danian, and Blake and Zinsmeister (1991) observed that, although the majority of modern marsupiate echinoids occur at higher latitudes, there are enough occurring in warm, lower latitude environments to make a one-to-one correlation of brooding with cool to cold water a risky interpretation. McNamara (1994a) reconsidered the entire question of the reasons for the evolution of marsupia and comes down in favor of high seasonality rather than merely low temperature. Roman (1983) described some European Miocene and Pliocene marsupiate echinoids that further emphasize this problem, as there is no evidence for very cold waters being involved.

Kruger (1997) described some French Lutetian marsupiate echinoids and some from the Danian in the Netherlands. Liao and Lin (1981) described a Pliocene marsupiate echinoid, *Tetradiella*, from Guangxi, in southwestern-most China, an undoubtedly warmwater region. Jeffery (1997) noted that non-planktotrophy in echinoids first appears in the late Campanian and that marsupia may have more to do with a seasonal food supply than with cold water. He also observed that female gonopore size and *c*-axis calcite plate orientations are useful in sexing echinoids. Van der Ham (1988) described some Early Paleocene marsupiate echinoids from the Maastricht region. Jagt and van der Ham (1994) described some Paleocene marsupiate regular echinoids from Belgium and concluded that they represent factors other than a coldwater environment. Néraudeau et al. (2003) described some Redonian marsupiate echinoids from northwestern France and made a case for correlation with coldwater environments. Kier (1969, 1987) illustrated an Eocene marsupiate from Georgia and provided an excellent discussion concerning fossil marsupiate echinoids (Kier, 1967). Roman (1976) described some Miocene marsupiate echinoids from Quatar. Poulia and Feral (1996) suggested that the presence of Antarctic marsupiates may reflect a more seasonal food supply rather than having anything to do with low temperatures.

Dudicourt et al. (2005) described some Pliocene marsupiate echinoids from southwestern France in another relatively warm region. Poulin et al. (2002) reviewed the Antarctic marsupiate echinoids and their geological history while making the novel suggestion that the marsupiates represent selection in the earlier Tertiary for coldwater conditions, and the abundant nonmarsupiates of one genus reflect more moderate temperature conditions relatively recently. All in all, these data suggest that more than one selective factor has been involved in the development of the marsupiate condition, as both coldwater and warm, temperate environments are involved.

Smith (1997) makes a good case for paternal care of the eggs, by watering them in their elevated site, in a German Late Jurassic belostomatid giant water bug (Figure 233), based on known sizes of varied instars and their eggs from Solnhofen; this water bug belongs to a group where the male brood the eggs on their backs and keep them moist. Vega et al. (2006) briefly described a poorly preserved, possible belostomatid with eggs on its back from the Albian of Chiapas, Mexico.

□ *Reliability*
Category 2B, functional morphology.

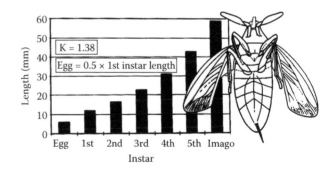

FIGURE 233 Smith (1997) made a good case that a Late Jurassic giant water bug, *Mesobelostomum leperditum*, from Solnhofen belongs to a group where the males brood the eggs on their backs and keep them moist. The histogram indicates that the adult, imago size of the *Mesobelostomum* is of the size where today the large eggs are brooded on the backs of the males. (From Smith, R.L., in J.C. Choe and B.J. Crespi, Eds., *The Evolution of Social Behavior in Insects and Arachnids*, Cambridge University Press, Cambridge, U.K. Reproduced with permission.

*VERTEBRATES

Haynes' work (1991; see also Section 5k, Acridid Aggregation) on Quaternary proboscidian aggregations provided good evidence from the preserved population structure favoring parental care in mammoths and mastodons similar to that in living elephants. Norell et al. (1994), while describing some Mongolian Cretaceous theropod dinosaur egg nests, provided evidence favoring parental care, but Ruben et al. (2003, pp. 157–158) refuted Norell's concept by pointing out the behaviors of modern crocodilians.

□ *Reliability*
Category 1, frozen behavior.

12 Depth Behavior

*12a. HETEROPODS

Bandel and Hemleben (1987) made a good case that masses of minute Early Jurassic gastropods from the Posidonienschiefer of southern Germany are actually the remains of planktonic heteropods little different from modern representatives of the same group. This is still more evidence of the relatively modern nature of the marine biota beginning with the Jurassic.

☐ *Reliability*
Category 2B, based on functional morphology.

*12b. PALEOECOLOGIC DISSONANCE

Woodring (1960) coined the term "paleoecologic dissonance" to cover those examples from the fossil record where an organism radically changed its environmental tolerances over evolutionary time. He used as an example *Astarte* (today, a very-cool-climate bivalve) from the Eocene London Clay, where it was associated with tropical/subtropical mangrove community organisms. Arkell (1956, p. 616) cited three Jurassic bivalve genera—*Trigonia*, *Astarte*, and *Pholadomya*—that commonly occur together in a shallow-water community and whose descendants have radically changed their temperature or depth tolerances. Today, *Astarte* is boreal, *Trigonia* only lives in warm Australian shallow waters, and *Pholadomya* is abyssal. Feldmann and Wilson (1988) cited a shallow-water Antarctic Eocene malacostracan example whose modern descendent is an abyssal organism. Cooper (1986) described a Cenozoic brachiopod example that has similarly changed its depth tolerances. Feldmann et al. (1993) summarized three Antarctic Late Cretaceous crustacean taxa that occupied inner-shelf depths during the Cretaceous, whereas their extant descendants are restricted to outer-shelf and bathyal depths. Wiedman et al. (1988) described the Seymour Island, Antarctica, brachiopods and observed that many of the 11 genera from this temperate, Eocene, shallow-water environment are represented today by bathyal, coldwater forms. They added the proviso that, because we know so little about Eocene bathyal brachiopods, there is the possibility that this might represent a "restriction" in the environmental tolerances of these genera rather than a change from shallow, temperate to deep, cold.

Consider also the photic-zone, late Middle Devonian monoplacophoran *Pilina* and its modern abyssal relative *Neopilina* (Runnegar, 1987). Studencka (1987) has documented the fact that the modern abyssal bivalve *Kelliella* occurs in the fossil condition in the photic-zone, shallow-water mid-Miocene of southern Poland. Nehm and Geary (1994) discussed a change from shallower to deeper water in the evolution of Neogene *Prunum* species. Potter and Boucot (1992) recognized that the Middle Ordovician–earlier Devonian brachiopod *Skenidioides* had a depth range of benthic assemblages 3 through 5 in the Ordovician but was restricted to mostly 5 and deeper 4 in the Silurian–Devonian (Boucot and Lawson, 1999), with none known from 3. Such examples of ecologic dissonance are very, very rare, the point being that most organisms at the generic level do not change their environmental tolerances over evolutionary time. Smith (2004) discussed the evidence for a number of holasteroids (Cretaceous and Cenozoic) being involved in environmental changes from shallow marine waters to relatively deep-sea environments; his evidence is convincing.

Racki (1982) discussed the apparently major environmental shift characteristic of the charophytes, or at least their oogonia, from the Silurian to the present. In the Paleozoic, they are apparently a fully marine to brackish water and freshwater group, whereas in the post-Paleozoic they are a strictly freshwater to brackish water group. Admittedly, the Paleozoic taxa at the generic to familial levels are distinct from those of the post-Paleozoic, which suggests a quantum evolution change near the Permian–Triassic boundary, but the distinction is impressive.

Rasnitsyn and Quicke (2002, p. 23) observed that stoneflies as a group have undergone several major environmental changes, resulting in a greater Mesozoic presence as contrasted with their much lower abundance in the Paleozoic and Cenozoic; they suggested that this probably results from a change in ecology for most taxa—more lotic in the Late Cretaceous and Cenozoic compared with lentic dominance earlier. Lake bed preservation is far better over time than is the case with fast-flowing water deposits.

Norris et al. (1996) discussed a foraminiferal lineage's environmental change through time from shallow to deep water. Boucot and Kříž (1999) reviewed the entire question of paleoecological dissonance, based on new information provided by Kříž's (1999) work on Silurian–Devonian bivalve communities. Kříž (1999) made a strong case for the ecological functioning of some of these communities, including phyletically descendant taxa, having changed through time. In other words, what had been an homologous series of communities evolved phyletically into an analogous community. This being the case, Boucot and Kříž proposed that it was necessary to modify the term "analogous community" with the modifiers "related" and "unrelated" to make clear that some examples involve taxa that, although functionally similar, are evolutionarily unrelated, in agreement with Boucot's (1981, pp. 360–361) original concept of analogous community, whereas others are closely related in a phyletically evolving manner. Woodring's (1960) original concept of ecological dissonance involved genera that had changed their ecological tolerances; that is, they are all examples of related analogous behavioral change insofar as the individual genera are concerned. (Note that, in his paper, nothing was said about the communities involved, and one assumes that only the named genera were involved in changing their behaviors through evolutionary time.) Kříž (1999) listed the genera that had changed their behaviors and represented examples of paleoecological dissonance.

Teichert (1991) documented the shift of the bizarre Permian crinoid *Calceolispongia* from coldwater, Gondwana Realm environments early in its history to warmwater, extra-Gondwana Realm "Tethyan" environments during the later Permian. Oleinik (1996) provided evidence that the warmwater marine gastropod *Arctomelon* of the Eocene gave rise to Neogene descendants that live in cold water. Park (1968) concluded that some subgenera of the Atlantic and Gulf Coast bivalve *Venericardia* underwent ecological changes during the Tertiary, whereas others did not. His evidence is convincing. Das (2002) suggested that Mesozoic pleurotomarian gastropods are shallow-water, soft-bottom forms, whereas a Cenozoic shift to bathyal, rocky bottoms occurred.

Oji (1996) made a good case that stalked crinoids as a group are essentially absent from shallow water since the earlier Jurassic, being present subsequently only in bathyal and deeper regions; however, because we know little about the bathyal and deeper region Triassic and Paleozoic stalked crinoid fauna, there is some uncertainty about whether the Jurassic to present situation indicates merely removal from the shelf environments or movement from the shelf to the bathyal and deeper regions, particularly as the Paleozoic crinoid orders became extinct by the end of the Permian.

Among the terrestrial vertebrates the movement of some proboscidians from warm to Arctic environments is notable (Maglio, 1973), as well as the presence of the wooly rhinoceros!

□ *Reliability*
Category 1, based on the paleoecologically depth-indicating information.

*12c. PELAGIC TRILOBITE DEPTH SELECTION

(See Figure 234 and Figure 235.) Boucot (1990a) commented on the widespread evidence of depth control being involved with benthic larval substrate behavior. In a very innovative

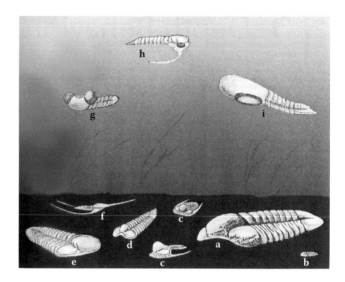

FIGURE 234 An atheloptic assemblage of blind, or nearly blind, bottom-dwelling Ordovician trilobites (a–f), and an overlying group with hypertrophied eyes (g–i) that probably lived in mesopelagic depths. (a) *Illaenopsis*, (b) *Shumardia*, (c) *Bergamia*, (d) *Colpocoryphe*, (e) *Orthamops*, (f) *Ampyx*, (g) *Pricyclopyge*, (h) *Fenniops*, and (i) *Degamella*. (Figure 13 in Fortey, R.A. and Owens, R.M., in *Evolutionary Trends*, Owens, R.M. and McNamara, K.J., Eds., University of Arizona Press, Tucson, 1990, pp. 121–142.)

paper, McCormick and Fortey (1998) indicated how the characteristics of the eyes in two pelagic Ordovician trilobite genera (*Carolinites*, epipelagic with large eyes, and *Pricyclopyge*, mesopelagic with *very* large eyes) concur with other information; for example, *Carolinites* occurs in shallow- to deep-water deposits, consistent with a relatively shallow depth, whereas *Pricyclopyge* occurs only in deeper water deposits, consistent with a deeper environment. This

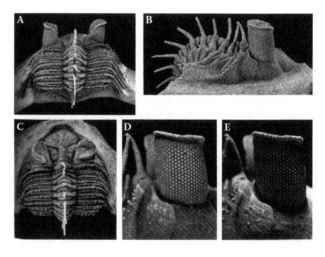

FIGURE 235 A unique Emsian, Early Devonian, Moroccan trilobite (*Erbenochile erbeni*) with an "eyeshade"; the headshield in Part C is 32 mm across. Part D is a side view of the right eye, showing lenses under optimum illumination, and Part E shows how the eyeshade cut out light from directly above. (Figure 1 in Fortey, R. and Chatterton, B., *Science*, 301, 1689, 2003. Reproduced with permission from AAAS.)

observation strongly substantiates the suggestion that these two trilobites lived at distinctly different water depths.

Fortey and Owens (1987) coined the term *atheloptic* ("shrunken eyes") for deeper water, blind to almost blind trilobites, possibly about benthic assemblage 6; the term may also be associated with cyclopygid trilobites with hypertrophied, large eyes, consistent with a mesopelagic life style (for good examples, see Crônier and Fortey, 2006; Fortey, 2006; Feist, 1995). Cotton (2001) discussed the blind Early and Middle Cambrian Conocoryphidae. Included in the atheloptic category are Feist and Clarkson's (1989) late Middle and early Late Devonian trilobites. Owens and Tilsley (1995) described a group of atheloptic trilobites from the Visean of Devon, while comparing them with similar age materials from Moravia; they also cited other British and German examples of Mississippian age. Vaněk (1995) described blind later Ordovician trilobites (a new species of *Colpocoryphe*, *Marrolithus*, *Pragolithus*, *Raphiophorus*). In all of these examples, the atheloptic condition originated convergently from different, unrelated groups that occupied similar deep-water, benthic environments where vision was clearly of little use. Fortey and Chatterton (2003) described a unique Devonian trilobite from Morocco that developed an "eye-shade" to improve its distance vision, unimpeded by light from above.

In line with the above is Benson's (1975) work showing that later Cenozoic psychrospheric deep-sea ostracodes were blind and that blindness in marine ostracodes is largely lacking in shelf forms, although it began to appear in deeper water. Oertli (1974) provided some similar evidence for the Albian from Deep Sea Drilling Project (DSPD) ostracods.

☐ *Reliability*

Category 2, based on functional morphology and synecology.

*12d. DEPTH DISTRIBUTIONS

Chen (2009) provided a method for determining the depth distributions (upper, lower, and maximum abundance) for pelagic organisms of the past.

13 Phoresy

One of the more gaudy examples of modern phoresy is Colwell's (1983) account of a mutualism among hummingbirds, the hummingbird flower mite (*Rhinoseuis colwelli*), and flowers. The mites are transported *inside* the nasal passages of the hummingbird, from which they emerge onto the flower while the bird is feeding. Colwell commented that, "For their size, they run about as fast as a cheetah—a useful ability when they have only 2 or 3 sec to disembark from a long-billed hummingbird." The mites then lay their eggs on the flowers. Archibald and Mathewes (2000) cited a possible example of phoresy involving a march fly and a beetle from the Early Eocene of British Columbia. Davis (1989) cited, but did not describe, a Dominican amber micropezid fly with phoretic mites on the ovipositor.

13a. PSEUDOSCORPIONS

(See Figures 236 through 239.) Poinar et al. (1998) surveyed the known examples of pseudoscorpion phoresy for both insects and chelicerates from the present and the fossil record. The modern and fossil examples make it very clear that the phoretic relations are very obligate; specific pseudoscorpion taxa have phoretic relations with specific insect and chelicerate taxa, and in one example even the position on the "host" is stereotypical. These represent a most elegant set of fixed coevolutionary examples, such as a Baltic amber specimen of *Chthonius* on the back of a moth. The Weitschat and Wichard (1998) example of phoresy with an opilionid became available too late to have been included in the

survey, but it further extends the conclusions just cited. Ross (1998, Figure 75) illustrated a snipe fly, Rhagionidae, with a phoretic pseudoscorpion hanging onto a leg. Kosmowska-Ceranowicz (2001, p. 27) illustrated a Baltic amber pseudoscorpion associated with a nematoceran fly that may be a

FIGURE 237 Pseudoscorpion attached to the leg of a long-legged fly (Dolichopodidae) in Baltic amber (Andrew Cholewinski collection); length of pseudoscorpion = ~1.5 mm.

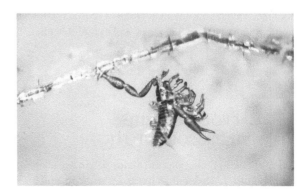

FIGURE 238 A pseudoscorpion attached to the leg of a harvestman in Baltic amber (Andrew Cholewinski collection); length of pseudoscorpion = ~1 mm.

FIGURE 236 Pseudoscorpion attached to a scolytid beetle in Mexican amber (unknown collection); length of beetle = 2 mm.

FIGURE 239   A pseudoscorpion (*Chthonius* sp.) on the back of a moth in Baltic amber (Poinar amber collection accession no. Sy-1-78); length of pseudoscorpion = 1.3 mm.

fungus gnat (Mycetophilidae); although it was not attached to the fly, it may represent phoresy. Poinar (1992b, Figure 113; see also Poinar and Poinar, 1999, Figure 60) illustrated a Dominican amber pseudoscorpion holding onto the abdomen of a platypodid beetle. Grimaldi (1996, p. 82) illustrated a platypodid beetle with a phoretic pseudoscorpion from an unspecified amber locality. Weitschat and Wichard (1998, Plate 11, Figures a–c) illustrated pseudoscorpions phoretic on a braconid wasp and on an opilionid.

☐ *Reliability*

Category 1, frozen behavior.

13b.   MITES WITH A MIDGE AND WITH A BARK BEETLE AND OTHER INSECTS

(See also frontispiece and Figures 40 and 45 through 50.) Poinar (1992b, Figure 135) illustrated a stingless bee, *Proplebeia dominicana*, with attached phoretic deutonymphs (hypopi) of the suborder Astigmati. Poinar (1992b, p. 246) also cited phoretic mites on adult platypodid beetles, adult chironomid midges, adult tipulid flies, and termites in Dominican amber. Poinar (1993, Figure 10) illustrated phoretic mites on the underside of a Dominican amber beetle belonging to the Leiodidae. Poinar and Poinar (1999, Figure 59) illustrated phoretic mites on the underside of a platypodid beetle.

☐ *Reliability*

Category 1, frozen behavior.

*13c.   MACROCHELID MITES AND DROSOPHILID FLIES

(See Figure 240.) Poinar and Grimaldi (1990; see also Poinar, 1992b, Figure 134) documented a Dominican amber, Late Eocene–Early Miocene drosophilid fly with phoretically attached macrochelid mites on its abdomen in the same manner as living examples. They pointed out that drosophilids from the same deposit were parasitized by allantonematid nematodes (Poinar, 1984), on which modern macrochelids

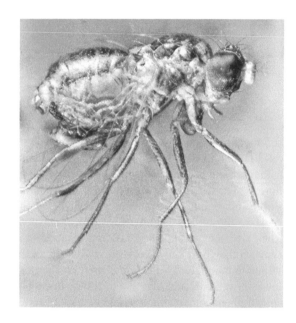

FIGURE 240   A drosophilid fly with three macrochelid mites attached to its abdomen in Dominican amber (Poinar amber collection accession no. Sy-1-2); length of fly = 2 mm.

feed, suggesting that a potentially complex relationship among mites, nematodes, and flies may have been established long ago.

☐ *Reliability*

Category 1, frozen behavior.

*13d.   MAMMALIAN HAIR EPIZOOCHORY

(See Figure 241.) Although not an example of phoresy in the normal sense, truly noteworthy is Poinar and Columbus's account (1992) of a Dominican amber, Eocene–Miocene

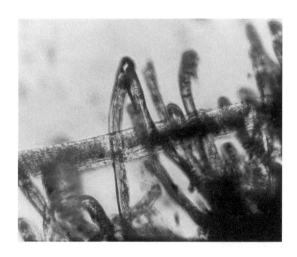

FIGURE 241   A rodent hair attached to hooks of a spikelet of the bamboo *Pharus primuncinatus* in Dominican amber provides the first fossil record of epizoochory, the transportation of seeds by animals (Poinar amber collection accession no. M-1-4); lengths of the hooked plant hairs grasping the rodent hair range from 0.22 to 0.56 mm.

mammal hair attached to specialized hooked hairs on the lemma of a grass seed, *Pharus*, itself the only known fossil example of the bamboo subfamily; this situation is termed *epizoochory* rather than phoresy (see also Poinar and Poinar, 1999, Figures 16, 17; Poinar and Judziewicz, 2005). Poinar (1998b, Figure 3B) illustrated this specimen and commented that the hair probably belongs to a carnivore. Angiosperm seeds commonly have various morphological features to aid dispersal, one of which is the presence of sticky, glandular hairs that adhere to animals. Two seeds in Burmese amber are covered with such glandular trichomes (see the back cover, Figure B).

☐ *Reliability*

Category 1, frozen behavior.

*13e.   FEMALE FIG WASPS AND NEMATODES

(See Figure 242.) Poinar and Poinar (1999, Figure 23, pp. 32, 34) illustrated and described a very specific nematode–parasitic fig wasp relationship wherein the nematodes use the wasp to get from fig flower to fig flower.

☐ *Reliability*

Category 1, frozen behavior.

*13f.   SWINGING SPRINGTAILS: PHORETIC BEHAVIOR IN FOSSIL COLLEMBOLA

by George O. Poinar, Jr.

The 6500 described species of Collembola, or springtails, comprise one of the most widespread and abundant groups of arthropods. These small animals rarely achieve lengths greater than 4 mm and obtain nourishment from plant (especially fungi) and animal matter (Hopkin, 1997). They are wingless and depend on crawling or jumping for locomotion. Wind movement, water runoff, and ocean currents are the major long-distance means of dispersal reported for members of this group (Christiansen, 1964; Hopkin, 1997).

There are no reports of phoretic associations involving springtails and other invertebrates even though many larger

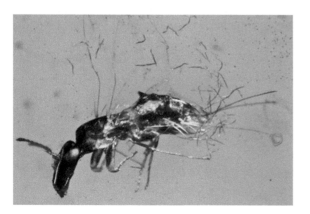

FIGURE 242   A Dominican amber fig wasp parasitized by nematodes (Poinar amber collection accession no. N-3-24); length of wasp = 3 mm.

FIGURE 243   Row of five individuals of *Sminthurus longicornis* adjacent to the leg of the harvestman *Dicranopalpus ramiger* in Baltic amber (Andrew Cholewinski amber collection). Note how most of the antennae of the springtails are curved toward the leg, suggesting that they were originally secured to this appendage. The length of each individual is about 1 mm.

winged and wingless forms that inhabit the same milieu as springtails serve as carriers for other flightless arthropods such as mites (Binns, 1982) and pseudoscorpions (Poinar et al., 1998). Inclusions in amber are well known for portraying examples of frozen behavior, including symbiotic associations between two or more species. A 40-million-year-old (late Eocene) piece of Baltic amber (Figure 243) from northern Russia (Kaliningrad region) contains an interesting association between a harvestman and five springtails that is both revealing and perplexing because it suggests a previously undescribed phoretic association (Poinar, 1992b). The specimen containing the springtails was repolished in order to better view details and was viewed under both a dissecting and compound microscope. The final piece of amber weighs 1 g and is approximately circular in outline, with a diameter of 15 mm and a greatest depth of 5 mm. It is maintained in the Andrew Cholewinski amber collection. The springtails have been identified as the extinct Baltic amber species *Sminthurus longicornis* Koch & Berendt 1854. The harvestman has been identified as a juvenile of *Dicranopalpus ramiger* Koch & Berendt 1854 (Phalangidae), also described from Baltic amber.

Five springtails are aligned in a row adjacent to the right second leg of the harvestman. One or both antennae of four springtails are completely or partially curved over the leg; the antennae of the fifth individual are extended adjacent to the harvestman's leg. Four of the springtails are facing away from the leg while the fifth is facing the leg. Of those facing away from the leg, three have the antennae curved backward toward or over the leg. The individual facing the leg has the antennae bent forward over the leg. The situation presented suggests that the springtails were holding onto the leg of the harvestman by their antennae in life and were carried into the resin when the harvestman walked or fell into the sticky substance. The springtails had time to only loosen their grip but no time to move away, thus remaining juxtaposed with the leg.

There are no reports of extant springtails being phoretic on any arthropod or displaying the type of behavior described here; however, Andrew Cholewinski informed the authors

TABLE 32

Phoretic Relations Preserved in the Fossil Record (Additions to Table 35 from Boucot, 1990a)

Behavioral Type	Taxonomic Level	Time Duration	Category	Refs.
13a. Pseudoscorpions	Genus–genus	Eocene–Holocene	1	Poinar et al. (1998a)
13b. Mites	Genus–genus	Oligocene–Holocene	1	Poinar (1992b, 1993)
13c. Macrochelid mites and drosophilid flies	Genus–genus	Miocene–Holocene	1	Poinar (1992b)
13d. Mammalian hair epizoochory	Genus–class	Miocene–Holocene	1	Poinar and Poinar (1999)
13e. Female fig wasps and nematodes	Genus–genus	Miocene–Holocene	1	Poinar and Poinar (1999)
13f. Swinging springtails	Genus	Eocene	2A	See text

(pers. comm., 2006) that he had found an additional harvestman in Baltic amber with two springtails attached by their antennae to one of its legs. Most springtails have four segmented, relatively short antennae that are not adapted to holding objects (Daly et al., 1998). However, it has been reported that the males of sminthurid springtails have slightly modified antennae capable of lifting females or being themselves lifted from the surface during mating rituals, thus illustrating the ability of the antennae to manipulate and support the animal (Massoud and Betsch, 1966). Also, the antennae of *Sminthurus longicornis* are composed of 18 segments; the terminal 15 segments are often considered as secondary segments of the final, fourth segment, but they are quite distinct in this species. The terminal segments are very flexible and capable of bending over structures, as demonstrated in this case. Such long antennae could be an adaptation for grasping objects.

Other soil arthropods, such as pseudoscorpions, are known to grasp the legs of harvestmen in phoretic associations in the present (Poinar et al., 1998a) and in Baltic amber (Weitschat and Wichard, 1998). In addition, mites have been observed attached to the legs of extant opilionids (James Cokendolpher, pers. comm.).

Not all associations in amber are natural; many are the result of random movements of organisms as they try to escape or are displaced after death by resin flows (dying or death assemblages). Reasons for not considering the harvestman–springtail association discussed here as a death assemblage, rather than a life assemblage, include: (1) All of the springtails are well preserved and juxtaposed to each other as well as to the harvestman's leg. In a death assemblage, they would be variously positioned in the amber. Springtails often occur in groups, sometimes comprising large aggregations (Hopkin, 1998), which would explain why five appear close together. (2) One or both antennae of all of the individuals are either curved over the leg (four examples) or extended adjacent to the leg (one example). (3) The harvestman is well preserved and in a natural position, indicating that it was quickly preserved after entering the resin.

Other cases of springtail behavior in the fossil record include a swarm of many individuals randomly positioned in Dominican amber (Poinar and Poinar, 1999) and stalked spermatophores with sperm of *Sminthurus longicornis* in Baltic amber (Poinar, 2000b). Aggregation behavior and stalked spermatophores are conditions found in extant springtails (Hopkin, 1997).

Regarding the reliability of fossil evidence in interpreting associations, the present association would fall between Category 1, where the evidence is obvious, and Category 2A, where organisms are preserved in close association but not actually in their original position.

Many flying insects inhabit the same habitat as springtails and could be potential carriers of the latter. Such associations could explain the appearance of springtails on newly formed islands as well as the widespread distribution of certain species. If the behavioral scenario suggested here is correct, why are there no reports of springtail phoresis? First, very few scientists study springtails, so this behavior could have gone unnoticed (especially if the springtails quickly release their hold when disturbed). Also, this behavior would be limited to those forms with elongate flexible antennae (as in *Sminthurus longicornis*), which represent a minority of species. The other possibility is that this behavior was only exhibited by certain clades of sminthurids that, like *S. longicornis*, have become extinct.

Summary

A number of generic level relations are available. (See Table 32.)

14 Defense

14a. OPERCULATE GASTROPODS

Zeng and Hu (1991) described some Late Cretaceous nonmarine operculae, and McLean (1981, Figure 14) illustrated an Australian Early Devonian, Lilydale Limestone *Liomphalus* with its operculum in place. Gubanov and Yochelson (1994) described an operculate Wenlockian *Australonema* from Siberia, and Kase and Maeda (1980) described the Early Cretaceous neritacean *Hayamia rex* (Plate 35, Figure 10) with its operculum in place. Majima (1984) described some Miocene and Mio–Pliocene naticid species (seven) with operculae in place, and Tomida and Ozawa (1996) described Late Miocene and Early Pliocene Japanese *Turbo* (*Lunatica*) operculae.

Lindström (1884), Perner (1903, 1907, 1911), and Knight (1941) illustrated a number of operculate Paleozoic gastropods, all of which Boucot (1990a) cited, either here or previously, based on the work of others. Rohr and Blodgett (2003) described some taxonomically unassigned marine opercula from the Alaskan Silurian and Devonian and noted the difficulty in discriminating operculae that belong to the Oriostomatidae from those belonging to the Omphalocirridae. Carrera (1999) described an Early Ordovician operculum from the San Juan Formation, Argentina. An immature prosobranchid snail of the Family Helicinidae in Dominican amber (Poinar and Roth, 1991) contains extruded soft tissue and an operculum at the base of the body whorl in front of the aperture (Figure 244). It is important to note that calcified operculae have convergently evolved in a number of unrelated gastropod groups over time. (See Table 33.)

14b. SERPULID WORM OPERCULAE

While describing varied Middle Miocene operculae from southern Poland, Radwańska (1994b) reviewed many aspects of serpulid operculae in the fossil record dating back to the Late Cretaceous; see also her companion paper on tube-dwelling polychaetes (Radwańska, 1994a). Bielokrys (1994) described a Maastrichtian serpulid operculum from the Crimea. Fleming (1971) reviewed the record of New Zealand serpulid tubes and also of operculae.

14c. CEPHALOPOD INK SACS

Garassino and Donovan (2000) described an Italian Lower Jurassic coleoid with its ink sac, and Donovan and Crane (1992) described a Jurassic coleoid with its ink sac. Frickhinger (1994, Figure 141, p. 92) illustrated a Solnhofen Tithonian *Plesioteuthis* preserved with its ink sac. Gekker and Gekker (1955) described ink sacs from Volgian, Late Jurassic Russian *Parabelopeltis*(?), and Donovan (2006) described *Phragmoteuthis* from the Sinemurian of Dorset with ink sacs. Fuchs et al. (2009) described Cenomanian, Upper Cretaceous coleoids from Lebanon with preserved ink sacs; see Fuchs (2006) for both Cenomanian and Santonian materials from Lebanon. Doguzhaeva et al. (2002) described ink sacs in Late Pennsylvanian orthocones from Nebraska; Doguzhaeva et al. (2003) described another one from the Late Pennsylvania of Oklahoma, while reviewing the fossil record of these coleoids; Doguzhaeva et al. (2004) reviewed the occurrences of coleoid ink sacs, as well as the morphology of the ink granules still preserved in the fossil condition; and Doguzhaeva et al. (2007) described Late Triassic coleoid ink remains associated with shells. The critical point of the Doguzhaeva et al. material is that modern cephalopod ink consists of a large grouping of micron-sized spherules of ink, including some kind of melanin, and that these masses of spherules are also recognized in fossilized material under scanning electron microscopy (SEM). It is worth noting that "melanin" is a very generic term for a complex class of pigments, with those restricted to cephalopods being distinct from many other different types (for an introduction to the literature on this topic, see Nicolaus et al., 1964).

FIGURE 244 An immature snail (Prosobranchia: Helicinidae) in Dominican amber with an operculum (arrow) at the base of the body whorl in front of the aperture (Costa Amber Museum, Puerto Plata, Dominican Republic); diameter of shell = 3 mm.

TABLE 33

Fossil Operculate Gastropoda (Additions to Table 36 from Boucot, 1990a)

Taxonomic Level	Time Duration	Category	Refs.
Ceratopea unguis	Early Ordovician (1 my)	2B	Rohr (2004); Yochelson and Wise (1972)
Teiichispira	Early Ordovician (1 my)	2B	Carrera (1999)
Teiichispira	Early Ordovician	2B	Wen (1993)
Maclurites	Middle Ordovician	2B	Rohr and Yochelson (1999)
Slehoferia excavata	Late Ordovician	2B	Rohr and Frýda (2001)
Oriostoma cf. *sculptus*	Mid–Silurian	1	Ebbestad (2007)
Australonema varvarae	Mid–Silurian (1 my)	2B	Gubanov and Yochelson (1994)
Tophicola linsleyi	Late Silurian (1 my)	2B	Horný and Peel (1995)
Tophicola linsleyi	Late Silurian (1 my)	1	Horný and Peel (1995)
Liomphalus northi	Early Devonian (1 my)	2B	McLean (1981)
Tychobrahea aerumnans	Early Devonian (1 my)	2B	Horný (1992b; Peel and Horný (1996)
Australonema cf. *guillieri*	Early Devonian (1 my)	2B	Horný (1998a)
Cyclospongia discus	Middle Devonian (Eifelian) (1 my)	2B	Solem and Nitecki (1968)
Hessonia	Middle Devonian (Givetian) (1 my)	1	Bandel and Heidelberger (2001)
Neritopsis	Late Triassic (1 my)	2B	Zardini (1978)
Neritopsis	Late Triassic (1 my)	2B	Zardini (1985)
Naticopsis	Late Triassic (1 my)	2B	Zardini (1978)
Eucycloscala(?)	Late Triassic (1 my)	2B	Zardini (1980)
Reesidella	Late Jurassic (1 my)	2B	Wang (1984)
Homalopoma	Jurassic (1 my)	1	McLean and Kiel (2007)
Petropoma	Early Cretaceous (1 my)	1	McLean and Kiel (2007)
Hayamia	Early Cretaceous (1 my)	1	Kase and Maeda (1980)
Reesidella, Bithynia, Mirolaminatus	Late Cretaceous (1 my)	2B	Zeng and Hu (1991)
Reesidella	Late Cretaceous (1 my)	1	Yen (1951)
Reesidella	Late Cretaceous	1	Tozer (1956)
Reesidella	Late Cretaceous, Paleocene	1	Stearns MacNeil (1939)
Reesidella	Late Cretaceous (1 my)	2B	Wen et al. (1990)
Metrioomphalus hupei	Late Cretaceous (1 my)	1	McLean and Kiel (2007)
Otostoma retzii	Cretaceous (30 my)	1	Jagt and Kiel (2008)
Sohlipoma	Late Cretaceous (1 my)	1	McLean and Kiel (2007)
Parafossarulus	Cenozoic	2B	Wang (1965)
Mirolaminatus, Pseudemmerica, Fluvinarita, Assiminea	Paleogene	2B	Wang (1980)
Natica hantoniensis and *N. semeri*	Oligocene (30 my)	1	von Koenen (1891)
Helicinid in amber	Dominican amber	1	Poinar and Roth (1991)
Neritilia neritinoides	Early Miocene	1	Lozouet (2004)
Pachylabra prisca	Miocene (20 my)	2B	Prashad (1924)
Bithynia sp.	Late Miocene	1	Rust (1997)
Seven naticid species	Miocene–Pliocene	1	Majima (1984)
Scalez	Pliocene	1	Hanna and Gaylord (1924)
Turbo (Lunatica)	Pliocene	2B	Tomida and Ozawa (1996)
Naticids (*Cryotonatica, Natica, Tectonatica, Cochlis*)	Pliocene to Recent	1	Pedriali and Robba (2005, 2008)

14d. CAMOUFLAGE

Gunji et al. (1999), Savazzi (1999b), and Parker (1999) provided a wealth of material referring to color patterns and color in invertebrates. Keep in mind, however, that color patterns in invertebrates are definitely not preserved in a random manner, as indicated by the fact that some taxa are greatly overrepresented in this regard whereas many other taxa are virtually never preserved with color patterns. It is obvious that some taxa, either due to the composition of their shell pigments or to something about their shell structure, are greatly overrepresented. Among the Paleozoic brachiopods, for example, there is good evidence for terebratuloids (Early Devonian to Permian) being overrepresented (Blodgett et al., 1988; Hoare, 1978) whereas many other orders are either missing or way underrepresented. It is also clear that the fossils with color patterns are essentially restricted to photic zone environments. The oldest known color-patterned representatives of major groups may be sampling artifacts as much as anything else.

CEPHALOPODS

Ruedemann (1921) described countershading in a Trenton Limestone, mid-Ordovician nautiloid from New York, the earliest occurrence of countershading, and provided a useful summary of the older fossil color pattern literature, chiefly North American and European; see also Hoare (1978). Lehmann (1990, pp. 74–77) described and illustrated non-countershading color patterns from several Mesozoic ammonites, suggesting that different behaviors were involved than those involving countershading. Mapes and Davis (1996) reviewed what is known of color patterns in ammonites. Engeser and Keupp (2002) cited color patterns in some Mesozoic aptychae. Mapes and Sneck (1987) described the color patterns of Lower Triassic ammonites from Nevada, while reviewing occurrences of color patterns elsewhere in the Mesozoic; none is known in the Paleozoic. Kobluck and Mapes (1989) reviewed the occurrences of color patterning in fossil nautiloids, including those where camouflage is a reasonable possibility. Mapes and Evans (1995) discussed a Cretaceous color-patterned nautiloid from South Dakota for which camouflage is a possibility.

ECHINODERMS

Beaver and Fabian (1996) discussed color patterns in blastoids, citing previous occurrences additional to their Mississippian materials. Although their work does not involve color patterns, O'Malley et al. (2008) reviewed the chemistry of some color residues preserved in crinoids from the Mississippian and Jurassic.

BRACHIOPODS

Blodgett et al. (1988; Sun et al., 1999) provided entry to the Devonian and younger record of color patterns in brachiopods. Modzalevskaya (2007) described what is the earliest known brachiopod with a color pattern from the Lochkovian of Novaya Zemlya.

BIVALVIA

Color patterns in fossil bivalve literature are summarized by Hoare (1978) and involve taxa as old as the Devonian.

GASTROPODS

Tichy (1980) provided an excellent account of color patterns preserved in Triassic gastropod shells, as well as useful comments about color patterns in general. Zardini (1978, 1980, 1985) provided extensive documentation on color patterns in Late Triassic gastropods and a few for bivalves from the Italian Saint Cassian Beds. There are Paleozoic examples as far back as the Silurian and Devonian (Kříž and Lukeš, 1974; Yochelson and Kříž, 1974).

MONOPLACOPHORANS

The earliest molluscan citation is for a Tremadocian shell from a Pechora Basin borehole (Gubanov and Bogolepova, 2004). This is one of the oldest known color patterns, exceeded only by that for some Cambrian trilobites.

LOBOPODIANS

Von Bitter et al. (2007) briefly cite and illustrate a Canadian Silurian lobopodian that appears to have well-preserved color bands encircling its body segments.

TRILOBITES

Harrington (1959, p. O107) listed some of the occurrences, beginning with the Middle Cambrian.

HYOLITHIDS

Color patterns in hyolithids were unknown until their description in an Early Devonian Bohemian species by Valent and Malinky (2008).

INSECTS

Poinar and Poinar (1999, Figure 56, p. 60) discussed protective cases constructed by varied beetle larvae, which even employed their own excreta as a means of camouflage, and a similar subterfuge for a psocid (illustrations of color-patterned insect wings, beginning in the Pennsylvanian, are widespread). The prevalence of color patterns in some groups is so pervasive, as in certain Late Pennsylvanian and Early Permian cockroach lineages (Schneider and Werneburg, 2006), that they have been used in nonmarine biostratigraphy.

BIRD FEATHERS

Kurochkin (1985) described a patterned Early Cretaceous bird feather from Mongolia, and Martill and Frey (1995) described an Aptian(?), Early Cretaceous patterned bird feather from northeastern Brazil. Kurochkin (2000) reviewed the fossil record of Late Jurassic and Early Cretaceous bird feathers from Mongolia in the former Soviet Union and mentioned the single patterned-feather occurrence. Schlee and Glöckner (1978, p. 63, Table 3) illustrated an Early Cretaceous Lebanese amber feather. Perrichot et al. (2008) described some French Albian feathers preserved in amber, and Perrichot (2004) listed a number of Cretaceous bird feather occurrences. Grimaldi and Case (1995) described a New Jersey Turonian feather and reviewed many prior occurrences of fossil feathers. Talent et al. (1966) described an Australian Koonwarra Early Cretaceous feather from Victoria, Australia. Christiansen and Brock (1981, p. 15) illustrated a Danish Eocene bird feather, and Franzen (2001) illustrated a Messel, Middle Eocene bird with feathers. Possibly the most extravagant example of patterned bird

FIGURE 245 Mayr (2006) described an Eocene bird from Messel that has well-preserved color pattern; scale bar = 5 mm. (From Mayr, G., *Paläontologisches Zeitschrift*, 80, 390–395, 2006. With kind permission from Springer Science and Business Media.)

feathers is Mayr's (2006) Messel Eocene messilirrisorid (see Figure 245), which also has possible wax preserved from a uropygial gland—the only example of this type of thing from the fossil record. Bonde (1987) illustrated a feather from the Danish Lower Eocene. Laybourne et al. (1994) described a Dominican amber feather assigned to the Picidae on good evidence, showing what can be done with this class of material for taxonomic purposes. Of late, a great deal has been written about the possibilities for feathered dinosaurs (e.g., Currie et al., 2004); however, Ruben et al. (2003) and Jones et al. (2000, 2001) make the case for disbelieving this possibility.

FISH

(See color plates 1A and 1B, after page 132.) Tischlinger (1998) provided the most comprehensive account of color patterns preserved in fossil fish, including an unpublished mention of a Devonian armored fish from Antarctica, while describing some Late Jurassic color patterning in *Thrissops* involving melanophores from the Solnhofen Plattenkalk of southern Germany. Gottfried (1989) described the earliest fossil evidence for protective pigmentation in an actinopterygian from near the Permian–Carboniferous boundary in southeastern Kansas. Bellwood (1999) and Bellwood and Sorbini, 1996) described color patterns preserved on earlier Eocene Monte Bolca reef fishes (see also Tischlinger, 1998, Figures 7, 8) that are similar to those present on modern reef fishes belonging to the same taxa. Sorbini (1972, Table XV, Figure 1) illustrated a specimen of the eel *Paranguilla tigrina* from Monte Bolca with well-preserved melanophores. These are the best vertebrate examples of camouflage where one can make comparisons with modern examples.

AMPHIBIA

Lund (1978) described a truly very rare item—an amphibian, an aïstopod, from the Mazon Creek Pennsylvanian with

its color pattern well preserved. Because this is an obviously carnivorous, snake-like form, the pattern might have been useful in securing prey.

REPTILIA

Sullivan et al. (1988) described color pattern on the carapace of a Paleocene turtle.

☐ *Reliability*
Category 2B, based on functional morphology.

CASES OF CAMOUFLAGE IN AMBER

by George O. Poinar, Jr.

Camouflage is a condition that allows organisms to remain hidden from view, generally by making it difficult to detect them against the background environment. The main purpose of camouflage is to help prey avoid detection by predators or to allow predators to approach their prey without being seen. Fossil evidence of camouflage is rare, mainly because color patterns fade and other devices are difficult to detect after fossilization. Some examples that occur in amber are presented below.

Camouflage can be accomplished by assuming a shape that blends into the background as done by walking sticks that assume the shape of a twig (Figure 246). A more common method of camouflage is by adapting a color pattern that matches the normal background of the organism, such as the color patterns on the back of some flat bugs (Aradidae: Hemiptera) (Figure 247), on the prothorax of some orthopterans that feed exposed on plant surfaces (Figure 248), or on the emesine bugs that stealthily creep up on their victim (Figure 249). Such patterns commonly occur on vertebrates, such as the dark areas on the back and legs of frogs (Figure 250) and on the head, body, and tail of geckos (Figure 251 and Figure 252).

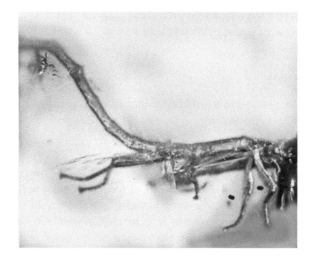

FIGURE 246 A walking stick (Phasmidia) in Dominican amber has a slender body that mimics a twig, thus providing protective camouflage (Morone amber collection, Torino, Italy); length is about 10 mm.

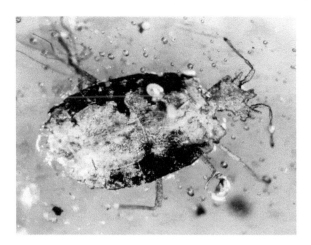

FIGURE 247 Color pattern on the back of an aradid bug (Hemiptera: Aradidae) in Dominican amber (Poinar amber collection accession no. HE-4-64); length of bug = 6 mm.

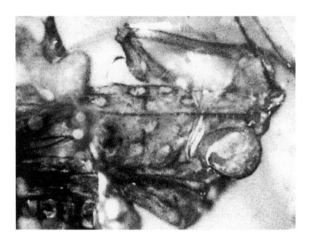

FIGURE 248 Light spots with dark borders symmetrically positioned on each side of the prothorax of the Burmese amber elcanid *Longioculus burmensis* (Orthoptera: Elcanidae) probably served as camouflage (Buckley amber collection accession no. AB-307, Florence, KY); length of insect = 7.1 mm.

Some insects, such as the predatory hemipteran in Figure 253, add debris to the surface of their bodies for concealment, while others, such as the bark louse shown in Figure 254, arrange debris in the form of a dense shield for protection. Tineid and psychodid moth larvae construct cases made out of fecal material and debris that surround their entire body (Figure 255); however, the addition of debris to an organism or residing in a debris-constructed case usually offers little protection once the occupant is detected.

A more elaborate form of protective camouflage is with hardened body cases that blend into the background. These cases, carried around by their builders, are stronger than debris cases and are entered when danger threatens. The entrance is generally blocked by the highly sclerotized head capsule of the occupant, thus physically prohibiting entry by predators and parasites. Such structurally secure cases are made by bagworm moths (Lepidoptera: Psychidae) that bind together portions of thick leaves and even pupate inside the

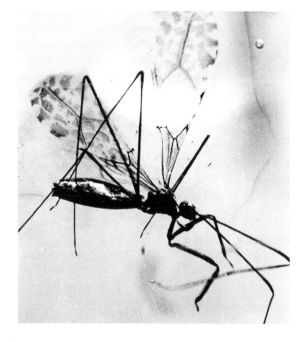

FIGURE 249 Color pattern on the wings of a predatory emesine bug (Hemiptera: Reduviidae) in Dominican amber (Poinar amber collection accession no. HE-4-6); length of bug = 6 mm.

FIGURE 250 Color pattern on back and legs of a frog (*Eleutherodactylus*) in Dominican amber (Poinar amber collection accession no. Am-5-6); length of frog = 5 mm.

FIGURE 251 Color pattern on the head of a gecko (*Sphaerodactylus*) in Dominican amber; length of head is about 10 mm.

FIGURE 252 Color patterns on the body and tail of a gecko (*Sphaerodactylus*) in Dominican amber (Morone amber collection, Torino, Italy); length is about 2.5 cm.

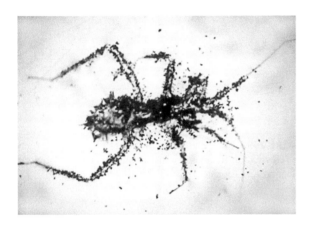

FIGURE 253 A predatory bug (Hemiptera) covered with debris in Mexican amber (Jim Work amber collection, Ashland, OR); (about 8 mm in length)

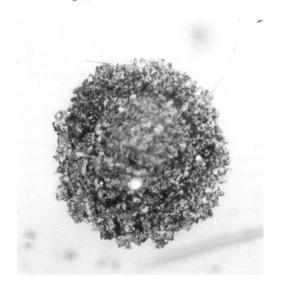

FIGURE 254 A bark louse (Psocoptera: *Blastopsocus* sp.) concealed under a protective shieldlike structure constructed from debris in Dominican amber—note one of the insect's antenna sticking out from under the case (Poinar amber collection accession no. CO-2-18); diameter of case = 2 mm.

chamber (Figure 256). Leaf beetle larvae (Chrysomelidae) use saliva to cement together their fecal matter into a thick-walled case that is carried around and looks simply like a fecal pellet when the larva is inside (Figure 148). In the

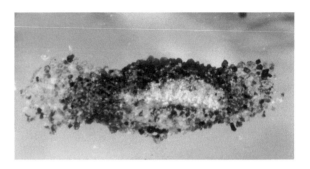

FIGURE 255 A moth larva (Lepidoptera: Tineidae) inside a loose case constructed with debris and fecal matter in Dominican amber (Poinar amber collection accession no. L-3-35); length of case = 2.3 mm.

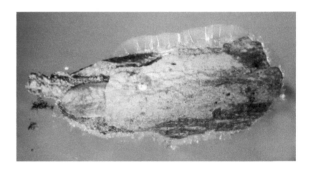

FIGURE 256 A case made by a moth larva (Lepidoptera: Psychidae) from leaves in Baltic amber—note the empty pupal case partially extended from one end (Poinar amber collection accession no. L-3-38); length of larva = 6 mm.

FIGURE 257 The large protuberances and dark color of the Dominican amber weevil *Velatis dominicana* Poinar and Voisin, 2003, camouflage it against the soil background (Poinar amber collection accession no. 7-387); length of weevil = 3.5 mm.

aquatic environment, caddisflies (Trichoptera) are well known for constructing cases with whatever occurs in their environments, be it stones or vegetable matter (see Figure 144 and Figure 145).

Another method of escaping detection is by bearing body structures that help the animal blend into the background (Figure 257). Tree hoppers (Hemiptera: Membracidae) often have prothoraxes that have developed into projections that resemble spines or thorns that occur on the branches of their host plants (Figure 258 and Figure 259). Some beetle larvae

FIGURE 259 A tree hopper (Hemiptera: Membracidae) with a single thorn-like projection on its thorax, in Dominican amber (Poinar amber collection accession no. HO-4-5B); length of insect = 6 mm.

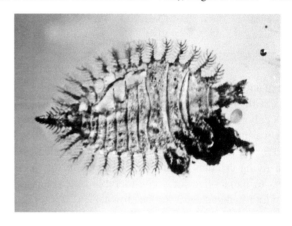

FIGURE 258 A tree hopper (Hemiptera: Membracidae) with three projections on the thorax resembling thorns located at the base of *Mimosa* twigs recovered from Dominican amber (Poinar amber collection accession no. HO-4-5A); length of insect = 5 mm.

possess lateral body projections that blend into the background (Figure 260) or long hairs on their dorsal surfaces that resemble structures that occur in their habitat (Figure 261). These are just a few examples of camouflage that occur in amber.

FIGURE 260 A beetle larva (Coleoptera: Brachypsectridae) with lateral body projections in Dominican amber (Cardoen collection); length is about 2 mm.

SUMMARY

Evidence is present for well-developed color patterns in both marine and nonmarine animals beginning with the Middle Cambrian (see Table 34).

14e. AUTOTOMY

ECHINODERMS

Lawrence and Vasquez (1996; for asteroid data, see Lawrence, 1992) provided a summary account of autotomy examples in many extant echinoderms. Baumiller and Ausich (1992) provided a very convincing account of autotomy in Fort Payne Chert, Mississippian crinoid stems; they compared the summary of similar autotomy in stems of living crinoids (from Emson and Wilkie, 1980) and cited somewhat similar data from the Triassic (Hagdorn, 1983). Emson and Wilkie (1980) pointed out that the actual process of stem autotomy in living crinoids has not actually been observed. Amemiya and Oji (1992) considered arm and crown autotomy in modern crinoids, with implications for the fossil record. Springer (1920, Plate LVI, Figures 11a–c, pp. 402–403) described a

FIGURE 261 Long hairs on the dorsal surface of a skin beetle (Coleoptera: Dermestidae) in Dominican amber (Poinar amber collection accession no. C-7-355); length of beetle = 1.3 mm.

TABLE 34

Evidence for Camouflage from the Fossil Record (Additions to Table 38 from Boucot, 1990a)

Behavioral Type	Taxonomic Level	Time Duration	Category	Refs.
Nautiloid countershading	Subclass	Middle Ordovician–Holocene	1	Ruedemann (1921)
Echinoderms	Class	Mississippian	1	Beaver and Fabian (1996)
Brachiopods	Class	Devonian–Holocene	1	Blodgett et al. (1988)
Bivalvia	Class	Devonian–Holocene	1	Hoare (1978)
Gastropoda	Class	Silurian–Holocene	1	Yochelson and Kříž (1974)
Monoplacophorans	Class	Early Ordovician	1	Gubanov and Bogolepova (2004)
Trilobita	Class	Middle Cambrian	1	Harrington (1959)
Hyolithids	Uncertain	Middle Cambrian	1	Valent and Malinky (2008)
Insecta	Class	Pennsylvanian–Holocene	1	Poinar and Poinar (1999); Schneider and Werneburg (2006)
Coleoptera	Order	Miocene–Holocene	1	Poinar and Poinar (1999)
Bird feathers	Class	Late Jurassic–Holocene	2B	Kurochkin (2000)
Fish	Class	Permo–Carboniferous	2B	Gottfried (1989)
	Genera	Eocene–Holocene	2B	Bellwood (1999)
Amphibia	Class	Pennsylvanian	2B	Lund (1978)
Reptilia	Class	Paleocene–Recent	2B	Sullivan et al. (1988)

case of autotomy in a Late Paleozoic crinoid in which a small calyx with six rays grew on top of a stem that had lost most of its calyx and another specimen with six rays where autotomy was not involved, possibly a genetic defect (Plate LVII, Figures 10a–c). Oji and Amemiya (1998) provided evidence from living crinoids regarding the capability of stem fragments to metabolize for a considerable time interval, even when totally detached from the crown.

Donovan and Pawson (1997; see also Donovan and Schmidt, 2001) provided still more evidence about the capability of stems totally detached from their crowns, additional to crowns totally detached from their stems, for continued life, with probable examples from the fossil record extending back to the Late Ordovician (Ausich and Baumiller, 1993). Gil Cid et al. (1998) described a Llandeilan Spanish crinoid with good evidence for autotomy; this is certainly just about the oldest evidence for autotomy among crinoids. Meyer and Ausich (1983) discussed many examples of regeneration in fossil crinoids. Baumiller (2001) suggested how oxygen and carbon isotope data may be useful in recognizing autotomy in fossil crinoids. Baumiller and Gahn (2004) provided a sample providing evidence for arm autotomy in Paleozoic crinoids and observed that less than 5% of Ordovician–Silurian crinoids show such evidence, as contrasted with more than 10% for Devonian–Pennsylvanian examples. They suggested that this indicates a major change in predation affecting crinoids between the Silurian and Devonian.

Aronson et al. (1997) discussed ophiuroid examples from the Late Eocene of Antarctica. Radwański (2002) commented extensively on autotomy in ophiuroids. Frickhinger (1994, Figure 371, p. 175) provided an outstanding illustration of a Solnhofen Tithonian ophiuroid (*Geocoma*) with two arms partly regenerated following autotomy. Lehmann (1951) described a number of German Bundenbach examples involving regeneration and summarized much of the literature on

starfish autotomy and arm regeneration from the fossil record, which goes back to the Ordovician. Tasnádi-Kubacska (1962) discussed autotomy examples in starfish present in the fossil record, including ones with regenerating arms.

Non-Insect Arthropods

Simpson and Middleton (1985) described good evidence for autotomy in some Early Cretaceous lobsters from England. Penney (2005) described some Dominican amber spiders with autotomized legs that exuded hemolymph droplets at the point of breakage—the first occurrence of this phenomenon. He prefers the term *autospasized* to *autotomy*. Some spiders also not only lose their legs readily but also have the ability to replace lost appendages. A spider in Baltic amber that regenerated parts of its patella is shown in color plate 4F after page 132 (Wunderlich, 2008). Although not exactly representing autotomy, Poinar and Poinar (1999, Figure 44, p. 50) illustrated and discussed a planthopper whose young produce long, waxy "tail" filaments at which predators may mistakenly strike, thus permitting the planthopper to escape. Rasnitsyn and Quicke (2002, p. 42) cited probably autotomized cranefly legs and suggested that the presence of numerous lepidopteran wing scales indicates possible escape from the resin.

Insects

Vickery and Poinar (1994) illustrated a cricket with an autotomized leg regrowing (Figure 262).

Vertebrates

Tischlinger (2005) used ultraviolet light on Solnhofen Limestone specimens to recognize autotomized lizard tails and autotomy septae, fracture planes (see also Boucot,

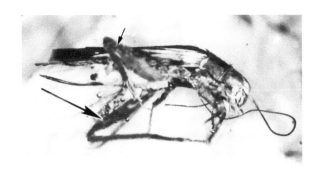

FIGURE 262 Evidence of autotomy and growing back of the right leg of the paratype of *Proanaxipha latoca* in Dominican amber (Poinar amber collection accession no. 0-2-16); length of cricket body = 6 mm. Photograph shows small right leg (small arrow) and large left leg (large arrow). (From Vickery, V.R. and Poinar, G.O., *The Canadian Entomologist*, 126, 13–22, 1994.)

1990a). Arnold (1988) discussed the basics of autotomy in reptiles. (See color plate 2A after page 132.)

Behavioral Implications from Lizards Preserved in Amber

Fossil lizards are rare, and those in amber are usually well preserved, with individual scales intact and original color patterns remaining. Of the handful of lizards found in amber, the majority have been from the Dominican Republic and are representatives of two groups: geckos and anoles. The actual age of Dominican amber is uncertain, with estimates ranging from 15 to 45 million years.

Lizards are thought to have evolved by the Upper Triassic, some 225 million years ago (Carroll, 1988). Although the Mesozoic was dominated by dinosaurs, lizards possessed certain attributes that allowed them to survive the K-T extinction event and survive to the present. The fossil record of geckos extends back to the Early Cretaceous, while that of the iguanids (which includes the anoles) began in the Eocene. The oldest unequivocal gecko, *Cretaceogekko burmae*, in Burmese amber with transverse lamellae bearing hair-like setae on its toes, reveals that the gecko's sophisticated dry adhesive mechanism evolved some 100 million years ago (Arnold and Poinar, 2008). Aside from some Recent anole bones from caves dated in thousands of years, the amber fossils represent the oldest known remains of *Anolis* and the only known fossil Least Gecko belonging to *Sphaerodactylus*.

The small size of Least Geckos probably explains why several have been recovered in Dominican amber. Large lizards would have been strong enough to free themselves from the sticky resin, but small ones would have had a hard time, especially if they had been injured by a predator. Like most geckos, Least Geckos have large lidless eyes and small, smooth scales on their heads. Over the eons, the upper and lower eyelids of these (and many other) geckos have fused, forming a transparent scale, or brille, that completely covers the eye. The geckos clean these "contact lenses" with their tongue without so much as a blink. Their lack of eyelids is why their stare is perpetual, even after millions of years (see color plate 1C after page 132). These shy, nocturnal creatures are unusual because they vocalize by emitting shrill chirps while foraging for invertebrates. Such sounds are probably used for guarding the male's territory and calling for mates. The unique adhesive toe pads of geckos (also seen on the fossils) facilitate climbing up the sides of trees and rocks in search of prey. Least Geckos lay hard-shelled eggs that look like miniature bird eggs. Both anole and gecko eggs are laid in rotting wood, under rocks, and in tree branches, which is why a few ended up in amber (see Eggs, Oviposition, and Maternal Care in Amber section and Figure 196 in Chapter 10).

Anoles are slightly larger than Least Geckos and have movable eyelids. Most anoles are arboreal, climbing swiftly with the aid of special toe pads (different from those found on geckos) to reach their perches from the shrub layer to the forest canopy. A fascinating character of anoles is their ability to change color, just like true chameleons. Depending on the background and circumstance, they can assume varying shades of gray, brown, yellow, and green. Male *Anolis* also possess prominent throat fans (dewlaps), which they expand when fighting or courting females. Males will defend territories that enclose one to several females and spend a fair amount of time on high perches advertising their presence. When an intruding male is detected, the defender will put on a display that usually involves extending and retracting his colorful dewlap while at the same time bobbing his head and body up and down. The bobbing is thought to help show off the large dewlap, which occurs on some of the Dominican amber anoles. If such tactics do not succeed in driving off the intruder, a fight might then ensue which could involve biting and chasing.

Resting on branches during the day and particularly displaying, in the case of males, make anoles an attractive meal to a number of forest dwellers. Snakes, birds (especially trogons and motmots), other lizards, and even monkeys dine on anoles. Having a host of enemies may explain why anoles residing high in the canopy have the spectacular ability to drop instantly from their branches and make remarkable recoveries after landing on the forest floor.

Many lizards, including both anoles and geckos, have the ability to lose their tails, usually at precarious moments. This tail shedding, which is a type of voluntary behavior, usually occurs when they are being attacked by a predator. The severed (autotomized) tail continues to wriggle convulsively for several minutes, thus distracting or confusing the predator, while the lizard makes its escape. The lost appendage is usually eaten by the predator or consumed later by invertebrate scavengers. The mechanism of voluntary tail loss is very complicated and depends on certain anatomical modifications of the caudal vertebrae, muscles, nerves, and blood sinuses. This evolutionary survival tactic probably took millions of years to develop. Special fracture planes consisting of bands of cartilage occur in the tail vertebrae (see Boucot, 1990a, 14e, Autotomy). As soon as the tail is amputated at these uniquely structured vertebrae, a sphincter muscle closes the fracture

FIGURE 263 Shed tail of a gecko in Dominican amber (Poinar amber collection accession no. R-3-10); length of shed tail = 12 mm. Such tails continue to wriggle for several minutes after being autotomized by their owners; the movements distract predators and allow the lizards to escape.

and prevents blood loss. The vertebra, skin, nerves, and blood vessels are all preconstricted at the same predetermined spot where the break will occur. The tail slowly starts to regenerate, providing us with one of the few instances where vertebrate appendages are replaced. The regeneration involves the posterior growth of a cartilaginous rod, accompanied by muscles, nerves, and blood vessels (Smith, 1946).

There are ways of recognizing if a lizard has a replaced tail. First, the new scales are smaller and different in color from those that were on the original tail and, second, there are no vertebrae in the reformed tail. Although the new tail may serve its bearer by functioning for balance and food storage as the original one did, it cannot regenerate if broken again. Some geckos, such as the New Zealand Green Gecko, utilize their tail as a prehensile organ. This function is obviously very important and is probably why these lizards rarely lose their tails, since a regrown tail has little grasping ability. Young lizards suffer when their mother loses her tail, as it is an important storage organ used for yolk production; females without tails have been known to produce smaller eggs and less vigorous offspring than those with normal tails. Most ground-dwelling geckos, such as those found in Dominican amber, readily lose their tails.

That tail shedding also occurred in the past is evident from the discovery of a cast-off tail in Dominican amber (Figure 263). Lizard tails have also been recovered from the more recent Madagascar copal (Figure 264). These appendages might have been lost on a branch of the amber-producing tree during the attack of a predator and fallen or wriggled off into a pool of resin. At least an isolated tail indicates the presence

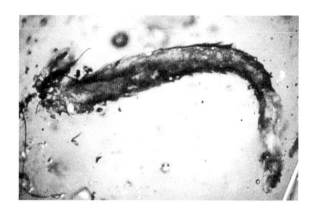

FIGURE 264 Shed tail of a lizard in Madagascar copal (Poinar amber collection accession no. R-3-11); length of shed tail = 10 mm.

of lizards, although it may take some detailed morphological or genetic analysis to determine just who the bearer was.

☐ *Reliability*
Category 2B, functional morphology.

SUMMARY

Evidence of autotomy is present from at least the Middle Ordovician for certain taxa (see Table 35).

14f. ENROLLMENT

Chatterton and Campbell (1993) provided additional insights into trilobite enrollment.

14g. SPINES AND THORNS

Dacqué (1921) offered the best survey of spinosity in marine invertebrate fossils. Storch and Richter (1994) illustrated and discussed the spines on a small Eocene insectivore from Messel that presumably were purely for defensive purposes (see the Camouflage section, earlier in this chapter, for several examples of spines being used by insects for camouflage). Insects have various types of morphological defensive mechanisms (e.g., areas of thickened cuticle, protective spines, and hairs).

☐ *Reliability*
Category 2B, functional morphology.

TABLE 35
Examples of Autotomy from the Fossil Record (Additions to Table 39 from Boucot, 1990a)

Behavioral Type	Taxonomic Level	Time Duration	Category	Refs.
Crinoids	Class	Middle Ordovician–Holocene	2B	Gil Cid et al. (1998)
Asteroidea	Class	Ordovician–Holocene	2B	Lehmann (1951)
Ophiuroidea	Class	Late Jurassic–Holocene	2B	Aronson et al. (1997); Frickhinger (1994)
Arthropods	Class	Cretaceous–Holocene	2B	Simpson and Middleton (1985)
Lizards	Order	Jurassic–Holocene	2B	Boucot (1990a); this work

14h.   BELEMNITE SWIMMING AND OTHER CEPHALOPODS

Baird et al. (1989) made a good case from the holoperipheral, aligned orientation of bryozoan epizoans on Late Ordovician and Middle Devonian orthocones that the orthocones were alive and actively swimming while the epizoans attached and grew.

☐ *Reliability*

Category 2B, functional morphology.

*14i.   TRILOBITES SHELTERED WITHIN NAUTILOID SHELLS AND CRUSTACEANS WITHIN AMMONITES

Davis et al. (2001) described varied trilobite occurrences within Ordovician to Devonian nautiloid conchs, which are easily interpreted for the most part as defensive shelters where they might have been recovering from ecdysis. Chatterton et al. (2003) described trilobites aligned in a manner that suggests sheltering in "worm" burrows.

☐ *Reliability*

Category 1, frozen behavior.

*14j.   STINGRAY SPINES AND OTHER VENOMOUS FISH SPINES

Halstead (1988) summarized information on modern venomous marine animals, including those fishes with spines involved in a venom delivery system (Figure 265 and Figure 266). Prominent among the modern venomous fish with a fossil record are the stingrays, members of the Myliobatoidea that have a fossil record extending back into the Early Cretaceous (Underwood et al., 1999b). Evans (1923) detailed much of the anatomy involved with the modern stingray venom delivery system, including its vasodentine caudal spine, and the nature of the holocephalan venom delivery system; he suggested that the pleuracanth sharks of the Late Paleozoic had spines designed for venom delivery, as did other various Paleozoic sharks (e.g., *Paleospinax*, *Sphenacanthus*, *Asteracanthus*, *Hybodus*). De Carvalho et al. (2004) illustrated a number of Early Eocene, Green River Formation freshwater stingrays with well-developed spines, and there are other localities for isolated spines (Figure 267). Siber (1982, p. 18) illustrated a spiny stinger still attached to the tail of a Green River *Heliobatis*. Cappetta (pers. comm., 2006) informed us that the oldest known actual stinger is one belonging to *Rhombodus* from the Maastrichtian of Morocco (Arambourg, 1952, Plate 30, Figures 16, 32), and that such stingers commonly accompany myliobatiform teeth. Well-preserved fossil stingrays are known only from the Eocene of the Green River Formation and Monte Bolca, whereas isolated stingray teeth and stingers make up the overwhelming bulk of the occurrences.

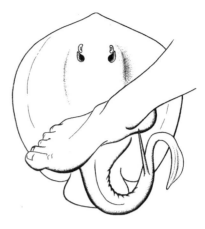

FIGURE 265   Drawing showing the method by which a stingray usually inflicts its sting. (Figure 33 in Halstead, B.W., *Poisonous and Venomous Marine Animals of the World*, 2nd ed., The Darwin Press, Princeton, NJ, 1988.)

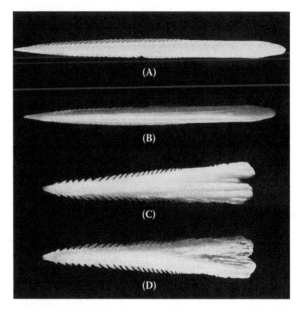

FIGURE 266   (A) Dorsal view of the spine of *Taeniura lymma*; length = 7 cm. (B) Ventral view of above spine. (C) Dorsal view of spine of *Gymnura marmorata*; length = 1.7 cm. (D) Ventral view of the spine of *G. marmorata*. (Figure 16b in Halstead, B.W., *Poisonous and Venomous Marine Animals of the World*, 2nd ed., The Darwin Press, Princeton, NJ, 1988.)

☐ *Reliability*

Category 2B, functional morphology, as no venomous fossil fish has ever been caught in the act of delivering its venom.

*14k.   ONYCHOPHORAN SLIME SECRETION

Poinar (2000c) described Dominican and Baltic amber onychophorans with the Dominican amber specimen showing emitted slime deposits adjacent to its oral papillae (Figure 268).

☐ *Reliability*

Category 1, frozen behavior.

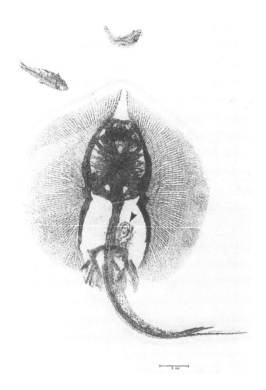

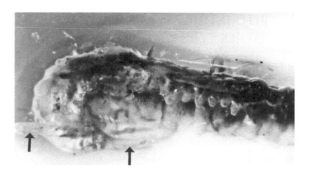

FIGURE 268 Evidence of defense secretions is shown by emitted slime deposits (arrows) adjacent to oral papillae of the fossil velvet worm, *Tertiapatus dominicana*, in Dominican amber (Poinar amber collection accession no. ON-1-2); length of velvet worm = 4.38 mm.

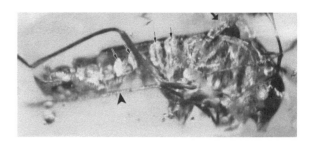

FIGURE 269 Lateral view of a soldier beetle (Coleoptera: Cantharidae) displaying a chemical defense response against a potential predator (arrowhead) in Burmese amber (Deniz Eren amber collection, Istanbul, Turkey); length of specimen = 4.5 mm. Four cuticular vesicles (small arrows) are everted. Large arrow shows chemical deposit.

FIGURE 267 *Asterotrygon maloneyi* female in ventral view with one unborn late-term fetus indicated by arrow—note the tail spine; this is a freshwater Eocene stingray from the Green River Formation. The two teleosts belong to the clupeid *Knightia*. (Figure F-2 in de Carvalho, M.R. et al., *Bulletin of the American Museum of Natural History*, 284, 1–136, 2004. Courtesy of the American Museum of Natural History.)

*14l. SOLDIER BEETLE

(See also Chapter 7, Communication.) The fossil record of chemical defense is rare and previously limited to ink sacs in Pennsylvanian and Jurassic cephalopods (Boucot, 1990a) and defensive secretions in amber termites and velvet worms (Poinar, 1998b, 2000c). The discovery of a 100-million-year-old soldier beetle in Burmese amber that employed a chemical defense response against a potential predator is very spectacular (see Figure 269). The beetle, a member of the family Cantharidae, whose modern descendants are known to use chemicals defensively (Eisner et al., 2005), had everted six pairs of cuticular vesicles with their associated gland reservoirs. A secretion released from one of these everted vesicles covered a portion of the antennae of a second insect species. While the actual insect that caused the response was missing, it was very likely a cockroach or orthopteran, based on the structure of the remaining antenna (Poinar et al., 2007b).

☐ *Reliability*
Category 1, frozen behavior.

14m. CRYPTORHYNCH WEEVIL LOCKING MECHANISM

Insects have various types of morphological defensive mechanisms (e.g., areas of thickened cuticle, protective spines and hairs). A Burmese amber weevil (Poinar, 2008c) possessed a unique femora interlocking mechanism consisting of a flange on the basal third of the prefemur that inserts into a groove along the basal portion of the mesofemur (see the back cover, Figure A). This is probably a defensive adaptation, although no modern weevil is known to possess this character.

☐ *Reliability*
Category 2B, functional morphology.

SUMMARY

Mostly at the family and higher levels but with some generic examples. (See Table 36.)

TABLE 36
Defense

Type	Time Duration	Category	Refs.
Gastropod operculae	Early Ordovician–Holocene	1	See text
Serpulid operculae	Late Cretaceous	1	See text
Cephalopod ink sacs	Pennsylvanian–Holocene	1	See text
Camouflage	Cambrian–Holocene	1	See text
Autotomy	Ordovician–Holocene	1	See text
Enrollment	Cambrian–Permian	1	See text
Spines and thorns	Cambrian–Holocene	1	See text
Cephalopod swimming	Ordovician–Holocene	1	See text
Trilobite inquilinism	Ordovician–Devonian	1	See text
Stingray spines	Late Cretaceous–Holocene	1	See text
Onychophoran slime	Miocene–Holocene	1	See text
Soldier beetle	Cretaceous–Holocene	1	See text

15 Carrier Shells

(See Figure 270.) Rohr (1993) illustrated a fine *Lytospira cerulus* from the earlier Middle Ordovician, Alaskan Telsitna Formation that shows excellent attachment scars; however, the nature of the attached objects is unknown. Rohr (1994) illustrated Whiterock-age *Lytospira yochelsoni* from the Antelope Valley Formation of Nevada. Ulrich and Scofield (1897, Plate LXXV, Figures 20, 21) illustrated depressions on a Middle Ordovician *Eccyliomphalus* due to adhering objects. Horný (1992b) described an Early Devonian *Lytospira* with attachment scars from the Barrandian area of the Czech Republic. El-Nakal and Bandel (1991) described specimens of a small gastropod, *Scaliola*, with attached sand grains from the modern Indo-Pacific fauna and the Paris Basin Eocene.

Dacqué (1921, Figure 323) reproduced spectacular figures from von Koenen (1892) of North German Oligocene *Xenophora* with attached pebbles. Perillat and Vega (2001) described a Late Cretaceous spiny xenophoran gastropod (*Acanthoxenophora*) from southern Mexico and reviewed the fossil record of Cretaceous xenophorans, with the oldest being a Late Cretaceous, Cenomanian item from Texas (there are no known pre-Late Cretaceous xenophorans). Squires and Saul (2001) described a spectacular Late Cretaceous (late Campanian–Maastrichtian) xenophoran from the San Diego region with attached pebbles, and they reviewed many of the other Late Cretaceous and some Cenozoic xenophorans.

Kiel and del Carmen Perrilliat (2001) described a Maastrichtian xenophoran (*Acanthoxenophora sinuosa*) from southern Mexico with an attached *Turritella* on one specimen and attachment scars on others. Darragh and Kendrick (1994) described *Xenophora* from the Carnarvon Basin Maastrichtian.

The fossil record demonstrates that the attachment of varied materials to gastropod shell exteriors is a convergent character that has independently evolved in many lineages, Paleozoic and post-Paleozoic.

□ *Reliability*
Category 2B, based on functional morphology.

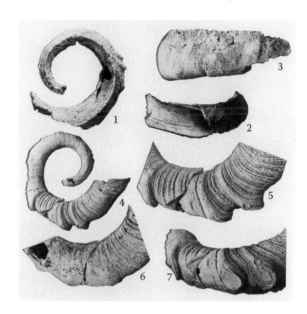

FIGURE 270 One of the oldest known carrier shells, *Lytospira gerulus*, from the Middle Ordovician of Alaska. (1) Top view and (2) interior detail of the specimen showing spiral ridges and groove (×1); (3) detail of repaired shell break (×0.7); (4) basal view, (5) detail of repaired break on base, and (7) oblique basal view of attachment scars (all, ×1); (6) basal view of repaired shell break (×0.7). (Figure 3 from Rohr, D.M., *Journal of Paleontology*, 67, 959–962, 1993. Reproduced with permission.)

16 Pollination Ecology

(See Figure 271.) Friis (1985) provided an excellent discussion of pollination ecology through time, based on a consideration of Late Cretaceous flowers, and concluded that beetles and flies were probably the major pre-Tertiary pollinators. Poinar (1998c) has described a euglossine bee from Dominican amber, which suggests that the obligate relation between most orchid bees and orchids was developed by at least the Miocene, and Poinar concluded that the two females might have become entrapped in amber while collecting resin for nest construction. Ramírez et al. (2007) described a stingless bee with orchid pollinia attached from Dominican amber.

Ren (1998) and Labandeira (1998b) discussed the time of origin and the insect–angiophyte taxa possibly involved in the initial appearance/development of insect pollination. Using Chinese data they ascribed a Tithonian, Late Jurassic age to the beds yielding flies with the appropriate mouthparts, but the beds have now been more reliably radiometrically dated as Early Cretaceous (Luo, 1999; Swisher et al., 1999). It is clear that the precise age and identity of the angiophyte groups involved are still uncertain. Labandeira (2000) reviewed the pollination story and, using biogeographic evidence from the present, concluded that the involvement of beetles with pre-Cretaceous conifers and also with cycads suggest that the pollination relationship with beetles at least is a pre-Cretaceous phenomenon. Norstog and Nicholls (1997; Norstog et al., 1995) reviewed the relationship between beetle pollination, curculionids, and cycad pollination. Delevoryas (1968) and Crepet (1974) provided evidence from the fossil record that cycads in the Cretaceous were subject to boring and chewing, presumably by insects.

FIGURE 271 The orchid bee, *Paleoeuglossa melissiflora*, in Dominican amber (Jim Work amber collection, Ashland, Oregon); length of bee = 17.7 mm.

17 Social Insects

Choe and Crespi (1997) provided an excellent account of the social, as well as parental, behaviors of many groups of insects and arachnids. Hasiotis et al. (1997), based on Late Triassic data from Arizona, suggested that trace fossils indicate the presence of social hymenopterans far earlier than body fossil evidence supports. Engel (2001) provided extensive evidence for rejecting the conclusion of Hasiotis et al. regarding the existence of late Triassic social hymenopterans. The alternative is a high level of behavioral convergence involving unrelated arthropod groups.

17a. STRATIGRAPHIC RANGES

WASPS

Carpenter and Rasnitsyn (1990) discussed the record for the Mesozoic Vespidae while describing Early Cretaceous taxa from Mongolia and Asiatic Russia as well as the Late Cretaceous of Kazakhstan. Poinar (1998b, Figure 1F) illustrated a piece of carton nest made by a vespid wasp from Dominican amber. Poinar and Poinar (1999, Figure 127) illustrated a Dominican amber paper wasp nest fragment.

ANTS

Brandão et al. (1989; see also Brandão, 1990) described the earliest putative ant from the Southern Hemisphere, from the Aptian of the Araripe region of eastern Brazil. Agosti et al. (1998) suggested that the Turonian *Sphecomyrma freyi* and *Baiburis* n. sp. are the oldest true ants, but Poinar et al. (2000a) disagreed, considering them to be nonformicine hymenopterans. Poinar et al. (2000a; see also Poinar and Milki, 2001, p. 62) presented evidence indicating that no Cretaceous modern ants have been proved to exist, with *Sphecomyrma* remaining in the uncertain category. All extant or "modern" ants are placed in a single family, the Formicidae, in the monotypic superfamily Formicoidea. Tertiary ants all belong to "modern" groups and can be accommodated within the Formicidae. The females and workers of Cretaceous ants are considered archaic or primitive ants and differ from "modern" ants by possessing a shortened scape (see the front cover, Figure B), bidentate mandibles, and long, flexible flagella. These archaic ants are best placed in a separate family, the Armaniidae (comprised of the subfamilies Armaniinae and Sphecomyrminae), within the Formicidae (Dlussky, 1999; Rasnitsyn and Quicke, 2002).

MESOZOIC BEES

Synopsis of Fossil Bees

The collection of pollen as a protein source for their young and the resulting pollination from these actions make bee–angiosperm associations an interesting and controversial topic. The distribution and types of flowering plants in our world were determined in large part by the past and current activities of bees. The fossil record of bees is based mainly on body fossils in amber and sedimentary deposits. Ichnofossils in the form of fossil nests in paleosols and petrified wood (Hasiotis and Dubiel, 1993) are quite interesting and some may indeed be the result of bee activity, but it is impossible to be certain without associated body fossils. Hymenopteran ichnofossil nests have been collected from the Upper Jurassic Morrison Formation (Utah), the Lower Cretaceous Dakota Formation (Rocky Mountain region), the Upper Cretaceous Two Medicine Formation (Montana), the Paleocene–Eocene Claron Formation (Utah), the Eocene Brian Head Formation (Utah), the Eocene Bridger Formation (Wyoming), the Miocene of Greece (Santorini), and Holocene dunes in Great Sandstone Dunes National Monument (Colorado) (Hasiotis et al., 1998a), to mention a few locales. These nests contain cells and cocoons that are very similar in shape and size to those of modern bees and wasps. Organic chemical analysis of the wall linings in Triassic nests produced straight-chain carbon molecules similar to those found in modern bees of the families Anthophoridae and Colletidae (Kay et al., 1997). The evidence that these nests belong to aculeate hymenopterans is impressive, however, because the shape and size of bee and wasp cells are similar, and it is difficult to determine if the nest material was constructed from wood fibers (wasps) or wax (bees); the makers of these fossil nests remain controversial (Michener, 2000). Such nests are found in amber (Figure 272) and even when additional details are preserved, there is still controversy over whether they were made by social bees or wasps.

The recently discovered Early Cretaceous Burmese amber bee, *Melittosphex burmensis* (Poinar and Danforth, 2006), which is morphologically intermediate between bees and wasps, provides new insights into the origin of bees (Figure 273 and Figure 274). This fossil possesses branched hairs, acutely pointed glossa, and other characters found in modern bees, as well as features of apoid wasps (paired midtibial spurs and a slender hind basitarsus); it represents a stage in the transition from predatory wasps to bees. Branched hairs

FIGURE 272 Controversy reigns over whether this nest in Dominican amber was formed by bees or social wasps. It is highly probable that these cells were constructed by a species of stingless bee in the genus *Melipona* (Apidae: Meliponini) (deposited in the amber collection of Hermann Dittrich, Santo Domingo, Dominican Republic). Cell diameter is approximately 10 mm.

are characteristic of all 11 families of extant bees and separates bees (in the series Apiformes) from the 9 extant families of sphecoid wasps (series Spheciformes) (Goulet and Huber, 1993; Prentice, 1998). Ancient eudicots obtained from amber at the same site as the bee show the types of flowering plants *Melittosphex* might have visited (Poinar and Chambers, 2005; Poinar et al., 2007d; Santiago-Blay et al., 2005).

Other Cretaceous bees have been seriously scrutinized. *Meliponorytes devictus* (Cockerell, 1921) was supposedly also in Early Cretaceous Burmese amber. In the year following its description, Cockerell (1922) commented that

the amber containing the fossil was pale and pellucid and originated from a group that had been artificially colored to "enhance their value." There was also a question about the origin of this material. Swinhoe, who had purchased the amber in Burma, "could not be sure that the material was really from Burma," as he had been informed that stained amber necklaces were imported into Burma from China. At any rate Cockerell (1922) concluded that this "light amber (or copal) is of very recent origin" and commented that his *M. devictus* did not differ at all from extant Indian *Trigona laeviceps* Smith. This conclusion was later supported by Zeuner and Manning (1976), who synonymized *M. devictus* with extant *Trigona iridipennis* Smith. It is obvious that the above bee was either naturally entombed or intentionally placed in a recent resin and not Burmese amber.

A similar scenario appears to have occurred with the New Jersey amber bee dated at 74 to 96 mya and originally described as *Trigona prisca* by Michener and Grimaldi (1988). There is a growing consensus that the amber (if it is amber and not copal) containing the New Jersey bee was incorrectly dated, as the specimen has quite modern characters (Rasnitsyn and Michener, 1991; Radchenko and Psenko, 1994; Rasnitsyn and Quicke, 2002). Although originally described in a present-day species group, it was later transferred into a new genus, *Cretotrigona*, by Engel (2000), who modified the age to 65 million years (which with the current geological time scale would place it at the beginning of the Tertiary). However, this action did not alter the fact that *C. prisca* is a representative of a derived genus belonging to a tribe (Meliponini) containing the most derived family of bees. Engel (2000) also pointed out that two of these bees were trapped in the same piece of amber. It is well known that trigonids collect resin for nest construction and, if these stingless bees were as common as indicated, then they would be expected to occur in other amber samples from the same location or from other Late Cretaceous deposits such as those in Alberta and Manitoba, Canada, or in Siberia. The amber piece containing the bees also held other advanced insects not expected to occur in Cretaceous deposits (Rasnitsyn and Michener, 1991).

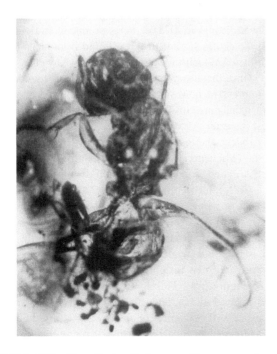

FIGURE 273 The Early Cretaceous Burmese amber bee *Melittosphex burmensis* (Apoidea: Melittosphecidae) possesses characters of both modern bees and aculeate wasps (Poinar amber collection accession no. B-Hy-7); length of bee = 2.95 mm.

FIGURE 274 Plumose body hairs (arrows) on *Melittosphex burmensis* (Apoidea: Melittosphecidae) in Early Cretaceous Burmese amber (Poinar amber collection accession no. B-Hy-7).

New Jersey amber occurs in a number of locations, and although some is Cretaceous (65 to 95 mya), with *Agathis* as a tree source (Lambert et al., 1990), Tertiary amber from the angiosperm *Liquidambar*, as well as from other unidentified botanical sources, occurs in at least six different localities within the state (Langenheim and Beck, 1968; Grimaldi et al., 1989). Perhaps the amber has been erroneously dated or is copal originating from a completely different region.

Some of the most interesting bees in Baltic amber are those in the genus *Electrapis* Cockerell, which are clearly the ancestors of our honey bee. These were studied by Zeuner and Manning (1976) and later by Engel (2001), who erected many new taxa.

Also of interest were the discoveries of the first fossil orchid bees, including an extinct genus, *Paleoeuglossa* (Poinar, 1998c,d) (see Figure 271 in Chapter 16) and a member of the extant genus *Euglossa* in Dominican amber (Engel, 1999). These bees provided indirect evidence of orchids in the Dominican Republic some 20 to 40 million years ago. Male orchid bees pollinate orchids while collecting aromatic compounds from the flowers apparently to convert them into female sex attractants. Each orchid has a different and unique method of attracting the bees, as well as attaching and removing pollinia from their bodies. These associations are so specialized that many orchids have come to depend on these bees for survival (Arditti, 1982). Putative orchid seeds were reported in Dominican amber (Poinar and Poinar, 1999), and Ramírez et al. (2007) discovered a Dominican amber stingless bee with an attached orchid pollinarium.

A list of described fossil bees, together with critical comments on their status, is provided elsewhere (Engel, 2000; Michener, 2000). Poinar (1998a) also cited stingless bee nests from Dominican amber and illustrated a stingless bee with resin balls (Figure 2A; see also Poinar, 1998b, Figure 3E) still attached to its corbiculae; these resin balls are employed in nest construction. Resin bugs (Hemiptera: Reduviidae) that specialize in predating other insects by attaching resin to their anterior legs are also present (Poinar, 1992a). Poinar (1998b, Figure 3C) illustrated a Dominican amber megachilid bee with a piece of resin adjacent to its mandibles that was probably being carried back to its nest for construction purposes.

TERMITES

Martínez-Delclos and Martinell (1995) described a basal Cretaceous termite worker from Montsec, Spain. Jarzembowski (1989) described a Spanish Early Cretaceous termite, and Engel et al. (2007) described additional Early Cretaceous termites. Poinar (2009b) described intestinal protozoa associated with an Early Cretaceous kalotermitid termite in Burmese amber, confirming that the mutualistic relations between termites and protists already existed at that time.

☐ *Reliability*
Category 2B, functional morphology.

17b. NEST BUILDING

Hasiotis (2003) extensively reviewed nest building in social insects, both ancient and modern, while also emphasizing the fact that social insect body fossils extend back only into the Early Cretaceous, whereas nests attributed by some to them go well back into the Triassic. Smith and Kitching (1997) briefly described some putative soil termite nests from South Africa, but in view of their Early Jurassic age it would be premature to accept their identity without considerably more work on the features, as well as obtaining a better understanding of whether these structures, if insect generated, might not reflect a convergent behavior by another group. Bordy et al. (2004) described in some detail Clarens Formation, Early Jurassic, southern African structures that they interpreted as termite nests. Hasiotis and Demko (1996) described some alleged Late Jurassic termite and ant nests from the Morrison Formation. Hasiotis and Dubiel (1995) described some probable soil nests recovered from the Late Triassic Chinle Formation of Arizona that resemble those of primitive termites.

There will undoubtedly be discussion about whether these represent termites, whose body fossil record currently extends only back into the Wealden, or another soil nest-building organism, but the structures are very tantalizing and are reported to be associated in the same beds with others that might be the work of bees, thus raising the parallel question.

Wheeler et al. (1994) cited termite nests with frass in Maastrichtian logs from Texas, the same horizon and region as the materials cited by Rohr et al. (1986), but with the nests in the outer parts of the trunk rather than in the central parts of a limb as in the specimens of Rohr et al. One of the two oldest known pieces of evidence of termite social behavior known is the Maastrichtian, Lambert Formation, drywood termite nest fragment from Vancouver Island, complete with frass (Ludvigsen, 1993; Ludvigsen and Beard, 1994, Figure 92; 1997, Figure 104).

Storch (1993, p. 82) observed "cemented wood of the same structure as hard carton nests of extant tree-dwelling termites" from the stomach contents of the mid-Eocene Messel anteater *Eurotamandua*, indicating still another very conservative character. Sombroek (1971) reported fossil termite mounds from several horizons in the Eocene and Late Cretaceous of northwestern Nigeria, and Zonneveld et al. (1971) provided some detail on their formation, using modern examples. Genise and Bown (1994b) described late Eocene–early Miocene soil termite nests from Egypt. Nel and Paicheler (1993, p. 173) provided references to varied nest occurrences.

Bown and Laza (1990) described a Miocene termite nest from southern Argentina, the first evidence of a fossil social insect from South America, and attributed it to an extant genus, *Syntermes*. Bown and Genise (1993) cited termite and ant nests from the Early Miocene of southern Ethiopia and the Late Miocene of Abu Dhabi. Vignaud et al. (2002, Figure 2) mentioned termite nests from the Late

Miocene of Chad. Crossley (1984) suggested that some carbonate mounds from the Old Stone Age of Malawi may represent termite mounds. Tessier (1959, 1965) discussed termite structures from Senegalese laterite, and Rozefelds (1990; see also Rozefelds and de Baar, 1991) provided an excellent example of Kalotermitidae borings and tunnels, with accompanying frass, from the mid-Tertiary rainforest of central Queensland.

TERMITE FRASS

(See also Section 9Bc, Termite Borings in Wood.) Noirot and Noirot-Timothée (1969) described the anatomical features of the termite anal region that give rise to the hexagonal cross-section of their frass pellets. Although not involving any actual borings, the illustration by Collinson (1990, Figure 46b) of a Late Eocene termite bit of frass from southern England is worth noting. Poinar (1998b, Figure 2B) illustrated a drywood termite (Kalotermitidae) actually excreting several drywood fecal pellets, presumably under the stress of being caught in the tree gum. Nel and Paicheler (1993) provided an account of the fossil record of drywood termite frass, as well as references to additional frass occurrences (pp. 123, 173) and termite eggs (p. 173). Genise (1995) described what he interpreted to be termite nests with frass in cycads from the Patagonian Late Cretaceous. Collinson (1990, p. 68P) reported drywood-type termite pellets from the Early Cretaceous Wealden (these are the oldest such termite coprolites reported) and from the British Paleocene. Hooker et al. (1995) reported similar termite pellets from the late Eocene of southern England.

□ *Reliability*

Category 2B, functional morphology, for the Cretaceous and younger termite nests, but Category 6, subject to confirmation for the pre-Cretaceous.

17c. WORKERS CARRYING LARVAE AND PUPAE

Weitschat and Wichard (1998, Plate 71, Figure e) illustrated a worker ant with a larva. Wagensberg et al. (1996) described a Dominican amber specimen containing numerous ant workers, some carrying larvae or pupae, plus eggs (the first known from the fossil record). Janzen (2002, Figures 184, 185) illustrated a Baltic amber ant carrying a thrip and an ant pupa. An ant holding a thrip in its mouth (see color plate 4B after page 132) in Burmese amber suggests a predator–prey behavior similar to present-day associations. Weaver ants (*Oecophylla smaragdina*, Hymeniptera: Formicidae) are used today as biocontrol agents for red-banded thrips (*Selenothrips rubrocinctus*, Thysanoptera: Theripidae) on mango trees in the Northern Territory of Australia (Peng and Christian, 2004).

□ *Reliability*

Category 1, frozen behavior.

17d. FUNGUS-GARDENING ANTS

It has been known for some time that New World attinid ants engage in fungus gardening. Currie et al. (1999), however, have discovered that these ants use an antibiotic-producing bacteria to control a virulent parasitic fungus, a uniquely coevolved relation of the mutualistic type. This is an example of a double mutualism between the ants and the fungus they farm plus the relation with the bacterium that prevents success by a parasitic fungus. The presence of fossil attinid ants suggests that the relations have considerable antiquity. Currie et al. (1999) repeated the comment about this symbiosis being at least 50 million years old, but this conclusion is based on the presence in Dominican amber of attinid ants, with the age of the amber (Eocene or late Oligocene–Miocene) still being contentious. Poinar and Poinar (1999, Figure 109) illustrated a Dominican amber fungus-gardening ant closely associated with a leaf fragment that it was presumably carrying back to the nest.

The fungus-growing ants are comprised of several genera that cultivate fungi and then use them as a food source (Wheeler, 1973). Members of the genus *Cyphomyrmex* construct small nests in various habitats, such as under stones or bark or in rotten wood. Some species have fungal gardens consisting of small roundish bodies known as bromatia, which range in diameter from 0.25 to 0.55 mm. When a queen leaves the old nest to start a new one, she will carry a bromatium as a starter culture. This was apparently the behavior with the queen *Cyphomyrmex* in Dominican amber whose bromatium is still present beneath her (see front cover, Figure C).

□ *Reliability*

Category 2B, functional morphology.

17e. SCALE AND ANT RELATION

Poinar and Poinar (1999, Figure 108) illustrated and discussed a Dominican amber ant with a scale insect in her mandibles that she was presumably transporting to a new site where the scale insect could produce honeydew. Although not an aphid, this example illustrates a similarly coevolved relation. Perkovsky (2006) described Eocene–Oligocene Saxonian amber with an association of ants and aphids that is suggestive of the typical ant–aphid relationship.

□ *Reliability*

Category 1, frozen behavior.

17f. TROPHALLAXIS

The great majority of worker ants from the Early Cretaceous possess a short first antennal segment (scape) (see front cover, Figure B) in contrast to extant worker ants. It has been suggested that, because this character would reduce their tactile and chemoreceptive abilities during trophallaxis, their eusocial status is questionable (Baroni Urbani, 1989).

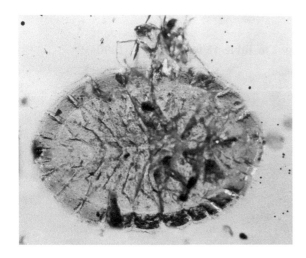

FIGURE 275 The flattened body of the termite bug *Termitaradus protera* (in Mexican amber) protected it against attack by termite soldiers (Poinar amber collection accession no. HE-4-39); length of bug = 7.10 mm.

17g. TERMITE NASUTES

Poinar (1998b, Figure 1D) illustrated an actual drop of sticky material coming out of the snout of a nasute from Dominican amber.

☐ *Reliability*
Category 1, frozen behavior.

17h. WASP NEST CONTROVERSY

The most recent word on the wasp nest controversy (see Boucot, 1990a) is that of Wenzel (1990), who accepted this Late Cretaceous specimen as that of a burrowing wasp.

*17i. TERMITE BUGS

(See Figure 275.) Poinar and Doyen (1992; see also Poinar, 1993, Figure 7) illustrated a Mexican amber termite bug (*Termitaradus protera*, Hemiptera: Termitaphididae), with a peculiarily flattened form, including loss of compound eyes, ocelli, and wings. Wilson (1971, Table 20-1) suggests the possibility that these bugs are coevolved to feed on fungi in termite nests. Poinar's specimen occurs in a block of amber adjacent to two termites, which strengthens this case. Poinar and Poinar (1999, Figure 119, p. 118) illustrated and discussed a termite bug from Dominican amber, a hemipteran known today only from termite nests.

☐ *Reliability*
Category 2B, functional morphology.

*17j. ANT MIMIC

Poinar (1993, Figure 8) illustrated and discussed a Dominican amber cerambycid beetle, *Tilloclytus*, that is an ant mimic today. The amber specimen possesses characters similar to the living forms and is preserved in the same amber block with an ant. Linsley (1959) reviewed the extensive evidence, including ant mimicry, among the Cerambycidae.

☐ *Reliability*
Category 2B, functional morphology.

*17k. TERMITE LARVAE

See Section 3r, Fossil Flatus: Indirect Evidence of Intestinal Microbes.

*17l. TERMITE NEST ASSOCIATES

Ross (1998, Figures 78, 79) illustrated a flightless scuttle fly, Phoridae, and a rove beetle, Staphylinidae, both known nest associates of termites in Dominican amber.

☐ *Reliability*
Category 2B, functional morphology.

*17m. ANT NEST BEETLES

Nagel (1987) has ably summarized the Baltic and Dominican amber occurrences of the coevolved ant nest beetles (Paussinae). For a brief account of the modern situation of these animals, see Wilson (1971); see also Poinar and Poinar (1999, Figures 112, 113) and Solórzano Kraemer (2006) for Mexican amber.

☐ *Reliability*
Category 2B, functional morphology.

*17n. ARMY ANT AND PREY

Poinar and Poinar (1999, Figure 104, pp. 98–99) discussed a Dominican amber army ant closely associated with a prey wasp pupa.

☐ *Reliability*
Just about Category 1, frozen behavior.

*17o. REPLETE ANT

Poinar and Poinar (1999, Figure 111, p. 109) discussed and illustrated a Dominican amber replete ant involved in storage of food for the colony.

☐ *Reliability*
Category 2B, functional morphology.

*17p. ORCHID BEES

(See Figure 271 in Chapter 16.) Poinar (1998c,d) and Poinar and Poinar (1999, Figures 131, 132, pp. 120–121) illustrated and described a Dominican amber orchid bee, a group which

TABLE 37
Insect Social Behaviors Leaving a Fossil Record (Additions to Table 41 from Boucot, 1990a)

Behavioral Type		Taxonomic Level	Time Duration	Category	Refs.
17a.	Stratigraphic ranges				
	Social bees	Superfamily	Paleocene–Holocene	2B	Rasnitsyn and Michener (1991)
	Social wasps	Superfamily	Early Cretaceous–Holocene	2B	Carpenter and Rasnitsyn (1990)
	Social ants	Superfamily	Tertiary–Holocene	2B	Poinar and Milki (2001)
17i.	Termite bugs	Family–superfamily	Miocene–Holocene	2B	Poinar (1993)
17j.	Ant mimic	Family–superfamily	Miocene–Holocene	2B	Poinar (1993)
17l.	Termite nest associates	Family–superfamily	Miocene–Holocene	2B	Ross (1998)
17m.	Ant nest beetles	Family–superfamily	Eocene–Holocene	2B	Nagel (1987)
17n.	Army ant and prey	Family–order	Miocene–Holocene	1	Poinar and Poinar (1999)
17o.	Replete ant	Superfamily	Miocene–Holocene	2B	Poinar and Poinar (1999)
17p.	Orchid bees	Family	Miocene–Holocene	2B	Poinar and Poinar (1999)
17q.	Bee pollen feeding	Superfamily	Miocene–Holocene	1	Poinar (1998b)
17r.	Weaver ants	Family	Eocene–Holocene	2B	Dlussky et al. (2008)

today is comprised of very obligate pollinators of specific orchid taxa representing a highly coevolved system.

□ *Reliability*

Category 2B, functional morphology.

*17q.   BEE POLLEN FEEDING

Poinar (1998b, Figure 2D) illustrated a bee fly (Bombyliidae) with a pollen grain attached to its body. Lutz (1993) described a Middle Eocene bee from the Eifel that still had attached pollen grains, as well as corbicula and other structures involved in pollen gathering.

□ *Reliability*

Category 1, frozen behavior.

*17r.   WEAVER ANTS

Dlussky et al. (2008) illustrated weaver ants from the Messel Eocene and summarized the fossil record of this leaf-nest-constructing group.

□ *Reliability*

Category 2B, functional morphology.

SUMMARY

The data are from the family to superfamily levels (see Table 37).

18 Long-Range Migration

(See also Section 5l, Mass Moth Migration.) Boles and Lohmann (2003) provided good evidence that spiny lobster migrations are guided by magnetic navigation capabilities. This is a remarkable discovery that may help to ultimately explain some of the other long-range migrational mysteries. Boucot (1981, pp. 242–243) reviewed some examples of migration for modern marine benthic invertebrates, including the spiny lobster (see also Lyons, 1980). Powell and Reynolds (1965) discussed annual movements of tagged Alaska king crabs, *Paralithodes camtschatica*. Shuster (1982) provided information about modern limuloid migration, Phillips (2006) provided extensive data on lobster migrations, and Crawford et al. (2008) summarized information about modern marine crab migrations. Carpenter et al. (2003) detail isotopic evidence for the migration of a fish from the Late Cretaceous of the North American interior seaway from brackish–estuarine waters into fully marine waters.

19 Molting

19a. TRILOBITES

Brandt (2002) extensively reviewed trilobite molting styles. Although not having anything to do with trilobites, Garcia-Bellido and Collins' (2004) specimen of the curious Middle Cambrian Burgess Shale arthropod *Marella splendens* actually molting is noteworthy.

19b. DECAPODS

(See Figure 276 and Figure 277.) Feldmann and McPherson, 1980, Plate 2, Figure 9) illustrated a Jurassic *Glyphea* from Arctic Canada preserved in a partially molted condition, which is as fine an example as the fossil record has provided to date. Feldmann and Gazdzicki (1998) illustrated a molted specimen of the Antarctic Miocene *Antarctidromia inflata*.

19c. INSECTS

Poinar and Poinar (1999, Figure 99, p. 93) illustrated and discussed a Dominican amber example of a larval mayfly molting into a subimago. Rasnitsyn and Quicke (2002, Figures 23, 24, pp. 20, 21) discussed and illustrated some insect molting examples.

FIGURER 276 Decapod molts: *Glyphea* from the Miocene of Arctic Canada (×0.5). (Plate 2, Figure 9, in Feldmann, R.M. and McPherson, C.B., *Fossil Decapod Crustaceans of Canada*, Paper 79-16, Geological Survey of Canada, Ottawa, Ontario, 1980. Reproduced with permission.)

FIGURE 277 *Antarctidromia inflata* from the Miocene of King George Island, Antarctica (×0.5). (Figure 3 in Feldmann, R.M. and Gazdzicki, A., *Acta Palaeontologica Polonica*, 43, 1–19, 1998. Reproduced with permission.)

20 Sensitive Plants

Dilcher et al. (1992, p. 39) mentioned that the mimosoid specimen illustrated earlier in Boucot (1990a, after Schlee, 1980) as an example of a sensitive plant with its leaves folded together is probably an example of "sleep posture" that can be induced in mimosoids by either drought conditions or night.

21 Reptilian and Mammalian Burrows and Dens

MAMMALIAN

(See Figure 278.) Hunt (1990, Figure 29, pp. 106–107) has further described some canine-relative beardog burrows from the Early Miocene of Nebraska and some mustelid burrows. See also Section 8CIIb for discussion of the Miocene beaver involved with the trace fossil *Daemonelix*.

Kurten's (1958) classic paper on European cave bears employed biometric methods on large collections from varied Pleistocene cave deposits to make a very strong case for hibernation, noting that the accompanying mortalities affected mostly the oldest members of the populations. His paper essentially cast doubt on earlier theories that blamed the extinction of these cave dwellers on genetic factors rather than the probability that habitat (cave) loss due to competition with humans was the real cause.

Szafer (1957) described some subrecent hamster burrows from Poland with stored seeds still preserved within. Hugueney and Escuillie (1995) described the structure and implications of a group of Early Miocene beavers (*Steneofiber*) from France that are consistent with their representing a family group, with three generations present, that may well have been catastrophically overwhelmed within their burrow. They commented on "the antiquity of life-history strategies in this family and that such fixed behavior can persist over long periods of time" (p. 225) while discussing the family group behaviors of the Miocene to recent aquatic beavers,

which are one more example of behavioral fixity during lengthy time intervals, this time at the family level.

Gee et al. (2003) provided a detailed description of an earlier Miocene Rhenish Browncoal region burrow system containing stored nuts. They also discussed the presence of some presumably insect-bored nuts and made comparisons with modern examples. They concluded that the seed hoard represents the work of a rodent, as is the case with many modern, similar examples, but such examples are very few from the fossil record.

Riggs (1945) described a Miocene carnivore (*Zodiolestes daimonelixensis*) coiled up in a typical, helically spiraled *Daimonelix* burrow. The beaver constructor was not present. Whether or not the little carnivore had predated a beaver was not apparent; the carnivore might have been merely resting.

REPTILIAN

(See Figure 279.) Botha-Brink and Modesto (2007) described a South African Permian "pelycosaur" aggregation involving an adult with young of the same species which implies

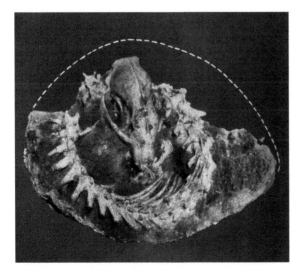

FIGURE 278 Riggs' (1945, Figure 45) *Zodiolester daimonelixensis* coiled up in a Miocene beaver burrow from Nebraska (×0.5). (For a discussion of the beaver burrow, see Boucot, 1990a.) (Courtesy of the Field Museum, Chicago.)

FIGURE 279 Reconstruction of a group of juvenile *Youngina* preserved in a Late Permian burrow from South Africa (about ×0.5). (Figure 5 in Smith, R.M.H. and Evans, S.E., *Palaeontology*, 39, 289–303, 1996. Reproduced with permission of Wiley-Blackwell.)

215

parental care and possibly a burrow environment. Smith and Evans (1995, 1996) described what must be a group of five Late Permian juvenile *Youngina* from South Africa, huddled together in such a manner that preservation in a burrow is the best explanation. Smith (1993) provided more information about helically spiraled Permian South African therapsid burrows, and Smith (1987) described a helical *Daemonelix* burrow from the South African Permian with a dicynodont preserved within. Damiani et al. (2003) described evidence of a cynodont burrow with an occupant (*Thrinaxodon*) from the Permo–Triassic boundary region in South Africa.

Groenewald et al. (2001) described an Early Triassic, Driekoppen Formation, Beaufort Group, South African burrow system with branches and widths suggesting that more than one therapsid (*Trirachodeon*, with skulls found in terminal portions at one locality) individual occupied the burrow system; this may represent a colonial therapsid in the same league, in a way, as the famous South African mole rats.

Smith (1987) described a therapsid burrow (*Diictodon*) from the South African Permian (Beaufort Group). Varricchio et al. (2007) described what appears to have been a Cenomanian burrow from Montana that included the partial skeletons of an adult and two juvenile hypsolophodont dinosaurs, from which it may be inferred that parental care was involved and that the climate was probably semiarid, as indicated by the presence of calcretes. Duvall (1986) provided a popular description of denning in rattlesnakes that provided an excellent background to the behavioral significance of snake dens in winter cold regions. Though not in the burrow or den category, Xu and Norell's (2004) account of an Early Cretaceous Laioning troodontid dinosaur skeleton in a sleeping position is suggestive of an avian-like posture.

AMPHIBIAN

Hembree et al. (2005) described a trace fossil genus, *Torridorefugium*, from the Kansas Early Permian that contains lysorophid amphibian remains of *Brachydectes*, probably the burrow maker in an environment interpreted as possibly being seasonally dry (for a citation of additional lysorophid burrow evidence, see Boucot, 1990a). Hembree et al. (2004) described some Early Permian lysorophid (*Brachydectes*) burrows, complete with occupant, from Kansas and reviewed similar occurrences elsewhere in North America.

☐ *Reliability*

Category 1, frozen behavior, for those examples with the maker still within the burrow.

22 Vertebrate Endocranial Casts

22a. MAMMALS AND PTEROSAURS

Edinger (1961) provided an excellent introduction to the whole subject of vertebrate endocasts and their significance, and Edinger (1955) provided a truly classic account of cetacean endocasts and their correlation with function while highlighting the relationship between acoustic problems and the reduction or loss of the olfactory lobes and hypertrophy of those regions. Edinger (1926) described Tertiary microchiropteran endocasts exhibiting greatly enlarged inferior colliculi that correlate very well with the expected greater auditory capability so characteristic of bats.

Radinsky (1968) summarized endocranial cast information indicating that both specialized vibrissae and handedness in early Pliocene to recent otters and in their Oligocene to Pliocene *Potamotherium* relatives are well documented in endocasts—one more example of behavioral conservatism. Dominguez Alonso et al.'s (2004) paper on *Archaeopteryx* endocast morphology demonstrated that this primitive bird had real flight capabilities as reflected in its endocast.

Edinger's classic monograph (1948) on horse brain evolution and its complement (1966) on camelid endocasts observed that the very primitive Eocene endocasts present in both families suggest that their behaviors must have differed very significantly from those of their Neogene descendents. The Paleogene ancestral types have endocasts more like those of opossums (*Didelphis*) and reptiles than those of typical placental mammals.

Edinger (1941) described pterosaur endocasts, emphasizing their convergence with bird brains and the fact that a bird-like condition was reached by them in the Cretaceous; they never approached the reptilian condition, and they show evidence of feeding in water as do some birds such as the Sphenisciformes. Witmer et al. (2003) described some pterosaur endocasts and semicircular canals, noting that they have large optic lobes and reduced olfactory lobes, as noted by others earlier (Edinger, 1941, described them as minute!), and that their enlarged cerebellar floccular lobes and semicircular canals may well be associated with flight specializations for flight control; these represent an excellent example of convergence. Kurochkin et al. (2007) mentioned a Cenomanian, Late Cretaceous bird endocast that still had relatively large olfactory lobes.

□ *Reliability*
Category 2B, functional morphology.

22b. HOMINID HANDEDNESS

Susman (1994) analyzed the implications of the functional morphology of the human thumb as contrasted with that of other higher anthropoids. He concluded that only in members of our genus is there reliable functional morphological evidence for the capability of using tools in a complex manner and that currently available evidence places the first appearance of the morphology back to 2.5 million years. Quade et al. (2004) push toolmaking back to 2.5 to 2.6 my.

*22c. HOMINIDS

(See also Section *36, Human Behavior.) Eccles (1989, p. 756) provided an excellent discussion of hominid endocranial casts, relative to behaviors, and commented, "Thus we have the revealing and provocative concurrence of three phenomena: the parts of the brain that govern spoken language become manifest at that stage where brain enlargement and marked encephalization first obtrude and soon after tools of hard materials first appear in the fossil record." Eccles concluded that the neurological basis for language was present in *Homo habilis* and *H. erectus* (pp. 94–96). Tobias (1987) concluded from *Homo habilis* endocasts that the "neurological basis of spoken language" was present. Goren-Inbar et al. (2004) discussed Israeli evidence for human use of fire from the Acheulian, more than 700,000 years ago. Belitsky et al. (1991) described a Middle Pleistocene wooden plank showing evidence of man-made polish. Bahn (1989) noted the age of a late Pleistocene female figurine from near Krems, dated at about 30,000 years, and discussed a still earlier, more controversial item from Israel dated at 230,000 to 800,000 years. Valladas et al. (1992) provided radiometric dates for some European prehistoric cave paintings that are between 1 and 2 millenia BP. Valladas et al. (1992) also cited radiocarbon evidence for dating some Spanish cave paintings as far back as 14,000 ± 400 years, with the caveat that stylistic dating is risky. Henshilwood et al. (2002) described South African evidence of "carved" bone with an age of about 77,000 years BP. Henshilwood et al. (2004) described bored snail shells used as beads from the South African Middle Stone age; they are about 75,000 years old and may be the currently oldest evidence of a typically "modern" human behavior (i.e., ornaments). The shells were large ones selected from a nearby intertidal source. See also d'Errico et al. (2005), who reviewed the bead evidence from other various localities. Vanbaeren et al. (2006) provided evidence suggesting that shell beads made by humans can be dated as far back as 100,100 to 135,000 years BP! Domínguez-Rodrigo et al. (2005) described Late Pliocene cut marks on ungulate bone that undoubtedly were made by humans of some kind, using sharp tools to remove flesh.

Marshak (1989) reviewed much of the evidence bearing on how early in the Acheulian, Mousterian, and Palaeolithic there is reliable evidence for truly human behaviors. Mania and Mania's (1988) description of the last interglacial engravings attributed to *Homo erectus* and the accompanying discussions provide a good sample of current thinking about data extending back for several hundred thousand years.

Brooks et al. (1995) and Yellen et al. (1995) described worked bone tools with minimum ages of about 90,000 years BP which they ascribe to the Middle Stone age, as a minimum estimate of very complex tools and associated hunting activities. Semaw et al. (1997) provided evidence of Early Pleistocene type of stone tools from Ethiopia that are dated at 2.5 my. Dennell (1997) and Thieme (1997) discussed the behavioral–cultural significance of a wooden throwing spear from a Palaeolithic, 400,000-year-old German locality and its implications for the presence of an advanced hunting technology far earlier than previously assumed.

Conroy (1990, pp. 351–357) summarized much of the data relating to cultural activity placing the use of tools back in the 2- to 2.5-million-year interval. Steele (1999) commented on the presence in Kenya of flint flake tools made by flint-knapping techniques that are about 2.5 million years old; they are the oldest known evidence for really hominid-type toolmaking behaviors. Steele (1999) also mentioned slightly older, reliable evidence from Ethiopia (Quade et al., 2004; Semaw et al., 1997). Roche et al. (1999) discussed the new Kenyan evidence in detail. Sieveking and Newcomer (1987; see also Potts, 2003) summarized the literature on human uses of flint, including chert, that go back into the Palaeolithic. Jacobs et al. (2008) suggested that Middle Stone Age evidence from Southern Africa, about 65,000 to 70,000 years old, suggests that modern humans were making stone tools and using beads and ocher, all relatively modern human behaviors.

Zhang et al. (1999) described early Neolithic musical instruments—flutes made from the ulnae of a crane—from China. They referred to a brief discussion (Slovenian Academy of Sciences, 1997) of a Mousterian (43,400–67,000 BC) flute fragment from Europe and some unreferenced citations of Magdalenian and Aurignacian flutes (Marcuse, 1975). Alperson-Afil (2008) described the evidence for Acheulian use of fire at a site in Israel of Early and early Middle Pleistocene age.

23 Preening

As briefly mentioned earlier (Boucot, 1990a; see also Habersetzer et al., 1992) evidence for preening in the fossil record is very rare. Evidence for hair is also very rare (see Boucot, 1990a, for citations and Sections 4AIIIc,d,h,j; 13d). Jungers et al. (2002) discussed the tooth comb present in some strepsirrhine primates that is used for grooming, among other things. Rose et al. (1981) reviewed the fossil record of tooth combs, making the point that they are present in several unrelated groups—lower primates, tree shrews, and arctocyonid condylarths. The arctocyonids have a mid-Paleocene and Early Eocene tooth comb record, whereas the lower primates go back to only the Late Miocene (now revised to the Late Eocene thanks to the discovery by Seiffert et al. (2003) of new material from the Fayum). In any event, the tooth comb is a convergent character in the unrelated groups. Ducrocq et al. (1992) described comb-like lower incisors in a Late Eocene Thai dermopteran. Rose et al. (1981) and Jungers et al. (2002) emphasized that the tooth comb is also used in feeding as well as in grooming.

□ *Reliability*

Category 2B, functional morphology.

24 Grain-Size Selectors

24b. MAGNETITE BALLAST GRAINS IN SAND DOLLARS ("WEIGHT BELTS")

Chen and Chen (1994) briefly described sand dollar juvenile and adult sand grain collecting, with the juveniles collecting relatively more than the adults, as in the fossils discussed earlier (Boucot, 1990a).

*25 The Seagrass Community Complex

Brasier (1975) pointed out the close relationship between the shallow-water seagrass taxa and larger Foraminifera in the present and employed the Foraminifera to extrapolate the seagrass community back into the early Tertiary and, by inference, even into the later Cretaceous. Eva (1980) extended this analysis into the Caribbean region and extrapolated it there into the Eocene. In a breakthrough paper, Ivany et al. (1990) recognized excellently preserved seagrasses in the Middle Eocene Avon Park Formation of Florida and provided extensive detail on the nature of the seagrass community complex. Selected Foraminifera can be used as indicators of the probable former presence of a seagrass community, as sirenians, known since the earlier Eocene, are obligate seagrass community feeders with a pantropical–subtropical distribution pattern. Domning (2001) reviewed what is known of sirenian occurrences in the Caribbean Cenozoic and drew conclusions about seagrass communities from this information. Voigt (1981) described some late Cretaceous seagrass immurations caused by attached bryozoans; the seagrasses are taxa distinct from those in the Cenozoic examples.

☐ *Reliability*

Category 1, frozen behavior.

*26 Shelter

Fraaye and Jäger (1995b) described four examples of Early Jurassic fishes from England and Germany preserved within ammonite conchs. These may be examples of feeding on debris from a dead ammonite or its scavengers, seeking shelter, or other possibilities. Fraaye and Jäger (1995a) reported on examples of decapods from Germany and England that are preserved in the same manner and may represent inquilinism.

☐ *Reliability*

Category 6, until the sample is considerably enlarged.

*27 Flying and Gliding Vertebrates

Storch et al. (1996) provided an excellent summary of gliding in rodents, including four independently evolved lineages: (1) Eomyidae (flying squirrels, Late Eocene–Pliocene), including the Late Oligocene flying squirrel specimen described by Storch et al.); (2) Sciuridae (sciurids, Early Oligocene–Holocene); (3) Anomaluridae (scaly-tailed flying squirrels, Late Eocene–Holocene); and (4) Gliridae (dormice, Early Oligocene–Holocene; see Carroll, 1988). The oldest goes back to the Oligocene, with gliding membranes (patagia) still preserved, as well as hair. Ducrocq et al. (1992) described a Late Eocene dermopteran (flying lemur) from Thailand. Bats are known well back into the Eocene. The earliest known gliding mammal is *Volaticotherium* from the mid-Mesozoic of Inner Mongolia (Meng et al., 2006), with a skin outline of the patagium well preserved. Birds are known back into the Late Jurassic. Carroll (1978) described a gliding reptile, *Daedalosaurus*, from the Madagascar Upper Permian and compared it to Late Triassic gliding reptiles as well as to living *Draco*. Fraser et al. (2007) described an Upper Triassic gliding reptile from Virginia and reviewed the morphology of other gliding reptiles known from the fossil record. Evans (1987) reviewed the morphology of an Upper Permian gliding reptile.

*28 Possible Genetic–Developmental Defects

(See also Section *4AIIIv, Tetrapod Osteomyelitis.)

*28a. TRANSPOSED BIVALVE HINGE LINES AND DOUBLE SIPHONAL GROOVES IN A GASTROPOD

Matsukuma (1996) described various bivalve taxa, both modern and ancient examples as far back as the Devonian and Permian, in which the hinge-line structures of the left valve have been transposed to the right valve and *vice versa*, with the additional complication in some taxa that not every feature of the hinge line need be transposed. This may be a genetically controlled developmental feature, although in the absence of experimental work on living bivalves it is difficult to evaluate the possibility. Cox (1969) discussed transposed bivalve hinges in some detail. Weaver (1963) described an Oligocene *Crassatella* from California with a transposed hinge. In the same vein is hinge transposition in various bivalves which involves different hinge-line elements, with first evidence involving Permian taxa. Boyd and Newell (1968) discussed Permian examples and cited a number of younger examples.

Leriche (1910) illustrated and described a unique Oligocene gastropod ("*Pleurotoma regularis*") from Belgium that has two well-developed siphonal grooves. Something clearly happened to generate the second groove as a replacement for the first. Whether or not it involved a genetic defect in a developmental pathway or a disease organism is unknown. Radwański (1977, Plate 12, Figure 15) illustrated a Miocene Polish gastropod with an additional siphonal groove that he interpreted as regrowth following a predation attempt. Following Radwanski's suggestion it is possible to interpret Leriche's specimen as also being the result of regrowth following a predation attempt. Radwanski has very generously provided an illustration (Figure 280) of a previously unpublished specimen of the same genus as he illustrated in 1977 from the same Korytnica Clays that even more vividly displays the two siphonal grooves. Bergman (1998) discussed the possibility that the reversal present in some scolecodonts is a genetically caused feature.

The presence in some echinoids of four or six, rather than the customary five, ambulacral regions may reflect a developmental defect, a genetic cause (see Tasnádi-Kubacska, 1962, for discussion). Springer (1920, Plates LVII and LVI, pp. 402–403) referred to some Late Paleozoic crinoids having six rays in some cases.

☐ *Reliability*
Category 2B, functional morphology.

*28b. RHINOCERATID TOOTH AND OTHER POSSIBLE GENETIC DEFECTS

Billia and Graovac (2001) described a genetic defect in a Pleistocene rhinoceratid tooth and cited a number of dental anomalies in rhinoceratids from the fossil record; theirs is one of the few known examples where a genetic defect can be indicated. Von Koenigswald et al. (2007) described a Pleistocene rhino mandible from near Wiesbaden having supernumerary teeth, which probably represent a developmental–genetic defect (Figure 281). Although not a rhinoceratid, hypodontia in a Middle Eocene lofiodontid was interpreted by Cuesta Ruíz-Colmenares et al. (2004) as a probable genetic defect. Palmqvist et al. (1999) described aberrant cranial morphology

FIGURE 280 (Left) *Pleurotoma regularis* with two siphonal grooves (×1) (Leriche, 1910, p. 343). (Right) A previously unpublished specimen (×1), generously provided by Radwański, of *Clavatula* from the Polish Miocene Korytnica Clays displaying two siphonal grooves; Radwański interprets the double grooves as being the result of regrowth following a failed predation attempt.

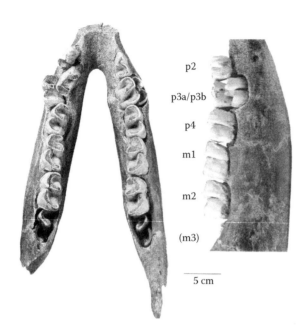

p2

p3a/p3b

p4

m1

m2

(m3)

5 cm

FIGURE 281 Lower jaw of a Middle Pleistocene rhino (*Stephanorhinus hundshelmensis*) from Germany with supernumerary teeth (bar = 5 cm). (Figure 1 in von Koenigswald, W. et al., *Paläontologische Zeitschrift*, 81, 416–428, 2007. http://www. schweizerbart.de. Reproduced with permission.)

and a missing right canine in a hunting dog from the Spanish Pleistocene that they suspect may represent a genetic defect compatible with pack behavior, including hunting. Wang and Rothchild (1992; see also Section 4AIIIv, Osteochondroma) suggested that osteochondroma in Oligocene *Hesperocyon* is probably a genetic defect.

☐ *Reliability*
Category 1, frozen behavior.

*29 Teratologies

In the absence of information about the diseases of most organisms, living and fossil, all one can do is to record in a systematic manner the varied irregularities present in skeletal materials of all sorts and try to consider whether or not predation or diseases of one kind or another might be responsible.

*MARINE INVERTEBRATES

*29a.   TRILOBITES

(See Figure 282.) In his review of trilobite abnormalities, Owen (1985) commented that some are probably the result of predation; see also Babcock and Robison (1989a,b) for some Cambrian trilobite predation data. Šnajdr (1978b) provided an excellent account of trilobite abnormalities with comment about potential causes; most examples were based on Bohemian paradoxitid material. Babcock (1993, 2003, 2007) provided additional data and reviewed further information concerning the basic problems involved with trilobite malformations and teratologies. Jell (1989) considered the difficult questions involved with trying to understand the causes of trilobite teratologies, citing and describing a number of examples while also referring to previously described items (Petr, 1983; Westergård, 1936); he made it clear how difficult it is to arrive at positive conclusions. Šnajdr (1978a) discussed a fine Barrandian *Bohemoharpes* example from the Late Silurian Kopanina Formation that he interpreted as a neoplasm.

Isidro et al. (1996; see also Capasso and Caramiello, 1996) provided more information about trilobite paleopathology. Rabano and Arbizu (1999) provided some Spanish Ordovician and Devonian examples of possible predation and parasitism. Lee et al. (2001) described a teratology in a Late Cambrian Korean trilobite while reviewing some earlier accounts (Chatterton, 1971; Conway-Morris and Jenkins, 1985; Henningsmoen, 1975; Lochman, 1941; Ludvigsen, 1977; Owen, 1980, 1983; Pratt, 1998; Ramskold, 1984; Rudkin, 1979, 1985; Schrank, 1969; Šnajdr, 1978b, 1979, 1981, 1990; Walcott, 1883; Whittington, 1956). Hughes (2001) illustrated and briefly described some Cambrian trilobite skeletal abnormalities. Jago and Haines (2002) illustrated and described a fine Australian Cambrian example of a healed injury and reviewed many earlier accounts of abnormalities.

☐ *Reliability*

Category 6, due to the overall uncertainties about causation.

*29b.   BIVALVES

(See Figure 283, Figure 284, and Figure 285.) Kříž (1979) described a number of Silurian cardiolid shell anomalies, although causation is elusive as always. Pek and Marek (1983) described a pectinid with what they considered a pathological

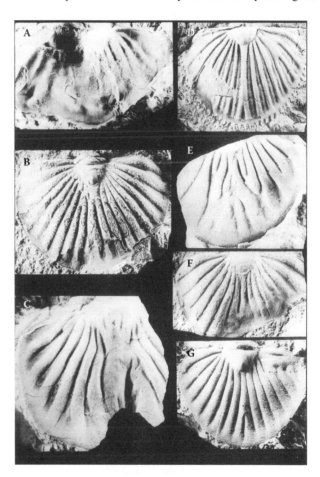

FIGURE 282 Šnajdr (1990) illustrated (all ×1) a number of Early Devonian trilobite pygidia from the Barrandian that suffered various deformities: (A) an incomplete *Radioscutellum intermixium* pygidium with a badly damaged posterior portion, possibly due to predator damage; (B) a teratological pygidium of *Spiniscutellum umbelliferum* with an excessive number of ribs; (C) a badly deformed *Bojoscutellum* that may be due to predator damage; (D) a *Thysanopeltis speciosa* pygidium showing some proximally fused ribs on the right side; (E) *Bojoscutellum obsoletum* with badly deformed ribs; (F) a *Spiniscutellum umbelliferum* with a right side possibly deformed by predator activity; (G) a *Bojoscutellum angusticeps* with a very deformed right side. In none of these examples can one be truly positive about causation. (From Šnajdr, M., *Bohemian Trilobites*, Czech Geological Survey, Prague, 1990. Reproduced with permission.)

FIGURE 283 Pek and Marek (1983) illustrated a Turonian *Lyropecten* valve (×1) that they suggested was first damaged by a predator, following which a prominent shell deformity developed that reflected mantle behavior. (Plates I and II in Pek, I. and Marek, J., *Vestnik Ustredniho Ustavu Geologického*, 58, 44–52, 1983. Reproduced with permission.)

condition from the Cretaceous of Czechoslovakia. Boucot (1981, pp. 283, 286) discussed varied molluscan examples while emphasizing that the only group about which much is known is the commercially important oysters. Boshoff (1968) considered some bivalve shell malformation situations from the present. This is a poorly understood area, as is unfortunately the case with most noneconomically important organisms. Čtyroký (1969) briefly described several Czech Miocene pectens with relatively uncommon teratologies that he ascribed to mantle injuries while noting that such conditions are rare.

☐ *Reliability*

Category 6, due to the overall uncertainties about causation.

*29c. BRACHIOPODS

Cooper and Grant (1972, 1975) discussed and illustrated a malformed ventral valve muscle scar on a Permian *Yakovlevia* (1975, Plate 473, Figure 17) and and the abnormal cardinal process of a large *Tropidelasma* (1972, Plate 52, Figure 28). Cooper and Grant (1975, Plate 474, Figures 15, 16) illustrated specimens of *Yakovlevia hessorum* with an extra anterior halteroid spine. Kaiser (1964) illustrated a *Cariniferella tioga* from the New York Late Devonian with a badly malformed ventral muscle field and another from a Devonian *Stropheodonta*. Alexander (1977, Plate 1, Figure 23) illustrated a Mississippian *Rhipidomella* from Utah that suffered some kind of mantle damage reflected in subsequent shell growth. Bergström (1968, Plate 6, Figures 3–6) illustrated valves of *Titanonema grandis* from the Swedish Late Ordovician that had clearly suffered mantle damage from an unknown cause. Wright (1974, Figures 3g,h) illustrated a "pathological" specimen of the Late Ordovician *Parastrophina angulosa* with deformed costae. Williams and Rowell (1965, pp. 73–75) described in some detail the nature of shell repair in several brachiopods.

FIGURE 284 Kříž (1979) illustrated and discussed a number of pathological Bohemian Silurian bivalves, including *Cardiolita* (Plate XXXIX, Figures 1–3, 5–11) and *Cardiola* (Plate XXXIX, Figure 4). They show distorted valves and shell damage of various kinds that was repaired. (Figures 1, 2, and 3, ×0.6; Figure 4, ×1.3; Figure 5, ×1; Figure 6, ×0.9; Figure 7, ×.7; Figure 8, ×1; Figure 9, ×6.1; Figure 10, ×0.9; Figure 11, ×1.8.) (From Kříž, J., *Sborník Geologických Věd, Paleontologie*, 22, 5–157, 1979. Reproduced with permission.)

Some deviations from normal bilateral symmetry in brachiopods are caused by the crowding, oyster fashion, present in some forms such as Silurian pentameroids that grew in blankets; these abnormalities reflect the ecology of the brachiopods rather than being evidence of disease of any sort. Fürsich and Palmer (1984) made a strong case for the "normal" type of asymmetry in some brachiopod taxa probably being a genetically neutral feature. Cooper and Grant (1972, p. 103) emphasized the malformations generated by crowding in some of their Permian brachiopods from the Glass Mountains.

*29d. ECHINOIDS

Radwańska and Radwański (2005, pp. 114–115) have thoroughly reviewed the varied, mostly poorly understood, teratologies reported in echinoids. It is clear that we have a lot to learn in this area.

FIGURE 285 Kříž (1979) illustrated (A) *Cardiola* showing a damaged shell (×4); (B) *Cardiola* showing pearl-like structures (×5); and (C) *Cardiola* showing impressions of an epibiont (×10.1). (From Kříž, J., *Sborník Geologických Věd, Paleontologie*, 22, 5–157, 1979. Reproduced with permission.)

*TERRESTRIAL INVERTEBRATES

Poinar and Poinar (1998, Figure 1) illustrated a Dominican amber mite with a necrotic spot on its idiosoma that was produced by an unknown pathogen. Tumor-like growths in a lepidopteran caterpillar in Mexican amber provide a rare example of tumors in insects (Poinar and Poinar, 2005; see color plate 3B after page 132). A viral infection was presented as a possible cause of the tumors. The same paper discusses viral, protozoan, and fungal infections of insects in Tertiary and Cretaceous ambers.

*30 Disease

Boucot (1990a) reviewed many examples of host–parasite relations, including fungal relations, but seldom dealt with other situations, except in the case of humans (4AIIIj), which are on a nonevolutionary time scale. Below we list a few of these and discuss them. The best treatment of skeletal paleopathology we have seen is provided by Rothschild and Martin (1993), who reviewed the principles involved and evaluated and provided a number of examples; their book appears to be far more reliable in its diagnoses and suggestions than prior efforts by others (see also Rothschild and Martin, 2006a).

*30a. METASTATIC CANCER IN THE JURASSIC

Rothschild et al. (1999a) provided convincing evidence for the presence of metastatic cancer in a Late Jurassic specimen of dinosaur bone from the Morrison Formation of Colorado. This is the oldest example known of metastatic cancer. Additional to this consideration of metastatic cancer, Rothschild et al. (1998) provided an example from the Utah Late Jurassic Morrison Formation of a hemangioma, a benign tumor, probably from a dinosaur. McWhinney et al. (2001a) discussed an example of periostitus in a dinosaur bone for which the nature of the teratology is uncertain. Although not pertaining to the Jurassic, the report by Helbling et al. (2001) of Cretaceous hemangiomas in hadrosaurs as contrasted with other dinosaurs is of interest. Also of interest here is Brothwell's (1967) account of human neoplasms of relatively modern age.

□ *Reliability*
Category 1, frozen behavior.

*30b. DISEASED ELEPHANT UPPER JAW AND DEFORMED TEETH

Klein (1971) provided an elegant description of a Pleistocene *Paraelephas trogontherii* upper jaw, from Rhine gravels near Manheim. A massive hole in the palate connects with the infected root of the left, third molar, with good evidence of bone resorption caused by the infection. Mayr (1992, Figure 3, p. 205) illustrated an upper jaw of *Palaeoloxodon trogontheria* from the Rhine Gravels that shows evidence of periostitus as a wide opening, presumably produced by an infection well within the jaw to the tooth eruptive region.

Kramer et al. (1996) cited various examples of skeletal pathology in a Late Pleistocene mammoth from Wisconsin. Rothschild and Rothschild (1996c) discussed inflammatory disease and pointed out that osteoarthritis is very rare in the fossil record; they cited two *Iguanodon* occurrences from Bernissart. Tanke and Rothschild (1997) and Rothschild (1990) showed that osteoarthritis is very rare in dinosaurs. Rothschild and Molnar (1988) described reasonable incidences of osteoarthritis from collections of Australian Pleistocene kangaroos and *Diprotodon*. Rothschild (1993b) indicated that skeletal evidence of arthritis is absent from sub-Recent higher primates. For reviews of skeletal pathology of rheumatic diseases, see Rothschild (1989, 1993a).

Bricknell (1987) reviewed varied vertebrate paleopathologic examples with an emphasis on Pleistocene proboscidian abnormalities, particularly teeth. Maschenko and Shpansky (2005) described a very abnormal mammoth tooth from the Pleistocene of the Tomsk region and reviewed other occurrences of aberrant mammoth teeth. Ducrocq et al. (1995) described some Tertiary dental anomalies and refer to many other accounts. Miles and Grigson (1990, pp. 314–315) briefly discussed and illustrated some badly misplaced elephant molars but said little about causation except for suggesting that the jaw size was too small. Hunter and Langston (1964) described a Pleistocene odontoma from the Yukon that was generated by a proboscidian; this type of aberrant dental structures is known from a number of mammals but they have not yet been recognized in the pre-Pleistocene. Isidro et al. (1996) provided data concerning some Miocene diseased mammalian cervical vertebrae, a sirenian mandible with a tumor-like lesion, and a rhinoceratid with a metapodial lesion.

*30c. BONE FRACTURES

In Boucot (1990a), Repenning provided a description of a marine mammal with unique anterior limb fractures; Section 4AIxb mentioned some dipnoan jaw fractures, and Section 6AIIh also cited some allosaur fractures. Warren and Ptasznik (2002) described the oldest known tetrapod fracture, which involved an Early Carboniferous, Australian amphibian with excellent development of bony callus. Lingham-Soliar (2004) described some evidence for fractured dentaries, with osteomyelitis involvement, in mosasaurs from the Late Cretaceous. MacDonald and Sibley (1969) described various Rancho La Brea mammalian examples. Dawson and Gottfried (2002) mentioned some healed rib fractures in a Miocene dolphin. Although modern fractured fish bones are known, we have not found any reports from the fossil record. Moodie (1923) devoted an entire chapter to evidence of fractures in fossil vertebrate materials, Permian to Pleistocene. Rothschild and Tanke (2005) reviewed the evidence of fractures in theropods, as well as bite marks. Sullivan et al. (2000) described a late Campanian impact fracture from an ornithomimid dinosaur. Tanke and Rothschild (1997) observed that healed bone fractures are common in the vertebrate fossil record. Rothschild et al. (2001b) discussed theropod stress fractures and tendon avulsions possibly related to activity. Although not necessarily involving fractures, the discussion by Lockley et al. (1994) addressed trackway evidence for limping gaits in a number of dinosaur examples from the Jurassic and Cretaceous. Rothschild (2003a) reported stress fractures in some Late Pleistocene mastodons from New York.

☐ *Reliability*
Category 1, frozen behavior.

*30d.   CERATOPSIAN STRESS FRACTURE

Rothschild (1988) described what he diagnosed as a stress fracture in a ceratopsian phalanx. He mentioned that such fractures have not previously been recognized in the fossil record and discussed their presence in humans, horses, and greyhounds today.

☐ *Reliability*
Category 1, frozen behavior.

*31 Marine Molluscan Larval Types and Their Behavior

Jablonski and Lutz (1980) discussed the nature of the larval gastropod protoconch and the bivalve prodissoconch, which can be useful in distinguishing between planktotrophic and nonplanktotrophic dispersal types. Beesley et al. (1998) provided a wealth of information regarding protoconch data. Haszprunar (1995) concluded that nonplanktotrophic larvae are probably the most primitive molluscan type.

The best example known to the authors of marine larval, presumably planktotrophic, bivalve behavior preserved in the fossil record is Palmer's (1989) suite of Late Jurassic bivalves involving three genera with four species from the South of England. The well-preserved prodissoconchs, including stages I and II, together with the adult dissoconchs, comprise a very convincing piece of evidence about larval behavior for the three bivalves.

Hansen (1978, 1980, 1983) discussed a number of correlates of planktonic vs. nonplanktonic protoconchs for gastropods. Scheltema (1978, 1979) reviewed similar material, as did Shuto (1974). Bouchet (1981), as well as Hansen, provided examples of planktotrophic ancestral forms giving rise to nonplanktotrophic descendents with no known examples of the reverse, *contra* Haszprunar (1995).

In a series of publications, Frýda et al. (2008; Frýda, 2004) used gastropod protoconch characters to indicate the presence of planktotrophic and nonplanktotrophic larvae back into the earlier Paleozoic, including data for many extinct higher taxa. Their data indicate that a high level of heterochrony has always been characteristic of gastropods. Changes from a planktotrophic to a nonplanktotrophic larval life style have evidently occurred repeatedly.

Page (1997) discussed protoconch development in *Haliotis*. Lutz et al. (1984) discussed protoconch morphology and its implications for hydrothermal vent gastropods and bivalves, and they addressed the problem of them having nonplanktotrophic protoconchs for the most part. Dzik (1994) reviewed a variety of earlier Paleozoic protoconch morphologies in gastropods. Bouchet (1981) discussed some Miocene to Recent protoconch characteristics in eastern Atlantic species of *Terebra*.

Williams (2003) summarized the information on "juvenile" shells of many linguliform brachiopods which presumably represent the larval condition, but due to our ignorance about the larval ecology of modern linguliforms it is not yet possible to interpret this potentially important data.

*32 Competition Involving Bryozoans

It is very difficult to prove competition in the fossil record, but McKinney (1995) makes a strong case that beginning in the Cretaceous cheilostome bryozoans have consistently overgrown cyclostomes. This is a taxonomically high-level behavior but is convincing.

□ *Reliability*
Category 1, frozen behavior.

*33 "Lost" Behaviors and Their Vestigial Evidence

Since 1859 the presence of vestigial organs has been of evolutionary importance. The fossil record provides evidence of changing behaviors involving a variety of organisms under the pressure of various selection pressures. Some of the most spectacular evidence comes from the vertebrates.

Gingerich and his colleagues (Gingerich, 1989, 2003; Gingerich et al., 2009) have produced a series of papers laying out the path taken by certain artiodactyls moving on the path from primitive, mid-Eocene, Lutetian whales with fully developed, functional posterior limbs to Oligocene whales lacking posterior limbs but possessing internal vestiges in the form of reduced pelvis and proximal limb bones. Gingerich (2003) provided an excellent summary of current understanding and problems while emphasizing that some Late Eocene whales have very reduced external limbs, the function of which was certainly not involved with locomotion (see also Gingerich et al., 2009). Andrews (1921) described some posterior, poorly developed external limbs on a modern humpback whale that obviously reflected a "mistake" in the developmental program.

Rieppel et al. (2003) summarized the problems involved with the relations of the earliest, mid-Cretaceous snakes and their ancestry. They noted that those early snake taxa possessed rudimentary, clearly vestigial posterior limbs. Apesteguía and Zahner (2006) described an Argentinian Cenomanian–Turonian snake in which a sacrum and small legs are preserved with the sacrum still attached to the vertebral column; this is a very transitional lizard to snake form that makes it clear that snakes were not derived from mosasaurs. The latest contribution (Palci and Caldwell, 2007) to the literature regarding the relationship of snakes with other lizards, as well as mosasaurs, makes it clear that this is still a complex problem for which more data are needed. In a very insightful paper, Wiens et al. (2006) employed a variety of morphological, developmental, biogeographic, paleontological, behavioral, and genetic data to make the point that the evolution of snake-like morphologies in lizards is a very complex business involving multiple examples of convergence; they observed that short-tailed forms tend to be burrowers, whereas long-tailed forms tend to be surface dwellers. The relatively sparse fossil record of this group does provide some help with understanding times of first appearance for some groups, but overall this summary is a caution against making undue generalizations about phylogeny based on inadequate paleontological information.

Other items, such as the loss of a bony tail in birds, are well illustrated by the stages represented by Late Jurassic *Archaeopteryx*, Early Cretaceous birds with very short tails, and more modern birds possessing only a pygostyle. Also instructive is the loss of a tail in the higher anthropoids (e.g., our own coccyx). MacFadden (1992) described some of the evidence concerning atavistic toes in modern horses, a polydactyly similar to that in humans and presumably due to a developmental failure possibly involving the *hox* gene complex.

☐ *Reliability*
Category 2B, functional morphology.

*34 Stunting

Evidence for stunting, a term preferred over dwarfing, has been widely discussed in the paleontological literature (Hallam, 1965) to explain aggregations of small shells. However, Hallam (1965) makes a good case that it is very difficult to determine, except in the cases of hypo- and hypersaline environments, that true stunting is involved. In the terrestrial environment, however, there are numerous instances of off-shore Pleistocene–Holocene islands of true dwarfing in many groups, as well as examples of gigantism (Azzaroli, 1982).

☐ *Reliability*
Category 1, frozen behavior.

*35 Oceanic vs. Neritic

The behavioral and coevolutionary evidence concerned with the selection of favored environments in which to live is well known in the present, and there is a vast literature concerning the many physical and biological variables involved on land and in the sea. The fossil record also provides a mass of evidence involving varied correlations of chiefly physical variables and the presence or absence of taxa on land and in the sea. For example, in the marine benthic environment, the massive documentation of community evolution through time correlates with variables of distinctive kinds; this is particularly true for the infaunal and epifaunal benthic plus demersal environments (see Table 38 in Chapter 37 for some examples). In the marine environment, there are varied correlations between water depth and the presence or absence of varied pelagic organisms. For example, the presence of abundant myctophid fish is good evidence of a bathypelagic environment (Table 38) and suggests that depth and temperature controls are involved to a large extent (see discussion in Gaudant, 1978). Additionally, the presence of certain decapods (Oji, 2001) is also good evidence for bathyal, benthic environments. The presence of certain radiolarian taxa in bedded chert, beginning with the Ordovician, provides evidence for oceanic vs. neritic plankton.

Problems arise, however, when considering the distinctive separation of neritic from oceanic organisms; for example, which variables control and determine the taxa present or absent in these two major marine environments? Clarke and Fitch (1979) described a mass of North American teuthoid (squid) statolith material from the Eocene to Pleistocene, all of neritic type, as are all modern teuthoids. Boucot (1990a) reviewed the distribution of the belemnites from an environmental viewpoint, concluding that concentrations of small specimens reflect inshore "nursery" environments, as contrasted with the more offshore neritic environments, with there being no evidence for oceanic environment material in deposits of Cretaceous age. There is good evidence for the presence of oceanic as contrasted with neritic plankton from the Cretaceous to present, while oceanic Radiolaria are present as far back as the Cambrian (Braun et al., 2007). In all of this information, though, there is still no good explanation for just how these neritic and oceanic organisms maintain their distinctive environments.

☐ *Reliability*
Category 1, frozen behavior.

*36 Human Behavior

(See also Section *22c, Hominids.) From what we know of basic human behaviors, it is clear that there has been no significant change since the beginnings of recorded history. Sandison (1967) reviewed the evidence of human sexual behaviors and concluded: "Human nature appears to have changed remarkably little over the millennia." The same conclusion is obvious when one considers warfare through time; the weapons and tactics have changed, but the basic behaviors remain the same. Consideration of other basic human behaviors permits one to conclude *nihil novum sub sol.* One need only read the daily newspaper to reach this conclusion. Courville (1950) discussed and described some Neanderthal skull injuries presumably caused by their fellows, with the implicit message that such lethal behavior within our genus has been with us for a long time, as well as more numerous Cro-Magnon examples. Zollikofer et al. (2002) described evidence for a weapon-produced injury on a French Neanderthal skull and reviewed other possible examples of Neanderthal intraspecific injuries caused by weapons, including Trinkaus's (1983) consideration of evidence from the Shanidar Cave in Iraq. In view of what we now know, it is possible that intraspecific violence using weapons by hominids against members of their own species is hard wired, as contrasted with the interspecific aggressive behaviors of the other higher anthropoids.

37 Summary and Conclusions

Before summarizing some of the available evidence about community "fixity" through evolutionary time, below, it might be good to say a few words about how to define a community. Boucot (1981, pp. 177–181) commented on the variety of ways to define a community. Boucot's view is that a community can most usefully be defined as a group of co-occurring organisms in which the varied taxa have abundances ranging from the very abundant to the rare, with the caveat that these abundances will vary within reasonable limits; one cannot define communities by using taxic abundances taken to the nearest decimal point.

When one considers the effects of short-term changes due to the vagaries of recruitment in the marine and nonmarine environments, the effects of varied predators and herbivores on a short-term basis, the effects of disease on the short-term decimation of one or another taxon, as well as the short-term climatic variations that affect many environments, it becomes clear that an overly restrictive, numerical community definition makes little biological sense. Examples of such variability from the shallower water marine environment include such things as Dayton's (1975) discussion of some Alaskan kelp communities where abundances of herbivores and carnivores interacting with the kelps produce wide abundance swings, and Scheibling's (1994) account of some Nova Scotia lobster, sea urchin, sea urchin epizootics, and kelp communities provides still more evidence about the wide abundance swings that are normal through time for a community.

Boucot (1990a) concluded that behavioral and coevolutionary relations were very, very conservative at the genus, family, and superfamily levels, as well as at higher levels, based on numerous examples drawn from the fossil record. In this treatment, we have added a large number of additional examples. The majority of the examples consist chiefly of two points in time: a younger one commonly but not always being the present and an older one from the fossil record. Few of these examples are based on a large number of points through time. Some sticklers might protest that almost none of these examples has any statistical validity. This is certainly true if one examines examples in isolation from each other; however, when one sums the results from all of the examples, the overall conclusion is obvious: overall behavioral and coevolutionary conservatism at varied taxonomic levels. It is foolish to insist on viewing each example in isolation from all the others.

1. COMMUNITY EVOLUTION BEHAVIORAL EVIDENCE

(See Table 38.) No attempt has been made to synthesize the massive data dealing with the relative fixity, climatic zone by climatic zone, of Cenozoic paleobotanical data from the evidence provided by megafossils and pollen plus spores. A good example that also includes similar information provided from fungal spores is that studied by Martha Sherwood (see Gray, 1985).

1a. COMMUNITY EVOLUTION: THE PLEISTOCENE PARADOX

Employing evidence from Holocene–latest Pleistocene pack-rat middens, Betancourt et al. (1990) pointed out that plant species during this time interval have moved in a relatively independent manner, as is the case with most terrestrial taxa during this time interval of rapidly changing global climatic gradient. Elias (1994) provided a wealth of information from terrestrial insects about the comings and goings of insects during the Pleistocene in response to the everchanging climates and consequent community changes. His conclusions parallel very well those provided earlier (Boucot, 1990a) for other organisms as well as insects. Elias (1994, p. 63) stated:

> Paleoentomologists are reasonably confident about the constancy of ecological requirements in Quaternary beetles because of the consistency of associations of insect species through time: species kept the same company in the past as they do today. In addition, certain well-studied regions, such as the British Isles, have yielded fossil insect assemblages of nearly identical composition during the different glacial, interglacial, and interstadial climatic episodes. In other words, the warm-adapted faunas of one interglacial are extremely similar to the warm-adapted faunas of other interglacials, even though these climatic episodes may have occurred hundreds of thousands of years apart. Likewise, the cold-adapted faunas found during one glacial stage have a great number of species in common with faunas found in previous and subsequent glaciations. If the physiological properties of the species in these faunas were changing through time, then those changes would have to be unidirectional and of constant rate, in order for insect species to assemble in these same communities time after time. The odds of an entire suite of species evolving new physiological requirements in this uniform a manner are too small to merit serious consideration.

TABLE 38

Community Fixity Examples

Time	Duration	Environment	Dominant Organisms	Refs.
Marine				
Invertebrates				
Pleistocene–Holocene	0.5 my	Rocky bottom	Mollusks	Lindberg and Lipps (1996)
Pleistocene–Holocene	1 my	Level bottom	Bivalves	Troitskiy (1974)
Early Pleistocene–Holocene	1–2 my	Outer shelf boulders	Bivalves	Dell (1951a)
Early Pleistocene–Holocene	1–2 my	Level bottom	Brachiopods, bivalves	Dell (1951b)
Early Pleistocene	1–2 my	Level bottom	Brachiopods	Neall (1970)
Pleistocene–Holocene	1–2 my	Shallow subtidal	Ostracods	Neale and Howe (1974)
Pleistocene–Holocene	0.1 my	Coral reef	Scleractinians	Pandolfi (1996)
Pleistocene–Holocene	0.5 my	Coral reef	Primarily scleractinians	Jackson (1992)
Pleistocene–Holocene	1–2 my	Coral reef	Scleractinians	Nakamori (1986)
Plio–Pleistocene	1–3 my	Level bottom	Mollusks	Taylor (1966)
Late Pliocene–Holocene	2–3 my	Level bottom	Mollusks	Powell (1937, 1939)
Pliocene–Holocene	4 my	Level bottom	Bivalves	Fleming (1952, 1953)
Pliocene–Late Miocene	6 my	Coral reef	Scleractinians	Jackson et al. (1996)
Late Miocene–Holocene	5 my	Hypersaline, shallow	Gastropods	Schreiber and Hsü (1980)
Miocene–Holocene	15 my	Level bottom	Crustacean, worm	MacGinitie and MacGinitie (1949)
Miocene–Holocene	15 my	Level bottom	Bivalves, gastropods	Watkins (1974c)
Miocene–Holocene	20 my	Shallow subtidal	Bivalves, gastropods	Chinzei (1978, 1984)[a]
Miocene–Holocene	20 my	Shallow subtidal	Hermatypic corals	Hayward (1977)
Miocene	5 my	Oceanic plankton	Radiolaria	Sachs and Hasson (1979)
Miocene–Holocene	15 my	Rocky and shelly substrates; intertidal to 800 m	Rhynchonellid, brachiopod, and various inverts	Lee (1978)
Miocene–Holocene	15 my	Deep-sea	Whale–falls and mollusca	Pyenson and Haasl (2007)
Mio–Pliocene–Holocene	5 my	Coldwater seep	Bivalves	Campbell (1992)
Miocene–Holocene	15 my	Soft bottom	Lunulitiform bryozoans	Di Geronimo et al. (1992)
Miocene–Holocene	15 my	Level and rocky bottoms	Mollusks	Eagle et al. (1995)
Miocene–Holocene	15 my	Soft bottom	Cheilostome, scleractinian	Cadée and McKinney (1994)
Miocene–Holocene	15 my	Soft and hard bottoms	Molluscs and other taxa	Eagle and Hayward (1992)
Miocene–Holocene	15 my	Rocky shore and coarse sediment	Mollusca, echinoids, other taxa	Eagle et al. (1994)
Miocene–Holocene	15 my	Rocky shore and enclosed bay	Mollusks and other taxa	Eagle et al. (1995)
Early Miocene–Holocene	15 my	Level bottom	Mollusks	Harzhauser and Kowalke (2001)
Late Oligocene	30 my	Level bottom	Bivalves, gastropods, echinoids	Baldi (1973)
Late Oligocene–Early Miocene	5 my	Level bottom	*Crassostrea* and varied invertebrate, algal, fungal taxa	Parras and Casadío (2006)
Oligocene–Holocene	30 my	Sargassum weed	*Planes* (decapod)	Glaessner (1965, p. 115)
Oligocene–Holocene	30 my	Sargassum weed	Portunidae (decapod)	Gekker and Merklin (1946)
Oligocene–Holocene	35 my	Level bottom	Bivalves, gastropods	Watkins (1974b)
Oligocene–Holocene	35 my	Level bottom	Mollusks and other taxa	Eagle and Hayward (1993)
Oligocene–Holocene	35 my	Deep-sea whale carcass	Bivalves	Squires et al. (1991)
Oligocene–Holocene	35 my	Rocky shore	Mollusks and other inverts	Lee et al. (1983)
Oligocene–Holocene	35 my	Marine caves	Gastropods	Lozouet (2004)
Late Eocene–Holocene	40 my	Methane seep	Bivalves	Squires and Goedert (1991)
Middle Eocene–Holocene	45 my	Seagrass environment	Bivalves	Brasier (1975); Ivany et al. (1990)
Eocene	50 my	Abyssal	Ostracodes	Guernet (1993)
Late Cretaceous–Paleocene to Holocene	70 my	Deep-sea benthos, spreading center	Tube worms	Haymon and Koski (1985); Haymon et al. (1984); Oudin and Constantinou (1984); Oudin et al. (1985)
Paleocene–Holocene	60 my	Level bottom	Bivalves, gastropods	Sorgenfrei (1965)
Paleocene–Holocene	60 my	Shallow subtidal	*Corbula* (bivalve) and *Turritella* (gastropod)	Hoffman (1977)
Paleocene–Holocene	60 my	Level bottom	Ahermatypic corals, brachiopods	Asgaard (1968)

TABLE 38 (cont.)
Community Fixity Examples

Time	Duration	Environment	Dominant Organisms	Refs.
Late Cretaceous–Holocene	90 my	Level bottom	Bivalves	Kauffman (1967)
Late Cretaceous–Holocene	90 my	Level bottom	Bivalves	Thomas (1975)
Late Cretaceous–Holocene	90 my	Cold seep	Acmaeid–vestimentiferan	Jenkins et al. (2007a,b)
Late Cretaceous–Holocene	90 my	Wood-falls	Varied mollusks (bivalves, gastropods)	Kiel et al. (2009)
Cretaceous	90 my	Level bottom	Bivalves	Sohl (1967)
Late Cretaceous–Holocene	90 my	Intertidal, level bottom	*Donax*	Sohl (1978)
Turonian–Holocene	90 my	Interstitial	Ostracodes	Danielopol and Wouters (1992)
Late Jurassic–Late Cretaceous	60–100 my	Level bottom	Bivalves	Fürsich (1977)
Middle Jurassic–Holocene	200 my	Level-bottom "reefs"	Brackish water oysters and associates (such as the annelid *Polydora* and the sponge *Cliona*)	Feldmann and Palubniak (1975); Gardner (1945); Hudson and Palmer (1976); Lawrence (1968); Wiedmann (1972)
Early Jurassic–Holocene	190 my	Plankton	Heteropod snails	Bandel and Hemleben (1987)
Jurassic–Holocene	200 my	Intertidal, level bottom	*Callianassa* (decapod)	DeWindt (1974); Sellwood (1971)
Jurassic–Holocene	200 my	Brackish to nearshore	Corbulids	Lewy and Samtleben (1979)
Jurassic	10 my	Intertidal	*Cochlearites* and *Lithiotis*	Lee (1983)
Middle Jurassic	2–4 my	Shallow subtidal	*Dacromya* (bivalve)	Fürsich and Sykes (1977)
Jurassic–Cretaceous	130 my	Neritic nekton	Belemnite shoals	Stevens (1965)
Permian	50 my	Level bottom	Brachiopods	Waterhouse (1973)
Late Pennsylvanian–Early Permian	50 my	Level bottom	Brachiopods, mollusks, Foraminifera, bryozoans	Typical examples in Mudge and Yochelson (1962)
Pennsylvanian	ca. 5–10 my	Level bottom	Brachiopods	Bambach and Bennington (1996)
Mississippian	20 my	Level bottom	Rugose corals	de Groot (1964); Hill (1938–41); Kullman (1966); Rowett (1975a,b)
Mississippian–Holocene	340 my	Hydrothermal vent	Tube "worms"	Banks (1985); Moore et al. (1986)
Late Devonian	350 my	Hydrothermal vent	Tube "worms," brachiopods	Poole (1988)
Middle Devonian	360 my	Hydrothermal vent	Tube "worms," bivalves	Kuznetsov et al. (1990); Zaykov and Maslennikov (1987)
Middle Devonian (Eifelian)	10 my	Marine	Rugose corals	Birenheide (1963)
Late Silurian–Mississippian, possibly to Permian	100 my	Marine	Pelagic myodocope ostracods	Gooday (1983); Siveter et al. (1991)
Silurian–Early Devonian	50 my	Level bottom	Ostracodes	Lundin (1978)
Silurian–Devonian	100 my	Level bottom	Brachiopods	Boucot (1975)
Silurian	30 my	Neritic	Brachiopods	Boucot and Lawson (1999)
Early Silurian	5 my	Subtidal benthic	Brachiopods	Johnson (1979)
Late Ordovician	3–5 my	Subtidal benthic	Trilobites	Bergström (1973)
Middle Ordovician	10 my	Subtidal benthic	Trilobites	Ludvigsen (1979)
Early Ordovician	5 my	Subtidal benthic	Trilobites	Sheldon (1987)
Late Cambrian–Ordovician	50 my	Oxygen-poor	Olenid trilobites	Fortey (2000)
Vertebrates				
Plio–Pleistocene	1–3 my	Freshwater	Fish	Smith (1975)
Pliocene–Holocene	5 my	Bathypelagic	Myctophid	Fitch (1969)
Late Miocene–Holocene	7 my	Bathypelagic	Fish	Arambourg (1927)
Middle Miocene–Holocene	15 my	Bathyal	Fish	Sato (1962)
Miocene–Holocene	15 my	Upper bathyal	Fish, benthic inverts	Ishida (1993); Mizuno (1993); Mizuno and Takeda (1993); Ohe (1993); Oji (2001); Yamaoka (1993)
Late Miocene–Holocene	7 my	Deep-water, bathypelagic, neritic	Fish	Gust (2000); Corsi et al. (1999); Morgan and Brown (1998); Wagner and Marks (1990); Crane (1966); Fitch (1969); Raschke (1989); Stadum (1997); Kalabis and Schultz (1974)

(continued)

TABLE 38 (cont.)
Community Fixity Examples

Time	Duration	Environment	Dominant Organisms	Refs.
Oligocene–Holocene	35 my	Bathypelagic	Fish	Arambourg (1967); Gregorová (2004)
Oligocene–Holocene	35 my	Bathypelagic	Fish	Danilchenko (1946, 1947, 1960)
Early Oligocene–Holocene	35 my	Bathypelagic	Fish	Nevesskaya (2002)
Middle Eocene–Holocene	45 my	Bathypelagic	Fish	Bannikov (1993)
Eocene–Holocene	40 my	Seagrass environment	Sirenians, Foraminifera	Eva (1980)
Eocene-Holocene	50 my	Bathypelagic	Fish	Carnevale (2003); Danilchenko (1962); Jerzmanska (1968); Jerzmanska and Kotlarczyk (1976, 1983); Prokofiev and Bannikov (2002)
Eocene–Holocene	10 my	Bathyal, abyssal, neritic	Fish	David (1943, 1956)
Eocene–Holocene	55 my	Deep-sea, demersal, bathypelagic	Macrurid fish	David (1946a,b); Fitch (1969); Grechina (1973); Prokofiev (2005)
Eocene–Holocene	50 my	Coral reef	Fish community	Bellwood (1996); Bellwood and Wainwright (2002); Landini and Sorbini (1996)
Maastrichtian	65 my	Bathypelagic	Myctophid	Uyeno and Matsui (1993)
Nonmarine				
Invertebrates				
Pleistocene–Late Miocene	6 my	Polar	Insects	Elias and Matthews (2002); Elias et al. (2006)
Miocene–Holocene	5 my	Terrestrial	Carabid beetles	Matthews (1979)
Miocene–Holocene	5 my	Lotic	Chironomids	Grund (2006)
Early Eocene	50 my	Level-bottom freshwater	Bivalves	Hanley (1976)
Middle Eocene–Holocene	45 my	Mangrove swamp complex	Verts, plants, inverts	Westgate and Gee (1990)
Paleocene–Holocene	60 my	Level bottom	*Elliptio* (bivalve) and by inference glochidia and fish	LaRoque (1960)
Mid–Cretaceous–Holocene	100 my	Litter	Varied insects	Perrichot (2004); Perrichot et al. (2007)
Permian–Carboniferous	100 my	Brackish–estuarine	Crustaceans	Schram (1981)
Carboniferous–Holocene	300 my	Ephemeral, shallow, freshwater bodies	Estherids	Tasch and Zimmerman (1961)
Vertebrates				
Miocene-Holocene	20 my	Swamp	Herpetofauna	Holman (1968)
Early Miocene	6 my	Equatorial rainforest	Mammals	Van Couvering (1980)
Late Cretaceous–Eocene	30 my	Freshwaters	Crocodilians, bony fish, amphibians, turtles	Estes and Berberian (1970); Jepsen (1963)
Permian	30 my	Nonmarine	Vertebrates	Olson (1980)
Early Permian	15 my	Lowland environments	Vertebrates	Olson (1952)
Land Plants				
Miocene–Holocene	15 my	Coastal swamp	Angiosperms	Anderson and Muller (1975)
Eocene–Holocene	40 my	Forests	Woody plants	Chaney (1947); Axelrod (1958)
Eocene–Holocene	55 my	Intertidal, level bottom	Mangroves, gastropods, bivalves	Gray (unpublished)

[a] Chinzei (1984) used the term "temporal parallelism" when referring to the stability at the generic level of communities through time.

Source: Boucot (1978; 1990a, Table 43) plus additional material accumulated since 1989.

Graham et al. (1996) analyzed Late Pleistocene North American mammal associations and arrived at conclusions similar to those of the palynologists, students of insects and small mammals: Late Pleistocene mammalian communities have been in a continual state of flux. They concluded that this supports a Gleasonian community concept without considering that the strongly changing Late Pleistocene climatic gradient is responsible for the effect. If they had also considered earlier Quaternary mammalian associations, they would have found correspondences in community terms to those similarly found with palynological data. Gasse and Van Campo (2001) provided late Quaternary pollen and diatom data from a Madagascar lake sequence that are compatible with community shifts back and forth through time in the southern tropics.

Buzas and Culver (1994), using benthic Tertiary Foraminifera from eastern North America, found no evidence for the concept of community evolution; however, their data consist of samples taken through a local "transect" that covers the entire Tertiary climatic change from Paleogene warmth to Neogene coolness, very different environments. They made no attempt to track a similar environment through time. In view of this fact, their conclusions have little validity.

Lopez Antonanzas and Cuenca Bescós (2002) described some Early to Middle Pleistocene small mammal associations from a Spanish site that further support the concept of back and forth climatic fluctuations during the Pleistocene being accompanied by corresponding animal changes.

1b. TIME SCALE FOR COMMUNITY CHANGE AND EVOLUTION

Cronin and Raymo (1997) made a good case for abyssal ostracodes in the later Pliocene changing their diversity through time on a Milankovich-correlated time scale; this is the first such example from the deep oceans. Raymo et al. (1998) provided data suggesting that short-term, millenial, Milankovitch-type climatic changes occurred about 1 million years ago in an overall warmer environment than in the later Pleistocene, which further supports the concept that Milankovitch-type climatic cycles have been with us since the beginning.

The Milankovitch-type cycles provide evidence that cyclic climatic fluctuations that will influence communities on land and sea have been with us since the beginning. Extensive and elegant Milankovitch data are provided by Park et al. (1993) from South Atlantic JOIDES core, Late Cretaceous–Paleocene data covering a million-year span. Olsen and Kent (1996) provided convincing data concerning the recognition of Milankovitch cyclicity in the Late Triassic–earliest Jurassic of the Newark Series lake beds of eastern North America. Bond et al. (1993) pointed out the difficulty of being certain as to whether or not all cyclic changes can be blamed on Milankovitch-type orbital forcing and considered some latest Quaternary data from Greenland ice cores. Willis et al. (1999) provided an excellent summary of some Hungarian Late Pliocene data supporting, from pollen diagrams covering several hundred thousand years, the repetition of plant communities through Milankovich and other perturbations. Mutterlose and Ruffell (1999) provided good evidence for a Milankovitch-type interpretation of cyclicly banded Early Cretaceous marine strata in eastern England and northern Germany. Using Lake Baikal cores, Kashiwaya et al. (2001) recognized Milankovitch cycles in the 400, 600, and 1000 kyr. Gong et al. (2001) provided an introduction to a mass of pre-Quaternary data dealing with Milankovitch cyclicity, including their own Late Devonian information; all of this supports the presence of Milankovitch cyclicity since the very beginning.

1c. LEVELS OF COEVOLUTION WITHIN COMMUNITIES

Although it does not involve coevolution, Brasier's (1992) paper used examples from the Mediterranean and eastern North American forests to illustrate how an exotic fungus can effectively attack local woody plant populations and either largely or totally destroy them. This example is of biogeographic importance in indicating how an exotic organism, coevolved in its native habitat (suspected to be the New Guinea–Celebes region) can wreak havoc elsewhere. Singer and Parmesan (1993) provided an elegant example of variations in plant–insect relations that emphasized the behavioral complexity that can be present; that is, one may not safely assume that all biotic relations remain relatively fixed, as there can be complex patterns of switching.

Allied to the question of coevolution within communities is that of rarity. Why are some taxa numerically dominant, whereas others are uncommon to rare? Why are the abundant genera also generally cosmopolitan and the rare genera endemic? Why do the abundant genera tend to be eurytopic, whereas the rare taxa tend to be stenotopic? Rabinowitz et al. (1984) reviewed some of the concepts advocated to explain the question of abundance vs. rarity in modern communities. They pointed out that some workers have even concluded that rare species are environmentally too feeble to persist for long with a community, which is at total variance with the paleontological evidence. Rabinowitz et al. (1984) pointed out that their own observational and experimental data with grass species made it very clear that the rare taxa are by no means feeble or unable to maintain their positions within the community. However, the essential question of why some species are rare and others abundant remains unanswered.

1d. QUATERNARY VS. PRE-QUATERNARY

Further support for the ecological back and forth shown by the Quaternary climatic fluctuations that is reflected in similar, co-occurring vegetational changes has been found from a 250,000-year record of pollen taken from marine beds to the northwest of the Sahara (Hooghiemstra et al., 1992). Tzedakis (1993) provided Late Pleistocene palynological data from Greece extending back over 400,000 years indicating the existence of at least three plant "biome" repetitions through time; thus, communities are continually forming and reforming under the influence of changing climatic conditions (for similar data from a Macedonian core, see Wijmstra and Groenhart, 1983). Elias (1996) provided further data from North American Pleistocene beetles. Kershaw and Whitlock (2000) assembled a number of papers making it very clear that later Quaternary glacial and interglacial floras at any one locality feature a back and forth change in pollen floras as one goes from warm to cold conditions. See Fuji (1988) for Lake Biwa, Japan, materials that further support the concept of recurring plant associations in the later Quaternary, as well as de Beaulieu and Reille (1989) and Hooghiemstra (1989), included in Kukla (1989), which address this question.

*1e. Rates of Species-Level Evolution within Evolving Community Groups and Subgroups Belonging to Ecological–Evolutionary Units and Subunits

Boucot has frequently used a diagram illustrating community evolution (Boucot, 1990a, Figure 409, is typical) which shows that there is an inverse relationship between the relative abundance of a genus within an evolving community and rate of evolution, invariably anagenetic. In other words, the abundant genera show little or no morphological change within any ecological–evolutionary unit or subunit, whereas the uncommon to rare genera show rates of anagenesis inversely related to their relative abundance.

Because rare genera are seldom collected from any community, unless very, very large samples are available through evolutionary time, it is the relatively uncommon genera that commonly supply the evidence cited below. The examples cited below are mostly drawn from Boucot's familiarity with Silurian–Devonian brachiopods; we suspect that workers familiar with other groups and time intervals could provide similar materials.

Boucot (1997, p. 15) listed the following:

1. *Eocoelia* species lineage of the middle to late Llandovery and earlier Wenlock
2. *Stricklandia lens* subspecies to *Costistricklandia* lineage of the early Llandovery to early Wenlock
3. *Microcardinalia–Plicostricklandia* lineage of the earlier Llandovery to mid-Wenlock
4. *Borealis–Pentamerus–Pentameroides* lineage of the middle to late Llandovery in Eurasia and the mid-Llandovery to mid-Wenlock of North America
5. Late Ludlow–Pridoli *Dayia* species lineage of the European Province
6. *Howellella vanuxemi–Howellella cyclopterus–Acrospirifer murchisoni* lineage of the Early Devonian Appohimchi Province
7. *Dalejina oblata–Discomyorthis* lineage of the Early Devonian Appohimchi Province
8. *Glypterina–Ptychopleurella* lineage of the Middle Ordovician through Early Devonian
9. *Leptostrophia (Leptostrophiella)–Leptostrophia (Leptostrophia)–Protoleptostrophia* lineage in the Early and Middle Devonian of the Appohimchi Province
10. Early Devonian–Eifelian age lineage from small *Nanothyris* to large *Nanothyris* to *Rensselaeria* to *Etymothyris* to *Amphigenia preparva* to small species of *Amphigenia* (species such as *curta*, *parva* and *chickasawensis*) to intermediate-size specimens assigned by Boucot (1959) to *Amphigenia elongata* to true, large Eifelian *Amphigenia elongata* (Boucot, 1959, 1997)
11. *Delthyris* to *Quadrifarius* lineage of the Wenlock through earlier Lochkovian of the Silurian European Province and successor Rhenish–Bohemian Region

Hurst (1975) described a lineage of *Resserella sabrinae* subspecies in the Welsh Wenlock. Sun and Boucot (1999) discussed Givetian *Stringocephalus* lineages in South China and western North America.

In the Uralian Region of the Early Devonian there is a good lineage for the species of the brachiopod *Karpinskia* (Zhivkovich and Chekhovich, 1985). Rozman (1999) discussed a good species lineage of the Silurian brachiopod *Tuvaella*, restricted to the Mongolo–Okhotsk Region of the Silurian in central Asia; there is a probable species lineage in the Late Jurassic–Early Cretaceous in Europe for the deep-water, "keyhole" brachiopod *Pygope* (Vogel, 1966, 1984). Smith (1984, pp. 115–118) discussed the *Infulaster–Hagenowia* Late Cretaceous urchin lineage from the northwestern European Chalk that is well documented by Gale and Smith (1982). Feist and Clarkson (1989) discussed an impressive sequence of tropidocoryphine trilobites that appear to belong in this category.

Truly rapid rates of evolution under locally very endemic, very small population, natural conditions include Lister's (1989) example of dwarfing in an offshore island population of red deer where the morphological changes took no more than a few thousand years. Additional to Lister's example are many additional examples known from the Quaternary of the Mediterranean islands and from additional offshore islands in other parts of the world, such as Agenbroad's (2003; see also Boucot, 1976) descriptions of *Mammuthus exilis* from the Channel Islands of Southern California and Vartanyan et al.'s (1993) Wrangel Island dwarf mammoths, not to mention the giant lizards and tortoises present on some islands, as well as very large flightless owls, or the recent excitement generated by the small-bodied hominin from the island of Flores, Indonesia (Brown et al., 2004).

1f. Host–Parasite Stability and Community Stability

Ebert (1994; for comment, see Gibbons, 1994) discussed evidence having to do with the marked changes that can occur in host–parasite systems that, on a geological time scale, can most easily be viewed as back and forth population changes within barriers erected by long-term stabilizing selection (i.e., small-scale noise). Frank (2002) emphasized the molecular basis of parasitism while also pointing out the basis for the many rapid changes possible in viruses, bacterial, and many parasitic protozoans.

1g. Community Ecologic, Behavioral, Coevolutionary, and Other Objections to the Concept of Punctuated Equilibrium

Maynard Smith (1988a,b) commented on the absence of any need for so-called species selection and suggested that Sheldon's work on gradualism in trilobites is very similar to the work done by Eldredge (1985) alleging punctuation. Lemen and Freeman (1989) noted that a cladistic analysis of data from small, monophyletic groups provided conclusions inconsistent with the concept of punctuated equilibrium. Goodfriend and Gould (1996) appear to have good evidence for hybridization in a post-Pleistocene, Bahamian land snail but lack any evidence for the changes they infer involve allopatry; that is, this is not an actual example of punctuated equilibrium. Nevo (1999) concluded that the evidence concerning the evolution of subterranean mammals during the Cenozoic is gradualistic. Samadi et al. (2000) provided a careful analysis of an African Rift Valley gastropod that allegedly displayed evidence favoring the concept of punctuated equilibrium but concluded that such "evidence" is ambiguous at best. Fortey (1985) provided evidence supporting the presence of gradualism in varied, pelagic earlier Ordovician trilobites, while Fortey (1974, p. 21) indicated that the morphological changes involving one species of a genus to another in some of his trilobites might involve relatively short intervals of environmental change followed by much longer intervals of stability featuring little or no morphological change.

2. BEHAVIORAL AND COEVOLUTIONARY CONCLUSIONS

2a. IS COEVOLUTION A NEVERENDING PROCESS?

Haldane (1949) provided a very thoughtful account of the role disease may play in regulating the evolutionary process. He pointed out that disease may even play an important role in major extinctions. Disease in his view could also play a very positive role in maintaining some groups as contrasted with other groups. He also considered the potential evolutionary importance of disease on abundant vs. uncommon species, with the former being more susceptible, other factors being equal. Herre (1993) provided some real data (see also Yoon, 1993) based on the nematodes parasitic on fig wasps indicating that for some cases disease *cum* parasitism becomes more intense and virulent if abundant hosts are provided, which also raises the prospect of switching, as well, in trying to explain widespread extinction events. Still, with the difficulty of finding enough paleopathology in the fossil record, as most disease and parasitism do not leave skeletal records, it will be very seldom that one can use this concept in a positive, well documented manner. The eminent British physician J.C.M. Given (1928) published a presidential address in *The Lancet* that encapsulates the problem rather well, ending with, "One cannot but feel certain that disease in its various manifestations has played a much greater part in the evolution of different forms of life in this world than the biologists have realized," particularly when musing about differing levels of immunity and resistance to bacterial, viral, and protozoan diseases.

2b. BEHAVIORAL CONSTANCY

Another aspect of behavioral constancy through evolutionary time might be labeled *use it or lose it*. By this we refer to the well-known fact that both structures and behaviors that fall into disuse are quickly lost. Darwin's emphasis on vestigial structures was the earliest solid discussion of the topic, but the well-known loss of both skin pigment and sight in obligate troglobites is still another, and there is a real wealth of examples in group after group. Paleontologically speaking, one can also discuss such things as the reduction and loss of the olfactory lobes in the wholly aquatic mammals, as well as such things as hind limb loss, as well as the reduction of the olfactory lobes with consequent enlargement of the optic lobes in such taxonomically unrelated, although functionally related, groups as birds, pterodactyls, and bats. Fortey and Owens (1990; see also Clarkson, 1997) carefully reviewed a number of examples demonstrating blindness (loss of eyes) in varied trilobite lineages; the actual causes of the eye loss can be attributed to such things as life in subphotic-zone, deeper water environments or activity within a soft substrate. Fortey and Owens (1990) also reviewed examples of the opposite trend—hypertrophied eyes in some forms concluded to have been pelagic in somewhat deeper waters with low light intensities.

It needs to be emphasized that behavioral characteristics should be routinely included with taxonomic descriptions, because behavior is an integral part of a taxon. It is also obvious that behaviors may be ranked taxonomically, from the phylum to the species level. Most of the behavioral evidence available in the fossil record is at the familial to superfamilial levels. Be that as it may, it is exciting to note that appropriate taxonomic level behaviors correlate very well with morphological characteristics, taxon by taxon. The basic conservatism of behavioral characteristics parallels that of morphologies at the appropriate taxonomic levels.

2c. NON-COENOLOGIC BEHAVIORAL AND COEVOLUTIONARY EVIDENCE

(See Table 39.) Schmidt-Kittler (1987) provided the following telling quote: "As a matter of fact, in Tertiary mammals gradual change of characters within lineages is frequent and in many cases well documented. In order to extract biochronological criteria from such morphoclines, particular evolutionary stages have to be defined as chronospecies (or subspecies) by designating type specimens. These stages can directly be used as points of reference within a biochronological concept based on reference levels (niveaux repères)." The volume edited by Martin and Barnosky (1993a) includes a number of contributions (especially Barnosky, 1993; Goodwin, 1993; Hulbert and Morgan, 1993; Martin, 1993; Martin and Barnosky, 1993b), making the point that species-level evolution may commonly include a mosaic of rate changes, affecting different characters, including such things as no change, gradual changes, abrupt changes, rapid changes ... all of which add up to a mosaic evolution rate picture affecting different parts of a structure or structures. Spaan et al. (1994) provided an excellent example from among the horses that appears to be a far more realistic view of species-level evolution than ones based merely on a single character through time. Most of the examples of evolutionary

TABLE 39
Examples of Phyletic Gradualism in the Fossil Record (Additions to Table 45 of Boucot, 1990a)

Refs.	Time	Example
Foraminifera		
Lazarus (2001)	Neogene	*Globorotalia*
Kelly et al. (1996)	Neogene	
Norris et al. (1996)[a]	Neogene	
Gourinard (1983)	Miocene	
Schaub (1963)	Eocene	
Lucas et al. (2001)	Paleocene	
Pearson (1993)	Paleogene	
Fondecave-Wallez et al. (1990)	Senonian	
Kucera and Malmgren (1996, 1998)	Late Cretaceous	
Bettenstaedt (1968)	Jurassic, Cretaceous, Tertiary	
Radiolaria		
Dumitrica and Dumitrica Jud (1995)	Early Cretaceous	*Aurisaturnalis*
Lazarus (2001)	Pliocene	*Pterocanium*
Bivalves		
Ward and Blackwelder (1975)	Miocene–Pliocene	*Chesapecten*
Walaszczyk and Cobban (2000)	Turonian–Lower Coniacian	Inoceramids
Kauffman and Harries (1996)	Late Cretaceous	Some *Mytiloides* lineages
Crampton and Gale (2005)	Late Cretaceous	*Actinoceramus*
Crampton and Gale (2009)	Early Cretaceous	Inoceramids
Johnson (1993)	Jurassic	*Gryphaea*
Johnson (1994)	Early Jurassic	*Gryphaea*
Fürsich and Werner (1986)	Jurassic	Several taxa
Brachiopods		
Johnson and Talent (1967)	Early Devonian	*Protocortezorthis, Cortezorthis*
Gastropods		
Nehm and Geary (1994)	Neogene	*Prunum* species
Geary (1990, 1992)	Late Miocene	*Melanopsis* species
Bouchet (1981)	Miocene–Holocene	Terebridae
Graptolites		
Springer and Murphy (1994)	Early Devonian	*Monograptus hercynicus*
Urbanek (1997)	Late Silurian	Linograptids
Ni (1997)	Late Silurian	*Colonograptus* species
Sudbury (1958)	Early Silurian	Triangulate monograptids
Conodonts		
Kozur (1992)	Late Permian	
Echinoderms		
Smith (1994); Smith and Bengtson (1991)	Late Cretaceous	*Mecaster* species
Crustaceans		
Feldmann et al. (1993)	Late Cretaceous	*Hoploparia stokesi*
Tschudy et al. (1998)	Late Cretaceous	*Hoploparia*
Olempska (1989)	Ordovician	*Mojczella* species (ostracodes)

rate changes have focused on change (or no change) within a single character, which eliminated any possibility for considering mosaic rate changes within the same lineage. With this concept in mind, it is now clear that many of the arguments over rate changes, or not, within the fossil record must be carefully reconsidered. The Norris et al. (1996) evidence discussed above fits into this picture very comfortably. Sondaar (1994) provided several excellent examples of these mosaic evolution changes from among Neogene offshore island mammals.

TABLE 39 (cont.)

Examples of Phyletic Gradualism in the Fossil Record (Additions to Table 45 of Boucot, 1990a)

Refs.	Time	Example
Insects		
Elias and Matthews (2002)	Pleistocene–Late Miocene	Various insect taxa
Trilobites		
Fortey and Owens (1990)	Paleozoic	Many taxa
Fortey (1985)	Early Ordovician	*Carolinites, Pricyclopyge*
Fortey (1975)	Earlier Ordovician	Nileidae
Sheldon (1987)	Ordovician	Various genera
Feist and Clarkson (1989)	Upper Devonian	Tropidocoryphines
Amphibians		
Werneburg et al. (2007)	Lower Permian	Three branchiosaurs
Mammals		
Hoffman and Fowler (1995)	Pleistocene	
Semken (1966)[b]	Pleistocene	
Lister (1993a)	Pleistocene	
Lister (1993b)	Pleistocene	
Neraudeau et al. (1995)	Plio–Pleistocene	
Chaline and Sevilla (1990)	Plio–Pleistocene	
Comte and Vianey-Liaud (1989)	Oligocene	
Schmidt-Kittler and Vianey-Liaud (1979)	Oligocene	
Schmidt-Kittler and Vianey-Liaud (1987)	Oligocene	
Vianey-Liaud (1989)	Oligocene	
Vianey-Liaud and Schmidt-Kittler (1987)	Oligocene	
Vianey-Liaud (1976)	Late Oligocene to Late Eocene	
Vianey-Liaud (1979)	Late Eocene	
Vianey-Liaud and Ringeade (1992)	Late Eocene	
Godinot (1985)	Late Eocene	
Gingerich (1989)	Early Eocene	
Simpson (1943)	Early Tertiary	
MacFadden (1992)	Tertiary	
Moissette et al. (2007)	Pliocene–Holocene	Seagrass community

[a] Norris et al. (1996) made the important point that within a morphologically gradualistic foraminiferal lineage there is good evidence for a relatively abrupt ecological change from shallow to deep water; that is, behavioral changes and what went with them evolved at very different rates from the gradualistic morphological changes.

[b] The late Bjorn Kurten (written comm., 1984) pointed out: "I enclose a Xerox of Semken's 1966 paper in which the muskrats are treated. The important point here is that, at that time, the stratigraphy was mixed up (because it was not realized that there were in fact three different 'Pearlette Ash' deposits). It is now known that the Borchers and Grandview material which was then placed in the 'Yarmouth' are really much older (late Blancan to early Irvingtonian) and so it predates the Kansan. That straightens out the apparent kink in the sequence. I think it's beautiful. You can get additional references from *Pleistocene Mammals of North America*. Spatial differentiation in latest Pleistocene muskrats may be observed: the Florida (Itchtucknee) molars are narrower than others. That was presumably an isolate, and the species is now extinct here. Otherwise, they may parsimoniously be referred to as a single lineage with the possible exception of the Port Kennedy specimen (early Irvingtonian) which might (or might not) represent a side-branch."

3. SPECIES EVOLUTION BETWEEN FAMILIES AND WITHIN FAMILIES

One of the closer approximations of paleontological evidence of quantum evolution is provided by Sondaar's (1994) analysis of Pleistocene morphological changes in Mediterranean island large mammals. Sondaar pointed out that the initial morphological changes in distal limb structure for such taxa as horses and hippos are geologically very rapid, approaching the quantum evolution level and involving very small, highly endemic populations, followed by much slower, phyletic-type dental changes.

4. CORRELATES OF RATES OF EVOLUTION

Martin et al. (1992) made it clear that rates of evolution of mitochondrial DNA in a subset of squalomorph sharks as contrasted with some primates are very different (much slower in the sharks); thus, one cannot assume similar rates of DNA evolution across major taxonomic boundaries even within a single class.

4a. Biogeographic–Distributional Variables

Sanfilippo and Riedel (1992) see a strongly positive correlation between stratigraphic range and geographic range of a sequence of oceanic radiolaria; that is, greater stratigraphic ranges correlate with greater geographic ranges and *vice versa*. Buzas and Culver (1989) pointed out the correlation between geographic range and stratigraphic range for continental margin benthic Foraminifera, as well as with their greater abundance.

4b. Abundance of Individuals as a Control

Boltovskoy and Boltovskoy (1988, p. 80; pers. comm., 1990) commented that "benthic Foraminifera is poor for stratigraphic purposes because of the high proportion of rare taxa which do not convey reliable information. The few abundant organisms, in turn, are usually too long ranging to allow chronological resolution of a medium to fine scale." Buzas and Culver (1989) noted the correlation between geographic range and stratigraphic range for continental margin benthic Foraminifera, as well as with their greater abundance. Lawton (1994) commented on the relationship among modern populations between greater geographic area occupied and greater numerical abundance locally. Kennett (pers. comm., 1985) also found that more geographically widespread species tend to be more abundant as individuals.

5. CLASSES OF PALEONTOLOGIC DATA AND RATES OF CHANGE

Although it is very unlikely that a bone protein such as osteocalcin, which is now recognized to occur in a Jurassic dinosaur bone (Moyzer et al., 1992), will add much to our knowledge of rates of evolution at the biochemical level, it is still comforting to learn that such basic building blocks can be shown to have been present from an early time. Nielsen-Marsh et al. (2002) found osteocalcin in sub-Recent bison bone dating back 55,000 years. Cano et al. (1992a) described a DNA fragment from a Dominican amber Miocene–Eocene stingless bee (*Proplebeia dominicana*), and DeSalle et al. (1992) did the same for an Oligocene–Miocene termite from Dominican amber, which raises the possibility of truly characterizing a few fossilized organisms under unusual preservation circumstances. Cano et al. (1993) extracted DNA fragments belonging to a weevil from Neocomian, Early Cretaceous, Lebanese amber, and Poinar et al. (1993b) reported chloroplast DNA from an Oligocene Dominican Republic leguminosean

preserved in amber. Poinar et al. (1994a) summarized the currently available evidence regarding DNA preserved in amber, and Cano et al. (1994) described a symbiotic bacillus DNA occurrence from a Dominican amber stingless bee. Stankiewicz et al. (1998a), however, commented about the absence of biochemical evidence for chitin in amber specimens, which makes the preservation of even more "delicate" DNA very unlikely. Noonan et al. (2005) used cave bear DNA data to place that taxon within a reasonable phylogeny of living bears.

Voss-Foucart's (1968) find of protein remnants, presumably derived from a protein or proteins based on the presence of varied amino acids, in Late Cretaceous dinosaurian eggshell is worthy of mention. Smith and Littlewood (1994) compiled many examples, including some not cited above (Cano et al., 1992b; Golenberg, 1991; Golenberg et al., 1990; Soltis et al., 1992), as well as a number of subrecent examples chiefly involving the relations of recently extinct mammals.

Poinar and Pääbo (2001) summarized the evidence of recovered DNA from the fossil record and concluded that the only reliable evidence is from the late Pleistocene to Recent, and chiefly from cool-temperature environments such as caves. Greenwood (2001) observed that even nuclear DNA from frozen Siberian mammoths has undergone a certain level of chemical change; if this is the case with our "best" preserved organic material, then it is clear that the very complex organic molecules will not remain unaltered. Krause et al. (2006) and Poinar et al. (2006) described mammoth DNA of later Pleistocene age. Fish et al. (2002) noted that RNA fragments from Late Silurian halite were the oldest they had recovered, and they suggested bacterial and haloarchaeal sources.

Willerslev et al. (2003) provided excellent data indicating that DNA from Siberian permafrost persisted for at least 400,000 years in taxonomically useful form and that several million years are feasible with samples preserved under the right conditions. Woodward et al. (1994) described DNA fragments from Late Cretaceous bone of unspecified taxonomic affinities. Herrmann and Hummel (1994) summarized much of what has been recovered in the way of "ancient DNA" from varied materials, mostly subrecent materials; it is clear that there are various possibilities for future investigation. Yang et al. (1996; for discussion of fossil DNA possibilities, see also Yang, 1997) reported on fossil DNA from the American mastodon.

Van Bergen et al. (1995) reviewed the record of resistant biomacromolecules in the fossil record, with an emphasis on fossil plant materials. Poinar (2000d) discussed preservation of fossil DNA, the type of damage that can destroy DNA and materials, and methods used to obtain fossil DNA. Nguyen Tu et al. (2003) discussed some higher land-plant chemotaxonomy, using fossil and extant data, and provided a useful review. Mitchell and Curry (1997) analyzed the amino acid content in a series of Pleistocene to recent shells from New Zealand and found that peptide bonds are progressively broken the further back one goes in time, thus emphasizing that different groups have somewhat different amino acid contents preserved.

Johns (1986) summarized a wealth of information about organic chemical "fragments" preserved in the strata and discussed the possible organic sources involved; this is a basic reference for anyone concerned with what evidence can be derived from organic chemicals about life of the past. Rowley and Rich (1986) discussed the identification of bone collagen in Australian Pleistocene birds and marsupials, as well as in a mid-Miocene mammals. Collins and Child-Gernaey (2001) reviewed the preservation of proteins from the fossil record. Evershed and Lockheart (2001) reviewed the preservation of lipids from the fossil record. Otto et al. (2002) described the recovery of terpenoids from Eocene and Miocene conifers that can be assigned to modern taxa. De Leeuw et al. (1995) discussed and documented the nature and significance of various organic molecules extracted from sedimentary rocks. Briggs and Eglinton (1994) provided a good review of some of the many biochemical items preserved in the fossil record as traces of once living organisms; they are dubious about the claims of DNA from the Clarkia Miocene plants. Stankiewicz et al. (1997) found good evidence for chitin in German Oligocene insect cuticle; this is currently the oldest known positive chemical evidence of chitin. Stankiewicz et al. (1998b; see also Jacob et al., 2007) pointed out that marine-sourced chitin is not preserved as contrasted with that in nonmarine rocks. Matheson and Brian (2003) pointed out the possibilities of molecular paleopathology for identifying disease in ancient materials.

*6. BIOLOGICAL EXTINCTION CAUSE POSSIBILITIES

Little has been written lately about the possibilities for biological causation being massively involved in any major extinction. Could this reflect the fact that most of the thinking about extinction events has been done by physically, rather than biologically, trained geologists? Brasier (1992), while discussing serious tree mortality ascribed to an exotic fungus, documented the possibility that the rapid spread of an exotic parasite by natural agencies could result in massive disruption of the ecosystem and extermination of a significant number of taxa. This example is thought provoking. Haldane (1949) commented previously on the possibility of disease playing an important role in the extinction process.

*7. ADAPTIVE RADIATION

Schluter (2000) extensively reviewed the ecological explanation of adaptive radiation, which does not really explain why some taxa adaptively radiate while others do not in varied adaptive radiation events, nor the fact that adaptive radiations mostly occur in varied unrelated taxa at about the same times; the problems raised by the fossil record were not considered. It is of concern that Simpson (1953), on whom Schluter (2000) evidently relied for the fossil record, did not try to employ the extensive marine invertebrate information because the terrestrial tetrapods do not provide adequate information

for considering this problem. As an example, the adaptive radiations involving such things as Darwin's finches and the Hawaiian avifauna are low-level items where allopatry can be called on; these are not major adaptive radiations. Brooks and McLennan (1993) discussed a number of aspects of the adaptive radiation problem, while pointing out our limited understanding in the area. As they noted, cladogenesis resulting from the simple imposition of barriers to reproductive communication, allopatric speciation, cannot be considered to represent adaptive radiation, as it is simply diacladogenesis (Boucot, 1978). The only exception that one can readily think of is the introduction during the Phanerozoic from time to time of sea-level continental glaciation, with the resulting generation of high-latitude, very cold marine environments, plus the occasional presence of high-latitude, cool-water, nonglacial environments that similarly permit adaptive radiations, such as that causing the Malvinokaffric Realm and Atlantic Realm faunas of higher southern latitudes during the Cambrian through mid-Middle Devonian. The cold marine environments—Hirnantian at the end of the Ordovician, Gondwanan near the end of the Pennsylvanian and in the Early Permian, and finally Oligocene to present in southern high latitudes—have generated the adaptive radiation of presumably cold-adapted marine invertebrates and vertebrates. The Pleistocene to present northern high-latitude, cold marine environment has not been present for a sufficient interval to permit much adaptive radiation to take place, although in the terrestrial environment adaptive radiations of relatively small-population plants and animals, such as *Betula nana*, polar bears, and snowshoe rabbits, did occur. Such things as the generation of various seed-eating, Amazon-region freshwater fish and of hymenopterans becoming specialized for feeding on pollen and nectar can be viewed as related to the availability of new, abundant resources.

So much for the little we can reasonably conclude about causation. The real mystery, as pointed out earlier (Boucot, 1990a) is what might be the causal factors involved in the restriction of marine, level-bottom adaptive radiations to a brief interval near the beginning of many of the ecological–evolutionary units. Why, during these brief intervals, should species belonging to varied higher taxa, within varied community groups, either adaptively radiate or not? Following this geologically relatively brief interval, adaptive radiation ceases within the level-bottom environment until another ecological–evolutionary unit or subunit occurs. Non-level-bottom adaptive radiations, such as those characterizing the reef community complex, commonly occur at times other than those affecting the level-bottom environment, but here, again, we have little understanding about causation. The reef community complex is present during some ecological–evolutionary intervals but missing from others, and, again, we have no real understanding of causation. It is clear, too, that adaptive radiations are random rather than regular in time, just as are extinctions (Boucot, 1994). Boucot earlier (1990b, Figure 1) itemized and ranked the relative intensity (numbers in brackets) of the major adaptive radiations—beginnings of the Ediacaran [1], Early Cambrian [1], Middle Cambrian [3],

Early Ordovician [3], Middle Ordovician [2], Mississippian [2], Middle Triassic [2], Jurassic [2]—and also briefly cited some of the mid-level and minor adaptive radiations.

*8. EVOLUTIONARY CONSERVATISM VS. EVOLUTIONARY RADICALISM

When one considers the fossil record overall from an evolutionary perspective it is clear that certain aspects are conservative whereas others are radical. *Conservative* is used here for the relative absence of radical changes in community ecology or adaptive radiations, whereas *radical* refers to the reverse. Extinctions in this sense are radical because they involve major changes in communities present following the event, including the restructuring of surviving communities and the presence of new community groups (community types; see Boucot, 1975, for definitions), and are commonly followed by adaptive radiations. Behavior and morphology are relatively conservative items, as they overall tend to remain very stable during both anagenesis and cladogenesis, unless affected by adaptive radiation. Adaptive radiations may be regarded as radical in this sense, although why some taxa adaptively radiate during a specific time interval whereas others do not is not understood. Biogeography overall may be largely regarded as conservative as it does not involve major changes in community groups present or in the presence of adaptive radiations, although exceptions to this statement are the adaptive radiations and community group changes generated when high-latitude freezing winter or continental glaciation first appear. Extinctions, of course, may be present during intervals of biogeographic constancy, such as the cosmopolitan mid-Late Devonian extinction, which does involve a radical set of changes.

References

Abdelhamid, A.M. (1999). Parasitism, abnormal growth and predation on Cretaceous echinoids from Egypt. *Revue de Paléobiologie*, 18, 69–83.

Abe, T., Bignell, D.E., and Higoshi, M., Eds. (2000). *Termites: Evolution, Sociality, Symbiosis, Ecology*. Kluwer Academic, Dordrecht.

Abel, O. (1935). *Vorzeitliche Lebenspuren*. Gustav Fischer, Jena, 644 pp.

Abletz, V.V. (1993). Sledi zhiznedeyatelnosti rachkov Acrothoracica (Cirripedia) iz eotsena Krivbassa. *Paleontologicheskii Zhurnal*, 2, 56–61.

Adamczak, F. (1968). Palaeocopa and Platycopa (Ostracoda) from Middle Devonian rocks in the Holy Cross Mountains, Poland. *Stockholm Contributions in Geology*, 17, 1–109.

Adamczak, F. (1991). Kloedenellids: morphology and relations to non-myodocopide ostracodes. *Journal of Paleontology*, 65, 255–267.

Adami-Rodrigues, K., Iannuzzi, R., and Pinto, I.D. (2004). Permian plant–insect interactions from a Gondwana flora of southern Brazil. *Fossils and Strata*, 51, 106–125.

Adang, M.J. (1991). *Bacillus thuringiensis* insecticidal crystal proteins: gene structure, action and utilization. In *Biotechnology for Biological Control of Pests and Vectors* (pp. 3–24), edited by K. Maramorosch. CRC Press, Boca Raton, FL.

Adkins, W.S. and Winton, W.M. (1919). Paleontological correlation of the Fredericksburg and Washita formations in North Texas. *University of Texas Bulletin*, No. 1945, 128 pp.

Afonin, S.A. (2000). Pollen grains of the genus *Cladaitina* extracted from the gut of the Early Permian insect *Tillyardembia* (Grylloblattida). *Paleontologicheskii Zhurnal*, 5, 105–109.

Agenbroad, L.D. (2003). New absolute dates and comparisons for California's *Mammuthus exilis*. *Deinsea*, 9, 1–16.

Agosti, D., Grimaldi, D., and Carpenter, J.M. (1998). Oldest known ant fossils discovered. *Nature*, 391, 447.

Aguilera, O.A., García, L., and Cozzuol, M.A. (2008). Giant-toothed white sharks and cetacean trophic interaction from the Pliocene Caribbean Paraguaná Formation. *Paläontologisches Zeitschrift*, 82, 204–208.

Aguirre, J. (1998). Bioconstrucciones de *Saccostrea cuccullata* Born, 1778 en el Plioceno superior de Cadiz (SO de España): implicaciones paleoambientales y paleoclimaticas. *Revista Española de Paleontologia*, 13, 27–36.

Aguirre-Urreta, M.B. and Olivero, E.B. (1992). A Cretaceous hermit crab from Antarctica: predatory activities and bryozoan symbiosis. *Antarctic Science*, 4, 207–214.

Aharon, P. (1994). Geology and biology of modern and ancient submarine hydrocarbon seeps and vents: an introduction. *Geo-Marine Letters*, 14, 69–73.

Ahlberg, P.E. (1992). The Palaeoecology and evolutionary history of the porolepiform sarcopterygians. In *Fossil Fishes as Living Animals* (pp. 71–90), edited by E. Mark-Kurik. Institute of Geology, Tallinn, Estonia.

Ahmadjian, V. (1995). Lichens—specialized groups of parasitic fungi. In *Pathogenesis and Host Specificity in Plant Diseases*. Vol. II. *Eucaryotes, Pathogenesis and Host Specificity in Plant Diseases* (pp. 277–286), edited by U. Singh. Pergamon Press, New York.

Akersten, W.A. (1985). Canine function in *Smilodon* (Mammalia; Felidae; Machairodontinae). *Contributions in Science, Natural History Museum of Los Angeles County*, No. 356, 22 pp.

Akpan, E.B. (1991). Palaeoecological significance of *Lithophaga* borings in Albian stromatolites, SE Nigeria. *Palaeogeography, Palaeoclimatology, Palaeoecology*, 88, 185–192.

Alcover, J.A., Perez-Obiol, R., Errikarta-Imanol, Y., and Bover, P. (1999). The diet of *Myotragus balearicus* Bate 1909 (Artiodactyla: Caprinae), an extinct bovid from the Balearic Islands: evidence from coprolites. *Biological Journal of the Linnean Society*, 66, 57–74.

Alekseev, A.S. and Endelman, L.G. (1989). Association of ectoparasitic prosobranch gastropods with Upper Cretaceous echinoid Galerites. In *Fossil and Recent Echinoderm Researches* (pp. 165–174). Academy of Sciences of the Estonian SSR, Tallinn.

Alessandrello, A., Pinna, G., and Teruzzi, G. (1988). Land planarian locomotion trail from the lower Permian of Lombardian pre-alps (Tricladida Terricola). *Atti della Società Italiana di Scienze Naturali e del Museo Civico di Storia Naturale di Milano*, 129, 139–145.

Alexander, J.P. and Burger, B.J. (2001). Stratigraphy and taphonomy of Grizzly Buttes, Bridger formation, and the Middle Eocene of Wyoming. In *Eocene Biodiversity: Unusual Occurrences and Rarely Sampled Habitats* (pp. 163–196), edited by G.F. Gunnell. Kluwer Academic, New York.

Alexander, R.E. (1994). Distribution of pedicle boring traces and the life habit of Late Paleozoic leiorhynchid brachiopods from dysoxic habitats. *Lethaia*, 27, 227–234.

Alexander, R.M. (1976). Estimates of speeds of dinosaurs. *Nature*, 261, 129–130.

Alexander, R.R. (1977). Growth, morphology and ecology of Paleozoic and Mesozoic opportunistic species of brachiopods from Idaho–Utah. *Journal of Paleontology*, 51, 1133–1149.

Alexander, R.R. (1981). Predation scars preserved in Chesterian brachiopods: probable culprits and evolutionary consequences for the articulates. *Journal of Paleontology*, 55, 192–203.

Alexander, R.R. (1986). Resistance to and repair of shell breakage induced by durophages in Late Ordovician brachiopods. *Journal of Paleontology*, 60, 273–285.

Alexander, R.R. and Dietl, G.P. (2003). The fossil record of shell-breaking predation on marine bivalves and gastropods. In *Predator–Prey Interactions in the Fossil Record* (pp. 141–176), edited by P.H. Kelley, M. Kowalewski, and T.A. Hansen. Kluwer Academic, New York.

Alexander, R.R. and Scharpf, C.D. (1990). Epizoans on Late Ordovician brachiopods from southeastern Indiana. *Historical Biology*, 4, 179–202.

Ali, M.S.M. (1982). Predation and repairing phenomena in certain clypeasteroid, echinoid, from the Miocene and Pliocene epochs of Egypt. *Journal of the Palaeontological Society of India*, 27, 7–8.

Allman. (1863). On a new fossil ophiuridian, from post-Pliocene strata of the Forth. *Proceedings of the Royal Society of Edinburgh*, 5, 101–104.

Allmon, W.D. (1993). Age, environment and mode of deposition of the densely fossiliferous Pinecrest Sand (Pliocene of Florida): implications for the rate of biological productivity in shell bed formation. *Palaios*, 8, 183–201.

Allmon, W.D., Nieh, J.C., and Norris, R.D. (1990). Drilling and peeling of turritelline gastropods since the Late Cretaceous. *Palaeontology*, 33, 595–611.

Alperson-Afil, N. (2008). Continual fire-making by hominins at Gesher Benot Ya'aqov, Israel. *Quaternary Science Reviews*, 27, 1733–1739.

Alt, K.W., Rosing, F.W., and Teschler-Nicola, M., Eds. (1998). *Dental Anthropology: Fundamentals, Limits, and Prospects*. Springer, New York, 564 pp.

Alvarez, F. and Taylor, P.D. (1987). Epizoan ecology and interactions in the Devonian of Spain. *Palaeogeography, Palaeoclimatology, Palaeoecology*, 61, 17–31.

Amano, K. (2006). Temporal pattern of naticid predation on *Glycymeris yessoensis* (Sowerby) during the Late Cenozoic of Japan. *Palaios*, 21, 369–375.

Amano, K. and Kiel, S. (2006). Fossil vesicomyid bivalves from the North Pacific region. *The Veliger*, 49, 270–293.

Amano, K. and Little, C.T.S. (2005). Miocene whale-fall community from Hokkaido, northern Japan. *Palaeogeography, Palaeoclimatology, Palaeoecology*, 215, 340–356.

Amano, K., Little, C.T.S., and Inoue, K. (2007). A new Miocene whale-fall community from Japan. *Palaeogeography, Palaeoclimatology, Palaeoecology*, 247, 236–242.

Ambrose, D.P. (1999). *Assassin Bugs*. Science Publishers, Enfield, NH, 337 pp.

Amemiya, S. and Oji, T. (1992). Regeneration in sea lilies. *Nature*, 357, 546–547.

Andersen, N.M. and Poinar, Jr., G.O. (1992). Phylogeny and classification of an extinct water strider genus (Hemiptera, Gerridae) from Dominican amber, with evidence of mate guarding in a fossil insect. *Zeitschrift für Systematik und Evolutionsforschung*, 30, 256–267.

Andersen, N.M. and Poinar, Jr., G.O. (1998). A marine water strider (Hemiptera: Veliidae) from Dominican amber. *Entomologica Scandinavica*, 29, 1–9.

Anderson, J.A.R. and Muller, J. (1975). Palynological study of a Holocene peat and a Miocene coal deposit form NW Borneo. *Review of Palaeobotany and Palynology*, 19, 291–351.

Anderson, J.L. and Feldmann, R.M. (1995). *Lobocarcinus lumacopius* (Decapoda: Cancridae), a new species of cancrid crab from the Eocene of Fayum, Egypt. *Journal of Paleontology*, 69, 922–932.

Anderson, O.R. (1996). The physiological ecology of planktonic sarcodines with applications to paleoecology: patterns in space and time. *Journal of Eukaryotic Microbiology*, 43(4), 261–274.

Anderson, W.F. (1946). Een Fossiele Parel in Nederland. *Publications of the Dutch Geological Society*, 1, 1–8.

Andrée, K. (1937). *Der Bernstein*. Grafe und Unzer, Königsberg, 219 pp.

Andrews, P. (1991). *Owls, Caves and Fossils*. University of Chicago Press, Chicago, IL, 231 pp.

Andrews, R.C. (1921). A remarkable case of external hind limbs in a humpback whale. *American Museum Novitates*, 9, 1–6.

Angel, J.L. (1966). Porotic hyperostosis, anemias, malaria and marshes in the prehistoric Eastern Mediterranean. *Science*, 153, 760–763.

Anisyutkin, L.N., Grachev, V.G., Ponomarenko, A.G., Rasnitsyn, A.P., and Vrsansky, P. (2008). Fossil insects in the Cretaceous Mangrove Facies of Southern Negev, Israel. In *Plant–Arthropod Interactions in the Early Angiosperm History: Evidence from the Cretaceous of Israel*, Part II (pp. 189–223), edited by V. Krassilov and A. Rasnitsyn. Pensoft/Brill, Boston, MA.

Anon. (1993). Webs woven indeed. *Geotimes*, January, p. 10.

Apesteguía, S. and Zahner, H. (2006). A Cretaceous terrestrial snake with robust hind limbs and a sacrum. *Nature*, 440, 1037–1040.

Arambourg, C. (1921). Sur un Scopélidé fossile à organes lumineux. *Bulletin de la Société géologique de France*, 4, XX.

Arambourg, C. (1927). Les poissons fossiles d'Oran. *Matériaux pour la Carte géologique de l'Algérie, 1 Série—Paleontologie*, 6, 1–218.

Arambourg, C. (1952). Les vertébrés fossiles des gisements de phosphates (Maroc-Algérie-Tunisie). *Notes et Mémoires du Service Géologique du Maroc (Rabat)*, 92, 1–372.

Arambourg, C. (1967). Résultats scientifiques de la mission C. Arambourg en Syrie et en Iran (1938–1939). II. Les poissons oligocènes de l'Iran. *Notes et Mémoires sur le Moyen-Orient*, 8, 1–210.

Araujo, A.J.G., Confalonieri, U.E.C., and Ferreira, L.F. (1982). Oxyurid (Nematoda) egg from Brazil. *Journal of Parasitology*, 68, 511–512.

Araujo, R. and Ramos, M.A. (2001). Life-history data on the virtually unknown *Margaritifera auricularia*. In *Ecological Studies*. Vol. 145. *Ecology and Evolution of the Freshwater Mussels Unionoida* (pp. 143–152), edited by G. Bauer and K. Wächtler. Springer-Verlag, Heidelberg.

Archibald, S.B. and Mathewes, R.W. (2000). Early Eocene insects from Quilchena, British Columbia, and their paleoclimatic implications. *Canadian Journal of Zoology*, 78, 1441–1462.

Arditti, J., Ed. (1982). *Orchid Biology*, Vol. 2. Cornell University Press, Ithaca, NY, 380 pp.

Arditti, J. (1992). *Fundamentals of Orchid Biology*. John Wiley & Sons, New York, 691 pp.

Arillo, A. (2007). Paleoethology: fossilized behaviours in amber. *Geologica Acta*, 5, 159–166.

Arillo, A. and Ortuno, V.M. (1997). The fossil Acrididae from the Oligocene of Izarra (Alava, Spain): the antiquity of gregarious behavior (Orthoptera, Caelifera). *Paleobios*, 30, 231–234.

Arkell, W.J. (1956). *Jurassic Geology of the World*. Oliver & Boyd, Edinburgh, 806 pp.

Arnold, D.C. (1956). A systematic revision of the fishes of the teleost family Carapidae (Percomorpha, Blennioidea), with descriptions of two new species. *Bulletin of the British Museum (Natural History)*, 4(6), 17–307.

Arnold, E.N. (1988). Caudal autotomy as a defense. In *Biology of the Reptila*. Vol. 16. *Ecology* (pp. 235–273), edited by C. Gans and R.B. Huey. Alan R. Liss, New York.

Arnold, E.N. and Poinar, Jr., G.O. (2008). A 100 million year old gecko with sophisticated toe pads, preserved in amber from Myanmar. *Zootaxa*, 1847, 62–68.

Aronson, R.B. and Blake, D.B. (1997). Evolutionary paleoecology of dense ophiuroid populations. *Paleontological Society Papers*, 3, 107–119.

Aronson, R.B. and Sues, H.-D. (1987). The paleoecological significance of an anachronistic community. In *Predation* (pp. 355–366), edited by W.C. Kerfoot and A. Sih. University Press of New England, Lebanon, NH.

Aronson, R.B., Blake, D.B., and Oji, T. (1997). Retrograde community structure in the late Eocene of Antarctica. *Geology*, 25, 903–906.

Arsenault, M. (1982). *Eusthenopteron foordi*, a predator on *Homalacanthus concinnus* from the Escuminac Formation, Miguasha, Quebec. *Canadian Journal of Earth Sciences*, 19, 2214–2217.

Arua, I. (1989). Gastropod predators and their dietary preferences in an Eocene molluscan fauna from Nigeria. *Palaeogeography, Palaeoclimatology, Palaeoecology*, 72, 283–290.

Arua, I. and Hoque, M. (1989). Predatory gastropod boreholes in an Eocene molluscan assemblage from Nigeria. *Lethaia*, 22, 49–59.

Asgaard, U. (1968). Brachiopod palaeoecology in Middle Danian limestones at Fakse, Denmark. *Lethaia*, 1, 103–121.

Asgaard, U. (2008). *Crania* at your service: the Antikythira shipwreck. *Fossils and Strata*, 54, 277–282.

Asgaard, U., Jensen, M., and Bromley, R.G. (1996). Two congeneric heart urchins having different architecture and trophic behavior: paleontologic and ichnologic implications. In *Sixth North American Paleontological Convention Abstracts of Papers* (p. 14), edited by J.E. Repetski, Special Publ. No. 8. Paleontological Society, Washington, D.C.

Ash, S. (1997). Evidence of arthropod–plant interactions in the Upper Triassic of the southwestern United States. *Lethaia*, 29, 237–248.

Ash, S. (1999). An upper Triassic *Sphenopteris* showing evidence of insect predation from Petrified Forest National Park, Arizona. *International Journal of Plant Science*, 160, 208–215.

Ash, S. (2000). Evidence of oribatid mite herbivory in the stem of a Late Triassic tree fern from Arizona. *Journal of Paleontology*, 74, 1065–1071.

Aufderheide, A.C. (2003). *The Scientific Study of Mummies*. Cambridge University Press, Cambridge, U.K., 608 pp.

Augusta, J. and Remes, M. (1947). *Úvod do Všeobecné Paleontologie*. Vydal Ferdinand Horký v Praze.

Ausich, W.I. and Baumiller, T.K. (1993). Column regeneration in an Ordovician crinoid (Echinodermata): paleobiologic implications. *Journal of Paleontology*, 67, 1068–1070.

Ausich, W.I. and Gurrola, R.A. (1979). Two boring organisms in a Lower Mississippian community of southern Indiana. *Journal of Paleontology*, 53, 335–344.

Autumn, K., Liang, Y.A., Hsieh, S.T., Zesch, W., Chan, W.P., Kenny, T.W., Fearing, R., and Full, R.J. (2000). Adhesive force of a single gecko foot-hair. *Nature*, 405, 681–685.

Avanzini, M., Mietto, P., Panarello, A., de Angelis, M., and Rolandi, G. (2008). The Devil's Trails: Middle Pleistocene human footprints preserved in volcanoclastic deposit of southern Italy. *Ichnos*, 15, 179–189.

Awramik, S.M., Schopf, J.W., and Walter, M.R. (1983). Filamentous fossil bacteria from the Archean of Western Australia. *Precambrian Research*, 20, 357–374.

Axelrod, D.L. (1958). Evolution of the Madro-Tertiary Geoflora. *Botanical Review*, 24, 433–509.

Azzaroli, A. (1982). Insularity and its effects on the terrestrial vertebrates: evolutionary and biogeographic aspects. In *Palaeontology, Essential of Historical Geology* (pp. 193–213), edited by E. Montanaro Gallitelli. STEM-Mucchi, Modena, Italy.

Babcock, L.E. (1993). Trilobite malformations and the fossil record of behavioral asymmetry. *Journal of Paleontology*, 67, 217–229.

Babcock, L.E. (2003). Trilobites in Paleozoic predator–prey systems, and their role in reorganization of Early Paleozoic ecosystems. In *Predator–Prey Interactions in the Fossil Record* (pp. 55–92), edited by P.H. Kelley, M. Kowalewski, and T.A. Hansen. Kluwer Academic, New York.

Babcock, L.E. (2007). Role of malformations in elucidating trilobite paleobiology: a historical synthesis. *New York State Museum Bulletin*, 507, 3–19.

Babcock, L.E. and Robison, R.A. (1989a). Preferences of Palaeozoic predators. *Nature*, 337, 695–696.

Babcock, L.E. and Robison, R.A. (1989b). Asymmetry of predation on trilobites. *28th International Geological Congress Abstracts*, 1, I-66.

Babcock, L.E., Wegweiser, M.D., Wegweiser, A.E., Stanley, T.M., and McKenzie, S.C. (1995). Horseshoe crabs and their trace fossils from the Devonian of Pennsylvania. *Pennsylvania Geology*, 26(2), 2–7.

Bachmayer, F. (1955). Die fossilen Asseln aus den oberjuraschichten von Ernstbrunn in Niederösterreich und von Stramberg in Mahren. *Sitzungsberichte der Österreichischen Akademie der Wissenschaften*, 1(164), 255–273.

Bachmayer, F. (1960). Eine fossile Cumaceenart (Crustacea, Malacostraca) aus dem Callovien von La Voulte-sur-Rhône (Ardèche). *Eclogae Geologicae Helvetiae*, 53, 422–426.

Bachmayer, F. and Binder, H. (1967). Fossile Perlen aus dem Wiener Becken. *Annalen der Naturhistorischen Museum zu Wien*, 71, 1–12.

Bachofen-Echt, A. (1949). *Der Bernstein und seine Einschlusse*. Springer-Verlag, Berlin, 204 pp.

Báez, A.M. (1981). Redescription of *Saltenia ibanezi*, a Late Cretaceous pipid frog from northwestern Argentina. *Ameghiniana*, 18, 127–154.

Báez, A.M. and Pugener, L.A. (2003). Ontogeny of a new Palaeogene pipid frog from southern South America and xenopodinomorph evolution. *Zoological Journal of the Linnean Society*, 139, 439–476.

Bahn, P.G. (1989). Age and the female form. *Nature*, 342, 345–346.

Baird, G.C., Brett, C.E., and Frey, R.C. (1989). "Hitchhiking" epizoans on orthoconic cephalopods: preliminary review of the evidence and its implications. *Senckenbergiana Lethaia*, 69, 439–465.

Baird, G.C., Brett, C.E., and Tomlinson, J.T. (1990). Host-specific acrothoracid barnacles on Middle Devonian platyceratid gastropods. *Historical Biology*, 4, 221–244.

Baker, J.M. (1963). Ambrosia beetles and their fungi, with particular reference to *Platypodus cylindrus* Fab. In *Symbiotic Associations* (pp. 232–265), edited by P.S. Nutman and B. Mosse. Cambridge University Press, Cambridge, U.K.

Baldi, T. (1973). Mollusc faunas of the Hungarian Upper Oligocene (Egerian). In *Studies in Stratigraphy, Palaeontology, Palaeogeography and Systematics*. Akademiai Kiado, Budapest.

Bałinski, A. (1993). A recovery from sublethal damage to the shell of a Devonian spiriferoid brachiopod. *Acta Palaeontologica Polonica*, 38, 111–118.

Bałuk, W. (1968). *Berthelina krachi* n. sp., a new bivalved gastropod from the Miocene of Poland. *Acta Palaeontologica Polonica*, 13, 291–302.

Bałuk, W. and Radwański, A. (1977). Organic communities and facies development of the Korytnica basin (Middle Miocene; Holy Cross Mountains, Central Poland). *Acta Geologica Polonica*, 27, 85–123.

Bałuk, W. and Radwański, A. (1979). Shell adaptations and ecological variability in the pelecypod species *Sphenia anatina* (Basterot) from the Korytnica basin (Middle Miocene; Holy Cross Mountains, Central Poland). *Acta Geologica Polonica*, 29, 269–286.

Bałuk, W. and Radwański, A. (1984). Creusioid cirripedes from the Korytnica Clays (Middle Miocene; Holy Cross Mountains, Central Poland). *Acta Geologica Polonica*, 34, 271–279.

Bałuk, W. and Radwański, A. (1985). Slipper-limpet gastropods (*Crepidula*) from the Eocene glauconitic sandstone of Kressenberg (Bavarian Alps, West Germany). *Neues Jahrbuch für Geologie und Paläontologie–Monatsheft*, H4, 237–247.

Bałuk, W. and Radwański, A. (1991). A new occurrence of fossil acrothoracican cirripedes: *Trypetesa polonica* sp. n. in her-mitted gastropod shells from the Korytnica Basin (Middle Miocene; Holy Cross Mountains, Central Poland), and its bearing on the behavioral evolution of the genus *Trypetesa*. *Acta Geologica Polonica*, 41, 1–36.

Bałuk, W. and Radwański, A. (1996). Stomatopod predation upon gastropods from the Korytnica Basin, and from other classical Miocene localities in Europe. *Acta Geologica Polonica*, 46, 279–304.

Bałuk, W. and Radwański, A. (1997). The micropolychaete *Josephella commensalis* sp. n. commensal to the scleractin-ian coral *Tarbellastraea reussiana* (Milne-Edwards & Haime, 1850) from the Korytnica Clays (Middle Miocene; Holy Cross Mountains, Central Poland). *Acta Geologica Polonica*, 47, 211–224.

Bambach, R.K. and Bennington, J.B. (1996). Do communities evolve? A major question in evolutionary paleoecology. In *Evolutionary Paleobiology* (pp. 123–160), edited by D. Jablonski, D.H. Erwin, and J.H. Lipps. University of Chicago Press, Chicago, IL.

Bandel, K. (1992). Platyceratidae from the Triassic St. Cassian Formation and the evolutionary history of the Neritomorpha (Gastropoda). *Paläontologisches Zeitschrift*, 66, 231–240.

Bandel, K. (1994). Triassic Euthyneura (Gastropoda) from St. Cassian formation (Italian Alps) with a discussion on the evo-lution of the Heterostropha. *Freiberger Forschungsheft*, C452, 79–100.

Bandel, K. and Heidelberger, D. (2001). The new family Nerrhenidae (Nerilimorpha, Gastropoda) from the Givetian of Germany. *Neues Jahrbuch für Geologie und Paläontologie–Monatsheft*, 12, 705–718.

Bandel, K. and Hemleben, C. (1987). Jurassic Heteropoda and their modern counterparts (Planktonic Gastropods, Mollusca). *Neues Jahrbuch für Geologie und Paläontologie–Abhandlungen*, 174, 1–22.

Bandel, K., Shinaq, R., and Weitschat, W. (1997). First insect inclusions from the amber of Jordan (Mid Cretaceous). *Mitteilungen aus dem Geologisch–Paläontologischen Institut der Universität Hamburg*, 80, 213–223.

Bandyopadhyay, S. (1988). A kannemeyeriid dicynodont from the Middle Triassic Yerrapalli Formation. *Philosophical Transactions of the Royal Society of London B*, 320, 185–233.

Banks, D.A. (1985). A fossil hydrothermal worm assemblage from the Tynagh lead-zinc deposit in Ireland. *Nature*, 313, 128–131.

Banks, H.P. and Colthart, B.J. (1993). Plant–animal–fungus interac-tions in Early Devonian trimerophytes from Gaspé, Canada. *American Journal of Botany*, 80, 992–1001.

Bannikov, A.F. (1993). The succession of the Tethys fish assem-blages exemplified by the Eocene localities of the southern part of the former USSR. *Kaupia*, 2, 241–246.

Barnes, D.K.A. (2001). Ancient homes for hard-up hermit crabs. *Nature*, 412, 785–786.

Barnes, E. (2005). *Diseases and Human Evolution*. University of New Mexico Press, Albuquerque, 484 pp.

Barnes, L.G. and McLeod, S.A. (1984). The fossil record and phyletic relationships of gray whales. In *The Gray Whale Eschrichtius robustus* (pp. 3–32), edited by M.L. Jones and S.L. Swartz. Academic Press, New York.

Barnosky, A.D. (1993). Mosaic evolution at the population level in *Microtus pennsylvanicus*. In *Morphological Change in Quaternary Mammals of North America* (pp. 24–59), edited by R.A. Martin and A.D. Barnosky. Cambridge University Press, Cambridge, U.K.

Baroni Urbani, C. (1989). Phylogeny and behavioural evolution in ants, with a discussion of the role of behaviour in evolutionary processes. *Ethology, Ecology and Evolution*, 1, 137–168.

Barron, G.L. (1977). *The Nematode-Destroying Fungi*, Topics in Mycobiology No. 1. Canadian Biological Publications, Guelph, Ontario, Canada, 140 pp.

Barthel, K.W., Swinburne, N.H.M., and Conway Morris, S. (1990). *Solnhofen, A Study in Mesozoic Palaeontology*. Cambridge University Press, Cambridge, U.K., 236 pp.

Basibuyuk, H.H., Rasnitsyn, A.P., Fitton, M.G., and Quicke, D.L.J. (2002). The limits of the family Evaniidae (Insecta: Hymenoptera) and a new genus from Lebanese Amber. *Insect Systematics and Evolution*, 32, 143–146.

Bassett, M.G. (1984). Life strategies of Silurian brachiopods, *Special Papers in Palaeontology*, 32, 237–263.

Bassett, M.G. and Bryant, C. (1993). The micromorphic rhyncho-nelloidean brachiopod *Lambdarina* from the type Dinantian. *Journal of Paleontology*, 67, 518–527.

Bassett, M.G., Popov, L.E., and Holmer, L.E. (2004). The oldest known metazoan parasite. *Journal of Paleontology*, 78, 1214–1216.

Bates, D.E.B. and Loydell, D.K. (2000). Parasitism on graptoloid graptolites. *Palaeontology*, 43, 1143–1152.

Batra, L.R. (1963). Ecology of ambrosia fungi and their trans-mission by beetles. *Transactions of the Kansas Academy of Science*, 66, 213–236.

Bauer, G. and Wächtler, K., Eds. (2001). *Ecology and Evolution of the Freshwater Mussels Unionida*. Springer, Heidelberg.

Baum, J. and Bar-Gal, G.K. (2003). The emergence and co-evo-lution of human pathogens. In *Emerging Pathogens: The Archaeology, Ecology, and Evolution of Infectious Disease* (pp. 67–78), edited by C.L. Greenblatt and M. Spigel. Oxford University Press, London.

Baumiller, T.K. (1990). Non-predatory drilling of Mississippian crinoids by platyceratid gastropods. *Palaeontology*, 33, 743–748.

Baumiller, T.K. (1993). Boreholes in Devonian blastoids and their implications for boring by platyceratids. *Lethaia*, 26, 41–47.

Baumiller, T.K. (1996). Boreholes in the Middle Devonian blastoid *Heteroschisma* and their implications for gastropod drilling. *Palaeogeography, Palaeoclimatology, Palaeoecology*, 123, 343–351.

Baumiller, T.K. (2001). Light stable isotope geochemistry of the crinoid skeleton and its use in biology and paleobiology. In *Echinoderms 2000* (pp. 107–112), edited by M. Barker. Swets & Zeitlinger, Lisse.

Baumiller, T.K. (2002). Multi-snail infestation of Devonian cri-noids and the nature of platyceratid-crinoid interactions. *Acta Palaeontologica Polonica*, 47, 133–139.

Baumiller, T.K. (2003). Evaluating the interaction between platy-ceratid gastropods and crinoids: a cost–benefit approach. *Palaeogeography, Palaeoclimatology, Palaeoecology*, 201, 199–209.

Baumiller, T.K. and Ausich, W.I. (1992). The broken-stick model as a null hypothesis for crinoid stalk taphonomy and as a guide to the distribution of connective tissue in fossils. *Paleobiology*, 18, 288–298.

Baumiller, T.K. and Bitner, M.A. (2004). A case of intense preda-tory drilling of brachiopods from the Middle Miocene of southeastern Poland. *Palaeogeography, Palaeoclimatology, Palaeoecology*, 214, 85–95.

Baumiller, T.K. and Gahn, F.J. (2002). Fossil record of parasitism on marine invertebrates with special emphasis on the platyc-eratid–crinoid interaction. *Paleontological Society Special Papers*, 8, 195–209.

Baumiller, T.K. and Gahn, F.J. (2003). Predation on crinoids. In *Predator–Prey Interactions in the Fossil Record* (pp. 263–278), edited by P.H. Kelley, M. Kowalewski, and T.A. Hansen. Kluwer Academic, New York.

Baumiller, T.K. and Gahn, F.J. (2004). Testing predator-driven evolution with Paleozoic crinoid arm regeneration. *Science*, 305, 1453–1455.

Baumiller, T.K. and Macurda, Jr., D.B. (1995). Borings in Devonian and Mississippian blastoids (Echinodermata). *Journal of Paleontology*, 69, 1084–1089.

Baumiller, T.K., Leighton, L.R., and Thompson, D.L. (1999). Boreholes in Mississippian spiriferid brachiopods and their implications for Paleozoic gastropod drilling. *Palaeogeography, Palaeoclimatology, Palaeoecology*, 147, 283–289.

Baumiller, T.K., Bitner, M.A., and Emig, C.C. (2006). High frequency of drill holes in brachiopods from the Pliocene of Algeria and its ecological implications. *Lethaia*, 39, 313–320.

Beamon, J.C. (2001). Possible prey acquisition behavior for the Cretaceous fish *Xiphactinus audax*. *Paleobios*, 21(2, Suppl.), 20.

Beauchamp, B. and Savard, M. (1992). Cretaceous chemosynthetic carbonate mounds in the Canadian Arctic. *Palaios*, 7, 434–452.

Beauchamp, B., Krouse, H.R., Harrison, J.C., Nassichuk, W.W., and Eliuk, L.S. (1989). Cretaceous cold-seep communities and methane derived carbonates in the Canadian Arctic. *Science*, 244, 53–56.

Beaver, H.H. and Fabian, A.J. (1996). Color patterns in blastoids. In *Sixth North American Paleontological Convention Abstracts of Papers* (p. 26), edited by J.E. Repetski, Special Publ. No. 8. Paleontological Society, Washington, D.C.

Bechly, G. and Wittmann, M. (2000). Two new tropical bugs (Insecta: Heteroptera: Thaumastocoridae-Xylastodorina and Hypsipterygidae) from Baltic amber. *Stuttgarter Beiträge zur Naturkunde, Serie B (Geologie und Paläontologie)*, 289, 1–11.

Beck, A.L. and Labandeira, C.C. (1998). Early Permian insect folivory on a gigantopterid-dominated riparian flora from north-central Texas. *Palaeogeography, Palaeoclimatology, Palaeoecology*, 142, 139–173.

Becker, M.A. and Chamberlain, Jr., J.A. (2006). Anomuran microcoprolites from the lowermost Navesink Formation (Maastrichtian), Monmouth County, New Jersey. *Ichnos*, 13, 1–9.

Beerbower, J.R. (1963). Morphology, paleoecology and phylogeny of the Permo–Pennsylvanian amphibian *Diploceraspis*. *Bulletin of the Museum of Comparative Zoology*, 130, 31–108.

Beesley, P.L., Ross, G.J.B., and Wells, A., Eds. (1998). *Fauna of Australia*. Vol. 5. *Mollusca: The Southern Synthesis*. CSIRO Publishing, Melbourne.

Belitzky, S., Goren-Imbar, N., and Erker, E. (1991). A Middle Pleistocene wooden plank with man-made polish. *Journal of Human Evolution*, 20, 349–353.

Bell, C.M. and Padian, K. (1995). Pterosaur fossils from the Cretaceous of Chile: evidence for a pterosaur colony on an inland desert plain. *Geological Magazine*, 132, 31–38.

Bell, Jr., G.L., Sheldon, M.A., Lamb, J.P., and Martin, J.E. (1996). The first direct evidence of live birth in Mosasauridae (Squamata): exceptional preservation in Cretaceous Pierre Shale of South Dakota. *Journal of Vertebrate Paleontology*, 16(3, Suppl.), 21A.

Bell, T. (1858/1913). *A Monograph of the Fossil Malacostracous Crustacea of Great Britain*. Palaeontographical Society, London, 44 and 40 pp.

Bell, W.J., Roth, L.M., and Nalepa, C.A. (2007). *Cockroaches, Ecology, Behavior, and Natural History*. The Johns Hopkins University Press, Baltimore, MD, 230 pp.

Bellomo, E. (1996). Predation traces on specimens of *Polinices lacteus* (Guilding, 1831) (Gastropoda, Prosobranchia, Naticidae) from last interglacial assemblage of Bovetto (Reggio Calabria, Southern Italy). *Bollettino della Societa Paleontologia Italiana*, 3, 9–17.

Bellwood, D.R. (1996). The Eocene fishes of Monte Bolca: the earliest coral reef fish assemblage. *Coral Reefs*, 15, 11–19.

Bellwood, D.R. (1999). Fossil pharyngognath fishes from Monte Bolca, Italy, with a description of a new pomacentrid genus and species. *Studi e Ricerche sui Giacimenti Terziari di Bolca*, 8, 207–217.

Bellwood, D.R. (2003). Origins and escalation of herbivory in fishes: a functional perspective. *Paleobiology*, 29, 71–83.

Bellwood, D.R. and Sorbini, L. (1996). A review of the fossil record of the Pomacentridae (Pisces: Labroidei) with a description of a new genus and species from the Eocene of Monte Bolca, Italy. *Zoological Journal of the Linnaean Society*, 117, 159–174.

Bellwood, D.R. and Wainwright, P.C. (2002). The history and biogeography of fishes on coral reefs. In *Coral Reef Fishes: Dynamics and Diversity in a Complex Ecosystem* (pp. 5–32), edited by P.F. Sale. Academic Press, San Diego, CA.

Beltan, L. (1977). La parturition d'un Actinoptérygien de l'Eotrias du nord-ouest de Madagascar. *Comptes Rendus de l'Academie des Sciences, Paris*, 284, 2223–2225.

Bengtson, S. (1968). The problematic genus *Mobergella* from the Lower Cambrian of the Baltic area. *Lethaia*, 1, 325–351.

Benner, J.S., Ekdale, A.A., and de Gibert, J.M. (2007). Enigmatic organisms preserved in early Ordovician macroborings, western Utah, USA. In *Current Developments in Bioerosion* (pp. 55–64), edited by M. Wisshak and L. Tapanila. Springer-Verlag, Heidelberg.

Bennett, G.F. and Peirce, M.A. (1988). Morphological form in the avian Haemoproteidae and an annotated checklist of the genus *Haemoproteus* (Kruse, 1890). *Journal of Natural History*, 22, 1683–1696.

Bennett, M.R., Harris, J.W.K., Richmond, B.G., Braun, D.R., Mbua, E., Kiura, P., Olago, D., Kibunjia, M., Omuombo, C., Behrensmeyer, A.K., Huddart, D., and Gonzalez, S. (2009). Early hominin foot morphology based on 1.5-million-year-old footprints from Ileret, Kenya. *Science*, 323, 1197–1201.

Bennett, S.C. (2003). A survey of pathologies of large pterodactyloid pterosaurs. *Palaeontology*, 46, 185–198.

Benson, R.H. (1975). Morphologic stability in Ostracoda. *Bulletins of American Paleontology*, 65, 13–46.

Berg, R.D. (1996). The indigenous gastrointestinal microflora. *Trends in Microbiology*, 4, 430–435.

Berger, L.R. and Clarke, R.J. (1995). Eagle involvement in accumulation of the Taung child fauna. *Journal of Human Evolution*, 29, 275–299.

Berger, W. (1953). Missbildungen an jungtertiären Laubblättern infolge Berlezung im Knospenzustand. *Neues Jahrbuch für Geologie und Paläontologie-Monatsheft*, 7, 322–323.

Bergman, C.F. (1998). Reversal in some fossil polychaete jaws. *Journal of Paleontology*, 72, 632–638.

Bergström, J. (1968). Upper Ordovician brachiopods from Västergötland, Sweden. *Geologica et Palaontologica*, 2, 1–35.

Bergström, J. (1973). Palaeoecologic aspects of an Ordovician *Tretaspis* fauna. *Acta Geologica Polonica*, 23, 179–206.

Berkyová, S., Frýda, J., and Lukeš, P. (2007). Unsuccessful predation on Middle Paleozoic plankton; shell injury and anomalies in Devonian dacryoconarid tentaculites. *Acta Palaeontologica Polonica*, 52, 407–412.

Berkowski, B. (2004). Monospecific rugosan assemblage from the Emsian hydrothermal vents of Morocco. *Acta Palaeontologica Polonica*, 49, 75–84.

Berman, D.S. (1976). Occurrence of *Gnathorhiza* (Osteichthyes: Dipnoi) in aestivation burrows in the Lower Permian of New Mexico with description of a new species. *Journal of Paleontology*, 50, 1034–1039.

Bernays, E.A. and Chapman, R.F. (1994). *Host–Plant Selection by Phytophagous Plants*. Chapman & Hall, London, 312 pp.

Berry, C., Hindley, J., and Oei, C. (1991). The *Bacillus sphaericus* toxins and their potential for biotechnological development. In *Biotechnology for Biological Control of Pests and Vectors* (pp. 35–51), edited by K. Maramorosch. CRC Press, Boca Raton, FL.

Berry, C.T. (1936). A Miocene pearl. *American Midland Naturalist*, 17, 464–470.

Berry, E.W. (1921). A insect-cut-leaf from the lower Eocene. *American Journal of Science*, 221, 301–303.

Berry, E.W. (1934). The Lower Lance Flora from Harding County, South Dakota. *U.S. Geological Survey Professional Paper*, 357, 1–145.

Berta, A. (1994). What is a whale? *Science*, 263, 180–181.

Bertling, M. (1992). *Arachnostega* n. ichnog.—burrowing traces in internal moulds of boring bivalves (Late Jurassic, Northern Germany). *Paläontologisches Zeitschrift*, 66, 177–185.

Bertrand, M., Coen-Aubert, M., Dumoulin, V., Preat, A., and Tourneur, F. (1993). Sedimentologie et paléoécologie de l'Emsien supérieur et de l'Eifelian inférieur des regions de Couvin et de Villers-la-Tour (bord sud du Synclinorium de Dinant, Belgique). *Neues Jahrbuch für Geologie und Paläontologie–Abhandlung*, 188, 177–211.

Beschin, C., Busulini, A., de Angeli, A., Tessier, G., and Ungaro, S. (1998). Crostacei eocenici di "Cava Rossi" presso Monte di Malo (Vicenza–Italia settentrionale). *Studi Trentini di Scienze Naturali–Acta Geologica*, 73, 7–34.

Bessels, R. (1869). Ueber fossile Selachier-Eier. *Jahreshefte des Vereins für vaterländische Naturkunde in Württemberg*, 25, 152–155.

Betancourt, J.L., Van Devender, T.R., and Martin, P.S. (1990). *Packrat Middens: The Last, 40,000 Years of Biotic Change*. University of Arizona Press, Tucson, 467 pp.

Bettenstaedt, F. (1968). Wechselbeziehungen zwischen angewandter Mikropaläontologie und Evolutionsforschung. *Beihefte zu den Berichten der Naturhistorischen Gesellschaft zu Hannover*, 5, 337–391.

Beu, A.G., Henderson, R.A., and Nelson, C.S. (1972). Notes on the taphonomy and paleoecology of New Zealand Tertiary Spatangoida. *New Zealand Journal of Geology and Geophysics*, 15, 275–286.

Bi, D.-C. (1986). Ecological explanation of some abnormal phenomena in fossil conchostracan shells. In *Selected Papers from the 13th and 14th Annual Conventions of the Palaeontological Society of China* (pp. 229–235). Anhui Science and Technology Publishing House, Hefei.

Biaggi, R.E., Leggitt, V.L., and Buchheim, H.P. (1999). Caddisfly (Insecta: Trichoptera) larvae mounds from the Eocene Tipton Member, Green River Formation, Wyoming. *Geological Society of America Abstracts with Programs*, 31(7), A-242.

Bianucci, G., Di Celma, C., Landini, W., and Buckeridge, J. (2006a). Palaeoecology and taphonomy of an extraordinary whale barnacle accumulation from the Plio–Pleistocene of Ecuador. *Palaeogeography, Palaeoclimatology, Palaeoecology*, 242, 326–342.

Bianucci, G., Landini, W., and Buckeridge, J. (2006b). Whale barnacles and Neogene cetacean migration routes. *New Zealand Journal of Geology and Geophysics*, 49, 115–120.

Bielokrys, L.S. (1994). Maloizvestnie serpulidi *Pyrgopolon* iz Maastrikhta Krima. *Paleontologicheskii Zhurnal*, 1, 44–50.

Bien, W.F., Wendt, J.M., and Alexander, R.R. (1999). Site selection and behavior of sponge and bivalve borers in shells of the Cretaceous oysters *Exogyra cancellata* and *Pycnodonte mutabilis* from Delaware, USA. *Historical Biology*, 13, 299–315.

Bigelow, P.K. (1994). Occurrence of a squaloid shark (Chondrichthyes: Squaliformes) with the pinniped *Allodesmus* from the Upper Miocene of Washington. *Journal of Paleontology*, 68, 680–684.

Bignell, D.E. (2000). Introduction to symbiosis. In *Termites: Evolution, Sociality, Ecology* (pp. 189–208), edited by T. Abe, D.E. Bignell, and M. Higashi. Kluwer Academic, Dordrecht.

Biknevicius, A.R. and Van Valkenburgh, B. (1996). Design for killing: craniodental adaptations of predators. In *Carnivore Behavior, Ecology, and Evolution*, Vol. 2 (pp. 393–428), edited by J.L. Gittleman. Cornell University Press, Ithaca, NY.

Billia, E.M.E. and Graovac, S.M. (2001). Amelogenesis Imperfecta on a deciduous molar of *Coelodonta antiquitatis* (Blumenbach, 1799) (Mammalia, Rhinoceratidae) from Late Pleistocene levels of Grotta di Furmane (Verona, Northern Italy): a rare case report. *Paleontologia i Evolucio*, 32–33, 93–98.

Binns, E.S. (1982). Phoresy as migration: some functional aspects of phoresy in mites. *Biological Reviews*, 57, 571–620.

Birenheide, R. (1963). Standortwechsel von Korallen aus dem Eifelmeer. *Natur und Museum*, 93, 405–409.

Bisconti, M. and Varola, A. (2006). The oldest eschrichtiid mysticete and a new morphological diagnosis of Eschrichtiidae (Gray whales). *Rivista Italiana di Paleontologia e Stratigrafia*, 112, 447–457.

Bishop, G.A. and Palmer, Sr., B.T. (2006). A new genus and species of crab from the bryozoan bioherms of the Eocene Santee Limestone: South Carolina, USA. *Revista Mexicana de Ciencias Geológicas*, 23, 334–337.

Black IV, W.C. (2003). Evolution of arthropod disease vectors. In *Emerging Pathogens: The Archaeology, Ecology, and Evolution of Infectious Disease* (pp. 49–63), edited by C.L. Greenblatt and M. Spigelman. Oxford University Press, London.

Blackwell, M. (2000). Terrestrial life: fungi from the start? *Science*, 289, 1884–1885.

Blair, W.N. and Armstrong, A.K. (1979). *Hualapei Limestone Member of the Muddy Creek Formation: The Youngest Deposit Predating the Grand Canyon, Southeastern Nevada and Northwestern Arizona*, U.S.G.S. Professionial Paper No. 1111. U.S. Geological Survey, Reston, VA, 14 pp.

Blake, D.B. and Aronson, R.B. (1998). Eocene stelleroids (Echinodermata) at Seymour Island, Antarctic Peninsula. *Journal of Paleontology*, 72, 339–353.

Blake, D.B. and Guensburg, T.E. (1994). Predation by the Ordovician asteroid *Promopalaester* on a pelecypod. *Lethaia*, 27, 235–239.

Blake, D.B. and Hagdorn, H. (2003). The Asteroidea (Echinodermata) of the Muschelkalk (Middle Triassic of Germany). *Paläontologisches Zeitschrift*, 77, 23–58.

Blake, D.B. and Zinsmeister, W.J. (1991). A new marsupiate cidaroid echinoid from the Maastrichtian of Antarctica. *Palaeontology*, 34, 629–635.

Bleckmann, H. (1986). Rose of the lateral line in fish behaviour. In *The Behaviour of Teleost Fishes* (pp. 177–202), edited by T.J. Pitcher. Croom Helm, London.

Blodgett, R.B., Boucot, A.J., and Koch II, W.F. (1988). New occurrences of color patterns in Devonian articulate brachiopods. *Journal of Paleontology*, 62, 46–51.

Blow, W.C. and Bailey, R.H. (2003). A Late Pliocene pea crab infestation of the slipper-shell *Crepidula fornicate* (Linnaeus, 1758) from the Yorktown Formation of southeastern Virginia. *Geological Society of America Abstracts with Programs*, 35(6), 58.

Blumer, L.S. (1979). Male parental care in the bony fishes. *Quarterly Review of Biology*, 54, 149–161.

Blunck, G. (1929). Bakterieneinschlüsse im Bernstein. *Centralblatt für Mineralogie, Geologie und Paläontologie*, 11, 554–555.

Bock, W. (1949). Triassic chimaeroid egg capsules from the Connecticut River Valley. *Journal of Paleontology*, 23, 515–517.

Boekschoten, G.J. (1967). Palaeoecology of some Mollusca from the Tielrode Sands (Pliocene, Belgium). *Palaeogeography, Palaeoclimatology, Palaeoecology*, 3, 311–362.

Boirat, J.-M. and Fouquet, Y. (1986). Découverte des tubes de vers hydrothermaux fossiles dans un amas sulfure de l'Eocène supérieur (Barlo, ophiolite de Zambales, Philippines). *Comptes Rendus de l'Academie des Sciences, Paris*, 302(15), 941–946.

Boles, I.C. and Lohmann, K.J. (2003). True navigation and magnetic maps in spiny lobsters. *Nature*, 421, 60–63.

Bolin, T. and Stanton, R. (1993). *Wind Breaks: Coming to Terms with Flatulence*. M. Gee Publishing, McMahons Point, Australia, 86 pp.

Boltovskoy, E. and Boltovskoy, D. (1988). Cenozoic deep-sea benthic Foraminifera: faunal turnovers and paleobiogeographic differences. *Revue de Micropaléontologie*, 31(2), 67–84.

Bonaparte, J. (1971). Descripción del Cráneo y Mandibulas de *Pterodaustro guiñazui* (Pterodactyloidea-Pterodaustriidae nov.) de la formación Lagarcito, San Luis, Argentina. *Publicaciones del Museo Municipal de Ciencias Naturales de Mar del Plata*, 1(9), 263–272.

Bond, G., Broecker, W., Johnsen, S., McManus, J., Labeyrie, L., Jouzel, J., and Bonani, G. (1993). Correlations between climate records from North Atlantic sediments and Greenland ice. *Nature*, 365, 143–147.

Bond, P.N. and Saunders, W.B. (1989). Sublethal injury and shell repair in Upper Mississippian ammonoids. *Paleobiology*, 15, 414–428.

Bonde, N. (1987). *Moler: Its Origin and Its Fossils Especially Fishes*. Skamol Skarrehage Moler-vaerk A/S, Morsø Bogtrykkeri, Nykøbing Mors, 47 pp.

Bongrain, M. (1995). Traces de bioerosion sur un Pectinidae (Bivalvia): du Miocène d'Aquitaine (SO France): un cas possible de commensalisme entre Pectinidae et Capulidae. *Géobios*, 28, 347–358.

Bordeaux, Y.L. and Brett, C.E. (1990). Substrate specific associations of epibionts on Middle Devonian brachiopods: implications for paleoecology. *Historical Biology*, 4, 203–220.

Bordy, E.M., Bumby, A.J., Catuneanu, O., and Eriksson, P.G. (2004). Advanced Early Jurassic termite (Insecta: Isoptera) nests: evidence from the Clarens Formation in the Tuli Basin, Southern Africa. *Palaios*, 19, 68–78.

Borkent, A. (1995). *Biting Midges in the Cretaceous Amber of North America (Diptera: Ceratopogonidae)*. Backhuys, Leiden, Germany, 237 pp.

Borkent, A. (2000). Biting midges (Ceratopogonidae: Diptera) from Lower Cretaceous Lebanese amber with a discussion of the diversity and patterns found in other ambers. In *Studies on Fossils in Amber, with Particular Reference to the Cretaceous of New Jersey* (pp. 355–451), edited by D. Grimaldi. Backhuys, Leiden, Germany.

Boshoff, P.H. (1968). A preliminary study on conchological physiopathology with special reference to Pelecypoda. *Annals of the Natal Museum*, 20, 199–216.

Botha-Brink, J. and Modesto, S.P. (2007). A mixed-age-classes "pelycosaur" aggregation from South Africa: earliest evidence of parental care in amniotes? *Philosophical Transactions of the Royal Society of London B*, 274, 2829–2834.

Böttcher, R. (1989). Uber die Nahrung eines *Leptopterygius* (Ichthyosauria, Reptilia) aus dem suddeutschen Posidonienschiefer (Unterer Jura) mit Bemerkungen über den Magen der Ichthyosaurier. *Stuttgarter Beiträge zur Naturkunde Serie B (Geologie und Paläontologie)*, 155, 1–19.

Böttcher, R. (1990). Neue Erkenntnisse über die Fortpflanzungsbiologie der Ichthyosaurier (Reptilia). *Stuttgarter Beiträge zur Naturkunde Serie B (Geologie und Paläontologie)*, 164, 1–51.

Botting, J.P. and Thomas, A.T. (1998). A pseudoplanktonic inarticulate brachiopod attached to graptolites and algae. *Palaeontology Newsletter*, 39, 27–28.

Botting, J.P. and Thomas, A.T. (1999). A pseudoplanktonic inarticulate brachiopod attached to graptolites and algae. *Acta Universitatis Carolinae–Geologica*, 43(1/2), 333–335.

Bouchet, P. (1981). Evolution of larval development in eastern Atlantic Terebridae (Gastropoda), Neogene to Recent. *Malacologia*, 21, 363–369.

Bouchet, P. (1989). A marginellid gastropod parasitizes sleeping fishes. *Bulletin of Marine Science*, 45, 76–84.

Bouchet, P. and Perrine, D. (1996). More gastropods feeding at night on parrotfishes. *Bulletin of Marine Science*, 59, 224–228.

Boucot, A.J. (1959). Brachiopods of the Lower Devonian rocks at Highland Mills, New York. *Journal of Paleontology*, 33, 727–769.

Boucot, A.J. (1975). *Evolution and Extinction Rate Controls*. Elsevier, Amsterdam, 427 pp.

Boucot, A.J. (1976). Rates of size increase and of phyletic evolution. *Nature*, 261, 694–696.

Boucot, A.J. (1978). Community evolution and rates of cladogenesis. *Evolutionary Biology*, 11, 545–655.

Boucot, A.J. (1981). *Principles of Benthic Marine Paleoecology*. Academic Press, San Diego, CA, 463 pp.

Boucot, A.J., Ed. (1990a). *Evolutionary Paleobiology of Behavior and Coevolution*. Elsevier, Amsterdam, 735 pp.

Boucot, A.J. (1990b). Phanerozoic extinctions: how similar are they to each other? In *Extinction Events in Earth History* (pp. 5–30), edited by E.G. Kauffman and O.H. Walliser, Lecture Notes in Earth History. Springer-Verlag, Heidelberg.

Boucot, A.J. (1994). The episodic, rather than the periodic nature of extinction events. *Revista de la Societa Mexicana Paleontologia*, 7, 15–35.

Boucot, A.J. (1997). Community Evolution. In *Evolutsiya Zhizin na Zemle*, edited by V.M. Podobina, First International Symposium on Evolution of Life on the Earth, Tomsk, pp. 13–15.

Boucot, A.J. and Chen, X. (in press). Graptolite depth zonation, *Palaeoworld*.

Boucot, A.J. and Kříž, J. (1999). Definition of the terms "homologous" and "analogous" community. In *Paleocommunities: A Case Study from the Silurian and Lower Devonian* (p. 32), edited by A.J. Boucot and J.D. Lawson. Cambridge University Press, Cambridge, U.K.

Boucot, A.J. and Lawson, J.D., Eds. (1999). *Paleocommunities: A Case Study from the Silurian and Lower Devonian*. Cambridge University Press, Cambridge, U.K., 895 pp.

Bourke, J.B. (1967). A review of the palaeopathology of the arthritic diseases. In *Diseases in Antiquity* (pp. 352–370), edited by D. Brothwell and A.T. Sandison. Charles C Thomas, Springfield, IL.

Bowman, T.E. (1971). *Palaega lamnae*, new species (Crustacea, Isopoda) from the Upper Cretaceous of Texas. *Journal of Paleontology*, 45, 540–541.

Bown, T.M. and Genise, J.F. (1993). Fossil nests and gallery systems of termites (Isoptera) and ants (Formicidae) from the early Miocene of southern Ethiopia and the late Miocene of Abu Dhabi Emirate, U.A.E. *Geological Society of America Abstracts with Programs*, 25, A58.

Bown, T.M. and Laza, J.H. (1990). A Miocene termite nest from southern Argentina and its paleoclimatological implications. *Ichnos*, 1, 73–79.

Boy, J. (1971). Zur Problematik der Branchiosaurier (Amphibia: Karbon-Perm). *Paläontologisches Zeitschrift*, 45, 107–119.

Boy, J. (1974). Die Larven der rhachitomen Amphibien (Amphibia: Temnospondyli: Karbon-Trias). *Paläontologisches Zeitschrift*, 48, 236–268.

Boy, J.A. and Sues, H.-D. (2000). Branchiosaurs: larvae, metamorphosis and heterochrony in temnospondyls and seymouriamorphs. In *Amphibian Biology*, Vol. 4 (pp. 150–1197), edited by H. Heatwole. Beatty & Sons, Chipping Norton, New South Wales.

Boyd, D.W. and Newell, N.D. (1968). Hinge grades in the evolution of Crassatellacean bivalves as revealed by Permian genera. *American Museum Novitates*, 2328, 1–52.

Bradley, J.S. (1956). A teratological *Parafusulina*. *Journal of Paleontology*, 30, 303–304.

Bradley, W.H. (1931). Origin and microfossils of the Oil Shale of the Green River Formation of Colorado and Utah. *U.S. Geological Survey Professional Paper*, 168, 1–57.

Brandão, C.R.F. (1990). Phylogenetic, biogeographic, and evolutionary inferences from the description of an Early Cretaceous South American Myrmeciinae. In *Proceedings of the 11th International Congress of the IUSSI* (pp. 313–314), Bangalore, India.

Brandão, C.R.F., Martin-Neto, R.G., and Vulcano, M.A. (1989). The earliest known fossil ant (first southern hemisphere Mesozoic record) (Hymenoptera: Formicidae: Myrmeciinae). *Psyche*, 96, 195–205.

Brandt, D.S. (2002). Ecdysial efficiency and evolutionary efficacy among marine arthropods: implications for trilobite survivorship. *Alcheringa*, 26, 399–421.

Brandt, D.S., Meyer, D.L., and Lask, P.B. (1995). *Isotelus* (Trilobita) "hunting burrow" from Upper Ordovician strata, Ohio. *Journal of Paleontology*, 69, 1079–1083.

Brasier, C.M. (1992). Oak tree mortality in Iberia. *Nature*, 360, 539.

Brasier, M.D. (1975). An outline history of seagrass communities. *Palaeontology*, 18, 681–702.

Brasier, M., Cotton, L., and Hiscocks, J. (2008). Earliest Cretaceous "firestorm amber" with spiders' webs, from the ~140 Ma dinosaur trackway beds of Bexhill. *The Palaeontological Association Newsletter*, 69, 18.

Braun, A., Chen, J.-Y., Waloszek, D., and Maas, A. (2007). First Early Cambrian Radioloria. In *The Rise and Fall of the Ediacaran Biota*, edited by P. Vickers-Rich and P. Komarower (pp. 143–149). Geological Society of London Special Publication 286. Geological Society, London.

Brecher, G. and Wigglesworth, V.B. (1944). The transmission of *Actinomyces rhodnii* Erikson in *Rhodnius prolixus* Stål (Hemiptera) and its influence on the growth of the host. *Parasitology*, 35, 220–224.

Breder, C.M. and Rosen, D.E. (1966). *Modes of Reproduction in Fishes*. Natural History Press, Garden City, NJ, 941 pp.

Bresciani, J., Haarlov, N., Nansen, P., and Moller, G. (1983). Head louse (*Pediculus humanus* subsp. *capitus* de Geer) from mummified corpses of Greenlanders, A.D., 1460 (+50). *Acta Entomologica Fennica*, 42, 24–27.

Bresciani, J., Dansgaard, W., Fredskild, B. et al. (1991). Living conditions. In *The Greenland Mummies* (pp. 151–167), edited by J.P.H. Hansen, J. Meldgaard, and J. Nordqvist. Smithsonian Institution Press, Washington, D.C.

Breton, G. (1992). Les Goniasteridae (Asteroidea, Echinodermata) jurassiques et crétacés de France: taphonomie, systématique, biostratigraphie, paléobiogéographie, évolution. *Bulletin Trimestriel de la Société Géologique Normandie et Amis Muséum du Havre, fascicule hors série*, 78(4), 590.

Brett, C.E. (2003). Durophagous predation in Paleozoic marine benthic assemblages. In *Predator–Prey Interactions in the Fossil Record* (pp. 401–432), edited by P.H. Kelley, M. Kowalewski, and T.A. Hansen. Kluwer Academic, New York.

Brett, C.E. and Bordeaux, Y.L. (1990). Taphonomy of brachiopods from a Middle Devonian shell bed: implications for the genesis of skeletal accumulations. In *Brachiopods Through Time* (pp. 219–226), edited by D.I. MacKinnon, D.E. Lee, and J.D. Campbell. Balkema Press, Rotterdam.

Brett, C.E. and Cottrell, J.F. (1982). Substrate specificity in the Devonian tabulate coral *Pleurodictyum*. *Lethaia*, 15, 247–262.

Brett, C.E. and Eckert, J.D. (1982). Paleoecology of a well-preserved crinoid colony from the Silurian Rochester Shale in Ontario. *Royal Ontario Museum Life Science Contributions*, 131, 1–20.

Brice, D. and Hou, H.-F. (1992). "Blisters" in a Famennian cyrtospiriferid brachiopod from Hunan (South China). *Palaeogeography, Palaeoclimatology, Palaeoecology*, 94, 253–260.

Brice, D. and Mistiaen, B. (1992). Epizoaires des brachiopodes Frasniens de Ferques (Boulonnais, Nord de la France). *Géobios, Mémoire Especial*, 14, 45–58.

Bricknell, I. (1987). Palaeopathology of Pleistocene proboscidians in Britain. *Modern Geology*, 11, 295–309.

Briggs, D.E.G. and Eglinton, G. (1994). Chemical traces of ancient life. *Chemistry in Britain*, 30, 907–912.

Briggs, D.E.G., Sutton, M.D., Siveter, D.J., and Siveter, D.J. (2005). Metamorphoses in a Silurian barnacle. *Philosophical Transactions of the Royal Society of London B*, 272, 2365–2369.

Bright, M. (2002). *Sharks*. Smithsonian Institution Press, Washington, D.C., 112 pp.

Brinkmann, W. (1996). Ein Mixosaurier (Reptilia, Ichthyosauria) mit Embryonen aus der Grenzbitumenzone (Mitteltrias) des Monte San Giorgio (Schweiz, Kanton Tessin). *Eclogae Geologicae Helvetiae*, 89, 1345–1362.

Britt, B.B., Scheetz, R.D., and Dangerfield, A. (2008). A suite of dermestid beetle traces on dinosaur bone from the Upper Jurassic Morrison Formation, Wyoming, USA. *Ichnos*, 15, 59–71.

Bromley, R.G. (1975). Comparative analysis of fossil and recent echinoid bioerosion. *Palaeontology*, 18, 725–739.

Bromley, R.G. (1981). Concepts in ichnotaxonomy illustrated by small round holes in shells. *Acta Geologica Hispanica*, 16, 55–64.

Bromley, R.G. (1993). Predation habits of octopus past and present and a new ichnospecies, *Oichnus ovalis*. *Bulletin of the Geological Society of Denmark*, 40, 167–173.

Bromley, R.G. (1994). The palaeoecology of bioerosion. In *The Palaeobiology of Trace Fossils* (pp. 134–154), edited by S.K. Donovan. Cambridge University Press, Cambridge, U.K.

Bromley, R.G. (1996). *Trace Fossils*. Chapman & Hall, London, 361 pp.

Bromley, R.G. (1999). Anomiid (bivalve) bioerosion on Pleistocene pectinid (bivalve) shells, Rhodes, Greece. *Geologie en Mijnbouw*, 78, 175–177.

Bromley, R.G. (2001). Tetrapod tracks deeply set in unsuitable substrates: recent musk oxen in fluid earth (East Greenland) and Pleistocene caprines in Aeolian sand (Mallorca). *Bulletin of the Geological Society of Denmark*, 48, 209–215.

Bromley, R.G. (2004). A stratigraphy of marine erosion. In *The Application of Ichnology to Palaeoenvironmental and Stratigraphic Analysis* (pp. 455–479), edited by D. McIlroy, Geological Society of London Special Publication 228. Geological Society, London.

Bromley, R.G. (2005). *Podichnus centrifugalis* Bromley & Surlyk, 1973 revisited: attachment scars of brachiopods. In *Fifth International Brachiopod Congress, Copenhagen 2005, Abstracts* (p. 9), edited by D.A.T. Harper, S.L. Long, and M. McCorry. Geological Survey of Denmark and Greenland, Copenhagen.

Bromley, R.G. (2008). Trace fossil *Podichnus obliquus*, attachment scar of the brachiopod *Terebratulina retusa*: Pleistocene, Rhodes, Greece. *Fossils and Strata*, 54, 227–230.

Bromley, R.G. and Asgaard, U. (1972). The burrows and microcoprolites of *Glyphea rosenkrantzei*, a Lower Jurassic palinuran crustacean from Jameson Land, East Greenland. *Grønlands Geologiske Undersøgelse Rapport*, 49, 15–21.

Bromley, R.G. and Asgaard, U. (1975). Sediment structures produced by a spatangoid echinoid: a problem of preservation. *Bulletin of the Geological Society of Denmark*, 24, 261–281.

Bromley, R.G. and Asgaard, U. (1979). Triassic freshwater ichnocoenoses from Carlsberg Fjord, East Greenland. *Palaeogeography, Palaeoclimatology, Palaeoecology*, 28, 39–80.

Bromley, R.G. and Asgaard, U. (1991). Ichnofacies: a mixture of taphofacies and biofacies. *Lethaia*, 24, 153–163.

Bromley, R.G. and Asgaard, U. (1993). Endolithic community replacement on a Pliocene rocky coast. *Ichnos*, 2, 93–116.

Bromley, R.G. and Heinberg, C. (2006). Attachment strategies of organisms on hard substrates: a palaeontological view. *Palaeogeography, Palaeoclimatology, Palaeoecology*, 232, 429–453.

Bromley, R.G. and Martinell, J. (1991). *Centrichnus*, new ichnogenus for centrically patterned attachment scars on skeletal substrates. *Bulletin of the Geological Society of Denmark*, 38, 243–252.

Bromley, R.G. and Schönberg, C.H.L. (2007). Borings, bodies and ghosts: spicules of the endolithic sponge *Aka akis* sp. nov. within the boring *Entobia cretacea*, Cretaceous, England. In *Current Developments in Bioerosion* (pp. 235–248), edited by M. Wisshak and L. Tapanila. Springer-Verlag, Heidelberg.

Bromley, R.G. and Surlyk, F. (1973). Borings produced by brachiopod pedicles, fossil and Recent. *Lethaia*, 6, 349–365.

Bromley, R.G., Pemberton, S.G., and Rahmani, R.A. (1984). A Cretaceous woodground: the *Teredolites* ichnofacies. *Journal of Paleontology*, 58, 488–498.

Bromley, R.G., Uchman, A., Gregory, M.R., and Martin, A.J. (2003). *Hillichnus lobosensis* igen. et isp. nov., a complex trace fossil produced by tellinacean bivalves, Paleocene, Monterey, California, USA. *Palaeogeography, Palaeoclimatology, Palaeoecology*, 192, 157–186.

Bromley, R.G., Wisshak, M., Glaub, I., and Botquelen, A. (2007). Ichotaxonomic review of dendriniform borings attributed to foraminiferans: *Semidendrina* gen. nov. In *Trace Fossils: Concepts, Problems, Prospects* (pp. 518–530), edited by W. Miller III. Elsevier, Amsterdam.

Brooks, A.S., Helgren, D.M., Cramer, J.S., Franklin, A., Hornyak, W. et al. (1995). Dating and context of three Middle Stone Age sites with bone points in the upper Semliki Valley, Zaire. *Science*, 268, 548–553.

Brooks, D.R. and McLennan, D.A. (1993). Comparative study of adaptive radiations with an example using parasitic flatworms (Platyhelminthes: Cercomeria). *American Naturalist*, 142, 755–778.

Brooks, M.A. (1963). The microorganisms of healthy insects. In *Insect Pathology: An Advanced Treatise*, Vol. 1 (pp. 215–250), edited by E.A. Steinhaus. Academic Press, London.

Brothwell, D. (1967). The evidence for neoplasms. In *Diseases in Antiquity* (pp. 320–345), edited by D. Brothwell and A.T. Sandison. Charles C Thomas, Springfield, IL.

Brown, P., Sutikna, T., Norwood, M.J., Soejono, R.P., Jatmiko, E. et al. (2004). A new small-bodied hominin from the Late Pleistocene of Flores, Indonesia. *Nature*, 425, 310–316.

Brown, R.W. (1940). Fossil pearls from the Colorado group of western Kansas. *Journal of the Washington Academy of Sciences*, 30, 365–374.

Brown, R.W. (1946a). Fossil egg capsules of chimaeroid fishes. *Journal of Paleontology*, 20, 261–266.

Brown, R.W. (1946b). A Pleistocene pearl from southern Maryland. *Journal of the Washington Academy of Sciences*, 36, 75–76.

Brown, R.W. (1950). Cretaceous fish egg capsule from Kansas. *Journal of Paleontology*, 24, 594–600.

Brown, R.W. (1962). Paleocene flora of the Rocky Mountains and Great Plains, *U.S. Geological Survey Professional Paper*, 375, 1–119.

Bruet, E. (1950). Le Loess de la République de l'Equateur et ses nids fossiles d'insectes. *Revue française d'entomologie*, 17, 280–283.

Brugerolle, G. and Lee, J.J. (2000a). Order Oxymonadidae. In *An Illustrated Guide to the Protozoa*, Vol. II, 2nd ed. (pp. 1186–1195), edited by J.J. Lee, G.F. Leedale, and P. Bradbury. Allen Press, Lawrence, KS.

Brugerolle, G. and Lee, J.J. (2000b). Phylum Parabasalia. In *An Illustrated Guide to the Protozoa*, Vol. II, 2nd ed. (pp. 1196–1250), edited by J.J. Lee, G.F. Leedale, and P. Bradbury. Allen Press, Lawrence, KS.

Brunton, C.H.C. and Mundy, D.J.C. (1988). Strophalosiacean and aulostegacean productoids (Brachiopoda) from the Craven Reef Belt (late Viséan) of North Yorkshire. *Proceedings of the Yorkshire Geological Society*, 47, 55–88.

Brunton, C.H.C. and Mundy, D.J.C. (1993). Productellid and Plicatiferid (Productoid) brachiopods from the Lower Carboniferous of the Craven Reef Belt, North Yorkshire. *Bulletin of the Natural History Museum London (Geology)*, 49(2), 99–119.

Brunton, H. (1966). Predation and shell damage in a Viséan brachiopod fauna. *Palaeontology*, 9, 355–359.

Bucher, W.H. (1938). A shell-boring gastropod in a *Dalmanella* bed of Upper Cincinnatian age. *American Journal of Science*, 36, 1–7.

Buehler, E.J. (1969). Cylindrical borings in Devonian shells. *Journal of Paleontology*, 43, 1291.

Buffetaut, E. (1983). Wounds on the jaw of an Eocene mesosuchian crocodilian as possible evidence for the antiquity of crocodilian intraspecific fighting behaviour. *Paläontologisches Zeitschrift*, 57, 143–145.

Buffetaut, E. and Mazin, J.-M., Eds. (2003). *Evolution and Palaeobiology of Pterosaurs*, Geological Society of London Special Publication 217. Geological Society, London, 347 pp.

Buffetaut, E. and Suteethorn, V. (1989). A sauropod skeleton associated with theropod teeth in the Upper Jurassic of Thailand: remarks on the taphonomic and palaeoecological significance of such associations. *Palaeogeography, Palaeoclimatology, Palaeoecology*, 73, 77–83.

Buffetaut, E., Martill, D., and Escuillié, F. (2004). Pterosaurs as part of a spinasaur diet. *Nature*, 430, 33.

Buffetaut, E., Grellet-Tinner, G., Suteethorn, V., Cuny, G., Tong, H. et al. (2005). Minute theropod eggs and embryo from the Lower Cretaceous of Thailand and the dinosaur–bird transition. *Naturwissenschaften*, 92, 477–482.

Buge, E. and Fischer, J.-C. (1970). *Atractosoecia incrustans* (d'Orbigny) (Bryozoa Cyclostomata) espèce bathonienne symbiotique d'un pagure. *Bulletin de la Société Géologique de France*, 7(12), 126–133.

Bukowski, F. and Bond, P. (1989). A predator attacks *Sphenodiscus*. *The Mosasaur*, 4, 69–74.

Bull, E.E. (1987). Upper Llandovery dendroid graptolites from the Pentland Hills, Scotland. *Palaeontology*, 30, 117–140.

Bullock, T.H., Northcutt, R.G., and Bodznick, D.A. (1982). Evolution of electroreception. *Trends in Neurosciences*, 5, 50–53.

Burns, F. (1899). Viviparous Miocene Turritellidae. *The Nautilus*, 13, 68–69.

Bürgin, T. (2000). *Euthynotus* cf. *incognitus* (Actinopterygia, Pachycormidae) als Mageninhalt eines Fischsauriers aus dem Posidonienschiefer Süddeutschlands (Unterer Jura, Lias epsilon). *Eclogae Geologicae Helvetiae*, 93, 491–496.

Busvine, J.R. (1980). The evolution and mutual adaptation of insects, microorganisms and man. In *Changing Disease Patterns and Human Behavior* (pp. 55–68), Academic Press, San Diego, CA.

Buzas, M.A. and Culver, S.J. (1989). Biogeographic and evolutionary patterns of continental margin benthic Foraminifera. *Paleobiology*, 15, 11–19.

Buzas, M.A. and Culver, S.J. (1994). Species pool and dynamics of marine paleocommunities. *Science*, 264, 1439–1441.

Cadée, G.C. (1999). Shell damage and shell repair in the Antarctic limpet *Nacella concinna* from King George Island. *Journal of Sea Research*, 41, 149–161.

Cadée, G.C. (2000). Herring gulls feeding on a recent invader in the Wadden Sea, *Ensis directus*. In *The Evolutionary Biology of the Bivalvia* (pp. 459–467), edited by E.M. Harper, J.D. Harper, and J.A. Crame. Geological Society of London Special Publication 177. Geological Society, London.

Cadée, G.C. (2008). *Hydrobia* as "Jonah in the Whale": shell repair after passing alive through shelducks. *The Palaeontological Association Newsletter*, 69, 47.

Cadée, G.C. and McKinney, F.K. (1994). A coral–bryozoan association from the Neogene of northwestern Europe. *Lethaia*, 27, 59–66.

Caetano, F.H. and Cruz-Landim, C. (1985). Presence of microorganisms in the alimentary canal of ants of the tribe Cephalotini (Myrmicinae): location and relationship with intestinal structure. *Naturalia (São Paulo)*, 10, 37–47.

Calcinai, B., Bavestrello, G., Cerrano, C., and Gaggero, L. (2007). Substratum microtexture affects the boring pattern of *Cliona albimarginata* (Clionaidae, Demospongiae). In *Current Developments in Bioerosion* (pp. 203–212), edited by M. Wisshak and L. Tapanila. Springer-Verlag, Heidelberg.

Caldas, E.B., Neto, R.G.M., and Filho, F.P.L. (1989). *Afropollis* sp. (pólem) no trato intestinal de vespa (Hymenoptera: Apocrita: Xyelidae), no Cretáceo da Bacia do Araripe. *XIII Simpósio de Geologia do Nordeste, Atas*, pp. 195–196.

Caldwell, M.W. (2003). "Without a leg to stand on": on the evolution and development of axial elongation and limblessness in tetrapods. *Canadian Journal of Earth Sciences*, 40, 573–588.

Caldwell, M.W. and Lee, M.S.Y. (2001). Live birth in Cretaceous marine lizards (Mosasauroids). *Proceedings of the Royal Society London B*, 268, 2397–2401.

Callen, E.O. and Cameron, T.W.M. (1960). A prehistoric diet revealed in coprolites. *The New Scientist*, 8, 35–40.

Callender, W.R., Staff, G.M., Powell, E.N., and MacDonald, I.R. (1990). Gulf of Mexico hydrocarbon seep communities. V. Biofacies and shell orientation of autochthonous shell beds below storm wave base. *Palaios*, 5, 2–14.

Calvo, J.O. (1994). Gastroliths in sauropod dinosaurs. *Gaia*, 10, 205–208.

Cambefort, Y. (1991). Evolution. In *Dung Beetle Ecology* (pp. 51–67), edited by I. Hanski and Y. Cambefort. Princeton University Press, Princeton, NJ.

Cameron, B. (1967). Oldest carnivorous gastropod borings found in Trentonian (Middle Ordovician) brachiopods. *Journal of Paleontology*, 41, 147–150.

Campbell, K.A. (1992). Recognition of a Mio–Pliocene cold seep setting from the northeast Pacific convergent margin, Washington, USA. *Palaios*, 7, 422–433.

Campbell, K.A. (2006). Hydrocarbon seep and hydrothermal vent paleoenvironments and paleontology: past developments and future research directions. *Palaeogeography, Palaeoclimatology, Palaeoecology*, 232, 362–407.

Campbell, K.A. and Bottjer, D.J. (1994). Brachiopods and chemosymbiotic bivalves in Phanerozoic cold seeps and hydrothermal vents. *Geological Society of America Abstracts with Programs*, 26, A53–A54.

Campbell, K.A. and Bottjer, D.J. (1995a). Brachiopods and chemosymbiotic bivalves in Phanerozoic hydrothermal vent and cold seep environments. *Geology*, 23, 321–324.

Campbell, K.A. and Bottjer, D.J. (1995b). *Peregrinella*: an Early Cretaceous cold-seep-restricted brachiopod. *Paleobiology*, 21, 461–478.

Campbell, K.A., Carlson, C., and Bottjer, D.J. (1993). Fossil cold seep limestones and associated chemosymbiotic macroinvertebrate faunas, Jurassic–Cretaceous Great Valley Group, California. In *Advances in the Sedimentary Geology of the Great Valley Group, Sacramento Valley, California* (pp. 37–50), edited by S.A. Graham and D.R. Lowe. Pacific Section, Society of Economic Paleontologists and Mineralogists, Los Angeles, CA.

Campbell, K.A., Peterson, D.E., and Alfaro, A.C. (2008). Two new species of *Retiskenea?* (Gastropoda: Neomphalidae) from Lower Cretaceous hydrocarbon-seep carbonates of northern California. *Journal of Paleontology*, 82, 140–153.

Campbell, L.D. (1993). *Pliocene Mollusks from the Yorktown and Chowan River Formation in Virginia.* Mineral Resources Department, Virginia Department of Mines, Minerals, and Energy, Charlottesville, 259 pp.

Canning, E.U. (1956). A new eugregarine of locusts, *Gregarina garnhami* n. sp., parasitic in *Schistocerca gregaria* Forsk. *Journal of Protozoology*, 3, 50–62.

Cannon, L.R.G. (1986). *Turbellaria of the World: A Guide to Families and Genera.* Poly-Graphics, Brisbane, 125 pp.

Cano, R.J. and Borucki, M.K. (1995). Revival and identification of bacterial spores in 25–40-million-year-old Dominican amber. *Science*, 268, 1060–1064.

Cano, R.J., Poinar, H.N., and Poinar, Jr., G.O. (1992a). Isolation and partial characterization of DNA from the bee *Proplebeia dominicana* (Apidae: Hymenoptera) in 25–40 million year old amber. *Medical Science Research*, 20, 249–251.

Cano, R.J., Poinar, H.N., Roubik, D.W., and Poinar, Jr., G.O. (1992b). Enzymatic amplification and nucleotide sequencing of portions of the 18s rRNA gene of the bee *Problebeia dominica* (Apidae: Hymenoptera) isolated from 25–40 million year old Dominican amber. *Medical Science Research*, 20, 619–622.

Cano, R.J., Poinar, H.N., Pieniazek, N.J., Acra, A., and Poinar, Jr., G.O. (1993). Amplification and sequencing of DNA from a 120–135-million-year-old weevil. *Nature*, 363, 536–538.

Cano, R.J., Borucki, M.K., Higby-Schweitzer, M., Poinar, H.N., Poinar, Jr., G.O., and Pollard, K.J. (1994). *Bacillus* DNA in fossil bees: an ancient symbiosis? *Journal of Applied and Environmental Microbiology*, 60, 2164–2167.

Cantrill, D.J. and Poole, I. (2005). A new Eocene *Araucaria* from Seymour Island, Antarctica: evidence from growth form and bark morphology. *Alcheringa*, 29, 341–350.

Capasso, L. (1998). Cranial pathology of *Ursus spelaeus* Rosenmüller & Heinroth from Chateau Pignon, Basque Territories (Spain). *International Journal of Osteoarchaeology*, 8, 107–115.

Capasso, L. and Caramiello, S. (1996). A healed injury in a Cambrian trilobite. *Journal of Paleopathology*, 8, 181–184.

Capasso, L., Bacchia, F., Rabottini, N., Rothschild, B.M., and Mariani-Costantini, R. (1996). Fossil evidence of intraspecific aggressive behavior of Devonian giant fishes. *Journal of Paleopathology*, 8, 153–160.

Cappetta, H. (1987). Chondrichthyes. II. Mesozoic and Cenozoic Elasmobranchii. In *Handbook of Paleoichthyology*, Vol. 3B (pp. 1–193), edited by H.-P. Schultze. Gustav Fischer Verlag, Stuttgart.

Cappetta, H. (1988). Les Torpédiniformes (Neoselachii, Batomorphii) des phosphates du Maroc. Observations sur la denture des genres actuels. *Tertiary Research*, 10, 21–32.

Carlson, K.J. (1968). The skull morphology and estivation burrows of the Permian lungfish *Gnathorhiza serrata. Journal of Geology*, 76, 641–663.

Carnevale, G. (2003). Redescription and phylogenetic relationships of *Argyropelecus logearti* (Teleostei: Stomiiformes: Sternoptychidae), with a brief review of fossil *Argyropelecus. Rivista Italiana di Paleontologia e Stratigrafia*, 109, 63–76.

Carpenter, F.M. (1992). Systematic descriptions of the superclass Hexapoda. In *Treatise on Invertebrate Paleontology.* Part R. *Arthropoda 4, Superclass Hexapoda* (pp. 1–655), edited by R.L. Kaesler. Geological Society of America/University Press of Kansas, Lawrence.

Carpenter, F.M. and Burnham, L. (1985). The geological record of insects. *Annual Review of Earth and Planetary Science*, 13, 297–314.

Carpenter, F.M. and Rasnitsyn, A.P. (1990). Mesozoic Vespidae. *Psyche*, 97, 1–20.

Carpenter, K. and Lindsey, D. (1980). The dentary of *Brachychampsa Montana* Gilmore (Alligatorinae; Crocodylidae), a Late Cretaceous turtle-eating alligator. *Journal of Paleontology*, 54, 1213–1217.

Carpenter, K., Hirsch, K.F., and Horner, J.R., Eds. (1994). *Dinosaur Eggs and Babies.* Cambridge University Press, Cambridge, U.K., 372 pp.

Carpenter, K., Sanders, F., McWhinney, L.A., and Wood, L. (2005). Evidence for predator–prey relationships: examples for *Allosaurus* and *Stegosaurus*. In *The Carnivorous Dinosaurs* (pp. 325–350), edited by K. Carpenter. Indiana University Press, Bloomington.

Carpenter, S.J., Erickson, J.M., and Holland, Jr., F.D. (2003). Migration of a Late Cretaceous fish. *Nature*, 423, 70–74.

Carranza, S., Baguñá, J., and Riutort, M. (1997). Are the Platyhelminthes a monophyletic primitive group? An assessment using 18S rDNA sequences. *Molecular Biology and Evolution*, 14, 485–497.

Carrera, M.G. (1999). Operculos de gastropodos en el Ordovicico Inferior de la Precordillera Argentina. *Ameghiniana*, 36, 91–94.

Carrera, M.G. (2000). Epizoan–sponge interactions in the Early Ordovician of the Argentine Precordillera. *Palaios*, 15, 261–272.

Carriker, M.R. and Yochelson, E.L. (1968). Recent gastropod boreholes and Ordovician cylindrical boreholes. *U.S. Geological Survey Professional Paper*, 593B, 1–26.

Carroll, E.J. (1962). Mesozoic fossil insects from Koonwarra, south Gippsland, Victoria. *Australian Journal of Science*, 25, 264–265.

Carroll, R.L. (1978). Permo-Triassic "lizards" from the Karoo System. Part II. A gliding reptile from the Upper Permian of Madagascar. *Palaeontographica Africana*, 21, 143–159.

Carroll, R.L. (1988). *Vertebrate Paleontology and Evolution.* W.H. Freeman, New York, 698 pp.

Carter, J.G. and Stanley, Jr., G.D. (2004). Late Triassic gastrochaenid and lithophaginid borings (Mollusca: Bivalvia) from Nevada (USA) and Austria. *Journal of Paleontology*, 78, 230–234.

Cary, S.C., Warren, W., Anderson, E., and Giovannoni, S.J. (1993). Identification and localization of bacterial endosymbionts in hydrothermal vent taxa with symbiont-specific polymerase chain reaction amplification and *in situ* hybridization techniques. *Molecular Marine Biology and Biotechnology*, 2(1), 51–62.

Castro, M.P. (1997). Huellas de actividad biologica sobre plantas del Estefaniense Superior de La Magdalena (Leon, España). *Revista Española de Paleontologia*, 12, 52–66.

Castro, P. (1976). Brachyuran crabs symbiotic with scleractinian corals: a review of their biology. *Micronesica*, 12, 99–110.

Cavin, L. (1999). Occurrence of a juvenile teleost, *Enchodus* sp., in a fish gut content from the Upper Cretaceous of Goulmima, Morocco. *Palaeontology*, 60, 57–72.

Ceranka, T. and Złotnik, M. (2003). Traces of cassid snail predation upon the echinoids from the Middle Miocene of Poland. *Acta Palaeontologica Polonica*, 48, 491–496.

Chaline, J. and Sevilla, P. (1990). Phyletic gradualism and developmental heterochronies in a European Plio/Pleistocene *Mimomys* lineage (Arvicolidae, Rodentia). In *International Symposium on the Evolution, Phylogeny, and Biostratigraphy of Arvicolids* (pp. 85–98), edited by O. Fejfar and W.D. Heinrich. Geological Survey, Prague.

Chaloner, W.G., Forey, P.L., Gardiner, B.G., Hill, A.J., and Young, V.T. (1980). Devonian fish and plants from the Bokkeveld Series of South Africa. *Annals of the South African Museum*, 81, 127–157.

Chaloner, W.G., Scott, A.C., and Stephenson, J. (1991). Fossil evidence for plant–arthropod interactions in the Palaeozoic and Mesozoic. *Philosophical Transactions of the Royal Society of London B*, 333(1267), 177–185.

Chandler, A.C. and Read, C.P. (1961). *Introduction of Parasitology*. John Wiley & Sons, New York, 822 pp.

Chaney, R.W. (1947). Tertiary centers and migration routes. *Ecological Monographs*, 17, 141–148.

Chang, M.-M., Chen, P.-J., Wang, Y.-Q., and Wang, Y., Eds. (2003). *The Jehol Biota*. Shanghai Scientific & Technical Publishers, Shanghai, China, 208 pp.

Charig, A.J. and Milner, A.C. (1997). *Baryonyx walkeri*, a fish-eating dinosaur from the Wealden of Surrey. *Bulletin of the Natural History Museum London (Geology)*, 53(1), 11–70.

Chatterton, B.D.E. (1971). Taxonomy and ontogeny of Siluro–Devonian trilobites from Yass, New South Wales. *Palaeontographica*, 137, 1–108.

Chatterton, B.D.E. and Campbell, M. (1993). Enrolling in trilobites: a review and some new characters. *Memoirs of the Association of Australasian Palaeontologists*, 15, 103–123.

Chatterton, B.D.E. and Whitehead, H.L. (1987). Predatory borings in the inarticulate brachiopod *Artiotreta* from the Silurian of Oklahoma. *Lethaia*, 20, 67–74.

Chatterton, B.D.E., Collins, D.H., and Ludvigsen, R. (2003). Cryptic behaviour in trilobites: Cambrian and Silurian examples from Canada, and other related occurrences. *Special Papers in Palaeontology*, 70, 157–173.

Chen, C.P. and Chen, B.Y. (1994). Diverticulum sand in a miniature sand dollar *Sinaechinocyamus mai*. In *Echinoderms Through Time* (p. 605), edited by B. David, A. Guille, J.-P. Feral, and M. Roux. Balkema, Rotterdam.

Chen, P.-J., Dong, Z.-M., and Zhen, S.-N. (1998). An exceptionally well preserved theropod dinosaur from the Yixian Formation of China. *Nature*, 391, 147–152.

Chen, X. (2009). Graptolite depth zonation. *Acta Palaeontologica Sinica*, 29, 507–526.

Cheng, T.C. (1964). *The Biology of Animal Parasites*. W.B. Saunders, Philadelphia, PA, 727 pp.

Cheng, Y.-N., Wu, X.-C., and Ji, Q. (2004). Triassic marine reptiles gave birth to young. *Nature*, 432, 383–386.

Cherns, L., Wheeley, J.R., and Karig, L. (2006). Tunneling trilobites: habitual infaunalism in an Ordovician carbonate seafloor. *Geology*, 34, 657–660.

Chevalier, A. and Chesnais, F. (1941). Nouvelles observations sur les domaties des feuilles des Juglandacées. *Comptes Rendus de l'Academie des Sciences, Paris*, 213, 389–392, 597–601.

Chiappe, L.M., Coria, R.A., Dingus, L., Jackson, F., Chinsamy, A., and Fox, M. (1998). Sauropod dinosaur embryos from the Late Cretaceous of Patagonia. *Nature*, 396, 258–261.

Chiappe, L.M., Kellner, A.W.A., Rivarola, D., Davila, S., and Fox, M. (2000). Cranial morphology of *Pterodaustro quinazui* (Pterosauria Pterodactyloidea) from the Lower Cretaceous of Argentina. *Contributions in Science*, 483, 1–19.

Chiappe, L.M., Codorniú, L., Grellet-Tinner, G., and Rivarola, D. (2004a). Argentinian unhatched pterosaur fossil. *Nature*, 432, 571.

Chiappe, L.M., Schmitt, J.G., Jackson, F.D., Garrida, A., Dingus, L., and Grettel-Tinner, G. (2004b). Nest structure of sauropods: sedimentary criteria for recognition of dinosaur nesting traces. *Palaios*, 19, 89–95.

Chin, K. and Gill, B.D. (1996). Dinosaurs, dung beetles, and conifers: participants in a Cretaceous food web. *Palaios*, 11, 280–285.

Chin, K., Tokartkm, T.T., Erickson, G.M., and Calk, L.C. (1998). A king-sized theropod coprolite. *Nature*, 393, 680–682.

Chinzei, K. (1978). Neogene molluscan faunas in the Japanese islands, an ecologic and zoogeographic synthesis. *The Veliger*, 21, 155–170.

Chinzei, K. (1984). Ecological parallelism in shallow water marine benthic associations of Neogene molluscan faunas of Japan. *Geobios Mémoire Spécial*, 8, 135–143.

Chipman, A.D. and Tschernov, E. (2002). Ancient ontogenies: larval development of the lower Cretaceous anuran *Shomranella jordanica* (Amphibia: Pipoidea). *Evolutionary Development*, 4, 86–95.

Chlupač, I. (1995). Lower Cambrian arthropods from the Paseky Shale (Barrandian area, Czech Republic). *Journal of the Czech Geological Society*, 40, 9–36.

Choe, J.C. and Crespi, B.J. (1997). *The Evolution of Social Behavior in Insects and Arachnids*. Cambridge University Press, Cambridge, U.K., 541 pp.

Christiansen, E.F. and Brock, V. (1981). *Biologforbundets blad Kaskelot*, 47, 33.

Christiansen, K. (1964). Bionomics of Collembola. *Annual Review of Entomology*, 9, 147–178.

Christiansen, P. (1996). The evidence for and implications of gastro-liths in sauropods (Dinosauria, Sauropoda). *Gaia*, 12, 1–7.

Chumakov, N.M. (1998). Stones scattered in Cretaceous deposits of South England. *Lithology and Mineral Resources*, 33, 313–326.

Cicimurri, D.J. and Everhart, M.J. (2001). An elasmosaur with stomach contents and gastroliths from the Pierre Shale (Late Cretaceous) of Kansas. *Transactions of the Kansas Academy of Science*, 104, 129–143.

Cigala-Fulgosi, F. (1990). Predation (or possible scavenging) by a great white shark on an extinct species of bottlenosed dolphin in the Italian Pliocene. *Tertiary Research*, 12, 17–36.

Cione, A.L. and Medina, F.A. (1987). A record of *Notidanodon pectinatus* (Chondrichthyes, Hexanchiformes) in the Upper Cretaceous of the Antarctic Peninsula. *Mesozoic Research*, 1, 79–88.

Clapp, W.F. and Kenk, R. (1967). *Marine Borers: An Annotated Bibliography*, ACR-74. Office of Naval Research, Department of the Navy, Washington, D.C., 1136 pp.

Clark, S.H.B., Poole, F.G., and Wang, Z.-C. (2004). Comparison of some sediment-hosted, stratiform barite deposits in China, the United States, and India. *Ore Geology Reviews*, 24, 85–101.

Clark, T.H. and Usher, J.L. (1948). The sense of *Climactichnites*. *American Journal of Science*, 246, 251–253.

Clark Sellick, J.T. (1994). Phasmida (stick insect) eggs from the Eocene of Oregon. *Palaeontology*, 37, 913–921.

Clarke, J.M. (1921). *Organic Dependence and Disease: Their Origin and Significance*. Yale University Press, New Haven, CT, 113 pp.

Clarke, M.R. and Fitch, J.E. (1979). Statoliths of Cenozoic teuthoid cephalopods from North America. *Palaeontology*, 22, 479–511.

Clarkson, E.N.K. (1997). The eye: morphology, function and evolution. In *Treatise on Invertebrate Paleontology*. Part O. *Arthropoda 1, Trilobita, Revised* (pp. 114–132), edited by R.L. Kaesler. Geological Society of America/University Press of Kansas, Lawrence.

Claus, H. (1958). Ein Neuer Splintkäfer (*Scolytus tiburtinus* n. sp.) aus dem Diluvialtravertin Nordwestthüringens (Coleoptera: Scolytidae). *Beiträge zur Entomologie*, 8, 710–716.

Clemen, A.J. (1956). Caries in the South African ape-man: some examples of undoubted pathological authenticity believed to be 800,000 years old. *British Dental Journal*, 101, 4–7.

Clopton, R.E. (2000). Order Eugregarinorida Leger (1900). In *An Illustrated Guide to the Protozoa*, Vol. I, 2nd ed. (pp. 205–288), edited by J.J. Lee, G.F. Leedale, and P. Bradbury. Allen Press, Lawrence, KS.

CoBabe, E.A., Chamberlain, K.R., Ivie, M.A., and Giersch, J.J. (2002). A new insect and plant Lagerstätte from Tertiary lake deposits along the Canyon Ferry Reservoir, southwestern Montana. *Rocky Mountain Geologist*, 37, 13–30.

Cockburn, T.A. (1971). Infectious diseases in ancient populations. *Current Anthropology*, 12, 45–62.

Cockburn, T.A. and Cockburn, E. (1980). *Mummies, Disease and Ancient Cultures*. Cambridge University Press, Cambridge, U.K., 340 pp.

Cockerell, T.D.A. (1917). Arthropods in Burmese amber. *Psyche*, 24, 40–42.

Cockerell, T.D.A. (1921). Fossil arthropods in the British Museum, Part VII. *Annals and Magazine of Natural History*, 8, 541–545.

Cockerell, T.D.A. (1922). Fossils in Burmese amber. *Nature*, 109, 713–714.

Codez, J. and de Saint-Seine, R. (1958). Révision des Cirripèdes Acrothoraciques Fossiles. *Bulletin de la Société Géologique de France*, 7, 699–719.

Codington, L.A. (1992). Fossil spider web from the Eocene of western Colorado. *Geological Society of America Abstracts with Programs*, 24, A344.

Coe, M.J., Dilcher, D.L., Farlow, J.O., Jarzen, D.M., and Russell, D.A. (1987). Dinosaurs and land plants. In *The Origins of Angiosperms and Their Biological Consequences* (pp. 225–258), edited by E.M. Friis, W.G. Chaloner, and P.R. Crane. Cambridge University Press, Cambridge, U.K.

Cohen, M.N. and Crane-Kramer, G. (2003). The state and future of paleoepidemiology. In *Emerging Pathogens: The Archaeology, Ecology, and Evolution of Infectious Disease* (pp. 79–91), edited by C. Greenblatt and M. Spigel. Oxford University Press, London.

Colbert, E.H. (1969). *Evolution of the Vertebrates*. John Wiley & Sons, New York, 535 pp.

Colbert, E.H. (1989). *The Triassic Dinosaur* Coelophysi, Bulletin Series 57. Museum of Northern Arizona Press, Flagstaff, 160 pp.

Collins, D. and Rudkin, D.M. (1981). *Priscanserminarus barnetti*, a probable lepadomorph barnacle from the Middle Cambrian Burgess Shale of British Columbia. *Journal of Paleontology*, 55, 1006–1015.

Collins, J.S.H. and Jakobsen, S.L. (2003). New crabs (Crustacea, Decapoda) from the Eocene (Ypresian/Lutetian) Lillebælt Clay Formation of Jutland, Denmark. *Bulletin of the Mizunami Fossil Museum*, 30, 63–96.

Collins, J.S.H. and Rasmussen, H.W. (1992). Upper Cretaceous–Lower Tertiary decapod crustaceans from West Greenland. *Grønlands Geologiske Undersøgelse Bulletin*, 162, 1–46.

Collins, J.S.H., Lee, C., and Noad, J. (2003). Miocene and Pleistocene crabs (Crustacea, Decapoda) from Sabah and Sarawak. *Journal of Systematic Palaeontology*, 1, 187–226.

Collins, M.J. and Child-Gearney, A.M. (2001). Ancient proteins. In *Palaeobiology II* (pp. 245–247), edited by D.E.G. Briggs and P. Crowther. Blackwell Scientific, Oxford.

Collinson, M.E. (1989). The fossil record of the Moraceae. In *Evolution, Systematics, and Fossil History of the Hamamelidae*, Vol. 2 (pp. 319–339), edited by P.R. Crane and S. Blackmore. Clarendon Press, London.

Collinson, M.E. (1990). Plant evolution and ecology during the Early Cainozoic diversification. *Advances in Botanical Research*, 17, 1–98.

Collinson, M.E. (1999a). Evolution of angiosperm fruit and seed morphology and associated functional biology: status in the Late Cretaceous and Palaeogene. In *The Evolution of Plant Architecture* (pp. 331–357), edited by M.H. Kurmann and A.R. Hemsley. Royal Botanic Garden, Kew.

Collinson, M.E. (1999b). Plants and animal diets. In *Fossil Plants and Spores: Modern Techniques* (pp. 316–319), edited by T.P. Jones and N.P. Rowe. Geological Society, London.

Collinson, M.E. and Hooker, J.J. (2000). Gnaw marks on Eocene seeds: evidence for early rodent behaviour. *Palaeogeography, Palaeoclimatology, Palaeoecology*, 157, 127–149.

Colwell, R.K. (1983). *Rhinoseius colwelli* (Acaro Floral del Colibri, Totolate Floral de Colibri, Hummingbird Flower Mite). In *Costa Rican Natural History* (pp. 767–768), edited by D.H. Janzen. University of Chicago Press, Chicago, IL.

Conroy, G.C. (1990). *Primate Evolution*. W.W. Norton & Company, New York, 492 pp.

Comte, B. and Vianey-Liaud, M. (1989). Eomyidae (Rodentia) de l'Oligocène d'Europe occidentale. *Palaeontographica*, A209, 33–91.

Conway-Morris, S. and Bengtson, S. (1994). Cambrian predators: possible evidence from boreholes. *Journal of Paleontology*, 68, 1–23.

Conway-Morris, S. and Jenkins, R.J.E. (1985). Healed injuries in Early Cambrian trilobites from South Australia. *Alcheringa*, 9, 167–177.

Cook, E.F. (1990). Fossil scatopsidae from Dominican amber. In *Evolutionary Paleobiology of Behavior and Coevolution* (pp. 389–390), edited by A.J. Boucot. Elsevier, Amsterdam.

Cook, T.L. and Stokes, D.S. (1995). Biogeological mineralization in deep-sea hydrothermal deposits. *Science*, 267, 1975–1979.

Coombs, M.C. (1975). Sexual dimorphism in chalicotheres (Mammalia, Perissodactylia). *Systematic Zoology*, 24, 55–62.

Coombs, S., Janssen, J., and Webb, J.F. (1988). Diversity of lateral line systems: evolutionary and functional considerations. In *Sensory Biology of Aquatic Animals* (pp. 553–593), edited by J. Atema, R.R. Fay, A.N. Popper, and W. Tavolga. Springer-Verlag, Heidelberg.

Coombs, S., Görner, P., and Munz, H., Eds. (1989). *The Mechanosensory Lateral Line: Neurobiology and Evolution*. Springer-Verlag, Heidelberg, 724 pp.

Coombs, Jr., W.P. (1989). Modern analogs for dinosaur nesting and parental behavior. In *Paleobiology of the Dinosaurs* (pp. 21–53), edited by J.O. Farlow. Geological Society of America, Boulder, CO.

Cooper, G.A. (1956). New Pennsylvanian brachiopods. *Journal of Paleontology*, 30, 521–530.

Cooper, G.A. (1975). Brachiopods from West African waters with examples of collateral evolution. *Journal of Paleontology*, 49, 911–927.

Cooper, G.A. and Grant, R.E. (1972). Permian Brachiopods of West Texas, Parts I and II. *Smithsonian Contributions to Paleobiology*, 14, 1–793.

Cooper, G.A. and Grant, R.E. (1975). Permian Brachiopods of West Texas, Part III. *Smithsonian Contributions to Paleobiology*, 19, 795–1921.

Cooper, M.R. (1986). A new species of the brachiopod genus *Agulhasia* (Terebratulidinae: Chlidonophoridae) from the Uloa Formation. *Suid-Afrikaanse Tydskrif vir Wetenskap*, 84, 35–38.

Coram, R.A. and Jarzembowski, E.A. (2002). Diversity and ecology of fossil insects in the Dorset Purbeck succession, southern England. *Special Papers in Palaeontology*, 68, 257–268.

Coria, R.A. (1994). On a monospecific assemblage of sauropod dinosaurs from Patagonia: implications for gregarious behavior. *Gaia*, 10, 209–213.

Corral, J.C., Pereda Suberbiola, X., and Bardet, N. (2004). Marcas de ataque atribuidas a un selacio en una vertebra de mosasaurio del Cretácico superior de Álava (region Vasco-Cantábrica). *Revista Española de Paleontología*, 19, 23–32.

Corsi, A., Landini, W., and Sorbini, C. (1999). A new ichthyofauna from the Upper Miocene of Ca'Matterella (Ravenna, Italy); paleoecologiccal and paleobiogeographical considerations. *Studi i Richerche sul giacimenti Terziari di Bolca*, 8, 59–76.

Cosma, T.N. and Baumiller, T.K. (2005). A trace fossil on a Silurian bivalve: evidence of predatory boring? *Ichnos*, 12, 135–139.

Cossey, P.J. and Mundy, D.J.C. (1990). *Tetrataxis*: a loosely attached limpet-like foraminifer from the Upper Palaeozoic. *Lethaia*, 23, 311–322.

Cotton, T.J. (2001). The phylogeny and systematics of blind Cambrian ptychoparioid trilobites. *Palaeontology*, 44, 167–207.

Cotton, W.D., Hunt, A.P., and Cotton, J.E. (1996). Cruising the dinosaur freeway. In *Sixth North American Paleontological Convention Abstracts of Papers* (p. 88), edited by J.E. Repetski, Special Publ. No. 8. Paleontological Society, Washington, D.C.

Courville, C.B. (1950). Cranial injuries in prehistoric man, with particular reference to the Neanderthals. *Bulletin of the Los Angeles Neurological Society*, 15, 1–21.

Courville, C.B. (1953). Cranial injuries in prehistoric animals. *Bulletin of the Los Angeles Neurological Society*, 18, 117–126.

Cox, L.R. (1969). Transposed hinges. In *Treatise on Invertebrate Paleontology*. Part N. *Mollusca 6, Bivalvia* (pp. 56–58), edited by R.C. Moore. Geological Society of America/University Press of Kansas, Lawrence.

Crampton, J.S. (1990). A new species of Late Cretaceous wood-boring bivalve from New Zealand. *Palaeontology*, 33, 981–992.

Crampton, J.S. (1996). *Inoceramid Bivalves from the Late Cretaceous of New Zealand*, Monograph 14. Institute of Geological and Nuclear Sciences, Lower Hutt, New Zealand, 188 pp.

Crampton, J.S. and Gale, A.S. (2005). A plastic boomerang: speciation and intraspecific evolution in the Cretaceous bivalve *Actinoceramus*. *Paleobiology*, 31, 559–577.

Crampton, J.S. and Gale, A.S. (2009). Taxonomy and biostratigraphy of the Late Albian *Actinoceramus sulcatus* lineage (Early Cretaceous Bivalvia, Inoceramidae). *Journal of Paleontology*, 83, 89–109.

Crane, Jr., J.M. (1966). Late Tertiary radiation of viperfishes (Chauliodontidae) based on a comparison of Recent and Miocene species. *Natural History Museum of Los Angeles County Contributions in Science*, 115, 1–37.

Crawford, R.S., Casadio, S., Feldmann, R.M., Griffin, M., Parras, A., and Schweitzer, C.E. (2008). Mass mortality of fossil decapods within the Monte León Formation (Early Miocene), southern Argentina: victims of Andean volcanism. *Annals of Carnegie Museum*, 77(2), 259–287.

Crepet, W.L. (1974). Investigations of North American cycadeoids: the reproductive biology of *Cycadeoidea*. *Palaeontographica B*, 148, 144–169.

Cressey, R. and Boxshall, G. (1989). *Kabatarina pattersoni*: a fossil parasitic copepod (Dichelesthiidae) from a Lower Cretaceous fish. *Micropaleontology*, 35, 150–167.

Crompton, R.H. and Pataky, T.C. (2009). Stepping out. *Science*, 323, 1174–1175.

Crônier, C. and Fortey, R.A. (2006). Morphology and ontogeny of an Early Devonian Phacopid trilobite with reduced sight from Southern Thailand. *Journal of Paleontology*, 80, 529–536.

Cronin, T.M. and Raymo, M.E. (1997). Orbital forcing of deep-sea benthic species diversity. *Nature*, 385, 624–627.

Cronin, T.W. (1988). Vision on aquatic invertebrates. In *Sensory Biology of Aquatic Animals* (pp. 403–418), edited by J. Atema, R.R. Fay, A.N. Popper, and W.N. Tavolga. Springer-Verlag, Heidelberg.

Crookall, R. (1928). The genus *Fayolia*. *The Naturalist*, 325–332.

Crookall, R. (1929). Further morphological studies in *Palaeoxyris*, etc. *Report of the Geological Survey of Great Britain and the Museum of Practical Geology, Summary of Progress*, 3, 8–36.

Cross, N.E. and Rose, E.P.E. (1994). Predation of the Upper Cretaceous spatangoid echinoid *Micraster*. In *Echinoderms Through Time* (pp. 607–612), edited by B. David, A. Guille, J.-P. Feral, and M. Roux. Balkema, Rotterdam.

Crossley, R. (1984). Fossil termite mounds associated with stone artifacts in Malawi, Central Africa. *Palaeoecology of Africa and of the Surrounding Islands and Antarctica*, 16, 397–401.

Cruikshank, R.D. and Ko, K. (2003). Geology of an amber locality in the Hukawng Valley, northern Myanmar. *Journal of Asian Earth Sciences*, 21, 441–455.

Čtyroký, P. (1969). The family Pectinidae in the Burdigalian of Czechoslovakia. *Sbornik Geologicky ved Paleontologia*, 10, 7–66.

Cuesta Ruíz-Colmenares, M.A., Jiménez Fuentes, E., and Pérez Pérez, P.J. (2004). Un caso de hipodoncia en un lofiodäntido (Perissodactyla, Mammalia) del Eocene Medio de la Cuenca del Duero (Castilla y Léon, España). Interpretación a luz de la agenesia dentaria humana. *Revista Española de Paleontología*, 19, 145–150.

Culver, S.J. and Lipps, J.H. (2003). Predation on and by Foraminifera. In *Predator–Prey Interactions in the Fossil Record* (pp. 7–32), edited by P.H. Kelley, M. Kowalewski, and T.A. Hansen. Kluwer Academic, New York.

Currah, R.S. and Stockey, R.A. (1991). A fossil smut fungus from the anthers of an Eocene angiosperm. *Nature*, 350, 698–699.

Currah, R.S., Stockey, R.A., and LePage, B.A. (1998). An Eocene tar spot on a fossil palm and its fungal hyperparasite. *Mycologia*, 90, 667–673.

Currie, C.R., Scott, J.A., Summerbell, R.C., and Malloch, D. (1999). Fungus-growing ants use antibiotic-producing bacteria to control garden parasites. *Nature*, 398, 701–704.

Currie, D.C. and Grimaldi, D. (2000). A new black fly (Diptera: Simuliidae) genus from mid Cretaceous (Turonian) amber of New Jersey. In *Studies on Fossils in Amber with Particular Reference to the Cretaceous of New Jersey* (pp. 473–485), edited by D. Grimaldi. Backhuys, Leiden, Germany.

Currie, P.J. (1980). Mesozoic vertebrate life in Alberta and British Columbia. *Mesozoic Vertebrate Life*, 1, 27–40.

Currie, P.J. and Dodson, P. (1984). Mass death of a herd of ceratopsian dinosaurs. In *Third Symposium on Mesozoic Terrestrial Ecosystems* (pp. 61–67), edited by W.E. Reif and F. Westphal. Attempto Verlag, Tübingen.

Currie, P.J. and Jacobsen, A.R. (1995). An azhdarchid pterosaur eaten by a velociraptorine theropod. *Canadian Journal of Earth Sciences*, 32, 922–925.

Currie, P.J., Koppelhus, E.B., and Fazal Muhammad, A. (1995). "Stomach" contents of a hadrosaur from the Dinosaur Park Formation (Campanian, Upper Cretaceous) of Alberta, Canada. In *Sixth Symposium on Mesozoic Terrestrial Ecosystems and Biota* (pp. 111–114), edited by A. Sun and Y. Wang. China Ocean Press, Beijing.

Currie, P.J., Koppelhus, E.B., Shugar, M.A., and Wright, J.L., Eds. (2004). *Feathered Dragons: Studies on the Transition from Dinosaurs to Birds*. Indiana University Press, Bloomington.

Cvancara, A.M. (1970). Teredinid (Bivalvia) pallets from the Palaeocene of North America. *Palaeontology*, 13, 619–622.

Dacqué, E. (1921). *Vergleichende biologische Formenkunde der fossilem niederen Tiere*. Geb. Borntraeger, Berlin, 777 pp.

Dahlström, A. and Brost, L. (1996). *The Amber Book*. Geosciences Press, Tucson, AZ, 134 pp.

Dal Sasso, C. and Pinna, G. (1996). *Besanosaurus leptorhynchus* n. gen, n. sp., a new shastosaurid ichthyosaur from the Middle Triassic of Besano (Lombardy, N. Italy). *Paleontologia Lombarda*, 4, 3–23.

Daley, A.C. (2008). Statistical analysis of mixed-motive shell borings in Ordovician, Silurian, and Devonian brachiopods from northern and eastern Canada. *Canadian Journal of Earth Sciences*, 45, 213–229.

Daley, G.M., Ostrowski, S., and Geary, D.H. (2007). Paleo-environmentally correlated differences in a classic predator–prey system: the bivalve *Chione elevata* and its gastropod predators. *Palaios*, 22, 166–173.

Dalla Vecchia, F.M., Muscio, G., and Wild, R. (1988). Pterosaur remains in a gastric pellet from the Upper Triassic (Norian) of Rio Seazza Valley (Udine, Italy). *Gortania*, 10, 121–132.

Daly, H.V., Doyen, A.H., and Purcell, A.H., Eds. (1998). *Introduction to Insect Biology and Diversity*, 2nd ed. Oxford University Press, London.

Daly, M. and Wilson, M. (1988a). Evolutionary social psychology and family homicide. *Science*, 242, 519–524.

Daly, M. and Wilson, M. (1988b). *Homicide*. Aldine de Gruyter, New York, 328 pp.

Damiani, R., Modesto, S., Yates, A., and Neveling, J. (2003). Earliest evidence of cynodont burrowing. *Proceedings of the Royal Society London B*, 270, 1747–1751.

Dando, P.R., Southward, A.J., Southward, E.C., Dixon, D.R., Crawford, A., and Crawford, M. (1992). Shipwrecked tube worms. *Nature*, 356, 667.

Daniel, J.F. (1934). *The Elasmobranch Fishes*. University of California Press, Berkeley, 332 pp.

Danielopol, D. and Wouters, K. (1992). Evolutionary (paleo)biology of marine interstitial Ostracoda. *Géobios*, 25, 207–211.

Danilchenko, P.G. (1946). Svetyashtsiesya ribi Semeistva *Gonostomidae* iz Tretichnikh otlozhenii Kavkaza i Krima. *Izvestiya Akademii Nauk Sooze SSR, Serie Biologicheskaiya Nauk*, 6, 639–646.

Danilchenko, P.G. (1947). Fishes of the family *Myctophidae* from the Caucasian Oligocene. *Fa parte di: Comptes Rendus (Doklady) de l'Académie des Sciences de l'URSS*, 2, 193–196.

Danilchenko, P.G. (1960). Bony fishes of the Maikop deposits of the Caucasus. *Transactions of the Paleontological Institute, Academy of Science, USSR*, 78, 1–248 (originally in Russian, translated by A. Mercado, Israel Program for Scientific Translations, Jerusalem, 1967).

Danilchenko, P.G. (1962). Ribi Dabakhanskoi Sviti Gruzii. *Paleontologicheskii Zhurnal*, 1, 111–126.

Darnell, J., Lodich, H., and Baltimore, D. (1990). *Molecular Cell Biology*. W.H. Freeman, New York, 1105 pp.

Darragh, T.A. and Kendrick, G.W. (1994). Maastrichtian Scaphopoda and Gastropoda from the Miria Formation, Carnarvon Basin, northwestern Australia. *Records of the Western Australian Museum*, 48(Suppl.), 1–76.

Darrell, J.G. and Taylor, P.D. (1989). Scleractinian symbionts of hermit crabs in the Pliocene of Florida. *Memoirs of the Association of Australasian Palaeontologists*, 8, 115–123.

Darrell, J.G. and Taylor, P.D. (1993). Macrosymbiosis in corals: a review of fossil and potentially fossilizable examples. *Courier Forschungsinstitut Senckenberg*, 164, 185–198.

Darteville, E. (1934). Les Perles Fossiles. *Journal de Conchyliologie*, 78(3), 169–195.

Darwin, C. (1851). *A Monograph on the Fossil Lepadidae*. Palaeontographical Society, London, 86 pp.

Das, S.S. (2002). Two new pleurotomariid (Gastropoda) species, including the largest *Bathrotomaria*, from the Berriasian (Early Cretaceous) of Kutch, western India. *Cretaceous Research*, 23, 99–109.

Dauphine, Y., Denys, C., and Kowalski, K. (1997). Analysis of accumulations of rodent remains: role of the chemical composition of skeletal elements. *Neues Jahrbuch für Geologie und Paläontologie*, 203, 295–315.

David, L.R. (1943). *Miocene Fishes of Southern California*, Special Paper 43. Geological Society of America, Boulder, CO, 193 pp.

David, L.R. (1946a). Use of fossil fish scales in micropaleontology. *Carnegie Institution of Washington Publication*, 551, 25–43.

David, L.R. (1946b). Some typical Upper Eocene fish scales from California. *Carnegie Institution of Washington Publication*, 551, 45–79.

David, L.R. (1956). Tertiary anacanthin fishes from California and the Pacific Northwest; their paleoecological significance. *Journal of Paleontology*, 30, 568–607.

Davidson, R.G. and Trewin, N.H. (2005). Unusual preservation of the internal organs of acanthodian and actinopterygian fish in the Middle Devonian of Scotland. *Scottish Journal of Geology*, 41, 129–134.

Davis, D.R. (1989). An exceptional fossil amber collection acquired by the Smithsonian Institution. *Proceedings of the Entomological Society of Washington*, 91, 545–550.

Davis, O.K., Mead, J.I., Martin, P.S., and Agenbroad, L.D. (1985). Riparian plants were a major component of the diet of mammoths of southern Utah. *Current Research in the Pleistocene*, 2, 81–82.

Davis, R.A., Furnish, W.M., and Glenister, B.F. (1969). Mature modification and dimorphism in Late Paleozoic ammonoids. In *Sexual Dimorphism in Fossil Metazoa and Taxonomic Implications* (pp. 101–110), edited by G.E.G. Westermann. Schweizerbart, Stuttgart.

Davis, R.A., Fraaye, R.H.B., and Holland, C.H. (2001). Trilobites within nautiloid cephalopods. *Lethaia*, 34, 37–45.

Dawson, S.D. and Gottfried, M.D. (2002). Paleopathology in a Miocene kentriodontid dolphin (Cetacea: Odontoceti). *Smithsonian Contributions to Paleobiology*, 93, 263–270.

Day, J.J., Norman, D.B., Upchurch, P., and Powell, H.P. (2002a). Dinosaur locomotion from a new trackway. *Nature*, 415, 494–495.

Day, J.J., Upchurch, P., Norman, D.B., Gale, A.S., and Powell, H.P. (2002b). Sauropod trackways, evolution, and behaviour. *Science*, 296, 1659.

Day, J.J., Norman, D.B., Gale, A.S., Upchurch, P., and Powell, H.P. (2004). A Middle Jurassic dinosaur trackway site from Oxfordshire, U.K. *Palaeontology*, 47, 319–348.

Datron, P.K. (1975). Experimental studies of algal canopy interactions in a sea otter-dominated kelp community at Amchitka Island, Alaska. *Fishery Bulletin*, 73(2), 230–237.

de Beaulieu, J.-L. and Reille, M. (1989). The transition from temperate phases to stadials in the long Upper Pleistocene sequence from Les Echets (France). *Palaeogeography, Palaeoclimatology, Palaeoecology*, 72, 147–159.

de Carvalho, M.R., Maisey, J.G., and Grande, L. (2004). Freshwater stingrays of the Green River Formation of Wyoming (Early Eocene), with the description of a new genus and species and an analysis of its phylogenetic relationships (Chondrichthyes: Myliobatiformes). *Bulletin of the American Museum of Natural History*, 284, 1–136.

de Gibert, J.M., Domeneck, R., Ekdale, A.A., and Steen, P.P. (2001). Cretaceous ray traces? An alternative interpretation for alleged dinosaur tracks of La Posa, Isona, NE Spain. *Palaios*, 16, 409–416.

de Gibert, J.M., Domènech, R., and Martinell, J. (2007). Bioerosion in shell beds from the Pliocene Roussillon Basin, France: implications for the (macro) bioerosion ichnofacies model. *Acta Palaeontologica Polonica*, 52, 783–798.

de Groot, G.E. (1964). Rugose corals from the Carboniferous of northern Palencia (Spain). *Leidse Geologische Mededelingen*, 29, 1–123.

De Klerk, W.K., Forster, C.A., Sampson, S.D., Chinsamy, A., and Ross, C.F. (2000). A new coelurosaurian dinosaur from the Early Cretaceous of South Africa. *Journal of Vertebrate Paleontology*, 20, 324–332.

De Leeuw, J.W., Frewin, N.L., Van Bergen, P.F., Sinninghe Damste, J.S., and Collinson, M.E. (1995). Organic carbon as a palaeoenvironmental indicator in the marine realm. In *Marine Palaeoenvironmental Indicators in the Marine Realm* (pp. 43–71), edited by D.W.J. Bosence and P.A. Allison, Geological Society Special Publication 83. Geological Society, London.

Dean, B. (1909). A chimaeroid egg-capsule from the North American Cretaceous: studies on fossil fishes (sharks, chimaeroids and arthrodires). *Memoir of the American Museum of Natural History*, 9, 209–287.

Dechaseaux, C. (1951). Contribution à la connaissance des esthéries fossils. *Annales de Paléontologie*, 37, 125–132.

Deelder, A.M., Miller, R.L., DeJonge, N., and Krijger, F.W. (1990). Detection of schistosome antigen in mummies. *The Lancet*, 335, 724–725.

Deeming, D.C. (2006). Ultrastructural and functional morphology of eggshells supports the idea that dinosaur eggs were incubated buried in a substrate. *Palaeontology*, 49, 171–185.

Deeming, D.C., Halstead, L.B., Manabe, M., and Unwin, D.M. (1995). An ichthyosaur embryo from the lower Lias (Jurassic: Hettangian) of Somerset, England, with comments on the reproductive biology of ichthyosaurs. In *Vertebrate Fossils and the Evolution of Scientific Concepts* (pp. 463–482), edited by W.A.S. Sarjeant. Gordon & Breach, New York.

Delance, J.H. and Emig, C.C. (2004). Drilling predation on *Gryphus vitreus* (Brachiopoda) off the French Mediterranean coasts. *Palaeogeography, Palaeoclimatology, Palaeoecology*, 208, 23–30.

Delevoryas, T. (1968). Investigations of North American cycadeoids: structure, ontogeny and phylogenetic considerations of cones of *Cycadeoidea. Palaeontographica B*, 121, 122–133.

Deline, B., Baumiller, T., Kaplan, P., Kowalewski, M., and Hoffmeister, A.P. (2003). Edge-drilling on the brachiopod *Perditocardinia* cf. *P. dubia* from the Mississippian of Missouri (USA). *Palaeogeography, Palaeoclimatology, Palaeoecology*, 201, 211–219.

Deline, D. (2008). The first evidence of predatory or parasitic drilling in stylophoran echinoderms. *Acta Palaeontologica Polonica*, 53, 739–743.

Dell, R.K. (1951a). Some animal communities of the sea bottom from Queen Charlotte Sound, New Zealand. *New Zealand Journal of Science and Technology*, B33, 19–29.

Dell, R.K. (1951b). A deep water molluscan fauna off Banks Peninsula. *Records of the Canterbury Museum*, 6, 53–60.

Denison, R.H. (1956). A review of the habitat of the earliest vertebrates. *Fieldiana Geology*, 11(8), 361–457.

Dennell, R. (1997). The world's oldest spears. *Nature*, 385, 767–768.

Denton, E.J. and Gray, J.A.B. (1988). Mechanical factors in the excitation of the lateral lines of fishes. In *Sensory Biology of Aquatic Animals* (pp. 595–617), edited by J. Atema, R.R. Fay, A.N. Popper, and W.N. Tavolga. Springer-Verlag, Heidelberg.

Denton, Jr., R.K., Dobie, J.L., and Parris, D.C. (1997). The marine crocodilian *Hyposaurus* in North America. In *Ancient Marine Reptiles* (pp. 375–397), edited by J.M. Callaway and E.L. Nicholls. Academic Press, San Diego, CA.

D'Errico, F., Henshilwood, C., Vanhaeren, M., and van Niekerek, K. (2005). *Nassarius kraussianus* shell beads from Blombos Cave: evidence for symbolic behaviour in the Middle Stone Age. *Journal of Human Evolution*, 48, 3–24.

DeSalle, R., Gatesy, J., Wheeler, W., and Grimaldi, D. (1992). DNA sequences from a fossil termite in Oligo-Miocene amber and their phylogenetic implications. *Science*, 257, 1933–1936.

DeVries, P.J. and Poinar, G.O. (1997). Ancient butterfly–ant symbiosis: direct evidence from Dominican amber. *Proceedings of the Royal Society London B*, 264, 1137–1140.

DeWindt, J.T. (1974). Callianassid burrows as indicators of subsurface beach trend, Mississippi River Delta regime. *Journal of Sedimentary Petrology*, 44, 1136–1139.

Di Geronimo, I., Rosso, A., and Sanfilippo, R. (1992). Bryozoans as sedimentary instability indicators. *Rivista Italiana di Paleontologia e Stratigrafia*, 98, 229–242.

Diéguez, C., Isidro, A., and Malgosa, A. (1996a). An introduction to zoo-paleopathology and an update on fossil phyto-paleopathology from Spain. *Journal of Paleopathology*, 8, 133–142.

Diéguez, C., Nieves-Aldrey, J.L., and Barrón, E. (1996b). Fossil galls from the Upper Miocene of La Cerdaña (Lérida, Spain). *Reviews of Palaeobotany and Palynology*, 94, 329–343.

Dieni, I. (2008). Coupling ampullinid gastropods: sexual behaviour frozen in Palaeogene deposits of northern Italy. *Rivista Italiana di Paleontologia e Stratigrafia*, 114, 505–514.

Dieppe, P. and Rogers, J.M. (1993). Skeletal paleopathology of rheumatic disorders. In *Arthritis and Allied Conditions*, 12th ed. (pp. 9–16), edited by D.J. McCarty and W.J. Koopmans. Lea & Febiger, Philadelphia, PA.

Dietl, G.P. (2003). Coevolution of a marine gastropod predator and its dangerous bivalve prey. *Biological Journal of the Linnean Society*, 80, 409–436.

Dietl, G.P. and Alexander, R.R. (2000). Post-Miocene shift in stereotypical naticid predation on confamilial prey from the Mid-Atlantic Shelf: coevolution with dangerous prey. *Palaios*, 15, 414–429.

Dietl, G.P. and Hendricks, J.R. (2006). Crab scars reveal survival advantage of left-handed snails. *Biology Letters*, 2, 439–442.

Dietl, G.P. and Kelley, P.H. (2006). Can naticid gastropod predators be identified by the holes they drill? *Ichnos*, 13, 103–108.

Dilcher, D.L., Herendeen, P.S., and Hueber, F. (1992). Fossil *Acacia* flowers with attached anther glands from Dominican Republic amber. In *Advances in Legume Systematics*. Vol. 4. *The Fossil Record* (pp. 33–42), P.S. Herendeen and D.L. Dilcher. The Royal Botanic Gardens, Kew.

Dillon, R.J. and Dillon, V.M. (2004). The gut bacteria of insects: nonpathogenic interactions. *Annual Review of Entomology*, 49, 71–92.

Distel, D.L., Baco, A.R., Chuang, E., Morrill, W., Cavanaugh, C., and Smith, C.R. (2000). Do mussels take wooden steps to deep-sea vents? *Nature*, 403, 725–726.

Dixson, A.F. (1995). Baculum length and copulatory behavior in carnivores and pinnipeds (Grand Order Ferae). *Journal of Zoology, London*, 235, 67–76.

Dlussky, G.M. (1999). The first find of Formicoidea (Hymenoptera) in the Lower Cretaceous of the Northern Hemisphere. *Paleontological Journal*, 33, 274–277.

Dlussky, G.M., Wappler, T., and Wedman, S. (2008). New middle Eocene formicid species from Germany and the evolution of weaver ants *Oecophylla*. *Acta Palaeontographica Polonica*, 53, 615–626.

Dockery III, D.T. (1996). Brooding in the late Cretaceous gastropod *Gyrodes*? *Mississippi Geology*, 17, 56–58.

Dodson, P. (1995). Dinosaur eggs: the inside scoop. *American Paleontologist*, 3(2), 1–2.

Doguzhaeva, L.A. (2002). Pre-mortem septal crowding and pathological shell wall ultrastructure of ammonite younglings from the lower Aptian of Central Volga (Russia). In *Aspects of Cretaceous Stratigraphy and Palaeobiogeography, Proceedings of the 6th International Cretaceous Symposium, Vienna, 2000* (pp. 171–184), edited by M. Wagreich. Austrian Academy of Sciences, Vienna.

Doguzhaeva, L.A., Mapes, R.H., Mutvei, H., and Pabian, R.K. (2002). The Late Carboniferous phragmocone-bearing orthoconic coleoids with ink sacs: their environment and mode of life. In *Geological Society of Australia Abstracts, First International Palaeontological Congress 6–10 July 2002*. Macquarie University, New South Wales, Australia.

Doguzhaeva, L.A., Mapes, R.H., and Mutvei, H. (2003). The shell and ink sac morphology and ulstrastructure of the Late Pennsylvanian cephalopod *Donovaniconus* and its phylogenetic significance. *Berliner Paläobiologische Abhandlungen*, 3, 61–78.

Doguzhaeva, L.A., Mapes, R.H., and Mutvei, H. (2004). Occurrence of ink in Paleozoic and Mesozoic coleoids (Cephalopoda). *Mitteilungen aus dem Geologisch–Paläontologichen Institut der Universität Hamburg*, 88, 145–156.

Doguzhaeva, L.A., Summesberger, H., Mutvei, H., and Brandstaetter, F. (2007). The mantle, ink sac, ink, arm hooks and soft body debris associated with the shells in Late Triassic coleoid cephalopod *Phragmoteuthis* from the Austrian Alps. *Palaeoworld*, 16, 272–284.

Dollo, L. (1887). Le hainosaure et les nouveaux vertébrés fossiles de Musée de Bruxelles. *Revue des Questions Scientifiques*, 21, 504–539.

Dollo, L. (1910). La Paléontologie Éthologique. *Bulletin de la Société belge de Géologie, de Paléontologie et d'Hydrologie, Mémoires*, XXIII, 377–421.

Dollo, L. (1923). Les allures des iguanodons, d'après les empreintes des pieds et de la queue. *Bulletin biologique de la France et de la Belgique*, IX, 1–12.

Dombrowski, H. (1963). Bacteria from Palaeozoic deposits. *Annals of the New York Academy of Science*, 108, 453–460.

Domenech, R., de Gibert, J.M., and Martinell, J. (2001). Ichnological features of a marine transgression: Middle Miocene rocky shores of Tarragona, Spain. *Géobios*, 34, 99–107.

Dominguez Alonso, P. and Coca Abia, M. (1998). Nidos de avispas minadoras en el Mioceno de Tegucigalpa (Honduras, Central America). *Coloquios de Paleontologia*, 49, 93–114.

Dominguez, Alonso, P., Milner, A.C., Ketcham, R.A., Cookson, M.J., and Rowe, T.B. (2004). The avian nature of the brain and inner ear of *Archaeopteryx*. *Naute*, 430, 666–669.

Domínguez-Rodrigo, M., Rayne Pickering, T., Semaw, S., and Rogers, M.J. (2005). Cutmarked bones from Pliocene archaeological sites at Gona, Afar, Ethiopia: implications for the function of the world's oldest stone tools. *Journal of Human Evolution*, 48, 109–121.

Domning, D.P. (2001). Sirenians, seagrasses, and Cenozoic ecological change in the Caribbean. *Palaeogeography, Palaeoclimatology, Palaeoecology*, 166, 27–50.

Donovan, D.T. (2006). Phragmoteuthida (Cephalopoda: Coleoidea) from the Lower Jurassic of Dorset, England. *Palaeontology*, 49, 673–684.

Donovan, D.T. and Crane, M.D. (1992). The type material of the Jurassic cephalopod *Belemnoteuthis*. *Palaeontology*, 35, 273–296.

Donovan, S.K. (1988). Palaeoecology and taphonomy of barnacles from the Plio–Pleistocene Red Crag of East Anglia. *Proceedings of the Geologists' Association*, 99, 279–289.

Donovan, S.K. and Gale, A.S. (1990). Predatory asteroids and the decline of the articulate brachiopods. *Lethaia*, 23, 77–86.

Donovan, S.K. and Harper, D.A.T. (2006). Palaeoecological implications of rare predatory borings in Pleistocene brachiopods from Antillean fore-reef palaeoenvironments. In *Proceedings of the 50th Annual Meeting of the Palaeontological Association, December 18–21, Sheffield, U.K.*

Donovan, S.K. and Hensley, C. (2006). *Gastrochaenolites* Leymerie in the Cenozoic of the Antillean Region. *Ichnos*, 13, 11–19.

Donovan, S.K. and Jagt, J.W.M. (2002). *Oichnus* Bromley borings in the irregular echinoid *Hemipneustes* Agassiz from the type Maastrichtian (Upper Cretaceous, The Netherlands and Belgium). *Ichnos*, 9, 67–74.

Donovan, S.K. and Jagt, J.W.M. (2004). Site selectivity of pits in the Chalk (Upper Cretaceous) echinoid *Echinocorys* Leske from France. *Bulletin of the Mizunami Fossil Museum*, 31, 21–24.

Donovan, S.K. and Pawson, D.L. (1997). Proximal growth of the column in bathycrinid crinoids (Echinodermata) following decapitation. *Bulletin of Marine Science*, 61, 571–579.

Donovan, S.K. and Pickerill, R.K. (2002). Pattern versus process or informative versus uninformative ichnotaxonomy: reply to Todd and Palmer. *Ichnos*, 9, 85–87.

Donovan, S.K. and Pickerill, R.K. (2004). Traces of cassid snails predation upon the echinoids from the Middle Miocene of Poland: comments on Ceranka and Złotnik (2003). *Acta Palaeontologica Polonica*, 49, 483–484.

Donovan, S.K. and Schmidt, D.A. (2001). Survival of crinoid stems following decapitation: evidence from the Ordovician and paleobiological implications. *Lethaia*, 34, 263–270.

Dörfelt, H. and Schmidt, A.R. (2006). An archaic slime mould in Baltic amber. *Palaeontology*, 49, 1013–1017.

Dorn, P. (1937). Fossile Perlen in Ostreen des Dogger Delta Schwabens nebst paläogeographischen Bemerkungen. *Neues Jahrbuch für Mineralogie, Geologie und Paläontologie*, B7, 295–304.

Dortangs, R.W., Schulp, A.S., Mulder, E.W.A., Jagt, J.W.M., Peeters, H.H.G., and de Graaf, D.T. (2002). A large new mosasaur from the Upper Cretaceous of the Netherlands. *Netherlands Journal of Geosciences/Geologie en Mijnbouw*, 81, 1–8.

Dover, C.L.V. (2000). *The Ecology of Deep-Sea Hydrothermal Vents*. Princeton University Press, Princeton, NJ, 424 pp.

Doyle, P. and MacDonald, D.I.M. (1993). Belemnite battlefields. *Lethaia*, 26, 65–80.

Doyle, P., Mather, A.E., Bennett, M.R., and Bussell, M.A. (1997). Miocene barnacle assemblages from southern Spain and their palaeoenvironmental significance. *Lethaia*, 29, 267–274.

Dreger-Jauffret, F. and Shorthouse, J.D. (1992). Diversity of gall-inducing insects and their galls. In *Biology of Insect-Induced Galls* (pp. 8–33), edited by J.D. Shorthouse and O. Rohfritsch. Oxford University Press, London.

Dreyfuss, M. (1933). Découverte de nodules phosphates à jeunes Ammonites dans le Toarcien de Creveney (Haute-Saone). *Bulletin de la Société Géologique de France, Comptes Rendu Sommaire des Séances*, 14, 224–226.

Druckenmiller, P.S. and Russell, A.P. (2008). Skeletal anatomy of an exceptionally complete specimen of a new genus of plesiosaur from the Early Cretaceous (Early Albian) of northeastern Alberta, Canada. *Palaeontographica Abteilung A*, 283(1–3), 1–33.

Druckenmiller, P.S., Daun, A.J., Skulan, J.L., and Pladziewicz, J.C. (1993). Stomach contents in the Upper Cretaceous shark *Squalicorax falcatus*. *Journal of Vertebrate Paleontology*, 13(3, Suppl.), 33A.

Drushchits, V.V. and Zevina, G.B. (1969). Novie predstaviteli usonogikh rakov iz nizhnemelovikh otlozhenii severnogo Kavkaza. *Paleontologicheski Zhurnal*, 2, 73–85.

Dubinin, V.B. (1948). The finding of pleistocene lice (Anoplura) and nematodes in the course of investigations on the bodies of fossil ground-squirrels. *Fa parte di: Comptes Rendus (Doklady) de l'Académie des Sciences de l'URSS*, 62(3), 417–420.

Ducrocq, S., Buffetaut, E., Buffetaut-Tong, H., Jaeger, J.-J., Jingkanjanasoontorn, Y., and Suteethorn, V. (1992). First flying lemur: a dermopteran from the Late Eocene of Thailand. *Palaeontology*, 35, 373–380.

Ducrocq, S., Jaeger, J.-J., and Sige, B. (1993). Un megachiroptère dans l'Eocène supérieur de Thailands. *Neues Jahrbuch für Geologie und Paläontologie–Monatsheft*, 9, 561–575.

Ducrocq, S., Chaimanee, Y., Suteethorn, V., and Jaeger, J.-J. (1995). Dental anomalies in Upper Eocene Anthracotheriidae: a possible case of inbreeding. *Lethaia*, 28, 355–360.

Duda, T.F., Kohn, A.J., and Palumbi, S.R. (2001). Origin of diverse feeding ecologies within *Conus*, a genus of venomous marine gastropods. *Biological Journal of the Linnean Society*, 73, 391–409.

Dudicourt, J.-C., Neraudeau, D., Nicolleau, P., Ceulemans, L., and Boutin, F. (2005). Une faune remarquable d'échinides marsupiaux dans le Pliocène di Vendée (Ouest de la France). *Bulletin de Société géologique de France*, 176, 545–557.

Duffin, C.J. (1993). Late Triassic shark teeth (Chondrichthyes, Elasmobranchii) from Saint-Nicolas-de-Port (north-east France). *Belgian Geological Survey, Professional Paper*, 264, 7–32.

Duffin, C.J. (1998). New shark remains from the British Rhaetian (latest Triassic). 1. The earliest basking shark. *Neues Jahrbuch für Geologie und Paläontologie–Monatsheft*, 1998(2), 157–181.

Dukashevich, E.D. (1995). First pupae of Eoptychopteridae and Ptychopteridae from the Mesozoic of Siberia (Insecta: Diptera). *Paleontological Journal*, 29(4), 164–171.

Dukashevich, E.D. and Mostovski, M.B. (2003). Nasekomie-Gematofagi v paleontologicheskoi detopisi. *Paleontologicheski Zhurnal*, 2, 48–56.

Dumitrica, P. and Dumitrica Jud, R. (1995). *Aurisaturnalis carinatus* (Foreman), an example of phyletic gradualism among saturnalid-type radiolarians. *Revue de Micropaléontologie*, 38, 195–216.

Duncan, I.J., Briggs, D.E.G., and Archer, M. (1998). Three-dimensionally mineralized insects and millipedes from the Tertiary of Riversleigh, Queensland, Australia. *Palaeontology*, 41, 835–851.

Duncan, K.W. (1981). The effect on *Orchestia hurleyi* (Amphipoda: Talitridae) of a whitey disease caused by *Bacillus subtilis*. *New Zealand Journal of Zoology*, 8, 517–528.

Dunham, A.E., Overall, K.L., Porter, W.P., and Forster, C.A. (1989). Implications of ecological energetics and biophysical and developmental constraints for life-history variation in dinosaurs. In *Paleobiology of the Dinosaurs* (pp. 1–19), edited by J. Farlow, Special Paper 238. Geological Society of America, Boulder, CO.

Dunlop, J.A., Anderson, L.I., Kerp, H., and Hass, H. (2003). Preserved organs of Devonian harvestmen. *Nature*, 425, 916.

Duringer, P., Brunet, M., Cambefort, Y., Beauvilain, A., Mackaye, H.T., Vignaud, P., and Schuster, M. (2000a). Des boules de bousiers fossiles et leurs terriers dans les sites à Australopitheques du Pliocène tchadien. *Bulletin de la Société Géologique de France*, 171(2), 259–269.

Duringer, P., Brunet, M., Cambefort, Y., Likens, A., Mackaye, H.T., Schuster, M., and Vignaud, P. (2000b). First discovery of fossil dung beetle brood balls and nests in the Chadian Pliocene Australopithecine levels. *Lethaia*, 33, 277–284.

Duvall, D. (1986). Snake, rattle, and roll. *Natural History*, 195(11), 66–73.

Dzik, J. (1994). Evolution of "small shelly fossil" assemblages of the Early Paleozoic. *Acta Palaeontologica Polonica*, 39, 247–313.

Dzik, J., Ivantsov, A.Y., and Deulin, Y.V. (2004). Oldest shrimp and associated phyllocarid from the Lower Devonian of northern Russia. *Zoological Journal of the Linnaean Society*, 142, 83–90.

Eagle, M.K. and Hayward, B.W. (1992). Paleontology and paleoecology of Early Miocene sequences in Hays and Tipakuri Streams, northern Hunua Ranges, Auckland. *Records of the Auckland Institute and Museum*, 29, 113–133.

Eagle, M.K. and Hayward, B.W. (1993). Oligocene paleontology and paleoecology of Waitete Bay, northern Coromandel Peninsula. *Records of the Auckland Institute and Museum*, 30, 13–26.

Eagle, M.K., Hayward, B.W., and Carter, G. (1994). Early Miocene rocky shore and coarse sediment fossil communities, Kawau Island, Auckland. *Records of the Auckland Institute and Museum*, 31, 187–204.

Eagle, M.K., Hayward, B.W., and Grant-Mackie, J.A. (1995). Early Miocene beach, rocky shore, and enclosed bay fossil communities, Waiheke Island, Auckland. *Records of the Auckland Institute and Museum*, 32, 17–44.

Eames, A.J. (1930). Report on ground sloth coprolite from Dona Ana County, New Mexico. *American Journal of Science*, 120, 353–356.

Ebbestad, J.O.R. (1998). Multiple attempted predation in the Middle Ordovician gastropod *Bucania gracillima*. *Geologiska Förenings I Stockholm Förhandlingar*, 120, 27–33.

Ebbestad, J.O.R. (2007). Gastropods. In *Silurian Fossils of the Pentland Hills* (pp. 109–122), edited by E.N.K. Clarkson, D.A.T. Harper, C.M. Taylor, and L.I. Anderson. Wiley-Blackwell, New York.

Ebbestad, J.O.R. and Hogstrom, A.E.S. (2000). Shell repair following failed predation in two Upper Ordovician brachiopods from central Sweden. *Geologiska Förenings I Stockholm Förhandlingar*, 122, 307–312.

Ebbestad, J.O.R. and Peel, J.S. (1997). Attempted predation and shell repair in Middle and Upper Ordovician gastropods from Sweden. *Journal of Paleontology*, 71, 1007–1019.

Ebbestad, J.O.R. and Stott, C.A. (2008). Failed predation in Late Ordovician gastropods (Mollusca) from Manitoulin Island, Ontario, Canada. *Canadian Journal of Earth Sciences*, 45, 231–241.

Ebert, D. (1994). Virulence and local adaptation of a horizontally transmitted parasite. *Science*, 265, 1084–1086.

Eccles, J.C. (1989). *Evolution of the Brain: Creation of the Self*. Routledge, London, 282 pp.

Edinger, T. (1926). Fossile Fliedermausgehirne. *Senckenbergiana*, 8(1), 1–6.

Edinger, T. (1941). The brain of *Pterodactylus*. *American Journal of Science*, 239, 665–682.

Edinger, T. (1948). Evolution of the horse brain. *Geological Society of America Memoir*, 25, 1–177.

Edinger, T. (1955). Hearing and smell in cetacean history. *Monatschrift für Psychiatrie und Neurologie*, 129, 37–58.

Edinger, T. (1961). Anthropocentric misconceptions in paleoneurology. *Proceedings of the Rudolf Virchow Medical Society in the City of New York*, 19, 55–107.

Edinger, T. (1966). Brains from 40 million years of camelid history. In *Evolution of the Forebrain* (pp. 153–161), edited by R. Hassler and H. Stephan. Plenum Press, New York.

Edwards, N. and Meco, J. (2000). Morphology and palaeoenvironment of brood cells of Quaternary ground-dwelling solitary bees (Hymenoptera, Apidae) from Fuerteventura, Canary Islands, Spain. *Proceedings of the Geologists' Association*, 111, 173–183.

Edwards, N., Jarzembowski, E.A., Pain, T., and Daley, B. (1998). Cocoon-like trace fossils from the lacustrine–palustrine Bembridge Limestone Formation (Late Eocene), southern England. *Proceedings of the Geologists' Association*, 109, 25–32.

Eeckhaut, I. and Améziane-Cominardi, N. (1994). Structural description of three myzostomes parasitic on crinoids and of the skeletal deformations they induce on their hosts. In *Echinoderms Through Time* (pp. 203–209), edited by B. David, A. Guille, J.-P. Feral, and M. Roux. Balkema, Rotterdam.

Eeckhaut, I., Parmentier, E., Becker, P., da Silva, S.G., and Jangoux, M. (2003). Parasites and biotic diseases in field and cultivated sea cucumbers. *Advances in Sea Cucumber Aquaculture and Management*, 463, 311–325.

Eisner, T., Eisner, M., and Siegler, M. (2005). *Secret Weapons: Defenses of Insects, Spiders, Scorpions, and Other Many-Legged Creatures*. Belknap Press, Cambridge, MA, 372 pp.

Ekdale, A.A. and Bromley, R.G. (2001a). Bioerosional innovation for living in carbonate hardgrounds in the Early Ordovician of Sweden. *Lethaia*, 34, 1–12.

Ekdale, A.A. and Bromley, R.G. (2001b). A day and a night in the life of a cleft-foot clam: *Protovirgularia–Lockeia–Lophosctenium*. *Lethaia*, 34, 119–124.

Eldredge, N. (1985). *Time Frames*. Simon & Schuster, New York.

El-Hawat, A.S. and Abdel Gawad, G.I. (1996). On the occurrence of globular bryozoan forms in Marada Formation, Middle Miocene, Sirt Basin, Libya. In *The Geology of Sirt Basin*, Vol. 1 (pp. 513–521), edited by M.J. Salem, A.J. Mouzughi, and O.S. Hammuda. Elsevier, Amsterdam.

Elias, R.J. and Lee, D.-J. (1993). Microborings and growth in Late Ordovician halysitids and other corals. *Journal of Paleontology*, 67, 922–934.

Elias, S.A. (1994). *Quaternary Insects and Their Environments*. Smithsonian Institution Press, Washington, D.C., 284 pp.

Elias, S.A. (1996). Biogeography of North American beetles: new perspectives from the Quaternary fossil record. *Sixth North American Paleontological Convention Abstracts of Papers*, 8, 114.

Elias, S.A. and Matthews, Jr., J.V. (2002). Arctic North American seasonal temperatures from the latest Miocene to the Early Pleistocene, based on mutual climatic range analyses of fossil beetle assemblages. *Canadian Journal of Earth Sciences*, 39, 911–920.

Elias, S.A., Kuzmina, S., and Kiselyov, S. (2006). Late Tertiary origins of the Arctic beetle fauna. *Palaeogeography, Palaeoclimatology, Palaeoecology*, 241, 373–392.

Elliott, D.K. and Bounds, S.D. (1987). Causes of damage to brachiopods from the Middle Pennsylvanian Naco Formation, central Arizona. *Lethaia*, 20, 327–335.

Elliott, D.K. and Brew, D. (1988). Cephalopod predation on a Desmoinesian brachiopod from the Naco Formation, central Arizona. *Journal of Paleontology*, 62, 145–147.

Elliott, D.K. and Nations, J.D. (1998). Bee burrows in the Late Cretaceous (Late Cenomanian) Dakota formation, northeastern Arizona. *Ichnos*, 5, 243–253.

Elliott, G.F. (1963). A Palaeocene teredinid (Mollusca) from Iraq. *Palaeontology*, 6, 315–317.

Ellis, D. (1914). Fossil microorganisms from the Jurassic and Cretaceous rocks of Great Britain. *Proceedings of the Royal Society of Edinburgh*, 35, 110–133.

Ellis, W.N. and Ellis-Adam, A.C. (1993). Fossil brood cells of solitary bees on Fuerteventura and Lanzarote, Canary Islands (Hymenoptera: Apoidea). *Entomologische Berichten*, 53(12), 161–173.

El-Nakal, H.A. and Bandel, K. (1991). Geographical distribution of the small gastropod genus *Scaliola*. *Micropaleontology*, 37, 423–424.

Emson, R.G. and Wilkie, I.C. (1980). Fission and autotomy in echinoderms. *Oceanography and Marine Biology Annual Review*, 18, 155–250.

Engel, M.S. (1999). The first fossil *Euglossa* and the phylogeny of orchid bees (Hymenoptera: Apidae: Euglossini). *American Museum Novitates*, 3272, 1–14.

Engel, M.S. (2000). A new interpretation of the oldest fossil bee (Hymenoptera: Apidae). *American Museum Novitates*, 3296, 1–11.

Engel, M.S. (2001). A monograph of the Baltic Amber bees and evolution of the Apoida (Hymenoptera). *Bulletin of the American Museum of Natural History*, 259, 1–192.

Engel, M.S. (2005). An Eocene ectoparasite of bees: the oldest definitive record of phoretic meloid triungulins (Coleoptera: Meloidae; Hymenoptera: Megachilidae). *Acta Zoologica Cracovensia*, 48B, 43–48.

Engel, M.S., Grimaldi, D.A., and Krishna, K. (2007). Primitive termites from the Early Cretaceous of Asia (Isoptera). *Stuttgarter Beiträge zur Naturkunde Serie B Geologie und Paläontologie*, 371, 1–32.

Engeser, T. (1990). Phylogeny of the fossil cephalopoda (Mollusca). *Berliner Geowissenschaftliche Abhandlungen A*, 124, 123–191.

Engeser, T. and Keupp, H. (2002). Phylogeny of the aptychi-possessing Neoammonoidea (Aptychophora nov., Cephalopoda). *Lethaia*, 35, 79–96.

Enos, P., Lehrmann, D.J., Wei Jiayang, Yu Youji, Xiao Jiafei, Chaikin, D.H., Minzoni, M., Berry, A.K., and Montgomery, P. (2006). *Triassic Evolution of the Yangtze Platform in Guizhou Province, People's Republic of China*, Special Paper 417. Geological Society of America, Boulder, CO, 105 pp.

Erickson, B.R. (1984). Chelonivorous habits of the Paleocene crocodile *Leidyosuchus formidabilis*. *Science Publications of the Science Museum of Minnesota*, 5(4), 3–9.

Erickson, B.R. and Sawyer, G.T. (1996). The estuarine crocodile *Gavialosuchus carolinensis* n. sp. (Crocodylia: Eusuchia) from the Late Oligocene of South Carolina, North America. *Monographs of the Science Museum of Minnesota*, 3, 1–47.

Erickson, G.M. and Olson, K.H. (1996). Bite marks attributable to *Tyrannosaurus rex*: preliminary description and implication. *Journal of Vertebrate Paleontology*, 16, 175–178.

Erickson, G.M., Van Kirk, S.D., Su, J., Levenston, M.E., Caler, W.E., and Carter, D.R. (1996). Bite-force estimation for *Tyrannosaurus rex* from tooth-marked bones. *Nature*, 382, 706–708.

Ernst, G. (1971). Stand und Zielsetzung der geologischen Forschungsarbeiten in der Lagerdorfer Schreibkreide. *Steinburger Jahrbuch*, 87–100.

Ernst, G., Kohring, R., and Rehfeld, U. (1996). Gastrolithe aus dem Mittel-Cenomanium von Baddeckenstedt (Harzvorland) und ihre paläogeographische Bedeutung fur eine prä-ilsedische Harzinsel. *Mitteilungen aus dem Geologisch–Paläontologischen Institut der Universität Hamburg*, 77, 503–543.

Erwin, D.M. and Schick, K.N. (2007). New Miocene oak galls (Cynipini) and their bearing on the history of cynipid wasps in western North America. *Journal of Paleontology*, 81, 568–580.

Esperante, R. and Brand, L. (2002). *Preservation of Baleen Whales in Tuffaceous and Diatomaceous Deposits of the Pisco Formation, Southern Peru*, First International Palaeontological Congress (IPC2002), Abstract No. 68. Geological Society of Australia, Sydney.

Espinosa-Arrubarrena, L. and Applegate, S.P. (1996). A paleoecological model of the vertebrate bearing beds in the Tlaycua Quarries, near Tepexi de Rodriguez, Puebla, Mexico. In *Mesozoic Fishes: Systematics and Paleoecology* (pp. 539–550), edited by G. Arratia and G. Viohl. Verlag Dr. Friedrich Pfeil, Munich.

Estes, J.A., Lindberg, D.R., and Wray, C. (2005). Evolution of large body size in abalones (*Haliotis*): patterns and implications. *Paleobiology*, 31, 591–606.

Estes, R. and Berberian, P. (1970). Paleoecology of a Late Cretaceous vertebrate community from Montana. *Breviora*, 343, 1–35.

Etches, S., Clarke, J., and Calloman, J. (2009). Ammonite eggs and ammonitellae from the Kimmeridge Clay formation (Upper Jurassic) of Dorset, England. *Lethaia*, 42, 204–217.

Ettensohn, F.R. (1978). Acrothoracic barnacle borings from the Chesterian of eastern Kentucky and Alabama. *Southeastern Geology*, 20, 27–31.

Etter, W. (2004). Redescription of *Opsepedon gracilis* Heer (Crustacea, Tanaidacea) from the Middle Jurassic of northern Switzerland and the palaeoenvironmental significance of tanaidaceans. *Palaeontology*, 47, 67–80.

Eva, A.N. (1980). Pre-Miocene seagrass communities in the Caribbean. *Palaeontology*, 23, 231–236.

Evans, A.C. (1996). Late stone-age coprolite reveals evidence of prehistoric parasitism. *South African Medical Journal*, 274–275.

Evans, F.J. (1998). Taphonomy of some Upper Palaeozoic actinopterygian fish from southern Africa. *Journal of African Earth Sciences*, 27(1A), 69–70.

Evans, H.M. (1923). The defensive spines of fishes, living and fossil and the glandular structure in connection therewith with observations on the nature of fish venoms. *Philosophical Transactions of the Royal Society of London B*, 212(391), 1–33.

Evans, S. (1999). Wood-boring bivalves and boring linings. *Bulletin of the Geological Society of Denmark*, 45, 130–134.

Evans, S.A. (1987). A review of the Upper Permian genera *Coelurasauravus*, *Weigeltisaurus* and *Gracilisaurus* (Reptilia: Diapsida). *Zoological Journal of the Linnaean Society*, 90, 275–303.

Everhart, M.J. (1999). Evidence of feeding on mosasaurs by the Late Cretaceous lamniform shark *Cretoxyrhina mantelli*. *Journal of Vertebrate Paleontology*, 17(3, Suppl.), 43A–44A.

Everhart, M.J. (2000). Gastroliths associated with plesiosaur remains in the Sharon Springs Member of the Pierre Shale (Late Cretaceous), western Kansas. *Kansas Academy of Science Transactions*, 103, 58–69.

Everhart, M.J. (2004a). Late Cretaceous interaction between predators and prey: evidence of feeding by two species of shark on a mosasaur. *PalArch, Vertebrate Palaeontology Series*, 1(1), 1–7.

Everhart, M.J. (2004b). Plesiosaurs as the food of mosasaurs: new data on the stomach contents of a *Tylosaurus proriger* (Squamata; Mosasauridae) from the Niobrara Formation of western Kansas. *The Mosasaur*, 7, 41–46.

Everhart, M.J. (2005a). Bite marks on an elasmosaur (Sauropterygia; Plesiosauria) paddle from the Niobrara Chalk (Upper Cretaceous) as probable evidence of feeding by the lamniform shark, *Cretoxyrhina mantelli*. *PalArch, Vertebrate Palaeontology Series*, 2(2), 14–24.

Everhart, M.J. (2005b). *Oceans of Kansas*. Indiana University Press, Bloomington, 322 pp.

Everhart, M.J. (2005c). Earliest record of the genus *Tylosaurus* (Squamata; Mosasauridae) from the Fort Hays Limestone (Lower Coniacian) of western Kansas. *Transactions of the Kansas Academy of Science*, 108, 149–155.

Everhart, M.J. and Hamm, S. (2005). A new nodosaur specimen (Dinosauria: Nodosauridae) from the Smoky Hill Chalk (Upper Cretaceous) of western Kansas. *Transactions of the Kansas Academy of Science*, 108(1/2), 15–21.

Everhart, M.J., Everhart, P.A., and Shimada, K. (1995). New specimen of shark bitten mosasaur vertebrae from the Smoky Hill Chalk (Upper Cretaceous) in western Kansas. *Transactions of the Kansas Academy of Science*, 14, 19.

Evershed, R.P. and Lockheart, M.J. (2001). Lipids. In *Palaeobiology*, Vol. II (pp. 247–253), edited by D.E.G. Briggs and P.R. Crowther. Blackwell Scientific, Oxford.

Ewing, H.E. (1926). A revision of the American lice of the genus *Pediculus*, together with a consideration of the significance of their geographical and host distribution. *Proceedings of the United States National Museum*, 68(1), 1–30.

Fage, L. (1937). Sur l'association d'un annélide polychète "Lumbriconereis flabellicola" n. sp. et d'un Madrépore Flabellum pavoninum distinctum E. et H. In *Comptes Rendus des Scéances du XII Congrès International de Zoologie, Lisbon, Portugal* (pp. 941–945), edited by A.R. Jorge.

Farlow, J.O. (1981). Estimates of dinosaur speeds from a new trackway site in Texas. *Nature*, 294, 747–748.

Fastovsky, D.E. and Smith, J.B. (2004). Dinosaur paleoecology. In *The Dinosauria* (pp. 614–626), edited by D.B. Weishampel, P. Dodson, and H. Osmalska. University of California Press, Berkeley.

Faulkner, C.T. (1991). Prehistoric diet and parasitic infection in Tennessee: evidence from the analysis of desiccated human paleofeces. *American Antiquity*, 56, 687–700.

Feibel, C.S. (1987). Fossil fish nests from the Koobi Fora Formation (Plio–Pleistocene) of northern Kenya. *Journal of Paleontology*, 61, 130–134.

Feist, R. (1995). Effect of paedomorphosis in eye reduction on patterns of evolution and extinction in trilobites. In *Evolutionary Change and Heterochrony* (pp. 225–244), edited by K.J. McNamara. John Wiley & Sons, New York.

Feist, R. and Clarkson, E.N.K. (1989). Environmentally controlled phyletic evolution, blindness and extinction in Late Devonian tropidocoryphine trilobites. *Lethaia*, 22, 359–373.

Fejfar, O. and Kaiser, T.M. (2005). Insect bone modification and paleoecology of Oligocene mammal-bearing sites in the Doupov Mountains, Northwestern Bohemia. *Palaeontologia Electronica*, 8(1), 1–11.

Feldman, H.R. and Brett, C.E. (1998). Epi- and endobiontic organisms on Late Jurassic crinoid columns from the Negev Desert, Israel: implications for co-evolution. *Lethaia*, 31, 57–71.

Feldmann, R.M. (1993). Additions to the fossil decapod crustacean fauna of New Zealand. *New Zealand Journal of Geology and Geophysics*, 36, 201–211.

Feldmann, R.M. (1998). Parasitic castration of the crab, *Tumidocarcinus giganteus* Glaessner, from the Miocene of New Zealand: coevolution within the Crustacea. *Journal of Paleontology*, 72, 493–498.

Feldmann, R.M. (2003). The Decapoda: new initiatives and novel approaches. *Journal of Paleontology*, 77, 1021–1039.

Feldmann, R.M. and Bearlin, R.K. (1988). *Linuparus korura* n. sp. (Decapoda: Palinura) from the Bortonian (Eocene) of New Zealand. *Journal of Paleontology*, 62, 245–250.

Feldmann, R.M. and de Saint Laurent, M. (2002). *Glyphea foresti* n. sp. (Decapoda) from the Cenomanian of Northern Territory, Australia. *Crustaceana*, 75, 359–373.

Feldmann, R.M. and Fordyce, R.E. (1996). A new cancrid crab from New Zealand. *New Zealand Journal of Geology and Geophysics*, 39, 509–513.

Feldmann, R.M. and Gazdzicki, A. (1998). Cuticular ultrastructure of fossil and living homolodromiid crabs (Decapoda: Brachyura). *Acta Palaeontologica Polonica*, 43, 1–19.

Feldmann, R.M. and May, W. (1991). Remarkable crayfish remains (Decapoda: Cambaridae) from Oklahoma-evidence of predation. *Journal of Paleontology*, 65, 884–886.

Feldmann, R.M. and McPherson, C.B. (1980). *Fossil Decapod Crustaceans of Canada*. Geological Survey of Canada, Ottawa, Ontario, 20 pp.

Feldmann, R.M. and Palubniak, D.S. (1975). Palaeoecology of Maestrichtian oyster assemblages in the Fox Hills Formation. *Geological Association of Canada Special Paper*, 13, 211–233.

Feldmann, R.M. and Wilson, M.T. (1988). Eocene decapod crustaceans from Antarctica. *Geological Society of America Memoir*, 169, 465–488.

Feldmann, R.M., Tshudy, D.M., and Thomson, M.R.A. (1993). Late Cretaceous and Paleocene decapod crustaceans from James Ross Basin, Antarctic Peninsula. *Paleontological Society Memoir*, 28, 1–41.

Feldmann, R.M., MacKinnon, D.L., Endo, K., and Chirino-Galvez, L. (1996a). *Pinnotheres laquei* Sakai (Decapoda: Pinnotheridae), a tiny crab commensal within the brachiopod *Laqueus rubellus* (Sowerby) (Terebratulida: Laqueidae). *Journal of Paleontology*, 70, 303–311.

Feldmann, R.M., Vega, F., Tucker, A.B., Garcia-Barrera, P., and Avendano, J. (1996b). The oldest record of *Lophoranina* (Decapoda: Raninidae) from the Late Cretaceous of Chiapas, southeastern Mexico. *Journal of Paleontology*, 70, 296–303.

Feldmann, R.M., Villamil, T., and Kauffman, E.G. (1999). Decapod and stomatopod crustaceans from mass mortality lagerstatten: Turonian (Cretaceous) of Colombia. *Journal of Paleontology*, 73, 91–101.

Feldmann, R.M., Schweitzer, C.E., and McLaughlan, D. (2006). Additions to the records for decapod Crustacea from Motunau and Glenafric Beaches, North Canterbury, New Zealand. *New Zealand Journal of Geology and Geophysics*, 49, 417–427.

Feng, R.-L. (1985). New discovery of fossil ophiuroids from Guizhou and southern Sichuan, China. *Acta Palaeontologica Sinica*, 24, 337–343.

Fenton, C.L. and Fenton, M.A. (1930–1931). Some snail borings of Paleozoic age. *American Midland Naturalist*, 7, 522–528.

Fenton, C.L. and Fenton, M.A. (1932). Orientation and injury in the genus *Atrypa*. *American Midland Naturalist*, 8, 63–74.

Fernald, R.D. (1988). Aquatic adaptations in fish eyes. In *Sensory Biology of Aquatic Animals* (pp. 435–466), edited by J. Atema, R.R. Fay, A.N. Popper, and W. Tavolga. Springer-Verlag, Heidelberg.

Ferreira, L.F., Araujo, A., Confalonieri, U., Chame, M., and Gomes, D.C. (1991). *Trichuris* eggs in animal coprolites dated from 30,000 years ago. *Journal of Parasitology*, 77, 491–493.

Ferreira, L.F., Araujo, A., Confalonieri, U., Chame, M., and Ribeiro, B. (1992). *Eimeria* oocysts in deer coprolites dated from 9,000 years ago. *Memórias do Instituto Oswaldo Cruz*, 87(Suppl. I), 105–106.

Ferreira, L.F., Araujo, A., and Duarte, A.N. (1993). Nematode larvae in fossilized animal coprolites from Lower and Middle Pleistocene sites, central Italy. *Journal of Parasitology*, 79, 440–442.

Ferrer, O. and de Gibert, J.M. (2005). Presencia de *Teredolites* en la Formació Arcillas de Morella (Cretácico Inferior, Castellón). *Revista Española de Paleontología*, X, 39–47.

Finger, T.E. (1988). Organization of chemosensory systems within the brains of bony fishes. In *Sensory Biology of Aquatic Animals* (pp. 339–363), edited by J. Atema, R.R. Fay, A.N. Popper, and W. Tavolga. Springer-Verlag, Heidelberg.

Fiore, L. and Ioale, P. (1973). Regulation of the production of subitaneous and dormant eggs in the turbellarian *Mesostoma ehrenbergii* (Focke), *Monitore Zoologico Italiano*, 7, 201–224.

Fiorillo, A.R. (1991a). Taphonomy and depositional setting of Careless Creek Quarry (Judith River Formation) Wheatland County, Montana, USA. *Palaeogeography, Palaeoclimatology, Palaeoecology*, 81, 157–166.

Fiorillo, A.R. (1991b). Prey bone utilization by predatory dinosaurs. *Palaeogeography, Palaeoclimatology, Palaeoecology*, 81, 281–311.

Fish, S.A., Shepherd, T.J., McGenity, T.J., and Grant, W.D. (2002). Recovery of 16S ribosomal RNA gene fragments from ancient halite. *Nature*, 417, 432–436.

Fisk, D.A. (1981). Sediment shedding and particulate feeding in two free-living sediment-dwelling corals (*Heteropsamia cochlea* and *Heterocyathus aequicostatus*) at Wistari Reef, Great Barrier Reef. *Proceedings of the Fourth International Coral Reef Symposium*, 2, 21–26.

Fitch, J.E. (1969). Fossil lanternfishes of California, with notes on fossil Myctophidae of North America. *Contributions in Science*, 173, 1–20.

Fleagle, J.G., Kay, R.F., and Simone, E.L. (1980). Sexual dimorphism in early anthropoids. *Nature*, 287, 328–330.

Fleming, C.A. (1952). A Foveaux Strait oyster bed. *New Zealand Journal of Science and Technology Series B*, 34, 73–85.

Fleming, C.A. (1953). The geology of Wanganui Subdivision. *New Zealand Geological Survey Bulletin*, 52, 1–362.

Fleming, C.A. (1971). A preliminary list of New Zealand fossil polychaetes. *New Zealand Journal of Geology and Geophysics*, 14, 742–756.

Flower, R.H. (1955). Trails and tentacular impressions of orthoconic cephalopods. *Journal of Paleontology*, 29, 857–867.

Folinsbee, K.E., Müller, J., and Reisz, R.R. (2007). Canine grooves: morphology, function, and relevance to venom. *Journal of Vertebrate Paleontology*, 27, 547–551.

Fondecave-Wallez, M.-J., Souquet, P., and Gourinard, Y. (1990). Sequence stratigraphy and grade-dating in the Senonian Series from the South-Pyrenees (Spain), the sedimentary record of eustasy and tectonics. In *Cretaceous Resources, Events and Rhythms* (pp. 63–74), edited by R.N. Ginsburg and B. Beaudoin. Kluwer Academic, Dordrecht.

Fordyce, R.E. and Jones, C.M. (1990). Penguin history and new fossil material from New Zealand. In *Penguin Biology* (pp. 419–446), edited by L.S. Davis and J.T. Darb. Academic Press, San Diego, CA.

Fordyce, R.E., Jones, C.M., and Field, B.D. (1986). The world's oldest penguin? *Geological Society of New Zealand Newsletter*, 74, 56–57.

Fornós, J.J., Bromley, R.G., Clemmensen, L.B., and Rodrígues-Perea, A. (2002). Tracks and trackways of *Myotragus balearicus* Bate (Artiodactyla, Caprinae) in Pleistocene aeolinites from Mallorca (Balearic Islands, Western Mediterranean). *Palaeogeography, Palaeoclimatology, Palaeoecology*, 180, 277–313.

Forrest, R. (2003). Evidence of scavenging by the marine crocodile *Metriorhynchus* on the carcass of a plesiosaur. *Proceedings of the Geologists' Association*, 114, 363–366.

Förster, R. (1969). Epökie, Entökie, Parasitismus und Regeneration bei fossilen Dekapoden. *Mitteilungen der Bayerische Staatssammlung für Paläontologie und Historische Geologie*, 9, 45–49.

Förster, R., Gadzicki, A., and Wrona, R. (1987). Homolodromid crabs from the Cape Melville Formation (Lower Miocene) of King George Island, West Antarctica. *Palaeontologica Polonica*, 49, 147–161.

Forteleoni, G. and Eliasova, H. (2000). I rapporti tra il bivalve *Lithophaga alpina* (Zittel, 1866) ed il corallo *Actinastea elongata* Alloiteau, 1954, nel Cretaceo superiore dell'Italia nord-orientale. *Bollettino della Societa Paleontologia Italiana*, 39, 47–54.

Fortey, R.A. (1974). The Ordovician Trilobites of Spitsbergen. I. Olenidae. *Norsk Polarinstitutt Skrifter*, 160, 1–80.

Fortey, R.A. (1975). The Ordovician trilobites of Spitsbergen. II. Asaphidae, Nileidae, Raphiophoridae and Telephinidae of the Valhallfonna Formation. *Norsk Polarinstitutt Skrifter*, 162, 1–125.

Fortey, R.A. (1985). Gradualism and punctuated equilibria as competing and complementary theories. *Special Papers in Palaeontology*, 33, 17–28.

Fortey, R.A. (2000). Olenid trilobites: the oldest known chemoautotrophic symbionts? *Proceedings of the National Academy of Sciences*, 97(12), 6574–6578.

Fortey, R.A. (2006). A new deep-water Upper Ordovician (Caradocian) trilobite fauna from South-West Wales. *Geological Journal*, 41, 243–253.

Fortey, R.A. and Chatterton, B. (2003). A Devonian trilobite with an eyeshade. *Science*, 301, 1689.

Fortey, R.A. and Owens, R.M. (1987). The Arenig Series in South Wales. *Bulletin of the British Museum (Natural History), Geology*, 41(3), 69–307.

Fortey, R.A. and Owens, R.M. (1990). Trilobites. In *Evolutionary Trends* (pp. 121–142), edited by K.J. McNamara. University of Arizona Press, Tucson.

Fortunato, H. (2007). Naticid gastropod predation in the Gatun Formationn (late Middle Miocene), Panama: preliminary assessment. *Paläontologische Zeitschrift*, 81, 356–364.

Fox, R.C. and Scott, C.S. (2005). First evidence of a venom delivery apparatus in extinct mammals. *Nature*, 435, 1091–1093.

Fraaije, R.H.B. (2003). The oldest *in situ* hermit crab from the Lower Cretaceous of Speeton, U.K. *Palaeontology*, 46, 53–58.

Fraaije, R.H.B. and Pennings, H.W.J. (2006). Crab carapaces preserved in nautiloid shells from the Upper Paleocene of Huesca (Pyrenees, Spain). *Revista Mexicana de Ciencias Geológicas*, 23, 361–363.

Fraaye, R.H.B. and Jäger, M. (1995a). Decapods in ammonite shells: examples of inquilinism from the Jurassic of England and Germany. *Palaeontology*, 38, 63–75.

Fraaye, R.H.B. and Jäger, M. (1995b). Ammonite inquilinism by fishes: examples from the Lower Jurassic of Germany and England. *Neues Jahrbuch für Geologie und Paläontologie–Monatsheft*, 9, 541–552.

Francis, J.E. (2000). Fossil wood from Eocene high latitude forests, McMurdo Sound, Antarctica. *Antarctic Research Series*, 76, 253–260.

Francis, J.E. and Harland, B.M. (2006). Termite borings in Early Cretaceous fossil wood, Isle of Wight, U.K. *Cretaceous Research*, 27, 773–777.

Frank, S.A. (2002). *Immunology and Evolution of Infectious Disease*. Princeton University Press, Princeton, NJ, 348 pp.

Franzén, C. (1974). Epizoans on Silurian–Devonian crinoids. *Lethaia*, 7, 287–301.

Franzen, J.L. (1992). The Messel horse show, and other odd-toed ungulates. In *Messel: An Insight into the History of Life and of the Earth* (pp. 241–247), edited by S. Schaal and W. Ziegler. Clarendon Press, London.

Franzen, J.L. (1997). Exponet des Monats Februar: Fossiler Paarhufer mit Embryo. *Natur und Museum*, 127, 61–62.

Franzen, J.L. (2000). *Europolemur kelleri* n. sp. von Messel und ein Nachtrag zu *Europolemur koenigswaldi* (Mammalia, Primates, Notharctidae, Cercamoniinae). *Senckenbergiana Lethaia*, 80, 275–287.

Franzen, J.L. (2001). Taphonomic analysis of the Messel Formation (Germany). In *Eocene Biodiversity: Unusual Occurrences and Rarely Sampled Habitats* (pp. 197–214), edited by G.F. Gunnell. Kluwer Academic/Plenum Press, New York.

Franzen, J.L. and Frey, E. (1993). *Europolemur* completed. *Kaupia*, 3, 113–130.

Franzen, J.L. and Richter, G. (1992). Primitive even-toed ungulates: loners in the undergrowth. In *Messel: An Insight into the History of Life and of the Earth* (pp. 251–256), edited by S. Schaal and W. Ziegler. Clarendon Press, London.

Fraser, N.C., Olsen, P.E., Dooley, Jr., A.C., and Ryan, T.R. (2007). A new gliding tetrapod (Diapsida: ?Archosauromorpha) from the Upper Triassic (Carnian) of Virginia. *Journal of Vertebrate Paleontology*, 27, 261–265.

Freeman, B.E. and Donovan, S.K. (1991). A reassessment of the ichnofossil *Chubutolithus gaimanensis* Bown & Ratcliffe. *Journal of Paleontology*, 65, 702–705.

Freess, W.B. (1991). Beiträge zur Kenntnis von Fauna und Flora des marinen Mitteloligozäns bei Leipzig. *Altenburger Naturwissenschaftliche Forschungen*, 6, 1–74.

Frenguelli, J. (1937). Sobre una perla fosil del Aonikense de Punta Norte. *Notas del Museo de La Plata, Paleontologia*, 2(11), 155–162.

Frey, D.G. (1964). Remains of animals in Quaternary lake and bog sediments and their interpretation. *Archiv für Hydrobiologie Ergebnisse der Limnologie*, 2, 1–114.

Frey, E., Martill, D.M., and Buchy, M.-C. (2003). A new crested ornithocheirid from the Lower Cretaceceous of northeastern Brazil and the unusual death of an unusual pterosaur. In *Evolution and Palaeobiology of Pterosaurs* (pp. 55–63), edited by E. Buffetaut and J.-M. Mazin. Geological Society of London Special Publication 217.

Frey, R.C. (1987). The occurrence of pelecypods in Early Paleozoic epeiric-sea environments: Late Ordovician of the Cincinnati, Ohio, area. *Palaios*, 2, 3–23.

Frey, S.W., Howard, J.D., and Hong, J.-S. (1986). Naticid gastropods may kill solenoid bivalves without boring: ichnologic and taphonomic consequences. *Palaios*, 1, 610–612.

Frickhinger, K.A. (1994). *The Fossils of Solnhofen*. Goldschneck-Verlag, Stuttgart, 336 pp.

Friis, E.M. (1985). Structure and function in Late Cretaceous angiosperm flowers. *Kongelige Danske Videnskabernes Selskab, Biologiske Skrifter*, 25, 1–37.

Fritzsch, B., Ryan, M.J., Wilczynski, W., Hetherington, T.E., and Walkowiak, W., Eds. (1988). *The Evolution of the Amphibian Auditory System*. Wiley-Interscience, New York, 705 pp.

Frizzell, D.L. and Exline, H. (1958). Crustacean gastroliths from the Claiborne Eocene of Texas. *Micropaleontology*, 4, 273–280.

Frizzell, D.L. and Horton, W.C. (1961). *Crustacean Gastroliths from the Jackson Eocene of Louisiana*, Bulletin No. 99. School of Mines and Metallurgy, University of Missouri, Columbia, pp. 3–6.

Fry, B.G., Vidal, N., Norman, J.A., Vonk, F.J., Scheib, H. et al. (2006). Early evolution of the venom system in lizards and snakes. *Nature*, 439, 584–588.

Fry, G.F. (1985). Analysis of fecal material. In *The Analysis of Prehistoric Diets* (pp. 127–154), edited by R.I. Gilbert and J.H. Mielke. Academic Press, San Diego, CA.

Frýda, J. (2004). Gastropods. In *Encyclopedia of Geology* (pp. 378–388), edited by R.C. Selley, L.R.M. Cocks, and I.R. Plimer. Elsevier, Amsterdam.

Frýda, J., Nützel, A., and Wagner, D.J. (2008a). Paleozoic gastropods. In *Phylogeny and Evolution of the Mollusca* (pp. 239–270), edited by W. Ponder and D.R. Lindberg. University of California Press, Berkeley.

Frýda, J., Racheboeuf, P.R., and Frýdova, B. (2008b). Mode of life of Early Devonian *Orthonychia protei* (Neritomorpha, Gastropoda) inferred from its post-larval shell ontogeny and muscle scars. *Bulletin of Geosciences*, 83, 491–502.

Frýda, J., Racheboeuf, P.R., Frýdova, B., Ferrová, L., Mergl, M., and Berkyová, S. (2009). Platyceratid gastropods: a stem group of patellogastropods, neritomorphs or something else? *Bulletin of Geosciences*, 84, 107–120.

Fryer, G. and Stanley, Jr., G.D. (2004). A Silurian porpitoid hydrozoan from Cumbria, England, and a note on porpitoid relationships. *Palaeontology*, 47, 1109–1119.

Fuchs, D. (2006). Morphology, taxonomy and diversity of vampyropod Coleoids (Cephalopoda) from the Upper Cretaceous of Lebanon. *Memorie della Società Italiana di Scienze Naturali e del Museo Civico di Storia Naturale di Milano*, 34, 1–28.

Fuchs, D., Bracchi, G., and Weis, R. (2009). New octopods (Cephalopods: Coleoidea) from the Late Cretaceous (Upper Cenomanian) of Hâkel and Hâdjoula, Lebanon. *Palaeontology*, 52, 65–81.

Fuji, N. (1988). Palaeovegetation and palaeoclimate changes around Lake Biwa, Japan during the last ca. 3 million years. *Quaternary Science Reviews*, 7, 21–28.

Fujioka, Y. and Yamazato, K. (1983). Host selection of some Okinawan coral associated gastropods belonging to the genera *Drupella*, *Coralliophila* and *Quoyula*. *Galaxea*, 2, 59–73.

Fujita, T. (1992). Dense beds of ophiuroids from the Paleozoic to the Recent: the significance of bathyal populations. *Otsuchi Marine Research Conference Report*, 18, 25–41.

Fujiyama, I. and Iwao, Y. (1974). Fossil insects from Togo, Kagoshima, Japan (Tertiary Insect Fauna of Japan, 5). *Bulletin of the National Science Museum*, 17, 87–96.

Fürsich, F.T. (1977). Corallian (Upper Jurassic) marine benthic associations from England and Normandy. *Palaeontology*, 20, 337–385.

Fürsich, F.T. (1979). Genesis, environments, and ecology of Jurassic hardgrounds. *Neues Jahrbuch für Geologie und Paläontologie–Abhandlungen*, 158, 1–63.

Fürsich, F.T. and Jablonski, D. (1984). Late Triassic naticid drill-holes: carnivorous gastropods gain a major adaptation but fail to radiate. *Science*, 224, 78–80.

Fürsich, F.T. and Palmer, T. (1984). Commissural asymmetry in brachiopods. *Lethaia*, 17, 251–265.

Fürsich, F.T. and Pandey, D.K. (1999). Genesis and environmental significance of Upper Cretaceous shell concentrations from the Cauvery Basin, southern India. *Palaeogeography, Palaeoclimatology, Palaeoecology*, 145, 119–139.

Fürsich, F.T. and Sykes, R.M. (1977). Palaeobiogeography of the European Boreal Realm during Oxfordian (Upper Jurassic) times: a quantitative approach. *Neues Jahrbuch für Geologie und Paläontologie–Abhandlungen*, 155, 137–161.

Fürsich, F.T. and Werner, W. (1986). Benthic associations and their environmental significance in the Lusitanian Basin (Upper Jurassic, Portugal). *Neues Jahrbuch für Geologie und Paläontologie–Abhandlungen*, 172, 271–329.

Fürsich, F.T., Palmer, T.J., and Goodyear, K.L. (1994). Growth and disintegration of bivalve dominated patch reefs in the Upper Jurassic of southern England. *Palaeontology*, 37, 131–171.

Gabbott, S.E. (1999). Orthoconic cephalopods and associated fauna from the Ordovician Soom Shale Lagerstätte, South Africa. *Palaeontology*, 42, 123–148.

Gahn, F.J. and Baumiller, T.K. (2003). Infestation of Middle Devonian (Givetian) camerate crinoids by platyceratid gastropods and its implications for the nature of their biotic interaction. *Lethaia*, 36, 71–82.

Gahn, F.J., Fabian, A., and Baumiller, T.K. (2003). Additional evidence for the drilling behavior of Paleozoic gastropods. *Acta Palaeontologica Polonica*, 48, 156.

Gaillard, C. and Pajaud, D. (1971). *Rioultina virdunensis* (Buv.) cf. *ornata* (Moore); Brachiopode thecideen de l'epifauna de l'Oxfordien Supérieur du Jura Meridional. *Géobios*, 4, 227–242.

Gaillard, C. and Rolin, Y. (1986). Paléobiocoenoses susceptibles d'être liées à des sources sous-marines en milieu sedimentaires. L'example pseudobiohermes des Terres Noires (S.E. France) et de Tepee Buttes de la Pierre Shale Formation (Colorado, USA). *Comptes Rendus de l'Academie des Sciences, France*, 303(II), 1503–1508.

Gaillard, C., Bourseau, J.P., Boudeville, M., Pailleret, P., Rio, M., and Roux, M. (1985). Les pseudo-biohermes de Beauvoisin (Drôme): un site hydrothermal sur la marge tethysienne à l'Oxfordien. *Bulletin de la Société Géologique de France*, 1, 69–78.

Gaillard, C., Rios, M., Rolin, Y., and Roux, M. (1992). Fossil chemosynthetic community related to vents or seeps in sedimentary basins: the pseudobioherms of southeastern France. *Palaios*, 7, 451–465.

Gaillard, C., Hantzpergue, P., Vannier, J., Margerard, A.-L., and Mazin, J.-M. (2005). Isopod trackways from the Crayssac Lagerstätte, Upper Jurassic, France. *Palaeontology*, 48, 947–962.

Gale, A.S. and Smith, A.B. (1982). The paleobiology of the Cretaceous irregular echinoids *Infulaster* and *Hagenowia*. *Palaeontology*, 25, 11–42.

Galil, B. (1986/87). Trapeziidae (Decapoda: Brachyura: Xanthoidea) of the Red Sea. *Israel Journal of Zoology*, 34, 159–182.

Galippe, V. (1920). Recherches sur la resistance des microzymes à l'action du temps et sur leur survivance dans l'ambre. *Comptes Rendus de l'Academie des Sciences, Paris*, 170, 856–858.

Gall, H. and Muller, D. (1975). *Balanus*-Rasen auf Brandungsgerollen der Oberen Meeresmolasse (Helvet) vom Dischingen Blockstrant. *Mitteilungen der Bayerische Staatssammlung für Paläontologie und Historische Geologie*, 15, 29–31.

Gall, J.C. and Grauvogel, L. (1966). Faune du Buntsandstein I. Pontes d'invertébrès du Buntsandstein supérieur. *Annales de Paléontologie, Invertébrès*, 52, 155–161.

Galle, A. and Parsley, R.L. (2005). Epibiont relationships on hyolithids demonstrated by Ordovician trepostomes (Bryozoa) and Devonian tabulates (Anthozoa). *Bulletin of Geosciences*, 80, 125–138.

Galle, A. and Plusquellec, Y. (2002). Systematics, morphology, and paleobiogeography of Lower Devonian tabulate coral epibionts: Hyostragulidae fam. nov. on hyolithids. *Coral Research Bulletin*, 7, 53–64.

Galle, A., Marek, L., Vannier, J., and Racheboeuf, P.R. (1994). Assemblage epibenthique à hyolithes, tabule epizoaire et ostracode Beyrichiacea du Dévonien inférieur du Maroc et d'Espagne. *Revue de Paléobiologie*, 13, 411–425.

Galton, P.M. (1973). On the anatomy and relationships of *Afraasia diagnostica* (Huene), n. gen., a prosauropod dinosaur (Reptilia: Saurischia) from the Upper Triassic of Germany. *Paläontologisches Zeitschrift*, 47, 229–255.

Gao, K.-Q. and Shubin, N.H. (2001). Late Jurassic salamanders from northern China. *Nature*, 410, 574–577.

Gao, K.-Q. and Shubin, N.H. (2003). Earliest known crown group salamanders. *Nature*, 422, 424–428.

Garassino, A. and Donovan, D.T. (2000). A new family of coleoids from the Lower Jurassic of Osteno, northern Italy. *Palaeontology*, 43, 1019–1038.

Garcia-Bellido, D.C. and Collins, D.H. (2004). Moulting arthropod caught in the act. *Nature*, 429, 40.

García-Robleda, C. and Staines, C.L. (2008). Herbivory in gingers from latest Cretaceous to present: is the ichnogenus *Cephaloleichnites* (Hispinae, Coleoptera) a rolled-leaf beetle? *Journal of Paleontology*, 82, 1035–10372.

Gardner, J. (1945). Mollusca of the Tertiary formations of Northeastern Mexico. *Geological Society of America Memoir*, 11, 1–332.

Garth, J.N. (1964). The Crustacea Decapoda (Brachyura and Anomura) of Eniwetok Atoll, Marshall Islands, with special reference to the obligate commensals of branching corals. *Micronesica*, 1, 137–144.

Gasse, F. and Van Campo, E. (2001). Late Quaternary envionmental changes from a pollen and diatom record in the southern tropics (Lake Tritrivakely, Madagascar). *Palaeogeography, Palaeoclimatology, Palaeoecology*, 167, 287–308.

Gatesey, S.M., Middleton, K.M., Jenkins, Jr., F.A., and Shubin, N.H. (1999). Three-dimensional preservation of foot movements in Triassic theropod dinosaurs. *Nature*, 399, 141–144.

Gaudant, J. (1978). Signification bathymétrique, paléoclimatique et paléogéographique de l'ichthyofaune marine du Miocène terminal de la Méditerranée occidentale. Remarques préliminaires. *Bulletin du Muséum national d'histoire naturelles, sciences de la terre*, 70, 137–148.

Gautier, A. (1974). Fossiele vliegenmaden (*Protophormia terraenovae* Robineau-Desvoidy, 1830) in een schedel van de wolharige neushoorn (*Coelodonta antiquatatis*) uit het Onder-Würm te Dendermonde (Oost-Vlaanderen, België. *Natuurwetenschappelijk Tijdschrift*, 56, 76–84.

Gautier, A. and Schumann, H. (1973). Puparia of the subarctic or black blowfly *Protophormia terraenovae* (Robineau-Desvoidy, 1830) in a skull of a Late Eemian (?) bison at Zemst, Brabant (Belgium). *Palaeogeography, Palaeoclimatology, Palaeoecology*, 14, 119–125.

Gay, M. and Rickards, B. (1989). *Pike*. The Boydell Press, Woodbridge, U.K.

Gay, R.J. (2002). The myth of cannibalism in *Coelophysis bauri*. *Journal of Vertebrate Paleontology*, 22, 57A.

Geary, D.H. (1990). Patterns of evolutionary tempo and mode in the radiation of *Melanopsis* (Gastropoda; Melanopsidae). *Paleobiology*, 16, 492–511.

Geary, D.H. (1992). An unusual pattern of divergence between two fossil gastropods: ecophenotypy, or hybridization? *Paleobiology*, 18, 93–109.

Geary, D.H., Allmon, W.D., and Reaka-Kudla, M.L. (1991). Stomatopod predation on fossil gastropods from the Plio–Pleistocene of Florida. *Journal of Paleontology*, 65, 355–360.

Gebhardt, M.D. (1987). Parental care: a freshwater phenomenon? *Environmental Biology of Fishes*, 19, 69–72.

Gee, C.T., Sandere, P.M., and Petzelberger, B.E.M. (2003). A Miocene rodent nut cache in coastal dunes of the Lower Rhine Embayment, Germany. *Palaeontology*, 46, 1133–1149.

Geib, K.W. (1952). Über eine fossile Perle aus dem mitteloligozänen Meeressand vom Welshberg bei Waldböckelheim (Nahebergland). *Notizblatt des Hessischen Landesamtes für Bodenforschung*, VI(3), 31–32.

Gekker, E.L. and Gekker, R.F. (1955). Ostatki Teuthoidea iz Verkhnei Oori i Nizhnego Mela Povolshya. *Voprosi Paleontologii*, 2, 36–44.

Gekker, R.F. (1935). Yavleniya prirastaniya i prikrepleniya sredi verkhnedevonskoi fauni i flori glavnogo polya. *Transactions of the Paleontological Institute, Academy of Science, USSR*, 4, 159–180.

Gekker, R.F. and Merklin, R.L. (1946). On the character of the embedding of fossil fishes in the Maikop Shales of North Ossetia (N. Caucasus). *Izvestiya Akademia Nauk SSSR, Seriya Biologicheska*, 6, 672–674.

Genise, J.F. (1995). Upper Cretaceous trace fossils in permineralized plant remains from Patagonia, Argentina. *Ichnos*, 3, 287–299.

Genise, J.F. and Bown, T.M. (1990). The constructor of the ichnofossil *Chubutolithes*. *Journal of Paleontology*, 64, 482–483.

Genise, J.F. and Bown, T.M. (1994a). New Miocene scarabeid and hymenopterous nests and Early Miocene (Santacrucian) paleoenvironments, Patagonia, Argentina. *Ichnos*, 3, 107–117.

Genise, J.F. and Bown, T.M. (1994b). New trace fossils of termites (Insecta: Isoptera) from the Late Eocene–Early Miocene of Egypt, and the reconstruction of ancient isopteran social behavior. *Ichnos*, 3, 155–183.

Genise, J.F. and Bown, T.M. (1996). *Uruguay* Roselli 1938 and *Rosellichnus*, n. ichnogenus: two ichnogenera for clusters of fossil bee cells. *Ichnos*, 4, 199–217.

Genise, J.F. and Laza, J.H. (1998). *Monesichnus ameghinoi* Roselli: a complex insect trace fossil produced by two distinct trace makers. *Ichnos*, 5, 213–223.

Genise, J.F., Sciutto, J.C., Laza, J.H., González, M.G., and Bellosi, E.S. (2002). Fossil bee nests, coleopteran pupal chambers and tuffaceous paleosols from the Late Cretaceous Laguna Palacios Formation, central Patagonia (Argentina). *Palaeogeography, Palaeoclimatology, Palaeoecology*, 177, 215–235.

Germonpré, M. and Leclercq, M. (1994). Des pupes de *Protophormia terraenova* associées à des mammifères pleistocènes de la Vallée flamande (Belgique). *Bulletin de l'Institut Royal des Sciences Naturelles de Belgique*, 64, 265–268.

Getty, P.R. and Hagadorn, J.W. (2008). Reinterpretation of *Climactichnites* Logan 1860 to include subsurface burrows, and erection of *Musculopodus* for resting traces of the trailmaker. *Journal of Paleontology*, 82, 1161–1172.

Geyer, G. and Kelber, K.-P. (1987). Flugelreste und Lebensspuren von Insekten aus dem Unteren Keuper Mainfrankens. *Neues Jahrbuch für Geologie und Paläontologie-Abhandlungen*, 174, 331–355.

Gibbons, A. (1994). Mistreating a long-time host. *Science*, 265, 1037.

Gibson, M.A. and Watson, J.B. (1989). Predatory and non-predatory borings in echinoids from the upper Ocala Formation (Eocene), north-central Florida, USA. *Palaeogeography, Palaeoclimatology, Palaeoecology*, 71, 309–321.

Gil Cid, M.D., Dominguez Alonso, P., and Silvan Pobes, S. (1998). *Coralocrinus sarachagae* gen. nov. sp. primer crinoide (Disparida, Inadunata) descrito en el Ordovicico medio de Sierra Morena. *Coloquios de Paleontologia*, 49, 115–128.

Gill, T. (1905). An interesting Cretaceous chimaeroid egg-case. *Science*, 22, 601–602.

Gilliam, M. (1992). Normal microorganisms associated with honey bees and bee mites. In *Asian Apiculture: Proceedings of the First International Symposium on Asian Honey Bees and Bee Mites* (pp. 732–736), edited by L.J. Connor, T.E. Rinderer, H.A. Sylvester, and S. Wongsiri. Wicwas Press, Cheshire, CN.

Gilliam, M., Buchmann, S.L., Lorenz, B.T., and Schmalzel, R.J. (1990). Bacteria belonging to the genus *Bacillus* associated with three species of solitary bees. *Apidologie*, 21, 99–105.

Gingerich, P.D. (1989). New earliest Wasatchian mammalian fauna from the Eocene of northwestern Wyoming: composition and diversity in a rarely sampled high-floodplain assemblage. *University of Michigan Papers on Paleontology*, 28, 1–97.

Gingerich, P.D. (2003). Land-to-sea transition in early whales: evolution of Eocene Archaeoceti (Cetacea) in relation to skeletal proportions and locomotion of living semiaquatic mammals. *Paleobiology*, 29, 429–454.

Gingerich, P.D., ul-Haq, M., von Koeinigswald, W., Sanders, W.J., Smith, B.H., and Zalmout, I.S. (2009). New protocetid whale from the Middle Eocene of Pakistan: birth on land, precocial development and sexual dimorphism. *PLoS ONE*, 4(2), 1–20.

Gischler, E., Sandy, M.R., and Peckmann, J. (2003). *Iberorhynchia contraria* (F.A. Romer 1850), an Early Carboniferous seep-related rhynchonellide brachiopod from the Harz Mountains, Germany: a possible successor to *Dzieduszyckia*? *Journal of Paleontology*, 77, 293–303.

Gittleman, M. and Van Valkenburgh, B. (1997). Sexual dimorphism in the canines and skulls of carnivores: effects of size, phylogeny, and behavioural ecology. *Journal of Zoology*, 242, 97–117.

Giusberti, L., Fantin, M., and Buckeridge, J. (2005). *Ovulaster protodecimae* n. sp. (Echinoidea, Spatangoida) and associated epifauna (Cirripedia, Verrucidae) from the Danian of northeastern Italy. *Rivista Italiana di Paleontologia e Stratigrafia*, 111, 455–465.

Given, J.C.M. (1928). Palaeopathology and evolution. *The Lancet*, 21, 164–166.

Givulescu, R. (1984). Pathological elements on fossil leaves from Chiuzbaia (galls, mines and other insect traces). *Dari de Seama ale Sedintelor (Paleontologie)*, 68, 123–133.

Glaessner, M.F. (1965). Vorkommen fossiler Dekapoden (Crustacea) in fisch-Schiefern. *Senckenbergiana Lethaea*, 46a, 111–122.

Glaessner, M.F. (1969). Decapoda. In *Treatise on Invertebrate Paleontology*. Part R. *Mollusca 6, Bivalvia* (pp. 399–566), edited by R.C. Moore. Geological Society of America/ University Press of Kansas, Lawrence.

Glaub, I. (1994). Mikrobohrspuren in ausgewahlten Ablagerungsraumen des europaischen Jura und der Unterkreide (Klassifikation und Palökologie). *Courier Forschungs-Institut Senckenberg*, 174, 1–324.

Glaub, I. and Bundschuh, M. (1997). Comparative study on Silurian and Jurassic/Lower Cretaceous microborings. *Courier Forschungsinstitut Senckenberg*, 201, 123–135.

Glaub, I. and Konigshof, P. (1997). Microborings in conodonts. *Courier Forschungsinstitut Senckenberg*, 201, 137–143.

Glaub, I., Golubic, S., Gektidis, M., Radke, G., and Vogel, K. (2007). Microborings and microbial endoliths: geological implications. In *Trace Fossils, Concepts, Problems, Prospects* (pp. 368–381), edited by W. Miller III. Elsevier, Amsterdam.

Glynn, P.W. (1976). Some physical and biological determinants of coral community structure in the Eastern Pacific. *Ecological Monographs*, 46, 431–456.

Glynn, P.W. (1983). Crustacean symbionts and the defense of corals: coevolution of the reef? In *Coevolution* (pp. 111–178), edited by M.H. Nitecki. University of Chicago Press, Chicago, IL.

Godefroid, J. (1999). *Invertrypa struvei*, a new atrypid brachiopod species from the Givetian of Morocco. *Senckenbergiana Lethaea*, 79, 267–272.

Godfrey, S.J. (1992). Fossilized eggs from Illinois. *Nature*, 356, 21.

Godfrey, S.J. (1995). Fossilized eggs from the Pennsylvanian of Illinois. *Ichnos*, 4, 71–75.

Godinot, M. (1985). Evolutionary implications of morphological changes in Palaeogene primates. *Special Papers in Palaeontology*, 33, 39–47.

Goedert, J.L. (1993). First Oligocene records of *Calyptogena* (Bivalvia: Vesicomyidae). *The Veliger*, 36, 72–77.

Goedert, J.L. and Benham, S.R. (1999). A new species of *Depressigyra*? (Gastropoda: Peltospiridae) from cold seep carbonates in Eocene and Oligocene rocks of western Washington. *The Veliger*, 42, 112–116.

Goedert, J.L. and Campbell, K.A. (1995). An Early Oligocene chemosynthetic community from the Makeh Formation, northwestern Olympic Peninsula, Washington. *The Veliger*, 38, 22–29.

Goedert, J.L. and Kaler, K.L. (1996). A new species of *Abyssochrysos* (Gastropoda: Loxonematoidea) from a Middle Eocene cold-seep carbonate in the Humptulips Formation, western Washington. *The Veliger*, 39, 65–70.

Goedert, J.L. and Squires, R.L. (1990). Eocene deep-sea communities in localized limestones formed by subduction-related methane seeps, southwestern Washington. *Geology*, 18, 1182–1185.

Goedert, J.L. and Squires, R.L. (1993). First Oligocene records of *Calyptogena* (Bivalvia: Vesicomyidae). *The Veliger*, 36, 72–77.

Goedert, J.L., Peckmann, J., and Reitner, J. (2000). Worm tubes in an allochthonous cold-seep carbonate from lower Oligocene rocks of western Washington. *Journal of Paleontology*, 74, 992–999.

Goedert, J.L., Thiel, V., Schmale, O., Rau, W.W., Michaelis, W., and Peckmann, J. (2003). The late Eocene "Whiskey Creek" methane-seep deposit (western Washington State). I. Geology, paleontology, and molecular geobiology. *Facies*, 48, 223–240.

Goldring, R. and Stephenson, D.G. (1972). The depositional environment of three starfish beds. *Neues Jahrbuch für Geologie und Paläontologie–Monatsheft*, 10, 611–624.

Golenberg, E.M. (1991). Amplification and analysis of Miocene plant fossil DNA. *Philosophical Transactions of the Royal Society of London B*, 333, 419–426.

Golenberg, E.M., Giannasi, D.E., Clegg, M.T., Smiley, C.J., Durbin, M., Henderson, D., and Zurawski, G. (1990). Chloroplast DNA sequence from a Miocene Magnolia species. *Nature*, 344, 656–658.

Golubic, S. and Radtke, G. (2007). The trace *Rhopalia clavigera* isp. n. reflects the development of its maker *Eugomontia sacculata* Kornmann, 1960. In *Current Developments in Bioerosion* (pp. 95–108), edited by M. Wisshak and L. Tapanila. Springer-Verlag, Heidelberg.

Gómez-Pérez, I. (2003). An Early Jurassic deep-water stromatolitic bioherm related to possible methane seepage (Los Molles Formation, Neuquén, Argentina). *Palaeogeography, Palaeoclimatology, Palaeoecology*, 201, 21–49.

Gong, Y.-M., Li, B.-H., Wang, C.-Y., and Wu, Y. (2001). Orbital cyclostratigraphy of the Devonian Frasnian–Famennian transition in South China. *Palaeogeography, Palaeoclimatology, Palaeoecology*, 168, 237–248.

Gooday, A.J. (1983). Entomozoacean ostracods from the lower Carboniferous of south-western England. *Palaeontology*, 26, 755–788.

Goodfriend, G.A. and Gould, S.J. (1996). Paleontology and chronology of two evolutionary transitions by hybridization in the Bahamian land snail *Cerion*. *Science*, 274, 1894–1897.

Goodwin, H.T. (1993). Patterns of dental variation and evolution in prairie dogs, genus *Cynomys*. In *Morphological Change in Quaternary Mammals of North America* (pp. 107–133), edited by R.A. Martin and A.D. Barnosky. Cambridge University Press, Cambridge, U.K.

Goren-Inbar, N., Alperson, N., Kisler, M.E., Simchoni, O., Melamed, Y., Ben-Nur, A., and Werker, E. (2004). Evidence of hominin control of fire at Gesher Benot Ya'aqol, Israel. *Science*, 304, 725–727.

Goth, K. and Wilde, V. (1992). Frass spuren in permischen Hölzern aus der Wetterau. *Senckenbergiana Lethaea*, 72, 1–6.

Gottfried, M.D. (1989). Earliest fossil evidence for protective pigmentation in an actinopterygian. *Historical Biology*, 3, 79–83.

Gougerot, L. (1969). Clefs de determination des petites espèces de gastéropodes de l'Éocène du Bassin parisien. IV. Le genre *Eulima* Risso. *Cahiers des Naturalistes*, 25, 117–126.

Goulet, H. and Huber, J.T., Eds. (1993). *Hymenoptera of the World: An Identification Guide to Families*, Publication 1894/E. Research Branch, Agriculture Canada, Ottawa, 688 pp.

Gourinard, Y. (1983). Quelques vitesses d'evolution observées dans les lignées de Foraminifères neogènes. Utilisations chronologiques. *Comptes Rendus de l'Academie des Sciences, Paris*, 297(3), 269–272.

Grabenhorst, H. (1985). Eine zweite Bremse (Tabanidae) zusammen mit ihrem Parasiten (Nematoda, Mermithoidae) aus dem oberpliozän von Willershausen, Krs. Osterode. *Osterode Aufschluss*, 36, 325–328.

Graham, R.W., Lundelius, Jr., E.L., Graham, M.A., Schroeder, E.K., Toomey III, R.S. et al. (1996). Spatial responses of mammals to Late Quaternary environmental fluctuation. *Science*, 272, 1601–1606.

Grandal-d'Anglade, A. and López-González, F. (2005). Sexual dimorphism and ontogenetic variation in the skull of the cave bear (*Ursus spelaeus* Rosenmüller) of the European Upper Pleistocene. *Géobios*, 38, 325–337.

Grande, L. (2001). An updated review of the fish faunas from the Green River Formation: the world's productive freshwater Lagerstätten. In *Eocene Biodiversity: Unusual Occurrences and Rarely Sampled Habitats* (pp. 1–38), edited by G.F. Gunnell. Kluwer Academic/Plenum Press, New York.

Grande, L. and Buchheim, R.H.P. (1994). Paleontological and sedimentological variation in early Eocene Fossil Lake. *Contributions to Geology, University of Wyoming*, 30, 33–56.

Grant, R.E. (1988). The family Cardiarinidae (late Paleozoic, rhynchonellid Brachiopoda). *Senckenbergiana Lethaea*, 69, 121–135.

Grauvogel-Stamm, L. and Kelber, K.-P. (1996). Plant–insect interactions and coevolution during the Triassic of western Europe. *Paleontologia Lombarda*, V, 5–23.

Gray, J. (unpublished). *The Mangrove Community Complex Through Time and Space*.

Gray, J. (1985). Interpretatiion of co-occurring megafossils and pollen: a comparative study with Clarkia as an example. In *Late Cenozoic History of the Pacific Northwest* (pp. 185–244), edited by C.J. Smiley. Pacific Division, American Association for the Advancement of Science, Washington, D.C.

Grechina, N.I. (1973). Novii vid roda *Coryphaenoides* (Teleostei) iz Oligotsena Kamchatki. *Paleontologicheskii Zhurnal*, 1, 116–118.

Greensill, N.A.R. (1902). Structure of leaf of certain species of *Coprosoma*. *Transactions and Proceedings of the New Zealand Institute*, 35, 342–355.

Greenwood, A.D. (2001). Mammoth biology: biomolecules, phylogeny, Numts, nuclear DNA, and the biology of an extinct species. *Ancient Biomolecules*, 3, 255–266.

Greenwood, M.T., Wood, P.J., and Monk, W.A. (2006). The use of fossil caddisfly assemblages in the reconstruction of flow environments from floodplain paleochannels of the River Trent, England. *Journal of Paleolimnology*, 35, 747–761.

Gregor, H.-J. (1982). *Pinus aurimontana* n.sp.—eine neue Kieferart aus dem Jungtertiar des Goldbergs (Ries). *Stuttgarter Beiträge zur Naturkunde Serie B Geologie und Paläontologie*, 83, 1–19.

Gregorová, R. (2004). A new Oligocene genus of lanternfish (family Myctophidae) from the Carpathian Mountains. *Revue de Paléobiologie, Genève*, 9, 81–97.

Greinert, J., Bollwerk, S.M., Derkachev, A., Bohrmann, G., and Suess, E. (2002). Massive barite deposits and carbonate mineralization in the Derugan Basin, Sea of Okhotsk: precipitation processes at cold seep sites. *Earth and Planetary Science Letters*, 203, 165–180.

Grellet-Tinner, G., Chiappe, L., Norell, M., and Bottjer, D. (2006). Dinosaur eggs and nesting behaviors: a paleobiological investigation. *Palaeogeography, Palaeoclimatology, Palaeoecology*, 232, 294–321.

Grignard, J.C. and Jangoux, M. (1994). Occurrence and effects of symbiotic pedunculate barnacles on echinoid tests. In *Echinoderms Through Time* (pp. 679–683), edited by B. David, A. Guille, J.-P. Feral, and M. Roux. Balkema, Rotterdam.

Grigorescu, D., Seclamen, M., Norman, D.B., and Weishampel, D.B. (1990). Dinosaur eggs from Romania. *Nature*, 346, 417.

Grimaldi, D.A. (1993). Forever in amber. *Natural History*, 6/93, 59–61.

Grimaldi, D.A. (1996). *Amber: Window to the Past*. Harry N. Abrams/American Museum of Natural History, New York, 216 pp.

Grimaldi, D.A. (2000). A diverse fauna of Neuropterodea in amber from the Cretaceous of New Jersey. In *Studies on Fossils in Amber, with Particular Reference to the Cretaceous of New Jersey* (pp. 259–303), edited by D. Grimaldi. Backhuys, Leiden, Germany.

Grimaldi, D.A. and Case, G.R. (1995). A feather in amber from the Upper Cretaceous of New Jersey. *American Museum Novitates*, 3126, 1–6.

Grimaldi, D.A. and Engel, M.S. (2005). *Evolution of the Insects*. Cambridge University Press, Cambridge, U.K., 755 pp.

Grimaldi, D.A., Beck, C.W., and Boon, J.J. (1989). Occurrences, chemical characteristics, and paleontology of the fossil resins from New Jersey. *American Museum Novitates*, 2948, 1–28.

Grimaldi, D.A., Kathirithamby, J., and Schawaroch, V. (2005). Strepsiptera and triungula in Cretaceous amber. *Insect Systematics and Evolution*, 36, 1–20.

Grimaldi, G., Shedrinsky, A., and Wampler, T.P. (2000). A remarkable deposit of fossiliferous amber from the Upper Cretaceous (Turonian) of New Jersey. In *Studies on Fossils in Amber, with Particular Reference to the Cretaceous of New Jersey* (pp. 1–76), edited by D. Grimaldi. Backhuys, Leiden, Germany.

Gripp, K. (1929). Uber Verletzungen an Seeigeln aus der Kreide Norddeutschlands. *Paläontologisches Zeitschrift*, 11, 238–245.

Groenewald, G.H., Welman, J., and MacEachern, A. (2001). *Cynognathus* zone burrow complexes from the Early Triassic Cynognathus zone (Driekoppen Formation, Beaufort Group) of the Karoo Basin, South Africa. *Palaios*, 16, 148–160.

Grogan, Jr., W.L. and Szadziewski, R. (1988). A new biting midge from Upper Cretaceous (Cenomanian) amber of New Jersey (Diptera, Ceratopogonidae). *Journal of Paleontology*, 62, 808–812.

Gross, M.R. and Sargent, R.C. (1985). The evolution of male and female parental care in fishes. *American Zoologist*, 25, 807–822.

Grund, M. (2006). Chironomidae (Diptera) in Dominican amber as indicators for ecosystem stability in the Caribbean. *Palaeogeography, Palaeoclimatology, Palaeoecology*, 241, 410–416.

Gubanov, A.P. and Bogolepova, O.K. (2004). The earliest record of a colour pattern in mollusks. *Erlanger geologische Abhandlungen, Sonderband*, 5, 39.

Gubanov, A.P. and Yochelson, E.L. (1994). A Wenlockian (Silurian) gastropod shell and operculum from Siberia. *Journal of Paleontology*, 68, 486–491.

Guernet, C. (1993). Ostracodes du Plateau d'Exmouth (Ocean Indien): Remarques systématiques et evolution des environnements oceaniques profonds au course du Cenozoique. *Géobios*, 26, 345–360.

Guerra, A. and Nixon, M. (1987). Crab and mollusc shell drilling by *Octopus vulgaris* (Mollusca: Cephalopoda) in the Ria de Vigo (north-west Spain). *Journal of Zoology*, 211, 515–524.

Guhl, F., Jaramillo, C., Yockteng, R., Vallejo, G.A., and Cárdenas-Arroyo, F. (1997). *Trypanosoma cruzi* DNA in human mummies. *The Lancet*, 349, 1370.

Guinot, D. and Breton, G. (2006). *Lithophylax trigeri* A. Milne-Edwards & Brocchi, 1879 from the French Cretaceous (Cenomanian) and placement of the family Lithophylacidae Van Straelen, 1936 (Crustacea, Decapoda, Brachyura). *Geodiversitas*, 28, 591–633.

Gundrum, L.E. (1979). Demosponges as substrates: an example from the Pennsylvanian of North America. *Lethaia*, 12, 105–119.

Gunji, Y.-P., Kusunoki, Y., and Ito, K. (1999). Pigmentation of molluscs: how does global synchronisation arise? In *Functional Morphology of the Invertebrate Skeleton* (pp. 37–55), edited by E. Savazzi. John Wiley & Sons, New York.

Guo, S.-X. (1991). A Miocene trace fossil of insect from Shanwang Formation in Linqu, Shandong. *Acta Palaeontologica Sinica*, 30, 739–742.

Gust, S. (2000). *Final Paleontological and Archaeological Monitoring Report for the Olinda Landfill Center Ridge Expansion Project, Brea, Orange County, California*, Project Number 97-1044. Prepared by RMW Paleo Associates for TRC Environmental Solutions, Mission Viejo, CA, 20 pp.

Guthrie, R.D. (1990). Frozen fauna of the mammoth steppe: the story of Blue Babe. University of Chicago Press, Chicago, IL.

Gwynne, D.T. (2003). Mating behaviors. In *Encyclopedia of Insects* (pp. 682–688), edited by V.H. Resh and R.T. Carde. Academic Press, San Diego, CA.

Habersetzer, J. and Storch, G. (1989). Ecology and echolocation of the Eocene Messel bats. In *European Bat Research 1987* (pp. 213–233), edited by V. Hanak, I. Horacek, and J. Gaisler. Charles University Press, Prague, Czech Republic.

Habersetzer, J., Richter, G., and Storch, G. (1992). Bats: already highly specialized insect predators. In *Messel: An Insight into the History of Life and of the Earth* (pp. 181–191), edited by S. Schaal and W. Ziegler. Clarendon Press, London.

Habersetzer, J., Richter, G., and Storch, G. (1994). Paleoecology of early Middle Eocene bats from Messel, FRG: aspects of flight, feeding and echolocation. *Historical Biology*, 8, 235–260.

Hachiya, K. (1993). Mammalia. In *Fossils from the Miocene Morozaki Group* (pp. 263–274), edited by F. Ohe, I. Nonogaki, T. Tanaka, K. Hachiya, Y. Mizuno et al. The Tokai Fossil Society, Nagoya, Japan.

Hagadorn, J.W. (2001). Bear Gulch: an exceptional Upper Carboniferous Plattenkalk. In *Exceptional Fossil Preservation* (pp. 167–183), edited by D.J. Bottjer, W. Etter, J.W. Hagadorn, and C.M. Tang. Columbia University Press, New York.

Hagdorn, H. (1983). *Holocrinus doreckae*, n. sp. aus dem Oberen Muschelkalk und die Entwicklung von Sollbruchstellen in Stiel der Isocrinida. *Neues Jahrbuch für Geologie und Paläontologie-Monatsheft*, 6, 345–368.

Hageman, S.A. and Kaesler, R.L. (2002). Fusulinids: predation damage and repair of tests from the Upper Pennsylvanian of Kansas. *Journal of Paleontology*, 76, 181–184.

Haldane, J.B.S. (1949). Disease and evolution. *La Ricerca Scientifica*, 1(Suppl.), 68–76.

Haldar, D.P. and Chakraborty, N. (1974). A new cephaline gregarine, *Phleobum gigantium*, n. gen., n. sp. (Protozoa: Sporozoa) from a grasshopper. *Proceedings of the Third International Congress of Parasitology, Munich*, 1, 8.

Hall, J. (1888). *Palaeontology of New York*. Vol. V, Part II. *Supplement Containing Descriptions and Illustrations of Pteropoda, Cephalopoda and Annelida*. Geological Survey of the State of New York, C. Van Benthuysen & Sons, Albany, NY, 42 pp.

Hall, J.P.W., Robbins, R.K., and Harvey, D.J. (2004). Extinction and biogeography in the Caribbean: new evidence from a fossil riodinid butterfly in Dominican amber. *Proceedings of the Royal Society London B*, 271, 797–801.

Hallam, A. (1963). Observations on the palaeoecology and ammonite sequence of the Frodingham Ironstone (Lower Jurassic). *Palaeontology*, 6, 554–574.

Hallam, A. (1965). Environmental causes of stunting in living and fossil marine benthonic invertebrates. *Palaeontology*, 8, 132–155.

Halstead, B.W. (1988). *Poisonous and Venomous Marine Animals of the World*, 2nd ed. The Darwin Press, Princeton, NJ, 1168 pp.

Hamamoto, T. and Horikoshi, K. (1994). Characterization of a bacterium isolated from amber. *Biodiversity and Conservation*, 3, 567–572.

Hamilton, A.G. (1896). On domatia in certain Australian and other plants. *Proceedings of the Linnaean Society of New South Wales*, 21, 758–792.

Han, N.-R. and Chen, X. (1994). Regeneration in *Cardiograptus*. *Lethaia*, 27, 117–118.

Hanger, R.A. (1992). *Podichnus centrifugalis* (Bromley and Surlyk, 1973) in the Cretaceous (Albian) Duck Creek Formation, Tarrant County, Texas. *Texas Journal of Science*, 44, 252–254.

Hanley, J.H. (1976). Paleosynecology of nonmarine mollusca from the Green River and Wasatch Formations (Eocene), southwestern Wyoming and northwestern Colorado. In *Structure and Classification of Paleo-communities* (pp. 235–262), edited by R.W. Scott and R.R. West. Dowden, Hutchinson & Ross, Stroudsburg, PA.

Hanna, G.D. and Gaylord, E.G. (1924). Description of a new genus and species of freshwater gastropod mollusk (*Scalex petrolia*) from the Etchegoin Pliocene of California. *Proceedings of the California Academy of Sciences*, 13(9), 147–149.

Hanna, R.R. (2002). Multiple injury and infection in a sub-adult theropod dinosaur *Allosaurus fragilis* with comparisons to allosaur pathology in the Cleveland–Lloyd dinosaur quarry collection. *Journal of Vertebrate Paleontology*, 22, 76–90.

Hansen, T.A. (1978). Larval dispersal and species longevity in lower Tertiary gastropods. *Science*, 199, 885–887.

Hansen, T.A. (1980). Influence of larval dispersal and geographic distribution on species longevity in neogastropods. *Paleobiology*, 6, 193–207.

Hansen, T.A. (1983). Modes of larval development and rates of speciation in early Tertiary neogastropods. *Science*, 220, 501–502.

Hansen, T.A., Graham, S.E., and Kelley, P.H. (1996). Does escalation occur in cycles: evidence from the Neogene. *Sixth North American Paleontological Convention Abstracts*, 8, 160.

Hanski, I. and Cambefort, Y., Eds. (1991). *Dung Beetle Ecology*. Princeton University Press, Princeton, NJ, 481 pp.

Häntzschel, W. (1962). Trace fossils and problematica. In *Treatise on Invertebrate Paleontology*. Part W. *Miscellanea* (pp. 177–245), edited by R.C. Moore. Geological Society of America/ University Press of Kansas, Lawrence.

Häntzschel, W. (1975). Trace fossils and problematica. In *Treatise on Invertebrate Paleontology*. Part W, Suppl. 1. *Miscellanea* (pp. 1–269), edited by C. Teichert. Geological Society of America/University Press of Kansas, Lawrence.

Harington, C.R. (1996). Paleoecology of a Pliocene beaver-pond in the Canadian Arctic Islands. *30th International Geological Congress Abstracts*, 2, 114.

Harmsworth, R.V. (1968). The developmental history of Blelham Tarn (England) as shown by animal microfossils, with special reference to the Cladocera. *Ecological Monographs*, 38, 223–241.

Harper, E.M. (1994). Are conchiolin sheets in corbulid bivalves primarily defensive? *Palaeontology*, 37, 551–578.

Harper, E.M. (2002). Plio–Pleistocene octopod drilling behavior in scallops from Florida. *Palaios*, 17, 292–295.

Harper, E.M. (2003). Assessing the importance of drilling predation over the Palaeozoic and Mesozoic. *Palaeogeography, Palaeoclimatology, Palaeoecology*, 210, 185–198.

Harper. E.M. (2005). Evidence of predation damage in Pliocene *Apletosia maxima* (Brachiopoda). *Palaeontology*, 48, 197–208.

Harper, E.M. (2006). Dissecting post-Palaeozoic arms races. *Palaeogeography, Palaeoclimatology, Palaeoecology*, 232, 322–343.

Harper, E.M. and Wharton, D.S. (2000). Boring predation and Mesozoic articulate brachiopods. *Palaeogeography, Palaeoclimatology, Palaeoecology*, 158, 15–24.

Harper, E.M., Forsythe, G.T.W., and Palmer, T. (1998). Taphonomy and the Mesozoic marine revolution: preservation state masks the importance of boring predators. *Palaios*, 13, 352–360.

Harper, E.M., Dulai, A., Forsythe, G.T.W., Fürsich, F.T., Kowalewski, M., and Palmer, T.J. (1999). A fossil record full of holes: the Phanerozoic history of drilling predation, discussion and reply. *Geology*, 27, 959–960.

Harries, P.J. and Ozanne, C.R. (1998). General trends in predation and parasitism upon inoceramids. *Acta Geologica Polonica*, 48, 377–386.

Harries, P.J. and Schopf, K.M. (2007). Late Cretaceous gastropod drilling intensities: data from the Maastrichtian Fox Hills Formation, Western Interior Seaway, USA. *Palaios*, 22, 35–46.

Harrington, H.J. (1959). General description of Trilobita: thoracic region. In *Treatise on Invertebrate Paleontology*. Part O. *Arthropoda 1* (pp. 33–117), edited by R.C. Moore. Geological Society of America, University of Kansas Press, Lawrence.

Harrison, W.R., Merbs, C.F., and Leathers, C.R. (1991). Evidence of coccidioidmycosis in the skeleton of an ancient Arizona Indian. *Journal of Infectious Diseases*, 164, 437–438.

Hart, M., Tewari, A., and Watkinson, M. (1996). *Teredolites*-infested log-grounds from the Cretaceous of the Cauvery Basin, southeast India. *Palaeontology Newsletter*, 32, xvi.

Hart, P.J.B. (1997). Foraging tactics. In *Behavioral Ecology of Teleost Fishes* (pp. 104–133), edited by J.-G. J. Godin. Oxford University Press, London.

Hary, A. (1987). Epifaune et endofaune de *Liogryphaea arcuata* (Lamarck). *Travaux Scientifiques du Musée d'Histoire naturelle de Luxembourg*, 10, 77.

Harzhauser, M. and Kowalke, T. (2001). Early Miocene brackish-water Mollusca from the Eastern Mediterranean and the Central Paratethys: a faunistic and ecological comparison by selected faunas. *Journal of the Czech Geological Society*, 46, 353–374.

Hasiotis, S.T. (1997). Abuzz before flowers. *Plateau Journal*, 1, 21–27.

Hasiotis, S.T. (2003). Complex ichnofossils of solitary and social soil organisms: understanding their evolution and roles in terrestrial paleoecosystems. *Palaeogeography, Palaeoclimatology, Palaeoecology*, 192, 259–320.

Hasiotis, S.T. and Bown, T.M. (1993). Ichnofossils of deposit-feeding echinoids, Lower Miocene Moghra Formation, Qattara Depression, northwestern Egypt. *Cordilleran/Rocky Mountain Annual Meeting Abstracts*, 25, A49.

Hasiotis, S.T. and Demko, T.M. (1996). Terrestrial and freshwater trace fossils, Upper Jurassic Morrison Formation, Colorado Plateau. In *The Continental Jurassic* (pp. 355–370), edited by M. Morales. Museum of Northern Arizona, Flagstaff.

Hasiotis, S.T. and Dubiel, R.F. (1993). Trace fossil assemblages in Chinle Formation alluvial deposits at the Tepees, Petrified Forest National Park, Arizona. In *The Nonmarine Triassic-Field Guidebook* (pp. G42–G43), edited by S.G. Lucas and M. Morales. New Mexico Museum of Natural History and Science, Albuquerque.

Hasiotis, S.T. and Dubiel, R.F. (1995). Termite (Insecta: Isoptera) nest ichnofossils from the Upper Triassic Chinle Formation, Petrified Forest National Park, Arizona. *Ichnos*, 4, 119–130.

Hasiotis, S.T. and Fiorillo, A.R. (1997). Dermestid beetle borings in dinosaur bones, Dinosaur National Monument, Utah: additional keys to bone bed taphonomy. *Geological Society of America Abstracts with Programs*, 29(2), 13.

Hasiotis, S.T. and Hannigan, R.E. (1991). Use of periodic acid Schiff (PAS) in the identification of possible lungfish burrows in the Upper Triassic Chinle Formation of southeastern Utah and western Colorado. *Geological Society of America Abstracts with Programs*, 23(1), 18590.

Hasiotis, S.T. and Mitchell, C.E. (1993). A comparison of crayfish burrow morphologies: Triassic and Holocene fossil, paleo- and neo-ichnological evidence, and the identification of their burrow signatures. *Ichnos*, 2, 291–314.

Hasiotis, S.T., Aslan, A., and Bown, T.M. (1993a). Origin, architecture, and paleoecology of the early Eocene continental ichnofossil *Scaphichnium hamatum*: integration of ichnology and paleopedology. *Ichnos*, 3, 1–9.

Hasiotis, S.T., Mitchell, C.E., and Dubiel, R.F. (1993b). Application of morphologic burrow interpretations to discern continental burrow architects: lungfish or crayfish? *Ichnos*, 2, 315–333.

Hasiotis, S.T., Bown, T.M., Kay, P.T., Dubiel, R.F., and Demko, T.M. (1996). The ichnofossil record of hymenopteran nesting behavior from Mesozoic and Cenozoic pedogenic and xylic substrates: examples of relative stasis. *Sixth North American Paleontological Convention Abstracts of Papers*, 8, 165.

Hasiotis, S.T., Dubiel, R.F., Kay, P.T., Demko, T.M., Kowalska, K., and McDaniel, D. (1998a). Research update on hymenopteran nests and cocoons, Upper Triassic Chinle Formation, Petrified Forest National Park, Arizona. In *National Park Service Paleontologic Research* (pp. 116–121), edited by V. Santucci and L. McClelland, Technical Report NPS/NRGRD/GRDTR-98/01. Natural Resources Publication Office, Fort Collins, CO.

Hasiotis, S.T., Kirkland, J.I., Windscheffel, G.W., and Safris, C. (1998b). Fossil caddisfly cases (Insecta: Trichoptera), Upper Jurassic Morrison Formation, Fruita Paleontological Area, Colorado. *Modern Geology*, 22, 493–502.

Hasiotis, S.T., Fiorillo, A.R., and Hanna, R.R. (1999). Preliminary report on borings in Jurassic dinosaur bones: evidence for invertebrate–vertebrate interactions. *Vertebrate Paleontology in Utah*, 1999, 193–200.

Hass, H., Taylor, T.N., and Remy, W. (1994). Fungi from the Lower Devonian Rhynie Chert: mycoparasitism. *American Journal of Botany*, 81, 29–37.

Haszprunar, G. (1988). Anatomy and relationships of the bone-feeding limpets *Cocculinella minutissima* (Smith) and *Osteoporella mirabilis* Marshall (Archaeogastropoda). *Journal of Molluscan Studies*, 54, 1–20.

Haszprunar, G. (1995). On the evolution of larval development in the Gastropoda, with special reference to larval planktotrophy. *Notiziario CISMA*, 16(1994), 5–13.

Hatai, K. (1951). A Lower Cretaceous *Teredo*. *Short Papers of the Institute of Geology and Paleontology, Sendai*, 3, 29–32.

Hattin, D.E. (1986). Carbonate substrates of the Late Cretaceous Sea, central Great Plains and Southern Rocky Mountains. *Palaios*, 1, 347–367.

Hattin, D.E. and Hirt, D.S. (1986). Paleoecology of scalpellomorph cirripeds in the Fairport Member, Carlile Shale (Middle Turonian) of central Kansas. *Palaios*, 6, 553–563.

Havlíček, V., Vanek, J., and Fatka, O. (1993). Floating algae of the genus *Krejciella* as probable hosts of epiplanktic organisms (Dobrotiva Series, Ordovician: Prague basin). *Journal of the Czech Geological Society*, 38, 79–88.

Haygood, M.G. (1993). Light organ symbioses in fishes. *Critical Reviews in Microbiology*, 19, 191–216.

Haymon, R.M. and Koski, R.A. (1985). Evidence of an ancient hydrothermal vent community: fossil worm tubes in Cretaceous sulfide deposits of the Samail Ophiolite, Oman. *Biological Society of Washington Bulletin*, 6, 57–65.

Haymon, R.M., Koski, R.A., and Sinclair, C. (1984). Fossils of hydrothermal vent worms from Cretaceous sulfide ores of the Samail Ophiolite, Oman. *Science*, 223, 1407–1409.

Haynes, G. (1991). *Mammoths, Mastodonts, and Elephants*. Cambridge University Press, Cambridge, U.K., 413 pp.

Hayward, B.W. (1977). Lower Miocene corals from Waitakere Ranges, North Auckland, New Zealand. *Journal of the Royal Society of New Zealand*, 7, 99–111.

Heads, S.W. (2006). A new caddisfly larval case (Insecta, Trichoptera) from the Lower Cretaceous Vectis Formation (Wealden Group) of the Isle of Wight, southern England. *Proceedings of the Geologists' Association*, 117, 307–310.

Heer, O. (1850). Der Insekten der Tertiärgebilde von Oeningen und von Radoboj in Croatien. II. Neue Denckschriften Allgem. *Schweizerischen Gesellschaft für die Gesammten Naturwissenschaften B*, XI, 1–254.

Heer, O. (1865). *Die Urwelt der Schweiz*. Friedrich Schultess, Zurich, 622 pp.

Heidelberger, D. and Amler, M.R.W. (2002). Devonian Gastropoda from the Dornap "Massenkalk" complex (Bergisches Land, Germany). *Paläontologisches Zeitschrift*, 76, 317–329.

Heie, O.E. and Penalver, E. (1999). *Palaeophylloxera* nov. gen., the first fossil specimen of the Family Phylloxeridae (Hemiptera, Phylloxeroidea): Lower Miocene of Spain. *Géobios*, 32, 593–597.

Heie, O.E. and Poinar, Jr., G.O. (1988). *Mindazerius dominicanus* nov. gen., nov. sp., a fossil aphid (Homoptera, Aphidoidea, Drespanosiphidae) from Dominican amber. *Psyche*, 95, 153–166.

Heimpel, A.M. (1966). A crystalliferous bacterium associated with a "blister disease" in the earthworm, *Eisenia foetida* (Savigny). *Journal of Invertebrate Pathology*, 8, 295–298.

Heinrich, A. (1988). Fliegenpuppen aus Eiszeitlichen Knochen. *Cranium*, 5(2), 82–83.

Helbling II, M., Rothschild, B., and Tanke, D. (2001). Tertiary neoplasia: a family affair. *Journal of Vertebrate Paleontology*, 21(3), 60A.

Helden, A.J. (2008). First extant records of mermithid nematode parasitism of Auchenorrhyncha in Europe. *Journal of Invertebrate Pathology*, 99, 351–353.

Hellmund, M. and Hellmund, W. (1991). Eiablageverhalten fossiler Kleinlibellen (Odonata, Zygoptera) aus dem Oberoligozän von Rott im Siebengebirge. *Stuttgarter Beiträge zur Naturkunde Serie B*, 177, 1–17.

Hellmund, M. and Hellmund, W. (1993). Neufund fossiler Eilogen (Odonata, Zygoptera, Coenagrionidae) aus dem Oberoligozän von Rott im Siebengebirge. *Decheniana*, 146, 348–351.

Hellmund, M. and Hellmund, W. (1996a). Zum Fortpflanzungsmodus fossiler Kleinlibellen (Insecta, Odonata, Zygoptera). *Paläontologisches Zeitschrift*, 70, 153–170.

Hellmund, M. and Hellmund, W. (1996b). Zur endophytischen Eiablage fossiler Kleinlibellen (Insecta, Odonata, Zygoptera), mit Beschreibung eines deuen Gelegetyps. *Mitteilungen der Bayerische Staatssammlung für Paläontologie und Historische Geologie*, 36, 107–115.

Hellmund, M. and Hellmund, W. (1998). Eilogen von Zygopteren (Insecta, Odonata, Coenagrionidae) in unteroligozänen Maarsedimenten von Hammerunterwiesenthal (Freistaat Sachsen). *Abhandlungen des Staatlichen Museums für Mineralogie und Geologie zu Dresden*, 43/44, 281–292.

Hellmund, M. and Hellmund, W. (2002). Neufunde und Ergänzungen zur Fortpflanzensbiologie fossiler Kleinlibellen (Insecta, Odonata, Zygoptera). *Stuttgarter Beiträge zur Naturkunde Serie B*, 319, 1–26.

Hembree, D.I., Martin, L.D., and Hasiotis, S.T. (2004). Amphibian burrows and ephemeral ponds of the Lower Permian Speiser Shale, Kansas. *Palaeogeography, Palaeoclimatology, Palaeoecology*, 203, 127–152.

Hembree, D.I., Hasiotis, S.T., and Martin, L.D. (2005). *Torridorefugium eskridgensis* (new ichnogenus and species): amphibian aestivation burrows from the Lower Permian Speiser Shale of Kansas. *Journal of Paleontology*, 79, 583–593.

Henderson, R.A., Kennedy, W.J., and Cobban, W.A. (2002). Perspectives of ammonite paleobiology from shell abnormalities in the genus *Baculites*. *Lethaia*, 35, 215–230.

Hengsbach, R. (1990). Die Paläoparasitologie, eine Arbeitsrichtung der Paläobiologie. Senckenbergiana *Lethaia*, 70, 439–461.

Hengsbach, R. (1991a). Studien zur Paläopathologie der Invertebraten. III. Parasitismus bei Ammoniten. *Paläontologisches Zeitschrift*, 65, 127–139.

Hengsbach, R. (1991b). Die Symmetropathie, ein Beiträg zur Erforschung sogenannter Anomalien. *Senckenbergiana Lethaia*, 71, 339–366.

Hengsbach, R. (1996). Ammonoid pathology. In *Ammonoid Paleobiology* (pp. 581–605), edited by N.H. Landman, K. Tanabe, and R.A. Davis. Plenum Press, New York.

Henningsmoen, G. (1975). Moulting in trilobites. *Fossils and Strata*, 4, 179–200.

Henshilwood, C.S., d'Errica, F., Yates, R., Jacobs, Z., Tribolo, C. et al. (2002). Emergence of modern human behavior: Middle Stone Age engravings from South Africa. *Science*, 295, 1278–1280.

Henshilwood, C.S., d'Errica, F., Venhaeren, M., van Niekerk, K., and Jacobs, Z. (2004). Middle Stone Age shell beads from South Africa. *Science*, 304, 404.

Hentschel, U. and Felbeck, H. (1993). Nitrate respiration in the hydrothermal vent tubeworm *Riftia pachyptila*. *Nature*, 366, 338–340.

Herbert, G.S. and Portell, R.W. (2004). First paleontological record of larval brooding in the calyptraeid gastropod genus *Crepidula* Lamarck, 1799. *Journal of Paleontology*, 78, 424–429.

Herbig, H.-G. (1993). First Upper Devonian crustacean coprolites: *Favreina prima* n. sp. from northern Morocco. *Journal of Paleontology*, 67, 98–103.

Herre, E.A. (1993). Population structure and the evolution of virulence in nematode parasites of fig wasps. *Science*, 259, 1442–1445.

Herrmann, B. and Hummel, S., Eds. (1994). *Ancient DNA*. Springer-Verlag, Heidelberg, 263 pp.

Hershkovitz, I.H., Kelley, J., Latimer, B., Rotschild, B.M., Simpson, S., Polak, J., and Rosenberg, M. (1997). Oral bacteria in Miocene *Sivapithecus*. *Journal of Human Evolution*, 33, 507–512.

Hess, H. (1960). Neubeschreibung von *Geocoma elegans* (Ophiuroidea) aus dem unteren Callovien von La Voulte-sur-Rhône (Ardèche). *Eclogae Geologicae Helvetiae*, 53, 335–385.

Hess, H., Ausich, W.I., Brett, C.E., and Simms, M.J. (1999). *Fossil Crinoids*. Cambridge University Press, Cambridge, U.K., 275 pp.

Hewitt, R.A. and Westermann, G.E.G. (1990). Mosasaur tooth marks on the ammonite *Placenticeras* from the Upper Cretaceous of Alberta, Canada. *Canadian Journal of Earth Sciences*, 27, 469–472.

Hibbett, D.S., Grimaldi, D., and Donoghue, M.J. (1997). Fossil mushrooms from Miocene and Cretaceous ambers and the evolution of Homobasidiomycetes. *American Journal of Botany*, 84, 981–991.

Higgins, R.P. and Thiel, H., Eds. (1988). *Introduction to the Study of Meiofauna*. Smithsonian Institution Press, Washington, D.C., 488 pp.

Hikida, Y., Suzuki, S., Togo, Y., and Ijiri, A. (2002). An exceptionally well-preserved seep community from the Cretaceous Yezo forearc basin in Hokkaido, northern Japan. *First International Palaeontological Congress Abstracts*, 68, 215–216.

Hikida, Y., Suzuki, S., Togo, Y., and Ijiri, A. (2003). An exceptionally well-preserved fossil seep community from the Cretaceous Yezo Group in the Nakagawa area, Hokkaido, northern Japan. *Paleontological Research*, 7, 329–342.

Hill, D. (1938–41). *Carboniferous Rugose Corals of Scotland*, Parts I, III, IV. Palaeontographical Society, London, pp. 1–78, 115–204, 205–213.

Hillson, S. (1996). *Dental Anthropology*. Cambridge University Press, Cambridge, U.K., 373 pp.

Hinegardner, R.T. (1958). The venom apparatus of the cone shell. *Hawaii Medical Journal*, 17, 533–536.

Hirsch, K.F. (1994). The fossil record of vertebrate eggs. In *The Palaeobiology of Trace Fossils* (pp. 269–294), edited by S.K. Donovan. Cambridge University Press, Cambridge, U.K.

Hirsch, K.F. (2001). Pathological amniota eggshell: fossil and modern. In *Mesozoic Vertebrate Life* (pp. 378–392), edited by D.H. Tanke and K. Carpenter. Indiana University Press, Bloomington.

Hirsch, K.F. and Harris, J. (1989). Fossil eggs from the Lower Miocene Legetet Formation of Koru, Kenya: snail or lizard? *Historical Biology*, 3, 61–78.

Hirsch, K.F., Krishtalka, L., and Stucky, R.K. (1987). Revision of the Wind River faunas, early Eocene of central Wyoming. Part 8. First fossil lizard egg (?Gekkonidae) and list of associated lizards. *Annals of Carnegie Museum*, 56, 223–230.

Hirsch, K.F., Stadtman, K.L., Miller, W.E., and Madsen, Jr., J.H. (1989). Upper Jurassic dinosaur egg from Utah. *Science*, 243, 1711–1713.

Hladilova, S. and Pek, I. (1998). Oysters attached to gastropod shells from Rudoltice, eastern Bohemia (Miocene, Lower Badenian). *Věstník Českého geologického ústavu*, 73, 137–142.

Hoagland, K.E. (1977). Systematic review of fossil and Recent *Crepidula* and discussion of evolution of the Calyptraeidae. *Malacologia*, 16, 353–420.

Hoagland, K.E. and Turner, R.E. (1981). Evolution and adaptive radiation of wood-boring bivalves (Pholadacea). *Malacologia*, 21, 111–148.

Hoare, R.D. (1978). Annotated bibliography on preservation of color patterns on invertebrate fossils. *The Compass of Sigma Gamma Epsilon*, 53, 39–63.

Hoare, R.D. (2003). *Brachiopods from the Maxville Limestone (Mississippian) of Ohio.* Ohio Geological Survey Report, Investigation 147. Department of Natural Resources, Columbus, 16 pp.

Hoare, R.D. and Steller, D.L. (1967). A Devonian brachiopod with epifauna. *The Ohio Journal of Sciences,* 67, 291–297.

Hoffman, A. (1977). Synecology of macrobenthic assemblages of the Korytnica Clays (Middle Miocene; Holy Cross Mountains, Poland). *Acta Geologica Polonica,* 27, 227–280.

Hoffman, E. and Fowler, M.E. (1995). *The Alpaca Book.* Clay City Press, Herald, CA, 255 pp.

Hoffmeister, A.P., Kowalewski, M., and Bambach, R.K. (2001). Evidence for predatory drilling in Late Paleozoic brachiopods and bivalve mollusks from West Texas. *7th North American Paleontological Convention Program and Abstracts,* 66–67.

Hoffmeister, A.P., Kowalewski, M., Bambach, R.K., and Baumiller, T.K. (2003). Intense drilling in the Carboniferous brachiopod *Cardiarina cordata* Cooper, 1956. *Lethaia,* 36, 107–117.

Hoffmeister, A.P., Kowalewski, M., Baumiller, T.K., and Bambach, R.K. (2004). Drilling predation on Permian brachiopods and bivalves from the Glass Mountains, West Texas. *Acta Palaontologica Polonica,* 49, 443–454.

Hogler, J.A. (1994). Speculations on the role of marine reptile deadfalls in Mesozoic deep-sea paleoecology. *Palaios,* 9, 42–47.

Hölder, H. (1972). Endo- und Epizoen von Belemniten-Rostren (*Megateuthis*) in nordwestdeutschen Bajocium (Mittlerer Jura). *Paläontologische Zeitschrift,* 46, 199–220.

Holland, C.H. (1971). Some conspicuous participants in Palaeozoic symbiosis. *Scientific Proceedings of the Royal Dublin Society Series A,* 4(2), 15–26.

Holland, C.H. (1988). The paper nautilus. *New Mexico Bureau of Mines and Mineral Resources Memoir,* 44, 109–114.

Hölldobler, B. (1976). The behavioral ecology of mating in harvester ants (Hymenoptera: Formicidae: *Pogonomyrmex*). *Behavior Ecology and Sociobiology,* 1, 405–423.

Hölldobler, B. and Haskins, C.P. (1977). Sexual calling behavior in primitive ants. *Science,* 195, 793–794.

Holliday, R.A. (1942). Some observations on Natal Onychophora. *Annals of the Natal Museum,* 10, 237–244.

Hollingworth, N.T.J. and Wignall, P.B. (1992). The Callovian–Oxfordian boundary in Oxfordshire and Wiltshore based on two new temporary sections. *Proceedings of the Geologists' Association,* 103, 15–30.

Holman, J.A. (1968). A small Miocene herpetofauna from Texas. *Quarterly Journal of the Florida Academy of Sciences,* 29, 267–275.

Holmer, L.E. (1989). Middle Ordovician phosphatic inarticulate brachiopods from Vastergotland and Dalarna, Sweden. *Fossils and Strata,* 26, 1–172.

Holmer, L.E., Popov, L.E., Koneva, S.P., and Bassett, M.G. (2001). Cambrian–Early Ordovician brachiopods from Malyi Karatau, the western Balkhash region, and Tien Shan, central Asia. *Special Papers in Palaeontology,* 65, 1–180.

Honigberg, B.M. (1970). Protozoa associated with termites and their role in digestion. In *Biology of Termites,* Vol. 2 (pp. 1–36), edited by K. Krishna and F.M. Wiesner. Academic Press, San Diego, CA.

Hooghiemstra, H. (1989). Quaternary and Upper-Pliocene glaciation and forest development in the tropical Andes: evidence from a long high-resolution pollen record from the sedimentary basin of Bogota, Colombia. *Palaeogeography, Palaeoclimatology, Palaeoecology,* 72, 11–26.

Hooghiemstra, H., Stalling, H., Agwu, C.O.C., and Dupont, L.M. (1992). Vegetational and climatic changes at the northern fringe of the Sahara 250,000–5,000 years BP: evidence from 4 marine pollen records located between Portugal and the Canary Islands. *Reviews of Palaeobotany and Palynology,* 74, 1–53.

Hooker, J.J., Collinson, M.E., van Bergen, P.F., Singer, R.L., de Leeuw, J.W., and Jones, P.T. (1995). Reconstruction of land and freshwater palaeoenvironments near the Eocene–Oligocene boundary, southern England. *Journal of the Geological Society,* 152, 449–468.

Hopfenberg, H.B., Witchey, L.C., and Poinar, Jr., G.O. (1988). Is the air in amber ancient? *Science,* 241, 717–724.

Hopkin, S.P. (1997). *Biology of the Springtails (Insects: Collembola).* Oxford University Press, New York.

Hopkins, C.S., Salva, E.W., and Feldmann, R.M. (1999). Re-evaluation of the genus *Xanthosia* Bell, 1863 (Decapoda: Brachyura: Xanthidae) and description of two new species from the Cretaceous of Texas. *Journal of Paleontology,* 73, 77–90.

Hopley, P.J. (2001). Plesiosaur spinal pathology: the first fossil occurrence of Schmorl's Nodes. *Journal of Vertebrate Paleontology,* 21, 253–260.

Horne, P. (1979). Head lice from an Aleutian mummy. *Paleopathology Newsletter,* 25, 7–8.

Hörnes, M. (1867). Die fossilen Mollusken des Tertiär-Beckens von Wien. *Jahrbuch der Kaiserlich-Königlichen Geologischen Reichsanstalt,* 17, 583–588.

Horný, R. (1992a). New Lower Devonian Gastropoda and Tergomya (Mollusca) of Bohemia. *Casopis Národniho Muzea v Praze. Rada Prirodovedná,* 159, 99–110.

Horný, R. (1992b). *Lytospira* Koken and *Murchisonia* (*Hormotomina*) Grabau et Shimer in the Lower Devonian of Bohemia. *Casopis Národniho Muzea v Praze. Rada Prirodovedná,* 160, 55–56.

Horný, R.J. (1997a). Ordovician Tergomya and Gastropoda (Mollusca) of the Anti-Atlas (Morocco). *Acta Musei Nationalis Pragae, Series B, Historia Naturalis,* 53, 37–78.

Horný, R.J. (1997b). Shell breakage and repair in explanate bellerophontoidean gastropods from the Middle Ordovician of Bohemia. *Bulletin of the Czech Geological Survey,* 72, 159–170.

Horný, R.J. (1998a). Two additional, isolated, paucispiral gastropod opercula from the Lower Devonian Koneprusy Limestone (Bohemia, Barrandian area). *Casopis Národniho Muzea v Praze. Rada Prirodovedná,* 167 (1–4), 91–94.

Horný, R.J. (1998b). A large injury and shell repair in *Boiotremus incipiens* (Mollusca, Gastropoda) from the Silurian of Bohemia. *Bulletin of the Czech Geological Survey,* 73, 343–345.

Horný, R. (2000). Mode of life of some Silurian and Devonian platyceratids. *Bulletin of the Czech Geological Survey,* 75, 135–143.

Horný, R.J. (2001). *Novakopteron,* a new *Beraunia*-like frilled genus from the Silurian of Bohemia (Mollusca, Gastropoda). *Casopis Národniho Muzea v Praze. Rada Prirodovedná,* 170(1–4), 37–41.

Horný, R.J. (2002). Anomalous development of apertural margin and failed predation in the Lower Devonian gastropod *Anarconcha pulchra* from the Barrandian (Czech Republic). *Casopis Národniho Muzea v Praze. Rada Prirodovedná,* 171, 1–6.

Horný, R.J. (2005). Muscle scars, systematics and mode of life of the Silurian Family Drahomiridae (Mollusca, Tergomya). *Acta Musei Nationalis Pragae, Series B, Historia Naturalis,* 61(1–2), 53–76.

Horný, R.J. (2006). The Middle Ordovician tergomyan mollusk: an obligatory epibiont on hyolithds. *Acta Musei Nationalis Pragae, Series B, Historia Naturalis*, 62, 81–95.

Horný, R.J. and Peel, J.S. (1995). A new Silurian gastropod from Bohemia with the operculum *in situ*. *Journal of the Czech Geological Society*, 40, 79–88.

Hotton, C.L., Hueber, F.M., and Labandeira, C.C. (1996). Plant–arthropod interactions from early terrestrial ecosystems: two Devonian examples. *Sixth North American Paleontological Convention Abstracts of Papers*, 8, 181.

Housa, V. (1963). Parasites of Tithonian decapod crustaceans (Stramberk, Moravia). *Sbornik Ustredniho ústavu Geologickeho*, 28, 101–114.

House, M.R. (1960). Abnormal growths in some Devonian goniatites. *Palaeontology*, 3, 129–136.

Houston, T.F. (1987). Fossil brood cells of stenotritid bees (Hymenoptera: Apoidea) from the Pleistocene of South Australia. *Transactions of the Royal Society of South Australia*, 111, 93–97.

Hu, C.-H. and Tao, H.-J. (1996). *Crustacean Fossils of Taiwan*. Ta-Jen Printers, Taipei, Taiwan, 228 pp.

Hua, H., Pratt, B.R., and Zhang, L.-Y. (2003). Borings in *Cloudina* shells: complex predatory–prey dynamics in the terminal Neoproterozoic. *Palaios*, 18, 454–459.

Hudson, J.D. and Palmer, T.J. (1976). A euryhaline oyster from the middle Jurassic and the origin of the true oysters. *Palaeontology*, 19, 79–94.

Huggett, J.M. and Gale, A.S. (1995). Palaeoecology and diagenesis of bored wood from the London Clay Formation of Sheppey, Kent. *Proceedings of the Geologists' Association*, 106, 119–136.

Hughes, N.C. (2001). Ecologic evolution of Cambrian trilobites. In *The Ecology of the Cambrian Radiation* (pp. 370–403), edited by A.Y. Zhuravlev and R. Riding. Columbia University Press, New York.

Hugueney, M. and Escuillie, F. (1995). K-strategy and adaptive specialization in *Steneofiber* from Montagiu-le-Blin (dept. Allier, France; Lower Miocene, MN, 2a, 23Ma): first evidence of fossil life-history strategies in castorid rodents. *Palaeogeography, Palaeoclimatology, Palaeoecology*, 113, 217–225.

Hugueney, M., Tachet, H., and Escuillie, F. (1990). Caddisfly pupae from the Miocene Indusial Limestone of Saint-Gérard-le-Puy, France. *Palaeontology*, 33, 495–502.

Hulbert, Jr., R.C. and Morgan, G.S. (1993). Quantitative and qualitative evolution in the giant armadillo *Holmesina* (Edentata: Pampatheridae) in Florida. In *Morphological Change in Quaternary Mammals of North America* (pp. 134–177), edited by R.A. Martin and A.D. Barnosky. Cambridge University Press, New York.

Hunt, Jr., R.M (1990). Taphonomy and sedimentology of Arikaree (lower Miocene) fluvial, eolian, and lacustrine paleoenvironments, Nebraska and Wyoming: a paleobiota entombed in fine-grained volcaniclastic rocks. In *Volcanism and Fossil Biotas* (pp. 69–111), edited by M.G. Lockley and A. Rice, Special Paper 244. Geological Society of America, Boulder, CO.

Hunter, H.A. and Langston, Jr., W. (1964). Odontoma in a northern mammoth. *Palaeontology*, 7, 674–781.

Hurst, J.M. (1974). Selective epizoan encrustacean of some Silurian brachiopods from Gotland. *Palaeontology*, 17, 423–429.

Hurst, J.M. (1975). *Resserella sabrinae* Bassett, in the Wenlock of Wales and the Welsh Borderland. *Journal of Paleontology*, 49, 316–328.

Hutchinson, J.H. and Frye, F.L. (2001). Evidence of pathology in early Cenozoic turtles. *PaleoBios*, 21(3), 12–19.

Hutchinson, J.H. and Harington, C.R. (2002). A peculiar new fossil shrew (Lipotyphla, Soricidae) from the High Arctic of Canada. *Canadian Journal of Earth Sciences*, 39, 439–443.

Huys, R. and Boxshall, G.A. (1991). *Copepod Evolution*. The Ray Society, London, 468 pp.

Hyman, L.H. (1940). *The Invertebrates: Protozoa Through Ctenophora*. McGraw-Hill, New York, 726 pp.

Hyman, L.H. (1951). *The Invertebrates: Platyhelminthes and Rhynchocoela, the Acoelomate Bilateria*, Vol. 2. McGraw-Hill, New York, 550 pp.

Insalaco, E. (1996). Upper Jurassic microsolenid biostromes of northern and central Europe: facies and depositional environment. *Palaeogeography, Palaeoclimatology, Palaeoecology*, 121, 169–194.

Iribarne, O.O., Pascual, M.S., and Zampatti, E.A. (1990). An uncommon oyster breeding system in a Late Tertiary Patagonian species. *Lethaia*, 23, 153–156.

Isakar, M. and Ebbestad, J.O.R. (2000). *Bucania* (Gastropoda) from the Ordovician of Estonia. *Paläontologisches Zeitschrift*, 74, 51–68.

Ishida, Y. (1993). Ophiuroidea. In *Fossils from the Miocene Morozaki Group* (pp. 123–140), edited by F. Ohe, I. Nonogaki, T. Tanaka, K. Hachiya, Y. Mizuno et al. The Tokai Fossil Society, Nagoya, Japan.

Ishida, Y. and Inoue, K. (1994). Paleoecology of fossil ophiuroids (*Ophiura sarsii* Lutken, 1854) from the Pleistocene Ichijuku formation (Kazusa Group), Chiba prefecture, Central Japan. In *Echinoderms Through Time* (pp. 437–440), edited by B. David, A. Guille, J.-P. Feral, and M. Roux. Balkema, Rotterdam.

Ishida, Y., Nagamori, H., and Narita, K. (1998). *Ophiura sarsii sarsii* (Echinodermata, Ophiuroidea) from the Late Miocene Ogawa Formation, Shinshushinmachi, Nagano Prefecture, central Japan. *Research Report of the Shinshushinmachi Fossil Museum*, 1, 9–16.

Ishida, Y., Nagasawa, K., and Tokairin, H. (1999). *Ophiura sarsii sarsii* (Echinodermata, Ophiuroidea) from the Late Miocene to Early Pliocene formations of Yamagata Prefecture, northern Japan. *Earth Science (Chikyu Kagaku)*, 53, 223–232.

Ishikawa, M. and Kase, T. (2007). Spionid bore hole *Polydorichnus subapicalis* new ichnogenus and ichnospecies: a new behavioral trace in gastropod shells. *Journal of Paleontology*, 81, 1466–1475.

Isidro, A., Malgosa, A., Belinchon, M., Vela, S., Alcala, L., Dieguez, C., Castellana, C., and Fernandez, S. (1996). Zoopaleopathology from Spain: a discussion of some cases. *Journal of Paleopathology*, 8, 143–152.

Itoigawa, J. (1963). Miocene rock- and wood-boring bivalves and their burrows from the Mizunami group, central Japan. *The Journal of Earth Sciences*, 11(1), 101–123.

Ivanov, V.D. (2006). Lichinki rucheinikov (Insecta: Trichoptera) iz Mezosoya Siberi. *Paleontologicheskii Zhurnal*, 2, 62–71.

Ivanova, E.A. (1949). Uslovya suschestvovania obraz zhyzni i istorya razvitia nekotorykh brakhiopod srednego karbona podmoskovnoi kotloviny. *Transactions of the Paleontological Institute, Academy of Science, USSR*, 21, 3–144.

Ivany, L.C., Portell, R.W., and Jones, D.S. (1990). Animal–plant relationships and paleobiogeography of an Eocene seagrass community from Florida. *Palaios*, 5, 244–258.

Jablonski, D. and Lutz, R.A. (1980). Molluscan larval shell morphology. In *Skeletal Growth of Aquatic Organisms* (pp. 323–377), edited by D.C. Rhoads and R.A. Lutz. Plenum Press, New York.

Jackson, J.B.C. (1992). Pleistocene perspectives on coral reef community structure. *American Zoologist*, 32, 719–731.

Jackson, J.B.C., Goreau, T.F., and Hartman, W.D. (1971). Recent brachiopod–coralline sponge communities and their paleoecological significance. *Science*, 173, 623–625.

Jackson, J.B.C., Budd, A.F., and Pandolfi, J.M. (1996). The shifting balance of natural communities. In *Evolutionary Paleobiology* (pp. 89–122), edited by D. Jablonski, D.H. Erwin, and J.H. Lipps. University of Chicago Press, Chicago, IL.

Jackson, J.W. (1909). On some fossil pearl growths. *Proceedings of the Malacological Society of London*, 8, 318–320.

Jacob, J., Paris, F., Monod, O., Miller, M.A., Tang, P., George, S.C., and Bény, J.-M. (2007). New insights into the chemical composition of chitinozoans. *Organic Geochemistry*, 38, 1782–1788.

Jacobs, M. (1966). On domatia: the viewpoints and some facts, Parts I–III. *Academie van Wetenschappen Amsterdam*, 69, 275–316.

Jacobs, Z., Roberts, R.G., Galbraith, R.F., Deacon, H.J., Grün, R., Mackay, A., Mitchell, P., Vogelsang, R., and Wadley, L. (2008). Age for the Middle Stone Age of Southern Africa: implications for human behavior and dispersal. *Science*, 322, 733–735.

Jacobsen, A.R. (1998). Feeding behaviour of carnivorous dinosaurs as determined by tooth marks on dinosaur bones. *Historical Biology*, 13, 17–26.

Jaekel, O. (1901). Ueber Jurassische Zahne und Eier von Chimäriden. *Neues Jahrbuch für Mineralogie, Geologie und Paläontologie*, 14, 540–564.

Jäger, M. (1991). Lias epsilon von Dotternhausen, 2. *Fossilien*, 8, 33–36.

Jäger, M. and Fraaye, R. (1997). The diet of the early Toarcian ammonite *Harpoceras falciferum*. *Palaeontology*, 40, 557–574.

Jago, J.B. and Haines, P.W. (2002). Repairs to an injured Early Middle Cambrian trilobite, Elkedra Area, Northern Territory. *Alcheringa*, 26, 19–21.

Jagt, J.W.M. and Collins, J.S.H. (1999). Log-associated late Maastrichtian cirripeds from northeast Belgium. *Paläontologisches Zeitschrift*, 73, 99–111.

Jagt, J.W.M. and Kiel, S. (2008). The operculum of *Otostoma retzii* (Nilsson, 1827) (Gastropoda, Neritidae; Late Cretaceous) and its phylogenetic significance. *Journal of Paleontology*, 82, 201–205.

Jagt, J.W.M. and van der Ham, R.W.J.M. (1994). Early Paleocene marsupiate regular echinoids from NE Belgium. In *Echinoderms Through Time* (pp. 725–729), edited by B. David, A. Guille, J.-P. Feral, and M. Roux. Balkema, Rotterdam.

Jagt, J.W.M., van Bakel, B.W.M., Fraaije, R.H.B., and Neumann, C. (2006). In *situ* hermit crabs (Paguroidea) from northwest Europe and Russia. Preliminary data on new records. *Revista Mexicana de Ciencias Geológicas*, 23, 364–369.

Jagt, J.W.M., Dortangs, R., Simon, E., and van Knippenberg, P. (2007). First record of the ichnofossil *Podichnus centrifugalis* from the Maastrichtian of northeast Belgium. *Bulletin de l'Institut Royal des Sciences Naturelles de Belgique, Sciences de la Terre*, 77, 95–105.

Jahnke, H. (1966). Beobachtungen an einem Hartgrund (Oberkants Terebratelbank mu gamma2 bei Göttingen). *Der Aufschluss*, 1, 2–5.

Jakobsen, S.L. and Collins, J.S.H. (1997). New Middle Danian species of anomuran and brachyuran crabs from Fakse, Denmark. *Bulletin of the Geological Society of Denmark*, 44, 89–100.

Jangoux, M. (1984). Diseases of echinoderms. *Helgoländer Meeresuntersuchungen*, 37, 207–216.

Jangoux, M. (1990). Diseases of Echinodermata. In Diseases of marine animals, Vol. III (pp. 439–567), edited by O. Kinne. Biologische Anstalt Helgoland, Oldendorf, Germany.

Jansson, H.-B. and Poinar, Jr., G.O. (1986). Some possible nematophagous fungi. *Transactions of the British Mycological Society*, 87, 471–474.

Jansson, I.-M., McLoughlin, S., and Vajda, V. (2008). Early Jurassic annelid cocoons from eastern Australia. *Alcheringa*, 32, 285–296.

Janvier, P. (1996). *Early Vertebrates*. Clarendon Press, London, 393 pp.

Janzen, J.-W. (2002). *Arthropods in Baltic Amber*. Ampyx Verlag, Halle, Germany, 167 pp.

Jarzembowski, E.A. (1989). A century plus of fossil insects. *Proceedings of the Geologists' Association*, 100, 433–449.

Jarzembowski, E.A. (1991). The Weald Clay of the Weald: report of 1988/89 field meetings. *Proceedings of the Geologists' Association*, 102, 83–92.

Jarzembowski, E.A. (1992). Fossil insects from the London Clay (Early Eocene) of Southern England. *Tertiary Research*, 13, 87–94.

Jarzembowski, E.A. (1995). Fossil caddis-flies (Insecta: Trichoptera) from the Early Cretaceous of southern England. *Cretaceous Research*, 16, 695–703.

Jarzembowski, E. and Ross, A. (1993). Time flies: the geological record of insects. *Geology Today*, 9, 218–223.

Jeffery, C.H. (1997). Dawn of echinoid nonplanktotrophy: coordinated shifts in development indicate environmental instability prior to the K-T boundary. *Geology*, 25, 991–994.

Jell, P.A. (1989). Some aberrant exoskeletons from fossil and living arthropods. *Memoirs of the Queensland Museum*, 27, 491–498.

Jeng, M.-S., Ng, N.K., and Ng, P.K.L. (2004). Hydrothermal vent crabs feast on sea 'snow'. *Nature*, 433, 969.

Jenkins, R.G., Kaim, A., and Hikida, Y. (2007a). Antiquity of the substrate choice among acmeaeid limpets from Late Cretaceous chemosynthesis-based communities. *Acta Palaeontologica Polonica*, 52, 369–373.

Jenkins, R.G., Kaim, A., Hikida, Y., and Tanabe, K. (2007b). Methane-flux-dependent lateral faunal changes in a Late Cretaceous chemosymbiotic assemblage from the Nakagawa area of Hokkaido, Japan. *Geobiology*, 5, 127–139.

Jepsen, G. (1963). Eocene vertebrates, coprolites, and plants in the Golden Valley formation of western North Dakota. *Geological Society of America Bulletin*, 74, 673–684.

Jerison, H.J. (1973). *Evolution of the Brain and Intelligence*. Academic Press, San Diego, CA, 482 pp.

Jerzmanska, A. (1968). Ichtyofauna des couches à Ménilite (Flysch des Karpathes). *Acta Palaeontologica Polonica*, XIII, 379–481.

Jerzmanska, A. and Kotlarczyk, J. (1976). The beginnings of the Sargasso Assemblage in the Tethys. *Palaeogeography, Palaeoclimatology, Palaeoecology*, 20, 297–306.

Jerzmanska, A. and Kotlarczyk, J. (1983). Ichthyofauna changes in the Tertiary of the Carpathians and of the Caucasus. In *Travaux du 12-éme congrès de l'association géologique carpatho-balkanique. Stratigraphie et Paléontologie* (pp. 191–198). Annuaire de l' Institut de Geålogie et de Geålophysique, Bucuresti.

Ji, Q., Luo, Z.-X., Wible, J.R., Zhang, J.-P., and Georgi, J.A. (2002). The earliest known eutherian mammal. *Nature*, 416, 816–822.

Ji, Q., Ji, S.-A., Cheng, Y.-N., You, H.-L., Lü, J.-C., Liu, Y.-Q., and Yuan, C.-X. (2004). Pterosaur egg with a leathery shell. *Nature*, 432, 542.

Ji, Q., Luo, Z.-X., Yuan, C.-X., and Tabrum, A.R. (2006). A swimming mammaliaform from the Middle Jurassic and ecomorphological diversification of early mammals. *Science*, 311, 1123–1127.

Jiménez-Fuentes, E., Martin de Jesus, S., and Alonso, E.M. (1987). Malformaciones y deformaciones patológicas en Tortugas fósiles. *Notas Informativas, Sala de las Tortugas*, 3, 1–3.

Johns, R.B., Ed. (1986). *Biological Markers in the Sedimentary Record*. Elsevier, Amsterstam, 361 pp.

Johnson, A.L.A. (1993). Punctuated equilibrium versus phyletic gradualism in European Jurassic *Gryphaea* evolution. *Proceedings of the Geologists' Association*, 104, 209–222.

Johnson, A.L.A. (1994). Evolution of European Lower Jurassic *Gryphaea* (*Gryphaea*) and contemporaneous bivalves. *Historical Biology*, 7, 167–186.

Johnson, J.E. and Borkent, A. (1998). *Chaoborus* Lichtenstein (Diptera: Chaoboridae) pupae from the middle Eocene of Mississippi. *Journal of Paleontology*, 72, 491–493.

Johnson, J.G. and Talent, J.A. (1967). Cortezorthinae, a new subfamily of Siluro-Devonian Dalmanellid Brachiopods. *Palaeontology*, 10, 142–170.

Johnson, M.E. (1979). Evolutionary brachiopod lineages from the Llandovery Series of eastern Iowa. *Palaeontology*, 22, 549–567.

Johnston, J.E. (1993). Insects, spiders, and plants from the Tallahatta Formation (Middle Eocene) in Benton County, Mississippi. *Mississippi Geology*, 14, 71–82.

Johnston, P.A., Eberth, D.A., and Anderson, P.K. (1996). Alleged vertebrate eggs from Upper Cretaceous redbeds, Gobi Desert, are fossil insect (Coleoptera) pupal chambers. *Canadian Journal of Earth Sciences*, 33, 511–525.

Jones, A.K.G. (1986). Parasitological investigation on Lindow Man. In *Lindow Man: The Body in the Bog* (pp. 136–139), edited by I.M. Stead, J.B. Bourke and D. Brothwell. Cornell University Press, Ithaca, NY.

Jones, B. (1982). Paleobiology of the Upper Silurian brachiopod *Atrypoidea*. *Journal of Paleontology*, 56, 912–923.

Jones, B. and Hurst, J.M. (1984). Autecology and distribution of the Silurian brachiopod *Dubaria*. *Palaeontology*, 27, 699–706.

Jones, B. and Pemberton, S.G. (1988). *Lithophaga* borings and their influence on the diagenesis of corals in the Pleistocene Ironshore Formation of Grand Cayman Island, British West Indies. *Palaios*, 3, 3–21.

Jones, T.D., Ruben, J.A., Martin, L.D., Kurochkin, E.N., Feduccia, A., Maderson, P.F.A., Hillenius, W.J., Geist, N.R., and Alifanov, V. (2000). Nonavian feathers in a Late Triassic Archosaur. *Science*, 288, 2202–2205.

Jones, T.D., Ruben, J.A., Maderson, P.F.A., and Martin, L.D. (2001). *Longisquama* fossil and feather morphology. *Science*, 291, 1899–1902.

Jonkers, H.A. (2000). Gastropod predation patterns in Pliocene and Recent pectinid bivalves from Antarctica and New Zealand. *New Zealand Journal of Geology and Geophysics*, 43, 247–254.

Jouy-Avantin, F., Combes, C., de Lumley, H., Miskovsky, J.-C., and Moné, H. (1999). Helminth eggs in animal coprolites from a middle Pleistocene site in Europe. *Journal of Parasitology*, 85, 376–379.

Joyce, W.G. (2000). The first complete skeleton of *Solnhofia parsonsi* (Cryptodira, Eurysternidae) from the Upper Jurassic of Germany and its taxonomic implications. *Journal of Paleontology*, 74, 684–700.

Joysey, K.A. (1959). Probable cirripede, phoronid, and echiurid burrows within a Cretaceous echinoid test. *Palaeontology*, 1, 397–400.

Jung, W., Selmeier, A., and Dernbach, U. (1992). *Araucaria: Petrified Cones and Petrified Wood of* Araucaria *from the Cerro Cuadrado, Argentina*. D'Oro Verlag, Heppenheim, Germany, 160 pp.

Jungers, W.L., Godfrey, L.R., Simons, E.L., Wunderlich, R.E., Richmond, B.G., and Chatrath, P.S. (2002). Ecomorphology and behavior of giant extinct lemurs from Madagascar. In *Reconstructing Behavior in the Primate Fossil Record* (pp. 371–411), edited by J.M. Plavcan, R.F. Kay, W.L. Jungers, and C.P. van Schaik. Kluwer Academic/Plenum Publishers, New York.

Kabat, A.R. (1990). Predatory ecology of naticid gastropods with a review of shell boring predation. *Malacologia*, 32, 155–193.

Kabat, A.R. and Kohn, A.J. (1986). Predation on early Pleistocene naticid gastropods in Fiji. *Palaeogeography, Palaeoclimatology, Palaeoecology*, 53, 255–269.

Kaddumi, H.F. (2006). A new genus and species of gigantic marine turtles (Chelonoidea: Cheloniidae) from the Maastrichtian of the Harrana Fauna–Jordan. *PalArch's Journal of Vertebrate Palaeontology*, 3(1), 1–14.

Kaim, A. (2006). The Middle Jurassic sunken wood association from Poland. In *Ancient Life and Modern Approaches: Abstracts of the Second International Palaeontological Congress* (pp. 482–483), edited by Q. Yang, Y. Wang, and E.A. Weldon. University of Science and Technology of China Press, Beijing.

Kaim, A., Jenkins, R.G., and Warén, A. (2008a). Provannid and provannid-like gastropods from the Late Cretaceous cold seeps of Hokkaido (Japan) and the fossil record of the Provannidae (Gastropoda: Abyssochrysoidea). *Zoological Journal of the Linnean Society*, 154, 421–436.

Kaim, A., Kobayashi, Y., Echizenya, H., Jenkins, R.G., and Tanabe, K. (2008b). Chemosynthesis-based associations on Cretaceous plesiosaurid carcasses. *Acta Paleontologica Polonica*, 53, 97–104.

Kaiser, H.E. (1964). Pathological conditions of the soft parts of a Devonian brachiopod species *Stropheodonta*. *Neues Jahrbuch für Mineralogie, Geologie und Paläontologie–Monatsheft*, 4, 196–198.

Kalabis, V. (1948). Sur les Poissons fossiles à organs lumineux du Paléogène (schistes ménilitiques) en Moravie (CSR). *Casopis Moravského Zemského Muzea*, 32, 131–234.

Kalabis, V. and Schultz, O. (1974). Die fischfauna der paläogenen Menilitschichten von Speitsch in Mähren, CSSR. *Annalen des Naturhistorischen Museums in Wien*, 78, 183–192.

Kalmijn, A.J. (1988a). Hydrodynamic and acoustic field detection. In *Sensory Biology of Aquatic Animals* (pp. 83–130), edited by J. Atema, R.R. Fay, A.N. Popper, and W. Tavolga. Springer-Verlag, Heidelberg.

Kalmijn, A.J. (1988b). Detection of weak electric fields. In *Sensory Biology of Aquatic Animals* (pp. 151–186), edited by J. Atema, R.R. Fay, A.N. Popper, and W. Tavolga. Springer-Verlag, Heidelberg.

Kalugina, N.S. (1991). New Mesozoic Simuliidae and Leptoconopidae and the origin of bloodsucking habit in the lower dipteran insects. *Paleontological Journal*, 1991(1), 66–77.

Kanie, Y. and Kuramochi, T. (1996). Description on possibly chemosynthetic bivalves from the Cretaceous deposits of the Obira-cho, northwestern Hokkaido. *Science Reports of the Yokosuka City Museum*, 44, 63–68.

Kanie, Y., Yoshikawa, Y., Sakai, T., and Kuramochi, T. (1996). Cretaceous chemosynthetic fauna from Hokkaido. *Science Reports of the Yokosuka City Museum*, 44, 69–74.

Kar, R.K., Sharma, N., Agarwal, A., and Kar, R. (2003). Occurrence of fossil-wood rotters (Polyporales) from the Lameta Formation (Maastrichtian), India. *Current Science*, 85, 37–40.

Kar, R.K., Sharma, N., and Kar, R. (2004). Occurrence of fossil fungi in dinosaur dung and its implications on food habit. *Current Science*, 87, 1053–1056.

Karasawa, H. (1993). Cenozoic Crustacea from Southwest Japan. *Mizunami Fossil Museum*, 20, 1–92.

Karasawa, H. (2002). Fossil uncinidean and anomalan Decapoda (Crustacea) in the Kitakyushu Museum and Institute of Natural History. *Bulletin of the Kitakyushu Museum of Natural History*, 21, 13–16.

Karim, T. and Westrop, S.R. (2002). Taphonomy and paleoecology of Ordovician trilobite clusters, Bromide Formation, South-central Oklahoma. *Palaios*, 17, 394–403.

Kase, T. and Ishikawa, M. (2003). Mystery of naticid predation history solved: evidence from a "living fossil" species. *Geology*, 31, 403–406.

Kase, T. and Maeda, H. (1980). Early Cretaceous gastropods from the Choshi District, Chiba Prefecture, Central Japan. *Transactions and Proceedings of the Palaeontological Society of Japan*, 118, 291–324.

Kase, T., Shigeta, Y., and Futakami, M. (1994). Limpet home depressions in Cretaceous ammonites. *Lethaia*, 27, 49–58.

Kase, T., Johnson, P.A., Seilacher, A., and Boyce, J.B. (1998). Alleged mosasaur bite marks on Late Cretaceous ammonites are limpet (patellogastropod) home scars. *Geology*, 26, 947–950.

Kashiwaya, K., Ochial, S., Sakai, H., and Kawai, T. (2001). Orbit-related long-term climate cycles revealed in a 12-myr continental record from Lake Baikal. *Nature*, 410, 71–74.

Kathrithamby, J. and Grimaldi, D.A. (1993). Remarkable stasis in some Lower Tertiary parasitoids: descriptions, new records and review of Streptsiptera in the Oligo–Miocene amber of the Dominican Republic. *Entomologica Scandinavica*, 24, 31–41.

Katinas, V. (1983). *Baltijos Gintaras*. Mokslas, Vilnius, Lithuania, 111 pp.

Kato, H. (2005). A new pinnotherid crab (Decapoda: Brachyura: Pinnotheridae) from the Miocene Niijukutoge Formation, northeast Japan. *Paleontological Research*, 9, 73–78.

Katsura, Y. (2004). Paleopathology of *Toyotomaphimeia machikanensis* (Diapsida, Crocodylia) from the Middle Pleistocene of Central Japan. *Historical Biology*, 16, 93–97.

Kauffman, E.G. (1967). Cretaceous *Thyasira* from the Western Interior of North America. *Smithsonian Miscellaneous Collection*, 152(1), 1–159.

Kauffman, E.G. (1972). *Ptychodus* predation upon a Cretaceous *Inoceramus*. *Palaeontology*, 15, 439–444.

Kauffman, E.G. (1990). Giant fossil Inoceramid bivalve pearls. In *Evolutionary Paleobiology of Behavior and Coevolution* (pp. 66–68), edited by A.J. Boucot. Elsevier, Amsterdam.

Kauffman, E.G. (2004). Mosasaur predation on Upper Cretaceous nautiloids and ammonites from the United States Pacific Coast. *Palaios*, 19, 96–100.

Kauffman, E.G. and Harries, P.J. (1996). The importance of crisis progenitors in recovery from mass extinction events. In *Biotic Recovery from Mass Extinction Events* (pp. 15–39), edited by M.B. Hart, Geological Society of London Special Publication 102. Geological Society, London.

Kauffman, E.G., Arthur, M.A., Howe, B., and Scholle, P.A. (1996). Widespread venting of methane-rich fluids in Late Cretaceous (Campanian) submarine springs (Tepee Buttes), Western Interior Seaway, USA. *Geology*, 24, 799–802.

Kay, P.T., King, D., and Hasiotis, S.T. (1997). Petrified Forest National Park Upper Triassic trace fossils yield biochemical evidence of phylogenetic link to modern bees (Hymenopteras: Apoidea). *Geological Society of America Abstracts with Programs*, 29, 102.

Kear, B.P. (2006). First gut contents in a Cretaceous sea turtle. *Biology Letters*, 2, 113–115.

Kear, B.P. and Godthelp, H. (2008). Inferred vertebrate bite marks on an Early Cretaceous unionoid bivalve from Lightning Ridge, New South Wales, Australia. *Alcheringa*, 32, 65–71.

Kear, B.P., Boles, W.E., and Smith, E.T. (2003). Unusual gut contents in a Cretaceous ichthyosaur. *Proceedings of the Royal Society London B*, 2(270, Suppl.), S206–S208.

Keen, A.M. and Smith, A.G. (1961). West American species of the bivalved gastropod *Berthelinia*. *Proceedings of the California Academy of Sciences*, 30(2), 47–66.

Keirans, J.E., Land, R.S., and Cauble, R. (2002). A series of larval *Amblyomma* species (Acari: Ixodidae) from amber deposits in the Dominican Republic. *International Journal of Acarology*, 28, 61–66.

Kelber, K.-P. (1990). Die versunkene Pflanzenwelt aus den Deltasümpfen Mainfrankens vor 230 Millionen Jahren. *Beringeria*, 1, 1–67.

Kelber, K.-P. and Geyer, G. (1989). Lebenspuren von Insekten an Pflanzen des Unteren Keupers. *Courier Forschungs-Institut Senckenberg*, 109, 165–174.

Keller, T. (1976). Magen- und Darminhalts von Ichthyosaurien des suddeutschen Posidonienschiefers. *Neues Jahrbuch für Geologie und Paläontologie–Monatsheft*, 5, 266–283.

Keller, T. and Schaal, S. (1992). Crocodiles: large ancient reptiles. In *Messel: An Insight into the History of Life and of the Earth* (pp. 109–118), edited by S. Schaal and W. Ziegler. Clarendon Press, London.

Kelley, P.H. (2007). Role of bioerosion in taphonomy: effect of predatory drillholes on preservation of mollusk shell. In *Current Developments in Bioerosion* (pp. 451–470), edited by M. Wisshak and L. Tapanila. Springer-Verlag, Heidelberg.

Kelley, P.H. and Hansen, T.A. (1993). Evolution of the naticid gastropod predator–prey system: an evaluation of the hypothesis of escalation. *Palaios*, 8, 358–375.

Kelley, P.H. and Hansen, T.A. (1996). Recovery of the naticid gastropod predator–prey system from the Cretaceous–Tertiary and Eocene–Oligocene extinctions. In *Biotic Recovery from Mass Extinction Events* (pp. 373–386), edited by M.B. Hart, Geological Society of London Special Publication 102. Geological Society, London.

Kelley, P.H. and Hansen, T.A. (2003). The fossil record of drilling predation on bivalves and gastropods. In *Predator–Prey Interactions in the Fossil Record* (pp. 113–139), edited by P.H. Kelley, M. Kowalewski, and T.A. Hansen. Kluwer Academic, New York.

Kelley, P.H. and Hansen, T.A. (2006). Comparisons of class- and lower taxon-level patterns in naticid gastropod predation, Cretaceous to Pleistocene of the U.S. Coastal Plain. *Palaeogeography, Palaeoclimatology, Palaeoecology*, 236, 302–320.

Kellogg, D.W. and Taylor, E.L. (2004). Evidence of oribatid mite detritivory in Antarctica during the Late Paleozoic and Mesozoic. *Journal of Paleontology*, 78, 1146–1153.

Kelly, D.C., Arnold, A.J., and Parker, W.C. (1996). Paedomorphosis and the origin of the Paleogene planktonic foraminiferal genus *Morozovella*. *Paleobiology*, 22, 266–281.

Kelly, S.R.A. (1988). Cretaceous wood-boring bivalves from western Antarctica with a review of the Mesozoic Pholadidae. *Palaeontology*, 31, 341–372.

Kelly, S.R.A. and Bromley, R.G. (1984). Ichnological nomenclature of clavate borings. *Palaeontology*, 27, 793–807.

Kelly, S.R.A., Blanc, E., Price, S.P., and Whitham, A.G. (2000). Early Cretaceous giant bivalves from seep-related limestone mounds, Wollaston Forland, Northeast Greenland. In *The Evolutionary Biology of the Bivalvia*, Vol. 177 (pp. 227–246), edited by E.M. Harper, J.D. Harper, and J.A. Crame, Geological Society of London Special Publication 177. Geological Society, London.

Kemp, A. (2001). Consequences of traumatic injury in fossil and Recent dipnoan dentitions. *Journal of Vertebrate Paleontology*, 21, 13–23.

Kemp, A. (2003a). Dental and skeletal pathology in lungfish jaws and tooth plates. *Alcheringa*, 27, 155–170.

Kemp, A. (2003b). Developmental anomalies in the tooth plates and jawbones of lungfish. *Journal of Vertebrate Paleontology*, 23, 517–531.

Kemp, A. (2005). New insights into ancient environments using dental characters in Australian Cenozoic lungfish. *Alcheringa*, 29, 123–149.

Kennedy, G.L. (1974). West American Cenozoic Pholadidae (Mollusca: Bivalvia). *San Diego Society of Natural History Memoir*, 8, 127 pp.

Kenward, H. (1999). Pubic lice (*Pthirus pubis* L.) were present in Roman and Medieval Britain. *Antiquity*, 73, 911–915.

Kerr, P.H. and Winterton, S.L. (2008). Do parasitic flies attack mites? Evidence in Baltic amber. *Biological Journal of the Linnean Society*, 93, 9–13.

Kershaw, A.P. and Whitlock, C., Eds. (2000). Palaeoecological records of the last glacial/interglacial cycle: patterns and causes of change. *Palaeogeography, Palaeoclimatology, Palaeoecology*, 155(1–2), 1–209.

Kershaw, S. (1987). Stromatoporoid–coral intergrowths in a Silurian biostrome. *Lethaia*, 20, 371–380.

Kesling, R.V. and Chilman, R.B. (1975). Strata and megafossils of the Middle Devonian Silica Formation. *University of Michigan Papers on Paleontology*, 8, 1–408.

Kesling, R.V. and Le Vasseur, D. (1971). *Strataster ohioensis*, a new early Mississippian brittle-star, and the paleoecology of its community. *University of Michigan Museum of Paleontology*, 23, 305–341.

Kesling, R.V., Hoare, R.D., and Sparks, D.K. (1980). Epizoans of the Middle Devonian brachiopod *Paraspirifer bownockerei*: their relationships to one another and to their host. *Journal of Paleontology*, 54, 1141–1154.

Keupp, H. (1985). Pathologische Ammoniten Kuriositäten oder paläobiologische dokumente?, Part 2. *Fossilien*, 1, 23–35.

Keupp, H. (1986). Perlen (Schalenkonkretionen) bei Dactylioceraten aus dem frankischen *Lias*. *Natur und Mensch*, 97–102.

Keyes, C.R. (1888). On the attachment of *Platyceras* to palaeocrinoids and its effects in modifying the form of the shell. *Proceedings of the American Philosophical Society*, XXV, 231–243.

Kiel, S. (2006). New records and species of mollusks from Tertiary cold-seep carbonates in Washington State, USA. *Journal of Paleontology*, 80, 121–137.

Kiel, S. (2008a). An unusual new gastropod from an Eocene hydrocarbon seep in Washington State. *Journal of Paleontology*, 82, 188–191.

Kiel, S. (2008b). Parasitic polychaetes in the Early Cretaceous hydrocarbon seep-restricted brachiopod *Peregrinella multicarinata*. *Journal of Paleontology*, 82, 1215–1217.

Kiel, S. and del Carmen Perrilliat, M. (2001). New gastropods from the Maastrichtian of the Mexicala Formation in Guerrera, southern Mexico. Part I. Stromboidea. *Neues Jahrbuch für Geologie und Paläontologie—Abhandlungen*, 222, 407–426.

Kiel, S. and Goedert, J.L. (2006a). A wood-fall association from Late Eocene deep-water sediments of Washington State, USA. *Palaios*, 21, 548–556.

Kiel, S. and Goedert, J.L. (2006b). Deep-sea food bonanzas (sunken wood and whales): munchies for mollusks. In *Proceedings of the 50th Annual Meeting of the Palaeontological Association, December 18–21, Sheffield, U.K.*

Kiel, S. and Goedert, J.L. (2006c). Deep-sea food bonanzas: Early Cenozoic whale-fall communities resemble wood-fall rather than seep communities. *Proceedings of the Royal Society London B*, 273, 2625–2631.

Kiel, S. and Little, C.T.S. (2006). Cold-seep mollusks are older than the general marine mollusk fauna. *Science*, 313, 1429–1431.

Kiel, S. and Peckmann, J. (2007). Chemosymbiotic bivalves and stable carbon isotopes indicate hydrocarbon seepage at four unusual Cenozoic fossil localities. *Lethaia*, 40, 345–357.

Kiel, S., Campbell, K.A., Eldeer, W.P., and Little, C.T.S. (2008). Jurassic and Cretaceous gastropods from hydrocarbon seeps in forearc and accretionary prism settings, California, Great Valley Group and Franciscan. *Acta Palaeontologica Polonica*, 53, 679–703.

Kiel, S., Amano, K., Hikida, Y., and Jenkins, R.G. (2009). Wood-fall associations from Late Cretaceous deep-water sediments of Hokkaido, Japan. *Lethaia*, 42, 74–82.

Kier, P.M. (1957). Tertiary Echinoidea from British Somaliland. *Journal of Paleontology*, 31, 839–902.

Kier, P.M. (1967). Sexual dimorphism in an Eocene echinoid. *Journal of Paleontology*, 41, 988–993.

Kier, P.M. (1969). Sexual dimorphism in fossil echinoids. In *Sexual Dimorphism in Fossil Metazoa and Taxonomic Implications* (pp. 215–222), edited by G.E.G. Westermann. Schweizerbart, Stuttgart.

Kier, P.M. (1987). Class Echinoidea. In *Fossil Invertebrates* (pp. 596–611), edited by R.S. Boardman, A.H. Cheetham, and A.J. Rowell. Blackwell Scientific, Oxford.

Kierst, J. and Wiesner, J. (1975). Fossile Fraßspuren an einer Conifere aus dem Dogger in Wolfsburg. *Aufschluss*, 26(6), 255–256.

Kieslinger, A. (1926). Untersuchungen an triadischen Nautiloideen. *Paläontologisches Zeitschrift*, VII, 101–122.

Kim, K.S., Kim, J.Y., Kim, S.H., Lee, C.Z., and Lim, J.D. (2009). Preliminary report on hominid and other vertebrate footprints from the Late Quaternary strata of Jeju Island, Korea. *Ichnos*, 16, 1–11.

King, J.E. and Saunders, J.J. (1984). Environmental insularity and the extinction of the American Mastodont. In *Quaternary Extinctions* (pp. 315–339), edited by P.S. Martin and R.G. Klein. University of Arizona Press, Tucson.

Kinzelbach, R. and Lutz, H. (1985). Stylopid larva from the Eocene: a spotlight on the phylogeny of the stylopids (Strepsiptera). *Annals of the Entomological Society of America*, 78, 600–602.

Kirchner, H. (1927). Perlbildung bei einem Ceratiten. *Centralblatt für Mineralogie und Paläontologie*, A, 148–150.

Kirkland, J.I., Zanno, L.E., Sampson, S.D., Clark, J.M., and DeBileux, D.D. (2005). A primitive therizinosauroid dinosaur from the Early Cretaceous of Utah. *Nature*, 435, 84–87.

Kitching, J.W. (1980). On some fossil arthropods from the Limeworks, Makapansgat, Potgietersrus. *Palaeontologica Africana*, 23, 63–68.

Kleemann, K.H. (1980). Korallenbohrende Muschel seit dem Mittleren Lias unverändert. *Beiträge zur Paläontologie von Österreich*, 7, 239–249.

Kleemann, K.H. (1983). Catalogue of Recent and fossil *Lithophaga* (Bivalvia). *Journal of Molluscan Studies*, 12(Suppl.), 1–46.

Kleemann, K.H. (1990a). Coral associations, biocorrosion, and space competition in *Pedum spondyloideum* (Gmelin) (Pectinacea, Bivalvia). *PSZN: Marine Ecology*, 11(1), 77–94.

Kleemann, K.H. (1990b). Evolution of chemically-boring Mytilidae (Bivalvia). In *The Bivalvia* (pp. 111–124), edited by B. Morton. Hong Kong University Press, Hong Kong.

Kleemann, K.H. (1994a). Mytilid bivalve in Upper Triassic coral *Pamiroseris* from Zlambach Beds compared with Cretaceous *Lithophaga alpina*. *Facies*, 30, 151–154.

Kleemann, K.H. (1994b). Associations of corals and boring bivalves since the Late Cretaceous. *Facies*, 31, 131–140.

Klein, H. (1971). Pathologische Veranderungen an einem Elefanten-Oberkiefer aus pleistozänen kiesen des Oberrheins (SW-Deutschland). *Neues Jahrbuch für Geologie und Paläontologie–Monatsheft*, 355–362.

Klembara, J. (1994). Electroreceptors in the Lower Permian tetrapod *Discosauriscus austriacus*. *Palaeontology*, 37, 609–626.

Kliks, M.M. (1990). Helminths as heirlooms and souvenirs: a review of New World paleoparasitology. *Parasitology Today*, 6, 93–100.

Klompen, H. and Grimaldi, D. (2001). First Mesozoic record of a parasitiform mite: a larval argesid tick in Cretaceous amber (Acarii: Ixodida: Argesidae). *Annals of the Entomological Society of America*, 94, 10–16.

Klug, C., Rücklin, M., Meyer-Berthaud, B., Soria, A., Korn, A., and Wendt, J. (2003). Late Devonian pseudoplanktonic crinoids from Morocco. *Neues Jahrbuch für Geologie und Paläontologie–Monatsheft*, 153–163.

Klug, S. and Kriwet, J. (2006). Anatomy and systematics of the Early Jurassic neoselachian *Synechodus smithwoodwardi* (Fraas, 1896) from southern Germany. *Neues Jahrbuch für Geologie und Paläontologie–Monatsheft*, 193–211.

Knight, J.B. (1941). *Paleozoic Gastropod Genotypes*, Special Paper 32. Geological Society of America, Boulder, CO, 510 pp.

Knight-Jones, E.W. (1953). Laboratory experiments on gregariousness during setting in *Balanus balanoides* and other barnacles. *Journal of Experimental Biology*, 30, 584–598.

Knox, L.W. and Miller, M.F. (1985). Environmental control of trace fossil morphology. In *Biogenic Structures: Their Use in Interpreting Depositional Environments* (pp. 167–176), edited by H.A. Curran. Society of Economic Mineralogists and Paleontologists, Tulsa, OK.

Knudsen, J.W. (1967). Trapezia and Tetralia (Decapoda, Brachyura, Xanthidae) as obligate ectoparasites of pocilloporid and acroporid corals. *Pacific Science*, 21, 51–57.

Kobayashi, Y., Lu, J.-C., Dong, Z.-M., Barsbold, R., Azuma, Y., and Tomida, Y. (1999). Herbivorous diet in an ornithomimid dinosaur. *Nature*, 402, 480–481.

Kobbert, M.J. (2005). *Bernstein Fenster in die Urzeit*. Planet Poster Press, Göttingen, Germany, 227 pp.

Kobluck, D.R. (1981). The record of cavity dwelling (coelobiontic) organisms in the Paleozoic. *Canadian Journal of Earth Sciences*, 18, 181–190.

Kobluck, D.R. and Mapes, R.H. (1989). The fossil record, function and possible origins of shell color patterns in Paleozoic marine invertebrates. *Palaios*, 4, 63–85.

Koch, C.L. and Berendt, G.C. (1854). *Die im Bernstein befindlichen Crustaceen, Myriapoden, Arachniden und Apteren der Vorwelt*. Berlin.

Kohn, A.J. (1990). Tempo and mode of evolution in Conidae. *Malacologia*, 32, 55–67.

Kohn, A.J. (2001). Maximal species richness in *Conus*: diversity, diet and habitat on reefs of northeast Papua New Guinea. *Coral Reefs*, 20, 25–38.

Kohn, A.J., Nishi, M., and Pernet, B. (1999). Snail spears and scimitars: a character analysis of *Conus* radular teeth. *Journal of Molluscan Studies*, 65, 461–481.

Kohring, R. (1990). Upper Jurassic chelonian eggshell fragments from the Guimarota Mine (Central Portugal). *Journal of Vertebrate Paleontology*, 10, 128–130.

Kohring, R. (1991). Lizard egg shells from the Lower Cretaceous of Cuenca Province, Spain. *Palaeontology*, 34, 237–240.

Kohring, R. and Hirsch, K.F. (1996). Crocodilian and avian eggshells from the Middle Eocene of the Geiseltal, eastern Germany. *Journal of Vertebrate Paleontology*, 16, 67–80.

Kolasa, J. (2001). Flatworms: Turbellaria and Nemertea. In *Ecology and Classification of North American Freshwater Invertebrates*, 2nd ed. (pp. 155–180), edited by J.H. Thorp and A.P. Covich. Academic Press, San Diego, CA.

Kolesnikov, C.M. (1973). Iskopaemy presnovody zhemchug. *Doklady Akademia Nauk SSSR*, 221(5), 1195–1197.

Kosmowska-Ceranowicz, B., Ed. (2001). *The Amber Treasure Trove*. The Tadeusz Giecewicz's Collection at the Museum of the Earth, Polish Academy of Sciences, Warsaw, 97 pp.

Kosmowska-Ceranowicz, B. and Konart, Z.T. (2005). *Tajemnice bursz Tynu*. Sport I Turstyka, Warsaw, 97 pp.

Koteja, J. (1998). Essays on coccids (Homoptera): sudden death in amber? *Polskie Pismo Entomologiczne*, 67, 185–218.

Koteja, J. (2000). Scale insects (Homoptera, Coccinea) from Upper Cretaceous New Jersey amber. In *Studies on Fossils in Amber, with Particular Reference to the Cretaceous of New Jersey* (pp. 147–229), edited by D. Grimaldi. Backhuys, Leiden, Germany.

Koteja, J. and Poinar, Jr., G.O. (2004). Scale insects (Coccinea) associated with mites (Acari) in the fossil record. In *Proceedings of the X International Symposium on Scale Insect Studies, April 19–23, 2004, Adana, Turkey*.

Kowalewski, M. (1993). Morphometric analysis of predatory drill holes. *Palaeogeography, Palaeoclimatology, Palaeoecology*, 102, 69–88.

Kowalewski, M. and Nebelsick, J.H. (2003). Predation on Recent and fossil echinoids. In *Predator–Prey Interactions in the Fossil Record* (pp. 279–302), edited by P.H. Kelley, M. Kowalewski, and T.A. Hansen. Kluwer Academic, New York.

Kowalewski, M., Dulai, A., and Fürsich, F.T. (1998). A fossil record full of holes: the Phanerozoic history of drilling predation. *Geology*, 26, 1091–1094.

Kowalewski, M., Simoes, M.G., Torello, F.F., Mello, L.H.C., and Ghilardi, R.P. (2000). Drill holes of Permian benthic invertebrates. *Journal of Paleontology*, 74, 532–543.

Kowalewski, M., Hoffmeister, A.P., Baumiller, T.K., and Bambach, R.K. (2005). Secondary evolutionary escalation between brachiopods and enemies of other prey. *Science*, 308, 1774–1777.

Kozur, H. (1992). Dzhulfian and Early Changxingian (Late Permian) Tethyan conodonts from the Glass Mountains, West Texas. *Neues Jahrbuch für Geologie und Paläontologie–Abhandlungen*, 187, 99–114.

Kramer, J.M., Overstreet, D.F., Aufderheide, A.C., and Krug, H.E. (1996). Mammoth pathologies and megafauna predation: evidence for opportunistic predation of the Hebior mammoth. In *Sixth North American Paleontological Convention Abstracts of Papers* (p. 222), edited by J.E. Repetski, Special Publ. No. 8. Paleontological Society, Washington, D.C.

Krassilov, V.A. (1982). Early Cretaceous flora of Mongolia. *Palaeontographica B*, 181, 1–77.

Krassilov, V.A. (1987). Palaeobotany of the Mesophyticum: state of the art. *Reviews of Palaeobotany and Palynology*, 50, 231–254.

Krassilov, V.A. (2003). *Terrestrial Palaeoecology and Global Change.* Pensoft, Sofia-Moscow, 464 pp.

Krassilov, V.A. and Bacchia, F. (2000). Cenomanian florule of Nammoura, Lebanon. *Cretaceous Research*, 21, 785–799.

Krassilov, V.A., and Makulbekov, N.M. (2003). Pervaya nekhodka gribov gasteromytsetov (Gasteromycetes) v Melovikh otlozheniyakh Mongoli. *Palaeontologicheski Zhurnal*, 4, 103–106.

Krassilov, V.A. and Rasnitsyn, A.P. (1997). Pollen in the guts of Permian insects: first evidence of pollenivory and its evolutionary significance. *Lethaia*, 29, 369–372.

Krassilov, V.A. and Rasnitsyn, A.P., Eds. (2008). *Plant–Arthropod Interactions in the Early Angiosperm History: Evidence from the Cretaceous of Israel.* Pensoft/Brill, Boston, MA.

Krassilov, V.A., Zherikhin, V.V., and Rasnitsyn, A.P. (1997a). *Classopolis* in the guts of Jurassic insects. *Palaeontology*, 40, 1095–1101.

Krassilov, V.A., Zherikhin, V.V., and Rasnitsyn, A.P. (1997b). Pollen in guts of fossil insects as evidence for coevolution. *Doklady Biological Sciences*, 354, 239–241.

Krassilov, V.A., Rasnitsyn, A.P., and Afonin, S.A. (1999). Pollen morphotypes from the intestine of a Permian booklouse. *Reviews of Palaeobotany and Palynology*, 106, 89–96.

Krassilov, V., Teklaeva, M., Meyer-Melikyan, N., and Rasnitsyn, A. (2003). New pollen morphotype from gut compression of a Cretaceous insect, and its bearing on palynomorphological evolution and palaeoecology. *Cretaceous Research*, 24, 149–156.

Krause, J., Dear, P.H., Pollack, J.L., Slatkin, M., Spriggs, H., Barnes, I., Lister, A.M., Eberberger, I., Pääbo, S., and Hofreiter, M. (2006). Multiplex amplification of the mammoth mitochondrial genome and the evaluation of Elephantidae. *Nature*, 439, 724–727.

Krell, F.T. (2000). The fossil record of Mesozoic and Tertiary scarabaeoidea (Coleoptera: Polyphaga). *Invertebrate Taxonomy*, 14, 871–905.

Krishtalka, L., Stucky, R.K., and Beard, K.C. (1990). The earliest fossil evidence for sexual dimorphism in primates. *Proceedings of the National Academy of Sciences USA*, 87, 5223–5226.

Kriwet, J. (2001). Feeding mechanisms and ecology of pycnodont fishes (Neopterygii, Pycnodontiformes). *Mitteilungen aus dem Museum für Naturkunde in Berlin, Geowissenschaftliche Reihe*, 4, 139–165.

Kříž, J. (1979). Silurian Cardiolidae (Bivalvia). *Sborník Geologických Věd, Paleontologie*, 22, 5–157.

Kříž, J. (1999). Bivalvia dominated communities of Bohemian type from the Silurian and Lower Devonian carbonate facies. In *Paleocommunities: A Case Study from the Silurian and Lower Devonian* (pp. 229–252), edited by A.J. Boucot and J.D. Lawson. Cambridge University Press, Cambridge, U.K.

Kříž, J. and Lukeš, P. (1974). Color patterns on Silurian *Platyceras* and Devonian *Merista* from Barrandian area, Bohemia. *Journal of Paleontology*, 48, 1974.

Kříž, J. and Mikuláš, R. (2006). Bivalve wood borings of the ichnogenus *Teredolites* Leymerie from the Bohemian Basin (Upper Cretaceous, Czech Republic). *Ichnos*, 13, 159–174.

Kröger, B. (2002a). On the efficiency of the buoyancy apparatus in ammonoids: evidences from sublethal shell injuries. *Lethaia*, 35, 61–70.

Kröger, B. (2002b). Antipredatory traits of the ammonoid shell: indications from the Jurassic ammonoids with sublethal injuries. *Paläontologisches Zeitschrift*, 76, 223–234.

Kröger, B. (2004). Large shell injuries in Middle Ordovician Orthocerida (Nautiloidea, Cephalopoda). *GFF*, 126, 311–316.

Kröger, B. and Keupp, H. (2004). A paradox survival: report of a repaired *syn vivo* perforation in a nautiloid phragmocone. *Lethaia*, 37, 439–444.

Kroh, A. and Nebelsick, J.H. (2006). Stachelige Leckerbissen. *Natur und Museum*, 136, 6–13.

Kruger, F.J. (1997). Brutpflege bei Seeigeln. *Natur und Museum*, 127, 101–112.

Krumbiegel, G. and Krumbiegel, B. (2001). *Faszination Bernstein.* Goldschneck-Verlag, Korb, 111 pp.

Krumbiegel, G. and Krumbiegel, B. (2005). *Bernstein.* Quelle & Meyer Verlag, Wiebelsheim, 112 pp.

Krumm, D.K. (1999). Bivalve bioerosion in Oligocene corals from Puerto Rico and Jamaica. *Bulletin of the Geological Society of Denmark*, 45, 179–180.

Krumm, D.K. and Jones, D.S. (1993). New coral–bivalve association (*Actinastrea-Lithophaga*) from the Eocene of Florida. *Journal of Paleontology*, 67, 945–951.

Krzeminska, E. and Krzeminski, W. (1992). *Les fantômes de l'ambre: insectes fossiles dans l'ambre de la Baltique.* Neuchâtel Musée d'Histoire Naturelle de Neuchâtel, 142 pp.

Kucera, M. and Malmgren, B.A. (1996). Gradual morphological evolution in a Late Cretaceous lineage of planktonic Foraminifera. In *Sixth North American Paleontological Convention Abstracts of Papers* (p. 225), edited by J.E. Repetski, Special Publ. No. 8. Paleontological Society, Washington, D.C.

Kucera, M. and Malmgren, B.A. (1998). Differences between evolution of mean form and evolution of new morphotypes: an example from Late Cretaceous planktonic Foraminifera. *Paleobiology*, 24, 49–63.

Kuhbandner, M. and Schleich, H.H. (1994). *Odontimyia*-Larven aus dem Randecker Maar (Insecta: Diptera, Stratiomydae). *Mitteilungen der Bayerischen Staatssammlung für Paläontologie und historische Geologie*, 34, 163–167.

Kukla, G., Ed. (1989). Long continental records of climate. *Palaeogeography, Palaeoclimatology, Palaeoecology*, 72, 1–225.

Kullman, J. (1966). Goniatiten-Korallen-Vergesellschaftungen im Karbon des kantabrischen Gebirges (Nordspanien). *Neues Jahrbuch für Geologie und Paläontologie–Abhandlungen*, 125, 443–466.

Kümel, F. (1935). Fossile Perlen im niederösterreichen Jungtertiär. *Verhandlungen der Geologische Bundesanstalt*, 7, 110–112.

Kurochkin, E.N. (1985). A true carinate bird from Lower Cretaceous deposits in Mongolia and other evidence of Early Cretaceous birds in Mongolia. *Cretaceous Research*, 6, 271–278.

Kurochkin, E.N. (2000). Mesozoic birds of Mongolia and the former USSR. In *The Age of Dinosaurs in Russia and Mongolia* (pp. 533–559), edited by M.J. Benton, M.A. Shiskin, D.M. Unwin, and E.N. Kurochkin. Cambridge University Press, Cambridge, U.K.

Kurochkin, E.N., Dyke, G.J., Saveliev, S.V., Perushov, E.N., and Popov, E.V. (2007). A fossil brain from the Cretaceous of European Russia and avian sensory evolution. *Biology Letters*, 3, 309–313.

Kurten, B. (1958). Life and death of the Pleistocene cave bear: a study in palaeoecology. *Acta Zoologica Fennica*, 95, 1–59.

Kurten, B. (1969). Sexual dimorphism in fossil mammals. In *Sexual Dimorphism in Fossil Metazoa and Taxonomic Implications* (pp. 226–227), edited by G.E.G. Westermann. Schweizerbart, Stuttgart.

Kuschel, G. and May, B.M. (1996). Discovery of Palophaginae (Coleoptera: Megalopodidae) on *Araucaria araucana* in Chile and Argentina. *New Zealand Entomologist*, 19, 1–13.

Kutassy, E. (1937). Die älteste fossile Perle und Verletzungsspuren an einem triadischen *Megalodus. Math. und Naturwiss. Anz. (Ung. Akad. Wiss.)*, 55(3), 1005–1023.

Kuznetsov, A.P., Maslennikov, V.V., Zaikov, V.V., and Zonenshain, L.P. (1990). Fossil hydrothermal vent fauna in Devonian sulfide deposits of the Uralian ophiolites. *Deep-Sea Newsletter (Denmark)*, 17, 9–10.

Kuznetsov, A.P., Zaikov, V.V., and Maslennikov, V.V. (1991). Ofioliti: "letopis" vulkanicheskikh, tektonicheskikh, fiziko-khimicheskikh i bioticheskikh sobitii formirovaniya zemnoi kori na dne paleookeanov. *Izvestiya Akademii Nauk SSSR, Seriya Biologicheskaya*, 2, 232–241.

Kuznetsov, A.P., Strizhov, V.P., and Galkin, S.V. (1994). Pozdnepaleozoiskie (Karbonovie) vestimentiferi (Vestimentifera, Obturata, Pogonophora) Severo-Vostochnoi Azii. *Izvestiya Akademii Nauk, Rossiiskaya Akademia Nauk, Seriya Biologicheskaya*, 6, 898–906.

Labandeira, C.C. (1991). Evidence for Pennsylvanian-age stem-mining and the early history of complete metamorphosis in insects. *Geological Society of America Abstracts with Programs*, 23, A405.

Labandeira, C.C. (1993). What's new with fossil insects. *American Paleontologist*, 1(4), 1–5.

Labandeira, C.C. (1996). The presence of a distinctive insect herbivore fauna during the Late Paleozoic. In *Sixth North American Paleontological Convention Abstracts of Papers* (p. 227), edited by J.E. Repetski, Special Publ. No. 8. Paleontological Society, Washington, D.C.

Labandeira, C.C. (1997a). Insect mouthparts: ascertaining the *paleobiology* of insect feeding strategies. *Annual Review of Ecology and Systematics*, 28, 153–193.

Labandeira, C.C. (1997b). Permian pollen eating. *Science*, 277, 1422–1423.

Labandeira, C.C. (1998a). Plant–insect associations from the fossil record. *Geotimes*, 43, 18–24.

Labandeira, C.C. (1998b). How old is the flower and the fly? *Science*, 280, 57–58.

Labandeira, C.C. (1998c). The role of insects in Late Jurassic to Middle Cretaceous ecosystems. In *Lower and Middle Cretaceous Terrestrial Ecosystems* (pp. 105–124), edited by S.G. Lucas, J.I. Kirkland, and J.W. Estep. New Mexico Museum of Natural History, Albuquerque.

Labandeira, C.C. (1998d). Early history of arthropod and vascular plant associations. *Annual Review of Earth and Planetary Sciences*, 26, 329–377.

Labandeira, C.C. (2000). The paleobiology of pollination and its precursors. In *Phanerozoic Terrestrial Ecosystems* (pp. 233–269), edited by R.A. Gastaldo and W.A. DiMichele. Paleontological Society, Baltimore, MD.

Labandeira, C.C. (2002a). The history of associations between plants and animals. In *Plant–Animal Interactions: An Evolutionary Approach* (pp. 26–74, 248–261), edited by C.M. Herrera and O. Pellmyr. Wiley-Blackwell, London.

Labandeira, C.C. (2002b). Paleobiology of middle Eocene plant-insect associations from the Pacific Northwest: a preliminary report. *Rocky Mountain Geology*, 37, 31–59.

Labandeira, C.C. (2007). Assessing the fossil record of plant–insect associations: ichnodata versus body-fossil data. In *Ichnology at the Crossroads: A Multidimensional Approach to the Science of Organism–Substrate Interactions*, edited by R. Bromley, L. Buatois, J. Genise, M.G. Mangano, and R. Melchor. Society for Sedimentary Geology, Tulsa, OK.

Labandeira, C.C. and Allen, E.G. (2007). Minimal insect herbivory for the Lower Permian Coprolite Bone Bed site of north-central Texas, USA, and comparison to other Late Paleozoic floras. *Palaeogeography, Palaeoclimatology, Palaeoecology*, 247, 197–219.

Labandeira, C.C. and Phillips, T.L. (1996a). A Carboniferous insect gall: insight into early ecologic history of the Holometabola. *Proceedings of the National Academy of Sciences USA*, 93, 8470–8474.

Labandeira, C.C. and Phillips, T.L. (1996b). Insect fluid-feeding on Upper Pennsylvanian tree ferns (Palaeodictyoptera, Mariattiales) and the early history of the piercing-and-sucking functional feeding groups. *Annals of the Entomological Society of America*, 89, 157–183.

Labandeira, C.C. and Phillips, T.L. (2002). Stem borings and petiole galls from Pennsylvanian tree ferns of Illinois, USA: implications for the origin of the borer and galler functional-feeding-groups and holometabolous insects. *Palaeontographica A*, 264, 1–84.

Labandeira, C.C., Dilcher, D.L., Davis, D.R., and Wagner, D.L. (1994). Ninety-seven million years of angiosperm-insect association: paleobiological insights into the meaning of coevolution. *Proceedings of the National Academy of Sciences USA*, 12278–12282.

Labandeira, C.C., Phillips, T.L., and Norton, R.A. (1997). Oribatid mites and the decomposition of plant tissues in Paleozoic coal-swamp forests. *Palaios*, 12, 319–353.

Labandeira, C.C., LePage, B.A., and Johnson, A.H. (2001). A *Dendroctonus* bark engraving (Coleoptera: Scolytidae) from a Middle Eocene *Larix* (Coniferales: Pinaceae): early or delayed colonization. *American Journal of Botany*, 88, 2026–2039.

Laird, M., Ed. (1981). *Blackflies*. Academic Press, London, 399 pp.

Lambers, P. and Boekschoten, G.J. (1986). On fossil and recent borings produced by acrothoracic cirripeda. *Geologie en Mijnbouw*, 65, 257–268.

Lambert, J. (1933). Echinoides de Madagascar. *Ann. Géol. Serv. Mines, Gouvernement Géneral da Madagascar et Dépendences*, 3, 7–49.

Lambert, J.B., Frye, J.S., and Poinar, Jr., G.O. (1990). Analysis of North American amber by carbon-13 NMR spectroscopy. *Geoarchaeology*, 5, 43–52.

Lancucka-Srodoniowa, M. (1964). Tertiary coprolites imitating fruits of the Araliaceae. *Acta Societatis Botanicorum Poloniae*, XXXIII(2), 469–473.

Landers, S.C. (2002). The fine structure of the gamont of *Pterospora floridiensis* (Apicomplexa: Eugregarinida). *Journal of Eukaryotic Microbiology*, 49, 220–226.

Landini, W. and Sorbini, L. (1996). Ecological and trophic relationships of Eocene Monte Bolca (Pesciara) fish fauna. Autoecology of selected fossil organisms: achievements and problems, edited by A. Cherchi. *Bolletino della Società Paleontologica Italiana*, 3, 105–112.

Lang, P.J., Scott, A.C., and Stephenson, J. (1995). Evidence of plant–arthropod interactions from the Eocene Branksome Sand Formation, Bournemouth, England. *Tertiary Research*, 15, 145–174.

Lange, R.T. (1978). Southern Australian Tertiary epiphyllous fungi, modern equivalents in the Australian region, and habitat indicator value. *Canadian Journal of Botany*, 56, 532–541.

Langenheim, J.H. and Beck, C.W. (1968). Catalogue of infrared spectra of fossil resin (ambers). 1. North and South America. *Botanical Museum Leaflets, Harvard University*, 22, 65–120.

Larew, H.G. (1987). Two cynipid wasp acorn galls preserved in the La Brea Tar Pits (Early Holocene). *Proceedings of the Entomological Society of Washington*, 89, 831–833.

Larew, H.G. (1992). Fossil galls. In *Biology of Insect-Induced Galls* (pp. 50–59), edited by J.D. Shorthouse and O. Rohfritsch. Oxford University Press, London.

LaRoque, A. (1960). Molluscan faunas of the Flagstaff Formation of central Utah. *Geological Society of America Memoir*, 78, 1–100.

Larsson, S.G. (1978). *Baltic Amber: A Palaeobiological Study*, Entomograph 1. Scandinavian Science Press, Klampenborg, Denmark, 192 pp.

Laub, R.S., Dufort, C.A., and Christensen, D.J. (1994). Possible mastodon gastrointestinal and fecal contents from the late Pleistocene of the Hiscock Site, western New York State. *New York State Museum Bulletin*, 481, 135–148.

Laudermilk, J.D. and Munz, P.A. (1934). Plants in the dung of *Nothrotherium* from Gypsum Cave, Nevada. *Carnegie Institution of Washington*, 483, 29–38.

Laudet, F. and Antoine, P.-O. (2004). Des chambres de pupation de Dermestidae (Insecta: Coleoptera) sur un os de mammifère tertiare (phosphorites de Quercy): implications taphonomiques et paléoenvironnementales. *Géobios*, 37, 376–381.

Laudet, F. and Selva, N. (2005). Ravens as small mammal bone accumulators: first taphonomic study on mammal remains in raven pellets. *Palaeogeography, Palaeoclimatology, Palaeoecology*, 226, 272–286.

Laurin, B. (1984). Un cas de grégarisme chez *Burmirhynchia decorate* (Schlotheim). *Géobios, Mémoire Special*, 8, 433–440.

Lavenberg, R.J. (1991). Megamania, the continuing saga of megamouth sharks. *Terra*, 30, 30–39.

Lawrence, D.L. (1968). Taphonomy and information losses in fossil communities. *Geological Society of America Bulletin*, 79, 1315–1350.

Lawrence, J.M. (1992). Arm loss and regeneration in Asteroidea (Echinodermata). In *Echinoderm Research 1991* (pp. 39–52), edited by L. Scalera-Liaci and C. Canicatti. Balkema, Rotterdam.

Lawrence, J.M. and Vasquez, J. (1996). The effect of sublethal predation on the biology of echinoderms. *Oceonologica Acta*, 19, 431–440.

Laws, R.R., Hasiotis, S.T., Fiorillo, A.R., Chure, D.J., Breithaupt, B.H., and Horner, J.R. (1996). The demise of a Jurassic Morrison dinosaur after death: three cheers for the dermestid beetle. *Geological Society of America Abstracts with Programs*, 28, 530A.

Lawton, J.H. (1994). Population dynamic principles. *Philosophical Transactions of the Royal Society of London B*, 344, 61–68.

Laybourne, R.C.D., Deedrick, D.W., and Hueber, F.M. (1994). Feather in amber is earliest New World fossil of Picidae. *Wilson Bulletin*, 106, 18–25.

Lazarus, D.B. (2001). Speciation and morphological change. In *Palaeobiology II* (pp. 133–137), edited by D.E.G. Briggs and P. Crowther. Blackwell Scientific, Oxford.

Le Renard, J., Sabelli, B., and Taviani, M. (1996). On *Candinia* (Sacoglossa: Juliidae), a new fossil genus of bivalved gastropod. *Journal of Paleontology*, 70, 230–235.

Leakey, L.S.B. (1952). Lower Miocene invertebrates from Kenya. *Nature*, 169, 624–625.

Lebold, J.G. (2000). Quantitative analysis of epizoans on Silurian stromatoporoids within the Brassfield Formation. *Journal of Paleontology*, 74, 394–403.

Lee, C.W. (1983). Bivalve mounds and reefs of the Central High Atlas, Morocco. *Palaeogeography, Palaeoclimatology, Palaeoecology*, 43, 153–168.

Lee, D.E. (1978). Aspects of the ecology and paleoecology of the brachiopod *Notosaria nigricans* (Sowerby). *Journal of the Royal Society of New Zealand*, 8, 395–417.

Lee, D.E., Carter, R.M., King, R.P., and Cooper, A.F. (1983). An Oligocene rocky shore community from Mt. Luxmore, Fiordland. *New Zealand Journal of Geology and Geophysics*, 26, 123–126.

Lee, J.G., Chol, D.K., and Pratt, B.R. (2001). A teratological pygidium of the Upper Cambrian trilobite *Eugenocare* (*Pseudeugenocare*) *bispinatum* from the Machari Formation, Korea. *Journal of Paleontology*, 75, 216–218.

Lefebvre, B. (2007). Early Palaeozoic palaeobiogeography and palaeoecology of stylophoran echinoderms. *Palaeogeography, Palaeoclimatology, Palaeoecology*, 245, 156–199.

Leggitt, V.L. and Cushman, Jr., R.A. (1999). Massive caddisfly bioherms from the Eocene Green River Formation. *Geological Society of America Abstracts with Programs*, 31(7), 242.

Leggitt, V.L. and Cushman, Jr., R.A. (2001). Complex caddisfly-dominated bioherms from the Eocene Green River Formation. *Sedimentary Geology*, 145, 377–396.

Leggitt, V.L. and Loewen, M.A. (2002). Eocene Green River Formation "*Oocardium tufas*" reinterpreted as complex arrays of calcified caddisfly (Insecta: Trichoptera) larval cases. *Sedimentary Geology*, 148, 139–146.

Leggitt, V.L., Biaggi, R.E., and Buchheim, H.P. (2007). Palaeoenvironments associated with caddisfly-dominated microbial-carbonate mounds from the Tipton Shale Member of the Green River Formation: Eocene Lake Gosiute. *Sedimentology*, 54, 661–699.

Legrand-Blain, M. and Poncet, J. (1991). Encroûtement et perforations de tests de brachiopodes dans le Carbonifère du Sahara algérien Implications pour les reconstitution de paléoenvironnements. *Bulletin de la Société Géologique de France*, 162(40), 775–789.

Lehane, M.J. (1991). *Biology of Blood-Sucking Insects*. Harper Collins Academic, London, 288 pp.

Lehmann, U. (1966). Dimorphismus bei Ammoniten der Ahrensburger Lias-Geschiebe. *Paläontologische Zeitschrift*, 41, 26–55.

Lehmann, U. (1981). *The Ammonites, Their Life and Their World*. Cambridge University Press, Cambridge, U.K., 246 pp.

Lehmann, U. (1990). *Ammonoideen*. Ferdinand Enke, Stuttgart, 257 pp.

Lehmann, W.M. (1951). Anomalien und regenerationserscheinungen an paläozoischen Asterozoan. *Neues Jahrbuch für Geologie und Paläontologie–Abhandlungen*, 93, 401–416.

Leighton, L.R. (2001). New examples of Devonian predatory boreholes and the influence of brachiopod spines on predator success. *Palaeogeography, Palaeoclimatology, Palaeoecology*, 165, 53–69.

Leighton, L.R. (2003a). Morphological response of prey to drilling predation in the Middle Devonian. *Palaeogeography, Palaeoclimatology, Palaeoecology*, 201, 221–234.

Leighton, L.R. (2003b). Predation on brachiopods. In *Predator–Prey Interactions in the Fossil Record* (pp. 215–237), edited by P.H. Kelley, M. Kowalewski, and T.A. Hansen. Kluwer Academic, New York.

Lemen, C.A. and Freeman, P.W. (1989). Testing macroevolutionary hypotheses with cladistic analysis: evidence against rectangular evolution. *Evolution*, 43, 1538–1554.

Leriche, M. (1910). Sur une coquille de *Pleurotoma regularis*, pourvue de deux siphons. *Annales de la Société Géologique du Nord*, 39, 343–344.

Lescinsky, H.L. (1997). Epibiont communities: recruitment and competition on North American Carboniferous brachiopods. *Journal of Paleontology*, 71, 34–53.

Lescinsky, H.L. (2001). Epibionts. In *Palaeobiology II* (pp. 460–464), edited by D.E.G. Briggs and P. Crowther. Blackwell Scientific, Oxford.

Lesnikowska, A.D. (1990). Evidence of herbivory in tree-fern petioles from the Calhoun Coal (Upper Pennsylvanian) of Illinois. *Palaios*, 5, 76–80.

Lespérance, P.J. and Sheehan, P.M. (1975). Middle Gaspe limestone communities on the Forillon Peninsula, Québec, Canada (Siegenian, Lower Devonian). *Palaeogeography, Palaeoclimatology, Palaeoecology*, 17, 309–326.

Lethiers, F., Damotte, R., and Whatley, R. (1997). Evidence of brooding in Permian non-marine Ostracoda. *Lethaia*, 29, 219–223.

Levett Smith, C., Rand, C.S., Schaeffer, B., and Atz, J.W. (1975). *Latimeria*, the living coelacanth is ovoviviparous. *Science*, 190, 1105–1106.

Levin, H.L. and Fay, R.O. (1964). Relationship between *Diploblastus kirkwoodensis* and *Platyceras* (*Platyceras*). *Oklahoma Geology Notes*, 24(2), 22–29.

Levin, L.A., James, D.W., Martin, C.M., Rathburn, A.E., Harris, L.H., and Michener, R.H. (2000). Do methane seeps support distinct macrofaunal assemblages? Observations on community structure and nutrition from the northern California slope and shelf. *Marine Ecology Progress Series*, 208, 21–39.

Lewin, R.A. (1999). *Merde, Excursions in Scientific, Cultural, and Sociohistorical Coprology*. Random House, New York, 187 pp.

Lewis, R.E. and Grimaldi, D.A. (1997). A pulicid flea in Miocene amber from the Dominican Republic (Insecta: Siphonoptera: Pulicidae). *American Museum Novitates*, 3205, 1–9.

Lewis, S.E. (1989). Eocene insect localities in the United States and Canada. *Occasional Papers in Paleobiology of St. Cloud State University*, 3(2), 1–38.

Lewis, S.E. (1992). Insects of the Klondike Mountain Formation, Republic, Washington. *Washington Geology*, 20(3), 15–19.

Lewis, S.E. (1994). Evidence of leaf-cutting bee damage from the Republic sites (middle Eocene) of Washington. *Journal of Paleontology*, 68, 172–173.

Lewis, S.E. and Carroll, M.A. (1991). Coeleoptera egg deposition on alder leaves from the Klondike Mountain Formation (Middle Eocene), northeastern Washington. *Journal of Paleontology*, 65, 334–335.

Lewis, S.E. and Carroll, M.A. (1992). Coleopterous egg deposition on an alder leaf from the John Day Formation (Oligocene), north-central Oregon. *Occasional Papers in Paleobiology of St. Cloud State University*, 6, 1–4

Lewis, S.E. and Heikes, P.M. (1990). Fossil caddisfly cases (Trichoptera), Miocene of northern Idaho, USA. *Ichnos*, 1, 143–146.

Lewis, S.E. and Heikes, P.M. (1991). A catalog of fossil sites from the Tertiary of the United States. *Occasional Papers in Paleobiology of St. Cloud State University*, 5(1), 1–487.

Lewy, Z. (1996). Octopods: nude ammonoids that survived the Cretaceous–Tertiary boundary mass extinction. *Geology*, 24, 627–630.

Lewy, Z. and Goldring, R. (2006). Campanian crustacean burrow system from Israel with brood and nursery chambers representing communal organization. *Palaeontology*, 49, 133–140.

Lewy, Z. and Samtleben, C. (1979). Functional morphology and palaeontological significance of the conchiolin layers in corbiculid pelecypods. *Lethaia*, 12, 341–351.

Liao, Y.-L. and Lin, C.-H. (1981). A new echinoid with sexual dimorphism from the Late Tertiary deposits of Beibuwan, Guangxi. *Acta Palaeontologica Sinica*, 20, 482–484.

Liddell, W.D. (1975). Ecology and Stratinomy of a Middle Ordovician Echinoderm Assemblage from Kirkfield, Ontario. M.S. thesis, University of Michigan.

Liljedahl, L. (1985). Ecological aspects of a silicified bivalve fauna from the Silurian of Gotland. *Lethaia*, 18, 53–66.

Liljedahl, L. (1994). Silurian nuculoid and modiomorphid bivalves from Sweden. *Fossils & Strata*, 33, 1–89.

Lincoln, G.A. (1994). Teeth, horns and antlers: the weapons of sex. In *The Differences Between the Sexes* (pp. 131–158), edited by R.V. Short and E. Balaban. Cambridge University Press, Cambridge, U.K.

Lincoln, R.J. and Boxshall, G.A. (1987). *The Cambridge Illustrated Dictionary of Natural History*. Cambridge University Press, Cambridge, U.K., 411 pp.

Lindberg, D.R. and Hedegaard, C. (1996). A deep water patellogastropod from Oligocene water-logged wood of Washington State, USA (Acmaeoidea; *Pectinodonta*). *Journal of Malacological Studies*, 62, 299–314.

Lindberg, D.R. and Lipps, J.H. (1996). Reading the chronicle of Quaternary temperate rocky shore faunas. In *Evolutionary Paleobiology* (pp. 161–182), edited by D. Jablonski, D.H. Erwin, and J.H. Lipps. University of Chicago Press, Chicago, IL.

Lindqvist, J.K. and Isaac, M.J. (1991). Silicified conifer forests and potential mining problems in seam M2 of the Gore Lignite Measures (Miocene), Southland, New Zealand. *International Journal of Coal Geology*, 17, 149–169.

Lindström, A. (2003). Shell breakage in two pleurotomarioid gastropods from the Upper Carboniferous of Texas, and its relation to shell morphology. *GFF*, 125, 39–46.

Lindström, A. and Peel, J.S. (1997). Failed predation and shell repair in the gastropod *Poleumita* from the Silurian of Gotland, Sweden. *Vestnik Ceského Geologického Ústavu*, 72, 115-126.

Lindström, A. and Peel, J.S. (2003). Shell repair and mode of life of *Praenatica gregaria* (Gastropoda) from the Devonian of Bohemia (Czech Republic). *Palaeontology*, 46, 623–633.

Lindström, A. and Peel, J.S. (2005). Repaired injuries and shell form in some Palaeozoic pleurotomarioid gastropods. *Acta Palaeontologica Polonica*, 50, 697–704.

Lindström, G. (1884). The Silurian Gastropoda and Pteropoda of Gotland. *Kongliga Svenska Vetenskaps-Akademiens Handlingar*, 19, 1–250.

Lingham-Soliar, T. (1992). The tylosaurine mosasaurs (Reptilidae, Mosasauridae) from the Upper Cretaceous of Europe and Africa. *Bulletin de l'Institut Royal des Sciences Naturelles de Belgique, Sciences de la Terre*, 62, 171–194.

Lingham-Soliar, T. (1998). Unusual death of a Cretaceous giant. *Lethaia*, 31, 308–310.

Lingham-Soliar, T. (2004). Palaeopathology and injury in the extinct mosasaurs (Lepidosauromorpha, Squamata) and implications for modern reptiles. *Lethaia*, 37, 255–262.

Linsley, E.G. (1959). Mimetic form and coloration in the Cerambycidae (Coleoptera). *Annals of the Entomological Society of America*, 52, 125–131.

Lister, A.M. (1989). Rapid dwarfing of red deer on Jersey in the last interglacial. *Nature*, 342, 539–542.

Lister, A.M. (1993a). Patterns of evolution in Quaternary mammal lineages. In *Evolutionary Patterns and Processes* (pp. 71–93), edited by D.R. Lees and D. Edwards, Linnaean Society Symposium Series 14, Academic Press, San Diego, CA.

Lister, A.M. (1993b). "Gradualistic" evolution: its interpretation in Quaternary large mammal species. *Quaternary International*, 19, 77–84.

Lister, A.M. (1993c). The Condover mammoth site: excavation and research (1986–93). *Cranium*, 10(1), 61–67.

Little, C.T.S. (1997). Fossil hydrothermal vent communities: an update. *BRIDGE Newsletter*, 13, 34–37.

Little, C.T.S. (2001). Ancient hydrothermal vent and cold seep faunas. In *Palaeobiology II* (pp. 447–451), edited by D.E.G. Briggs and P. Crowther. Blackwell Scientific, Oxford.

Little, C.T.S. (2002). The fossil record of hydrothermal vent communities. *Cahiers du Biologie Marine*, 43, 313–316.

Little, C.T.S., Herrington, R.J., Maslennikov, V.V., Morris, N.J., and Zaykov, V.V. (1997). Silurian hydrothermal-vent community from the southern Urals, Russia. *Nature*, 385, 146–148.

Little, C.T.S., Herrington, R.J., Maslennikov, V.V., and Zaykov, V.V. (1998a). The fossil record of hydrothermal vent communities. In *Modern Ocean Floor Processes and the Geological Record* (pp. 259–270), edited by R.A. Mills and K. Harrison, Geological Society of London Special Publication 148. Geological Society, London.

Little, C.T.S., Danelian, T., Herrington, R.J., and Haymon, R.M. (1998b). Jurassic vent. *Palaeontology Newsletter*, 39, 15.

Little, C.T.S., Maslennikov, V.V., Morris, N.J., and Gubanov, A.P. (1999). Two Palaeozoic hydrothermal vent communities from the southern Ural Mountains, Russia. *Palaeontology*, 42, 1043–1078.

Little, C.T.S., Campbell, K.A., and Herrington, R.J. (2002). Why did ancient chemosynthetic seep and vent assemblages occur in shallower water than they do today? [comment]. *International Journal of Earth Sciences (Geol. Rundsch.)*, 91, 149–153.

Little, C.T.S., Danelian, T., Herrington, R.J., and Haymon, R.M. (2004). Early Jurassic hydrothermal vent community from the Franciscan Complex, California. *Journal of Paleontology*, 78, 542–559.

Little, C.T.S., Magalashvili, A.G., and Banks, D.A. (2007). Neotethyan Late Cretaceous volcanic arc hydrothermal vent fauna. *Geology*, 35, 835–838.

Lochman, C. (1941). A pathological pygidium from the Upper Cambrian of Missouri. *Journal of Paleontology*, 15, 324–325.

Lockley, M.G. (2001). Trackways: dinosaur locomotion. In *Palaeobiology II* (pp. 408–412), edited by D.E.G. Briggs and P. Crowther. Blackwell Scientific, Oxford.

Lockley, M.G. and Matsukawa, M. (1999). Some observations on trackway evidence for gregarious behavior among small bipedal dinosaurs. *Palaeogeography, Palaeoclimatology, Palaeoecology*, 150, 25–31.

Lockley, M.G. and Meyer, C. (2000). *Dinosaur Tracks and Other Fossil Footprints of Europe*. Columbia University Press, New York, 323 pp.

Lockley, M.G., Hunt, A.P., Moratalla, J., and Matsukawa, M. (1994a). Limping dinosaurs? Trackway evidence for abnormal gaits. *Ichnos*, 3, 193–202.

Lockley, M.G., Meyer, C.A., and dos Santos, V.F. (1994b). Trackway evidence for a herd of juvenile sauropods from the Late Jurassic of Portugal. *Gaia*, 10, 27–35.

Lockley, M., Schulp, A.S., Meyer, C.A., Leonardi, G., and Mamani, D.K. (2002). Titanosaurid trackways from the upper Cretaceous of Bolivia: evidence for large manus, wide-gauge locomotion and gregarious behavior. *Cretaceous Research*, 23, 383–400.

Loewen, M.A. (1999). Morphologic variation in caddisfly (Trichoptera) larval case architecture from the Eocene Green River and Wasatch formations of Wyoming. *Geological Society of America Abstracts with Programs*, 31(7), A-470.

Long, C., Wings, O., Xiaohong, C., and Sander, P.M. (2006). Gastroliths in the Triassic Ichthyosaur *Panjiangsaurus* from China. *Journal of Paleontology*, 80, 583–588.

Long, J.A. (1991). Arthrodire predation by *Onychodus* (Pisces, Crossopterygii) from the Late Devonian Gogo Formation, Western Australia. *Records of the Western Australian Museum*, 15, 479–481.

Long, J.A. (1995). *The Rise of Fishes*. The Johns Hopkins University Press, Baltimore, MD, 223 pp.

Long, J.A., Trinajstic, K., Young, G.C., and Senden, T. (2008). Live birth in the Devonian period. *Nature*, 453, 650–652.

Lopes, S.G.B., Domaneschi, O., de Moraes, D.T., Morita, M., and Meserani, G. de L.C. (2000). Functional anatomy of the digestive system of *Neoteredo reynei* (Bartsch, 1920) and *Psiloteredo healdi* (Bartsch, 1931) (Bivalvia: Teredinidae). In *The Evolutionary Biology of the Bivalvia* (pp. 257–271), edited by E.M. Harper, J.D. Harper, and J.A. Crame. Geological Society of London Special Publication 177. Geological Society, London.

Lopez Antonanzas, R. and Cuenca Bescós, G. (2002). The Gran Dolina site (lower to middle Pleistocene, Atapuerca, Burgos, Spain); new palaeoenvironmental data based on the distribution of small mammals. *Palaeogeography, Palaeoclimatology, Palaeoecology*, 186, 311–334.

Loydell, D.K., Zalasiewicz, J., and Cave, R. (1998). Predation on graptolites: new evidence from the Silurian of Wales. *Palaeontology*, 41, 423–427.

Lozouet, P. (1997). Commensalisme chez *Crepidula unguis* et *Bicatillus deformis* (Gastropoda: Calyptraeidae) du Miocène inférieur d'Aquitaine (Sud-Ouest de la France). *Cossmanniana*, 4(1–2), 15–19.

Lozouet, P. (2004). The European Neritiliidae (Molluscs, Gastropoda, Neritopsina): indicators of tropical submarine cave environments and freshwater faunas. *Zoological Journal of the Linnean Society*, 140, 447–467.

Lozouet, P. and Renard, P. (1998). Les Coralliophilidae, Gastropoda d'Oligocène et du Miocène inférieur d'Aquitaine (Sud-Ouest de la France): Systématique et coraux hotes. *Géobios*, 31, 171–184.

Lucas, S.G. (2000). Pathological aetosaur armor from the Upper Triassic of Germany. *Stuttgarter Beiträge zur Naturkunde Serie B*, 281, 1–6.

Lucas, S.G., Libed, S.A., and Kondrashov, P.E. (2001). Species-level evolution of *Periptychus*, a Paleocene "condylarth" from the western United States. *Journal of Vertebrate Paleontology*, 21(3), 75A.

Ludvigsen, R. (1977). Rapid repair of traumatic injury by an Ordovician trilobite. *Lethaia*, 10, 205–207.

Ludvigsen, R. (1979). *Fossils of Ontario*. Part 1. *The Trilobites*. Royal Ontario Museum Life Sciences Miscellaneous Publications, Toronto, 96 pp.

Ludvigsen, R. (1993). A cryptic fossil of importance. *The Gulf Islands Guardian*, Spring, pp. 6–8.

Ludvigsen, R. and Beard, G. (1994). *West Coast Fossils*. Whitecap Books, Toronto, 195 pp.

Ludvigsen, R., and Beard, G. (1997). *West Coast Fossils*, rev. ed. Harbour Publishing, Maderia Park, British Columbia, 216 pp.

Lukashevich, E.D. and Mostovski, M.B. (2003). Hematophagous insects in the fossil record. *Paleontological Journal*, 37(2), 153–161.

Lund, R. (1977). A new petalodont (Chondrichthyes, Bradyodonti) from the Upper Mississippian of Montana. *Annals of Carnegie Museum*, 46, 129–155.

Lund, R. (1978). Anatomy and relationships of the Family Phlegethontiidae (Amphibia, Aïstopoda). *Annals of Carnegie Museum*, 47(4), 53–79.

Lund, R. (1982). *Harpagofututo volsellorhinus* new genus and species (Chondrichthyes, Chondrenchelyiformes) from the Namurian Bear Gulch limestone, *Chondrenchelys problemastica* Traquair (Visean) and their sexual dimorphism. *Journal of Paleontology*, 56, 938–958.

Lund, R. (1985). The morphology of *Falcatus falcatus* (St. John & Worthen), a Mississippian stethacanthid chondrichthyan from the Bear Gulch Limestone of Montana. *Journal of Vertebrate Paleontology*, 5, 1–19.

Lund, R. (1990). Chondrichthyan life history styles as revealed by the 320 million years old Mississippian of Montana. *Environmental Biology of Fishes*, 27, 1–19.

Lund, R. and Lund, W.L. (1985). Coelacanths from the Bear Gulch Limestone (Namurian) of Montana and the evolution of the Coelacanthiformes. *Bulletin of the Carnegie Museum of Natural History*, 25, 74 pp.

Lundin, R. (1978). Letter. *Evolutionary Biology*, 11, 554–555.

Lundstrom, A.N. (1887). Pflanzenbiologische studien. II. Die anpassungen der Pflanzen an Thiere. *Nova Acta Regiae Societatis Scientiarum Uppsaliensis*, 2, 1–88.

Luo, Z.-X. (1999). A refugium for relicts. *Nature*, 400, 23–25.

Luo, Z.-X., Ji, Q., Wible, J.R., and Yuan, C.-X. (2003). An Early Cretaceous tribosphenic mammal and metatherian evolution. *Science*, 302, 1934–1940.

Luther III, G.W., Rozan, T.F., Taillefert, M., Nuzzio, D.B., Di Meo, C., Shank, T.M., Lutz, R.A., and Cary, S.C. (2001). Chemical speciation drives hydrothermal vent ecology. *Nature*, 410, 813–816.

Lutz, A.I. and Herbst, R. (1992). Saprophytic fungi in Upper Permian ferns from Paraguay. *Courier Forschungs-Institut Senckenberg*, 147, 163–169.

Lutz, H. (1989). Die fossile Insektenfauna von Rott. In *Fossillagerstätte Rott bei Hennef am Siebengebirge* (pp. 33–46), edited by W. von Koenigswald. Rheinlandia Verlag, Siegburg.

Lutz, H. (1990). Systematische und palökologische Untersuchungen an Insekten aus dem Mittel-Eozän der Grube Messel bei Darmstadt. *Courier Forschungsinstitut Senckenberg*, 124, 1–165.

Lutz, H. (1992). Giant ants and other rarities: the insect fauna. In *Messel: An Insight into the History of Life and of the Earth* (pp. 55–67), edited by S. Schaal and W. Ziegler. Clarendon Press, London.

Lutz, H. (1993). *Eckfeldapis electrapoides* nov. gen. n. sp., eine "Honigbiene" aus dem Mittel-Eozän des "Eckfelder Maares" bei Manderscheid/Eifel, Deutschland (Hymenoptera: Apidae, Apinae). *Mainzer Naturwissenschaftliches Archiv*, 31, 177–199.

Lutz, R.A., Jablonski, D., and Turner, R.D. (1984). Larval development and dispersal at deep-sea hydrothermal vents. *Science*, 226, 1451–1454.

Lyons, W.G. (1980). Possible sources of Florida's spiny lobster population. In *Proceedings of the 33rd Annual Gulf and Caribbean Fisheries Institute, November 12–14, 1980, San Jose, Costa Rica*, pp. 253–266.

MacDonald, J.R. and Sibley, G. (1969). Paleopathological ponderings or how to tell a sick saber-tooth. *Los Angeles County Museum of Natural History Quarterly*, 8(2), 26–30.

MacEachern, J.A., Pemberton, S.G., Gingras, M.K., and Bann, K.L. (2007). The ichnofacies paradigm: a fifty-year perspective. In *Trace Fossils, Concepts, Problems, Prospects* (pp. 52–77), edited by W. Miller III. Elsevier, Amsterdam.

MacFadden, B.J. (1992). *Fossil Horses: Systematics, Paleobiology, and Evolution of the Family Equidae*. Cambridge University Press, Cambridge, U.K., 369 pp.

MacGinitie, G.E. and MacGinitie, N. (1949). *Natural History of Marine Animals*. McGraw-Hill, New York, 473 pp.

Machado, C.A., Jousselin, E., Kjellberg, F., and Crompton, S.C. (2001). Phylogenetic relationships, historical biogeography and character evolution of fig-pollinating wasps. *Proceedings of the Royal Society London B*, 268, 685–694.

MacKay, M.R. (1969). Microlepidopterous larvae in Baltic amber. *The Canadian Entomologist*, 101, 1173–1180.

MacKinnon, D.I. and Biernat, G. (1970). The probable affinities of the trace fossil *Diorygma atrypophilia*. *Lethaia*, 3, 163–172.

Mackness, B. and Sutton, R. (2000). Possible evidence for infraspecific aggression in a Pliocene crocodile from north Queensland. *Alcheringa*, 24, 55–62.

MacNaughton, R.B. and Pickerill, R.K. (1995). Invertebrate ichnology of the nonmarine Lepreau Formation (Triassic), southern New Brunswick, eastern Canada. *Journal of Paleontology*, 69, 160–171.

MacRae, C. (1999). *Life Etched in Stone: Fossils of South Africa*. The Geological Society of South Africa, Johannesburg, 305 pp.

MacSwain, J.W. (1956). *A Classification of the First Instar Larvae of the Meloidae (Coleoptera)*, Vol. 12. University of California Publications in Entomology, Berkeley, 182 pp.

Mader, D. (1999). *Geologische und Biologische Entemoökologie der Rezenten Seidenbiene* Colletes. Logabook, Köln, 807 pp.

Maes, P. and Jangoux, M. (1984). The bald sea-urchin disease: a biopathological approach. *Helgoländer Meeresuntersuchungen*, 37, 217–224.

Magallon-Puebla, S. and Cevallos-Ferriz, S.R.S. (1993). A fossil earthstar (Geasteraceae: Gasteromycetes) from the Late Cenozoic of Puebla, Mexico. *American Journal of Botany*, 80, 1162–1167.

Maglio, V.J. (1973). Origin and evolution of the elephantoidea. *Transactions of the American Philosophical Society*, 63, 3, 3–149.

Mahon, A.R., Amsler, C.D., McClintock, J.B., Amsler, M.O., and Baker, B.J. (2003). Tissue-specific palatability and chemical defenses against macropredators and pathogens in the common articulate brachiopod *Liothyrella uva* from the Antarctic Peninsula. *Journal of Experimental Marine Biology and Ecology*, 290, 197–210.

Maisey, J.G. (1994). Predator relationships and trophic level reconstruction in a fossil fish community. *Environmental Biology of Fishes*, 40, 1–22.

Maisey, J.G. (1996). *Discovering Fossil Fishes*. Henry Holt & Company, New York, 223 pp.

Maisey, J.G. and de Carvalho, M. da G.P. (1995). First records of fossil sergestid decapods and fossil brachyuran crab larvae (Arthropoda, Crustacea), with remarks on some supposed Palaemonid fossils, from the Santana Formation (Aptian-Albian, NE Brazil). *American Museum Novitates*, 3132, 1–20.

Majima, R. (1984). Observations on occurrences of Japanese Neogene naticids (Gastropoda) bearing calcareous opercula. *Transactions and Proceedings of the Palaeontological Society of Japan*, 134, 361–373.

Majima, R., Nobuhara, T., and Kitazaki, T. (2005). Review of fossil chemosynthetic assemblages in Japan. *Palaeogeography, Palaeoclimatology, Palaeoecology*, 227, 86–123.

Makowski, H. (1962). Problem of sexual dimorphism in ammonites. *Palaeontologia Polonica*, 12, 92 pp.

Małkowski, K. (1976). Regeneration of some brachiopod shells. *Acta Geologica Polonica*, 26, 439–442.

Malone, K.M. and Nolan, R.A. (1978). Aerobic bacterial flora of the larval gut of the black fly *Prosimulium mixtum* (Diptera: Simulidae) from Newfoundland, Canada. *Journal of Medical Entomology*, 14, 641–645.

Malzahn, E. (1968). Uber neue Funde von *Janassa bituminosa* (Schloth.) im niederrheinischen Zechstein. *Geologisches Jahrbuch*, 85, 67–96.

Mamay, S.H. (1994). Fossil eggs of probable piscine origin preserved on Pennsylvanian *Sphenopteridium* foliage from the Kinney Quarry, central New Mexico. *Journal of Vertebrate Paleontology*, 14, 320–326.

Mamay, S.H., Hook, R.W., and Hotton III, N. (1998). Amphibian eggs from the Lower Permian of north-central Texas. *Journal of Vertebrate Paleontology*, 18, 80–84.

Manceñido, M.O. and Damborenea, S.E. (1990). Corallophilous micromorphic brachiopods from the Lower Jurassic of west central Argentina. In *Brachiopods Through Time* (pp. 89–96), edited by D.I. MacKinnon, D.E. Lee, and J.D. Campbell. Balkema, Rotterdam.

Manceñido, M.O. and Gourvennec, R. (2008). A reappraisal of feeding current systems inferred for spire-bearing brachiopods. *Earth and Environmental Science Transactions of the Royal Society of Edinburgh*, 98, 345–356.

Manchester, S.R. (1987). The fossil history of the Juglandaceae. *Monographs in Systematic Botany from the Missouri Botanical Garden*, 21, 1–137.

Mandic, O., Harzhauser, M., and Roetzel, R. (2008). Benthic mass-mortality events on a Middle Miocene incised-valley tidal-flat (North Alpine Foredeep Basin). *Facies*, 54, 343–359.

Mania, D. and Mania, U. (1988). Deliberate engravings on bone artifacts of *Homo erectus*. *Rock Art Research*, 5, 91–107.

Manning, P.L. (2004). A new approach to the analysis and interpretation of tracks: examples from the dinosauria. In *The Application of Ichnology to Palaeoenvironmental and Stratigraphic Analysis* (pp. 93–123), edited by D. McIlroy, Geological Society of London Special Publication 228. Geological Society, London.

Manning, P.L., Margetts, L., Leng, J.M., and Smith, I.M. (2006). Parallel 3D finite element analysis of dinosaur track formation. In *Proceedings of the 50th Annual Meeting of the Palaeontological Association, December 18–21, Sheffield, U.K.*

Manum, S.B. (1996). Clitellate cocoons. In *Palynology: Principles and Applications* (pp. 361–364), edited by J. Jansonius and D.C. McGregor. AASP Foundation, College Station, TX.

Manum, S.B., Bose, M.N., and Sawyer, R.T. (1991). Clitellate cocoons in freshwater deposits since the Triassic. *Zoologica Scripta*, 20, 347–366.

Manum, S.B., Bose, M.N., and Sawyer, R.T. (1992). Seeds (*Burejospermum* Krassilov) and palynomorphs (*Dictyothylakos* Horst) with a netted wall structure reinterpreted: clitellate cocoons. *Courier Forschungs-Institut Senckenberg*, 147, 399–404.

Manum, S.B., Bose, M.N., Sayer, R.T., and Bostrom, S. (1994). A nematode (*Captivonemus cretacea* gen. et sp. n.) preserved in a clitellate cocoon wall from the Early Cretaceous. *Zoologica Scripta*, 23, 27–31.

Mapes, R.H. and Chaffin, D.T. (2003). Predation on cephalopods. In *Predator–Prey Interactions in the Fossil Record* (pp. 177–213), edited by P.H. Kelley, M. Kowalewski, and T.A. Hansen. Kluwer Academic, New York.

Mapes, R.H. and Davis, R.A. (1996). Color patterns in ammonoids. In *Ammonoid Paleobiology* (pp. 103–127), edited by N. Landman, K. Tanabe, and R.A. Davis. Plenum Press, New York.

Mapes, R.H. and Evans, T.S. (1995). The color pattern on a Cretaceous nautiloid from South Dakota. *Journal of Paleontology*, 69, 785–786.

Mapes, R.H. and Sneck, D.A. (1987). The oldest ammonoid "colour" patterns: description, comparison with *Nautilus*, and implications. *Palaeontology*, 30, 299–309.

Mapes, R.H., Fahrer, T.R., and Babcock, L.E. (1989). Sublethal and lethal injuries of Pennsylvanian conularids from Oklahoma. *Journal of Paleontology*, 63, 34–37.

Mapes, R.H., Sims, M.S., and Boardman II, D.R. (1995). Predation on the Pennsylvanian ammonoid *Gonioloboceras* and its implications for allochthonous vs. autochthonous accumulations of goniatites and other ammonoids. *Journal of Paleontology*, 69, 441–446.

Marcuse, S.A. (1975). *A Survey of Musical Instruments*. Harper & Row, New York, 555 pp.

Marek, L. and Galle, A. (1976). The tabulate coral *Hyostragulum*, an epizoan with bearing on hyolithids ecology and systematics. *Lethaia*, 9, 51–64.

Mark-Kurik, E. (1992). The inferognathal in the Middle Devonian arthrodire *Homostius*. *Lethaia*, 25, 173–178.

Mark-Kurik, E. and Carls, P. (2004). *Tityosteus*, a marine fish (Arthrodira, Homostiidae) from the Emsian of Aragón, Spain, and its distribution. *Revista Española de Paleontología*, 19, 139–144.

Marron, A. and Moore, J. (2006). Evidence of frugivory in terrestrial chelonians of the Scenic Member, Brule Formation, South Dakota., In *Proceedings of the 50th Annual Meeting of the Palaeontological Association, December 18–21, Sheffield, U.K.*

Marshak, A. (1989). Evolution of human capacity: the symbolic evidence. *Yearbook of Physical Anthropology*, 32, 1–34.

Marshall, A.G. (1987). Nutritional ecology of ectoparasitic insects. In *Nutritional Ecology of Insects, Mites, Spiders, and Related Invertebrates* (pp. 721–739), edited by F. Slansky, Jr., and J.G. Rodriguez. John Wiley & Sons, New York.

Marshall, B.A. (1985a). Recent and Tertiary deep-sea limpets of the genus *Pectinodonta* Dall (Mollusca: Gastropoda) from New Zealand and New South Wales. *New Zealand Journal of Zoology*, 12, 273–282.

Marshall, B.A. (1985b). Recent and Tertiary Cocculinidae and Pseudococculinidae (Mollusca: Gastropoda) from New Zealand and New South Wales. *New Zealand Journal of Zoology*, 12, 505–546.

Marshall, B.A. (1987). Osteopeltidae (Mollusca: Gastropoda): a new family of limpets associated with whale bone in the deepsea. *Journal of Molluscan Studies*, 53, 121–127.

Marshall, B.A. (1994). Deep-sea gastropods from the New Zealand region associated with Recent whale bones and an Eocene turtle. *The Nautilus*, 108, 1–8.

Marshall, J., Kelley, W.P., Rubakhin, S.S., Bingham, J.-P., Sweedler, J.V., and Gilly, W.F. (2002). Anatomical correlates of venom production in *Conus californicus*. *Biological Bulletin, Hopkins Marine Station*, 203, 27–41.

Martill, D.M. (1985). The preservation of marine vertebrates in the Lower Oxford Clay. *Philosophical Transactions of the Royal Society of London B*, 311, 155–165.

Martill, D.M. (1986a). The diet of *Metriorhynchus*, a Mesozoic marine crocodile. *Neues Jahrbuch für Geologie und Paläontologie–Monatsheft*, 621–625.

Martill, D.M. (1986b). The stratigraphic distribution and preservation of fossil vertebrates in the Oxford Clay of England. *Mercian Geologist*, 10, 161–188.

Martill, D.M. (1988). *Leedsichthys problematicus*, a giant filter-feeding teleost from the Jurassic of England and France. *Neues Jahrbuch für Geologie und Paläontologie–Monatsheft*, 670–680.

Martill, D.M. (1990). Predation on *Kosmoceras* by semionotid fish in the Middle Jurassic Lower Oxford Clay of England. *Palaeontology*, 33, 739–742.

Martill, D.M. (1992). Pliosaur stomach contents from the Oxford Clay. *Mercian Geologist*, 13, 37–42.

Martill, D.M. and Barker, M.J. (2006). A paper nautilus (Octopoda, *Argonauta*) from the Miocene Pakhna Formation of Cyprus. *Palaeontology*, 49, 1035–1041.

Martill, D.M. and Davis, P.G. (1998a). Did dinosaurs come up to scratch? *Nature*, 396, 528–529.

Martill, D.M. and Davis, P. (1998b). Ectoparasites on a Cretaceous feather. *Palaeontology Newsletter*, 39, 17.

Martill, D.M. and Frey, E. (1995). Colour patterning preserved in Lower Cretaceous birds and insects: the Crato formation of N.E. Brazil. *Neues Jahrbuch für Geologie und Paläontologie–Monatsheft*, 118–128.

Martill, D.M., Taylor, M.A., and Duff, K.L. (1994). The trophic structure of the biota of the Peterborough Member, Oxford Clay formation (Jurassic), U.K. *Journal of the Geological Society*, 151(1), 173–194.

Martin, A.P. and Naylor, G.J.P. (1994). Independent origins of filter-feeding in megamouth and basking sharks (Order Lamniformes) inferred from phylogenetic analysis of Cytochrome *b* gene sequences. In *Biology of the Megamouth Sharks* (pp. 39–50), edited by K. Yano, J.F. Morrissey, Y. Yabumoto, and K. Nakaya. Tokai University Press, Tokyo.

Martin, J.E. and Bjork, P.R. (1987). Gastric residues associated with a mosasaur from the Late Cretaceous (Campanian) Pierre shale in South Dakota. *Dakoterra*, 3, 68–72.

Martin, J.E. and Fox, J.E. (2004). Molluscs in the stomach contents of Globidens, a shell-crushing mosasaur, from the Late Cretaceous Pierre Shale, Big Bend area of the Missouri River, central South Dakota. *Geological Society of America Abstracts with Programs*, 36(4), 80.

Martin, J.E. and Kennedy, L.E. (1988). A plesiosaur with stomach contents from the Late Cretaceous (Campanian) Pierre Shale of South Dakota: a preliminary report. *Proceedings of the South Dakota Academy of Science*, 67, 76–79.

Martin, J.G., Naylor, J.P.J., and Palumbi, S.R. (1992). Rates of mitochondrial DNA evolution in sharks are slow compared with mammals. *Nature*, 357, 153–155.

Martin, L.D. (2003). Earth history, disease, and the evolution of primates. In *Emerging Pathogens: The Archaeology, Ecology, and Evolution of Infectious Disease* (pp. 13–24), edited by C. Greenblatt and M. Spigel. Oxford University Press, London.

Martin, L.D. and Rothschild, B.M. (1989). Paleopathology and diving mosasaurs. *American Scientist*, 77, 460–467.

Martin, L.D., Rothschild, B.M., Greenblatt, C., and Lev, G. (1999). DNA verification of tuberculosis in Pleistocene artiodactyls. *Journal of Vertebrate Paleontology*, 19, 61A.

Martin, L.D. and West, D.L. (1995). The recognition and use of dermestid (Insecta, coleoptera) pupation chambers in paleoecology. *Palaeogeography, Palaeoclimatology, Palaeoecology*, 113, 303–310.

Martin, R.A. (1993). Patterns of variation and speciation in Quaternary rodents. In *Morphological Change in Quaternary Mammals of North America* (pp. 226–280), edited by R.A. Martin and A.D. Barnosky. Cambridge University Press, New York.

Martin, R.A. and Barnosky, A.D., Eds. (1993a). *Morphological Change in Quaternary mammals of North America*. Cambridge University Press, New York, 415 pp.

Martin, R.A. and Barnosky, A.D. (1993b). Quaternary mammals and evolutionary theory: introductory remarks and historical perspective. In *Morphological Change in Quaternary Mammals of North America* (pp. 1–12), edited by R.A. Martin and A.D. Barnosky. Cambridge University Press, New York.

Martin, R.A., Nesbitt, E.A., and Campbell, K.A. (2007). Carbon stable isotopic composition of benthic Foraminifera from Pliocene cold methane seeps, Cascadia accretionary margin. *Palaeogeography, Palaeoclimatology, Palaeoecology*, 246, 260–277.

Martin, R.D., Willner, L.A., and Dettling, A. (1994). The evolution of sexual size dimorphism in primates. In *The Differences Between the Sexes* (pp. 159–200), edited by R.V. Short and E. Balaban. Cambridge University Press, Cambridge, U.K.

Martínez-Delclòs, X. and Martinell, J. (1995). The oldest known record of social insects. *Journal of Paleontology*, 69, 594–599.

Martínez-Delclòs, X., Briggs, D.E.G., and Peñalver, E. (2004). Taphonomy of insects in carbonates and amber. *Palaeogeography, Palaeoclimatology, Palaeoecology*, 203, 19–64.

Marty, P. (1894). De l'ancienneté de la *Cecidomyia fagi*. *Feuille des naturalistes revue mensuelle d'histoire naturelle*, 24, 173.

Marwick, J. (1922). Fossil pearls in New Zealand. *The New Zealand Journal of Science and Technology*, 5, 202.

Marwick, J. (1971). An ovoviviparous gastropod (Turritellidae, *Zeocolpus* [sic.]) from the upper Miocene of New Zealand. *New Zealand Journal of Geology and Geophysics*, 14, 66–70.

Maschenko, E.N. and Shpansky, A.V. (2005). Abnormal dental morphology in the mammoth *Mammuthus primigenius* Blumenbach, 1799. *Paleontological Journal*, 39(1), 93–100.

Massare, J.A. (1987). Tooth morphology and prey preference of Mesozoic marine reptiles. *Journal of Vertebrate Paleontology*, 7, 121–137.

Massare, J.A. and Callaway, J.M. (1988). Live birth in ichthyosaurs: evidence and implications. *Journal of Vertebrate Paleontology*, 3(Suppl.), 21A.

Masse, P. and Vachard, D. (1996). A crustacean coprolite, *Palaxius salataensis* in the Upper Carboniferous of the southern Urals. *Neues Jahrbuch für Geologie und Paläontologie–Monatsheft*, 490–494.

Massin, C. (1988). Boring Coralliophilidae (Mollusca, Gastropoda): coral–host relationship. *Proceedings of the Sixth International Coral Reef Symposium*, 3, 177–184.

Massoud, Z. and Betsch, J.M. (1966). Considérations sur l'antenne des Sminthuridinae et description de deux nouvelles espèces de Collemboles interstitials du genre *Sminthurides* Boerner, 1900 (Symphypléones). *Bulletin du Muséum national d'histoire naturelles, sciences de la terre*, 2(38), 574–585.

Masuda, K. (1968). Sandpipes penetrating igneous rocks in the environs of Sendai, Japan. *Transactions and Proceedings of the Palaeontological Society of Japan*, 72, 351–362.

Masuda, K. and Noda, H. (1969). Pliocene boring shells and their burrows from the environs of Sendai, Japan. *Transactions and Proceedings of the Palaeontological Society of Japan*, 75, 130–135.

Matheson, C.D. and Brian, D. (2003). The molecular taphonomy of biological molecules and biomarkers of disease. *The Archaeology, Ecology, and Evolution of Infectious Disease* (pp. 127–142), edited by C. Greenblatt and M. Spigelman. Oxford University Press, London.

Matsukuma, A. (1978). Fossil boreholes made by shell-boring predators or commensals. I. Boreholes of capulid Gastropoda. *Venus: Japanese Journal of Malacology*, 37(1), 29–45.

Matsukuma, A. (1996). Transposed hinges: a polymorphism of bivalve shells. *Journal of Molluscan Studies*, 62, 415–431.

Matsumoto, T., Obata, I., Okazaki, Y., and Kani, Y. (1982). An interesting occurrence of a fossil reptile in the Cretaceous of the Obira Area, Hokkaido. *Proceedings of the Japanese Academy Series B*, 58(5), 109–113.

Matthews, Jr., J.V. (1979). Late Tertiary carabid fossils from Alaska and the Canadian Archipelago. In *Carabid Beetles: Their Evolution, Natural History, and Classification* (pp. 425–445), edited by T.I. Erwin et al. Dr. W. Junk, the Hague.

Maugh, T.H. (1975). Paleontology facing a choice between fossils and trash. *Science*, 189, 985–986.

Maxwell, E.E. and Caldwell, M.W. (2003). First record of live birth in Cretaceous ichthyosaurs: closing an 80 million year gap. *Proceedings of the Royal Society London B*, 270(1, Suppl.), S104–S107.

Maynard Smith, J.M. (1988a). *Did Darwin Get It Right: Essays on Games, Sex and Evolution*. Chapman & Hall, London, 264 pp.

Maynard Smith, J.M. (1988b). Punctuation in perspective. *Nature*, 332, 311–312.

Mayr, G. (2004). Old World fossil record of modern-type hummingbirds. *Science*, 304, 861–864.

Mayr, G. (2006). New specimens of the Eocene Messelirrisoridae (Aves: Bucerotes), with comments on the preservation of uropygial gland waxes in fossil birds from Messel and the phylogenetic affinities of the Bucerotes. *Paläontologisches Zeitschrift*, 80, 390–405.

Mayr, H. (1992). *A Guide to Fossils*. Princeton University Press, Princeton, NJ, 256 pp.

Mazin, J.-M. and de Buffrenil, V., Eds. (2001). *Secondary Adaptation of Tetraods to Life in Water*. Verlag Dr. Friedrich Pfeil, Munich, 367 pp.

McAlpine, J.F. (1970). First record of calypterate flies in the Mesozoic era (Diptera: Calliphoridae). *Canadian Entomologist*, 102, 342–346.

McArthur, A.G. and Tunnicliffe, V. (1998). Relics and antiquity revisited in the modern vent fauna. In *Modern Ocean Floor Processes and the Geological Record* (pp. 271–291), edited by R.A. Mills and K. Harrison. Geological Society of London Special Publication 148. Geological Society, London.

McCormick, T. and Fortey, R.A. (1998). Independent testing of a paleobiological hypothesis: the optical design of two Ordovician pelagic trilobites reveals their relative paleobathymetry. *Paleobiology*, 24, 235–255.

McDonald, H.G. (2003). Sloth remains from North American Caves and associated karst features. In *Ice Age Cave Faunas of North America* (pp. 1–16), edited by B.W. Schubert, J.I. Mead, and R.W. Graham. Indiana University Press, Bloomington.

McFall-Ngai, M.J. (1991). Luminous bacterial symbiosis in fish evolution: adaptive radiation among leiognathid fishes. In *Symbiosis as a Source of Evolutionary Innovation* (pp. 381–408), edited by L. Margulis and R. Fester. MIT Press, Boston.

McGhee, Jr., G.R. and Richardson, Jr., E.S. (1982). First occurrence of the problematical fossil *Vetacapsula* in North America. *Journal of Paleontology*, 56, 1295–1296.

McHenry, C.R., Cook, A.G., and Wroe, S. (2005). Bottom-feeding plesiosaurs. *Science*, 310, 75.

McHenry, H.M. (1991). Sexual dimorphism in *Australopithecus afarensis*. *Journal of Human Evolution*, 20, 21–32.

McKee, J.W.A. (1987). The occurrence of the Pliocene penguin *Tereingaomis moisleyi* (Sphenisciformes; Spheniscidae) at Hawera, Taranaki, New Zealand. *New Zealand Journal of Zoology*, 14, 557–561.

McKeown, K.C. (1937). New fossil insect wings (Protohemiptera, Family Mesotitanidae). *Records of the Australian Museum*, 20, 31–37.

McKinney, F.K. (1968). A bored ectoproct from the Middle Mississippian of Tennessee. *Southeastern Geology*, 9, 165–170.

McKinney, F.K. (1995). One hundred years of competitive interactions between bryozoan clades: asymmetrical but not escalating. *Biological Journal of the Linnean Society*, 56, 465–481.

McKinney, F.K., Broadhead, T.W., and Gibson, M.A. (1990). Coral-bryozoan mutualism: structural innovation and greater resource exploitation. *Science*, 248, 466–468.

McKinney, F.K., Taylor, P.D., and Lidgard, S. (2003). Predation on bryozoans and its reflection in the fossil record. In *Predator–Prey Interactions in the Fossil Record* (pp. 239–261), edited by P.H. Kelley, M. Kowalewski, and T.A. Hansen. Kluwer Academic, New York.

McLean, J.H. (1981). The Galapagos Rift limpet *Neomphalus*: Relevance to understanding the evolution of a major Paleozoic-Mesozoic radiation. *Malacologia*, 21, 323–346.

McLean, J.H. and Kiel, S. (2007). Cretaceous and living Colloniidae of the redefined Subfamily Petropomatinae, with two new genera and one new species, with notes on opercular evolution in turbinoideans and the fossil record of Liotiidae (Vetigastropods; Turbinoidea). *Palaeontologische Zeitschrift*, 81, 254–266.

McLoughlin, S., Tosolini, A.-M.P., Nagalingum, N.S., and Drinnan, A.N. (2002). The Early Cretaceous (Neocomian) flora and fauna of the lower Strzelecki Group, Gippsland Basin, Victoria, Australia. *Association of Australasian Palaeontologists Memoirs*, 26, 1–144.

McNamara, K.J. (1991). Murder and mayhem in the Miocene. *Natural History*, 8/91, 40–45.

McNamara, K.J. (1994a). Diversity of Cenozoic marsupiate echinoids as an environmental indicator. *Lethaia*, 27, 257–268.

McNamara, K.J. (1994b). The significance of gastropod predation to patterns of evolution and extinction in Australian Tertiary echinoids. In *Echinoderms Through Time* (pp. 785–793), edited by B. David, A. Guille, J.-P. Feral, and M. Roux. Balkema, Rotterdam.

McNamara, K.J. and Barrie, D.J. (1992). A new genus of marsupiate spatangoid echinoid from the Miocene of South Australia. *Records of the South Australian Museum*, 26, 139–147.

McNamara, M.E., Orr, P.J., Kearns, S.L., Alcala, L., Anadón, P., and Peñalver-Molla, E. (2006). Taphonomy of exceptionally preserved tadpoles from the Miocene Libros fauna, Spain: ontogeny, ecology and mass mortality. In *Proceedings of the 50th Annual Meeting of the Palaeontological Association, December 18–21, Sheffield, U.K.*

McRoberts, C.A. and Stanley, Jr., G.D. (1989). A unique bivalve-algae life assemblage from the Bear Gulch Limestone (Upper Mississippian) of central Montana. *Journal of Paleontology*, 63, 578–581.

McWhinney, L., Carpenter, K., and Rothschild, B. (2001a). Dinosaurian humeral periostitus: a case of a juxtacortical lesion in the fossil record. In *Mesozoic Vertebrate Life* (pp. 364–392), edited by D.H. Tanke and K. Carpenter. Indiana University Press, Bloomington.

McWhinney, L.A., Rothschild, B.M., and Carpenter, K. (2001b). Posttraumatic chronic osteomyelitis in *Stegosaurus* dermal spikes. In *The Armored Dinosaurs* (pp. 141–156), edited by K. Carpenter. Indiana University Press, Bloomington.

Mead, A.J. (2000). Sexual dimorphism and paleoecology in *Teleoceras*, a North American Miocene rhinoceras. *Paleobiology*, 26, 689–706.

Mead, J.I. and Agenbroad, L.D. (1989). Pleistocene dung and the extinct herbivores of the Colorado Plateau, southwestern USA. *Cranium*, 6(1), 29–44.

Meehan, T.J. (1994). Sediment analysis of the Middle Whitney Member: climatic implications for the Upper Oligocene of Western Nebraska, *TER-QUA Symposium Series*, 2, 57–87.

Meek, F.B. and Worthen, A.H. (1866). Radiata, Echinodermata, Crinoidea. *Proceedings of the Academy of Natural Sciences, Philadelphia*, 251–275.

Meek, F.B. and Worthen, A.H. (1868). Notes on some points in the structure and habits of the Paleozoic Crinoidea. *Proceedings of the Academy of Natural Sciences, Philadelphia*, 323–334.

Mehl, J. (1978). Ein Koprolith mit Ammoniten-Aptychen aus den Solnhofer Plattenkalken. *Wetterauische Gesellschaft für die gesamte Naturkunde*, 85–89.

Mehl, J. (1986). Die fossile Dokumentation der Orchideen. *Jahresberichte des Naturwissenschaftlichen Vereins in Wuppertal*, 39, 121–133.

Melchor, R., Genise, J.F., and Miquel, S.E. (2002). Ichnology, sedimentology and paleontology of Eocene calcareous paleosols from a palustrine sequence, Argentina. *Palaios*, 17, 16–35.

Meng, J. and Wyss, A.R. (1997). Multituberculate and other mammal hair recovered from Palaeogene excreta. *Nature*, 385, 712–714.

Meng, J., Hu, Y., Wang, Y., Wang, X., and Li, C. (2006). A Mesozoic gliding mammal from northeastern China. *Nature*, 444, 886–893.

Meng, Q., Liu, J., Varricchio, D.J., Huang, T., and Gao, C. (2004). Parental care in an ornithiscian dinosaur. *Nature*, 431, 148.

Menge, A. (1856). Lebenszeichen vorweltlicher, im bernstein eingeschlossener thiere. *Programm der Petrischule, Danzig*, 32 pp.

Menge, A. (1866). Ueber ein Rhipidopteron und einige andere im Bernstein eingeschlossene tiere. *Naturforschenden Gesellschaft Danzig*, 1, 1–8.

Merkt, J. (1966). Über Austern und Serpeln als Epöken auf Ammonitengehäusen. *Neues Jahrbuch für Geologie und Paläontologie-Abhandlungen*, 125, 467–479.

Merriam, C.W. (1921). Notes on a brittle star limestone from the Miocene of California. *American Journal of Science*, 221, 304–310.

Merrill, G.K. (1979). Unusual substrate adaptation in Late Paleozoic acrothoracic barnacles. *Journal of Paleontology*, 53, 1433–1435.

Meyer, C.A. (1988a). Subtidal lagoon communities of a late Jurassic turtle deposit from northern Switzerland. *Museo Regionale di Scienze Naturli di Torino*, 107–121.

Meyer, C.A. (1988b). Paléoécologie d'une communauté d'ophiures du Kimmeridgien Supérieur de la region Havraise (Seine-Maritime). *Bulletin Trimestriel de la Société géologique de Normandie et des amis du Muséum du Havre*, 75, 25–35.

Meyer, D.L. (1990). Population paleoecology and comparative taphonomy of two edrioasteroid (Echinodermata) pavements: Upper Ordovician of Kentucky and Ohio. *Historical Biology*, 4, 155–178.

Meyer, D.L. and Ausich, W.I. (1983). Biotic interactions among recent and among fossil crinoids. In *Biotic Interactions in Recent and Fossil Benthic Communities* (pp. 377–427), edited by M.J. Tevesz and S.L. McCall. Plenum Press, New York.

Meyer, H.W. (2003). *The Fossils of Florissant*. Smithsonian Books, Washington, D.C., 258 pp.

Meyer, R.C. (1999). Helical burrows as a palaeoclimate response: *Daimonelix* by *Palaeocastor*. *Palaeogeography, Palaeoclimatology, Palaeoecology*, 147, 291–298.

Michalik, J. (1977). Systematics and ecology of *Zeilleria bayle* and other brachiopods in the uppermost Triassic of the west Carpathians. *Geologicky Zbornik-Geologica Carpathica*, 28, 323–346.

Michaux, J., Hutterer, R., and Lopez-Martinez, N. (1991). New fossil faunas from Fuerteventura, Canary Islands: evidence for a Pleistocene age of endemic rodents and shrews. *Comptes Rendus de l'Academie des Sciences, Paris*, 312(II), 801–806.

Michener, C.D. (2000). *The Bees of the World*. The Johns Hopkins University Press, Baltimore, MD, 913 pp.

Michener, C.D. and Grimaldi, D. (1988). A *Trigona* from Late Cretaceous amber of New Jersey. *American Museum Novitates*, 2917, 1–10.

Michener, C.D. and Grimaldi, D.A. (1989). The oldest fossil bee: apoid history, evolutionary stasis, and antiquity of social behavior. *Proceedings of the National Academy of Science USA*, 85, 6424–6426.

Mikhailov, K.E. (1994). Yaitsa teropodovikg i prototseratopsovikh dinozavrov iz Melovikh otlozhenii Mongolii i Kazakhstana. *Paleontologicheskii Zhurnal*, 2, 81–96.

Mikhailov, K.E. (1996). Yaitsa ptitsa v verkhnem Meli Mongolii. *Paleontologicheskii Zhurnal*, 1, 119–121.

Mikhailov, K.E. (2000). Eggs and eggshells of dinosaurs and birds from the Cretaceous of Mongolia. In *The Age of Dinosaurs in Russia and Mongolia* (pp. 560–572), edited by M.J. Benton, M.A. Shiskin, D.M. Unwin, and E.N. Kurochkin. Cambridge University Press, Cambridge, U.K.

Mikuláš, R. (1990). The ophiuroid *Taeniaster* as a tracemaker of *Asteriacites*, Ordovician of Czechoslovakia. *Ichnos*, 1, 133–137.

Mikuláš, R. (1993). Teredolites from the Upper Cretaceous near Prague (Bohemian Cretaceous basin, Czechoslovakia). *Vestnik Ceského Geologického Ústavu*, 68, 7–10.

Mikuláš, R., Pek, I., and Zimák, J. (1995a). *Teredolites clavatus* from the Cenomanian near Maletín (Bohemian Cretaceous basin), Moravia, Czech Republic. *Vestnik Ceského Geologického Ústavu*, 70, 51–57.

Mikuláš, R., Petr, V., and Prokop, R.J. (1995b). The first occurrence of a "brittlestar bed" (Echinodermata, Ophiuroidea) in Bohemia (Ordovician, Czech Republic). *Vestnik Ceského Geologického Ústavu*, 70/3, 15–24.

Mikuláš, R., Dvořák, Z., and Pek, I. (1998). *Lamniporichnus vulgaris* igen. et isp. nov.: traces of insect larvae in stone fruits of hackberry (*Celtis*) from the Miocene and Pleistocene of the Czech Republic. *Journal of the Czech Geological Society*, 43(4), 277–280.

Miles, A.E.W. and Grigson, C. (1990). *Colyer's Variations and Diseases of the Teeth of Animals*. Cambridge University Press, Cambridge, U.K., 672 pp.

Miles, R.S. and Westoll, S.T. (1968). The placoderm fish *Coccosteus cuspidatus* Miller ex Agassiz from the Middle Old Red Sandstone of Scotland. Part I. Descriptive morphology. *Transactions of the Royal Society of Edinburgh*, 67(9), 373–476.

Miller, R.H. and Sundberg, F.A. (1984). Boring Late Cambrian organisms. *Lethaia*, 17, 185–190.

Miller, R.L., Armelagos, G.J., Ikram, S., De Jonge, N., Krijger, F.W., and Deelder, A.M. (1992). Palaeoepidemiology of *Schistosoma* infections in mummies. *British Medical Journal*, 304, 355–356.

Miller, R.L., Ikraum, S., Armelagos, G.J., Walker, R., Harer, W.B., Shiff, C.J., Baggett, D., Carrigan, M., and Marel, S.M. (1994). Diagnosis of *Plasmodium falciparum* in mummies using the rapid manual ParaSight™-F test. *Transactions of the Royal Society of Tropical Medicine and Hygiene*, 88, 31–32.

Miller III, W. and Brown, N.A. (1979). The attachment scars of fossil balanids. *Journal of Paleontology*, 53, 208–210.

Milner, A.R. (1982). Small temnospondyl amphibians from the Middle Pennsylvanian of Illinois. *Palaeontology*, 25, 635–664.

Mistiaen, B. (1984). Comments on the caunopore tubes: stratigraphic distribution and microstructure. *Palaeontographica Americana*, 54, 501–508.

Mitchell, L. and Curry, G.B. (1997). Diagenesis and survival of intracrystalline amino acids in fossil and Recent mollusc shells. *Palaeontology*, 40, 855–874.

Mitchell, P. and Wighton, D. (1979). Larval and adult insects from the Paleocene of Alberta, Canada. *Canadian Entomologist*, 111, 777–782.

Mitrović-Petrović, J. (1964). Les apparitions des irregularities et des anomalies sur le squelette des echinides du Miocène Moyen, comme la conséquence du parasitisme et des lésions biotiques. *Geologiski anali Balkanskogo poluostrova*, 31, 135–145.

Mizuno, Y. (1993). Echinoidea. In *Fossils from the Miocene Morozaki Group* (pp. 141–155), edited by F. Ohe, I. Nonogaki, T. Tanaka, K. Hachiya, Y. Mizuno, T. Momoyama, and T. Yamaoka. The Tokai Fossil Society, Nagoya, Japan.

Mizuno, Y. and Takeda, M. (1993). Crustacea. In *Fossils from the Miocene Morozaki Group* (pp. 77–90), edited by F. Ohe, I. Nonogaki, T. Tanaka, K. Hachiya, Y. Mizuno, T. Momoyama, and T. Yamaoka. The Tokai Fossil Society, Nagoya, Japan.

Modzalevskaya, T.L. (2007). The earliest terebratuloids. *Palaeontology*, 50, 869–882.

Moissette, P. and Saint Martin, J.-P. (1990). Cirripèdes Pyrgomatidae actuels et fossiles d'Oranie (Algérie). *Revue de Paléobiologie*, 9, 37–47.

Moissette, P., Koskeridou, E., Cornée, J.-J., Guilocheau, F., and Lécuyer, C. (2007). Spectacular preservation of seagrasses and seagrass-associated communities from the Pliocene of Rhodes, Greece. *Palaios*, 22, 200–211.

Molnar, R.E. (2001). Theropod paleopathology: a literature survey. In *Mesozoic Vertebrate Life* (pp. 337–363), edited by D.H. Tanke and K. Carpenter. Indiana University Press, Bloomington.

Molnar, R.E and Clifford, H.T. (2001). An ankylosaur cololite from the Lower Cretaceous of Queensland, Australia. In *The Armored Dinosaurs* (pp. 394–412), edited by K. Carpenter. Indiana University Press, Bloomington.

Monks, N. (2000). Mid-Cretaceous heteromorph ammonite shell damage. *Journal of Molluscan Studies*, 6, 283–285.

Montgomery de Merette, L. (1984). L'ambre de St. Dominique. *Monde et Minéraux*, 10, 36–37, 40–41.

Moodie, R.L. (1923). *Paleopathology: An Introduction to the Study of Ancient Evidences of Disease*. University of Chicago Press, Chicago, IL, 567 pp.

Moore, D.W., Young, L.E., Modene, J.S., and Plahuta, J.T. (1986). Geological setting and genesis of the Red Dog zinc-lead-silver deposit, western Brooks Range, Alaska. *Economic Geology*, 81, 1696–1727.

Moore, R.C., Ed. (1956). *Treatise on Invertebrate Paleontology*. Part F. *Coelenterata*. Geological Society of America/University Press of Kansas, Lawrence.

Moosleitner, G. (2000). *Rastellum rectangulare* (Roemer), une petite huître servant d'espace vital à des organismes sessiles et foreurs. *Minéraux & Fossiles*, 289, 5–15.

Moran, N.A. (1989). A 48-million-year-old aphid–host plant association and complex life cycle: biogeographic evidence. *Science*, 245, 173–175.

Morard, A. (2002). Post-pathological keel-loss compensation in ammonoid growth. *Lethaia*, 35, 21–31.

Morgan, G.S. (1994). Whither the giant white shark? *American Paleontologist*, 2(3), 1–2.

Morgan, M. and Brown, J. (1998). *Paleontological and Archaeological Monitoring and Mitigation Report for the Alpha-Olinda Landfill, Orange County, California, Phase, 2, Center Ridge Vertical Expansion and Northwest Perimeter Storm Drain/Access Road*, Service Contract 97-175-003. Prepared by RMW Paleo Associates, Mission Viejo, CA.

Morris, J. (1851). Palaeontological notes. *The Annals and Magazine of Natural History*, 8(2), 85–90.

Morris, P.J., Linsley, R.M., and Cottrell, J.F. (1991). A Middle Devonian symbiotic relationship involving a gastropod, a trepostomatous bryozoan, and an inferred secondary occupant. *Lethaia*, 24, 55–67.

Morris, R.W. and Felton, S.H. (1993). Symbiotic association of crinoids, platyceratid gastropods, and *Cornulites* in the Upper Ordovician (Cincinnatian) of the Cincinnati, Ohio, region. *Palaios*, 8, 465–476.

Morris, R.W. and Felton, S.H. (2003). Paleoecologic associations and secondary tiering of *Cornulites* on crinoids and bivalves in the Upper Ordovician (Cincinnatian) of southwestern Ohio, southeastern Indiana, and northern Kentucky. *Palaios*, 18, 546–558.

Morris, R.W. and Rollins, H.B. (1971). The distribution and paleoecological interpretation of *Cornulites* in the Waynesville Formation (Upper Ordovician) of southwestern Ohio. *The Ohio Journal of Science*, 71, 159–170.

Morris, S.F. (1993). The fossil arthropods of Jamaica. *Geological Society of America Memoir*, 182, 115–124.

Morris, S.F. and Collins, J.S.H. (1991). Neogene crabs from Brunei, Sabah and Sarawak. *Bulletin of the British Museum (Natural History) (Geology)*, 47, 1–33.

Morton, B. (1990). Corals and their bivalve borers: the evolution of a symbiosis. In *The Bivalvia* (pp. 11–46), edited by B. Morton. Hong Kong University Press, Hong Kong.

Motani, R. (2002a). Scaling effects in caudal fin propulsion and the speed of ichthyosaurs. *Nature*, 415, 309–312.

Motani, R. (2002b). Swimming speed estimation of extinct marine reptiles: energetic approach revisited. *Paleobiology*, 28, 251–262.

Motani, R., Rothschild, B.M., and Wahl, Jr., W. (1999a). Large eyeballs in diving ichthyosaurs. *Nature*, 402, 747.

Motani, R., Rothschild, B.M., and Wahl, Jr., W. (1999b). What to do with a 10-inch eyeball? Evolution of vision in ichthyosaurs. *Journal of Vertebrate Paleontology*, 19, 65A.

Mottequin, B. and Sevastopoulo, G. (2007). Predatory boreholes in Tournaisian (Lower Carboniferous) spiriferid brachiopods. *The Palaeontological Association Newsletter*, 66, 84.

Mourer-Chauviré, C. (1994). A large owl from the Palaeocene of France. *Palaeontology*, 37, 339–348.

Moyzer, G., Sandberg, P., Knepen, M.H.J., Vermeer, C., Collins, M., and Westbroek, P. (1992). Preservation of the bone protein osteocalcin in dinosaurs. *Geology*, 20, 871–874.

Mudge, M.R. and Yochelson, E.L. (1962). Stratigraphy and paleontology of the uppermost Pennsylvanian and lowermost Permian rocks in Kansas. *U.S. Geological Survey Professional Paper*, 323, 1–213.

Mueller-Töwe, I.J., Sander, P.M., Schüller, H., and Thies, D. (2002). Hatching and infilling of dinosaur eggs as revealed by computed tomography. *Palaeontographica A*, 267, 119–168.

Muir-Wood, H.M. (1965). Productina. In *Treatise on Invertebrate Paleontology*. Part H. *Brachiopoda* (pp. 439–521), edited by R.C. Moore. Geological Society of America/University Press of Kansas, Lawrence.

Mukhopadhyay, S.K. (2003). Plastogamy and its early morphological indication in *Nummulites boninensis* Hanzawa from the Middle Eocene of Cambay Basin, India. *Revue de Paléobiologie*, 22, 231–242.

Mulder, E.W.A. (2002). Co-ossified vertebrae of mosasaurs and cetaceans: implications for the mode of locomotion of extinct marine reptiles. *Paleobiology*, 27, 724–734.

Müller, A.H. (1957). *Lehrbuch der Paläozoologie*. Vol. 1. *Allgemeine Grundlagen*. Fischer-Verlag, Jena, 322 pp.

Müller, A.H. (1969a). Ammoniten mit "Eierbeutel" und die Frage nach dem Sexualdimorphismus der Ceratiten (Cephalopoda). *Deutsche Akademie der Wissenschaften zu Berlin Monatsberichte*, 11, 411–420.

Müller, A.H. (1969b). Uber Raubschneckenbefall und Okologie fossiler ditrupinen (Polychaeta: Sedentaria). *Monatsberichte der Deutschen Akademie der Wissenschaften zu Berlin*, 11(7), 517–525.

Müller, A.H. (1978a). Uber *Palaeoxyris* und andere Eikapseln fossiler Knorpelfische (Chondrichthyes). *Freiberger Forschungsheft C*, 342, 7–28.

Müller, A.H. (1978b). Zur Oologie fossiler Tiere. *Biologische Rundschau*, 16, 155–174.

Müller, A.H. (1979). Fossilization (Taphonomy). In *Treatise on Invertebrate Paleontology*. Part A. *Introduction* (pp. 2–78), edited by R.A. Robison and C. Teichert. Geological Society of America/University Press of Kansas, Lawrence.

Müller, P. (1975). Trapezia (Crustacea. Decapoda) dans l'Eocène et le Miocène de Hongrie. *Bulletin of the Hungarian Geological Society*, 105, 516–523.

Müller, P. (1984). Decapod crustacean of the Badenian. *Geologica Hungarica, Series Palaeontologica, Fasciculus*, 42, 1–317.

Müller, P. and Collins, J.S.H. (1991). Late Eocene coral-associated decapods (Crustacea) from Hungary. *Contributions to Tertiary and Quaternary Geology*, 28, 47–92.

Mumcuoglu, Y.K. and Zias, J. (1988). Head lice, *Pediculus humanus capitus* (Anoplura: Pediculidae), from hair combs excavated in Israel and dated from the first century B.C. to the eighth century A.D. *Journal of Medical Entomology*, 25, 545–547.

Munk, W. and Sues, H.-D. (1993). Gut contents of *Parasaurus* (Pareisauria) and *Protosaurus* (Archosauromorpha) from the Kupferschiefer (Upper Permian) of Hessen, Germany. *Paläontologisches Zeitschrift*, 67, 169–176.

Münster, G. Graf Zu (1840). Ueber die Balanen. *Beiträge zur Petrefacten-Kunde*, III, 27–32.

Mustoe, G.E. (1993). Eocene bird tracks from the Chuckanut Formation, northwest Washington. *Canadian Journal of Earth Sciences*, 30, 1205–1208.

Mutterlose, J. and Ruffell, A. (1999). Milankovitch-scale palaeoclimate changes in pale-dark bedding rhythms from the Early Cretaceous (Hauterivian and Barremian) of eastern England and northern Germany. *Palaeogeography, Palaeoclimatology, Palaeoecology*, 154, 133–160.

Naehr, T.H., Stakes, D.S., and Moore, W.S. (2000). Mass wasting, ephemeral fluid flow, and barite deposition on the California continental margin. *Geology*, 28, 315–318.

Nagel, P. (1987). Fossil ant nest beetles. *Entomologische Arbeiten aus dem Museum G. Frey*, 35/36, 137–170.

Nakamori, T. (1986). Community structures of Recent and Pleistocene hermatypic corals in the Ryukyu Islands. *Science Reports of the Tohoku University, 2nd Series (Geology)*, 56, 71–133.

Neale, J.W. and Howe, H.V. (1974). The marine ostracoda of Russian Harbour, Novaya-Zemlya and other high latitude faunas. *Geoscience and Man*, 6, 81–98.

Neall, V.E. (1970). Notes on the ecology and paleoecology of *Neothyris*, an endemic New Zealand brachiopod. *New Zealand Journal of Marine and Freshwater Research*, 4, 117–125.

Nebelsick, J.H. and Kroh, A. (2002). The stormy path from life to death assemblages: the formation and preservation of mass accumulations of fossil sand dollars. *Palaios*, 17, 378–393.

Nehm, R.H. and Geary, D.H. (1994). A gradual morphologic transition during a rapid speciation event in marginellid gastropods (Neogene: Dominican Republic). *Journal of Paleontology*, 68, 787–795.

Nekvasilová, O. (1975). The etching traces produced by pedicles of Upper Cretaceous brachiopods from Bohemia (Czechoslovakia). *Casopis pro mineralogii a geologii*, 20, 69–74.

Nekvasilová, O. (1976). The etching traces produced by pedicles of Lower Cretaceous brachiopods from Stramberk (Czechoslovakia). *Casopis pro mineralogii a geologii*, 21, 405–408.

Nel, A. (1994). Traces d'activités d'insectes dans des bois et fruits fossiles de la formation de Nkondo (Mio-Pliocène du Rift Occidental, Ouganda). In *Geology and Palaeobiology of the Albertine Rift Valley, Uganda-Zaire*. Vol. II. *Palaeobiology* (pp. 47–57). CIFEG Occasional Publications, Orleans, France.

Nel, A. and Paicheler, J.-C. (1993). Les Isoptera fossils: état actuel des connaissances, implications paléoécologiques et paléoclimatiques [Insecta, Dictyoptera]. In *Essai de révision des Aeschinioidea (Insecta, Odonata, Anisoptera) / Les Isoptera fossiles (Insecta, Dictyoptera)* (pp. 103–179), edited by A. Nel, X. Martínez-Delclòs, and J-C. Paicheler. CNRS Editions [Cahiers de Paléontologie], Paris.

Nel, A., Waller, A., and de Ploëg, G. (2004). The oldest palm bug in the lowermost Eocene amber of the Paris Basin (Heteroptera: Cimicomorpha: Thaumastocoridae). *Geologica Acta*, 2(1), 51–55.

Nelson, C.H. and Johnson, K.R. (1987). Whales and walruses as tillers of the sea floor. *Scientific American*, 256, 112–117.

Nentwig, W. and Heimer, S. (1987). Ecological aspects of spider webs. In *Ecophysiology of Spiders* (pp. 211–225), edited by W. Nentwig. Springer-Verlag, Heidelberg.

Néraudeau, D., Viriot, L., Chaline, J., Laurin, B., and van Kolfschoten, T. (1995). Discontinuity in the Plio–Pleistocene Eurasian water vole lineage. *Palaeontology*, 38, 77–85.

Néraudeau, D., Barbe, S., Mercier, D., and Roman, J. (2003). Signatures paléoclimatiques des échinides du Néogène final atlantique à faciès redonien. *Annales de Paléontologie*, 89, 153–170.

Nerini, M. (1984). A review of gray whale feeding ecology. In *The Gray Whale* Eschrichtius robustus (pp. 423–450), edited by M.L. Jones, S.L. Swartz, and S. Leatherwood. Academic Press, San Diego, CA.

Nesbitt, E.A. (2005). A novel trophic relationship between cassid gastropods and mysticete whale carcasses. *Lethaia*, 38, 17–25.

Nesbitt, S.J., Turner, A.H., Erickson, E.M., and Norell, M.A. (2006). Prey choice and cannibalistic behaviour in the theropod *Coelophysis*. *Biology Letters*, 22, 611–614.

Neto de Carvalho, C., Viegas, P.A., and Cachão, M. (2007). *Thalassinoides* and its producer: populations of *Mecochirus* buried within their burrow systems, Boca do Chapim Formation (Lower Cretaceous), Portugal. *Palaios*, 22, 104–109.

Neuman, B.B.E. (1988). Some aspects of life strategies of Early Palaeozoic rugose corals. *Lethaia*, 21, 97–114.

Neumann, C. (2000). Evidence of predation on Cretaceous sea stars from northwest Germany. *Lethaia*, 33, 65–70.

Neumann, C. and Wisshak, M. (2006). A foraminiferal parasite on the sea urchin *Echinocorys*: ichnological evidence from the Late Cretaceous (Lower Maastrichtian, northern Germany). *Ichnos*, 13, 185–190.

Neumann, C., Wisshak, M., and Bromley, R. (2007). Boring a mobile domicile: an alternative to the conchicolous life habit. In *Current Developments in Bioerosion* (pp. 307–328), edited by M. Wisshak and L. Tapanila. Springer-Verlag, Heidelberg.

Nevesskaya, L.A., Ed. (2002). Biogeography of the Late Eocene to the Early Miocene. Part 2. Early Oligocene. *Paleontology Journal*, 2(36, Suppl.), S185–S259.

Nevo, E. (1979). Adaptive convergence and divergence of subterranean mammals. *Annual Review of Ecology and Systematics*, 10, 269–308.

Nevo, E. (1999). *Mosaic Evolution of Subterranean Mammals*. Oxford University Press, London, 413 pp.

Newman, W.A. (1985). The abyssal hydrothermal vent invertebrate fauna: a glimpse of antiquity? *Bulletin of the Biological Society of Washington*, 6, 231–247.

Newman, W.A., Zullo, V.A., and Withers, T.H. (1969). Cirripedia. In *Treatise on Invertebrate Paleontology*. Part R. *Arthropoda 4* (pp. 206–295), edited by R.C. Moore. Geological Society of America/University of Kansas Press, Lawrence.

Newton, R.B. (1908). Fossil pearl growths. *Proceedings of the Malacological Society of London*, 8, 128–139.

Nguyen Tu, T.T., Derenne, S., Largeau, C., Mariotti, A., and Bocherens, H. (2003). Comparison of leaf lipids from a fossil ginkgoalean plant and its extant counterpart at two degradation stages: diagenetic and chemotaxonomic implications. *Reviews of Palaeobotany and Palynology*, 124, 63–78.

Ni, Y.-N. (1997). Late Homerian (Wenlock, Silurian) graptolites from Shidian, Western Yunnan, China. *Acta Palaeontologica Sinica*, 36, 310–320.

Nicolaus, R.A., Piatellei, M., and Fattorusso, E. (1964). The structure of melanins and melanogenesis, Part IV. *Tetrahedron*, 20, 1163–1172.

Niebuhr, B. and Wilmsen, M. (2005). First record of the hydroid *Protulophila gestroi* Rovereto, 1901, a serpulid symbiont, from the Middle Cenomanian *primus* Event, northern Germany. *Neues Jahrbuch für Geologie und Paläontologie–Monatsheft*, 219–232.

Nield, E.W. (1986). *Liljevallis gotlandica*: encrustacean patterns in the earliest cemented articulate brachiopod and their implications for its larval behaviour. *Palaeogeography, Palaeoclimatology, Palaeoecology*, 56, 277–290.

Nielsen-Marsh, C.M., Ostrom, P.H., Gandhi, H., Shapiro, B., Cooper, A., Hauschke, P.V., and Collins, M.J. (2002). Sequence preservation of osteocalcin protein and mitochondrial DNA in bison bones older than 55 Ka. *Geology*, 30, 1099–1102.

NIAMS. (2006). *Questions and Answers About Osteonecrosis (Avascular Necrosis)*, NIH Publ. No. 06-485716. National Institute of Arthritis and Musculoskeletal and Skin Diseases, National Institutes of Health, Washington, D.C. (http://www.niams.nih.gov/Health_Info/Osteonecrosis/default.asp).

Nikolayev, G.V. (1993). The taxonomic placement in the subfamily Aphodiinae (Coleoptera, Scarabaeidae) of the new genus of Lower Cretaceous scarabid beetles from Transbaykal. *Paleontological Journal*, 27, 1–8.

Nobuhara, T. (2002). Pliocene chemosynthetic carbonate mounds composed of *Calyptogena kawamurai* (Bivalvia: Vesicomyidae) from the upper to middle slope deposits in the Sagara-Kakegawa area, central Japan. *First International Palaeontological Congress Abstracts*, 68, 249–250.

Nobuhara, T. (2003). Cold seep carbonate mounds with *Vesicomya (Caltyptogena) kawamurai* (Bivalvia: Vesicomyidae) in slope-mud facies of the Pliocene forearc basin of the Sagara-Kakegawa area, central Japan. *Paleontological Research*, 7, 313–328.

Noda, H. (1991). Fossil homing scar on the gastropod *Hipponix (Malluvium) lissus* from the Pliocene Shinzato Formation in Okinawa Prefecture, southwestern Japan. *Annual Report of the Institute of Geoscience, University of Tsukuba*, 17, 43–47.

Noda, H. and Lee, Y.-G. (1989). Wood-boring bivalve *Martesia striata* from the Middle Miocene Sinhyeon Formation in the Ulsan Basin, Korea. *Annual Report of the Institute of Geoscience, University of Tsukuba*, 15, 61–67.

Noirot, C. and Noirot-Timothée, C. (1969). The digestive system. In *Biology of Termites*, Vol. I (pp. 49–80), edited by K. Krishna and F.M. Weesner. Academic Press, San Diego, CA.

Nolfe, D. (1985). *Otolithi piscium*. In *Handbook of Paleoichthyology*, Vol. 10 (pp. 1–145), edited by H.-P. Schultze. Gustav Fischer Verlag, Stuttgart.

Nomura, M., Hatanaka, O., Nishimoto, H., Karasawa, H., and Nanao Nojiriko Group. (1991). *Megasqualus serriculus* Jordan and Hannibal (Squalidae: Squaliformes: Elasmobranchii) from the Middle Miocene Nanao Calcareous Sandstone, Nanao City, Noto Peninsula, Central Japan. *Bulletin of the Mizunami Fossil Museum*, 18, 33–45.

Noonan, J.P., Hofreiter, M., Smith, D., Priest, J.R., Rohloand, N., Rabeder, G., Krause, J., Detter, J.C., Pääbo, S., and Rubin, E.M. (2005). Genomic sequences of Pleistocene cave bears. *Science*, 309, 597–600.

Norell, M.A., McKenna, M.C., and Novacek, M.J. (1992). *Estesia mongoliensis*, a new fossil varanoid from the Late Cretaceous Barun Goyot Formation of Mongolia. *American Museum Novitates*, 3045, 24 pp.

Norell, M.A., Clark, J.M., Demberelyin, D., Rhinchen, B., Chiappe, L.M., Davidson, A.R., McKenna, M.C., Altangerel, P., and Novacek, M.J. (1994). A theropod dinosaur embryo and the affinities of the Flaming Cliffs dinosaur eggs. *Science*, 266, 779–782.

Norell, M.A., Clark, J.M., Chiappe, L.M., and Dashzeveg, D. (1995). A nesting dinosaur. *Nature*, 378, 774–776.

Norris, R.D., Corfield, R.M., and Cartlidge, J. (1996). What is gradualism? Cryptic speciation in globorotaliid Foraminifera. *Paleobiology*, 22, 386–405.

Norstog, K.J. and Nicholls, T.J. (1997). *The Biology of Cycads.* Cornell University Press, Ithaca, NY, 363 pp.

Norstog, K.J., Fawcett, P.K.S., Nicholls, T.J., Vovides, A.P., and Espinosa, E. (1995). Insect-pollination of cycads: evolutionary and ecological considerations. In *Proceedings of the Third International Conference on Cycad Biology* (pp. 265–285), edited by P. Vorster. Cycad Society of South Africa, Pretoria.

Norton, S.F. (1988). Role of gastropod shell and operculum in inhibiting predation by fishes. *Science*, 241, 92–94.

Nummela, S. Thewissen, J.G.M., Bajpai, S., Hussain, S.T., and Kumar, K. (2004). Eocene evolution of whale hearing. *Nature*, 430, 776–778.

Nury, D. and Schreiber, B.C. (1997). The Paleogene Basin of Southern Provence. In *Sedimentary Deposition in Rift and Foreland Basins in France and Spain* (pp. 240–300), edited by G. Busson and B.C. Schreiber. Columbia University Press, New York.

Nützel, A. and Frýda, J. (2003). Paleozoic plankton revolution: evidence from early gastropod ontogeny. *Geology*, 31, 829–831.

Nye, Jr., O.B., Brower, J.C., and Wilson, S.E. (1975). Hitchhiking clams in the Marcellus Sea. *Bulletins of American Paleontology*, 67, 287–297.

Oakley, K.P. (1966). Some pearl-bearing Ceramoporidae (Polyzoa). *Bulletin of the British Museum (Natural History) (Geology)*, 14 (1), 1–20.

Obruchev, D.V., Ed. (1967). *Fundamentals of Palaeontology*. Vol. XI. *Agnatha, Pisces*. Israel Program for Scientific Translation, Jerusalem, 825 pp.

O'Dea, A., Jackson, J.B.C., Taylor, P.D., and Rodríguez, F. (2008). Modes of reproduction in Recent and fossil cupuladrid bryozoans. *Palaeontology*, 51, 847–864.

O'Dowd, D.J. and Willson, M.F. (1989). Leaf domatia and mites on Australasian plants: ecological and evolutionary implications. *Biological Journal of the Linnaean Society*, 37, 191–236.

O'Dowd, D.J., Brew, C.R., Christophel, D.C., and Norton, R.A. (1991). Mite–plant associations from the Eocene of southern Australia. *Science*, 252, 99–101.

Oertli, H.J. (1974). Lower Cretaceous and Jurassic ostracods from DSDP Leg 27: a preliminary account. In *Initial Reports of the Deep Sea Drilling Project*, Vol. 27 (pp. 947–965), edited by J.J. Veevers, J.R. Heirtzler et al. U.S. Government Printing Office, Washington, D.C.

Ogawa, Y., Fujioka, K., Fujikura, K., and Iwabuchi, Y. (1996). En echelon patterns of *Calyptogena* colonies in the Japan Trench. *Geology*, 24, 807–810.

Ohe, F. (1993). Osteichthyes. In *Fossils from the Miocene Morozaki Group* (pp. 169–262), edited by F. Ohe, I. Nonogaki, T. Tanaka, K. Hachiya, Y. Mizuno, T. Momoyama, and T. Yamaoka. The Tokai Fossil Society, Nagoya, Japan.

Oji, T. (1993). Echinodermata, Crinoidea. In *Fossils from the Miocene Morozaki Group* (pp. 103–108), edited by F. Ohe, I. Nonogaki, T. Tanaka, K. Hachiya, Y. Mizuno, T. Momoyama, and T. Yamaoka. The Tokai Fossil Society, Nagoya, Japan.

Oji, T. (1996). Is predation intensity reduced with increasing depth? Evidence from the west Atlantic stalked crinoid *Endoxocrinus parrae* (Gervais) and implications for the Mesozoic marine revolution. *Paleobiology*, 22, 339–351.

Oji, T. (2001). Deep-sea communities. In *Palaeobiology II* (pp. 444–447), edited by D.E.G. Briggs and P. Crowther. Blackwell Scientific, Oxford.

Oji, T. and Amemiya, S. (1998). Survival of crinoid stalk fragments and its taphonomic implications. *Paleontological Research*, 2, 67–70.

Okada, H. and Cadet, J.-P., Eds. (1989). Geology, geochemistry and biology of subduction zones. *Palaeogeography, Palaeoclimatology, Palaeoecology*, 71, 1–203.

O'Leary, M.A., Lucas, S.G., and Williamson, T.E. (2000). A new specimen of *Ankalagon* (Mammalia, Mesonychia) and evidence of sexual dimorphism in mesonychians. *Journal of Vertebrate Paleontology*, 20, 387–393.

Oleinik, A.E. (1996). Genus *Arctomelon* (Gastropoda, Volutidae) in the Tertiary of the north-western Pacific: evolution and adaptations. *Journal of Paleontology*, 70, 236–246.

Olempska, E. (1989). Gradual evolutionary transformations of ontogeny in an Ordovician ostracod lineage. *Lethaia*, 22, 159–168.

Oliveira, C.E.M., Santucci, R.M., de Andrade, M.B., Basílio, J.A.F., and Benton, M.J. (2008). New crocodyloid eggs and eggshells from the Upper Cretaceous of Brazil (Bauru Group). *The Palaeontological Association Newsletter*, 69, 72.

Oliver, J.S., Slattery, P.N., Silberstein, M.A., and O'Connor, E.F. (1984). Gray whale feeding on dense ampeliscid amphipod communities near Bamfield, British Columbia. *Canadian Journal of Zoology*, 62, 41–49.

Olivero, E.B. and Aguirre-Urreta, M.B. (1994). A new tube-builder hydractinian symbiotic with hermit crabs, from the Cretaceous of Antarctica. *Journal of Paleontology*, 68, 1169–1182.

Olmi, M. (2003/2004). A revision of the world Sclerogibbidae (Hymenoptera Chrysedoidea). *Frustula Entomologica*, 26–27, 46–193.

Olmi, M. and Bechly, G. (2001). New parasitic wasps from Baltic amber (Insecta: Hymenoptera: Dryinidae). *Stuttgarter Beiträge zur Naturkunde B*, 306, 58 pp.

Olsen, P.E. and Kent, D.V. (1996). Milankovitch climate forcing in the tropics of Pangaea during the Late Triassic. *Palaeogeography, Palaeoclimatology, Palaeoecology*, 122, 1–26.

Olson, E.C. (1952). The evolution of a Permian vertebrate chronofauna. *Evolution*, 6, 181–196.

Olson, E.C. (1969). Sexual dimorphism in extinct amphibians and reptiles. In *Sexual Dimorphism in Fossil Metazoa and Taxonomic Implications* (pp. 223–225), edited by G.E.G. Westermann, Schweizerbart, Stuttgart.

Olson, E.C. (1980). Taphonomy: its history and role in community evolution. In *Fossils in the Making* (pp. 5–19), edited by A.K. Behrensmeyer and A.P. Hill. University of Chicago Press, Chicago, IL.

Olson, E.C. (1985). A larval specimen of a trematopsid (Amphibia: Temnospondyli). *Journal of Paleontology*, 59, 1173–1180.

Olu-LeRoy, K., Sibuet, M., Fiala-Médioni, Gofas, S., Salas, C., Mariotti, A., Foucher, J.-P., and Woodside, J. (2004). Cold seep communities in the deep eastern Mediterranean Sea: composition, symbiosis and spatial distribution on mud volcanoes. Deep-Sea Research, Part 1. *Oceanographic Research Papers*, 51(12), 1915–1936.

O'Malley, C.E., Ausich, W.I., and Chin, Y.-P. (2008). Crinoid biomarkers (Borden Group, Mississippian): implications for phylogeny. In *Echinoderm Paleobiology* (pp. 291–306), edited by W.I. Ausich and G.D. Webster. Indiana University Press, Bloomington.

Orr, C.M., Delezene, L.K., Scott, J.E., Tocheri, M.W., and Schwartz, G.T. (2007). The comparative method and the inference of venom delivery systems in fossil mammals. *Journal of Vertebrate Paleontology*, 27, 541–546.

Osgood, R.G. (1970). Trace fossils of the Cincinnati area. *Palaeontographica Americana*, 6(41), 281–444.

O'Sullivan, J.B., McConnaughey, R.R., and Huber, M.E. (1987). A blood-sucking snail: the Cooper's Nutmeg *Cancellaria cooperi* Gabb, parasitizes the California electric ray, *Torpedo californica* Ayres. *Biological Bulletin*, 172, 362–366.

Otto, A., White, J.D., and Simonett, B.R.T. (2002). Natural product terpenoids in Eocene and Miocene conifer fossils. *Science*, 297, 1542–1545.

Oudin, E. and Constantinou, G. (1984). Black smoker chimney fragments in Cyprus sulphide deposits. *Nature*, 308, 349–353.

Oudin, E., Bouladon, J., and Paris, J.-P. (1985). Vers hydrothermaux fossiles dans une mineralisation sulfurée des ophiolites de Nouvelle Caledonie. *Comptes Rendus de l'Academie des Sciences, Paris Séries 2*, 301, 157–162.

Owen, A.W. (1980). An abnormal cranidium of the trilobite *Calyptaulax norvegicus*. *Norsk Geologisk Tidsskrift*, 60, 87–88.

Owen, A.W. (1983). Abnormal cephalic fringes in the Trinucleidae and Harpetidae. *Special Papers in Palaeontology*, 30, 241–247.

Owen, A.W. (1985). Trilobite abnormalities. *Transactions of the Royal Society of Edinburgh*, 76, 255–272.

Owens, R.M. and Tilsley, J.W. (1995). An atheloptic trilobite assemblage from the Carboniferous of North Devon. *Geological Magazine*, 132, 713–728.

Ozanne, C.R. and Harries, P.J. (2002). Role of predation and parasitism in the extinction of the inoceramid bivalves: an evaluation. *Lethaia*, 35, 1–19.

Packard, M.J. and Hirsch, K.F. (1989). Structure of shells from eggs of the geckos *Gecko gecko* and *Phelsuma madagascarensis*. *Canadian Journal of Zoology*, 67, 746–758.

Page, L.R. (1997). Ontogenetic torsion and protoconch form in the archaeogastropod *Haliotis kamschatkana*: evolutionary implications. *Acta Zoologica*, 78, 227–245.

Paicheler, J.-C., Broin, D. de, Gaudant, J., Mourer-Chauviré, C., Rage, J.-C., and Vergnaud-Grazzine, C. (1978). Le basin lacustre Miocène de Bes-Konak (Anatolie, Turquie): géologie et introduction à la paleontology des vertébrés. *Géobios*, 11, 43–65.

Paik, I.S. (2000). Bone chip-filled burrows associated with bored dinosaur bone in floodplain paleosols of the Cretaceous Hasandong Formation, Korea. *Palaeogeography, Palaeoclimatology, Palaeoecology*, 157, 213–225.

Pajaud, D. (1974). Écologie des Thécidées. *Lethaia*, 7, 203–218.

Pajaud, D. (1977). Les brachiopods du Pliocène de la region d'Aquiles (Sud d'Almeria, Espagne). *Annales de Paléontologie (Invertébrés)*, 63, 59–71.

Palci, A. and Caldwell, M.W. (2007). Vestigial forelimbs and axial elongation in a 95-million-year-old non-snake squamate. *Journal of Vertebrate Paleontology*, 27, 1–7.

Palmer, C.P. (1989). Larval shells of four Jurassic bivalve molluscs. *Bulletin of the British Museum (Natural History) (Geology)*, 45(1), 57–69.

Palmer, K.V.W. (1958). Viviparous *Turritella pilsbryi* Gardner. *Journal of Paleontology*, 32, 210–213.

Palmer, K.V.W. (1961). Additional note on ovoviparous *Turritella*. *Journal of Paleontology*, 35, 633.

Palmer, T.J. and Fürsich, F.T. (1974). The ecology of a Middle Jurassic hardground and crevice fauna. *Palaeontology*, 17, 507–524.

Palmqvist, P., Arribas, A., and Martínez-Navarro, B. (1999). Ecomorphological study of large canids from the lower Pleistocene of southeastern Spain. *Lethaia*, 32, 75–88.

Panagiotakopulu, E. (2001). Fleas from Pharaonic Amarna. *Antiquity*, 75, 499–500.

Panagiotakopulu, E. and Buckland, P.C. (1999). *Cimex lectularius* L., the common bed bug from Pharaonic Egypt. *Antiquity*, 73, 908–911.

Pandolfi, J.M. (1996). Limited membership in Pleistocene reef coral assemblages from the Huon Peninsula, Papua New Guinea: constancy during global change. *Paleobiology*, 22, 152–176.

Papp, A., Zapfe, H., Bachmayer, F., and Tauber, A.F. (1947). Lebenspuren mariner Krebse. *Sitzungsberichte, Akademie der Wissenschaften in Wien, Abteilung I*, 155, 281–317.

Park, I.S., Huh, M., and Kim, H.J. (2004). Dinosaur egg-bearing deposits (Upper Cretaceous) of Boseone, Korea: occurrence, palaeoenvironments, taphonomy, and preservation. *Palaeogeography, Palaeoclimatology, Palaeoecology*, 205, 155–168.

Park, J., D'Hondt, S.L., King, J.W., and Gibson, C. (1993). Late Cretaceous precessional cycles in double time: a warm-earth Milankovitch response. *Science*, 261, 1431–1434.

Park, R.A. (1968). Paleoecology of *Venericardium sensu lato* (Pelecypoda) in the Atlantic and Gulf Coastal Province: an application of paleosynecologic methods. *Journal of Paleontology*, 42, 955–986.

Parker, A.R. (1999). Invertebrate structural colours. In *Functional Morphology of the Invertebrate Skeleton* (pp. 65–90), edited by E. Savazzi. John Wiley & Sons, New York.

Parras, A. and Casadío, S. (2006). The oyster *Crassostrea? hatcheri* (Ortmann, 1897), a physical ecosystem engineer from the Upper Oligocene–Lower Miocene of Patagonia, southern Argentina. *Palaios*, 21, 168–186.

Paterson, J.R., Hughes, N.C., and Chatterton, B.D.E. (2008). Trilobite clusters: what do they tell us? A preliminary investigation. In *Advances in Trilobite Research* (pp. 313–318), edited by I. Rábano, R. Gozalo, and D. Garcia-Bellido. Cuadernos del Museo Geominero, 9, Instituto Geológico y Minero de España, Madrid.

Patterson, C. (1975). The braincase of pholidophorid and leptolepid fishes, with a review of the Actinopterygian braincase. *Philosophical Transactions of the Royal Society of London B*, 269, 275–579.

Paulian, R. (1976). Three fossil dung beetles (Coleoptera: Scarabaeidae) from the Kenya Miocene. *Journal of the East Africa Natural History Society and National Museum*, 31, 1581–1584.

Pawlowska, A.M., Palińska, K.A., and Piekarek-Jankowska, H. (2007). Colonisation and bioerosion of marine bivalve shells from the Baltic Sea by euendolithic cyanobacteria: an experimental study. In *Current Developments in Bioerosion* (pp. 109–122), edited by M. Wisshak and L. Tapanila. Springer-Verlag, Heidelberg.

Pearson, P.N. (1993). A lineage phylogeny for the Paleogene planktonic Foraminifera. *Micropaleontology*, 39, 193–232.

Peckmann, J., Thiel, V., Michaelis, M., Clari, P., Gaillard, G., Martire, L., and Reitner, J. (1999). Cold seep deposits of Beauvoisin (Oxfordian; southeastern Fance) and Marmorito (Miocene; northern Italy): microbially induced authigenic carbonates. *International Journal of Earth Sciences*, 88, 60–75.

Peckmann, J., Gischler, E., Oschmann, W., and Reitner, J. (2001). An Early Carboniferous seep community and hydrocarbon-derived carbonates from the Harz Mountains, Germany. *Geology*, 29, 271–274.

Peckmann, J., Goedert, J.L., Thiel, V., Michaelis, W., and Reitner, J. (2002). A comprehensive approach to the study of methane-seep deposits from the Lincoln Creek Formation, western Washington State, USA. *Sedimentology*, 49, 855–873.

Peckmann, J., Campbell, K.A., Walliser, O.H., and Reitner, J. (2007a). A Late Devonian hydrocarbon-seep deposit dominated by dimerelloid brachiopods, Morocco. *Palaios*, 22, 114–122.

Peckmann, J., Senobari-Daryan, B., Birgel, D., and Goedert, J.L. (2007b). The crustacean ichnofossil *Palaxius* with callianassid body fossils in an Eocene methane-seep limestone, Humptulips Formation, Olympic Peninsula, Washington. *Lethaia*, 40, 273–280.

Pedriali, L. and Robba, E. (2005). A revision of the Pliocene naticids of Northern and Central Italy. I. The subfamily Naticinae except *Tectonatica*. *Rivista Italiana di Paleontologia e Stratigrafia*, 111, 109–179.

Pedriali, L. and Robba, E. (2008). A revision of the Pliocene naticids of northern and central Italy. II. Subfamily Naticinae: additions to *Cochlis*, *Tanea* and *Tectonatica*. *Rivista Italiana di Paleontologia e Stratigrafia*, 114, 77–117.

Peel, J.S. (1984). Attempted predation and shell repair in *Euomphalopterus* (Gastropoda) from the Silurian of Gotland. *Bulletin of the Geological Society of Denmark*, 32, 163–168.

Peel, J.S. and Horný, R.J. (1996). Sinistral hyperstrophic coiling in a Devonian gastropod from Bohemia with an *in situ* operculum. *Palaeontology*, 39, 709–718.

Peel, J.S., Ebbestad, J.O.R., and Lindström, A. (1996). Shell repair and failed predation in Lower Palaeozoic gastropods from Sweden. *Sixth North American Paleontological Convention Abstracts of Papers*, 8, 305.

Pek, I. (1977). Agnostid trilobites of the Central Bohemian Ordovician. *Sbornik geologickych ved, paleontologie*, 1, 7–44.

Pek, I. and Marek, J. (1983). Pathological effect in *Lyropecten* (*Aequipecten*?) *ternatus* (Muenster, 1833) from the Cretaceous of Czechoslovakia. *Vestnik Ustredniho Ustavu Geologického*, 58, 49–52.

Pellmyr, O. and Leebens-Mack, J. (1999). Forty million years of mutualism: evidence for Eocene origin of the yucca–yucca moth association. *Proceedings of the National Academy of Sciences USA*, 96, 9178–9183.

Pemberton, R.W. and Turner, C.E. (1989). Occurrence of predatory and fungivorous mites in leaf domatia. *American Journal of Botany*, 76, 105–112.

Peñalver, E. and Grimaldi, D. (2006). Assemblages of mammalian hair and blood-feeding midges (Insecta: Diptera: Phlebotominae) in Miocene amber. *Transactions of the Royal Society of Edinburgh, Earth Sciences*, 96, 177–195.

Peñalver, E., Engel, M.S., and Grimaldi, D.A. (2006). Fig wasps in Dominican amber (Hymenoptera: Agaonidae). *American Museum Novitates*, 3541, 1–16.

Peñalver, E., Grimaldi, D.A., and Delclòs, X. (2006). Early Cretaceous spider web with its prey. *Science*, 312, 176.

Peng, R.K. and Christian, K. (2004). The weaver ant, *Oecophylla smaragdina* (Hymenoptera: Formicidae), an effective biological control agent of the red-banded thrips. *International Journal of Pest Management*, 50, 107–114.

Penney, D. (2005). Fossil blood droplets in Miocene Dominican amber yield clues to speed and direction of resin secretion. *Palaeontology*, 48, 925–927.

Pérez, P.J. (1996). Resultados de las investigaciones paleopatologicas en hominidos fosiles. *Revista Española de Paleontologia*, 256–268.

Perillat, M.C. and Vega, F.J. (2001). A new genus and species of Late Cretaceous xenophorid gastropod from southern Mexico. *The Veliger*, 44, 73–78.

Perkins, S. (2002). Sea dragons. *Science News*, 162, 122–124.

Perkovsky, E.E. (2006). Vstrechaemost sininklozov muravev (Hymentoptera, Formicidae) i trei (Homoptera, Aphidinea) v Saksonskom i Rovenskom yantaryakh. *Paleontologicheskii Zhurnal*, 2, 72–74.

Perner, J. (1903). Gasteropodes. Système Silurien du Centre de la Bohême, 1ère Partie. *Recherches Paléontologiques*, 4, 1–164.

Perner, J. (1907). Gasteropodes. Système Silurien du Centre de la Bohême, Tome II. *Recherches Paléontologiques*, 4, 1–380.

Perner, J. (1911). Gasteropodes. Système Silurien du Centre de la Bohême, Tome III. *Recherches Paléontologiques*, 4, 1–390.

Perrichot, V. (2004). Early Cretaceous amber from south-western France: insight into the Mesozoic litter fauna. *Geologica Acta*, 2, 9–22.

Perrichot, V., Nel, A., and Néraudeau, D. (2007). Schizopterid bugs (Insecta: Heteroptera) in mid-Cretaceous ambers from France and Myanmar (Burma). *Palaeontology*, 50, 1367–1374.

Perrichot, V., Marion, L., Néraudeau, D., Vullo, R., and Tafforeau, P. (2008). The early evolution of feathers: fossil evidence from Cretaceous amber of France. *Proceedings of the Royal Society London B*, 275, 1197–1202.

Perrone, Jr., M. and Zaret, T.M. (1979). Parental care patterns of fishes. *American Naturalist*, 113, 351–361.

Perry, C.T. and Bertling, M. (2000). Spatial and temporal patterns of macroboring within Mesozoic and Cenozoic coral reef systems. In *Carbonate Platform Systems: Interactions and Processes* (pp. 33–50), edited by E. Insalaco, P. Skelton, and T. Palmer. Geological Society Special Publication 178. Geological Society, London.

Peters, D.S. (1992). Messel birds: a land-based assemblage. In *Messel: An Insight into the History of Life and of the Earth* (pp. 137–151), edited by S. Schaal and W. Ziegler. Clarendon Press, London.

Peters, S.E. and Bork, K.B. (1998). Secondary tiering on crinoids from the Waldron shale (Silurian: Wenlockian) of Indiana. *Journal of Paleontology*, 72, 887–894.

Pether, J. (1995). *Belichnus* new ichnogenus, a ballistic trace on mollusc shells from the Holocene of the Benguela Region, South Africa. *Journal of Paleontology*, 69, 171–181.

Petr, V. (1983). Teratological pygidium of the trilobite species *Radioscutellum intermixtum* (Hawle et Corda) from the Koneprusy Limestone (Lower Devonian, Pragian) deposited in the district museum of Beroun. *Cesky Kras*, 8, 56–59.

Petr, V. (1989). Revision of morphology and ecology of *Bohemura jahni* Jaeckel, 1903 (Ophiuroidea, Protasteridae) from Bohemian Middle Ordovician. *Sborník Národního Muzea v Praze*, 45B(1), 1–20.

Petuch, E.J. (1994). *Atlas of Florida Fossil Shells*. Chicago Spectrum Press, Chicago, IL, 394 pp.

Philippe, M. (1983). Deformation d'une scutelle (Echinoidea, Clypeasteroida) miocène due à la fixation d'une balane. Hypothèse paléoécologique. *Géobios*, 16, 371–374.

Phillips, B., Ed. (2006). *Lobsters: Biology, Management, Aquaculture and Fisheries*. Blackwell Publishing, Chichester, U.K., 506 pp.

Phipps, C.J. and Rember, W.C. (2004). Epiphyllous fungi from the Miocene of Clarkia, Idaho: reproductive structuring. *Reviews of Palaeobotany and Palynology*, 129, 67–79.

Phipps, C.J. and Taylor, T.N. (1996). Mixed arbuscular mycorrhizae from the Triassic of Antarctica. *Mycologia*, 88, 707–714.

Phipps, K.J. (2008). Evidence of predation on *Gryphaea* (*Bilobissa*) *lituola* Lamarck, 1819, from the Oxford Clay Formation of South Cave Station Quarry, Yorkshire. *Proceedings of the Geological Association*, 119, 277–285.

Pickerill, R.K. and Donovan, S.K. (1998). Ichnology of the Pliocene Bowden Shell Bed, southeast Jamaica. *Contributions to Tertiary and Quaternary Geology*, 35, 161–175.

Pickford, M. (1996). Fossil crocodiles (*Crocodylus lloydi*) from the Lower and Middle Miocene of southern Africa. *Annales de Paléontologie (Vertébrés-Invertébrés)*, 82, 235–250.

Pierce, W.D. (1960). Silicified turbellaria from Calico Mountains nodules. *Bulletin of the Southern California Academy of Sciences*, 59, 138–143.

Pierce, W.D. (1964). The Strepsiptera are a true order, unrelated to Coleoptera. *Annals of the Entomological Society of America*, 57, 603–605.

Pike, A.W. (1967). The recovery of parasite eggs from ancient cesspit and latrine deposits: an approach to the study of early parasite infections. In *Diseases in Antiquity* (pp. 184–188), edited by D. Brothwell and A.T. Sandison. Charles C Thomas, Springfield, IL.

Pinna, G., Arduini, P., Pesarini, C., and Teruzzi, G. (1985). Some controversial aspects of the morphology and anatomy of *Ostenocaris cypriformis* (Crustacea, Thylococephala). *Transactions of the Royal Society of Edinburgh, Earth Sciences*, 76, 373–379.

Pirrie, D., Feldmann, R.M., and Buatois, L.A. (2004). A new decapod trackway from the Upper Cretaceous, James Ross Island, Antarctica. *Palaeontology*, 47, 1–12.

Pitcher, T.J., Ed. (1986). *The Behaviour of Teleost Fishes*. Croom Helm, Baltimore, MD, 553 pp.

Platt, B.F. and Hasiotis, S.T. (2006). Newly discovered sauropod dinosaur tracks with skin and foot-pad impressions from the Upper Jurassic Morrison Formation, Bighorn Basin, Wyoming, USA. *Palaios*, 21, 249–261.

Platt, S.G., Rainwater, T.R., Finger, A.G., Thorbjarnarson, J.B., Anderson, T.A., and McMurray, S.T. (2006). Food habits, ontogenetic dietary partitioning and observations of foraging behaviour of Morelet's crocodile (*Crocodylus moreletii*) in northern Belize. *Herpetological Journal*, 16, 281–290.

Plavcan, J.M., Kay, R.F., Jungers, W.L., and van Schaik, C.P. (2002). *Reconstructing Behavior in the Primate Fossil Record*. Springer-Verlag, New York, 437 pp.

Pohl, H. and Kinzelbach, R. (2001). First record of a female stylopid (Strepsiptera: ?Myrmecolacidae) parasite of a prionomyrmecine ant (Hymenoptera: Formicidae) in Baltic amber. *Insect Systematics and Evolution*, 32, 143–146.

Poinar, Jr., G.O. (1977). Fossil nematodes from Mexican amber. *Nematologica*, 23, 232–238.

Poinar, Jr., G.O. (1983). *The Natural History of Nematodes*. Prentice Hall, Englewood, NJ, 323 pp.

Poinar, Jr., G.O. (1984). First fossil record of parasitism by insect parasitic Tylenchida (Allantonematidae: Nematoda). *Journal of Parasitology*, 70, 306–308.

Poinar, Jr., G.O. (1985). Nematode parasites and infectious diseases of Tabanidae (Diptera). *Myia*, 3, 599–616.

Poinar, Jr., G.O. (1991a). The mycetophagous and entomophagous stages of *Iotonchium californicum* n. sp. (Iotonochiidae: Tylenchida). *Revue de Nématologie*, 14, 565–580.

Poinar, Jr., G.O. (1991b). *Praecoris dominicana* gen. n., sp. n. (Hemiptera: Reduviidae: Holoptilinae) from Dominican amber, with an interpretation of past behavior based on functional morphology. *Entomologica Scandinavica*, 22, 193–199.

Poinar, Jr., G.O. (1992a). Fossil evidence of resin utilization by insects. *Biotropica*, 24, 466–468.

Poinar, Jr., G.O. (1992b). *Life in Amber*. Stanford University Press, Palo Alto, CA, 350 pp.

Poinar, Jr., G.O. (1993). Insects in amber. *Annual Reviews of Entomology*, 38, 145–149.

Poinar, Jr., G.O. (1994). The range of life in amber: significance and implications in DNA studies. *Experientia*, 50, 2164–2167.

Poinar, Jr., G.O. (1995a). Fleas (Insecta: Siphonoptera) in Dominican amber. *Medical Science Research*, 23, 789.

Poinar, Jr., G.O. (1995b). First fossil soft ticks, *Ornithodoros antiquus* n. sp. (Acari: Argasidae) in Dominican amber with evidence of their mammalian host. *Experienta*, 51, 384–387.

Poinar, Jr., G.O. (1995c). *Discovering the Mysteries of Amber*. Geofin, Udine, 67 pp.

Poinar, G., Jr. (1998a). Trace fossils in amber: a new dimension for the ichnologist. *Ichnos*, 6, 47–52.

Poinar, Jr., G.O. (1998b). Fossils explained 22: palaeontology of amber. *Geology Today*, 14(4), 154–160.

Poinar, Jr., G.O. (1998c). *Paleoeuglossa melissiflora* gen. n., sp. n. (Euglossinae: Apidae), fossil orchid bees in Dominican amber. *Journal of the Kansas Entomological Society*, 71, 29–34.

Poinar, Jr., G.O. (1998d). Orchid bees. *Orchids*, 67, 578.

Poinar, Jr., G.O. (1999a). *Paleochordodes protus* n. g., n. sp. (Nematomorpha: Chordodidae), parasites of a fossil cockroach, with a critical examination of other fossil hairworms and helminths of extant cockroaches (Insecta: Blattaria). *Invertebrate Biology*, 188, 109–115.

Poinar, Jr., G.O. (1999b). Chrysomelidae in fossilized resin: behavioural inferences. In *Advances in Chrysomelidae Biology*, Vol. 1 (pp. 1–16), edited by M.L. Cox. Backhuys, Leiden, 1–16.

Poinar, Jr., G.O. (1999c). A fossil palm bruchid, *Caryobruchus dominicanus* sp. n. (Pachymerini: Bruchidae) in Dominican amber. *Entomologica Scandinavica*, 30, 219–224.

Poinar, Jr., G.O. (2000a). *Heydenius araneus* n. sp. (Nematoda: Mermithidae), a parasite of a fossil spider, with an examination of helminths from extant spiders (Arachnida: Araneae). *Invertebrate Biology*, 119, 388–393.

Poinar, Jr., G.O. (2000b). First fossil record of stalked spermatophores with sperm (Collembola: Hexapoda). *Historical Biology*, 14, 229–234.

Poinar, Jr., G.O. (2000c). Fossil onychophorans from Dominican and Baltic amber *Tertiapatus dominicanus* n. g., n. sp. (Tertiapatidae n. fam.) and *Succinipatopsis balticus* n. g., n. sp. (Succinipatopsidae n. fam.) with a proposed classification of the subphylum Onychophora. *Invertebrate Biology*, 199, 104–109.

Poinar, Jr., G.O. (2001a). Fossil puffballs (*Gasteromycetes*: Lycoperdales) in Mexican amber. *Historical Biology*, 15, 219–221.

Poinar, Jr., G.O. (2001b). Dominican amber. In *Palaeobiology II* (pp. 362–364), edited by D.E.G. Briggs and P. Crowther. Blackwell Scientific, Oxford.

Poinar, Jr., G.O. (2001c). *Heydenius brownii* sp. n. (Nematoda: Mermithidae) parasitizing a planthopper (Homoptera: Achilidae) in Baltic amber. *Nematology*, 3(8), 753–757.

Poinar, Jr., G.O. (2002a). First fossil record of nematode parasitism of ants; a 40 million year tale. *Parasitology*, 125, 457–459.

Poinar, Jr., G.O. (2002b). Fossil palm flowers in Dominican and Mexican amber. *Botanical Journal of the Linnean Society*, 138, 57–61.

Poinar, Jr., G.O. (2002c). Fossil palm flowers in Dominican and Baltic amber. *Botanical Journal of the Linnean Society*, 139, 361–367.

Poinar, Jr., G.O. (2003a). Trends in the evolution of insect parasitism by nematodes as inferred from fossil evidence. *Journal of Nematology*, 35, 129–132.

Poinar, Jr., G.O. (2003b). Coelomycetes in Dominican and Mexican amber. *Mycological Research*, 107, 117–122.

Poinar, Jr., G.O. (2003c). A rhabdocoel turbellarian (Platyhelminthes, Typhloplanoidea) in Baltic amber with a review of fossil and sub-fossil platyhelminthes. *Invertebrate Biology*, 122, 308–312.

Poinar, Jr., G.O. (2003d). Fossil evidence of phorid parasitism (Diptera: Phoridae) by allonematid nematodes (Tylenchida: Allantonematidae). *Parasitology*, 127, 589–592.

Poinar, Jr., G.O. (2003e). Amber. In *Encyclopedia of Insects* (pp. 9–12), edited by V. Resh and R.T. Carde. Academic Press, San Diego, CA.

Poinar, Jr., G.O. (2004a). Fossil evidence of spider egg parasitism by ichneumonid wasps. In *Fossil Spiders in Amber and Copal* (pp. 1874–1877), edited by J. Wunderlich. Author.

Poinar, Jr., G.O. (2004b). Fossil evidence of scale phoresy on spiders. In *Fossil Spiders in Amber and Copa* (pp. 1878–1880), edited by J. Wunderlich. Author.

Poinar, Jr., G.O. (2004c). Behaviour and development of *Elasmosoma* sp. (Neoneurinae: Braconidae: Hymenoptera), an endoparasite of *Formica* ants (Formicidae: Hymenoptera). *Parasitology*, 128, 521–531.

Poinar, Jr., G.O. (2004d). Evidence of parasitism by Strepsiptera in Dominican amber. *BioControl*, 49, 239–244.

Poinar, Jr., G.O. (2004e). *Programinis burmitis* gen. et sp. nov. and *P. laminatus* sp. nov., Early Cretaceous grass-like monocots in Burmese amber. *Australian Systematic Botany*, 17, 497–504.

Poinar, Jr., G.O. (2004f). *Palaeomyia burmitis* (Diptera: Phlebotomidae), a new genus and species of Cretaceous sand flies with evidence of blood-sucking habits. *Proceedings of the Entomological Society of Washington*, 106, 598–605.

Poinar, Jr., G.O. (2005a). *Plasmodium dominicana* n. sp. (Plasmodiidae: Haemospororida) from Tertiary Dominican amber. *Systematic Parasitology*, 61, 47–52.

Poinar, Jr., G.O. (2005b). *Triatoma dominicana* sp. n. (Hemiptera: Reduviidae: Triatominae), and *Trypanosoma antiquus* sp. n. (Stercoraria: Trypanosomatidae), the first fossil evidence of a Triatomine–Trypanosomatid vector association. *Vector-Borne and Zoonotic Diseases*, 5(1), 72–81.

Poinar, Jr., G.O. (2005c). *Culex malariager*, n. sp. (Diptera: Culicidae) from Dominican amber: the first fossil mosquito vector of *Plasmodium*. *Proceedings of the Entomological Society of Washington*, 107, 548–553.

Poinar, Jr., G.O. (2005d). A Cretaceous palm bruchid, *Mesopachymerus antiqua*, n. gen., n. sp. (Coleoptera: Bruchidae: Pachymerini) and biogeographical implications. *Proceedings of the Entomological Society of Washington*, 197, 392–397.

Poinar, Jr., G.O. (2006). Retracing the long journey of the insects (book reviews of evolution of the insects and history of insects). *American Scientist*, 94, 376–378.

Poinar, Jr., G.O. (2008a). *Lutzomyia adiketis* sp. n. (Diptera: Phlegbotomidae), a vector of *Paleoleishmania neotropicum* sp. n. (Kinoplastida: Trypanosomatidae) in Dominican amber. *Parasites and Vectors*, 1, 1–22 (http://www.parasitesandvec-tors.com/content/1/1/22).

Poinar, Jr., G.O. (2008b). *Leptoconops nosopheris* sp. n. (Diptera: Ceratopogonidae) and *Paleotryapanosoma burmanicus* gen. n., sp. n. (Kinetoplastidae: Trypanosomatidae), a biting midge–trypanosoma vector association from the Early Cretaceous. *Memorias Instituto Oswaldo Cruz*, 103, 468–471.

Poinar, Jr., G.O. (2008c). *Palaeocryptorhynchus burmanus*, a new genus and species of Early Cretaceous weevils (Coleoptera: Curculionidae) in Burmese amber. *Cretaceous Research*, 30, 587–591.

Poinar, Jr., G.O. (2008d). *Palaeosiro burmanicum* n. gen., n. sp., a fossil cyphophthalmi (Arachnida: Opiliones: Sironidae) in Early Cretaceous Burmese amber. In *Advances in Arachnology and Developmental Biology* (pp. 267–274), edited by S.E. Makarov and R.N. Dimitrijevic. Institute of Zoology Monographs, Belgrade.

Poinar, Jr., G.O. (2009a). *Meloe dominicanus* n. sp. (Coleoptera: Meloididae) phoretic on the bee *Proplebia dominicana* (Hymenoptera: Apidae) in Dominican amber. *Proceedings of the Entomological Society of Washington*, 111, 145–150.

Poinar, Jr., G.O. (2009b). Description of an Early Cretaceous termite (Isoptera: Kalotermitidae) and its associated intestinal protozoa, with comments on their coevolution. *Parasites and Vectors*, 2, 12 (doi: 10.1186/1756-3305-2-12; www.parasite-sandvectors.com/content/2/1/12).

Poinar, Jr., G.O. (2009c). Early Cretaceous protist flagellates (Parabasalia: Hypermastigia: Oxymonada) of cockroaches (Insecta: Blattaria) in Burmese amber. *Cretaceous Research*, 30, 1066–1072.

Poinar, Jr., G.O. and Anderson, N.H. (2005). Hymenopteran parasites of Trichoptera: the first fossil record. *Proceedings of the 11th International Symposium on Trichoptera* (pp. 343–346), edited by K. Tanida and A. Rossiter. Tokai University Press, Tokyo.

Poinar, Jr., G.O. and Boucot, A.J. (2006). Evidence of intestinal parasites of dinosaurs. *Parasitology*, 133, 245–249.

Poinar, Jr., G.O. and Brown, A.E. (2002). *Hymanaea mexicana* sp. nov. (*Leguminosae*: Caesalpinioideae) from Mexican amber indicates Old World connections. *Botanical Journal of the Linnean Society*, 139, 125–132.

Poinar, Jr., G.O. and Brown, A.E. (2003a). A non-gilled hymenomycete in Cretaceous amber. *Mycological Research*, 107, 763–768.

Poinar, Jr., G.O. and Brown, A. (2003b). A new genus of hard ticks in Cretaceous Burmese amber (Acari: Ixodida: Ixodidae). *Systematic Parasitology*, 54, 199–205.

Poinar, Jr., G.O. and Brown, A.E. (2005). New Aphidoidea (Hemiptera: Sternorrhyncha) in Burmese amber. *Proceedings of the Entomological Society of Washington*, 107, 835–845.

Poinar, Jr., G.O. and Brown, A.E. (2006a). The enigmatic *Dacochile microsoma* Poinar & Brown: Tanyderidae or Bruchomyiinae? *Zootaxa*, 1162, 19–31.

Poinar, Jr., G.O. and Brown, A.E. (2006b). Remarks on *Paraverrucosa annulata* (=*Verrucosa annulata* Poinar and Brown, 2005) (Hemiptera: Sternorrhyncha: Aphidoidea). *Proceedings of the Entomological Society of Washington*, 108, 734–735.

Poinar, Jr., G.O. and Buckley, R. (2006). Nematode (Nematoda: Mermithidae) and hairworm (Nematomorpha: Chorodidae) parasites in Early Cretaceous amber. *Journal of Invertebrate Pathology*, 93, 36–41.

Poinar, Jr., G.O. and Buckley, R. (2007). Evidence of mycoparasitism and hypermycoparasitism in Early Cretaceous amber. *Mycological Research*, 111, 503–506.

Poinar, Jr., G.O. and Buckley, R. (2008a). *Compluriscutula vetulum* (Acari: Ixodida: Ixodidae), a new genus and species of hard tick from Lower Cretaceous Burmese amber. *Proceedings of the Entomological Society of Washington*, 110, 445–450.

Poinar, Jr., G.O. and Buckley, R. (2008b). *Cretacifilix fungiformis* gen. and sp. nov., an eupolypod fern (Polypodiales) in early Cretaceous Burmese amber. *Journal of the Botanical Research Institute of Texas*, 2, 1175–1182.

Poinar, Jr., G.O. and Buckley, R. (2009). *Palaeoleptus burmanicus* n. gen., n. sp., an Early Cretaceous shore bug (Hemiptera: Palaeoleptidae n. fam.) in Burmese amber. *Cretaceous Research*, 30, 1000–1004.

Poinar, Jr., G.O. and Chambers, K.L. (2005). *Palaeoanthella huangii* gen. and sp. nov., an early Cretaceous flower (Angiospermae) in Burmese amber. *Sida*, 21, 2087–2092.

Poinar, Jr., G.O. and Columbus, J.T. (1992). Adhesive grass spikelet with mammalian hair in Dominican amber: first fossil evidence of epizoochory. *Experientia*, 48, 906–908.

Poinar, Jr., G.O. and Danforth, B.N. (2006). A fossil bee from Early Cretaceous Burmese amber. *Science*, 314, 614.

Poinar, Jr., G.O. and Doyen, J.T. (1992). A fossil termite bug, *Termitaradus protera*, new species (Termitaphidae: Hemipters) from Mexican amber. *Entomologica Scandinavica*, 23, 89–93.

Poinar, Jr., G.O. and Grimaldi, D.A. (1990). Fossil and extant macrochelid mites (Ascari: Macrochelidae) phoretic on drosophilid flies (Diptera: Drosophilidae). *Journal of the New York Entomological Society*, 98, 88–92.

Poinar, Jr., G.O. and Jansson, H.B., Eds. (1988). *Diseases of Nematodes*, Vols. 1 and 2. CRC Press, Boca Raton, FL.

Poinar, Jr., G.O. and Judziewicz, E.J. (2005). *Pharus primuncinatus* (Poacae: Pharoideae: Phareae) from Dominican amber. *Sida*, 21(4), 2095–2103.

Poinar, Jr., G.O. and Milki, R. (2001). *Lebanese Amber: The Oldest Insect Ecosystem in Fossilized Resin*. Oregon State University Press, Corvallis, 96 pp.

Poinar, Jr., G.O. and Miller, J.C. (2002). First fossil record of endoparasitism of adult ants (Formicidae: Hymenoptera) by Braconidae (Hymenoptera). *Annals of the Entomological Society of America*, 95, 41–43.

Poinar, Jr., G.O. and Monteys, V.S. (2008). Mermithids (Nematoda: Mermithidae) of biting midges (Diptera: Ceratopogonidae): *Heleidomermis cataloniense* n. sp. from *Culicoides circumscriptus* Kieffer in Spain and a species of *Cretacimermis* Poinar, 2001 from a ceratopogonid in Burmese amber. *Systematic Parasitology*, 69, 13–21.

Poinar, Jr., G.O. and Poinar, R. (1994). *The Quest for Life in Amber*. Addison-Wesley, New York, 219 pp.

Poinar, Jr., G.O. and Poinar, R. (1998). Parasites and pathogens of mites. *Annual Review of Entomology*, 43, 449–469.

Poinar, Jr., G.O. and Poinar, R. (1999). *The Amber Forest*. Princeton University Press, Princeton, NJ, 270 pp.

Poinar, Jr., G.O. and Poinar, R. (2004a). *Paleoleishmania proterus* n. gen., n. sp. (Trypanosomatidae: Kinetoplastida) from Cretaceous Burmese amber. *Protist*, 155, 305–310.

Poinar, Jr., G.O. and Poinar, R. (2004b). Evidence of vector-borne disease of Early Cretaceous reptiles. *Vector-Borne and Zoonotic Diseases*, 4, 281–284.

Poinar, Jr., G.O. and Poinar, R. (2005). Fossil evidence of insect pathogens. *Journal of Invertebrate Pathology*, 89, 243–250.

Poinar, Jr., G.O. and Poinar, R. (2008). *What Bugged the Dinosaurs? Insects, Disease and Death in the Cretaceous*. Princeton University Press, Princeton, NJ, 264 pp.

Poinar, Jr., G.O. and Roth, B. (1991). Terrestrial snails (Gastropoda) in Dominican amber. *The Veliger*, 34, 253–258.

Poinar, Jr., G.O. and Santiago-Blay, J.A. (1997). *Paleodoris lattini* gen. n., sp. n., a fossil palm bug (Hemiptera: Thaumastocoridae, Xylastodorinae) in Dominican amber, with habits discernible by comparative functional morphology. *Entomologica Scandinavica*, 28, 307–310.

Poinar, Jr., G.O. and Telford, Jr., S.R. (2005). *Paleohaemoproteus burmacis* gen. n., sp. n. (Haemosporida: Plasmodiidae) from an Early Cretaceous biting midge (Diptera: Ceratopogonidae). *Parasitology*, 131, 79–84.

Poinar, Jr., G.O. and Thomas, G.M. (1978). *Diagnostic Manual for the Identification of Insect Pathogens*. Plenum Press, New York, 218 pp.

Poinar, Jr., G.O. and Thomas, G.M. (1984). *Laboratory Guide to Insect Pathogens and Parasites*. Plenum Press, New York, 392 pp.

Poinar, Jr., G.O. and Voisin, J.-F. (2003). A Dominican amber weevil, *Velatis dominicana* gen. n., sp. n. and key to the genera of the Anchonini (Molytinae, Curculionidae). *Nouvelle Revue d'Entomologie*, 19, 373–381.

Poinar, Jr., G.O., Treat, A.E., and Southcott, R.V. (1991). Mite parasitism of moths: examples of paleosymbiosis in Dominican amber. *Experientia*, 47, 210–212.

Poinar, Jr., G.O., Pike, E.M., and Krantz, G.W. (1993a). Animal–animal parasitism. *Nature*, 361, 307–308.

Poinar, Jr., G.O., Waggoner, B.M., and Bauer, U.-C. (1993b). Terrestrial soft-bodied protists and other microorganisms in Triassic amber. *Science*, 259, 222–224.

Poinar, Jr., G.O., Acra, A., and Acra, F. (1994a). Earliest fossil nematode (Mermithidae) in Cretaceous Lebanese amber. *Fundamental and Applied Nematology*, 17, 475–477.

Poinar, Jr., G.O., Acra, A., and Acra, F. (1994b). Animal–animal parasitism in Lebanese amber. *Medical Science Research*, 22, 159.

Poinar, Jr., G.O., Poinar, H.N., and Cano, R.J. (1994c). DNA from amber inclusions. In *Ancient DNA: Recovery and Analysis of Genetic Material from Paleontological, Archaeological, Museum, Medical, and Forensic Specimens* (pp. 92–103), edited by B. Herrmann and S. Hummel. Springer-Verlag, Heidelberg.

Poinar, Jr., G.O., Krantz, G.W., Boucot, A.J., and Pike. T.N. (1997). A unique Mesozoic parasitic association. *Naturwissenschaften*, 84, 321–322.

Poinar, Jr., G.O., Curcic, B.P.M., and Cokendolpher, J.C. (1998a). Arthropod phoresy involving pseudoscorpions in past and present. *Acta Arachnologica*, 47, 79–96.

Poinar, Jr., G.O., Baroni Urbani, C., and Brown, A. (2000a). The oldest ants are Cretaceous, not Eocene [reply]. *The Canadian Entomologist*, 132, 695–696.

Poinar, Jr., G.O., Peterson, E.B., and Platt, J.L. (2000b). Fossil *Parmelia* in New World amber. *Lichenologist*, 32, 263–269.

Poinar, Jr., G.O., Zavortink, T.J., Pike, T., and Johnston, P.A. (2000c). *Paleoculis minutus* (Diptera: Culicidae) n. gen., n. sp., from Cretaceous Canadian amber, with a summary of described fossil mosquitoes. *Acta Geologica Hispanica*, 35, 119–128.

Poinar, Jr., G.O., Jacobson, R.L., and Eisenberger, C.L. (2006a). Early Cretaceous phlebotomine sand fly larvae (Diptera: Psychodidae). *Proceedings of the Entomological Society of Washington*, 108, 785–792.

Poinar, Jr., G.O., Lachaud, J.-P., Castillo, A., and Infante, F. (2006b). Recent and fossil nematode parasites (Nematoda: Mermithidae) of Neotropical ants. *Journal of Invertebrate Pathology*, 91, 19–26.

Poinar, Jr., G.O., Lambert, J.B., and Wu, Y. (2007a). Araucarian source of fossiliferous Burmese amber: spectroscopic and anatomical evidence. *Journal of the Botanical Research Institute of Texas*, 1, 449–455.

Poinar, Jr., G.O., Marshall, C., and Buckley, R. (2007b). One hundred million years of chemical warfare by insects. *Journal of Chemical Ecology*, 33, 1663–1665.

Poinar, Jr., G.O., Voisin, C., and Voisin, J.-F. (2007c). Bird eggshell in Dominican amber. *Palaeontology*, 50(6), 1381–1383.

Poinar, Jr., G.O., Chambers, K.L., and Buckley, R. (2007d). *Eoëpigynia burmensis* gen. and sp. nov., an Early Cretaceous eudicot flower (Angiospermae) in Burmese amber. *Journal of Botatnical Research, Institute of Texas*, 1, 91–96.

Poinar, Jr., G.O., Kerp, H., and Hass, H. (2008). *Palaeonema phyticum* gen. n., sp. n. (Nematoda: Palaeonematidae fam. n.), a Devonian nematode associated with early land plants. *Nematology*, 10, 9–14.

Poinar, H.N. (2000d). *Preservation of DNA in the Fossil Record*. Verlag Helmar Wodtke and Katharina Stegbauer Gbr, Leipzig, 111 pp.

Poinar, H.N. and Pääbo, S. (2001). DNA. In *Palaeobiology II* (pp. 241–245), edited by D.E.G. Briggs and P. Crowther. Blackwell Scientific, Oxford.

Poinar, H.N., Cano, R.J., and Poinar, Jr., G.O. (1993c). DNA from an extinct plant. *Nature*, 363, 677.

Poinar, H.N., Hofreiter, M., Spaulding, W.G., Martin, P.S., Stankiewicz, B.A., Bland, H., Evershed, R.P., Possnert, G., and Pääbo, S. (1998b). Molecular coproscopy: dung and diet of the extinct ground sloth *Nothrotheriops shastensis*. *Science*, 281(5375), 402–406.

Poinar, H.N., Schwarz, C., Qi, J., Shapiro, B., MacPhee, R.D., Buigues, B., Tikhonov, A., Huson, D.H., Tomsho, L.P., Auch, A., Rampp, M., Miller, W., and Schuster, S.C. (2006c). Metagenomics to paleogenomics: large-scale sequencing of mammoth DNA. *Science*, 311, 392–394.

Pojeta, Jr., J. and Palmer, T.J. (1976). The origin of rock boring in mytilacean pelecypods. *Alcheringa*, 1, 167–179.

Pokorny, V. (1989). Pussella and Saipanetta (Ostracoda, Crustacea) in the Lower Turonian of Bohemia, Czechoslovakia. *Casopis pro mineralogii a geologii*, 34, 225–237.

Pollard, J.E. (1968). The gastric contents of an ichthyosaur from the lower Lias of Lyme Regis, Dorset. *Palaeontology*, 11, 376–388.

Poole, F.G. (1988). Stratiform barite in Paleozoic rocks of the Western United States. In *Proceedings of the Seventh International Quadrennial IAGOD Symposium* (pp. 309–319), edited by E. Zachrisson. IAGOD, Stuttgart.

Poole, F.G. and Dutro, Jr., J.T. (1988). Late Devonian fossils in seafloor hydrothermal-vent barites of Nevada (USA) and Sonora (Mexico) [extended abs.]. *Symposium on Barite and Barite Deposits Abstracts*, 51–53.

Poole, F.G., Madrid, R.J., and Oliva-Becerril, J.F. (1991). Geological setting and origin of stratiform barite in central Sonora, Mexico. In *Geology and Ore Deposits of the Great Basin* (pp. 517–522), edited by G.L. Raines, R.E. Lisle, R.W. Schaeffer, and W.H. Wilkinson. Geological Society of Nevada, Reno.

Poole, J.H. (1994). Sex differences in the behaviour of African elephants. In *The Differences Between the Sexes* (pp. 331–346), edited by R.V. Short and E. Balaban. Cambridge University Press, Cambridge, U.K.

Por, F.D. (1970). Boring species of *Aspidosiphon* (Sipuncula) on the coasts of Israel. In *Proceeding of the International Symposium on the Biology of Sipuncula and Echiura*, Vol. I (pp. 301–304), edited by M.E. Rice and M. Todorovic. Institute for Biological Research and National Museum of Natural History, Smithsonian Institution, Washington, D.C.

Potter, A.W. and Boucot, A.J. (1992). Middle and late Ordovician brachiopod benthic assemblages of North America. In *Global Perspectives on Ordovician Geology: Proceedings of the Sixth International Symposium on the Ordovician System* (pp. 307–323), edited by B.D. Webby and J.R. Laurie, Balkema, Rotterdam.

Potts, R. (2003). Early human predation. In *Predator–Prey Interactions in the Fossil Record* (pp. 359–376), edited by P.H. Kelley, M. Kowalewski, and T.A. Hansen. Kluwer Academic, New York.

Poulia, E. and Feral, J.-P. (1996). Why are there so many species of brooding Antarctic echinoids? *Evolution*, 50, 820–830.

Poulin, E., Palma, A.T., and Féral, J.-C. (2002). Evolutionary versus ecological success in Antarctic benthic invertebrates. *Trends in Ecology & Evolution*, 17, 218–222.

Powell, A.W.B. (1937). Animal communities of the sea-bottom in Auckland and Manukau Harbours. *Royal Society of New Zealand Bulletin*, 66, 354–401.

Powell, A.W.B. (1939). Note of the importance of recent animal ecology as a basis of paleoecology. *Pacific Science Congress Proceedings*, 6, 607–617.

Powell, G.C. and Nickerson, R.B. (1965). Aggregations among juvenile king crabs (*Paralithodes camtschatica*, Tilesius), Kodiak, Alaska. *Animal Behaviour*, 13, 374–380.

Powell, G.C. and Reynolds. R.E. (1965). *Movements of Tagged King Crabs* Paralithodes camschatica *(Tilesius) in the Kodiak Island–Lower Cook Inlet Region of Alaska (1954–1963)*. Department of Fish and Game, Juneau, AK, 10 pp.

Powers, B.G. and Ausich, W.I. (1990). Epizoan associations in a Lower Mississippian paleocommunity (Borden Group, Indiana, USA). *Historical Biology*, 4, 245–265.

Prakfalvi, P. (1992). A matraszelei (Nograd Megye) eggenburgi koru, ophioroideas retegek foldtani viszonyai. *A Magyar Allami Foldtani Intezet Évi Jelentese az 1990. évröl*, pp. 481–494.

Prantl, F. (1948). The genus *Conchicolites* Nicholson, 1872 (Serpulimorpha) in the Ordovician of Bohemia. *Vestnik Kralovske Ceske Spolecnosti Nauk*, 9, 1–7.

Prasad, V., Strömberg, C.A.E., Alimohammadian, H., and Sahni, A. (2005). Dinosaur coprolites and the early evolution of grasses and grazers. *Science*, 310, 1177–1180.

Prashad, B. (1924). On a fossil ampulariid from Poonch, Kashmir. *Records of the Geological Survey of India*, 56, 210–212.

Pratt, B.R. (1996). Lower Cambrian *Rusophycus* from Arctic Canada: ichnofossil of a predatory, non-trilobite arthropod. In *Sixth North American Paleontological Convention Abstracts of Papers* (p. 312), edited by J.E. Repetski, Special Publ. No. 8. Paleontological Society, Washington, D.C.

Pratt, B.R. (1998). Probable predation on Upper Cambrian trilobites and its relevance for the extinction of soft-bodied Burgess Shale-type animals. *Lethaia*, 31, 73–88.

Pratt, H.L. and Carrier, J.C. (2005). Elasmobranch courtship and mating behavior. In *Reproductive Biology and Phylogeny of Chondrichthyes* (pp. 129–169), edited by W.C. Hamlett. Science Books Publishers, Enfield, NH.

Prentice, M.A. (1998). The Comparative Morphology and Phylogeny of Apoid Wasps (Hymenoptera: Apoidea). Ph.D. dissertation, University of California, Berkeley, 1439 pp.

Preston, S.J., Roberts, D., and Montgomery, W.I. (1996). Crab predation as a selective agent on shelled gastropods: a case study of *Calliostoma zizyphinum* (Linnaeus) (Prosobranchia: Trochidae). In *Origin and Evolutionary Radiation of the Mollusca* (pp. 313–325), edited by J. Taylor. Oxford University Press, London.

Prokofiev, A.M. (2005). Periopisanie i rodstvennie otlosheniya *Beckerophotus gracilis* (Daniltshenko, 1962) (Teleostei: Myctophiformes: Neoscopelidae) iz Srednego Eotsena Gruzii. *Paleontologicheskii Zhurnal*, 6, 65–72.

Prokofiev, A.M. and Bannikov, A.F. (2002). Novii rod Stomiformes (Pisces, Actinopterygii) iz Eotsena Severnogo Kavkaza. *Paleontologicheskii Zhurnal* 2, 43–48.

Prothero, D.R. (1994). *The Eocene–Oligocene Transition*. Columbia University Press, New York.

Pufahl, P.K. and James, N.P. (2006). Monospecific Pliocene oyster buildups, Murray Basin, South Australia: brackish water end member of the reef spectrum. *Palaeogeography, Palaeoclimatology, Palaeoecology*, 233, 11–33.

Pugaczewska, H. (1970). Traces of the activity of bottom organisms on the shells of the Jurassic ostreiform pelecypods of Poland. *Acta Palaeontologica Polonica*, 15(4), 425–440.

Pugaczewska, H. (1985). Les organisms sédentaires sur les rostres des belemnites du Crétacé Supérieur. *Acta Palaeontologica Polonica*, 10, 73–95.

Purdy, R.W., Schneider, V.P., Applegate, S.P., McLellan, J.H., Meyer, R.L., and Slaughter. B.H. (2001). The Neogene sharks, rays, and bony fishes from Lee Creek Mine, Aurora, North Carolina. *Smithsonian Contributions to Paleobiology*, 90, 71–202.

Pyenson, N.D. and Haasl, D.M. (2007). Miocene whale-fall from California demonstrates that cetacean size did not determine the evolution of modern whale-fall communities. *Biology Letters, Palaeontology*, 3, 709–711.

Quade, J., Levin, N., Semaw, S., Stout, D., Renne, P., Rogers, M., and Simpson, S. (2004). Paleoenvironments of the earliest stone toolmakers, Gona, Ethiopia. *Geological Society of America Bulletin*, 116, 1529–1544.

Quenstedt, F.A. (1858). *Der Jura*. Laupp & Siebeck, Tübingen, 842 pp.

Rabano, I. and Arbizu, M. (1999). Casos de malformaciones en trilobites de España. *Revista Española de Paleontologia*, 109–113.

Rabinowitz, D., Rapp, J.K., and Dixon, P.M. (1984). Competitive abilities of sparse grass species: means of persistence or cause of abundance. *Ecology*, 65, 1144–1154.

Racheboeuf, P.R. and Lespérance, P.J. (1995). Revision of Silurian and Devonian Chonetoidean brachiopods from Québec and northern new Brunswick (Canada). *Documents des Laboratoires de Géologie Lyon*, 136, 7–99.

Racki, G. (1982). Ecology of the primitive charophyte algae: a critical review. *Neues Jahrbuch für Geologie und Paläontologie–Abhandlungen*, 162, 388–399.

Radchenko, V.G. and Pesenko, Y.A. (1994). *Biologiya pchel (Biology of Bees)*. Russian Academy of Sciences, St. Petersburg.

Radinsky, L.B. (1968). Evolution of somatic specialization in otter brains. *Journal of Comparative Neurology*, 134, 495–505.

Radwańska, U. (1994a). Tube-dwelling polychaetes from the Korytnica Basin (Middle Miocene; Holy Cross Mountains, southern Poland). *Acta Geologica Polonica*, 44, 35–81.

Radwańska, U. (1994b). A new group of microfossils: Middle Miocene (Badenian) opercular caps (*calottae*) of the tube-dwelling polychaetes *Vermiliopsis* Saint-Joseph, 1894. *Acta Geologica Polonica*, 44, 83–96.

Radwańska, U. (1996). Tube-dwelling polychaetes from some Upper Cretaceous sequences of Poland. *Acta Geologica Polonica*, 46, 61–80.

Radwańska, U. (1999). Lower Kimmeridgian echinoids of Poland. *Acta Geologica Polonica*, 49, 287–364.

Radwańska, U. (2004). Tube-dwelling polychaetes from the Upper Oxfordian of Wapienno/Bielawy, Couiavia region, north-central Poland. *Acta Geologica Polonica*, 54, 35–52.

Radwańska, U. (2005). Lower Kimmeridgian comatulid crinoids of the Holy Cross Mountains, Central Poland. *Acta Geologica Polonica*, 55, 269–282.

Radwańska, U. and Radwański, A. (2005). Myzostomid and copepod infestation of Jurassic echinoderms: a general approach, some new occurrences, and/or re-interpretation of previous reports. *Acta Geologica Polonica*, 55, 109–130.

Radwański, A. (1972). Isopod-infected prosoponids from the Upper Jurassic of Poland. *Acta Geologica Polonica*, 22, 499–506.

Radwański, A. (1977). Present-day types of trace in the Neogene sequence: their problems of nomenclature and preservation. In *Trace Fossils*, Vol. 2 (pp. 227–264), edited by T.P. Crimes and J.C. Harper. Seel House Press, Liverpool.

Radwański, A. (1996). The predation upon, and the extinction of, the latest Maastrichtian populations of the ammonite species *Hoploscaphites constrictus* (J. Sowerby, 1817) from the Middle Vistula Valley, Central Poland. *Acta Geologica Polonica*, 46, 117–135.

Radwański, A. (2002). Triassic brittlestar beds of Poland: a case of *Aspiduriella ludeni* (v. Hagenow, 1846) and *Arenorbis squamosus* (E. Picard, 1858). *Acta Geologica Polonica*, 42, 395–410.

Radwański, A. and Wysocka, A. (2001). Mass aggregations of Middle Miocene spine-coated echinoids (echinocardium) and their integrated eco-taphonomy. *Acta Geologica Polonica*, 51, 295–316.

Radwański, A. and Wysocka, A. (2004). A farewell to Swiniary sequence of mass-aggregated, spine-coated echinoids *Psammechinus* and their associates (Middle Miocene; Holy Cross Mountains, Central Poland). *Acta Geologica Polonica*, 54, 381–399.

Radwański, A., Kin, A., and Radwański, U. (2009). Queues of blind phacopid trilobites Trimerocephalus: a case of frozen behaviour of Early Famennian age from the Holy Cross Mountains, Central Poland. *Acta Geologica Polonica*, 59, 459–481.

Rajchel, J. and Uchman, A. (1998). Insect borings in Oligocene wood, Kliwa Sandstones, outer Carpathians, Poland. *Annales Societatis Geologorum Poloniae*, 68, 219–224.

Ramírez, S.R., Gravendee, B., Singer, R.B., Marshall, C.R., and Pierce, N.E. (2007). Dating the origin of the Orchidaceae from a fossil orchid and its pollinator. *Nature*, 448, 1042–1045.

Ramskold, L. (1984). Silurian odontopleurid trilobites from Gotland. *Palaeontology*, 27, 239–264.

Ranson, G. (1959). Mollusques perliers et perles (Bibliographie). *Bulletin de l'Institut Océanographique, Monaco*, 1140, 1–43.

Raschke, R. (1989). *Final Report on Paleontological Monitoring at Tract 13295, Rincon Village, West Chino Hills, California*. RMW Paleo Associates, Mission Viejo, CA.

Rasmussen, H.W. (1972). Lower Tertiary Crinoidea, Asteroidea and Ophiuroidea from Northern Europe and Greenland. *Det Kongelige Danske Videnskabernes Selskab, Biologiske Skrifter*, 19(7), 1–83.

Rasnitsyn, A.P. (1980). Nadotryad Vespidea. *Transactions of the Paleontological Institute, Academy of Science, USSR*, 175, 122–127.

Rasnitsyn, A.P. (1992). *Strashila incredibilis*, a new enigmatic mecopteroid insect with possible siphonapteran affinities from the Upper Jurassic of Siberia. *Psyche*, 99, 323–333.

Rasnitsyn, A.P. and Krassilov, V.A. (1996). First find of pollen grains in the gut of Permian insects: *Paleontologicheski Zhurnal*, 3, 119–124.

Rasnitsyn, A.P. and Krasilov, V.A. (2000). Pervoe Podtverzhdenie filofagii domelovikh nasekomish: listovie tkani v Kishechnike verkhneoorskikh nasekomikh iz oozhnogo Kazakhstana. *Paleontologicheski Zhurnal*, 3, 73–81.

Rasnitsyn, A.P. and Michener, C.D. (1991). Miocene fossil bumblebee from the Soviet Far East with comments on the chronology and distribution of fossil bees. *Annals of the Entomological Society of America*, 84, 583–589.

Rasnitsyn, A.P. and Quicke, D.L.J., Eds. (2002). *History of Insects*. Kluwer Academic, Dordrecht, 517 pp.

Rathbun, M.J. (1916). Description of a new genus and species of fossil crab from Port Townsend, Washington. *American Journal of Science*, 41, 344–346.

Raymo, M.E., Ganley, K., Carter, S., Oppe, D.W., and McManus, J (1998). Millenial-scale climate instability during the early Pleistocene epoch. *Nature*, 392, 699–702.

Rayner, J.M.V. (1991). Complexity and a coupled system: flight, echolocation and evolution in bats. In *Constructional Morphology and Evolution* (pp. 173–191) edited by N. Schmidt-Kittler and K. Vogel. Springer-Verlag, Heidelberg.

Redecker, D., Kodner, R., and Graham, L.E. (2000). Glomalean Fungi from the Ordovician. *Science*, 289, 1920–1924.

Reimer, D. (1989). Phosphatisierte Fossilien aus dem unteren Ordoviz von Südschweden. *Berliner Geowissenschaftliche Abhandlungen (A)*, 106, 9–l9.

Reinhard, K.J., Helvy, R.H., and Anderson, G.A. (1987). Helminth remains from prehistoric Indian coprolites on the Colorado Plateau. *Journal of Parasitology*, 73, 630–639.

Reitner, J. (2005). Mikrobielles Leben an Methan-Quellen ("Cold Seeps") des Schwarzen Meeres–Einblicke in ursprüngliche anaerobe Lebenswelten der Tiefen-Biosphäre. *Jahrbuch der Akademie der Wissenschaften zu Göttingen*, 101–110.

Remy, W., Taylor, T.N., and Hass, H. (1994). Early Devonian fungus: a blastocladalean fungus with sexual reproduction. *American Journal of Botany*, 81, 690–702.

Ren, D. (1998). Flower-associated Brachycera flies as fossil evidence for Jurassic Angiosperm origins. *Science*, 280, 85–88.

Renault, B. (1896). Recherches sur les Bactericées fossils. *Annales des Sciences Naturelles, Botanique, Séries VIII*, 3, 275.

Renders, E. (1984). The gait of *Hipparion* sp. from fossil footprints in Laetoli, Tanzania. *Nature*, 308, 179–181.

Renouf, D., Ed. (1991). *Behaviour of Pinnipeds*. Chapman & Hall, London, 410 pp.

Repenning, C.A. and Packard, E.I. (1990). Locomotion of a desmostylian and evidence of ancient shark predation. In *Evolutionary Paleobiology of Behavior and Coevolution* (pp. 199–203), edited by A.J. Boucot. Elsevier, Amsterdam.

Rex, G.M. and Galtier, J. (1986). Sur l'évidence d'interactions animal-végétal dans le Carbonifère inférieur français. *Comptes Rendus de l'Academie des Sciences, Paris*, 2, 1623–1626.

Reyment, R.A. (1967). Paleoethology and fossil drilling gastropods. *Transactions of the Kansas Academy of Science*, 70, 33–50.

Reyment, R.A. (1999). Drilling gastropods. In *Functional Morphology of the Invertebrate Skeleton* (pp. 197–204), edited by E. Savazzi. John Wiley & Sons, New York.

Reyment, R.A. and Elewa, A.M.T. (2003). Predation by drills on Ostracoda. In *Predator–Prey Interactions in the Fossil Record* (pp. 93–111), edited by P.H. Kelley, M. Kowalewski, and T.A. Hansen. Kluwer Academic, New York.

Reyment, R.A., Reyment, E.R., and Honigstein, A. (1987). Predation by boring gastropods on Late Cretaceous and Early Palaeocene ostracods. *Cretaceous Research*, 8, 189–209.

Reynoso, V.-H. (2005). Possible evidence of venom apparatus in a Middle Jurassic sphenodontian from the Huizachal red beds of Tamaulipas, Mexico. *Journal of Vertebrate Paleontology*, 25, 646–654.

Ricci Lucchi, F. and Vai, G.B. (1994). A stratigraphic and tectono-facies framework of the eastern "calcari a *Luchina*" in the Apennine Chain, Italy. *Geo-Marine Letters*, 14, 210–218.

Richards, R.P. (1972). Autecology of Richmondian brachiopods (Late Ordovician of Indiana and Ohio). *Journal of Paleontology*, 46, 386–405.

Richards, R.P. (1974). Ecology of the Cornulitidae. *Journal of Paleontology*, 48, 514–523.

Richards, R.P. and Shabica, C.W. (1969). Cylindrical living burrows in Ordovician dalmanellid brachiopod beds. *Journal of Paleontology*, 43, 838–841.

Richardson, J.R. (1987). Brachiopods from carbonate sands of the Australian Shelf. *Proceedings of the Royal Society of Victoria*, 99, 37–50.

Richter, G. (1993). Proof of feeding specialism in Messel bats? *Kaupia*, 3, 107–112.

Richter, G. and Baszio, S. (2001a). First proof of planctivory/insectivory in a fossil fish: *Thaumaturus intermedius* from the Eocene Lake Messel (FRG). *Palaeogeography, Palaeoclimatology, Palaeoecology*, 173, 75–85.

Richter, G. and Baszio, S. (2001b). Traces of a limnic food web in the Eocene Lake Messel: a preliminary report based on fish coprolite analyses. *Palaeogeography, Palaeoclimatology, Palaeoecology*, 166, 345–368.

Richter, G. and Wedmann, S. (2005). Ecology of the Eocene Lake Messel revealed by analysis of small fish coprolites and sediments from a drilling core. *Palaeogeography, Palaeoclimatology, Palaeoecology*, 223, 147–161.

Rieppel, O. (1985). Die Gattung *Saurichthys* (Pisces, Actinopterygii) aus der Mittleren Trias des Monte San Giorgio, Kanton Tessin. *Schweizerische paläontologische Abhandlungen*, 108, 1–85.

Rieppel, O., Zaher, H., Tchernov, E., and Polcyn, M.J. (2003). The anatomy and relationships of *Haasiophis terrasanctus*, a fossil snake with well-developed hind limbs from the Mid-Cretaceous of the Middle East. *Journal of Paleontology*, 77, 536–558.

Rigby, J.K. and Goedert, J.L. (1996). Fossil sponges from a localized cold-seep limestone in Oligocene rocks of the Olympic Peninsula, Washington. *Journal of Paleontology*, 70, 900–908.

Riggs, E.S. (1945). Some Early Miocene carnivores. *Fieldiana: Geology*, 9(3), 69–114.

Rikkinen, J. and Poinar, Jr., G.O. (2000). A new species of resinicolous *Chaenothecopsis* (Mycocaliciaceae, Ascomycota) from 20 million year old Bitterfeld amber, with remarks on the biology of the resinicolous fungi. *Mycological Research*, 104, 7–15.

Rikkinen, J. and Poinar, Jr., G.O. (2002). Fossilized *Anzia* (Lecanorales, lichen-forming Ascomycota) from European Tertiary amber. *Mycological Research*, 106, 984–990.

Risk, M.J. and Szczuczko, R.B. (1977). A method for staining trace fossils. *Journal of Sedimentary Petrology*, 47, 855–859.

Roberts, E.M., Rogers, R.R., and Foreman, B.Z. (2007). Continental insect borings in dinosaur bone: examples from the Late Cretaceous of Madagascar and Utah. *Journal of Paleontology*, 81, 201–208.

Robertson, D. (1883). On the post-Tertiary beds of Garvel Park, Greenock. *Transactions of the Geological Society of Glasgow*, 7, 1–37.

Robinson, J.H. (2005). Brachiopod pedicle traces: redefinition of *Podichnus centrifugalis* to differentiate three types of trace. In *Fifth International Brachiopod Congress, Copenhagen 2005, Abstracts* (p. 26), edited by D.A.T. Harper, S.L. Long, and M. McCorry. Geological Survey of Denmark and Greenland, Copenhagen.

Robinson, J.H., and Lee, D.E. (2008). Brachiopod pedicle traces: recognition of three separate types of trace and redefinition of *Podichnus centrifugalis* Bromley & Surlyk, 1973. *Fossils and Strata*, 54, 219–225.

Robinson, J.H., Ahlberg, P.E., and Koentges, G. (2005). The braincase and middle ear region of *Dendrerpeton acadianum* (Tetrapoda: Temnospondyla). *Zoological Journal of the Linnean Society*, 143, 577–597.

Robson, S.P. and Pratt, B.R. (2007). Predation of late Marjumian (Cambrian) linguliformean brachiopods from the Deadwood Formation of South Dakota, USA. *Lethaia*, 40, 19–32.

Roček, Z. (2003). Larval development in Oligocene palaeobatrachid frogs. *Acta Palaeontographica Polonica*, 48, 595–607.

Roček, Z. and Rage, J.-C. (2000). Tertiary anura of Europe, Africa, Asia, North America, and Australia. In *Amphibian Biology*. Vol. 4. *Palaeontology* (pp. 1332–1337), edited by H. Heatwole and R.L. Carroll. Surrey Beatty & Sons, Chipping Norton, Australia.

Roček, Z. and Van Dijk, E. (2006). Patterns of larval development in Cretaceous pipid frogs. *Acta Palaeontologica Polonica*, 51, 111–126.

Roček, Z., Böttcher, R., and Wassersug, R. (2006). Gigantism in tadpoles of the Neogene frog *Palaeobatrachus*. *Paleobiology*, 32, 666–675.

Roche, H., Delagnes, A., Brugal, J.-P. et al. (1999). Early hominid stone tool production and technical skill 2.34 myr ago in West Turkana, Kenya. *Nature*, 399, 57–60.

Rodda, P.U. and Fisher, W.L. (1962). Paleozoic acrothoracic barnacles from Texas. *Texas Journal of Science*, 14, 460–479.

Rodriguez, J. and Gutschick, R.C. (1977). Barnacle borings in live and dead hosts from the Louisiana Limestone (Famennian) of Missouri. *Journal of Paleontology*, 52, 718–724.

Roger, J. (1944). Essai d'interprétation d'une forme curieuse de *Flabellum* du Pliocène ancien de Dar bel Hamri (Maroc). *Bulletin du Muséum national d'Histoire naturelle, 2e Série*. 16, 245–254.

Rogers II, J.V. (2000). A complete crocodiloid egg from the Lower Cretaceous (Albian) Glen Rose Formation. *Journal of Vertebrate Paleontology*, 20, 780–783.

Rogers, P.H. and Cox, M. (1988). Underwater sound as a biological stimulus. In *Sensory Biology of Aquatic Animals* (pp. 131–149), edited by J. Atema, R.R. Fay, A.N. Popper, and W. Tavolga. Springer-Verlag, Heidelberg.

Rogers, R.R. (1992). Non-marine borings in dinosaur bones from the Upper Cretaceous Two Medicine Formation, northwestern Montana. *Journal of Vertebrate Paleontology*, 12, 528–531.

Rogers, R.R. (1998). Sequence analysis of the Upper Cretaceous Two Medicine and Judith River formations: nonmarine response to the Claggett and Bearpaw marine cycles. *Journal of Sedimentary Research*, 68, 615–631.

Rogers, R.R., Krause, D.W., and Rogers, K.C. (2003). Cannibalism in the Madagascan dinosaur *Majungatholus atopuc*. *Nature*, 422, 515–518.

Rohr, D.M. (1976). Silurian predator borings in the brachiopod *Dicaelosia* from the Canadian Arctic. *Journal of Paleontology*, 50, 1175–1179.

Rohr, D.M. (1991). Borings in the shell of an Ordovician (Whiterockian) gastropod. *Journal of Paleontology*, 65, 687–688.

Rohr, D.M. (1993). Middle Ordovician carrier shell *Lytospira* (Mollusca, Gastropoda) from Alaska. *Journal of Paleontology*, 67, 959–962.

Rohr, D.M. (1994). Ordovician (Whiterockian) gastropods of Nevada: Bellerophontoidea, Macluritoidea, and Euomphaloidea. *Journal of Paleontology*, 68, 473–486.

Rohr, D.M. (2004). Life association of shell and operculum of *Ceratopea* Ulrich 1911 (Ordovician, Gastropoda). *Journal of Paleontology*, 78, 218–220.

Rohr, D.M. and Blodgett, R.B. (2003). Gastropod opercula from the Silurian and Devonian of Alaska. In *Short Notes on Alaska Geology*, Professional Report 120 (pp. 83–85), edited by K.H. Clautice and P.K. Davis. Division of Geological and Geophysical Surveys, Fairbanks.

Rohr, D.M. and Frýda, J. (2001). A new Ordovician gastropod and operculum from the Czech Republic. *Journal of Paleontology*, 75, 461–462.

Rohr, D.M. and Yochelson, E.L. (1999). Life association of shell and operculum of Middle Ordovician gastropod *Maclurites*. *Journal of Paleontology*, 73, 1078–1080.

Rohr, D.M., Boucot, A.J., Abbott, M., and Miller, J.M. (1986). Oldest termite nest from the Upper Cretaceous of West Texas. *Geology*, 14, 87–88.

Rolfe, W.D.I. (1985). Form and function in Thylococephala, Conchyliocarida and Cancavicarida (?Crustacea): a problem in interpretation. *Transactions of the Royal Society of Edinburgh, Earth Sciences*, 76, 391–399.

Rolin, Y., Gaillard, C., and Roux, M. (1990). Ecologie des pseudo-biohermes des Terres Noires Jurassiques liés à des paléo-sources sous-marines. Le Site Oxfordien de Beauvoisen (Drôme, Bassin du Sud-Est, France). *Palaeogeography, Palaeoclimatology, Palaeoecology*, 80, 79–105.

Roman, J. (1976). Echinides Eocènes et Miocène du Quatar (Golfe Persique). *Annales de Paléontologie (Invertébrès)*, 62, 49–85.

Roman, J. (1983). Echinides 'marsupiaux' (Fam. Temnopleuridae) dans le Neogène de l'Ouest Européen. *Annales de Paléontologie*, 69, 13–42.

Röper, M., Leich, H., and Rothgaenger, M. (1999). *Die Plattenkalke von Pfalzpaint*. Eichendorf-Verlag, Eichendorf, 120 pp.

Rose, K.D., Walker, A., and Jacobs, L.L. (1981). Function of the mandibular tooth comb in living and extinct mammals. *Nature*, 289, 583–585.

Rosen, B.R., Aillud, G.S., Bosellini, F.R., Clack, N.J., Insalaco, E., Valldeoeras, F.X., and Wilson, M.E.J. (2000). Platy coral assemblages: 200 million years of functional stability in response to the limiting effects of light and turbidity. *Proceedings of the Ninth International Coral Reef Symposium*, 1, 1–10.

Roskam, J.C. (1992). Evolution of the gall-inducing guild. In *Biology of Insect-Induced Galls* (pp. 34–49), edited by J.D. Shorthouse and O. Rohfritsch. Oxford University Press, London.

Ross, A. (1964). Cirripedia from the Yorktown Formation (Miocene) of Virginia. *Journal of Paleontology*, 38, 71–72.

Ross, A. (1998). *Amber*. The Natural History Museum, London, 73 pp.

Ross, D.M. (1983). Symbiotic relations. In *The Biology of Crustacea*. Vol. 7. *Behavior and Ecology* (pp. 193–195), edited by F.J. Vernberg and W.B. Vernberg. Academic Press, San Diego, CA.

Rossi, W., Kotrba, M., and Triebel, D. (2005). A new species of Stigmatomyces from Baltic amber, the first fossil record of Laboulbeniomycetes. *Mycological Research*, 109, 271–274.

Rössler, R. and Schneider, J.W. (1997). Eine bemerkenswerte Paläobiocoenose im Unterkarbon Mitteleuropas: Fossilführung und Paläoenvironment der Hainichen-Subgruppe (Erzegebirge-Becken). *Veröffentlichungen des Museums für Naturkunde Chemnitz*, 20, 5–44.

Roth, L.M. and Willis, E.R. (1960). The biotic associations of cockroaches. *Smithsonian Miscellaneous Collection*, 141, 1–470.

Rothe, P. and Klemmer, K. (1991). Fossil eggs of terrestrial tortoises (Family Testudinidae) from Pliocene calcarenites of Fuenteventura (Canary Islands, Spain). *Senckenbergiana Lethaia*, 71, 307–317.

Rothschild, B.M. (1987). Decompression syndrome in fossil marine turtles. *Annals of the Carnegie Museum*, 56, 253–258.

Rothschild, B.M. (1988). Stress fracture in a ceratopsian phalanx. *Journal of Paleontology*, 62, 302–303.

Rothschild, B.M. (1989). Rheumatic diseases in the fossil record. In *Arthritis and Allied Conditions: A Textbook of Rheumatology*, 11th ed. (pp. 3–7), edited by D.J. McCarty. Lea & Febiger, Philadelphia, PA.

Rothschild, B.M. (1990). Radiologic assessment of osteoarthritis in dinosaurs. *Annals of Carnegie Museum*, 59, 295–301.

Rothschild, B.M. (1991). Stratophenetic analysis of avascular necrosis in turtles: affirmation of decompression syndrome hypothesis. *Comparative Biochemistry and Physiology A*, 100, 529–535.

Rothschild, B.M. (1993a). Skeletal paleopathology of rheumatic diseases: the sub-Homo connection. In *Arthritis and Allied Conditions: A Textbook of Rheumatology*, 12th ed. (pp. 3–7), edited by D.J. McCarty. Lea & Febiger, Philadelphia, PA.

Rothschild, B.M. (1993b). Arthritis in non-human African primates: transition from the Miocene to the Holocene. *Journal of Vertebrate Paleontology*, 13, 55A.

Rothschild, B.M. (2003a). Infectious processes around the dawn of civilization. In *The Archaeology, Ecology, and Evolution of Infectious Disease* (pp. 103–116), edited by C. Greenblatt and M. Spigelman. Oxford University Press, London.

Rothschild, B.M. (2003b). Pathology in Hiscock Site vertebrates, and its bearing on hyperdisease among North American mastodons. In "The Hiscock Site: Late Pleistocene and Holocene paleoecology and archaeology of western New York State," edited by R.S. Laub. *Bulletin of the Buffalo Society of Natural Sciences*, 37, 171–175.

Rothschild, B.M. (2003c). Infectious processes around the dawn of civilization. In *Emerging Pathogens: The Archaeology, Ecology, and Evolution of Infectious Disease* (pp. 103–124), edited by C. Greenblatt and M. Spigel. Oxford University Press, London.

Rothschild, B.M. (2005). History of syphilis. *Clinical Infectious Diseases*, 40, 1454–1463.

Rothschild, B.M. (2006). Diving disease in marine reptiles from the Cretaceous to the present. *Geological Society of America Abstracts with Programs*, 38(4), 62.

Rothschild, B.M. and Helbling, M.H. (2001). Documentation of hyperdisease in the Late Pleistocene: validation of an early 20th century hypothesis. *Journal of Vertebrate Paleontology Abstracts of Papers*, 94A.

Rothschild, B.M. and Laub, R. (2006). Hyperdisease in the late Pleistocene: validation of an early 20th century hypothesis. *Naturwissenschaften*, 93, 557–564.

Rothschild, B.M. and Martin, L.D. (1987). Avascular necrosis: occurrence in diving Cretaceous mosasaurs. *Science*, 236, 75–77.

Rothschild, B.M. and Martin, L.D. (1993). *Paleopathology: Disease in the Fossil Record*. CRC Press, Boca Raton, FL, 386 pp.

Rothschild, B.M. and Martin, L.D. (2005). Mosasaur ascending: the phylogeny of bends. *Netherlands Journal of Geosciences/ Geologie en Mijnbouw*, 84, 341–344.

Rothschild, B.M. and Martin, L.D. (2006a). *Skeletal Impact of Disease*, Bulletin 33. New Mexico Museum of Natural History & Science, Albuqueque, 226 pp.

Rothschild, B.M. and Martin, L.D. (2006b). Did ice-age bovids spread tuberculosis? *Naturwissenschaften*, 93, 565–569.

Rothschild, B.M. and Molnar, R.E. (1988). Osteoarthritis in fossil marsupial populations of Australia. *Annals of Carnegie Museum*, 57, 155–158.

Rothschild, B.M. and Rothschild, C. (1995). Treponemal disease revisited: skeletal discriminators for yaws, bejel, and venereal syphilis. *Clinical Infectious Diseases*, 20, 1402–1408.

Rothschild, B.M. and Rothschild, C. (1996a). Treponemal disease in the New World. *Current Anthropology*, 37, 555–561.

Rothschild, B.M. and Rothschild, C. (1996b). Treponematoses: origins and 1.5 million years of transition. *Human Evolution*, 11, 225–232.

Rothschild, B.M. and Rothschild, C. (1996c). Trans-mammalian pandemic of inflammatory arthritis (spondyloarthropathy variety): persistence since the Pleistocene. In *Sixth North American Paleontological Convention Abstracts of Papers* (p. 330), edited by J.E. Repetski, Special Publ. No. 8. Paleontological Society, Washington, D.C.

Rothschild, B.M. and Rothschild, C. (1996d). Post-Paleocene inflammatory arthritis. *Journal of Vertebrate Paleontology*, 16, 61A.

Rothschild, B.M. and Rothschild, C. (1998a). Skeletal examination-based recognition of treponematoses: a four continent odyssey of denouement, transition and spread. *Bulletin et Mémoires de la Société d'Anthropologie de Paris*, 10, 29–40.

Rothschild, B.M. and Rothschild, C. (1998b). Horse and rhinoceros evolutionary evidence for protective effect of spondyloarthropathy? *Arthritis and Rheumatism*, 41, S286.

Rothschild, B.M. and Saunders, A. (1997). Cetacean pathology: evolution of avascular necrosis/bends. *Journal of Vertebrate Paleontology*, 17, 72A.

Rothschild, B.M. and Storrs, G.W. (2003). Decompression syndrome in plesiosaurs (Sauropterygia: Reptilia). *Journal of Vertebrate Paleontology*, 23, 324–328.

Rothschild, B.M. and Tanke, D. (1992). Paleopathology of vertebrates: insights to lifestyles and health in the geological record. *Geoscience Canada*, 19, 73–82.

Rothschild, B.M. and Tanke, D. (1997). Thunder in the Cretaceous: interspecies conflict as evidence for ceratopsian migration? In *Dinofest International: Proceedings of a Symposium Sponsored by Arizona State* (pp. 77–81), edited by D.L. Wolberg, E. Stump, and G.D. Rosenberg. The Academy of Natural Sciences, Philadelphia, PA.

Rothschild, B.M. and Tanke, D. (2005). Theropod paleopathology: state-of-the-art review. In *The Carnivorous Dinosaurs* (pp. 351–365), edited by K. Carpenter. Indiana University Press, Bloomington.

Rothschild, B.M. and Thillaud, P.L. (1991). Oldest bone disease. *Nature*, 399, 288.

Rothschild, B.M., Wang, X., and Shoshani, J. (1994). Spondyloarthropathy in proboscidians. *Journal of Zoo and Wildlife Medicine*, 25, 360–366.

Rothschild, B.M., Hershkovitz, I., and Rothschild, C. (1995). Origin of yaws in Pleistocene East Africa: *Homo erectus. Nature*, 378, 343–344.

Rothschild, B.M., Molnar, R.E., and Sebes, J.I. (1997a). Mycobacteriosis in the Pliocene. *Annals of Internal Medicine*, 127, 168–169.

Rothschild, B.M., Tanke, D., and Carpenter, K. (1997b). Tyrannosaurs suffered from gout. *Nature*, 387, 357.

Rothschild, B.M., Tanke, D., and Carpenter, K. (1997c). Spheroid erosions in tyrannosaurids: Mesozoic gout. *Journal of Vertebrate Paleontology*, 17, 72A.

Rothschild, B.M., Sebes, J.I., and Rothschild, C. (1998a). Antiquity of arthritis: spondyloarthropathy identified in the Paleocene of North America. *Clinical and Experimental Rheumatology*, 16, 573–575.

Rothschild, B.M., Tanke, D., Hershkovitz, I., and Schultz, M. (1998b). Mesozoic neoplasia: origins of haemangioma in the Jurassic age. *The Lancet*, 351, 1862.

Rothschild, B.M., Motani, R., and Wahl, Jr., W. (1999a). Repeated diving was not for all ichthyosaurs. *Journal of Vertebrate Paleontology*, 19, 71A.

Rothschild, B.M., Witzke, B.J., and Hershkovitz, I. (1999b). Metastatic cancer in the Jurassic. *The Lancet*, 351, 398.

Rothschild, B.M., Calderon, F.L., Coppa, A., and Rothschild, C. (2000). First European exposure to syphilis: the Dominican Republic at the time of Columbian contact. *Clinical Infectious Diseases*, 31, 936–941.

Rothschild, B.M., Martin, L.D., Lev, G., Bercovier, H., Bar-gal, G.K., Greenblatt, C., Donoghue, H., Spigelman, M., and Brittain, D. (2001a). *Mycobacterium tuberculosis* complex DNA from an extinct Bison dated 17,000 years before the present. *Clinical Infectious Diseases*, 33, 305–311.

Rothschild, B., Tanke, D.H., and Ford, T.L. (2001b). Theropod stress fractures and tendon avulsions as a clue to activity. In *Mesozoic Vertebrate Life* (pp. 331–336), edited by D.H. Tanke and K. Carpenter. Indiana University Press, Bloomington.

Rothschild, B.M., Tanke, D.H., Helbling II, M., and Martin, L.P. (2003). Epidemiologic study of tumors in dinosaurs. *Naturwissenschaften*, 90(11), 495–500.

Rothschild, B.M., Martin, L.D., and Schulp, A.S. (2005). Sharks eating mosasaurs, dead or alive? *Netherlands Journal of Geosciences/Geologie en Mijnbouw*, 84, 335–340.

Rowett, C.L. (1975a). Stratigraphic distribution of Permian corals in Alaska. *Drevnie Cnidaria*, 2, 105–112.

Rowett, C.L. (1975b). Stratigraphic distribution of Permian corals in Alaska. *U.S. Geological Survey Professional Paper*, 823D, 59–75.

Rowley, M.J. and Rich, P.V. (1986). Immunoreactive collagen in avian and mammalian fossils. *Naturwissenschaften*, 73, 620–622.

Rozefelds, A.C. (1990). A mid-Tertiary rainforest flora from Capella, central Queensland. In *Proceedings of the Third International Organization of Palaeobotany Conference, Melbourne, 1988*, pp. 123–136.

Rozefelds, A.C. and de Baar, M. (1991). Silicified Kalotermitidae (Isoptera) frass in conifer wood from a mid-Tertiary rainforest in central Queensland, Australia. *Lethaia*, 24, 439–442.

Rozman, K.S. (1999). Silurian brachiopod communities of Mongolia. In *Paleocommunities: A Case Study from the Silurian and Lower Devonian* (pp. 164–176), edited by A.J. Boucot and J.D. Lawson. Cambridge University Press, Cambridge, U.K.

Ruben, J.A. (1990). Evidence of convergent behavioral patterns in male crocodilians and phytosaurs. In *Evolutionary Paleobiology of Behavior and Coevolution* (pp. 427–428), edited by A.J. Boucot. Elsevier, Amsterdam.

Ruben, J.A., Jones, T.D., and Geist, N.R. (2003). Respiratory and reproductive paleophysiology of dinosaurs and early birds. *Physiological and Biochemical Zoology*, 76, 141–164.

Rudkin, D.M. (1979). Healed injuries in *Ogygopsis klotzi* (Trilobita) from the Middle Cambrian of British Columbia. *Royal Ontario Museum, Life Sciences Occasional Paper*, 32, 1–8.

Rudkin, D.M. (1985). Exoskeletal abnormalities in four trilobites. *Canadian Journal of Earth Sciences*, 22, 479–483.

Ruedemann, R. (1921). On color bands in Orthoceras. *New York State Museum Bulletin*, 227–228, 79–88.

Ruffer, M.A. (1910). Note on the presence of *Bilharzia haematobium* in Egyptican mummies of the Twentieth Dynasty (1250–1000 BC). *British Medical Journal*, 1, 16.

Ruiz, G.M. (1991). Consequences of parasitism to marine invertebrates: host evolution? *American Zoologist*, 31, 831–838.

Ruiz, G.M. and Lindberg, D.R. (1989). A fossil record for trematodes: extent and potential uses. *Lethaia*, 22, 431–438.

Rundle, A.J. and Cooper, J. (1970). Occurrence of a fossil insect larva from the London Clay of Herne Bay, Kent. *Proceedings of the Geological Association*, 82, 293–296.

Runnegar, B. (1983). A *Diprotodon* ulna chewed by the marsupial lion *Thylacoleao carnifex*. *Alcheringa*, 7, 23–25.

Runnegar, B. (1987). Class Monoplacophora. In *Fossil Invertebrates* (pp. 297–304), edited by R.S. Boardman, A.H. Cheetham, and A.J. Rowell. Blackwell Scientific, Oxford.

Russell, D.A. (1967). Systematics and morphology of American mosasaurs. *Peabody Museum of Natural History, Yale University, Bulletin*, 23, 1–241.

Russell, M.P., Huelsenbeck, J.P., and Lindberg, D.R. (1992). Experimental taphonomy of embryo preservation in a Cenozoic brooding bivalve. *Lethaia*, 25, 353–359.

Russell, R.D. (1929). Fossil pearls from the Chico formation of Shasta County, California. *American Journal of Science*, 18, 416–428.

Rust, J. (1997). Evolution, Systematik, Paläoökologie und stratigraphischer Nutzen Neogener Süss- und Brackwasser-Gastropoden im Nord-Ägäis-Raum. *Palaeontographica, Abteilung A*, 243, 37–180.

Rust, J. (2000). Fossil record of mass moth migration. *Nature*, 405, 530–531.

Rust, J., Stumpner, A., and Gottwald, J. (1999). Singing and hearing in a Tertiary bushcricket. *Nature*, 399, 650.

Rutot, M.A. (1879). Nouvelles découvertes faites dans le Tongrien inférieur du Limbourg. *Annales de la Société Malacologique de Belgique*, 14, 77–78.

Ryan, K.J. (1990). *Corynebacterium* and other aerobic and facultative gram-positive rods. In *Medical Microbiology* (pp. 313–323), edited by J.C. Sherris. Elsevier, New York.

Rybczynski, N. (2008). Woodcutting behavior in beavers (Castoridae, Rodentia): estimating ecological performance in a modern and a fossil taxon. *Paleobiology*, 34, 389–402.

Rybczinski, N. and Reicz, R.R. (2001). Earliest evidence for efficient oral processing in a terrestrial herbivore. *Nature*, 411, 684–686.

Rydell, J., Hammarlund, J., and Seilacher, A. (2001). Trace fossil associations in the Swedish Mickwitzia sandstone (Lower Cambrian): did trilobites really hunt for worms? *GFF*, 123, 247–250.

Sachs, H.M. and Hasson, P.F. (1979). Comparison of species vs. character description for very high resolution biostratigraphy using cannartid radiolarians. *Journal of Paleontology*, 53, 1112–1120.

Sadler, J.P. (1990). Records of ectoparasites on humans and sheep from Viking Age deposits in the former Western Settlement on Greenland. *Journal of Medical Entomology*, 27, 628–631.

Sahni, A. and Khosla, A. (1994). Paleobiological, taphonomical and paleoenvironmental aspects of Indian Cretaceous sauropod nesting sites. *Gaia*, 10, 215–223.

Saint-Seine, P. de (1949). Les poissons des calcaires lithographiques de Cerin (Ain). *Nouvelles Archives du Muséum d'Histoire Naturelle de Lyon*, II, 1–357

Saint-Seine, R. de (1950). Lésions et regeneration chez le *Micraster*. *Bulletin de la Société Géologique de France, Série 5*, 20, 309–315.

Sainte-Seine, R. de (1951). Un Cirripède acrothoracique du Crétacé; *Rogersella lecointrei*, n. g., n. sp. *Comptes Rendus de l'Academie des Sciences, Paris*, 223, 1051–1053.

Saint-Seine, R. de (1954). Existence de Cirripèdes acrothoraciques des Lias: *Zapfella pattei* nov. gen., nov. sp. *Bulletin de la Société Géologique de France, Série 6*, 4, 447–451.

Sainte-Seine, R. de (1955). Les Cirripèdes Acrothoraciques echinicoles. *Bulletin de la Société Géologique de France, Série 6*, 5, 299–303.

Saint-Seine, R. de (1959). Sur un cas de commensalisme d'une annélide et d'un *Clypeaster* Miocène. *Bulletin de la Société Géologique de France, Série 6*, 8, 581–584.

Sala Burgos, N., Cuevas González, J., and López Martínez, N. (2007). Estudio paleopatológico de una hemimandíbula de *Tethytragus* (Artiodactyla, Mammalia) del Mioceneo Medio de Somosaguas (Pozuelo de Alarcón, Madrid). *Coloquios de Paleontología*, 57, 7–14.

Samadi, S., David, P., and Jarne, P. (2000). Variation of shell shape in the clonal snail *Melanoides tuberculata* and its consequences for the interpretation of fossil series. *Evolution*, 54, 492–502.

Sampson, F.B. and McLean, J. (1965). A note on the occurrence of domatia on the underside of leaves in New Zealand plants. *New Zealand Journal of Botany*, 3, 104–112.

Sander, P.M. (1988). A fossil reptile embryo from the Middle Triassic of the Alps. *Science*, 239, 780–783.

Sander, P.M., Peitz, C., Jackson, F., and Chiappe, L. (2008). Upper Cretaceous titanosaur nesting sites and their implications for sauropod dinosaur reproductive biology. *Palaeontographica A*, 284, 69–107.

Sanders, F., Manley, K., and Carpenter, K. (2001). Gastroliths from the Lower Cretaceous sauropod *Cedarosaurus weiskopfae*. In *Mesozoic Vertebrate Life* (pp. 166–180), edited by D.H. Tanke and K. Carpenter. Indiana University Press, Bloomington.

Sanderson, S.L. and Wassersug, R. (1993). Convergent and alternative designs for vertebrate suspension feeding. In *The Skull*. Vol. 3. *Functional and Evolutionary Mechanisms* (pp. 37–112), edited by J. Hanken and B.K. Hall. University of Chicago Press, Chicago, IL.

Sandison, A.T. (1967). Sexual behavior in ancient societies. In *Diseases in Antiquity* (pp. 734–755), edited by D. Brothwell and A.T. Sandison. Charles C Thomas, Springfield, IL.

Sandison, A.T. (1980). Diseases in ancient Egypt. In *Mummies, Disease, and Ancient Cultures* (pp. 29–44), edited by A. Cockburn and E. Cockburn. Cambridge University Press, Cambridge, U.K.

Sando, W.J. (1984). Significance of epibionts on horn corals from the Chainman Shale (Upper Mississippian) of Utah. *Journal of Paleontology*, 58, 185–196.

Sandy, M.R. (1996). Oldest record of peduncular attachment of brachiopods to crinoid stems, Upper Ordovician, Ohio, USA (Brachiopoda; Atrypida: Echinodermata; Crinoidea). *Journal of Paleontology*, 70, 532–534.

Sandy, M.R. and Campbell, K.A. (1994). New rhynchonellid brachiopod genus from Tithonian (Upper Jurassic) cold seep deposits of California and its paleoenvironmental setting. *Journal of Paleontology*, 68, 1243–1252.

Sanfilippo, A. and Riedel, W.R. (1992). The origin and evolution of Pterocorythidae (Radiolaria): a Cenozoic phylogenetic study. *Micropaleontology*, 38, 1–36.

Santiago-Blay, J.A., Anderson, S.R., and Buckley, R.T. (2005). Possible implications of two new Angiosperm flowers from Burmese amber (Lower Cretaceous) for well-established and diversified insect–plant associations. *Entomological News*, 116, 341–346.

Santos, A. and Mayoral, E. (2008). Colonization by barnacles on fossil *Clypeaster*: an exceptional example of larval settlement. *Lethaia*, 41, 317–332.

Santos, A., Mayoral, E., Muñiz, F., Bajo, I., and Adriaensens, O. (2003). Bioerosion en erizoz Irregulares (Clypeasteroidea) del Mioceno Superior en el sector suroccidental de La Cuenca del Guadalquivir (Provincia de Sevilla). *Revista Española de Paleontología*, 18, 131–141.

Santos, A., Mayoral, E., and Muñiz, F. (2005). Bioerosion scars of acorn barnacles from the south-western Iberian Peninsula, Upper Neogene. *Rivista Italiana di Paleontologia e Stratigrafia*, 111, 181–189.

Sanz, J.L., Chiappe, L.M., Fernandez-Jalvo, Y., Ortega, F., Sanchez-Chillon, B., Poyato-Ariza, F.J., and Perez-Moreno, B.P. (2001). An Early Cretaceous pellet. *Nature*, 409, 998–1000.

Sarycheva, T.G. (1949). O prizhyznennykh povrezhdeniakh rakovina kammenougolnykh produktid. *Transactions of the Paleontological Institute, Academy of Science, USSR*, 20, 280–292.

Sarzetti, L.C., Labandeira, C.C., and Genise, J.F. (2008). A leafcutter bee trace fossil from the Middle Eocene of Patagonia, Argentina, and a review of megachilid (Hymenoptera) ichnology. *Palaeontology*, 51, 933–941.

Sass, D.B. and Condrate, R.A. (1985). Destruction of a Late Devonian ophiuroid assemblage: a victim of changing ecology at the Catskill delta front. In *The Catskill Delta* (pp. 237–246), edited by D.L. Woodrow and W.D. Sevon, Special Paper 201. Geological Society of America, Boulder, CO.

Sato, J. (1962). Miocene fishes from the western area of Shizukuishi Basin, Iwate Prefecture, northeastern Japan. *Earth Science, Journal of the Association for Geological Collaboration of Japan*, 59, 1–29.

Sato, T. and Tanabe, K. (1998). Cretaceous plesiosaurs ate ammonites. *Nature*, 394, 629–630.

Sato, T., Hasegawa, Y., and Manare, M. (2006). A new elasmosaur from the Upper Cretaceous of Fukushima, Japan. *Palaeontology*, 49, 467–484.

Sauer, E.G.F. (1972). Ratite eggshells and phylogenetic questions. *Bonner Zoologische Beiträge*, 22, 3–48.

Saul, L.R. and Stadum, C.J. (2005). Fossil Argonauts (Mollusca: Cephalopoda: Octopodida) from Late Miocene siltstones of the Los Angeles Basin, California. *Journal of Paleontology*, 79, 520–531.

Saul, L.R., Squires, R.L., and Goedert, J.L. (1996). A new genus of cryptic lucinid? bivalve from Eocene cold seeps and turbidite-influenced mudstone, western Washington. *Journal of Paleontology*, 70, 788–794.

Savazzi, E. (1991). Constructional morphology of strombid gastropods. *Lethaia*, 24, 311–331.

Savazzi, E. (1994). Functional morphology of boring and burrowing invertebrates. In *The Palaeobiology of Trace Fossils* (pp. 43–82), edited by S.K. Donovan. Cambridge University Press, Cambridge, U.K.

Savazzi, E. (1995). Parasite-induced teratologies in the Pliocene bivalve *Isognomon maxillatus*. *Palaeogeography, Palaeoclimatology, Palaeoecology*, 116, 131–139.

Savazzi, E. (1998). Constructional morphology of the bivalve *Pedum*. In *Bivalves: An Eon of Evolution* (pp. 413–421), edited by P.A. Johnston and J.W. Haggart. University of Calgary Press, Calgary, Alberta, Canada.

Savazzi, E., Ed. (1999a). *Functional Morphology of the Invertebrate Skeleton*. John Wiley & Sons, New York, 706 pp.

Savazzi, E. (1999b). Three-dimensional accretional colour patterns. In *Functional Morphology of the Invertebrate Skeleton* (pp. 57–63), edited by E. Savazzi. John Wiley & Sons, New York.

Savazzi, E. (1999c). Soft-bottom dwellers and the Leaning Tower of Pisa: adaptive exploitation of unstable life positions. In *Functional Morphology of the Invertebrate Skeleton* (pp. 123–128), edited by E. Savazzi. John Wiley & Sons, New York.

Savazzi, E. (1999d). Cemented and embedded gastropods. In *Functional Morphology of the Invertebrate Skeleton* (pp. 183–195), edited by E. Savazzi. John Wiley & Sons, New York.

Savazzi, E. (1999e). Boring, nestling and tube-dwelling bivalves. In *Functional Morphology of the Invertebrate Skeleton* (pp. 205–237), edited by E. Savazzi. John Wiley & Sons, New York.

Savrda, C.E. (1990). *Teredolites*, wood substrates and sea-level dynamics. *Geology*, 19, 905–908.

Savrda, C.E. and King, Jr., D.T. (1993). Log-ground and *Teredolites* lagerstatte in a transgressive sequence, Upper Cretaceous (Lower Campanian) Moorville Chalk, central Alabama. *Ichnos*, 3, 69–77.

Savrda, C.E. and Smith, M.W. (1996). Behavioral implications of branching and tube-lining in *Teredolites*. *Ichnos*, 4, 191–198.

Savrda, C.E., Ozalas, K., Demko, T.H., Huchison, R.A., and Schiewe, T.D. (1993). Log-ground and the ichnofossil *Teredolites* in the transgressive deposits of the Clayton Formation (Lower Paleocene), western Alabama. *Palaios*, 8, 311–324.

Savrda, C.E., Counts, J., McCormick, O., and Urash, R. (2005). Log-grounds and *Teredolites* in transgressive deposits, Eocene Tallahatta Formation (Southern Alabama, USA). *Ichnos*, 12, 47–57.

Sawyer, G.T. and Erickson, B.R. (1987). Injury and diseases in fossil animals. *Field Museum of Natural History Bulletin*, 58, 20–25.

Sawyer, G.T. and Erickson, B.R. (1998). Paleopathology of the Paleocene crocodile *Leidyosuchus* (=*Borealosuchus*) *formidabilis*. *Science Museum of Minnesota Paleontology Monograph*, 4, 1–38.

Schaarschmidt, F. (1988). Der Wald, fossile Pflanzen als Zeugen eines warmen Klimas. In *Messel: Ein Schaufenster in die Geschichte der Erde und des Lebens* (pp. 27–53), edited by S. Schaal and W. Ziegler. Waldemar Kramer, Frankfurt-am-Main.

Schaarschmidt, F. (1992). The vegetation: fossil plants as witnesses of a warm climate. In *Messel: An Insight into the History of Life and of the Earth* (pp. 29–52), edited by S. Schaal and W. Ziegler. Clarendon Press, London.

Schaarschmidt, F. and Wilde, V. (1986). Palmenblüten und –blätter aus dem Eozän von Messel: Fossilfundstelle Messel Nr. 51. *Courier Forschungsinstitut Senckenberg*, 86, 177–202.

Schaub, H. (1963). Uber einige Entwicklungseihen von *Nummulites* und *Assilina* und ihre stratigraphisches Bedeutung. In *Evolutionary Trends in Foraminifera* (pp. 282–297), edited by G.H.R. von Koenigswald, J.D. Emets, W.L. Buning, and C.W. Wager. Elsevier, Amsterdam.

Schaumberg, G. (1979). Neue Nachweise von Bryozoan und Brachiopoden als Nahrung des permischen Holocephalen *Janassa bituminosa* (Schlotheim). *Philippia*, 4(1), 3–11.

Scheibling, R.E. (1994). Interactions between lobsters, sea urchins and kelp in Nova Scotia, Canada. In *Echinoderms Through Time* (pp. 865–870), edited by B. David, A. Guille, J.-P. Feral, and M. Roux. Balkema, Rotterdam.

Scheltema, R.S. (1978). On the relationship between dispersal of pelagic veliger larvae and the evolution of marine Prosobranch gastropods. In *Marine Organisms: Genetics, Ecology, and Evolution* (pp. 303–322), edited by J.C. Beardmore and B. Battaglia. Plenum Press, New York.

Scheltema, R.S. (1979). Dispersal of pelagic larvae and the zoogeography of Tertiary marine benthic gastropods. In *Historical Biogeography, Plate Tectonics, and the Changing Environment* (pp. 391–397), edited by D.M. Gray and A.J. Boucot. Oregon State University Press, Corvallis.

Scheven, J. (2004). *Bernstein-Einschlüsse: Eine untergegangene welt bezeugt die schöpfung. Erinnerungen an die welt vor der sintflut.* Kuratorium Lebendige Vorwelt e.V. Hofheim a.T., Germany, 160 pp.

Schierning, V.-H. (2005). Possible evidence of a venom apparatus in a Middle Jurassic sphenodontian from the Huizachal red beds of Tamaulipas, Mexico. *Journal of Vertebrate Paleontology*, 25, 646–654.

Schindel, D.E., Vermeij, G.J., and Zipser, E. (1982). Frequencies of repaired shell fractures among the Pennsylvanian gastropods of north-central Texas. *Journal of Paleontology*, 56, 729–740.

Schlaudt, C.M. and Young, K. (1960). Acrothoracic barnacles from the Texas Permian and Cretaceous. *Journal of Paleontology*, 34, 199–120.

Schlee, D. and Glöckner, W. (1978). Bernstein. *Stuttgarter Beiträge zur Naturkunde Serie C*, 8, 1–72.

Schlee, E.I. (1980). *Bernstein-Raritäten: Farben, Strukturen, Fossilen, Handwerk.* Staatliches Museum für Naturkunde, Stuttgart, 88 pp.

Schlinger, E.I. (1987). The biology of Acroceridae (Diptera): true endoparasitoids of spiders. In *Ecophysiology of Spiders* (pp. 319–327), edited by W. Nentwig. Springer-Verlag, Heidelberg.

Schlögl, J., Michalik, J., Zágoršek, K., and Atrops, F. (2008). Early Tithonian serpulid-dominated cavity-dwelling fauna, and the recruitment pattern of the serpulid larvae. *Journal of Paleontology*, 82, 351–361.

Schluter, D. (2000). *The Ecology of Adaptive Radiation.* Oxford University Press, London, 288 pp.

Schlüter, T. (1978). Zur Systematik und Palökologie harzkonservierter Arthropoda einer Taphozönose aus dem Cenomanium von NW-Frankreich. *Berliner Geowissenschaftliche Abhandlungen, Reihe A, Geologie und Paläontologie*, 9, 1–150.

Schmid, D.U. and Leinfelder, R.R. (1996). The Jurassic *Lithocodium aggregatum–Troglotella incrustans* foraminiferal consortium. *Palaeontology*, 39, 21–52.

Schmid, R. and Schmid, M.J. (1977). Fossil history of the orchidaceae. *Orchid Biology*, 1, 25–45.

Schmidt, A.R. and Schäfer, U. (2005). *Leptotrichites resinatus* new genus and species: a fossil sheathed bacterium in Alpine Cretaceous amber. *Journal of Paleontology*, 79, 175–184.

Schmidt, A.R., von Eynatten, H., and Wagreich, M. (2001). The Mesozoic amber of Schliersee (southern Germany) is Cretaceous in age. *Cretaceous Research*, 22, 423–428.

Schmidt, A.R., Dörfelt, H., and Perrichot, V. (2008). *Palaeoanellus dimorphus* gen. et sp. nov. (Deuteromycotina): a Cretaceous predatory fungus. *American Journal of Botany*, 95, 1328–1334.

Schmidt, G.D. and Roberts, L.S. (1989). *Foundations of Parasitology*, 4th ed. Mosby, St. Louis, MO, 750 pp.

Schmidt, G.D., Duszynski, D.W., and Martin, P.S. (1992). Parasites of the extinct Shasta ground sloth *Nothrotheriops shastensis*, in Rampart Cave, Arizona. *Journal of Parasitology*, 78, 811–816.

Schmidt, H. (1951). Aufgeprägte Form bei Tieren. *Natur und Volk*, 81, 84–89.

Schmidt-Kittler, N. (1987). Comments of the editor. *International Symposium on Mammalian Biostratigraphy and Paleoecology of the European Paleogene, Münchner Geowissenschaftliche Abhandlungen*, 10, 15–20.

Schmidt-Kittler, N. and Vianey-Liaud, M. (1979). Evolution des Aplodontidae Oligocènes Européens. *Palaeovertebrata*, 9, 34–82.

Schmidt-Kittler, N. and Vianey-Liaud, M. (1987). Morphometric analysis and evolution of the dental pattern of the genus *Issiodoromys* (Theridomyidae, Rodentia) of the European Oligocene as a key to its evolution. *Proceedings of the Koninklijke Nederlandse Akademie van Wetenschappen, Series B*, 90(3), 281–306.

Schneider, C.L. (2003). Hitchhiking on Pennsylvanian echinoids: epibionts on *Archaeocidaris*. *Palaios*, 18, 435–444.

Schneider, C.L., Sprinkle, J., and Ryder, D. (2005). Pennsylvanian (Late Carboniferous) echinoids from the Winchell Formation, north-central Texas, USA. *Journal of Paleontology*, 79, 745–762.

Schneider, J. and Reichel, W. (1989). Chondrichthyer-Eikapseln aus dem Rotliegenden (Unterperm) Mitteleuropas: Schlussfolgerungen zur Paläobiologie paläozoischer Süsswasserhaie. *Freiberger Forschungshefte C*, 436, 58–69.

Schneider, J.W. and Werneburg, R. (2006). Insect biostratigraphy of the Euramerican Continental Late Pennsylvanian and Early Permian. In *Non-Marine Permian Biostratigraphy and Biochronology* (pp. 325–336), edited by S.G. Lucas, G. Cassinis, and J.W. Schneider, Geological Society of London Special Publication 265. Geological Society, London.

Schneider, S., Hochleitner, R., and Janssen, R. (2008). A new bivalved gastropod, *Candinia lakoniae* n. sp. (Sacoglossa: Juliidae) from the Upper Pliocene of Greece (Glykovrysi, SE Peloponnese and Lardoa, Rhodes), with a short survey of the geological range and geographic distribution of the family. *Neues Jahrbuch für Geologie und Paläontologie–Abhandlungen*, 247, 79–91.

Schneider, V.P. and Fierstine, A.L. (2004). Fossil tuna vertebrae punctured by istiophorid billfishes. *Journal of Vertebrate Paleontology*, 24, 253–255.

Schönberg, C.H.L. (2007). A history of sponge erosion: from past myths and hypotheses to recent approaches. In *Current Developments in Bioerosion* (pp. 165–202), edited by M. Wisshak and L. Tapanila. Springer-Verlag, Heidelberg.

Schönberg, C.H.L. and Shields, G. (2007). Micro-computed tomography for studies on *Entobis*: transparent substrate versus modern technology. In *Current Developments in Bioerosion* (pp. 147–164), edited by M. Wisshak and L. Tapanila. Springer-Verlag, Heidelberg.

Schormann, J. (1987). Bissspuren an Seeigeln. *Arbeitskreis Paläontologie*, 15(4), 73–75.

Schram, F.R. (1975). A Pennsylvanian lepadomorph barnacle from the Mazon Creek area, Illinois. *Journal of Paleontology*, 49, 928–930.

Schram, F.R. (1981). Paleozoic crustacean communities. *Journal of Paleontology*, 55, 126–137.

Schram, F.R. and Newman, W.A. (1980). *Verruca withersi* n. sp. (Crustacea: Cirripedia) from the Middle of the Cretaceous of Colombia. *Journal of Paleontology*, 54, 229–233.

Schrank, E. (1969). Odontopleuriden (Trilobita) aus silurischen Geschieben. *Berichte der Deutschen Geologischen Gesellschaft für Geologische Wissenschaften, Reihe A, Geologie und Paläontologie*, 14, 705–726.

Schreiber, B.C. and Hsü, K.J. (1980). Evaporites. In *Developments in Petroleum Geology*, Vol. 2 (pp. 87–138), edited by G.D. Hobson. Applied Science Publishers, London.

Schuh, R.T. and Slater, J.A. (1995). *True Bugs of the Word (Hemiptera: Heteroptera)*. Cornell University Press, Ithaca, NY, 336 pp.

Schultz, A.H. (1967). Notes on diseases and healed fractures of wild apes. In *Diseases in Antiquity* (pp. 47–55), edited by D. Brothwell and A.T. Sandison. Charles C Thomas, Springfield, IL.

Schumacher, B.A. and Everhart, M.J. (2005). A stratigraphic and taxonomic review of plesiosaurs from the old "Fort" Benton Group of central Kansas: a new assessment of old records. *Paludicola*, 5, 33–54.

Schumann, D. (1967). Die Lebensweise von *Mucrospirifer* Grabau, 1931 (Brachiopoda). *Palaeogeography, Palaeoclimatology, Palaeoecology*, 3, 381–392.

Schwammer, H.M. (1989). Bald-sea-urchin disease: record of incidence in irregular echinoids, *Spatangus purpureus*, from the SW coast of Krk (Croatia–Jugoslavia). *Zoologischer Anzeiger*, 223, 100–106.

Schwanke, C. and Kellner, A.W. (1999). Presence of insect? Borings in synapsid bones from the terrestrial Triassic Santa Maria Formation, southern Brazil. *Journal of Vertebrate Paleontology*, 19, 74.

Schwarzhans, W. (2007). Otoliths from casts from the Eocene Lillebælt Clay Formation of Trelde Næs near Fredericia (Denmark) with remarks on the diet of stomatopods. *Neues Jahrbuch für Geologie und Paläontologie–Abhandlungen*, 69–81.

Schweitzer, C.E. (2003). Utility of proxy characters for classification of fossils: an example from the fossil Xanthoidea (Crustacea: Decapoda: Brachyura). *Journal of Paleontology*, 77, 1107–1128.

Schweitzer, C.E. (2005a). The genus *Xanthilites* Bell, 1858 and a new xanthoid family (Crustacea: Decapoda: Brachyura: Xanthoidea): new hypotheses on the origin of the Xanthoidea MacLeay, 1938. *Journal of Paleontology*, 79, 277–295.

Schweitzer, C.E. (2005b). The Trapeziidae and Domeciidae (Decapoda: Brachura: Xanthoidea) in the fossil record and a new Eocene genus from Baja California, Sur, Mexico. *Journal of Crustacean Biology*, 25, 625–636.

Schweitzer, C.E. and Feldmann, R.M. (2000). *Callichirus? Symmetricus* (Decapoda: Thalassinoidea) and associated burrows, Eocene, Antarctica. *Antarctic Research Series*, 76, 335–347.

Schweitzer, C.E. and Feldmann, R.M. (2008). New Eocene hydrocarbon seep decapod crustacean (Anomura: Galatheidae: Shinkaiinae) and its paleobiology. *Journal of Paleontology*, 82, 1021–1029.

Schweitzer, C.E., Feldmann, R.M., Fam, J., Hessin, W.A., Hetrick, S.W., Nyborg, T.G., and Ross, R.L.M. (2003). *Cretaceous and Eocene Decapod Crustaceans from Southern Vancouver Island, British Columbia, Canada*. NRC Research Press, Ottawa, 66 pp.

Schweitzer, M.H., Jackson, F.D., Chiappe, L.M., Schmitt, J.G., Calvo, J.O., and Rubilar, D.E. (2002). Late Cretaceous avian eggs with embryos from Argentina. *Journal of Vertebrate Paleontology*, 22, 191–195.

Schweitzer Hopkins, C. and Feldmann, R.M. (1997). Sexual dimorphism in fossil and extant species of *Callianopsis* de Saint Laurent. *Journal of Crustacean Biology*, 17, 236–252.

Schwimmer, D.R., Stewart, J.D., and Williams, G.D. (1997). Scavenging by sharks of the genus *Squalicorax* in the Late Cretaceous of North America. *Palaios*, 12, 71–83.

Scott, A.C. (1992). Trace fossils of plant-arthropod interactions. In *Trace Fossils, Short Courses in Paleontology No. 5* (pp. 197–223), edited by C.G. Maples and R.R. West. Paleontological Society, University of Tennessee, Knoxville.

Scott, A.C. and Taylor, T.N. (1983). Plant/animal interactions during the Upper Carboniferous. *Botanical Review*, 49, 259–307.

Scott, A.C., Chaloner, W.G., and Paterson, S. (1985). Evidence of pteridophyte–arthropod interactions in the fossil record. *Proceedings of the Royal Society of Edinburgh*, 86B, 133–140.

Scott, A.C., Stephenson, J., and Chaloner, W.G. (1992). Interaction and correlation of plants and arthropods during the Palaeozoic and Mesozoic. *Proceedings of the Royal Society London B*, 335, 129–165.

Scott, A.C., Stephenson, J., and Collinson, M.E. (1994). The fossil record of leaves with galls. In *Plant Galls: Organisms, Interactions, Populations* (pp. 447–470), edited by M.A.J. Williams, Systematics Association Publ. No. 49. Clarendon Press, London.

Scott, A.C., Anderson, J.M., and Anderson, H.M. (2004). Evidence of plant–insect interactions in the Upper Triassic Molteno Formation of South Africa. *Journal of the Geological Society, London*, 161, 401–410.

Scott, S.D., Chase, R.L., Hannington, M.D., Michael, P.J., McConachy, T.F., and Shea, G.T. (1990). Sulphide deposits, tectonics and petrogenesis of Southern Explorer Ridge, Northeast Pacific Ocean. In *Ophiolites: Oceanic Crustal Analogues, Symposium "Troodos, 1987," Nicosia* (pp. 719–733), edited by J. Malpas, E.M. Moores, A. Panayiotou, and C. Xenophontos. Geological Survey Department, Ministry of Agriculture and Natural Resources, Nicosia, Cyprus.

Scott, W.B. and Jepsen, G.L. (1936). The mammalian fauna of the White River Oligocene. Part 1. Insectivora and Carnivora, *Transactions of the American Philosophical Society*, 28, 1–153.

Scrutton, C.T. (1975). Hydroid-serpulid symbiosis in the Mesozoic and Tertiary. *Palaeontology*, 18, 255–274.

Scrutton, C.T. (1998). The Palaeozoic corals. II. Structure, variation and palaeoecology. *Proceedings of the Yorkshire Geological Society*, 52, 1–57.

Seeley, H. (1861). Notes on Cambridge palaeontology. *The Annals and Magazine of Natural History, Third Series*, 7, 116–124.

Seguenza, G. (1876). Ricerche paleontologische intorno ai Cirripedi Terziarii della Provincia di Messina. II. Terza famiglia Lepadidi Darwin. *Atti della Accademia Pontaniana*, 10, 369–481.

Seibertz, E. and Spaeth, C. (2005). Cretaceous belemnites of Mexico II. Mexican *Mesohiblites* and paleobiogeographic implications on the distribution of the genus (Albian, Lower Cretaceous). *Neues Jahrbuch für Geologie und Paläontologie–Abhandlungen*, 236, 95–113.

Seiffert, E.R., Simons, E.L., and Attia, Y. (2003). Fossil evidence for an ancient divergence of lorises and galagos. *Nature*, 422, 421–424.

Seilacher, A. (1968). Swimming habits of belemnites, recorded by boring barnacles. *Palaeogeography, Palaeoclimatology, Palaeoecology*, 4, 279–285.

Seilacher, A. (1969). Paleoecology of boring barnacles. *American Zoologist*, 9, 705–719.

Seilacher, A. (1998). Mosasaurs, limpets or diagenesis: how *Placenticeras* shells got punctured. *Mitteilungen des Museums für Naturkunde in Berlin—Geowissenschaftliche Reihe*, 1, 93–102.

Seilacher, A. (2005). Whale barnacles: exaptational access to a forbidden paradise. *Paleobiology*, 31(2), 27–35.

Seilacher, A. (2007). *Trace Fossil Analysis*. Springer-Verlag, Heidelberg, 226 pp.

Seilacher, A., Drozdzewski, G., and Houde, R. (1968). Form and function of the stem in a pseudoplanktonic crinoid (*Seirocrinus*). *Palaeontology*, 11, 275–287.

Selden, P.A. (1990). Lower Cretaceous spiders from the Sierra de Montsech, north-east Spain. *Palaeontology*, 33, 257–285.

Selden, P.A. and Nudds, J. (2004). *Evolution of Fossil Ecosystems*. University of Chicago Press, Chicago, IL, 160 pp.

Selden, P.A., Shear, W.A., and Bonamo, P.M. (1991). A spider and other arachnids from the Devonian of New York, and reinterpretations of Devonian Araneae. *Palaeontology*, 34, 241–281.

Sellwood, B.W. (1971). A *Thalassinoides* burrow containing the crustacean *Glyphaea udressieri* (Meyer) from the Bathonian of Oxfordshire. *Palaeontology*, 14, 589–591.

Selmeier, A. (1984). Fossile Bohrgange von *Anobium* sp. in einem jungtertiären Lorbeerholz aus Egweil (Südliche Frankenalb). *Archaeopteryx*, 2, 13–29.

Semaw, S., Renne, P., Harris, J.W.K., Feibel, C.S., Bernor, R.L., Fesseha, N., and Mowbray, K. (1997). 2.5-million-year-old stone tools from Gona, Ethiopia. *Nature*, 385, 333–336.

Semken, Jr., H.A. (1966). Stratigraphy and paleontology of the McPherson *Equus* beds (Sandahl local fauna), McPherson County, Kansas. *Contributions from the Museum of Paleontology, The University of Michigan*, 20 (6), 121–178.

Senter, P. (2005). Function of the stunted forelimbs of *Mononykus olecranus* (Theropoda), a dinosaurian anteater. *Paleobiology*, 31, 373–381.

Sequeira, A.S. and Farrell, B.D. (2001). Evolutionary origins of Gondwanan interactions: how old are *Araucaria* beetle herbivores? *Biological Journal of the Linnean Society*, 74, 459–474.

Sheehan, P.M. and Lespérance, P.J. (1978). Effect of predation on the population dynamics of a Devonian brachiopod. *Journal of Paleontology*, 52, 812–817.

Sheldon, P.R. (1987). Parallel gradualistic evolution of Ordovician trilobites. *Nature*, 330, 561–563.

Shen, S., Fan, B., Zhang, C., and Zhang, X. (1994). A new species of permianellids (Brachiopoda): taxonomic and palaeoecologic significance. *Géobios*, 27, 477–485.

Sherwood-Pike, M. (1990). Fossil evidence for fungus–plant interactions. In *Evolutionary Paleobiology of Behavior and Coevolution* (pp. 118–123), edited by A.J. Boucot. Elsevier, Amsterdam.

Shields, J.D., Stephens, F.J., and Jones, B. (2006). Pathogens, parasites, and other symbionts. In *Lobsters: Biology, Management, Aquaculture, and Fisheries* (pp. 146–204), edited by B. Philips. Wiley-Blackwell, New York.

Shimada, K. (1997). Paleoecological relationships of the Late Cretaceous lamniform shark *Cretoxyrhinus mantelli* (Agassiz). *Journal of Paleontology*, 71, 926–933.

Shimada, K. (2007). Mesozoic origin for megamouth shark (Lamniformes: Megachasmidae). *Journal of Vertebrate Paleontology*, 27, 512–516.

Shimada, K. and Everhart, M.J. (2004). Shark-bitten *Xiphactinus audax* (Teleostei: Ichthyodectiformes) from the Niobrara Chalk (Upper Cretaceous) of Kansas. *The Mosasaur*, 7, 35–39.

Shimada, K. and Hooks III, G.E. (2004). Shark-bitten protostegid turtles from the Upper Cretaceous Mooreville Chalk, Alabama. *Journal of Paleontology*, 78, 205–210.

Shimada, K., Everhart, M.J., and Hooks, G.E. (2002). Ichthyodectid fish and protostegid turtle bitten by the Late Cretaceous lamniform shark, *Cretoxyrhinus mantelli*. *Journal of Vertebrate Paleontology*, 22(3, Suppl.), 106A.

Shine, R. (1985). The evolution of viviparity in reptiles: an ecological analysis. In *Biology of the Reptiles* (pp. 605–694), edited by C. Gans and F. Billett. John Wiley & Sons, New York.

Shirley, J. and Lambert, C.A. (1922). On *Coprosoma baueri* Endl. *Proceedings of the Royal Society of Victoria*, 35, 19–23.

Shone, R.W. (1978). Giant *Cruziana* from the Beaufort Group. *Transactions of the Geological Society of South Africa*, 81, 327–329.

Shpanskaya, A.Y., Maslennikov, V.V., and Little, K.T.S. (1999). Trubki vestimentifer iz rannesiluriiskikh i srednedevonskikh prigidrotermalnikh biot Uralskogo Paleookeana. *Paleontologischeskii Zhurnal*, 3, 12–16.

Shuster, Jr., C.N. (1982). A pictorial review of the natural history and ecology of the horseshoe crab *Limulus polyphemus*, with reference to other Limulidae. In *Physiology and Biology of Horseshoe Crabs: Studies on Normal and Environmentally Stressed Animals* (pp. 1–52), edited by J. Bonaventura et al. Alan E. Liss, New York.

Shuto, T. (1974). Larval ecology of prosobranch gastropods and its bearing on biogeography and paleontology. *Lethaia*, 7, 239–256.

Siber, H.J. (1982). *Green River Fossilien*. Siber & Siber AG, Aathal, Switzerland, 81 pp.

Sieveking, G. de G. and Newcomer, M.H., Eds. (1987). *The Human Uses of Flint and Chert*. Cambridge University Press, Cambridge, U.K., 263 pp.

Sigé, B. (1991). Rhinolophoidea et Vespertilionoidea (Chiroptera) du Chambi (Eocène inférieur de Tunisie). Aspects biostratigraphique, biogéographique et paléoécologique de l'origine des chiroptères modernes. *Neues Jahrbuch für Geologie und Paläontologie–Abhandlungen*, 355–376.

Sigé, B., Habersetzer, J., and Storch, G. (1998). The deciduous dentition and dental replacement in the Eocene bat *Palaeochiropteryx tupaiodon* from Messel: the primitive condition and beginning of specialization of milk teeth among Chiroptera. *Lethaia*, 31, 349–358.

Signor III, P.W. and Brett, C.E. (1984). The mid-Paleozoic precursor to the Mesozoic marine revolution. *Paleobiology*, 10, 229–245.

Simmons, N.B., Seymour, K.L., Habersetzer, J., and Gunnell, G.F. (2008). Primitive Early Eocene bat from Wyoming and the evolution of flight and echolocation. *Nature*, 451, 818–821.

Simões, M.G., Rodrigues, S.C., and Kowalewski, M. (2007). Comparative analysis of drilling frequencies in Recent brachiopod-mollusk associations from the southern Brazilian shelf. *Palaios*, 22, 143–154.

Simon, L., Bousquet, J., Levesque, R.C., and Lalonde, M. (1993). Origin and diversification of endomycorrhizal fungi and coincidence with vascular plants. *Nature*, 363, 67–69.

Simon, R.A., Kelly, E.B., Price, S.P., and Whitham, A.G. (2000). Early Cretaceous giant bivalves from seep-related limestone mounds, Wollaston Forland, Northeast Greenland. *U.S. Geological Society Special Publication*, 177, 227–247.

Simonsen, A.H. and Cuffey, R.J. (1980). Fenestrate, pinnate and ctenostome bryozoans and associated barnacle borings in the Wreford Megacyclothem (Lower Permian) of Kansas, Oklahoma and Nebraska. *University of Kansas Paleontological Contributions*, 101, 11–38.

Simpson, G.G. (1943). Criteria for genera, species, and subspecies in zoology and paleozoology. *Annals of the New York Academy of Sciences*, 44, 145–178.

Simpson, G.G. (1953). *The Major Features of Evolution*. Columbia University Press, New York, 434 pp.

Simpson, M.I. and Middleton, R. (1985). Gross morphology and the mode of life of two species of lobster from the Lower Cretaceous of England: *Meyeria ornata* (Phillips) and *Meyerella magna* (M'Coy). *Transactions of the Royal Society of Edinburgh, Earth Sciences*, 76, 203–215.

Singer, M.C. and Parmesan, C. (1993). Sources of variations in patterns of plant–insect association. *Nature*, 361, 251–253.

Sivak, J.G. (1988). Optics of amphibious eyes in vertebrates. In *Sensory Biology of Aquatic Animals* (pp. 466–485), edited by J. Atema, R.R. Fay, A.N. Popper, and W. Tavolga. Springer-Verlag, Heidelberg.

Siveter, D.J., Briggs, D.E.G., Siveter, D.J., Sutton, M.D., and Fortey, R.A. (2008). Horns, eggs and legs: exceptionally preserved new arthropods from the Herefordshire (Silurian) Lagerstätte. In *Advances in Trilobite Research* (pp. 371–374), edited by I. Rábano, R. Gozalo, and D. Garcia-Bellido. Cuadernos del Museo Geominero, 9, Instituto Geológico y Minero de España, Madrid.

Siveter, D.J., Vannier, J.M.C., and Palmer, D. (1991). Silurian Myodocopes: pioneer pelagic ostracods and the chronology of an ecological shift. *Journal of Micropalaeontology*, 10, 151–173.

Siveter, D.J., Siveter, D.J., Sutton, M.D., and Briggs, D.E.G. (2006). Brood care in a Silurian ostracod. In *Proceedings of the 50th Annual Meeting of the Palaeontological Association, December 18–21, Sheffield, U.K.*

Skovsted, C.B., Brock, G.A., Lindström, A., Peel, J.S., Paterson, J.R., and Fuller, M.K. (2007). Early Cambrian record of failed durophagy and shell repair in an epibenthic mollusk. *Biology Letters*, 3(3), 314–317.

Slovenian Academy of Sciences (1997). Early music. *Science*, 276(5310), 203–205.

Smith, A.B. (1984). *Echinoid Palaeobiology*. Allen & Unwin, London, 190 pp.

Smith, A.B. (1994). *Systematics and the Fossil Record*. Blackwell Scientific, Oxford, 223 pp.

Smith, A.B. (2004). Phylogeny and systematics of holasteroid echinoids and their migration into the deep sea. *Palaeontology*, 47, 123–150.

Smith, A.B. and Bengtson, P. (1991). Cretaceous echinoids from north-eastern Brazil. *Fossils and Strata*, 31, 1–88.

Smith, A.B. and Jeffery, C.H. (2000). Changes in the diversity, taxic composition and life-history patterns of echinoids over the past 145 million years. In *Biotic Response to Global Change: The Last 145 Million Years* (pp. 181–194), edited by S.J. Culver. Cambridge University Press, Cambridge, U.K.

Smith, A.B. and Littlewood, D.T.J. (1994). Paleontological data and molecular phylogenetic analysis. *Paleobiology*, 20, 259–273.

Smith, C.R. and Baco, A.R. (2003). Ecology of whale falls at the deep-sea floor. *Oceanography and Marine Biology: Annual Review*, 41, 311–354.

Smith, C.R., Kukert, H., Wheatcroft, R.A., Jumars, P.A., and Deming, J.W. (1989). Vertebrate fauna on whale remains. *Nature*, 341, 27–28.

Smith, G.R. (1975). Fishes of the Pliocene Glenns Ferry Formation, Southwest Idaho. *University of Michigan Paleontology Paper*, 141, 1–68.

Smith, H.R. (1946). *Handbook of Lizards*. Comstock Publishing, Ithaca, NY, 557 pp.

Smith, R.L. (1997). Evolution of paternal care in the giant water bugs (Heteroptera: Belostomatidae). In *The Evolution of Social Behavior in Insects and Arachnids* (pp. 116–149), edited by J.C. Choe and B.J. Crespi. Cambridge University Press, Cambridge, U.K.

Smith, R.M.H. (1987). Helical burrow cast of therapsid origin from the Beaufort Group (Permian) of South Africa. *Palaeogeography, Palaeoclimatology, Palaeoecology*, 60, 155–170.

Smith, R.M.H. (1993). Vertebrate taphonomy of Late Permian floodplain deposits in the southwestern Karoo Basin of South Africa. *Palaios.*, 8, 45–67.

Smith, R.M.H. and Evans, S.E. (1995). An aggregation of juvenile *Youngina* from the Beaufort Group, Karoo Basin, South Africa. *Palaeontographica Africana*, 32, 45–49.

Smith, R.M.H. and Evans, S.E. (1996). New material of *Youngina*: evidence of juvenile aggregations in Permian diapsid reptiles. *Palaeontology*, 39, 289–303.

Smith, R.M.H. and Kitching, J.W. (1997). Sedimentology and vertebrate taphonomy of the *Tritylodon* Acme Zone: a reworked palaeosol in the Lower Jurassic Elliot Formation, Karoo Supergroup, South Africa. *Palaeogeography, Palaeoclimatology, Palaeoecology*, 131, 29–50.

Smith, S.A., Thayer, C.W., and Brett, C.E. (1985). Predation in the Paleozoic: gastropod-like drillholes in Devonian brachiopods. *Science*, 230, 1033–1035.

Šnajdr, M. (1978a). Pathological neoplasms in the fringe of Bohemoharpes (Trilobita). *Vestnik Ustredniho Ústavu Geologického*, 53, 301–304.

Šnajdr, M. (1978b). Anomalous carapaces of Bohemian paradoxid trilobites. *Sborník Geologickych Ved Paleontologie*, 20, 7–31.

Šnajdr, M. (1979). Two trinucleid trilobites with repair of traumatic injury. *Vestnik Ustredniho Ústavu Geologického*, 54, 49–50.

Šnajdr, M. (1981). Bohemian Proetidae with malformed exoskeletons (Trilobita). *Sborník Geologickych Ved Paleontologie*, 24, 37–60.

Šnajdr, M. (1990). *Bohemian Trilobites*. Czech Geological Survey, Prague, 265 pp.

Sohl, N.F. (1967). Upper Cretaceous gastropod assemblages of the Western Interior of the United States. In *Paleoenvironments of the Cretaceous Seaway: A Symposium* (pp. 1–37), edited by E.G. Kauffman and H.C. Kent. Colorado School of Mines, Golden.

Sohl, N.F. (1969). The fossil record of shell boring by snails. *American Zoologist*, 9, 725–734.

Sohl, N.F. (1978). Letter. *Evolutionary Biology*, 11, 553.

Soja, C.M., Gobetz, K.E., Thibeau, J., Zavala, E., and White, B. (1996). Taphonomy and paleobiological implications of Middle Devonian (Eifelian) nautiloid concentrates, Alaska. *Palaios*, 11, 422–436.

Solem, A. and Nitecki, M.H. (1968). *Cyclospongia discus* Miller, 1891: a gastropod operculum, not a sponge. *Journal of Paleontology*, 42, 1007–1013.

Solórzano Kraemer, M.M. (2006). The first fossil Paussinae (Coleoptera: Carabidae) from Mexican amber. *Paläontologische Zeitschrift*, 80, 107–111.

Solórzano Kraemer, M.M. (2007). Systematic, palaeoecology, and paleobiogeography of the insect fauna from Mexican amber. *Palaontographica Abteilung A*, 282(1–6), 1–133.

Soltis, P.S., Soltis, D.E., and Smiley, C.J. (1992). An rbcL sequence from Miocene *Taxodium* [bald cypress]. *Proceedings of the National Academy of Sciences USA*, 89, 449–451.

Sombroek, W.G. (1971). Ancient levels of plinthisation in N.W. Nigeria. In *Paleopedology: Origin, Nature, and Dating of Paleosols* (pp. 329–337), edited by D.H. Yaalon. International Society of Soil Science and Israel Universities Press, Jerusalem.

Sondaar, P.Y. (1976). Insularity and its effect on mammalian evolution. In *Major Patterns in Vertebrate Evolution* (pp. 671–707), edited by M.K. Hecht, P.C. Goody, and B.M. Hecht. Plenum Press, New York.

Sondaar, P.Y. (1994). Paleoecology and evolutionary patterns in horses and island mammals. *Historical Biology*, 8, 1–13.

Sorauf, J.E. (2001). External morphology and paleobiology of *Heliophyllum halli* (Zoantharia, Rugosa) from the Middle Devonian, Hamilton Group of New York State. *Journal of Paleontology*, 75, 24–33.

Sorbini, L. (1972). *I Fossili di Bolca*. Edizioni Corev, Verona, 132 pp.

Sørensen, A.M. and Surlyk, F. (2008). A brachiopod boring (*Podichnus*) in a Late Cretaceous oyster from a mangrove-like environment, Skåne, Sweden. *Lethaia*, 41, 295–298.

Sorgenfrei, T. (1965). Some trends in the evolution of European molluscan faunas. In *Proceedings of the First European Malacological Congress* (pp. 69–78), edited by L.R. Cox and J.F. Peake. Conchological Society of Great Britain and Ireland and the Malacological Society of London.

Southward, A.J. (1995). Occurrence in the English Channel of a warm water cirripede, *Soilidobalanus fallax*. *Journal of the Marine Biological Association of the United Kingdom*, 75, 199–210.

Spaan, A., Sondaar, P.Y., and Hartman, W. (1994). The structure of the evolutionary process. *Géobios*, 27, 385–390.

Sparks, D.K., Hoare, R.D., and Kesling, R.V. (1980). Epizoans on the brachiopod *Paraspirifer bownockeri* (Stewart) from the Middle Devonian of Ohio. *University of Michigan Paleontology Paper*, 23, 1–105.

Speyer, S.E. (1990). Gregarious behavior and reproduction in trilobites. In *Evolutionary Paleobiology of Behavior and Coevolution* (pp. 405–409), edited by A.J. Boucot. Elsevier, Amsterdam.

Spjeldnaes, N. (1976). Silurian bryozoans which grew in the shade. In *Bryozoa 1974* (pp. 415–424), edited by S. Pouyet. Université Claude Bernard, Lyon.

Spjeldnaes, N. (1984). Epifauna as a tool in autecological analysis of Silurian brachiopods. *Special Papers in Palaeontology*, 32, 225–235.

Spoor, F., Bajpai, S., Hussain, S.T., Kumar, K., and Thewissen, J.G.M. (2002). Vestibular evidence for the evolution of aquatic behaviour in early cetaceans. *Nature*, 417, 163–166.

Sprague, V. (1941). Studies on *Gregarina blattarum* with particular reference to the chromosome cycle. *Illinois Biological Monograph*, 18, 1–57.

Springer, F. (1920). *The Crinoidea Flexibilia*. The Smithsonian Institution, Washington, D.C., 486 pp.

Springer, K.B. and Murphy, M.A. (1994). Punctuated stasis and collateral evolution in the Devonian lineage of *Monograptus hercynicus*. *Lethaia*, 27, 119–128.

Squires, R.L. and Goedert, J.L. (1991). New Late Eocene mollusks from localized limestone deposits formed by subduction-related methane seeps, southwestern Washington. *Journal of Paleontology*, 65, 412–416.

Squires, R.L. and Goedert, J.L. (1995). An extant species of *Leptochiton* (Mollusca: Polyplacophora) in Eocene and Oligocene cold-seep limestones, Olympic Peninsula, Washington. *The Veliger*, 38, 47–53.

Squires, R.L. and Goedert, J.L. (1996). A new species of *Thalassonerita*? (Gastropoda: Neritidae?) from a Middle Eocene cold-seep carbonate in the Humptulips Formation, western Washington. *The Veliger*, 39, 270–272.

Squires, R.L. and Gring, M.P. (1996). Late Eocene chemosynthetic? Bivalves from suspect cold seeps, Wagonwheel Mountains, central California. *Journal of Paleontology*, 70, 63–73.

Squires, R.L. and Saul, L.R. (2001). New Late Cretaceous gastropods from the Pacific Slope of North America. *Journal of Paleontology*, 75, 46–65.

Squires, R.L., Goedert, J.L., and Barnes, L.G. (1991). Whale carcasses. *Nature*, 349, 574.

Srivasta, A.K. (1996). Plant/animal relationship in the Lower Gondwanas of India. In *Gondwana Nine: Ninth International Gondwana Symposium*, Vol. I (pp. 549–555), edited by Geological Survey of India. Balkema, Rotterdam.

Sroka, S.D. and Richardson, Jr., E.S. (1997). Problematica. In *Richardson's Guide to the Fossil Fauna of Mazon Creek* (pp. 270–280), edited by C.W. Shabica and A.A. Hay. Northeastern Illinois University, Chicago.

Stadum, C.J. (1997). *Paleontological Monitoring and Salvage Report, Olinda Landfill Landslide Remediation Retention Basins A and B*, Project Number, 95-1103. RMW Paleo Associates, Mission Viejo, CA.

Stankiewicz, B.A., Briggs, D.E.G., Evershed, R.P., Flannery, M.B., and Wuttke, M. (1997). Preservation of chitin in 25-million-year-old fossils. *Science*, 276, 1541–1543.

Stankiewicz, B.A., Briggs, D.E.G., Evershed, R.P., Miller, R.F., and Bierstedt, A. (1998a). The fate of chitin in Quaternary and Tertiary strata. In *Nitrogen-Containing Macromolecules in the Bio- and Geosphere* (pp. 211–224), edited by B.A. Stankiewicz and P.F. van Bergen, American Chemical Society Symposium Series No. 707. American Chemical Society, Washington, D.C.

Stankiewicz, B.A., Poinar, H.N., Briggs, D.E.G., Evershed, R.P., and Poinar, Jr., G.O. (1998b). Chemical preservation of plants and insects in natural resins. *Proceedings of the Royal Society London B*, 265, 641–647.

Stanley, Jr., G.D. and Stürmer, W. (1987). A new fossil ctenophore discovered by X-rays. *Nature*, 327, 61–63.

Stanley, Jr., G.D. and Swart, P.K. (1995). Evolution of the coral-zooxanthellae symbiosis during the Triassic: a geochemical approach. *Paleobiology*, 21, 179–199.

Stearns MacNeil, F. (1939). Fresh-water invertebrates and land plants of Cretaceous age from Eureka, Nevada. *Journal of Paleontology*, 13, 355–360.

Stebich, M. (2006). *Epipactis palustris*, Eine fossile Orchidee aus dem Weimarer Ilmtal-Travertinen. *Natur und Museum*, 136, 270–271.

Steele, J. (1999). Stone legacy of skilled hands. *Nature*, 399, 24–25.

Steineck, P.L., Maddocks, R.F., Turner, R.D., Coles, G., and Whatley, R. (1990). Xylophile Ostracoda in the deep sea. In *Ostracoda and Global Events* (pp. 307–319), edited by R. Whatley and C. Maybury. Chapman & Hall, London.

Steinich, G. (1965). Die artikulaten Brachiopoden der Rügener. Schreibkreide (Unter-Maastricht). *Paläontologische Abhandlungen*, 2(1), 1–220.

Stephenson, J. and Scott, A.C. (1992). The geological history of insect-related plant damage. *Terra Nova*, 4, 542–552.

Stevcic, Z. (1971). Laboratory observations on the aggregations of the spiny spider crab (*Maja squinado* Herbst). *Animal Behaviour*, 19, 18–25.

Stevens, G.R. (1965). The Jurassic and Cretaceous belemnites of New Zealand and a review of the Jurassic and Cretaceous belemnites of the Indo-Pacific Region. *New Zealand Geological Survey Paleontological Bulletin*, 36, 1–231.

Stewart, J.D. (1990). Niobrara Formation symbiotic fish in inoceramid bivalves. In *1990 Society of Vertebrate Paleontology Niobrara Chalk Excursion Guidebook* (pp. 31–41), edited by S.C. Bennett. Museum of Natural History and the Kansas Geological Survey, Lawrence.

Stewart, J.D. (1991). Fossil teeth fill in part of the story. *Terra*, 30, 34–35.

Stewart, J.E. (1980). Diseases. In *The Biology and Management of Lobsters* (pp. 301–342), edited by J.S. Cobb and B.F. Phillips. Academic Press, San Diego, CA.

Stilwell, J.D., Levy, R.H., Feldmann, R.M., and Harwood, D.M. (1997). On the rare occurrence of Eocene callianassid decapods (Arthropoda) preserved in their burrows, Mount Discovery, East Antarctica. *Journal of Paleontology*, 71, 284–287.

Stinnesbeck, W., Ifrim, C., Schmidt, H., Rindfleisch, A., Buchy, M.-C. et al. (2005). A new lithographic limestone deposit in the Upper Cretaceous Austin Group at El Rosario, county of Múzquiz, Coahuila, northeastern Mexico. *Revista Mexicana de Ciencias Geológicas*, 22, 401–418.

Stokes, W.L. (1957). Pterodactyl tracks from the Morrison Formation. *Journal of Paleontology*, 31, 952–954.

Stokes, W.L. (1987). Dinosaur gastroliths revisited. *Journal of Paleontology*, 61, 1242–1246.

Stolarski, J., Zibrowius, H., and Löser, H. (2001). Antiquity of the scleractinian–sipunculan symbiosis. *Acta Palaeontologica Polonica*, 46, 309–330.

Stone, H.M.I. (1998). On predator deterrence by pronounced shell ornament in epifaunal bivalves. *Palaeontology*, 41, 1051–1068.

Storch, G. (1993). "Grube Messel" and African-South American faunal connections. In *The Africa–South America Connection* (pp. 76–86), edited by W. George and R. Lavocat. Clarendon Press, London.

Storch, G. (2001). Paleobiological implications of the Messel mammalian assemblage. In *Eocene Biodiversity: Unusual Occurrences and Rarely Sampled Habitats* (pp. 215–235), edited by G.F. Gunnell. Kluwer Academic/Plenum Publishers, New York.

Storch, G. and Richter, G. (1992a). Pangolins: almost unchanged for 50 million years. In *Messel: An Insight into the History of Life and of the Earth* (pp. 203–207), edited by S. Schaal and W. Ziegler. Clarendon Press, London.

Storch, G. and Richter, G. (1992b). The ant-eater *Eurotamandua*: a South American in Europe. In *Messel: An Insight into the History of Life and of the Earth* (pp. 211–215), edited by S. Schaal and W. Ziegler. Clarendon Press, London.

Storch, G. and Richter, G. (1994). Zur Paläobiologie Messeler Igel. *Natur und Museum*, 124, 81–90.

Storch, G., Engesser, B., and Wuttke, M. (1996). Oldest fossil record of gliding in rodents. *Nature*, 379, 439–441.

Storch, G., Sigé. B., and Habersetzer, J. (2002). *Tachypteron franzeni* n. gen., n. sp., earliest emballonurid bat from the Middle Eocene of Messel (Mammalia, Chiroptera). *Paläontologische Zeitschrift*, 76, 223–234.

Størmer, L. (1963). Gigantoscorpio willsi *a New Scorpion from the Lower Carboniferous of Scotland and Its Associated Preying Microorganisms*. Oslo University Press, Oslo, Norway, 171 pp.

Størmer, L. (1969). Sexual dimorphism in eurypterids. In *Sexual Dimorphism in Fossil Metazoa and Taxonomic Implications* (pp. 201–214), edited by G.E.G. Westermann. Schweizerbart, Stuttgart.

Storrs, G.W. (1995). A juvenile specimen of ?*Plesiosaurus* from the Lias (Lower Jurassic, Pliensbachian) near Charmouth, Dorset, England. *Proceedings of the Dorset Natural History and Archaeological Society*, 116, 71–76.

Strauch, F. and Pockrandt, W. (1985). Ein *Encrinaster*-Vorkommen (Ophiuroides) aus der Unterdevon der Eifel. *Paläontologische Zeitschrift*, 59, 125–145.

Strouhal, E. (1987). La tuberculose vertébrale en Égypte et Nubie anciennes. *Bulletins et mémoires de la Société d'Anthropologie de Paris*, 14, 261–270.

Strouhal, E. (1998). Paleopathological evidence of jaw tumors. In *Dental Anthropology: Fundamentals, Limits, and Prospects* (pp. 277–292), edited by K.W. Alt, F.W. Rosing, and M. Teschler-Nicola. Springer, New York.

Struve, W. (1964). Strömungs-Orientierung bei bodenverwachsen, schlosstragenden Brachiopoden. *Natur und Museum*, 94(12), 515–529.

Struve, W. (1978). Fixo-sessile Brachiopoden aus dem Rheinischen Devon: *Schuchertellopsis* (*Krejcigrafella*) und *Auchmerella* (Strophomenida). *Senckenbergiana Lethaea*, 59, 93–115.

Struve, W. (1980). Zur Paläökologie fixo-sessile articulater Brachiopoden aus dem Rheinischen Gebirge. *Senckenbergiana Lethaea*, 60, 399–433.

Stubblefield, J.W. and Seger, J. (1994). Sexual dimorphism in the Hymenoptera. In *The Differences Between the Sexes* (pp. 71–103), edited by R.V. Short and E. Balaban. Cambridge University Press, Cambridge, U.K.

Stubblefield, S.P. and Taylor, T.N. (1986). Wood decay in silicified gymnosperms from Antarctica. *Botanical Gazette*, 147, 116–125.

Stubblefield, S.P. and Taylor, T.N. (1988). Recent advances in palaeomycology. *New Phytologist*, 108, 3–25.

Stubblefield, S.P., Taylor, T.N., and Beck, C.B. (1985). Studies of Paleozoic fungi. IV. Wood-decaying fungi in *Callixylon newberryi* from the Upper Devonian. *American Journal of Botany*, 72, 1765–1774.

Studencka, B. (1987). The occurrence of the genus *Kelliella* (Bivalvia, Kelliellidae) in shallow-water, Middle Miocene deposits of Poland. *Acta Palaeontologica Polonica*, 32, 73–81.

Suchy, D.R. and West, R.R. (1988). A Pennsylvanian cryptic community associated with laminar chaetetid colonies. *Palaios*, 3, 404–412.

Sudbury, M. (1958). Triangulate monograptids from the *Monograptus gregarious* zone (Lower Llandovery) of the Rheidol Gorge (Cardiganshire). *Philosophical Transactions of the Royal Society of London B*, 241, 485–555.

Sues, H.-D. (1991). Venom-conducting teeth in a Triassic reptile. *Nature*, 351, 141–143.

Sues, H.-D. (1996). A reptilian tooth with apparent venom canals from the Chinle Group (Upper Triassic) of Arizona. *Journal of Vertebrate Paleontology*, 16, 571–572.

Sues, H.-D., Ed. (2000). *Evolution of Herbivory in Terrestrial Vertebrates: Perspectives from the Fossil Record*. Cambridge University Press, Cambridge, U.K., 256 pp.

Sues, H.-D., Olsen, P.E., Carter, J.G., and Scott, D.M. (2003). A new crocodylomorph archosaur from the Upper Triassic of North Carolina. *Journal of Vertebrate Paleontology*, 23, 329–343.

Sukacheva, I.D. (1989). Kainozoiskie rucheiniki Primorskogo Kraya. In *Kainozoi Dalnego Vostoka* (pp. 151–160). Biologo-Pochvennogo Instituta, Vladivostok.

Sukacheva, I.D. (1990). Rucheiniki. Phryganeida. In "Pozdne-Mezozoiskie nasekomie vostochnogo Zabaikalya," edited by A.P. Rasnitsyn. *Transactions of the Paleontological Institute, Academy of Science, USSR*, 94–122.

Sukacheva, I.D. (1994). Oorskie domiki rucheinikov (Insecta, Trichoptera) Mongolii. *Paleontologicheskii Zhurnal*, 4, 76–85.

Sukacheva, I.D. (2005). Nakhoda lichinochnogo domika rucheinika *Folindusia poroda* Acrindusia (Trichoptera) v verkhnem Melu Amurskoi Oblasti. *Paleontologicheskii Zhurnal*, 5, 47–49.

Sullivan, C., Reisz, R.R., and Smith, R.M.H. (2003). The Permian mammal-like herbivore *Diictodon*, the oldest known example of sexually dimorphic armament. *Proceedings of the Royal Society London B*, 270, 173–178.

Sullivan, R.M., Lucas, S.G., Hunt, A.P., and Fritts, T.H. (1988). Color pattern on the selmacryptodiran turtle *Neurankylus* from the Early Paleocene (Puercan) of the San Juan Basin, New Mexico. *Natural History Museum of Los Angeles County Contributions in Science*, 401, 1–9.

Sullivan, R.M., Tanke, D.H., and Rothschild, B.M. (2000). An impact fracture in an ornithomimid (Ornithomimisauria: Dinosauria) metatarsal from the Upper Cretaceous (Late Campanian) of New Mexico. In *Dinosaurs of New Mexico* (pp. 109–111), edited by S.G. Lucas and A.B. Heckert, Bulletin No. 17. New Mexico Museum of Natural History and Science, Albuquerque.

Sulser, H. and Meyer, C. (1998). Taxonomy and palaeoecology of terebratulid brachiopods (*Sellithyris subsella*-group) from the Late Jurassic of northwestern Switzerland. *Eclogae Geologicae Helvetiae*, 91, 439–451.

Sumrall, C.D., Sprinkle, J., and Bonem, R.M. (2006). An edrioasteroid-dominated echinoderm assemblage from a Lower Pennsylvanian marine conglomerate in Oklahoma. *Journal of Paleontology*, 80, 229–244.

Sun, B. (1995). *Shanwang Fossils*. Science Press, Beijing, 77 pp.

Sun, Y.-L. and Boucot, A.J. (1999). Ontogeny of *Stringocephalus gubiensis* and origin of *Stringocephalus*. *Journal of Paleontology*, 73, 860–871.

Sun, Y.L., Boucot, A.J., Blodgett, R.B., and Ran, W.Z. (1999). Color pattern on a martiniid brachiopod from South China. *Journal of Paleontology*, 73, 973–976.

Sung, G.-H., Poinar, Jr., G.O., and Spatafora, J.M. (2008). The oldest fossil evidence of fungal–arthropod symbioses. *Molecular Phylogenetics and Evolution*, 49, 495–502.

Surlyk, F. (1972). Morphological adaptations and population structures of the Danish chalk brachiopods (Maastrichtian, Upper Cretaceous). *Det Kongelige Danske Videnskabernes Selskab, Biologiske Skrifter*, 19(2), 1–57.

Susman, R.L. (1994). Fossil evidence for early hominid tool use. *Science*, 265, 1570–1573.

Sutton, A.H. (1935). Ovoviviparous reproduction of Miocene Turritellidae. *American Midland Naturalist*, 16, 107–109.

Swift, C.C. and Barnes, L.G. (1996). Stomach contents of *Basilosaurus cetoides*: implications for the evolution of cetacean feeding behavior, and evidence for vertebrate fauna of epicontinental Eocene seas. In *Sixth North American Paleontological Convention Abstracts of Papers* (p. 380), edited by J.E. Repetski, Special Publ. No. 8. Paleontological Society, Washington, D.C.

Swindler, D.R., Ryan, D.P., and Rothschild, B.M. (1995). Dental remains from the Valley of the Kings, Luxor. Egypt. New Kingdom (1550 to 1070 B.C.). In *Aspects of Dental Biology, Palaeontology, Anthropology and Evolution* (pp. 365–372), edited by J. Moggi-Cecci. International Institute for the Study of Man, Florence, Italy.

Swisher III, C.C., Wang, Y.-Q., Wang, X.-L., Lu, X., and Wang, Y. (1999). Cretaceous age for the feathered dinosaurs of Laioning, China. *Nature*, 400, 58–61.

Sylvester-Bradley, P.C. (1969). Comparative and functional sex in ostracods and cephalopods. In *Sexual Dimorphism in Fossil Metazoa and Taxonomic Implications* (pp. 242–250), edited by G.E.G. Westermann. Schweizerbart, Stuttgart.

Szadziewski, R. (1999). Amber in 1999. *Meganeura*, Summer/Autumn, 4.

Szafer, W. (1957). Subfosylny biotop chomika w Mielniku nad Bugient. *Acta Societatis Botanicorum Poloniae*, 26, 105–130.

Sze, H.C. (1954). Description and discussion of a problematic organism from Lingwu, Kansu, northwestern China. *Acta Palaeontologica Sinica*, 2, 315–322.

Szidat, L. (1944). Über die Erhaltungsfähigkeit von Helmintheneiern in Vor-und Frühgeschtlichen Moorleichen. *Zeitschrift für Parasitenkunde*, 13, 265–274.

Taddei Ruggiero, E. (1990). A study of damage evidence in brachiopod shells. In *Brachiopoda Through Time* (pp. 203–210), edited by D.I. MacKinnon, D.E. Lee and J.D. Campbell. Balkema, Rotterdam.

Taddei Ruggiero, E. (2001). Brachiopods of the Isca submarine cave: observations during ten years. In *Brachiopods Past and Present* (pp. 261–267), edited by C.H. Brunton, L.R.M. Cocks, and S.L. Long. Taylor & Francis, London.

Taddei Ruggiero, E. and Annunziata, G. (2002). Bioerosion on a *Terebratula scillae* population from the Lower Pleistocene of Lecce area (Southern Italy). *Acta Geologica Hispanica*, 37, 43–51.

Taddei Ruggiero, E. and Bitner, M.A. (2008). Bioerosion on brachiopod shells: a Cenozoic perspective. *Transactions of the Royal Society of Edinburgh, Earth and Environmental Sciences*, 98, 369–378.

Taddei Ruggiero, E., Buono, G., and Raia, P. (2006). Bioerosion on brachiopod shells of a thanatocoenosis of Alboràn Sea (Spain). *Ichnos*, 13, 175–184.

Talent, J.A., Duncan, P.M., and Handry, P.L. (1966). Early Cretaceous feathers from Victoria. *The Emu*, 66(2), 81–86.

Tambussi, C.P., Reguero, M.A., Marenssi, S.A., and Santillana, S.N. (2005). *Crossvallia unienwillia*, a new Spheniscidae (Sphenisciformes, Aves) from the Late Paleocene of Antarctica. *Géobios*, 38, 667–675.

Tampa, A. (1976). Fossil eggs of the land snail genus *Vallonia* (Pulmonata: Valloniidae). *The Nautilus*, 90, 5–7.

Tanke, D.H. and Rothschild, B.M. (1997). Paleopathology. In *Encyclopedia of Dinosaurs* (pp. 525–530), edited by P.J. Currie and K. Padian. Academic Press, San Diego, CA.

Tankersley, S.L., Nagy, F., Rupert, L., Tankersley, K.O., and Tankersley, K.B. (1994). Detection of *Giardia lamblia* in Early woodland Indian paleofecal specimens from Salts Cave, Mammoth Cave National Park, Kentucky. In *Abstracts of the 69th Annual Meeting of the American Society of Parasitologists, August 9–13, Fort Collins, CO, 1994*, abstract 90 (supplement to *Journal of Parasitology*).

Tapanila, L. (2004). The earliest *Helicosalpinx* from Canada and the global expansion of commensalism in Late Ordovician sarcinulid corals (Tabulata). *Palaeogeography, Palaeoclimatology, Palaeoecology*, 215, 99–110.

Tapanila, L. (2007). The medium is the message: a complex microboring (*Pyrodendrina cupra* igen. n., isp. n.) from the early Paleozoic of Anticosti Island, Canada. In *Current Developments in Bioerosion* (pp. 123–146), edited by M. Wisshak and L. Tapanila. Springer-Verlag, Heidelberg.

Tarlo, L.B. (1959). *Pliosaurus brachyspondylus* (Owen) from the Kimmeridge Clay. *Palaeontology*, 1, 283–291.

Tasch, P. (1961). Valve injury and repair in living and fossil conchostracans. *Transactions of the Kansas Academy of Sciences*, 64, 144–149.

Tasch, P. and Zimmerman, J.R. (1961). Fossil and living conchostracan distribution in Kansas–Oklahoma across a 200-million-year time gap. *Science*, 133, 584–586.

Tasnádi-Kubacska, A. (1962). *Paläopathologie*. VEB Gustav Fischer Verlag, Jena, 269 pp.

Taviani, M. (1994). The "Calcari a *Lucina*" macrofauna reconsidered: deep-sea faunal oases from Miocene-age cold vents in the Romagna Apennines, Italy. *Geo-Marine Letters*, 14, 185–191.

Taylor, B.J. (1965). Aptian cirripedes from Alexander Island. *Bulletin of the British Antarctic Survey*, 7, 37–42.

Taylor, D.W. (1966). Summary of North American Blancan nonmarine mollusks. *Malacologia*, 4, 1–172.

Taylor, E.L. (1955). Parasitic Helminthes in Mediaeval remains. *Veterinary Record*, 67, 216–218.

Taylor, J.D., Morris, N.J., and Taylor, C.N. (1980). Food specialization and the evolution of predatory prosobranch gastropods. *Palaeontology*, 23, 375–409.

Taylor, J.D., Cleevely, S.J., and Morris, N.J. (1983a). Predatory gastropods and their activities in the Blackdown Greensand (Albian) of England. *Palaeontology*, 26, 521–553.

Taylor, L.R., Compagno, L.J.V., and Struksaker, P.J. (1983b). Megamouth: a new species, genus, and family of lamnoid shark (*Megachasma pelagios*, Family Megachasmidae) from the Hawaiian Islands. *Proceedings of the California Academy of Sciences*, 43A, 87–110.

Taylor, M.A. (1993). Stomach stones for feeding or buoyancy? The occurrence and function of gastroliths in marine tetrapods. *Philosophical Transactions of the Royal Society of London B*, 341, 163–175.

Taylor, P.D. (1979). Palaeoecology of the encrusting epifauna of some British Jurassic bivalves. *Palaeogeography, Palaeoclimatology, Palaeoecology*, 28, 241–262.

Taylor, P.D. (1991). Observations on symbiotic associations of bryozoans and hermit crabs from the Otago Shelf of New Zealand. In *Bryozoaires Actuels et Fossiles: Bryozoa Living and Fossil* (pp. 487–495), edited by F.P. Bigey and J.-L. d'Hondt. Société des Sciences Naturelles de l'Ouest, France.

Taylor, P.D. (1994). Evolutionary palaeoecology of symbioses between bryozoans and hermit crabs. *Historical Biology*, 9, 157–205.

Taylor, P.D. and Schindler, K.S. (2004). A new Eocene species of the hermit-crab symbiont *Hippoporidra* (Bryozoa) from the Ocala Limestone of Florida. *Journal of Paleontology*, 78, 790–794.

Taylor, P.D. and Vinn, O. (2006). Convergent morphology in small spiral worm tubes ('*Spirorbis*') and its palaeoenvironmental implications. *Journal of the Geological Society, London*, 163, 225–228.

Taylor, P.D., Gordon, D.P., and Batson, P.B. (2004a). Bathymetric distribution of modern populations of some common Cenozoic Bryozoa from New Zealand, and paleodepth estimation. *New Zealand Journal of Geology and Geophysics*, 47, 57–69.

Taylor, T.N. and Osborn, J.M. (1992). The role of wood in understanding saprophytism in the fossil record. *Courier Forschungs-Institut Senckenberg*, 147, 147–153.

Taylor, T.N. and Scott, A.C. (1983). Interactions of plants and animals during the Carboniferous. *Bioscience*, 33, 488–493.

Taylor, T.N. and Taylor, E.L. (1993). *The Biology and Evolution of Fossil Plants*. Prentice-Hall, Englewood Cliffs, NJ, 982 pp.

Taylor, T.N. and White, J.F. (1989). Triassic fungi with suggested affinities to the Endogonales (Zygomycotina). *Review of Palaeobotany and Palynology*, 61, 53–61.

Taylor, T.N., Remy, W., and Hass, H. (1992a). Parasitism in a 400-million-year-old green alga. *Nature*, 357, 493–494.

Taylor, T.N., Remy, W., and Hass, H. (1992b). Fungi from the Lower Devonian Rhynie Chert: Chytridiomyces. *American Journal of Botany*, 79, 1233–1241.

Taylor, T.N., Hass, H., and Remy, W. (1992c). Devonian fungi interactions with the green alga *Palaeonitella*. *Mycologia*, 84, 901–910.

Taylor, T.N., Remy, W., and Hass, H. (1994). *Allomyces* in the Devonian. *Nature*, 367, 601.

Taylor, T.N., Hass, H., and Kerp, H. (1997). A cyanolichen from the Lower Devonian Rhynie Chert. *American Journal of Botany*, 84, 992–1004.

Taylor, T.N., Hass, H., Krings, M., Klavins, S.D., and Kerp, H. (2004b). Fungi in the Rhynie Chert: a view from the dark side. *Transactions of the Royal Society of Edinburgh, Earth Sciences*, 94, 457–473.

Taylor, W.L. and Brett, C.E. (1996). Taphonomy and paleoecology of echinoderm *Lagerstätten* from the Silurian (Wenlockian) Rochester Shale. *Palaios*, 11, 118–140.

Tedford, R.H. and Harington, C.R. (2003). An arctic mammal fauna from the Early Pliocene of North America. *Nature*, 425, 388–390.

Teerink, B.J. (1991). *Hair of West-European Mammals: Atlas and Identification Key*. Cambridge University Press, Cambridge, U.K., 224 pp.

Teichert, C. (1991). New data on the Permian crinoid family Calceolispongiidac in Western Australia. *Journal of the Royal Society of Western Australia*, 73, 113–121.

Telford, Jr., S.R. (1984). Haemoparasites of reptiles. In *Diseases of Amphibians and Reptiles* (pp. 385–517), edited by G.L. Hoff, F.L. Frye, and E.R. Henderson. Plenum Press, New York.

Teller, L. (1998). Abnormalities in the development of colonies in some Monograptidae. In *Proceedings of the Sixth International Graptolite Conference of the Graptolite Working Group, International Palaeontological Association, June 19–22, Madrid, Spain* (pp. 264–265), edited by J.C. Gutiérrez-Marco and I. Rábano. University of Kansas, Lawrence.

Terzi, C., Aharon, P., Ricci Lucchi, F., and Vai, G.B. (1994). Petrography and stable isotope aspects of cold-vent activity imprinted on Miocene-age "calcari a *Lucina*" from Tuscan and Romagna Apennines, Italy. *Geo-Marine Letters*, 14, 177–184.

Tessier, F. (1959). Termitières fossiles dans la latérite de Dakar (Senegal); Remarques sur les structures latéritiques. *Annales de la Faculté des sciences Université de Dakar*, 4, 91–132.

Tessier, F. (1965). Les niveaux latéritique du Senegal. *Annales de la Faculté des sciences de Marseille*, 37, 221–237.

Tester, M., Smith, S.E., and Smith, F.A. (1987). The phenomenon of "nonmycorrhizal" plants. *Canadian Journal of Botany*, 65, 419–431.

Thackray, G.D. (1994). Fossil nest of sweat bees (Halictinae) from a Miocene paleosol, Rusinga Island, western Kenya. *Journal of Paleontology*, 68, 795–800.

Thewissen, J.G.M. and Hussain, S.T. (1993). Origin of underwater hearing in whales. *Nature*, 361, 444–445.

Thewissen, J.G.M., Hussain, S.T., and Arif, M. (1994). Fossil evidence for the origin of aquatic locomotion in Archaeocete whales. *Science*, 263, 210–212.

Thewissen, J.G.M., Roe, L.J., O'Neil, J.R., Hussain, S.T., Sahni, A., and Bajpal, S. (1996). Evolution of cetacean osmoregulation. *Nature*, 381, 379–380.

Thieme, H. (1997). Lower Palaeolithic hunting spears from Germany. *Nature*, 385, 807–810.

Thies, D. (1985). Bissspuren an Seeigel-Gehäusen der Gattung *Echinocorys* Leske, 1778 aus dem Maastrichtium von Hemmoor (NW-Deutschland). *Mitteilungen aus dem Geologisch–Paläontologichen Institut der Universität Hamburg*, 59, 71–82.

Thomas, G.M. and Poinar, Jr., G.O. (1973). Report of diagnoses of diseased insects, 1962–1971. *Hilgardia*, 42, 1–359.

Thomas, R.D.K. (1975). Functional morphology, ecology, and evolutionary conservatism in the Glycymerididae (Bivalvia). *Palaeontology*, 18, 217–254.

Thompson, R.S., Van Devender, T.R., Martin, P.S., Foppe, T., and Long, A. (1980). Shasta ground sloth (*Nothrotheriops shastense* Hoffstetter) at Shelter Cave, New Mexico: environment, diet, and extinction. *Quaternary Research*, 14, 360–376.

Thomson, K.S., Sutton, M., and Thomas, B. (2003). A larval Devonian lungfish. *Nature*, 426, 833–834.

Thorne, P. (1973). Records of fossil pearls. *The Canadian Rockhound*, 17(4), 7–32.

Thornhill, R. and Alcock, J. (1983). *The Evolution of Insect Mating Systems*. Harvard University Press, Cambridge, MA, 547 pp.

Thornton, M.L. and Rasmussen, D.T. (2001). Taphonomic interpretation of Gnat-Out-of-Hell, an Early Uintan small mammal locality in the Uinta Formation, Utah. In *Eocene Biodiversity: Unusual Occurrences and Rarely Sampled Habitats* (pp. 299–316), edited by G.F. Gunnell. Plenum Press, New York.

Thulborn, R.A. (1982). Speeds and gaits of dinosaurs. *Palaeogeography, Palaeoclimatology, Palaeoecology*, 38, 227–256.

Thulborn, T. and Turner, S. (1993). An elasmosaur bitten by a pliosaur. *Modern Geology*, 18, 489–501.

Thulborn, T. and Turner, S. (1995). An elasmosaur bitten by a pliosaur. In *Vertebrate Fossils and the Evolution of Scientific Concepts* (pp. 499–511), edited by W.A.S. Sarjeant. Gordon & Breach, New York.

Tichy, G. (1980). Uber die erhaltung von Farben und Farbmustern an triassichen Gastropoden-Gehausen. *Verhandlung der Geologischen Bundesanstalt*, 2, 175–217.

Tillier, A.-M., Arensburg, B., Raky, Y., and Vandermeersch, B. (1995). Middle Palaeolithic dental caries: new evidence from Kebara (Mount Carmel, Israel). *Journal of Human Evolution*, 29, 189–192.

Tintori, A. (1990). The actinopterygian fish *Prohalectites* from the Triassic of northern Italy. *Palaeontology*, 33, 155–174.

Tischlinger, H. (1998). Erstnachweis von Pigmentfarben bei Plattenkalk-Teleosteern. *Archaeopteryx*, 16, 1–18.

Tischlinger, H. (2005). Ultraviolet light investigations of fossils from the Upper Jurassic plattenkalks of southern Franconia. *Zitteliana*, B26, 26.

Tobias, P.V. (1987). The brain of *Homo habilis*: a new level of organization in cerebral evolution. *Journal of Human Evolution*, 16, 741–761.

Tobien, H. (1965). Insekten Frasspuren an tertiären und pleistozänen Säugetier-Knochen. *Senckenbergiana Lethaia*, 46a, 441–451.

Tobien, H. (1977). Ein Gewollrest mit Megacricetodon (Rodentia, Mammalia) aus dem Obermiozän von Ohningen (Baden-Wurttemberg). *Berichte der Naturforschenden Gesellschaft zu Freiburg i. Br. (Pfannenstiel Gedenkband)*, 67, 359–369.

Tomida, S. and Ozawa, T. (1996). Occurrence of *Turbo* (*Lunatica*) species (Gastropoda: Turbinidae) in the upper Neogene of Japan and their implications for Neogene marine climates. *Tertiary Research*, 17, 65–71.

Tomlinson, J.T. (1963). Acrothoracican barnacles in Paleozoic myalinids. *Journal of Paleontology*, 37, 164–166.

Tampa, A. (1976). The fossil eggs of the land snail genus *Vallonia* (Pulmonata: Valloniidae). *The Nautilus*, 90, 5–7.

Torres, M.E., McManus, J., and Huh, C.-A. (2002). Fluid seepage along the San Clemente Fault scarp: basin-wide impact on barium cycling. *Earth and Planetary Science Letters*, 203, 181–194.

Torres, M.E., Bohrmann, G., Dubé, T.E., and Poole, F.G. (2003). Formation of modern and Paleozoic stratiform barite at cold methane seeps on continental margins. *Geology*, 31, 897–900.

Tozer, E.T. (1956). Uppermost Cretaceous and Paleocene nonmarine molluscan fauna of western Alberta. *Geological Survey of Canada Memoir*, 280, 1–125.

Trapani, J., Sanders, W.J., Mitani, J.C., and Heard, A. (2006). Precision and consistency of the taphonomic signature of predation by Crowned Hawk-Eagles (*Stephanoaetus coronatus*) in Kibale National Park, Uganda. *Palaios*, 21, 114–131.

Traquair, R.H. (1879). Evidence as to the predaceous habits of the larger Palaeoniscidae. *Proceedings of the Royal Physical Society of Edinburgh*, 5, 128–130.

Trewin, N.H. (1986). Palaeoecology and sedimentology of the Achanarras fish bed of the Middle Old Red Sandstone, Scotland. *Transactions of the Royal Society of Edinburgh, Earth Sciences*, 77, 21–46.

Tribollet, A. (2007). The boring microflora in modern coral reef ecosystems: a review of its role. In *Current Developments in Bioerosion* (pp. 67–94), edited by M. Wisshak and L. Tapanila. Springer-Verlag, Heidelberg.

Trinkaus, E. (1983). *The Shandidar Neanderthals*. Academic Press, San Diego, CA, 502 pp.

Troitskiy, S.L. (1974). Subarctic Pleistocene molluscan fauna. In *Marine Geology and Oceanography of the Arctic Seas* (pp. 257–270), edited by Y. Herman. Springer-Verlag, Heidelberg.

Tshudy, D.M., Feldmann, R.M., and Ward, P.D. (1989). Cephalopods: biasing agents in the preservation of lobsters. *Journal of Paleontology*, 63, 621–626.

Tshudy, D., Baumiller, T.K., and Sorhannus, U. (1998). Morphologic change in the clawed lobster *Hoploparia* (Nephropsidae) from the Cretaceous of Antarctica. *Paleobiology*, 24, 64–73.

Tsujita, C.J. and Westermann, G.E.G. (2001). Were limpets or mosasaurs responsible for the perforations in the ammonite *Placenticeras*? *Palaeogeography, Palaeoclimatology, Palaeoecology*, 169, 245–270.

Tucker, A.B., Feldmann, R.M., and Powell II, C.L. (1994). *Speocarcinus berglundii* n. sp. (Decapoda: Brachyura), a new crab from the Imperial Formation (late Miocene-late Pliocene) of southern California. *Journal of Paleontology*, 68, 800–807.

Tunnicliffe, V. (1992). The nature and origin of the modern Hydrothermal vent fauna. *Palaios*, 7, 338–350.

Tunnicliffe, V., Juniper, S.K., and Sibuet, M. (2003). Reducing environments of the deep-sea floor. In *Ecosystems of the World: The Deep Sea* (pp. 81–110), edited by P.A. Tyler. Elsevier, Amsterdam.

Turek, V. (1990). Comments to upper Wenlock zonal subdivisions in the Silurian of Central Bohemia. *Casopis pro mineralogii a geologii*, 35, 337–354.

Turner, R.F. (1973). Occurrence and implications of fossilized burrowing barnacles (Cirripedia: Order Acrothoracica). *Geological Society of America Abstracts with Programs*, 5(2), 230–231.

Twitchett, R.J., Feinberg, J.M., O'Connor, D.D., Alvarez, W., and McCollum, L.B. (2005). Early Triassic ophiuroids: their paleoecology, taphonomy, and distribution. *Palaios*, 20, 213–223.

Tyler, M.J. (1977). Frog pellets, etc. *Transactions of the Royal Society of South Australia*, 101, 85–89.

Tyler, M.J. (1992). Early Holocene frogs from the Tantanoola Cave, South Australia. *Transactions of the Royal Society of South Australia*, 116, 153.

Tyler, M.J., Leong, A. S.-Y., and Godthelp, H. (1994). Tumors of the ilia of modern and Tertiary Australian frogs. *Journal of Herpetology*, 28, 528–529.

Tzedakis, P.C. (1993). Long-term tree populations in northwest Greece through multiple Quaternary climatic cycles. *Nature*, 364, 437–440.

Uchman, A. and Alvaro, J.J. (2000). Non-marine invertebrate trace fossils from the Tertiary Calatayud–Teruel Basin, NE Spain. *Revista Española de Paleontologia*, 15, 203–218.

Uchman, A. and Pervesler, P. (2006). Surface lebensspuren produced by amphipods and isopods (Crustaceans) from the Isonzo Delta tidal flat, Italy. *Palaios*, 21, 384–390.

Uhen, M.D. (2004). Form, function, and anatomy of *Dorudon atrox* (Mammalia, Cetacea): an Archaeocete from the Middle to Late Eocene of Egypt. *University of Michigan Paleontology Paper*, 34, 1–222.

Ulrich, E.O. and Scofield, W.H. (1897). The Lower Silurian Gastropoda of Minnesota. *Final Report Geological and Natural History Survey of Minnesota*, 3, 813–1081.

Ulrich, H. and Schmelz, R.M. (2001). Enchytraeidae as prey of Dolichopodidae, recent and in Baltic amber (Oligochaeta; Diptera). *Bonner Zoologische Beiträge*, 50, 89–101.

Unal, E. and Zinsmeister, W.J. (2006). Earliest record of sheltering strategy for predator avoidance from the Early Cambrian of the Marble Mountains, California. *Geological Society of America Abstracts with Programs*, 38(7), 550.

Underwood, C.J., Mitchell, S.F., and Veltkamp, C.J. (1999a). Microborings in mid-Cretaceous fish teeth. *Proceedings of the Yorkshire Geological Society*, 52, 269–274.

Underwood, C.J., Mitchell, S.F., and Veltkamp, C.J. (1999b). Shark and ray teeth from the Hauterivian (Lower Cretaceous) of north-east England. *Palaeontology*, 42, 287–302.

Unwin, D.M. (1997). Pterosaur tracks and the terrestrial ability of pterosaurs. *Lethaia*, 29, 373–386.

Upeniece, I. (1999). Fossil record of parasitic helminths in fishes. In *Abstracts of the Fifth International Symposium on Fish Parasites*, August 9–13, 1999, Ceské Budejovice, Czech Republic, p. 154.

Upeniece, I. (2001). The unique fossil assemblage from the Lode Quarry (Upper Devonian, Latvia). *Mitteilungen des Museums für Naturkunde in Berlin—Geowissenschaftliche Reihe*, 4, 101–119.

Urbanek, A. (1997). The emergence and evolution of linograptids. *Palaeontologia Polonica*, 56, 233–269.

Ushatinskaya, G.T. (2003). Ostatki mikroorganizmov i sledov ikh zhiznedeyatelnosti v rakovinakh drevnikh brachiopod. *Paleontologicheski Zhurnal*, 2, 40–43.

Uyeno, T. and Matsui, N. (1993). Late Cretaceous fish fossils from Nemuro, Hokkaido, Japan. *Memoirs of the National Science Museum*, 26, 39–46.

Vadasz, M.E. (1914). Regenerationserscheinungen an fossilen Echinoiden. *Centralblatt für Mineralogie, Geologie und Paläontologie*, 283–288.

Vahldiek, B.-W. and Schweigert, G. (2007). Ältester Nachweis Holz Bohrender Muscheln. *Neues Jahrbuch für Geologie und Paläontologie–Abhandlungen*, 244, 261–271.

Vajda, V. (2005). A new Maastrichtian water fern, *Azolla*, a hardy survivor of extreme environmental conditions. *GFF*, 127, 58.

Vakrameev, V.A. and Pushcharovskii, Y.M. (1954). The geological part of the Verkhoyansk marginal flexure in the Mesozoic. *Voprosy Geologii Azii*, 1, 588–628.

Valent, M. and Malinky, J.M. (2008). Early Devonian (Emsian) hyolith *Ottomarites discors* (Barrande, 1867) with colour pattern. *Bulletin of Geoscience*, 83, 503–506.

Valladas, H., Cachier, H., Maurice, P., Bernaldo de Quiros, Clottes, J., Cabrera Valdes, V., Uzquiano, P., and Arnold, M. (1992). Direct radiocarbon dates for prehistoric paintings at the Altamira, El Castillo and Niaux caves. *Nature*, 357, 68–70.

Van Bakel, B.W.M., Fraaije, R.H.B., Jagt, J.W.M., and Artai, P. (2008). An unexpected diversity of late Jurassic hermit crabs (Crustacea, Decapoda, Anomura) in Central Europe. *Neues Jahrbuch für Geologie und Paläontologie*, 250, 137–156.

Van Bergen, P.F., Collinson, M.E., Briggs, D.E.G., de Leeuw, J.W., Scott, A.C., Evershed, R.P., and Finch, P. (1995). Resistant biomacromolecules in the fossil record. *Acta Botanica Neerlandica*, 44, 319–342.

Van Couvering, J.A.H. (1980). Community evolution in East Africa during the Late Cenozoic. In *Fossils in the Making* (pp. 272–298), edited by A.K. Behrensmeyer and A.P. Hill. University of Chicago Press, Chicago, IL.

van der Ham, R.W.J.M. (1988). Echinoids from the Early Palaeocene (Danian) of the Maastricht area (NE Belgium, SE Netherlands): preliminary results. *Mededelingen van de Werkgroep voor Tertiaire en Kwartaire Geologie*, 25, 127–161.

Van Devender, T.R., Martin, P.S., Foppe, T., and Long, A. (1980). Shasta ground sloth (*Nothrotheriops shastense* Hoffstetter) at Shelter Cave, New Mexico: environment, diet, and extinction. *Quaternary Research*, 14, 360–376.

Van Dover, C.L., German, C.R., Speer, K.G., Parson, L.M., and Vrijenhoek, R.C. (2002). Evolution and biogeography of deep-sea vent and seep invertebrates. *Science*, 295, 1253–1257.

van Konijnenburg-van Cittert, J.H.A. and Schmeissner, S. (1999). Fossil insect eggs on Lower Jurassic plant remains from Bavaria (Germany). *Palaeogeography, Palaeoclimatology, Palaeoecology*, 152, 215–223.

Van Valkenburgh, B. (1985). Locomotor diversity within past and present guilds of large predatory mammals. *Paleobiology*, 11, 406–428.

Van Valkenburgh, B. (1988). Incidence of tooth breakage among large predatory mammals. *American Naturalist*, 131, 291–302.

Van Valkenburgh, B. and Hertel, F. (1993). Tough times at La Brea: tooth breakage in large carnivores of the Late Pleistocene. *Science*, 456–459.

Van Valkenburgh, B. and Ruff, C.B. (1987). Canine tooth strength and killing behaviour in large carnivores. *Journal of Zoology*, 212, 379–397.

Van Valkenburgh, B. and Sacco, T. (2002). Sexual dimorphism, social behavior and intrasexual competition in large Pleistocene carnivores. *Journal of Vertebrate Paleontology*, 22, 164–169.

Van Valkenburgh, B., Teaford, M.F., and Walker, A. (1990). Molar microwear and diet in large carnivores: inferences concerning diet in the sabretooth cat, *Smilodon fatalis*. *Journal of Zoology*, 222, 319–340.

Van Valkenburgh, B., Wang, X., and Damuth, J. (2004). Cope's rule, hypercarnivory, and extinction in North American canids. *Science*, 306, 101–104.

Vanbaeren, M., d'Errico, F., Stringer, C., James, S.L., Todd, J.A., and Mierrj, H.K. (2006). Middle Paleolithic shell beads in Israel and Algeria. *Science*, 312, 1785–1788.

Vander Wall, S.B. (1990). *Food Hoarding in Animals*. University of Chicago Press, Chicago, IL, 445 pp.

Vaněk, J. (1995). New deeper-water trilobites in the Ordovician of the Prague Basin (Czech Republic). *Palaeontologia Bohemiae*, 1, 1–12.

Vannier, J. and Chen, J.-Y. (2005). Early Cambrian food chains: new evidence from fossil aggregates in the Maotianshan Shale biota, SW China. *Palaios*, 20, 3–26.

Vannier, J., Thiéry, A., and Racheboeuf, P.R. (2003). Spinicaudatans and ostracods (Crustacea) from the Montceau Lagerstätte (Late Carboniferous, France): morphology and palaeoenvironmental significance. *Palaeontology*, 46, 999–1030.

Vannier, J., Steiner, M., Renvoisé, E., Hu, S.-X., and Casanova, J.-P. (2007). Early Cambrian origin of modern food webs: evidence from predator arrow worms. *Philosophical Transactions of the Royal Society of London B*, 274, 627–633.

Varricchio, D.J. (2001). Gut contents from a Cretaceous tyrannosaurid: implications for theropod dinosaur digestive tracts. *Journal of Paleontology*, 75, 401–406.

Varricchio, D.J., Jackson, F., and Trueman, C.N. (1999). A nesting trace with eggs for the Cretaceous theropod *Troodon formosus*. *Journal of Vertebrate Paleontology*, 19, 91–100.

Varricchio, D.J., Martin, A.J., and Katsura, Y. (2007). First trace and body fossil evidence of a burrowing, denning dinosaur. *Proceedings of the Royal Society London B*, 274, 1361–1368.

Varricchio, D.J., Sereno, P.C., Zhao Xijin, Tan Lin, Wilson, J.A., and Lyon, G.H. (2008). Mud-trapped herd captures evidence of distinctive dinosaur sociality. *Acta Palaeontologica Polonica*, 53, 567–579.

Vartanyan, S.L., Garutt, V.E., and Sher, A.V. (1993). Holocene dwarf mammoths from Wrangel Island in the Siberian Arctic. *Nature*, 362, 337–340.

Vasileiadou, K., Hooker, J.J., and Collinson, M.E. (2007). Taphonomic evidence of a Paleogene mammalian predator–prey interaction. *Palaeogeography, Palaeoclimatology, Palaeoecology*, 243, 1–22.

Vasilenko, D.V. (2005). Damages on Mesozoic plants from the Transbaikalian locality, Chernovskie Kopi. *Paleontological Journal*, 39(6), 628–633.

Vasilenko, D.V. (2006). Kraevie pobrezhdeniya listev khvoinikh i ginkgovokh iz Mezozoya Zabailkalya. *Paleontologicheskii Zhurnal*, 3, 53–55.

Vasilenko, D.V. (2007). Povrezhdeniya rastenii iz verkhnepermskikh otlozhenii po R. Sukhone. *Paleontologicheskii Zhurnal*, 2, 87–90 (plate with figures, No. XII, missing from journal).

Vasilenko, D.V. (2008). Kladki yaits nasekomikh na listyakh pravaootsshikh rastenii *Quereuxia* iz verkhnego Mela Priamurya. *Palaeontologicheskii Zhurnal*, 5, 60–66.

Vega, F.J., Cosma, T., Countiño, M.A., Feldmann, R.M., Nyborg, T.G., Schweitzer, C.E., and Waugh, D.A. (2001). New Middle Eocene decapods (Crustacea) from Chiapas, México. *Journal of Paleontology*, 75, 929–946.

Vega, F.J., Garcia-Barreta, P., Perilliat, M. del C., Coutiño, M.A., and Mariño, R. (2006). El Espinal, a new plattenkalk facies locality from the Lower Cretaceous Sierra Madre Formation, Chiapas, southeastern Mexico. *Revista Mexicana de Ciencias Geológicas*, 23, 323–333.

Velcescu, M. (2000). Epizoans on the Rupelian bivalves *Pycnodonta* (*Pycnodonta*) *gigantica gigantica* (Sol.) level: their relationships to their host. *Acta Palaeontologica Romaniae*, 2, 483–488.

Vénec-Peyré, M.-T. (1996). Bioeroding Foraminifera: a review. *Marine Micropaleontology*, 28, 19–30.

Vermeij, G.J. (1998). *Sabia* on shells: a specialized Pacific-type commensalism in the Caribbean Neogene. *Journal of Paleontology*, 72, 465–472.

Vetter, R.D. (1991). Symbiosis and the evolution of novel trophic strategies: thiotrophic organisms at hydrothermal vents. In *Symbiosis as a Source of Evolutionary Innovation* (pp. 219–245), edited by L. Margulis and R. Fester. MIT Press, Boston.

Vianey-Liaud, M. (1976). Les Issiodoromyinae (Rodentia, Theridomyidae) de l'Eocène supérieur à l'Oligocène supérieur en Europe occidentale. *Palaeovertebrata*, 7(1–2), 1–149.

Vianey-Liaud, M. (1979). Evolution des rongeurs à l'Oligocène en Europe occidentale. *Palaeontographica A*, 166, 136–236.

Vianey-Liaud, M. (1989). Parallelisme chez les Theridomyinae (Rodentia) de l'Oligocène: étude de deux nouvelles espèces des genres *Theridomys* et *Blainvillimys*. *Neues Jahrbuch für Geologie und Paläontologie–Abhandlungen*, 178, 203–241.

Vianey-Liaud, M. and Lopez-Martinez, N. (1997). Late Cretaceous dinosaur eggshells from the Tremp Basin, southern Pyrenees, Lleida, Spain. *Journal of Paleontology*, 71, 1157–1171.

Vianey-Liaud, M. and Ringeade, M. (1992). La radiation des Theridomyidae (Rodentia) hypsodontes à l'Eocène supérieur. *Géobios*, 26, 455–495.

Vianey-Liaud, M. and Schmidt-Kittler, N. (1987). Biostratigraphie de l'Oligocène d'Europe: importance des lignées-guides de rongeurs Theridomyidae, et particulièrement des *Issiodoromys*. *Münchner Geowissenschaftliche Abhandlungen*, 10, 211–216.

Vianey-Liaud, M., Hirsch, K., Sahni, A., and Sige, B. (1997). Late Cretaceous Peruvian eggshells and their relationships with Laurasian and Eastern Gondwanan materials. *Géobios*, 30, 75–90.

Vickery, V.R. and Poinar, Jr., G.O. (1994). Crickets (Grylloptera: Grylloidea) in Dominican amber. *The Canadian Entomologist*, 126, 13–22.

Vignaud, P., Duringer, P., Mackaye, H.T., Likius, A., Blondel, C. et al. (2002). Geology and palaeontology of the Upper Miocene Toros-Menalla hominid locality, Chad. *Nature*, 418, 152–155.

Vinn, O. (2006). Early cornulitids from the Ordovician of China. *The Palaeontological Association Newsletter*, 62, 109–111.

Vinn, O. and Mutvei, H. (2005). Observations on the morphology and affinities of cornulitids from the Ordovician of Anticosti Island and the Silurian of Gotland. *Journal of Paleontology*, 79, 726–737.

Vinogradov, A.V. (1996). New fossil freshwater bryozoans from the Asiatic part of Russia and Kazakhstan. *Paleontological Journal*, 30(3), 284–292.

Viohl, G. (1990). Piscivorous fishes of the Solnhofen lithographic limestone. In *Evolutionary Paleobiology of Behavior and Coevolution* (pp. 287–303), edited by A.J. Boucot. Elsevier, Amsterdam.

Vishnu-Mittre (1957). Fossil galls on some Jurassic conifer leaves. *Current Science*, 26(7), 210–211.

Vishnyakova, V.N. (1980a). Otryad Blattida. In "Istorischeskoe razvitie klasse nasekomykh," edited by B.B. Rohdendorf and A.P. Rasnitsyn. *Transactions of the Paleontological Institute, Academy of Science, USSR*, 175, 138–145.

Vishnyakova, V.N. (1980b). Earwigs from the Upper Jurassic of the Karatau range. *Paleontological Journal*, 1, 78–95.

Vogel, K. (1966). Eine funktionsmorphologische Studie an der Brachiopodengattung *Pygope* (Malm bis Unterkreide). *Neues Jahrbuch für Geologie und Paläontologie–Abhandlungen (Festband Schindewolf)*, 125, 423–442.

Vogel, K. (1984). *Lebensweise und Umwelt fossiler Tiere, Biologische Arbeitsbucher*, Vol. 39. Quelle & Meyer, Heidelberg, 171 pp.

Vogel, K. (1987). Bohrorganismen und Fazies im Mitteldevon des Staates New York, USA. *Natur und Museum*, 117(7), 207–216.

Vogel, K. (1993). Bioeroders in fossil reefs. *Facies*, 28, 109–114.

Vogel, K., Golubic, S., and Brett, C.E. (1987). Endolith associations and their relations to facies distribution in the Middle Devonian of New York State, USA. *Lethaia*, 20, 263–290.

Voigt, E. (1959). *Endosacculus moltkiae* n. g. n. sp., ein vermutlicher fossiler Ascothoracide (Entomostr.) als Cystenbildner bei Oktokoralle *Moltkeia minuta*. *Paläontologisches Zeitschrift*, 33, 211–233.

Voigt, E. (1967). Über einen neuen acrothoraciden Cirripedier aus dem Essener Grünsand (Cenoman). *Abhandlungen und Verhandlungen des Naturwissenschaftlichen Vereins in Hamburg*, 11, 117–121.

Voigt, E. (1981). Upper Cretaceous bryozoan–seagrass association in the Maastrichtian of the Netherlands. In *Recent and Fossil Bryozoa* (pp. 281–298), edited by G.P. Larwood and C. Nielsen. Olsen & Olsen, Fredensborg, Denmark.

Vokes, H.E. (1955). Cenozoic pearls from the Atlantic Coastal Plain. *Journal of the Washington Academy of Sciences*, 45, 260–262.

von Bitter, P.H., Scott, S.D., and Schenk, P.E. (1990). Early Carboniferous low-temperature hydrothermal vent communities from Newfoundland. *Nature*, 344, 145–148.

von Bitter, P.H., Scott, S.D., and Schenk, P.E. (1992). Chemosynthesis: an alternate hypothesis for Carboniferous biotas in bryozoan/microbial mounds, Newfoundland, Canada. *Palaios*, 7, 466–484.

von Bitter, P.H., Purnell, M.A., Tetreault, D.K., and Stott, C.A. (2007). Exceptionally preserved soft-bodied biotas with shallow-marine shelly and bioturbating organisms (Silurian, Ontario, Canada). *Geology*, 35, 879–882.

von Koenen, A. (1891). Das Norddeutsche Unter-Oligocän und seine Mollusken-Fauna. *Abhandlungen zur geologischen Specialkarte von Preussen und den Thuringischen Staaten*, X(3), 575–818.

von Koenen, A. (1892). Das Norddeutsche Unter-Oligocän und seine Mollusken-Fauna. *Abhandlungen zur geologischen Specialkarte von Preussen und den Thuringischen Staaten*, X(4), 819–1004.

von Koenigswald, W. (1987). Die Fauna des Ölschiefers von Messel. In *Fossilien der Messel-Formation* (pp. 73–142), edited by P.R. Heil, W. von Koenigswald, H.G. Lippmann, D. Graner, and C. Heunisch. Hessisches Landesmuseum, Darmstadt.

von Koenigswald, W. (1990). Die Paläobiologie der Apatemyiden (Insectivora s.l.) und die Ausdeutung der Skelettfunds von *Heterohyus nanus* aus dem Mitteleozän von Messel bei Darmstadt. *Palaeontographica Abteilung A*, 210, 41–77.

von Koenigswald, W. and Schierning, H.-P. (1987). The ecological niche of an extinct group of mammals, the early Tertiary apatemyids. *Nature*, 326, 595–597.

von Koenigswald, W., Storch, G., and Richter, G. (1992a). Primitive insectivores, extraordinary hedgehogs, and long-fingers. In *Messel: An Insight into the History of Life and of the Earth* (pp. 161–177), edited by S. Schaal and W. Ziegler. Clarendon Press, London.

von Koenigswald, W., Storch, G., and Richter, G. (1992b). Rodents: at the start of a great career. In *Messel: An Insight into the History of Life and of the Earth* (pp. 219–222), edited by S. Schaal and W. Ziegler. Clarendon Press, London.

von Koenigswald, W., Braun, A., and Pfeiffer, T. (2004). Cyanobacteria and seasonal death: a new taphonomic model for the Eocene Messel lake. *Paläontologische Zeitschrift*, 78, 417–424.

von Koenigswald, W., Rose, K.D., Grande, L., and Martin, R.D. (2005). First Apatemyid skeleton from the Lower Eocene Fossil Butte Member, Wyoming (USA), compared to the European Apatemyid from Messel, Germany. *Palaeontographica Abteilung A*, 272, 149–169.

von Koenigswald, W., Smith, B.H., and Keller, T. (2007). Supernumerary teeth in a subadult rhino mandible (*Stephanorhinus hundsheimensis*) from the middle Pleistocene of Mosbach in Wiesbaden (Germany). *Paläontologische Zeitschrift*, 81, 416–428.

Voorhies, M.R. (1981). Pompeii of animals in Nebraska. *National Geographic Magazine*, 159(1), 66–75.

Voronets, N.S. (1952). First findings of the egg capsules of chimaeras in the USSR. *Doklady Akademia Nauk SSSR*, 84, 587–589.

Voss-Foucart, M.F. (1968). Paléoproteines des coquilles fossiles d'oeufs de Dinosauriens de Crétace supérieur de Provence. *Comparative Biochemistry and Physiology*, 24, 31–36.

Waage, J.K. (1976). Insect remains from ground sloth dung. *Journal of Paleontology*, 50, 991–994.

Wagensberg, J., Brandão, C.R.F., and Baroni Urbani, C. (1996). Le mystère de la chambre jaune. *La Recherche*, 288, 54–59.

Waggoner, B.M. (1993). Fossil actinomycetes and other bacteria in Eocene amber from Washington State, USA. *Tertiary Research*, 14, 155–160.

Waggoner, B.M. (1994). Fossil microorganisms from Upper Cretaceous amber of Mississippi. *Review of Palaeobotany and Palynology*, 80, 75–84.

Waggoner, B.M. (1996). Bacteria and protists from Middle Cretaceous amber of Ellsworth County, Kansas. *Paleobios*, 17, 20–26.

Waggoner, B.M. (1999). Fossil oak leaf galls from the Stinking Water paleoflora of Oregon (Middle Miocene). *PaleoBios*, 19(3), 8–14.

Waggoner, B.M. and Poteet, M.F. (1996). Unusual oak leaf galls from the Middle Miocene of northwestern Nevada. *Journal of Paleontology*, 70, 1080–1084.

Wagner, H.M. and Marks, E. (1990). *Final paleontologic report Laband Village, Tract 13651, Phase II (Areas 11–16), Chino Hills, San Bernardino County, California*, Project No. 87119-20. Zeiser Geotechnical, Costa Mesa, CA.

Wahl, W. (1998). Description and identification of fossil coleoid hooks from the Upper Redwater Shale (Lower Oxfordian) of the Sundance formation (Jurassic) of Wyoming. In *Fifth International Symposium on the Jurassic System, International Union of Geological Sciences, August 17–20, 1998, Vancouver, Abstracts and Program*, p. 93.

Walaszczyk, I. and Cobban, W.A. (2000). Inoceramid faunas and biostratigraphy of the Upper Turonian–Lower Coniacian of the Western Interior of the United States. *Special Papers in Palaeontology* 64, 1–118.

Walcott, C.D. (1883). Injury sustained by the eye of a trilobite at the time of the moulting of the shell. *American Journal of Science*, 62, 302.

Waldron, T. (1994). *Counting the Dead: The Epidemiology of Skeletal Populations*. John Wiley & Sons, New York, 109 pp.

Walker, S.E. (1992). Criteria for recognizing marine hermit crabs in the fossil record using gastropod shells. *Journal of Paleontology*, 66, 535–558.

Walker, S.E. (2001). Paleoecology of gastropods preserved in turbiditic slope deposits from the Upper Pliocene of Ecuador. *Palaeogeography, Palaeoclimatology, Palaeoecology*, 166, 141–163.

Walker, S.E. (2007). Traces of gastropod predation on molluscan prey in tropical reef environments. In *Trace Fossils: Concepts, Problems, Prospects* (pp. 324–344), edited by W. Miller III. Elsevier, Amsterdam.

Walker, S.E. and Yamada, S.B. (1993). Implications for the gastropod fossil record of mistaken crab predation on empty mollusk shells. *Palaeontology*, 36, 735–741.

Walker, S.E., Holland, S.M., and Gardiner, L. (2003). *Coenobichnus currani* (new ichnogenus and ichnospecies): fossil trackway of a land hermit crab, early Holocene, San Salvador, Bahamas. *Journal of Paleontology*, 77, 576–582.

Wallace, F.G. (1966). The trypanosomatid parasites of insects and Arachnids. *Experimental Parasitology*, 18, 124–193.

Walossek, D. and Müller, K.J. (1990). Upper Cambrian stem-lineage crustaceans and their bearing on the monophyletic origin of Crustacea and the position of *Agnostus*. *Lethaia*, 23, 409–427.

Walossek, D. and Müller, K.J. (1994). Pentastomid parasites from the Lower Palaeozoic of Sweden. *Transactions of the Royal Society of Edinburgh, Earth Sciences*, 85, 1–37.

Walsh, S.A. and Hume, J.P. (2001). A new Neogene avian assemblage from north-central Chile. *Journal of Vertebrate Paleontology*, 21, 484–491.

Walter, H. and Werneburg, R. (1988). Uber Liegespuren (Cubichnia) aquatischer Tetrapoden (Diplocauliden, Nectridea) aus den Rotteroder Schichten (Rotliegendes, Thuringer Wald/DDR). *Freiberger Forschungshefte (Leipzig)*, C419, 96–106.

Wang, H.-J. (1965). Note on some opercula of the Cenozoic Gastropoda from Kiangsu Province. *Acta Palaeontologica Sinica*, 13, 618–628.

Wang, H.-J. (1980). Palaeogene gastropod opercula from China. *Acta Palaeotologica Sinica*, 19, 311–320.

Wang, H.-J. (1984). Two Upper Jurassic gastropod opercula in China. *Acta Palaeontologica Sinica*, 23, 369–372.

Wang, X. and Rothschild, B.M. (1992). Multiple hereditary osteochondroma in Oligocene *Hesperocyon* (Carnivora: Canidae). *Journal of Vertebrate Paleontology*, 12, 387–394.

Wang, X. and Zhou, Z. (2004). Pterosaur embryo from the Early Cretaceous. *Nature*, 429, 621.

Wang, X., Hagdorn, H., and Wang, C. (2006). Pseudoplanktonic lifestyle of the Triassic crinoid *Traumatocrinus* from Southwest China. *Lethaia*, 39, 187–193.

Wang, X., Bachmann, G.H., Hagdorn, H., Sanders, P.M., Cuny, G. et al. (2008). The Late Triassic black shales of the Guanling Area, Guizhou Province, South-West China: a unique marine reptile and pelagic crinoid fossil lagerstätte. *Palaeontology*, 51, 27–61.

Wang, Z.-Q. (1997). Permian *Supaia* fronds and an associated *Autunia* fructification from Shanxi, China. *Palaeontology*, 40, 245–277.

Wappler, T. and Ben-Dov, Y. (2008). Preservation of armoured scale insects on angiosperm leaves from the Eocene of Germany. *Acta Palaeontologica Polonica*, 53, 627–634.

Wappler, T. and Engel, M.S. (2003). The Middle Eocene bee faunas of Eckfeld and Messel, Germany (Hymenoptera: Apoidea). *Journal of Paleontology*, 77, 908–921.

Wappler, T., Smith, V.S., and Dalgleish, R.C. (2004). Scratching an ancient itch: an Eocene bird louse fossil. *Proceedings of the Royal Society London B*, 271(Suppl.), S255–S258.

Ward, D.J. and Hollingworth, N.T.J. (1990). The first record of a bitten ammonite from the Middle Oxford Clay (Callovian, Middle Jurassic) of Bletchley, Buckinghamshire. *Mesozoic Research*, 2, 153–161.

Ward, L.W. and Blackwelder, B.W. (1975). *Chesapecten*, a new genus of Pectinidae (Mollusca: Bivalvia) from the Miocene and Pliocene of eastern North America. *U.S. Geological Survey Professional Paper*, 861, 1–24.

Waren, A. (1980). Revision of the genera *Thyca, Stilifer, Scalenostoma, Mucronalia* and *Echineulima* (Mollusca, Prosobranchia, Eulimidae). *Zoologica Scripta*, 9, 187–210.

Waren, A. (1981). Bloodsucking snails: the Eulimidae. *La Conchiglia*, 146–147, 7–10.

Waren, A. (1983). A generic revision of the family Eulimidae (Gastropoda, Prosobranchia). *The Journal of Molluscan Studies*, 3(Suppl.), 1–96.

Waren, A. and Bouchet, P. (2001). Gastropods and Monoplacophora from hydrothermal vents and seeps: new taxa and records. *The Veliger*, 44, 116–231.

Warren, A. and Ptasznik, R. (2002). The earliest fractured tetrapod bone. *Alcheringa*, 26, 459–463.

Warren, P.S. (1947). Chimaeroid egg capsules from Alberta. *Geological Society of America Bulletin*, 58, 1238.

Wassersug, R.J. and Wake, D.B. (1995). Fossil tadpoles from the Miocene of Turkey. *Alytes*, 12, 145–157.

Waterhouse, J.B. (1973). Communal hierarchy and significance of environmental parameters for brachiopods: the New Zealand Permian model. *Royal Ontario Museum, Life Science Contributions*, 92, 1–49.

Watkins, R. (1974a). Carboniferous brachiopods from northern California. *Journal of Paleontology*, 48, 304–325.

Watkins, R. (1974b). Palaeobiology of an offshore molluscan fauna from the California Oligocene. *Palaeogeography, Palaeoclimatology, Palaeoecology*, 15, 245–266.

Watkins, R. (1974c). Molluscan *paleobiology* of the Miocene Wimer Formation, Del Norte County, California. *Journal of Paleontology*, 48, 1264–1282.

Watkins, R. (1981). Epizoan ecology in the type Ludlow Series (Upper Silurian), England. *Journal of Paleontology*, 55, 29–32.

Watkins, R. (1990). Pliocene channel deposits of oyster shells in the Salton Trough region, California. *Palaeogeography, Palaeoclimatology, Palaeoecology*, 79, 249–262.

Watson, D.M.S. (1927). The reproduction of the coelacanth fish *Undina*. *Proceedings of the Zoological Society of London*, 2, 453–457.

Waugh, D.A., Feldmann, R.M., Crawford, R.S., Jakobsen, S.L., and Thomas, K.B. (2004). Epibiont preservational and observational bias in fossil and marine decapods. *Journal of Paleontology*, 78, 961–972.

Weaver, D.W. (1963). Transposed hinge of a crassatellid. *Journal of Paleontology*, 37, 294–298.

Weaver III, J.S. and Morse, J.C. (1986). Evolution of feeding and case-making behaviour in Trichoptera. *Journal of the North American Benthological Society*, 5, 150–158.

Weaver, L., McLoughlin, S., and Drinnan, A.N. (1997). Fossil woods from the Upper Permian Bainmedart Coal Measures, northern Prince Charles Mountains, East Antarctica. *Journal of Australian Geology and Geophysics*, 16, 655–676.

Wedmann, S. and Richter, G. (2007). The ecological role of immature phantom midges (Diptera: Chaoboridae) in the Eocene Lake Messel, Germany. *African Invertebrates*, 48, 59–70.

Weedon, M.J. (1991). Microstructure and affinity of the enigmatic Devonian tubular fossil *Trypanopora*. *Lethaia*, 24, 227–234.

Wehr, W.C. (1995). Early Tertiary flowers, fruits and seeds of Washington State and adjacent area. *Washington Geology*, 23(3), 3–16.

Weidner, H. (1958). Einige interessante Insektenlarven aus der Bernsteininklusen-Sammlung des Geologischen Staatsinstituts Hamburg (Odonata, Coleoptera, Megaloptera, Plannipennia). *Mitteilungen aus dem Geologischen Staatsinstitut in Hamburg*, 27, 50–68.

Weitschat, W. (1980). Leben im Bernstein. *Geologisch–Paläontologisches Institut der Universität Hamburg*, 48 pp.

Weitschat, W. and Wichard, W. (1998). *Atlas der Pflanzen und Tieren im Baltischen Bernstein*. Verlag Dr. Friedrich Pfeil, München, 256 pp.

Welles, S.P. and Bump, J.D. (1949). *Alzadasaurus pembertoni*, a new elasmosaur from the Upper Cretaceous of South Dakota. *Journal of Paleontology*, 23, 521–535.

Wellnhofer, P. (1978). *Handbuch der Paläoherpetologie*. Teil 19. *Pterosauria*. Verlag Gustav Fischer, Stuttgart, 82 pp.

Wellnhofer, P. (1991). *The Illustrated Encyclopedia of Pterosaurs*. Salamander Books, London, 192 pp.

Wen, Y. (1993). Early Ordovician gastropods from the Canning Basin, Western Australia. *Records of the Western Australian Museum*, 16, 437–458.

Wen, Y., Gu, H.-L., and Zhang, X.-Q. (1990). Assemblage sequence of Late Cretaceous and Early Tertiary non-marine gastropods from Nanxiong Basin, Guangdong. *Acta Palaeontologica Sinica*, 160–182.

Wenzel, J.W. (1990). A social wasp's nest from the Cretaceous Period, Utah, USA, and its biogeographical significance. *Psyche*, 97, 21–29.

Wermelinger, E.D. and Zanuncio, J.C. (2001). Development of *Lutzomyia intermedia* and *Lutzomyia longipalpis* (Diptera: Psychodidae: Phlebotominae) larvae in different diets. *Brazilian Journal of Biology*, 61, 405–408.

Werneburg, R. (1986). Die Stegocephalen (Amphibia) der Goldlauterer Schichten (Unterrotliegendes, Perm) des Thüringer Waldes. Teil I. *Apateon flagrifer* (Whittard). *Freiberger Forschungshefte*, C410, 87–100.

Werneburg, R. (1988). Die Stegocephalen der Goldlauterer Schichten (Unterrotliegendes, Unterperm). Teil II. *Apateon kontheri* n. sp., *Melanerpeton eisfeldi* n. sp. des Thüringer Waldes und andere. *Freiberger Forschungshefte*, C427, 7–29.

Werneburg, R. (1989). Die Amphibienfauna der Manebacher Schichten (Unterrotliegendes, Unterperm) des Thüringer Waldes. *Veröffentlichungen des Naturhistorischen Museum Schloss Bertholdsburg, Schleusingen*, 4, 55–68.

Werneburg, R. (2004). First record of postcranial lateral lines in fossil amphibians (Branchiosauridae; Rotliegend, Upper Carboniferous–Lower Permian). *Neues Jahrbuch für Geologie und Paläontologie–Abhandlungen*, 232, 365–373.

Werneburg, R., Ronchi, A., and Schneider, J.W. (2007). The Early Permian branchiosaurids (Amphibia) of Sardinia (Italy): systematic palaeontology, palaeoecology, biostratigraphy and palaeobiogeographic problems. *Palaeogeography, Palaeoclimatology, Palaeoecology*, 252, 383–404.

Westergård, A.H. (1936). *Paradoxides oelandicus* Beds of Öland with the account of a diamond boring through the Cambrian at Mossberga. *Sveriges Geologiska Undersökning*, C394, 1–66.

Westermann, G.E.G., Ed. (1969). *Sexual Dimorphism in Fossil Metazoa and Taxonomic Implications*. Schweizerbart, Stuttgart, 251 pp.

Westgate, J.W. and Gee, C.T. (1990). Paleoecology of a middle Eocene mangrove biota (vertebrates, plants, and invertebrates) from southwest Texas. *Palaeogeography, Palaeoclimatology, Palaeoecology*, 78, 163–177.

Wetzel, W. (1960). Nachtrag sum Fossilarchiv der Quiriquina-Schichten. *Neues Jahrbuch für Geologie und Paläontologie–Monatsheft*, 439–446.

Wheeler, E.A., Lehman, T.M., and Gasson, P.E. (1994). *Javelinoxylon*, an Upper Cretaceous dicotyledonous tree from Big Bend National Park, Texas, with presumed malvalean affinities. *American Journal of Botany*, 81, 703–710.

Wheeler, W.M. (1973). *The Fungus-Growing Ants of North America*. Dover, New York, 136 pp.

Whitfield, R.P. (1885). Brachiopoda and Lamellibranchiata of the Raritan clays and Greensand marls of New Jersey. *Monographs of the U.S. Geological Survey*, 9, 1–269.

Whittington, H.B. (1956). Silicified Middle Ordovician trilobites: the Odontopleuridae. *Bulletin of the Museum of Comparative Zoology, Harvard University*, 114, 155–288.

Whittington, H.B. (1978). The lobopod animal *Aysheaia pedunculata* Walcott, Middle Cambrian, Burgess Shale, British Columbia. *Philosophical Transactions of the Royal Society of London B*, 284, 165–197.

Whittington, H.B. (1985). *The Burgess Shale*. Yale University Press, New Haven, CT, 151 pp.

Whittle, C.H. and Everhart, M.J. (2000). Apparent and implied evolutionary trends in lithophagic vertebrates from New Mexico and elsewhere. In *Dinosaurs of New Mexico* (pp. 75–82), edited by S.G. Lucas and A.B. Heckert, Bulletin No. 17. New Mexico Museum of Natural History and Science, Albuquerque.

Wichard, W. and Weitschat, W. (2004). *Im Bernsteinwald*. Gerstenberg Verlag, Hildesheim, 168 pp.

Wiedman, L.A., Feldmann, R.M., Lee, D.E., and Zinsmeister, W.J. (1988). Brachiopods from the La Meseta Formation (Eocene), Seymour Island, Antarctica. *Geological Society of America Memoir*, 169, 449–457.

Wiedmann, H.U. (1972). Shell deposits and shell preservation in Quaternary and Tertiary estuarine sediments in Georgia, USA. *Sedimentary Geology*, 7, 103–125.

Wieland, G.R. (1909). Revision of the Protosteganidae. *American Journal of Science, 4th Series*, 27, 101–130.

Wiens, J.J., Brandley, M.C., and Reeder, T.W. (2006). Why does a trait evolve multiple times within a clade? Repeated evolution of snakelike body form in squamate reptiles. *Evolution*, 60, 123–141.

Wiggins, G.B. (2004). *Caddisflies: The Underwater Architects*. University of Toronto Press, Toronto, 292 pp.

Wignall, P.B. and Simms, M.J. (1990). Pseudoplankton. *Palaeontology*, 33, 359–378.

Wijmstra, T.A. and Groenhart, M.C. (1983). Record of 700,000 years of vegetational history in eastern Macedonia (Greece). *La Revista de la Academia Colombiana de Ciencias Exactas, Fisicas y Naturales*, 15, 87–98.

Wilby, P.R. and Martill, D.M. (1992). Fossil fish stomachs: a microenvironment for exceptional preservation. *Historical Biology*, 6, 25–36.

Wilf, P. and Labandeira, C.C. (1999). Response of plant–insect associations to Paleocene–Eocene warming. *Science*, 284, 2153–2156.

Wilf, P., Labandeira, C.C., Kress, W.J., Staines, C.L., Windsor, D.M., Allen, A.L., and Johnson, K.R. (2000). Timing the radiation of leaf beetles: hispines on gingers from latest Cretaceous to Recent. *Science*, 289, 291–294.

Wilf, P., Labandeira, C.C., Johnson, K.R., and Ellis, B. (2006). Decoupled plant and insect diversity after the end-Cretaceous extinction. *Science*, 313, 1112–1115.

Wilke, P.J. and Hall, H.J. (1975). *Analysis of Ancient Feces: A Discussion and Annotated Bibliography*. Archaeological Research Facility, Department of Anthropology, University of California, Berkeley, 47 pp.

Willerslev, E., Hansen, A.J., Binladen, J., Brand, T.B., Gilbert, M.T.P.G., Shapiro, B., Bunce, M., Wiuf, C., Gilichinsky, D.A., and Cooper, A. (2003). Diverse plant and animal records from Holocene and Pleistocene sediments. *Science*, 300, 791–795.

Willey, P.R., Wilkinson, I.P., and Riley, N.J. (2006). Late Carboniferous scavenging ostracods: feeding strategies and taphonomy. *Transactions of the Royal Society of Edinburgh, Earth Sciences*, 96, 309–316.

Williams, A. (2003). Microscopic imprints on the juvenile shells of Palaeozoic linguliform brachiopods. *Palaeontology*, 46, 67–92.

Williams, A. and Rowell, A.J. (1965). Morphology. In *Treatise on Invertebrate Paleontology*. Part H. *Brachiopoda 1* (pp. 57–138), edited by R.L. Kaesler. Geological Society of America, Boulder, CO.

Williams, M.E. (1990). Feeding behavior in Cleveland Shale fishes. In *Evolutionary Paleobiology of Behavior and Coevolution* (pp. 273–287), edited by A.J. Boucot. Elsevier, Amsterdam.

Williamson, T.E. (1996). ?*Brachychampsa sealeyi*, sp. nov. (Crocodylia, Alligatoroidea) from the Upper Cretaceous (lower Campanian) Menefee Formation, northwestern New Mexico. *Journal of Vertebrate Paleontology*, 16, 421–431.

Willis, C. (1996). *Yellow Fever, Black Goddess*. Addison-Wesley, New York, 324 pp.

Willis, K.J., Kieczkowski, A., and Crowhurst, S.J. (1999). 124,000-year periodicity in terrestrial vegetation change during the late Pliocene epoch. *Nature*, 397, 685–688.

Williston, S.W. (1899). Some additional characters of the mosasaurs. *Kansas University Quarterly*, 8, 39–41.

Wills, L.U. (1963). *Cyprilepas holmi* Wills 1962, a pedunculate cirripede from the Upper Silurian of Oesel, Esthonia. *Palaeontology*, 6, 161–165.

Wills, M.A. (1993). Miscellania. In *The Fossil Record*, Vol. 2 (pp. 555–560), edited by M.J. Benton. Chapman & Hall, London.

Wilson, E.O. (1971). *The Insect Societies*. Harvard University Press, Cambridge, MA, 548 pp.

Wilson, H.M. (2006). Aggregation behaviour in juvenile millipedes from the Upper Carboniferous of Mazon Creek, Illinois. *Palaeontology*, 49, 733–740.

Wilson, M.A. (1986). Coelobites and spatial refuges in a Lower Cretaceous cobble-dwelling hardground fauna. *Palaeontology*, 29, 691–703.

Wilson, M.A. and Palmer, T.J. (1998). The earliest *Gastrochaenolites* (Early Pennsylvanian, Arkansas, USA): an Upper Paleozoic bivalve boring? *Journal of Paleontology*, 72, 769–772.

Wilson, M.A. and Taylor, P.D. (2001). Palaeoecology of hard substrate faunas from the Cretaceous Qahlah Formation of the Oman Mountains. *Palaeontology*, 44, 21–41.

Wilson, M.A. and Taylor, P.D. (2006). Predatory drill holes and partial mortality in Devonian colonial metazoans. *Geology*, 34, 565–568.

Wilson, M.I. and Daly, M. (1996). Male sexual proprietariness and violence against women. *Current Directions in Psychological Science*, 5(1), 2–7.

Wings, O. (2007). A review of gastrolith function with implications for fossil vertebrates and a revised classification. *Acta Palaeontologica Polonica*, 52, 1–16.

Wings, O. and Martin Sander, P. (2007). No gastric mill in sauropod dinosaurs: new evidence from analysis of gastrolith mass and function in ostriches. *Transactions of the Royal Society of London B*, 274, 635–640.

Winkler, D.A. and Murry, P.A. (1989). Paleoecology and hypsilophodontid behavior at the Proctor Lake dinosaur locality (Early Cretaceous), Texas. *Geological Societey of America Memoir*, 238, 55–61.

Wisshak, M. (2007). Two new dwarf *Entobia* ichnospecies in a diverse aphotic ichnocoenosis (Pleistocene; Rhodes, Greece). In *Current Developments in Bioerosion* (pp. 213–234), edited by M. Wisshak and L. Tapanila. Springer-Verlag, Heidelberg.

Wisshak, M. and Neumann, C. (2006). A symbiotic association of a boring polychaete and an echinoid from the Late Cretaceous of Germany. *Acta Palaeontographica Polonica*, 51, 589–597.

Wisshak, M. and Tapanila, L., Eds. (2007). *Current Developments in Bioerosion*. Springer-Verlag, Heidelberg, 499 pp.

Wisshak, M., Volohonsky, E., and Blomeier, D. (2004). Acanthodian fish trace fossils from the Early Devonian of Spitsbergen. *Acta Palaeontologica Polonica*, 49, 629–634.

Wisshak, M., Seuss, B., and Nützel, A. (2007). Evolutionary implications of exceptionally preserved Carboniferous microboring assemblage in the Buckhorn Asphalt lagerstätte (Oklahoma, USA). In *Current Developments in Bioerosion* (pp. 21–54), edited by M. Wisshak and L. Tapanila. Springer-Verlag, Heidelberg.

Withers, T.H. (1924). The fossil cirripedes of New Zealand. *New Zealand Geological Survey, Paleontological Bulletin*, 10, 1–47.

Withers, T.H. (1928). *Catalogue of fossil Cirripedia in the Department of Geology. I. Triassic and Jurassic.* British Museum of Natural History, London.

Withers, T.H. (1935). *Catalogue of Fossil Cirripedia in the Department of Geology. II. Cretaceous.* British Museum of Natural History, London.

Withers, T.H. (1953). *Catalogue of Fossil Cirripedia in the Department of Geology. III. Tertiary.* British Museum of Natural History, London.

Witmer, L.M., Chatterjee, S., Franzosa, J., and Rowe, T. (2003). Neuroanatomy of flying reptiles and implications for flight, posture and behaviour. *Nature*, 425, 950–953.

Wittlake, E.B. (1969). Fossil phylloxerid plant galls from the Lower Eocene. *Arkansas Academy of Science Proceedings*, 23, 164–167.

Wittlake, E.B. (1981). Fossil galls. In *Neoplasms: Comparative Pathology of Growth in Animals, Plants, and Man* (pp. 729–731), edited by H.E. Kaiser. Williams & Wilkins, Baltimore, MD.

Witzmann, F. (2006). Cranial morphology and ontogeny of the Permo–Carboniferous temnospondyl *Archegosaurus decheni* Goldfuss, 1847 from the Saar-Nahe Basin, Germany. *Transactions of the Royal Society of Edinburgh, Earth Sciences*, 96, 131–162.

Woodring, W.P. (1960). Paleoecologic dissonance: *Astarte* and *Nipa* in the Early Eocene London Clay. *American Journal of Science*, 258A, 418–419.

Woodward, S.R., Weyand, N.J., and Bunnell, M. (1994). DNA sequences from Cretaceous Period bone. *Science*, 266, 1229–1232.

Worthy, T.H. and Holdaway, R.N. (1994). Scraps from an owl's table: predator activity as a significant taphonomic process newly recognized from New Zealand Quaternary deposits. *Alcheringa*, 18, 229–245.

Wright, A.D. (1968). The brachiopod *Dicoelosia biloba* (Linnaeus) and related species. *Arkiv för Zoologi, Series 2*, 20(14), 261–319.

Wright, A.D. (1974). A revision of the Upper Ordovician brachiopod "*Pentamerus angulosus* Törnquist." *Geologiska Föreningens i Stockholm Förhandlingar*, 96, 237–246.

Wright, K. (1991). Dinosaur doctors: tracing modern disease to the ancient reptiles. *Discover*, November, 46–51.

Wu, R. (1996). *Secrets of a Lost World: Dominican Amber and Its Inclusions.* Calle Puerto Rico, Santo Domingo, 222 pp.

Wunderlich, J. (2004). Fossil spiders in amber and copal: conclusions, revisions, new taxa and family diagnoses of fossil and extant taxa. *Beiträge zur Araneologie*, 3A(1), 1–848; 3B(2), 1–1060 pp.

Wunderlich, J. (2008). Fossil and extant spiders (Areneae). *Beiträge zur Araneologie*, 5(1), 1–870.

Wuttke, M. (1989). Die Frösche von Rott und Orsberg. In *Fossillagerstätte* (pp. 53–58), edited by W. von Koenigswald. Rheinlandia Verlag, Siegburg.

Wuttke, M. (1992a). Amphibia at Lake Messel: salamanders, toads, and frogs. In *Messel: An Insight into the History of Life and of the Earth* (pp. 95–98), edited by S. Schaal and W. Ziegler. Clarendon Press, London.

Wuttke, M. (1992b). Conservation—dissolution—transformation: on the behaviour of biogenic materials during fossilization. In *Messel: An Insight into the History of Life and of the Earth* (pp. 265–275), edited by S. Schaal and W. Ziegler. Clarendon Press, London.

Wyse Jackson, P.N., Taylor, P.D., and Tilsley, J.W. (1999). The "Balladoole Coral" from the Lower Carboniferous of the British Isles, reinterpreted as the unusual cystoporate bryozoan *Meekoporella* Moore and Dudley, 1944. *Proceedings of the Yorkshire Geological Society*, 52, 257–268.

Xu, X. and Norell, M.A. (2004). A new troodontid dinosaur from China with avian-like sleeping posture. *Nature*, 431, 838–841.

Yamaguchi, T. (1977). Taxonomic studies on some fossil and Recent Japanese Balanoidea, Part 2. *Transactions and Proceedings of the Palaeontological Society of Japan*, 108, 161–201.

Yamaguchi, T. (1980). A new species belonging to the *Balanus amphitrite* Darwin group (Cirripedia, Balanomorpha) from the Late Pleistocene of Japan: an example of peripheral speciation. *Journal of Paleontology*, 54, 1084–1101.

Yamaoka, M. (1993). Geology of the Morozaki Group. In *Fossils from the Miocene Morozaki Group* (pp. 100–122), edited by F. Ohe, I. Nonogaki, T. Tanaka, K. Hachiya, Y. Mizuno, T. Momoyama, and T. Yamaoka. The Tokai Fossil Society, Nagoya, Japan.

Yang, H. (1997). DNA sequences for paleontologists: new challenges from old molecules. *Palaios*, 12(3), 1–2.

Yang, H. and Yang, S. (1994). The Shanwang fossil biota in eastern China: a Miocene *Konservat-Lagerstätte* in lacustrine deposits. *Lethaia*, 27, 345–354.

Yang, H., Golenberg, E.M., and Shoshani, J. (1996). Phylogenetic resolution within the elephantidae using fossil DNA sequence from the American mastodon (*Mammut americanum*) as an outgroup. *Proceedings of the National Academy of Sciences USA*, 93, 1190–1194.

Yang, S.-Y., Lockley, M.G., Greben, R., Erickson, B.E., and Lim, S.-K. (1995). Flamingo and duck-like bird tracks from the Late Cretaceous and Early Tertiary: evidence and replication. *Ichnos*, 4, 21–34.

Yeh, H.K. (1981). Third note on fossil bird from Miocene of Linqu, Shandong. *Vertebrata PalAsiatica*, 18(2), 149–155.

Yellen, J.E., Brooks, A.S., Cornellisser, E., Mehlman, M.J., and Stewart, K. (1995). A Middle Stone Age worked bone industry from Kelande, upper Semliki Valley, Zaire. *Science*, 268, 553–556.

Yen, T.-C. (1951). Fresh-water mollusks of Cretaceous age from Montana and Wyoming. *U.S. Geological Survey Professional Paper*, 233-A, 20 pp.

Yochelson, E.L. and Fedonkin, M.A. (1993). *Paleobiology of Climactichnites, an enigmatic Late Cambrian fossil.* *Smithsonian Contributions to Paleobiology*, 74, 1–74.

Yochelson, E.L. and Kříž, J. (1974). Platyceratid gastropods from the Oriskany Sandstone (Lower Devonian) near Cumberland, Maryland: synonymies, preservation and color markings. *Journal of Paleontology*, 48, 474–483.

Yochelson, E.L. and Wise, Jr., O.A. (1972). A life association of shell and operculum in the Early Ordovician gastropod *Ceratopea unguis*. *Journal of Paleontology*, 46, 681–684.

Yokoyama, M. and Kodaira, R. (1921). Fossil nutshells of *Juglans sieboldiana*, Maximiovosz in the lignite of Asahiyama, near Nagano City, Province of Shinano. *Chishitsugaku zasshi*, 28, 1–5.

Yonge, C.M. (1970). A note on mutualism between sipunculans and scleractinian corals. In *Proceeding of the International Symposium on the Biology of Sipuncula and Echiura*, Vol. I (pp. 305–311), edited by M.E. Rice and M. Todorovic. Institute for Biological Research and National Museum of Natural History, Smithsonian Institution, Washington, D.C.

Yoon, C.K. (1993). What might cause parasites to become more virulent. *Science*, 259, 1402.

Yoshikawa, H. (2000). *Juglans ailanthifolia* Carr. gnawed by large Japanese field mice, *Apodemus speciosus* (Temminck), from the Middle Pleistocene Sahama Mud in Hanamatsu City, Shizuoka Prefecture, central Japan. *Science Report of the Toyohashi Museum of Natural History*, 10, 23–30.

You, H.-L., Lamanna, M.C., Harris, J.D., Chiappe, L.M., O'Connor, J. et al. (2006). A nearly modern amphibious bird from the Early Cretaceous of northwestern China. *Science*, 312, 1640–1643.

Yuan, C., Zhang, H., Li, M., and Ji, X. (2004). Discovery of a Middle Jurassic fossil tadpole from Daohugou Region, Ningcheng, Inner Mongolia, China. *Acta Geologica Sinica*, 78, 145–148.

Yuan, X., Xiao, S., and Taylor, T.N. (2005). Lichen-like symbionts 600 million years ago. *Science*, 308, 1017–1020.

Zakharenko, G.V. (2008). Vozmozhnoe svidetelstvo khishnichestva sredi plakoderm (Pisces) evlanovskogo basseina Tsentralnoi Rossi. *Paleontologicheskii Zhurnal*, 5, 67–70.

Zakharov, V.A. (1966). Pozdneiurskie i rannemelovye dvustvorchatye molluski severa Siberi i usloviya ikh sushtsestvovaniya. *Akademiya Nauk SSSR, Sibirskoe Otdelenie, Institut geologii i geofiziki, Trudy*, 113, 1–189.

Zangerl, R. and Case, G.R. (1976). *Cobelodus aculeatus* (Cope), an anacanthous shark from Pennsylvanian black shales of North America. In *Palaeontographia* (pp. 107–157), edited by H.K. Erben and W. Haas. Schweizerbart, Stuttgart.

Zapfe, H. (1936). Spuren bohrender Cirripedier in Gastropoden-Gehäusen des Miozäns. *Senckenbergiana*, 18, 130–134.

Zardini, R. (1978). *Fossili Cassiani (Trias medio-superiore): Atlante dei Gasteropodi della Formazione di S. Cassiano raccolti nella regione Dolomitica Attorno a Cortina d'Ampezzo*. Edizioni Ghedina, Cortina d'Ampezzo, Italy, 41 pp.

Zardini, R. (1980). *Fossili Cassiani (Trias medio-superiore): Primo aggiornamento all'Atlante dei Gasteropodi della formazione di S. Cassiano raccolti nella regione Dolomitica attorno a Cortina d'Ampezzo*. Edizioni Ghedina, Cortina d'Ampezzo, Italy, 16 pp., 6 plates.

Zardini, R. (1985). *Fossili Cassiani (Trias medio-superiore): Primo aggiornamento all'Atlante dei bivalvi e secondo aggiornamento all'Atlante dei gasteropodi con illustrazioni dei gusci che hanno conservato la pigmentazione originaria; fossili raccolti nella formazione di S. Cassiano della regione Dolomitica attorna a Cortina d'Ampezzo*. Edizioni Ghedina, Cortina d'Ampezzo, Italy, 17 pp., 10 plates.

Zaton, M. and Salamon, M.A. (2008). Durophagous predation on Middle Jurassic molluscs, as evidenced from shell fragmentation. *Palaeontology*, 51, 63–70.

Zaton, M., Salamon, M.A., Boczarowski, A., and Sitek, S. (2008). Taphonomy of dense ophiuroid accumulations from the Middle Triassic of Poland. *Lethaia*, 41, 47–58.

Zavortink, T.J. and Poinar, Jr., G.O. (2000). *Anopheles* (*Nyssorhynchus*) *dominicanus* sp. n. (Diptera: Culicidae) from Dominican amber. *Annals of the Entomological Society of America*, 93, 1230–1235.

Zaykov, V.V. and Maslennikov, V.V. (1987). Sea-bottom sulfide structures in massive sulfide deposits of the Urals. *Doklady Akademia Nauk SSSR*, 293, 60–62 (translation).

Zelenitsky, D.K. and Hirsch, K.F. (1997). Fossil eggs: identification and classification. In *Dinofest International: Proceedings of the Symposium* (pp. 279–286), edited by D.L. Wolberg and E. Stump. The Academy of Natural Sciences, Philadelphia, PA.

Zelenitsky, D.K. and Thierrien, F. (2008). Phylogenetic analysis of reproductive traits of maniraptoran theropods and its implications for egg parataxonomy. *Palaeontology*, 51, 807–816.

Zeng, D.-M. and Hu, J.-M. (1991). Gastropod opercula from Late Cretaceous of Nanxiong Basin, Guangdong. *Acta Palaeontologica Sinica*, 30, 406–412.

Zeuner, F. and Manning, F.J. (1976). A monograph on fossil bees (Hymenoptera: Apoidea). *Bulletin of the British Museum (Natural History) (Geology)*, 27(2), 3149–3268.

Zhan, R. and Vinn, O. (2007). Cornulitid epibionts on brachiopod shells from the Late Ordovician (middle Ashgill) of East China. *Estonian Journal of Earth Sciences*, 56, 101–108.

Zhang, J., Harbottle, G., Wang, C., and Kong, Z. (1999). Oldest playable musical instruments found at Jiahu early Neolithic site in China. *Nature*, 401, 366–368.

Zhang, M.-M., Ed. (2001). *Jehol Biota*. Shanghai Science and Technology Publishing, Shanghai, China, 150 pp.

Zhantiev, R.D. (2006). Novie vidi zhukov-kozheedov (Coleoptera, Dermestidae) iz Rovenskogo i Baltiiskogo Yantarei. *Paleontologicheskii Zhurnal*, 5, 87–89.

Zhao Qi, Barrett, P.M., and Eberth, D.A. (2006). A multi-generational herd of the basal ceratopsian dinosaur *Psittacosaurus*. In *Proceedings of the 50th Annual Meeting of the Palaeontological Association, December 18–21, Sheffield, U.K.*

Zhao Qi, Barrett, P.M., and Eberth, D.A. (2007). Social behaviour and mass mortality in the basal dinosaur *Psittacosaurus* (Early Cretaceous), People's Republic of China. *Palaeontology*, 50, 1023–1029.

Zhivkovich, A.E. and Chekhovich, P.A. (1985). Paleozoiskie formatsii i tektonika Ufimskogo Amfiteatra. *Izdatelskvo "Nauka," Moskva, Akademia Nauk SSSR, Institut Litosferi*, 1–183.

Zhou, Z. and Zhang, B. (1989). A sideritic *Protocupressinoxylon* with insect borings and frass from the Middle Jurassic, Henan, China. *Reviews of Palaeobotany and Palynology*, 59, 133–143.

Zhou, Z. and Zhang, F. (2002). A long-tailed, seed-eating bird from the Early Cretaceous of China. *Nature*, 418, 405–409.

Zhou, Z. and Zhang, F. (2003). Anatomy of the primitive bird *Sapeornis chaoyangensis* from the Early Cretaceous of Liaoning, China. *Canadian Journal of Earth Sciences*, 40, 731–747.

Zhou, Z. and Zhang, F. (2004). A precocial avian embryo from the Lower Cretaceous of China. *Science*, 306, 653.

Zhu, M.-Y., Vannier, J., Van Iten, H., and Zhao, Y.-L. (2004). Direct evidence for predation on trilobites in the Cambrian. *Proceedings of the Royal Society, Biological Sciences*, 7(271, Suppl. 5), S277–S280.

Zhuravlev, K.I. (1943). Nakhodki ostatkov verkhneoorskikh reptilii v savelevskom slantsevom rudnike. *Izvestiya Akademii Nauk Sooza SSR, Otdelenie Biologich Nauk*, 5, 293–306.

Zibrowius, H. (1997). A new type of symbiosis: *Heterocyathus japonicus* (Cnidaria: Scleractinia) living on *Fissidentalium vernedei* (Mollusca: Scaphopoda). *Zoologische Verhandelingen*, 323, 319–340.

Zibrowius, H., Southward, E.C., and Day, J.H. (1975). New observations on a little-known species of *Lumbrineris* (Polychaeta) living on various cnidarians, with notes on its Recent and fossil scleractinian hosts. *Journal of the Marine Biological Association of the United Kingdom*, 55, 83–108.

Ziegelmeier, E. (1954). Beobachtungen über den Nahrungserwerb bei der Naticide *Lunatia nitida* Donovan (Gastropoda Prosobranchia). *Helgoländer Wissenschaftliche Meeresunters*, 5, 1–33.

Ziegler, B. (1992). Guide to the Lowentor Museum. *Stuttgarter Beiträge zur Naturkunde Serie C*, 27(E), 1–100.

Zilch, A. (1936). Unsre (sic. Unsere) Kenntnis von fossilen Perlen. *Archiv für Molluskenkunde*, 68, 238–252.

Zinsmeister, W.J. (1980). Observations on the predation of the clypeasteroid echinoid *Monophoraster darwini*, from the Upper Miocene Entrerrios Formation, Patagonia, Argentina. *Journal of Paleontology*, 54, 910–912.

Žítt, J. and Mikuláš, R. (2006). Substrate of bivalve borers as recorded on phosphatic fills of *Gastrochaenolites*, Palaeoenvironmental context (Bohemian Cretaceous Basin, Czech Republic). *Ichnos*, 13, 191–198.

Zlotnik, M. (2001). Size-related changes in predatory behaviour of naticid gastropods from the Middle Miocene Korytnica Clays, Poland. *Acta Palaeontologica Polonica*, 46, 87–97.

Zollikofer, C.P.E., Ponce de León, M.S., Vandermeersch, B., and Lévêque, F. (2002). Evidence for interpersonal violence in the St. Césaire Neanderthal. *Proceedings of the National Academy of Sciences USA*, 99(9), 6444–6448.

Zonneveld, I.S., de Leeuw, P.N., and Sombroek, W.G. (1971). An ecological interpretation of aerial photographs in a savanna region in northern Nigeria. *International Institute for Aerial Survey and Earth Sciences, Series B*, 63, 1–41.

Zschokke, S. (2003). Spider-web silk from the Early Cretaceous. *Nature*, 424, 636–637.

Zullo, V.A. (1966). A new species of *Balanus* (Cirripedia, Thoracica) from the Late Eocene Cowlitz Formation of southern Washington, USA. *Crustaceana*, 11, 198–204.

Zullo, V.A. and Chivers, D.D. (1969). Pleistocene symbiosis: pinnotherid crabs in pelecypods from Cape Blanco, Oregon. *The Veliger*, 12, 72–73.

Zumpt, F. (1965). *Myiasis in Man and Animals in the Old World*. Buttersworths, London, 267 pp.

Addendum

1. FUNCTIONAL MORPHOLOGY

SLENDER STRAIGHT-SNOUTED WEEVILS

Some straight-snouted weevils enter the galleries of bark beetles for oviposition and possible predation. In order to navigate these narrow tunnels these beetles have evolved a narrow, elongate body form (Figure A1). The straight-snouted weevil, *Dominibrentus leptus*, from Dominican amber, has such a narrow body and also possesses body concavities for the reception of its legs. However, in contrast to extant straight-snouted weevils that have cavities for just the front two pairs of legs, *Dominibrentus* has cavities for the reception of all three pairs of legs (Poinar, 2009a). Apparently, the fossil weevil also used its tarsal spines, tibial pegs, and hardened hair tufts to move through the tunnels of platypodids and scolytids as it deposited its eggs after ejecting or devouring the inhabitants.

MODIFICATIONS FOR VISUAL DETECTION OF ENEMIES

Many insects have two to three ocelli on the top of their heads to detect movements of approaching predators. An Early Cretaceous unicorn fly of the Bibionomorpha (a group that includes today's flower flies or bibionids) has three ocelli placed on the top of a horn-like protuberance arising from the middle of its head (Figure A2) (Poinar, 2009b). The reduced mouthparts and pollen grains associated with its tarsi indicate that *Cascoplecia insolitis* probably pollinated primitive angiosperm flowers in the Burmese amber forest. The ocelli raised on the "horn" would have allowed it to better detect approaching danger while foraging among flower clusters.

FIGURE A1 The elongate body of *Dominibrentus leptus* (Poinar, 2009a) in Dominican amber facilitated entry into the narrow tunnels of bark beetles for oviposition and possible predation (Ron Cauble amber collection); length of beetle = 9.5 mm.

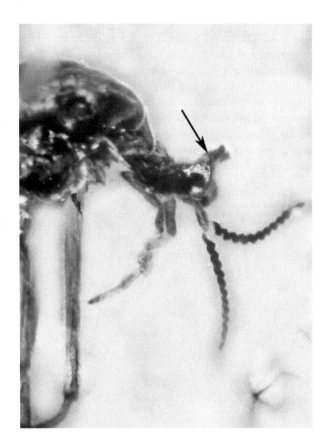

FIGURE A2 Head of *Cascoplecia insolitis* (Poinar, 2009b) with horn-like protuberance (arrow) in Early Cretaceous Burmese amber (Poinar amber collection accession no. B-D-68); length of horn-like protuberance = 210 µm.

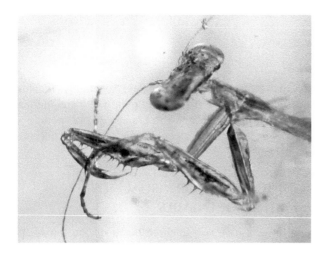

FIGURE A3 Praying mantid in Dominican amber. Note spines on lower edges of femora and tibia (Poinar amber collection accession no. O-2-12); width of head = 2 mm.

FORELEG MODIFICATIONS ON INSECTS

Insects can have foreleg modifications, especially for catching and holding prey. These features are so constant and unique to genera and even species that they are often used in classifying the bearers. The most well known insects with raptorial forelegs are praying mantids. The forelegs of these large insects contain a series of spines on the lower edges of the femora and tibia (Figure A3). These two segments are brought together to trap the victim.

A similar type of trap arrangement has evolved independently in the mantisflies (Neuroptera: Mantispidae), as is shown on the foreleg of the Dominican amber *Feroseta priscus* (Figure A4). This species has a large downward-pointing spine on its femora that would impale potential prey, providing an extra security measure.

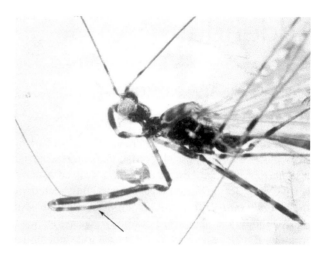

FIGURE A5 A thread-legged bug (Heteroptera: Reduviidae) in Dominican amber with raptorial front legs. Arrow shows tibia folded against the femur (Poinar amber collection accession no. HE-4-37); length of femur = 1 mm.

Thread-legged bugs (Heteroptera: Reduviidae: Emesinae) appear so delicate that one wonders how they can subdue prey with their thin raptorial forelegs; however, the legs are much stronger than they appear (Figure A5). Again, it is the tibia appressed against the femur that grasps the prey even though there is only a row of very fine spines along the surfaces of these leg segments. Their color patterns provide good camouflage as these bugs crawl over vegetation in search of prey.

Another group of small predatory bugs are the gnat bugs (Heteroptera: Enicocephalidae). They have compact bodies with the femur and tibia on the front legs larger than those on the other legs (Figure A6). However, the raptorial portion of the forelegs consists of the one-segmented tarsus that folds back against the enlarged apical portion of the tibia (Figure A7). These bugs search for prey on bark and other flat surfaces.

FIGURE A4 Raptorial foreleg of the mantisfly *Feroseta priscus* (Poinar, 2006) in Dominican amber. Arrow shows tibia folded back over femur. Note large femoral spine which would impale the prey (Poinar amber collection accession no. NU-4-16); length of femur = 4 mm.

FIGURE A6 A gnat bug (Heteroptera: Enicocephalidae) in Dominican amber. Note enlarged tibiae (arrows) on front legs (Poinar amber collection accession no. HE-4-7D); length of specimen = 4 mm.

FIGURE A7 Detail of forelegs of the gnat bug shown in Figure A6. Arrows show single segmented tarsus terminated with a long claw. When catching prey the tarsus is folded down against the widened end of the tibia (Poinar amber collection accession no. HE-4-7D); length of tibia and tarsus = 0.7 mm.

Dragonflies and damselflies have spiny legs for perching and holding prey that they catch on the wing. The damselfly *Diceratobasis worki* (Poinar, 1996) was probably looking for ovipositional sites when it was captured in Dominican amber. The spines on the tibia and tarsal segments are clearly visible (Figure A8).

Dance flies of the Empididae are small to medium size predatory flies that prey on smaller insects, especially other flies. Some dance flies have their front legs modified to capture and hold prey as shown in a specimen in Madagascar copal (Figure A9). The forelegs bear a series of forward pointing setae or spines for securing prey (Figure A10).

Some insects have forelegs modified for digging or grasping. In the case of cicada larvae, which spend most of their lives in the soil feeding on plant roots, the tibia and tarsal

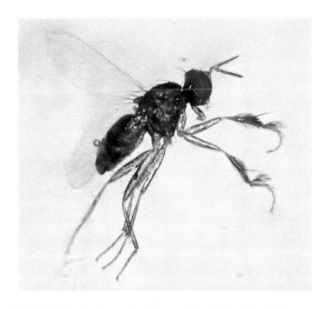

FIGURE A9 A dance fly (Diptera: Empididae) in Madagascar copal with front legs modified for holding prey (Alex Brown amber collection); length of fly = 3 mm.

segments are enlarged. These modifications can be seen on newly hatched larvae that usually fall to the ground after leaving the egg (Figure A11). The same modifications probably assist the mature larvae as they crawl back up the tree trunk after a prolonged period of feeding (Figure A12).

Dung beetles (Coleoptera: Scarabaeidae), such as the one in Burmese amber shown in Figure A13, often have spines or teeth on their front legs to assist in digging as they bury food for their future brood.

FEEDING TUBES ON APHIDS

As with other members of the Heteroptera, aphids obtain nourishment through their beak or feeding tube. In plant parasites, the beak is inserted into plant tissue and the

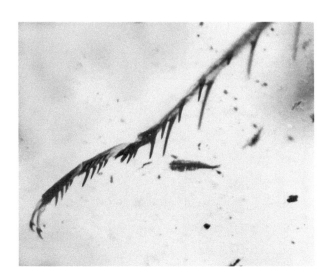

FIGURE A8 Spines on the foreleg of the damselfly *Diceratobasis worki* (Poinar, 1996) in Dominican amber (Jim Work amber collection); length of tarsus = 1 mm.

FIGURE A10 Detail of front legs of dance fly in Figure A9, showing setae and spines for holding prey (Alex Brown amber collection); length of front leg = 3 mm.

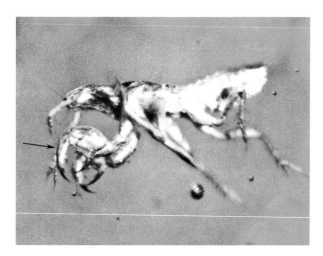

FIGURE A11 A cicada larva in Dominican amber with its front legs modified for digging and climbing. Arrow shows enlarged tarsal segment with large spine at tip (Poinar amber collection accession no. HO-4-16); length of specimen = 1.5 mm.

withdrawn juices pass directly into the insect's alimentary tract. Most extant aphids have beaks that do not surpass their body length; however, some extinct aphids had beaks that greatly surpassed their body length. In members of the Baltic amber genus *Germaraphis*, the beak can be greater than two body lengths (Figure A14) (Heie, 1967). The long beak indicates that the aphid fed on trees with thick bark and needed to extend its beak some distance through the woody tissue

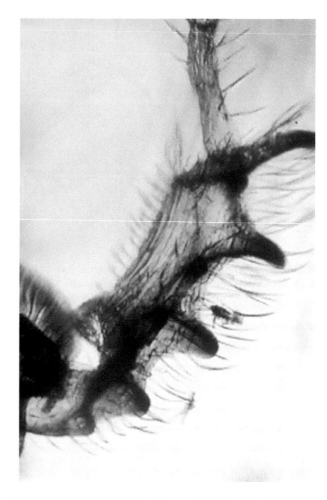

FIGURE A13 The "teeth" on the modified front leg of this Early Cretaceous Burmese amber scarab beetle are for burying dung balls or dead organisms that would have served as food for the future brood (Poinar amber collection accession no. B-C-13); length of beetle = 5 mm.

before reaching a feeding site. Some Cretaceous aphids also had beaks longer than their bodies, probably for the same reason (Figure A15).

FIGURE A12 A cicada larva after feeding underground for a prolonged period; note front legs modified for digging and climbing (private collection).

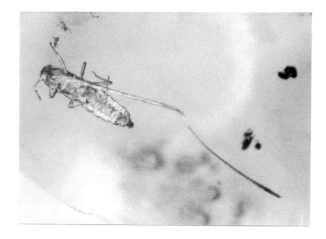

FIGURE A14 The beak (feeding tube) of this Baltic amber aphid (*Germaraphis* sp.) is over twice as long as its body (location of specimen unknown).

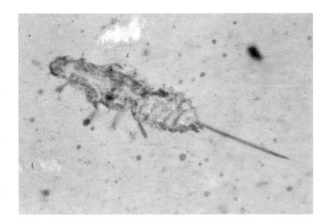

FIGURE A15 An aphid larva in Early Cretaceous Burmese amber with its beak extending posteriorly beyond the body (Poinar amber collection accession no. B-HE-21); length of aphid = 2 mm.

TERMINAL ABDOMINAL MODIFICATIONS IN PARASITOID WASPS

Female parasitoid wasps often have elongated ovipositors for depositing eggs in or on hosts concealed in various substrates (Figure A16). Rarely, male wasps have elongated abdomens that superficially resemble ovipositors. For example, males of the Dominican amber braconid *Aivalykus dominicanus* (Zuparko and Poinar, 1997) have a greatly elongated and extremely narrowed abdominal terminus (Figure A17). This feature allows them to mate through substrates with females that are still in their emergence burrows.

HALTERES AND HAMULOHALTERES

Winged flies (Diptera) have only a single pair of functional wings attached to the mesothorax, while the metathoracic wings are represented by reduced, stub-like appendages called *halteres* (Figure A18). Halteres have been considered

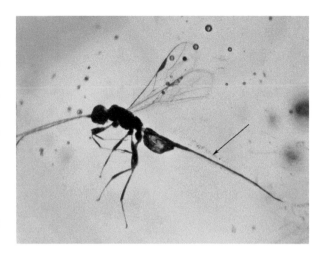

FIGURE A17 A male *Aivalykus dominicanus* (Zuparko and Poinar, 1997) with elongated abdominal terga (arrow) superficially resembling an ovipositor in Dominican amber (Poinar amber collection accession no. Hy-10-93); length of wasp = 4.0 mm.

balancing or equilibrium organs because, when removed, the power of flight is lost or greatly reduced in most flies. One exception appears to be horseflies (Tabanidae), which can still fly without halteres. Early experimenters showed that some dehaltered flies became airborne again if a thread of cotton was attached to the tip of the insect's abdomen (Wigglesworth, 1965). Halteres consist of a basal scabellum, a stalk-like pedicel, and a terminal expanded knob-like capitellum. They contain hemolymph and at least one tracheole (Imms, 1948). Whereas flies have the metathoracic pair of wings reduced to halteres, it is the mesothoracic wings that are represented by halteres in male Strepsiptera (Figure A19). Again, if these are removed, the insects cannot fly (Wigglesworth, 1965).

Another insect group that has only a single pair of wings are male scale insects (Hemiptera: Sternorrhyncha: Coccoidea). In coccids, the second pair of wings are reduced and often represented by hamulohalteres, which are similar

FIGURE A16 A female torymid wasp parasitoid with her exserted ovipositor and sheath (arrow) in Dominican amber (Poinar amber collection accession no. St-1-4); length of wasp (with ovipositor) = 7.5 mm.

FIGURE A18 The Burmese amber cranefly *Drinosa prisca* (Podenas and Poinar, 2009), showing the left halter (arrow) attached to the metathorax (Poinar amber collection no. B-D-10U); length of fly = 1.9 mm.

FIGURE A19   Male strepsipteran in Dominican amber showing the right halter (arrow) attached to the mesothorax (Poinar amber collection accession no. St-1-11); length of specimen = 1.52 mm.

in appearance to the halteres of flies, except that the former structures possess small hooks or hamuli at the tip of the capitellum (Figure A20). These hooks attach to an area on the base of the front wing, presumably providing stability during flight.

Extant aphids, which belong to the same taxonomic group as scale insects (Sternorrhyncha), all have two pairs of wings. The hind pair is smaller than the front, but they are membranous and normally contain veins. However, some Early Cretaceous aphid clades possessed vestigial hind wings reduced to hamulohalteres. One of these lineages is represented by the Burmese amber *Paraverrucosa annulata* (Poinar and Brown, 2005, 2006), whose hind wings were represented by hamulohalteres (Figure A21).

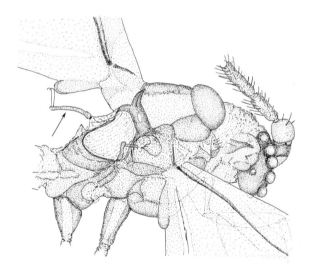

FIGURE A20   Hamulohalter (arrow) attached to the metathorax of the Cretaceous coccid *Kukaspis usingeri* (Koteja and Poinar, 2001) in Alaskan amber. Note the paired hami at the tip of the hamulohalter (Poinar amber collection accession no. HO-4-30); length of hamulohalter = 120 μm. (From Koteja, J. and Poinar, Jr., G.O., *Proceedings of the Entomological Society of Washington*, 103, 356–363, 2001. With permission.)

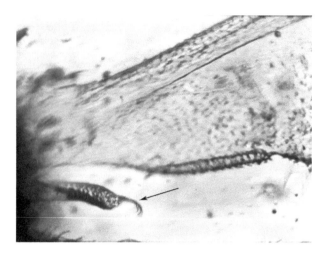

FIGURE A21   Hamulohalter with paired hami (arrow) representing the reduced metathoracic wing of the primitive aphid *Parvaverrucosa annulatus* (Poinar and Brown, 2005, 2006) in Early Cretaceous Burmese amber (Poinar amber collection accession no. B-He-13A); length of hamulohaltere = 86 μm.

SEAHORSES

One of the most bizarre taxa among the fishes is the seahorse (Syngnathidae: Hippocampinae); the males brood eggs deposited by the females within a male brood pouch. These fish live in shallow waters among various marine plants. Žalohar et al. (2009) described the first fossil seahorses from the Slovenian Miocene, and the fossils differ in no significant regard from extant taxa; they are an elegant example of morphological fixity through time and its behavioral implications.

2.   SPECIALIZED, POTENTIALLY INTERACTING BIOLOGICAL SUBSTRATES

2A.   MARINE INVERTEBRATE BENTHOS

2Ac.   PLATYCERATIDS

Lane (1979) considered varied Crawfordsville, Indiana, Mississippian crinoid–platyceratid associations.

2Ae.   PRODUCTID BRACHIOPOD SPINES

The life orientations of productid brachiopods are complex. Many are cemented to a hard substrate by their ventral valves, whereas many others are free living on their ventral valves on what are likely to have been poorly cemented or uncemented substrates or employ a pedicle buried in the substrate. Among those commonly cemented to the substrate, mostly in the ventral posterior region, are the Strophalosioidea. The Richthofenioidea are commonly attached to hard substrates by spines on the ventral valve or cemented directly by that valve to the substrate. The Lyttonioidea are commonly cemented to the substrate in the ventral posterior region. The Productoidea are commonly free living, but Sun and Baliński (2008) illustrated silicified specimens of young *Argentiproductus* from the Tournaisian, Early Mississippian of Guizhou, South China, with well-developed clasping rings that in life served as attachments wrapping around crinoid stems. Brunton and Mundy (1993) illustrated Visean young of *Plicatifera pseudoplicatilis*, another member of the Productoidea, which were presumably attached by means of clasping spines in their young stage. Grant (1963) illustrated and described *Linoproductus*, of the Linoproductoidea, from the Texas Permian, that was attached to crinoid columnals throughout life, not just in the young stage. Shen et al. (1994) illustrated Late Permian *Laterispina*, of the Permianelloidea, with well-developed attachment spines present in the adult stage that presumably were attached to crinoid stems.

Fragmentary as these findings are, with much of the data depending on well-preserved, silicified early growth stages, the evidence suggests repeated convergent evolution, as well as the development of complex larval attachment behaviors in many unrelated groups. It is interesting that the shelly claspers on mangrove oysters, ranging from the Eocene to present, and wrapped around mangrove "roots," perform the same function over time.

2Ag. PALEOZOIC HOST-SPECIFIC, PIT-FORMING CRINOID EPIZOANS

(Previously 2Al in Boucot, 1990.) Sprinkle and Sumrall (2008) illustrated and described an Ordovician parablastoid with numerous *Tremichnus* pits and with "raised reaction rims" around the pits. As with the previously described cases of *Tremichnus* infesting crinoids, this case is of interest as broadening the group of affected echinoderms.

2Am. LEPADOMORPH BARNACLES AND EURYPTERIDS, OTHER SUBSTRATES, AND BALANOIDS

Balanomorphs

Nomura and Maeda (2008) described a species of *Concava* from the Japanese Miocene that attached to cobbles and shells, presumably in a very high-energy environment, which is a most unusual, rare occurrence (most of the others featuring attached balanids are from relatively quiet water environments).

Ascothoracicans

Boucot (1990) provided some information on the traces left by ascothoracican borings, and Lane (1979) provided examples from Indiana Mississippian platyceratid–crinoid associations where the platyceratid was bored by a barnacle.

2Aza. *LUMBRINERIS FLABELLICOLA*– SCLERACTINIAN RELATIONSHIP

Martinell and Domènech (2009) described a number of eunicid polychaete bioerosion traces on western Mediterranean Pliocene solitary corals.

3. MUTUALISM

3g. TUBE WORMS, BIVALVES, AND RHYNCHONELLID BRACHIOPODS FROM DEEP SEA VENTS, PLUS CRUSTACEANS

The problems provided by deep sea vent faunas, cold seeps, whale falls, etc., all deal in part with the questions arising from consideration of the age of the deep sea fauna overall. Smith and Stockley (2005) addressed this question in terms of deep sea echinoid taxa, concluding that there is no evidence for any being of pre-Mesozoic age. The deep sea vent faunas, and allied faunas such as those of whale falls, all suggest nothing older than the Cretaceous at the generic level. The enigmatic "tubes" that first appear in the Silurian hot vent environment are unhelpful, as their very simple morphology precludes certainty that they all belong to the same or related taxa through time. The Devonian *Pilina* and the closely related modern *Neopilina* do, however, raise the possibility that some taxa may have considerable antiquity.

COLD SEEPS

Kiel et al. (2008) described some Japanese Cretaceous cold-seep deposits from Hokkaido that include bivalves.

BARITE DEPOSIT VENTS

Baoshun et al. (1994) described some modern barite "chimneys" from the Gulf of Mexico and their implications. Bogoch and Shirav (1978) described some Senonian, presumably sedimentary, barite deposits from Israel, while considering the origins of sediment-associated barite deposits known elsewhere.

3s. TERMITE AND COCKROACH GUT MUTUALISTS

Two Early Cretaceous Burmese amber cockroaches contain protists related to mutualistic flagellates occurring in extant *Cryptocercus* cockroaches and lower termites. Representatives of the orders Trichomonadida, Trichonymphida, and Oxymonadida were represented (Poinar, 2009c). One flagellate belonging to the order Trichonymphida was recovered from a member of the cockroach family Blattellidae (Figure A22). Although extant cockroaches contain a diverse assemblage of intestinal microbiota, none, other than members of the genus *Cryptocercus* of the Cryptocercidae, contains mutualists belonging to the Trichonymphida and Oxymonadida.

It is intriguing to find representatives of the Trichonymphida and Oxymonidida in lineages of Early Cretaceous cockroaches that do not belong to the cryptocercid clade. Apparently, symbiotic flagellates were more widespread among diverse cockroaches in the Early Cretaceous than at present. It is highly probable that these fossil flagellates were true mutualists and digested cellulose, as lower termites and cryptocercid cockroaches do today. This would also imply

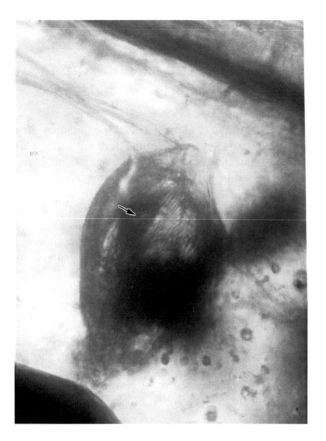

FIGURE A22 The Early Cretaceous trichonymphid flagellate *Burmanymphus cretacea* from the gut of a blattellid cockroach in Burmese amber (Poinar, 2009c). The body contains what appears to be a wood particle (arrow) undergoing digestion. Several long flagella extend outward from the rostrum (Poinar amber collection accession no. B-OR-1D); length of specimen = 170 μm.

that xylophagous cockroaches were more widespread in the past. As the diet of cockroaches shifted from xylophagy to omnivory, their flagellate fauna apparently disappeared. The Early Cretaceous flagellates probably formed spores that were excreted in fecal matter and then reacquired by subsequent generations of blattids, as coprophagy is characteristic of extant cockroaches.

4. HOST–PARASITE AND HOST–PARASITOID RELATIONSHIPS AND DISEASE

4A. ANIMAL–ANIMAL

4AI. MARINE

4AIb. Bopyrid Isopod–Decapod
See Table 2B.

4AIII. TERRESTRIAL

4AIIIp. Vertebrate Pathology Other than Human
Ellingson (2009) called attention to a late Pleistocene bison from western Oregon that had what is apparently a "large bone cancer on the metatarsals."

TABLE 2B
Bopyrid Isopod–Decapod Examples (Additional)

Isopods	Time	Refs.
Gastrosaccus	Late Jurassic	Förster (1969)
Cyclothyreus	Late Jurassic	Schweitzer and Feldmann (2009)
Prosopon	Late Jurassic	Schweitzer and Feldmann (2009)
Longodromites	Late Jurassic	Müller et al. (2000)
Cyphonotus	Early Cretaceous	Förster (1969)
Homolopsis	Late Cretaceous	Bishop (1983)

4AIIIzk. Tyrant Dinosaur Mandibles
Wolff et al. (2009) correlated erosive lesions on the mandibles of 10 out of a total of 61 tyrannosaurid specimens examined with similar lesions on the mandibles of modern birds caused by the protozoan flagellate *Trichomonas gallinae*. They concluded that a species of *Trichomonas* similar to *T. gallinae* caused the erosive lesions on the dinosaurs. Based on the lethal effects of *T. gallinae* on birds, especially doves and pigeons (one parasitologist considered *T. gallinae* to be the possible cause of the final extinction of the passenger pigeon), the authors concluded that this microbe could have eliminated many tyrant dinosaurs in the Late Cretaceous.

4B. ANIMAL–PLANT RELATIONSHIP

In 2008, there appeared two important accounts of Cretaceous arthropod-damaged plant parts, probably due chiefly to insects. The first (Krassilov and Rasnitsyn, 2008) for the first time provided a trace fossil ichnotaxonomic nomenclature for the varied leaf mines, galls, ovipositing scars, etc., abounding in the later Mesozoic and Cenozoic fossil record. This nomenclature (Krassilov's "phyllostigmas") is very important, as it provides a robust ichnotaxonomic basis for morphologically classifying the various types of plant damage, in contrast to previous works by many others that described varied traces but did not make it easy to sort them out into morphological groups. The second paper (Krassilov and Shuklina, 2008) provided a succinct account of the time sequence of plant damage types, emphasizing the earlier Cretaceous, pre-angiosperm level of far less varied damage, followed by an explosive adaptive radiation of damage types (phyllostigmas). For anyone concerned with insect evolution, as well as its ecological consequences, these are two very thought-provoking accounts. Krassilov (2008) for the first time recognized predation scars on leaf mines and galls, thus providing a further complication in organism relations from the present that can now be recognized in the past; he used more Cretaceous material from Israel.

4Ba. CRAB–BROMELIAD ASSOCIATION

Fossil terrestrial crabs are rare, so discovering them in amber is of some interest. A report of a small crab, approximately 4 mm in width, in Mexican amber (Vega et al., 2009) raised the question of how a terrestrial freshwater crab could have ended up in a piece of fossilized resin. The specimen was determined to be a member of the genus *Sesarma* of the family Grapsidae. This genus is restricted to the New World tropics and subtropics; most of the 12 described species occur along the seashore, in mangrove swamps, in estuaries, or near other water sources and sometimes in terrestrial habitats as high as 300 m in elevation (Rodriguez, 1981). At least one extant species, *Sesarma angustipes* Dana, occurs in water collected at the base of bromeliad leaves, and it is likely that the fossil *Sesarma* also occurred in such a habitat. This would explain its occurrence in amber, as other life forms such as tadpoles and aquatic beetles captured in Dominican amber were considered to have lived in the same habitat (Poinar and Poinar, 1999).

Another crab in Mexican amber is slightly larger (approximately 6 mm in width) (Figure A23). This specimen is also a grapsid and resembles the extant Jamaican bromeliad crab *Metopalulias depressus* Rathbun. The latter species completes its entire lifecycle in water-filled bases of bromeliads (Abele, 1972).

5. DENSITY AND SPACING

5a. BENTHIC SHELL EXAMPLES

ECHINOIDS "SEA URCHINS" AND EDRIOASTEROIDS

Sumrall et al. (2009) described some Late Silurian, McKenzie member edrioasteroids from central Pennsylvania that occur as a population on a clast that is similar to populations in other parts of the Paleozoic. What must be among the most

gaudy examples in the edrioasteroid category is Prokop's (1965) illustration (Figure A24) and discussion of a Czech Middle Ordovician trilobite carapace (*Selenopeltis*) literally covered with edrioasteroids (mostly *Hemicystites*), presumably during the life of the trilobite, because the edrioasteroids avoided those parts of the carapace that would have involved movement.

5r. CHAIN-LIKE CAMBRIAN ARTHROPOD ASSOCIATION

Hou et al. (2009) described a unique Early Cambrian arthropod from the Chengjiang Lagerstätte that occurs chiefly in chain-like associations, head to tail. The behavioral significance of this behavior is very uncertain, being unknown in any other arthropod living or fossil.

Chatterton and Fortey (2008) described a number of Ordovician trilobites in linear arrays within burrows excavated by another organism from the Lower Ordovician of southern Morocco. The behavioral utility of this material is uncertain, but protection against predation is one possibility. Some of the linear arrays involve more than a single taxon.

☐ *Reliability*
Category 1, frozen behavior.

FIGURE A24 Czech Middle Ordovician trilobite carapace (*Selenopeltis*) literally covered with edrioasteroids (mostly *Hemicystites*); bar = 1 cm. (From Prokop, R., *Casopis Národního Muzea—Oddil Prírodovèdny*, 134, 30–34, 1965. With permission.)

FIGURE A23 Grapsid crab in Mexican amber. Arrow shows stalked eye (private collection); width of crab approximately 6 mm.

6. PREDATION AND FEEDING BEHAVIORS

It is obvious from first principles that feeding, including predation and herbivory of all sorts, has been present since the beginning. The fossil record is replete with functional morphological examples of taxa suited for both predation and herbivory, as well as less common examples of visceral-region contents that demonstrate predation or herbivory. It is also obvious that the global ecosystem, since the beginning, has been capable of recycling organic materials; otherwise, the environment, both aquatic and terrestrial, would be swamped in a morass of its own garbage. The presence of fungal taxa and bacteria, as well as the obvious possibilities for physical recycling processes, oxidation, etc., make recycling clear. What is not so obvious is the extent to which one can estimate varying levels of predation and herbivory through time from evidence preserved in the fossil record. We would like to learn whether or not there have been time intervals characterized by higher levels of one type of predation as contrasted with time intervals with lower levels. After reviewing the varied types of predation evidence, it becomes clear that we lack such a capability.

First off one is tempted to use the evidence of functional morphology to tabulate the changing numbers of taxa capable of predation and herbivory. For example, within the marine environment one can count the changing numbers through time of taxa with obvious predatory morphologies, such as fish and reptiles having sharp, pointed teeth. Such compilations are probably in no small part heavily influenced by varied sampling artifacts. With predacious fish one faces the problem that taxonomically varied faunas are almost completely restricted to those localities and horizons where favorable preservation environments are present, which are relatively few in number through time. Additionally, there is the fact that not only are numbers of taxa involved but also the problem that some taxa are relatively abundant whereas others are relatively rare, with a complete abundance spectrum in between. It is clear that changing numbers of predaceous taxa have little chance of providing a reliable estimate of predation intensities through time. An additional problem concerns which predators were omnivorous as contrasted with those that were specialists; we have little information on this score with few exceptions, except for some marine Mesozoic reptiles and sharks, both of which show signs of omnivory.

Complicating the problem of estimating levels of predation through time is the fact that most marine invertebrates are soft bodied; see Whittington (1985), who made it clear that, for the Cambrian, soft-bodied megafaunal taxa significantly outnumbered well-skeletonized taxa, as is also the case today. Obviously, trying to estimate levels of predation that involve soft-bodied predators is next to impossible, as most of them are either unknown or poorly known through time, with the rare exceptions of taxa such as the morphologically bizarre, soft-bodied *Anomalocaris* in the Cambrian. Can one assume that relative numbers and abundances of soft-bodied predators will be similar to those of well-skeletonized predators?

The near impossibility of identifying the predators responsible for coprolites containing the remains of prey underlines the problem; see Underwood (1993) for an account of varied graptolite fragments preserved in coprolites. The difficulty of dealing with coprolites is underlined by the comment by Vijaya et al. (2009, p. 91) regarding some pollen and spores in nonmarine, Late Triassic Indian coprolites: "The coprolite-producing animals (spiral and non-spiral groups) possibly ingested gymnospermous and pteridophyte plant remains passively along with water or their herbivorous prey."

There is evidence, beginning in the Cambrian, of marine plankton, although this evidence involves skeletonized taxa such as Radiolaria for the most part, with our knowledge of soft-bodied plankton being very fragmentary. The presence of potentially planktiferous jellyfish (scyphozoans, siphonophores, ctenophores), beginning in the Cambrian, makes it clear that there were available plankton as well as probable soft-bodied, unknown pelagic taxa. In the post-Paleozoic, the presence of fish featuring gill rakers suited for planktivory begins at least by the Jurassic, as with the clupeid fishes, but again quantitative estimates are not possible. The mid-Cenozoic arrival of the whalebone whales clearly indicates planktivory, but, again, quantitative estimates are not possible.

Turning to the marine benthic invertebrates, one finds a mass of data concerning predation. Great effort has gone into trying to understand the effects of predatory gastropods that bore holes into their prey bivalves and gastropods. Some have even suggested that such predatory gastropods, as evidenced by naticid and muricid boreholes, mostly in the later Mesozoic and Cenozoic, have increased their activities through evolutionary time. However, the extensive efforts of Kelley and Hansen (2006) have negated this possibility; their work on Gulf Coast and Coastal Plain Late Cretaceous and Tertiary borehole frequencies make it clear that they go both up and down several times during this interval. Additionally, using modern East Coast examples, Kelley and Hansen (2007) have shown that there is a biogeographic component to borehole frequency that further complicates the issue. Turning to the Paleozoic record, one finds numerous examples of boreholes associated with skeletonized, presumed prey taxa, such as brachiopods, but no real understanding about the identity of the predators. For the circular, cylindrical Paleozoic boreholes, it has been assumed that predaceous gastropods are involved, but there is no really positive evidence. One can, with adequate samples, determine which prey taxa in a sample are most favored by boring invertebrates as contrasted with those that are less favored, but this alone tells us nothing about overall levels of predation through time.

In addition to the boreholes generated by predators are the "healed" scars on gastropod shells that have been unsuccessfully attacked by predatory crabs. The problem with this class of evidence is that there is still too little of it and that the evidence and frequency of successful predatory crab attacks is minimal. In addition, such scars left behind by unsuccessful attempts on Paleozoic gastropods are little reported and difficult to ascribe to the predatory activities of any particular organism.

The overall difficulty of the problem is illustrated by Carter's (1968) account of the varied modern molluscan predators. His summary made it very clear how difficult it is to try to transfer data from the present to interpret the past. Further complicating the problem is Chattopadhyay's (2009) suggestion that drilling-type predation decreases in the presence of crab predation. Schiffbauer et al. (2009) suggested that microtraces of radular teeth on borehole surfaces will aid in identifying the predator, which is a serious problem when considering Paleozoic boreholes.

The boreholes left behind by Mesozoic and Tertiary cephalopods have been very infrequently noted, although they undoubtedly would be reported far more frequently if searched for, even in museum collections. We have noted (see Section 6AIi, Octopus Boreholes) the few references to octopus boreholes into fossil mollusks, merely back to the Pliocene, as well as one into brachiopods from the Polish Miocene, but these few records clearly reflect more about the lack of attention by paleontologists to the presence of octopus boreholes than they do to their true stratigraphic range.

There is the problem that not all predatory efforts directed at shelly marine benthos results in shell damage. For example, with the aid of a little suction and their leathery lips, walruses from the Miocene to present merely suck out the flesh of bivalves, leaving behind the undamaged shells. There are other predators that successfully deal with marine invertebrate prey without leaving any traces of their activities on prey shells.

Finally, there is the problem of how to interpret shell fragments in the fossil record. Are they mostly the results of predation or are physical factors largely responsible? There is no easy solution to this problem.

It is clear that trying to estimate global levels of predation through geological time is currently not practical. We can provide overwhelming evidence for the presence of predation since at least the Cambrian in the marine realm, and from at least the Late Silurian in the non-marine, but providing solid, reliable evidence for levels of global predation through time, changing or unchanging, is beyond our present capabilities.

6A. MARINE

6AI. INVERTEBRATE

6AIn. Paleozoic Predation on Gastropods
Ebbestad et al. (2009) provided the most comprehensive account to date of failed predation attempts, as evidenced by healed shell damage, on Paleozoic gastropods and bellerophontids. They dealt with Ordovician through Carboniferous examples.

6AIs. Boreholes in Brachiopods and Predation on Brachiopods in General
Mundy (1982) described evidence of predation affecting some British Dinantian (Mississippian) brachiopods with an attempt to decide the identities of the predators. Mottequin and Sevastopulo (2009) described rather random boreholes, largely restricted to the ventral valves of a Mississippian *Crurithyris*, while other associated taxa (their Table 1) had few boreholes (i.e., prey selectivity). Małkowski (1976) described some Polish Albian terebratuloid brachiopods with evidence of shell repair following failed predatory attacks.

6AIzh. Ammonoid Feeding
Lehmann and Weitschat (1973) summarized what is known about ammonite feeding as determined from crop contents that include Foraminifera, ostracodes, small ammonites, and *Saccocomma* (stemless crinoids).

6AII. VERTEBRATE

Fish
Urlichs et al. (1994, Figure 84, p. 72) illustrated and briefly cited a Posidonienschiefer, Jurassic bony fish (*Saurostomus*) with a coleoid (*Loligosepia*) in its visceral region, complete with ink sac and tentacles.

6AIIg. Plankton Feeding
While describing some newly discovered *Megachasma* teeth from the Belgian Neogene, De Schutter (2009) reviewed the known occurrences of the genus, with the oldest being from the Late Oligocene, and there being some question about the form described from the Colorado Cenomanian.

6C. TERRESTRIAL

6CI. INVERTEBRATES
Labandeira (2007) reviewed the plant–insect associations from the fossil record while making the point that trace fossil evidence commonly precedes body fossil evidence for many of the categories.

6CIa. Spider Web, Spinnerets, and Bundled Prey
A piece of 140-mya amber discovered in alluvial soils of the Ashdown Formation near Hastings, England, contained some interesting microfossils. Along with vascular tissue, tracheid cells, putative resin ducts, soil microbe remains resembling actinobacterial colonies, and putative fungal or cyanobacterial filaments were the earliest examples of spider silk (Brasier et al., 2009).

6CIb. Dung Beetles
Sánchez and Genise (2009) described some earlier Cenozoic Patagonian dung beetle brood balls that show evidence of the presence of a kleptoparasitic accompanying beetle or fly that also fed on the nutrients provided by the dung balls.

□ *Reliability*
Category 2B, frozen behavior.

6CIc. Flesh-Eating Insects
Chin and Bishop (2007) described some Jurassic Morrison Formation bored bone contained in coprolites that they interpreted as dinosaur bone within theropod coprolites that had

been bored by "ancestors of modern dermestids." They summarized a number of possible dermestid boring reports from the fossil record.

6CIza. Dung-Feeding Gastropods

Chin et al. (2009) described an association of Upper Cretaceous Two Medicine Formation possible dinosaur coprolites from Montana with associated gastropods interpreted as having fed on dung nutrients.

☐ *Reliability*

Category 6, speculative.

6CIzb. Herbivory at the Rhynie Chert Site

Taylor et al. (2009) provided a wide overview of fossil plants (excluding algae), which includes discussions of symbiotic fungi, lichens, and past interactions between plants and animals. One of the controversial issues raised by the authors is the earliest record of herbivory in the terrestrial environment. Discoveries of animal life at several Silurian–Devonian terrestrial sites have produced equivocal results. Even from the well-studied Rhynie Chert deposits, the previously discovered mites, springtails, and arachnids all now appear to be either carnivores or detritivores (Selden and Nudds, 2004). There is no evidence that the mites inside the empty sporangia of the primitive vascular plants *Aglaophyton major* were actually feeding on plant tissues rather than simply using the microhabitats for protection against predators, such as the trigonotarbids. Selden and Nudds (2004) considered that the assemblage of animals in the Rhynie Chert, as well as at other investigated Silurian–Devonian terrestrial sites, was composed of carnivores and detritivores, and that the trophic system at that time period predated the evolution of herbivory. However, more recently, nematodes were discovered breeding within substomatal cavities and intercellular spaces and inside cortex cells of the primitive terrestrial plant *Aglaophyton major* (Poinar et al., 2008). The type of plant parasitism exhibited by the nematodes is similar to that noted in several extant genera of soil nematodes. Thus, it would appear that *Palaeonema phyticum* is the earliest known terrestrial herbivore.

6CIzc. Scorpionfly Feeding Behavior

A recent study of the mouthparts of 11 species of Jurassic and Cretaceous scorpionflies (Mecoptera) concluded that all taxa had elongate, siphonate (tubular) proboscids (Ren et al., 2009). Based on this conclusion, the authors speculated that these early scorpionflies fed on ovulate secretions of extinct gymnosperms and were early pollinators. This is an interesting theory, especially because modern scorpionflies are all carnivores and the authors were unable to obtain any evidence of pollen associated with the fossils. One of the scorpionfly groups they investigated was the Pseudopolycentropodidae, of which some species occur in Burmese amber (Figure A25); however, these forms have stylate, rather than siphonate, mouthparts. These Burmese amber pseudopolycentropodids have rows of fine serrations along the tip and lower edge of their stylets (Figure A26). This armature strongly

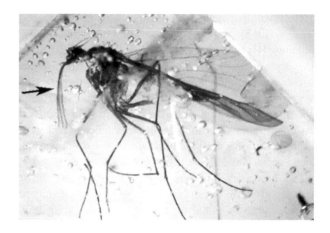

FIGURE A25 A member of the scorpionfly family Pseudopolycentropodidae in Burmese amber. Arrow shows stylet (Poinar amber collection accession no. B-M-1); body length = 4 mm.

suggests that they were primarily blood feeders and not phytophagous. They may have imbibed some plant secretions on the side, similar to that of many extant blood-feeding Diptera that supplement their diet with energy providing sugar solutions. The Ren et al. study provides detailed information on the mouthparts of extinct scorpionfly groups.

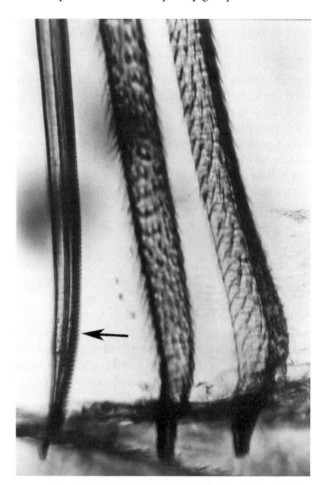

FIGURE A26 Serrations (arrow) along the inner edge of the stylet of the pseudopolycentropodid specimen shown in Figure A25 (Poinar amber collection accession no. B-M-1); stylet length = 2 mm.

6CII. VERTEBRATES

6CIIj. Venom-Conducting Reptilian Teeth

Gong et al. (2009) described teeth from a Chinese Cretaceous small carnivorous dinosaur, *Sinornithosaurus*, that bear grooves indicating that it was venomous and a space on the maxillary bone suggesting the presence of a venom pouch that led to the fangs.

6CIII. Frozen Pleistocene Mammals

Harrington (2007) provided an excellent summary account of the known Northern Hemisphere Pleistocene mammals preserved with soft tissues, including stomach contents in some cases, and a good set of references.

*6CIIzb. Snake Predation of a Hatchling Dinosaur

Wilson et al.'s (2010) detailed description of a Lameta Formation, Maastrichtian, Late Cretaceous snake in the act of predating a hatchling sauropod from Gujarat, western India, is a truly remarkable discovery. The snake (estimated length, 3.5 m) is a nonmacrostomatan that had just crushed the sauropod egg in a nest with associated uncrushed eggs, wrapped around the hatchling with parts of the hatchling skeleton preserved although not yet swallowed.

☐ *Reliability*
Category 1, frozen behavior.

7. COMMUNICATION

VISION

Although positive evidence of color vision is absent from the fossil record, it is clear from the consideration of color vision of different types in living animals when combined with information about their presence in the fossil record that color vision was undoubtedly present very early (Jacobs and Rowe, 2004; Maximov, 2000; Regan et al., 2001; Yokoyama, 2002; Yokoyama and Yokoyama, 1996). Schoenemann et al. (2009) reviewed the nature of light-perceiving and image-perceiving organs in terms of what the fossil record provides. See later discussion on bird feathers regarding the existence of color vision in birds at least by the Early Cretaceous.

CEPHALOPOD STATOLITHS

Octopods, sepioids, and teuthoids have calcified, aragonitic statoliths. The functions of these statoliths are uncertain, although they may be involved with motion in some manner, and they have a fossil record extending back to the Jurassic (Engeser, 1990). Clarke and Fitch (1979) provided an account of some North American statoliths from the Cenozoic, extending back to the Eocene, while Clarke et al. (1980) described a Miocene European example. They commonly occur together with fish otoliths (see Weiler, 1968), with most fish otolith studies concentrating on Cenozoic forms.

8. TRACE FOSSILS AND THEIR FORMERS

8A. MARINE

Getty and Hagadorn (2009) provided a comprehensive, well-documented account of *Climatichnites* leading to the conclusion that a mollusk or mollusk-like organism was responsible for this unique trace.

8Ad. CRAB BURROWS

Richards (1975) briefly described some Late Cretaceous Vancouver Island crab burrows with the exuviae of *Longusorbis* still in them, probably where the crab molted.

8C. TERRESTRIAL

8CII. VERTEBRATE

8CIIa. Vertebrate Tracks

Mazin et al. (2009) discovered and described an amazing set of Late Jurassic, French pterosaur hind-limb landing tracks, followed by purely terrestrial hind- and anterior-limb walking tracks.

9. SPECIALIZED SUBSTRATES

9A. MARINE

9Af. ROCK- AND WOOD-BORING BIVALVES

Teredinid and Pholad Bored Wood

Noda and Lee (1989) described some fine specimens of Korean Miocene *Martesia* borings into woody branches, with the bivalves still preserved within the wood.

10. SEXUAL BEHAVIOR

10A. TERRESTRIAL

10AII. INVERTEBRATES

*10AIIr. Pseudoscorpion Egg Sac

Female pseudoscorpions carry their eggs on the outer surface of a sac attached to the genital opening between the second and third abdominal sterna. The first fossil evidence of this behavior is shown in a Dominican amber pseudoscorpion with at least 12 eggs attached to her egg sac (Figure A27).

10B. AQUATIC INVERTEBRATES

10Bc. AMMONOID EGG SACS

Etches et al. (2009) described an occurrence of ammonite eggs and ammonitellae from the Kimmeridge Clay, Upper Jurassic, of Dorset. Mapes and Nützel (2009) provided evidence suggesting that at least some ammonoid and bactritoid cephalopod eggs in the later Paleozoic were laid either high in the water column or attached to floating materials, as the bottom environment in their sample was probably anoxic

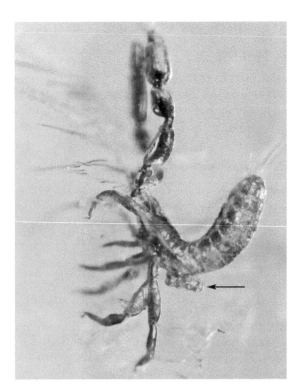

FIGURE A27   A female pseudoscorpion in Dominican amber carrying her egg sac with at least 12 attached eggs (Poinar amber collection accession no. Sy-1-185); length of pseudoscorpion = 2.2 mm.

or dysaerobic; the associated very small gastropods in their sample probably represent planktonic larval types, for the most part, that died on reaching an anoxic bottom.

☐ *Reliability*
Category 1, frozen behavior.

10Bd.   TRILOBITE AND CRAB CLUSTERS

Paterson et al. (2008) reviewed the trilobite cluster evidence and the varied possibilities for interpreting the evidence. This is clearly a complex set of questions.

10C.   VERTEBRATES

10Ca.   VERTEBRATE EGGS AND EGG CASES

Sabath (1991) provided a very comprehensive account of Cretaceous eggs from the Gobi Desert, emphasizing dinosaurs and birds, as well as discussing how eggs may be best described and studied paleontologically.

*10Cl.   PROTOCETID WHALE DELIVERY ATTITUDE

Gingerich et al. (2009) described a most unique early Middle Eocene protocetid whale from Pakistan that includes a gravid female as well as a male specimen. The female is unique in that it encloses the remains of a well-developed fetus, probably near term. The fetus is in a head-posterior position, as is the case with all terrestrial mammals and in contrast with marine

FIGURE A28   Female Lower Jurassic ichthyosaur *Stenopterygius* with extruded amniotic sac (arrow) situated near the cloacal region; length of skeleton = 2.8 m. (From Urlichs, M. et al., *Stuttgarter Beiträge zur Naturkunde, Serie C*, 36, 1–95, 1994. Copyright Staatliches Museum für Naturkunde Stuttgart. Photograph by H. Lumpe. With permission.)

FIGURE A29   Enlargement of amniotic sac shown in Figure A28 indicating the skull of the embryo (arrow); skull length = 7.9 cm. (From Urlichs, M. et al., *Stuttgarter Beiträge zur Naturkunde, Serie C*, 36, 1–95, 1994. Copyright Staatliches Museum für Naturkunde Stuttgart. Photograph by H. Lumpe. With permission.)

mammals and with marine Mesozoic reptiles (see Section 10Ce, Ichthyosaur and Mosasaur Birth Delivery Attitude, for data on ichthyosaurs and mosasaurs). The presence of a single fetus is consistent with the presence in living whales and porpoises of only a single fetus. The skeleton of the adult protocetids indicates an ability to move on land, as contrasted with the more advanced cetaceans that lack this capability. This unique specimen provides more evidence about the evolutionary intermediate stages in the progression from the wholly terrestrial condition to the wholly aquatic condition.

☐ *Reliability*
Category 1, frozen behavior.

10Cm.   ICHTHYOSAUR AMNIOTIC SAC

Urlichs et al. (1994, Figures 92 and 93) illustrated and briefly cited a Posidonienschiefer ichthyosaur closely associated with an amniotic sac and contained embryo (Figure 28 and Figure 29).

12.   DEPTH BEHAVIOR

12c.   PELAGIC TRILOBITE DEPTH SELECTION

Holloway and de Carvalho (2009) described a second Devonian trilobite, from a second trilobite family, with an "eyeshade."

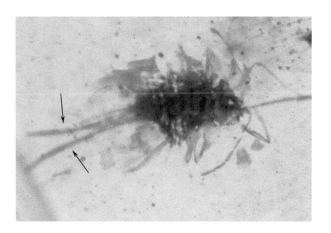

FIGURE A30 Mealybug producing wax in Burmese amber. Note pair of wax extensions (arrows) protruding from the rear of the insect (Poinar amber collection accession no. B-HE-21); length of body without extensions = 1.5 mm.

12d. DEPTH DISTRIBUTIONS

Boucot and Chen (2009) provided a scheme for determining the depth distribution of fossil plankton, upper and lower depth limits, and depth of maximum abundance.

14. DEFENSE

14d. CAMOUFLAGE

Parker (1998) described examples of color preserved in the Middle Cambrian arthropods *Marrella* and *Wiwaxia* due to diffraction gratings.

WAX PRODUCTION IN INSECTS

Waxes, composed of alcohols, fatty acids, and paraffin, are produced by a variety of terrestrial and aquatic insects in the form of layers, powder, threads, or plates. The most common use of insect waxes is to modify the permeability of the cuticle. Surface wax produced from epidermal glands on the surface serves to waterproof insects, which is why drops of water run off the back of a cockroach without wetting it. Body waxes also serve to retard evaporation, lubricate joints, provide camouflage, form protective coverings, and supply nest material.

Two groups of insects that produce copious supplies of wax are honeybees and scale insects. Worker honeybees force plates of wax from openings between their ventral abdominal segments that are used for comb construction.

In scale insects or coccids, wax is produced from minute openings in the cuticle that are sometimes extended as cuticular tubes. In the group of scale insects known as mealybugs, the wax often extends from the sides of the insect in a series of short filaments with two longer wax filaments protruding out from the posterior end of the insect. A mealybug from Burmese amber possessed these wax extensions, showing that the use of wax in this group occurred some 100 million years ago (Figure A30 and Figure A31).

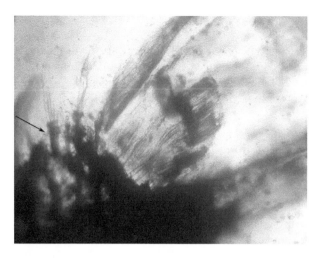

FIGURE A31 Detail of wax produced by mealybug shown in Figure 30. Arrow shows a cuticular tube extruding wax filaments (Poinar amber collection accession no. B-HE-21); length of cuticular tube = 63 μm.

BIRD FEATHERS

Vinther et al. (2008, 2009) have extensively described the presence in Early Cretaceous Brazilian bird feathers and Eocene German bird feathers of eumelanin-containing melanosomes that were capable of producing varied colors. The German melanosomes were earlier mistakenly identified as bacteria. The Brazilian specimens are patterned feathers of the type previously considered by Martill and Frey (1995). Vinther et al. pointed out that the presence of color-producing melanosomes in feathers is consistent with color vision in birds from at least the Early Cretaceous, as well as with the use of color patterning and feather color for sexual display and camouflage. Davis and Briggs (1995, table) compiled occurrences of fossil feathers known at the time, a very useful summary.

☐ *Reliability*

Category 1, frozen behavior.

*SHELTERED ENVIRONMENTS

(See also Sections 2As, Decapod Inquilinism within Ammonite Shells; 14i, Trilobites Sheltered within Nautiloid Shells and Crustaceans within Ammonites; and 26, Shelter.) Stewart and Carpenter (1990) carefully described a case of Late Cretaceous inquilinism involving an inoceramid and small halecostome fish in what was probably a commensal relation with the fishes using the bivalve as a shelter. Vullo et al. (2009) described a Kimmeridgian example from France wherein a fish used an abandoned ammonite conch as a shelter. The hermit crab–gastropod or ammonite shelter environment discussed in Section 9Ac (Bryozoan–Snail–Hermit Crab Complex and Hydrozoan–Gastropod Complex) is an obligate relation as opposed to merely being a shelter.

14e. AUTOTOMY

Buffetaut (1985; see also Urlichs et al., 1994, Figure 100) described a Toarcian crocodilian that suffered amputation of its tail, followed by the growth of a cartilaginous regenerated tip, as in some modern crocodilians (Figure A32).

☐ *Reliability*
Category 1, frozen behavior.

15. CARRIER SHELLS

Pavlovec (1964) briefly described an Eocene *Onustus* from Croatia to which are attached *Nummulites* and *Assilina* specimens. Harzhauser (2009) described an Aquitanian, Early Miocene *Xenophora* from coastal Tanzania while discussing some Miocene occurrences of the genus elsewhere in the Indopacific region. Harzhauser et al. (2009) described *Xenophora* from Kutch in northwestern India, and reviewed occurrences of the genus elsewhere in the western Indopacific region. The concept of carrier shells is commonly used when discussing gastropods, but Sartori and Harper (2009) have shown that a few anomalodesmatan bivalves also have the capability, through the production of arenophilic secretion, of also cementing debris to shell exteriors.

21. REPTILIAN AND MAMMALIAN BURROWS AND DENS

MAMMALIAN

Vizcaíno et al. (2001) carefully described a number of large burrows, with meter widths and corresponding heights, plus many meters lengths, from the Pleistocene of the Mar del Plata area. They made a convincing case that the most likely burrowers were large ground sloths, similar to those whose remains are found in both North and South American caves.

22. VERTEBRATE ENDOCRANIAL CASTS

22c. HUMAN BEHAVIOR

Conard (2009) described a well-dated human figurine from a basal Aurignacian German cave dated as far back as at least 35,000 years BP that displays the exaggerated female secondary sexual characteristics known for some time from significantly younger figurines such as the famous Venus of Willendorf. The behavioral significance of these figurines is not known, although they may have had something to do with fertility rites (this latest example has a ring instead of a head which would have enabled the owner to wear it around the neck on a thong).

FIGURE A32 Regenerated tail end of the Lower Jurassic marine crocodile *Steneosaurus*; length of the regenerated segment (arrow) = 12 cm. (From Buffetaut, E. et al., *Stuttgarter Beiträge zur Naturkunde, Serie B (Geologie und Paläontologie)*, 113, 1–7, 1985. Copyright Staatliches Museum für Naturkunde Stuttgart. Photograph by H. Lumpe. With permission.)

30. DISEASE

30b. DISEASED ELEPHANT UPPER JAW AND DEFORMED TEETH

Adam (1994) described some anomalously displaced molars in a German Late Pleistocene lower jaw of *Elephas primigenius* and discussed their significance.

34. STUNTING

Schöne (1999) made a good case, using some German Middle Devonian brachiopod examples, that stunting may also be recognized in situations where lower oxygen levels are probably the causative agent. In all of the cases of stunting, there is the problem of whether or not the causal factor, such as hyper- or hyposalinity or low oxygen, was actually significantly lowered nutrient levels rather than factors correlated with nutrient levels.

Urbanek (1993), after considering a mid-Late Silurian graptolite extinction event, an overall lower level extinction event, pointed out that the post-extinction graptolites were significantly smaller within lineages than were the pre-extinction event forms. He coined the term "Lilliput Effect" for this phenomenon. Twitchett (2001, 2007) noted, using Italian samples, that during the Permian–Triassic extinction event the post-Permian, earlier Triassic descendant taxa are significantly smaller than their pre-extinction forebears, and suggested that lower nutrient levels in the earlier Triassic were causal. Twitchett also referred to some Frasnian–Famennian age size changes within certain conodonts that appeared to follow this path. All of this involves sampling problems. The mere existence of one or a few localities consisting of very small individuals with reasonable phyletic relations to slightly older, significantly larger taxa does not by itself demonstrate a widespread phenomenon rather than a local abundance of small specimens reflecting a locally high small-specimen mortality rate. However, the Early Triassic evidence summarized by Twitchett does appear convincing. Bosetti et al. (submitted) described a Middle Devonian, post Kačak Event Brazilian (Paraná Basin) example of the Lilliput Effect that is very convincing, although additional examples of the same age from elsewhere will be needed before it can be established as a widespread phenomenon at this time. Whether other extinction horizons, large to small, will also involve immediately post-extinction small specimens with pre-extinction larger precursors remains to be studied.

Poinar (1979) discussed the abnormal condition of stunting in female steinernematid nematodes that develop in the body cavity of various insects. Such pygmy females, which were about 1/5 the size of normal females, were first noted by Bovien (1937) in *Steinernema bibionis* (Bovien). Similar pygmy females were subsequently noted in several other species of *Steinernema* (Poinar, 1979). Even though these pygmies occur along with normal females, they probably are the result of suboptimal developmental conditions related either directly or indirectly to obtaining adequate nourishment.

Benton et al. (in press) have made a very strong case for the presence of latest Cretaceous (Maastrichtian) dwarfed dinosaurs on an island in the Tethys (their Haţeg Island) in present-day western Romania. They provide convincing evidence for the dwarfed condition and also include a comprehensive review of dwarfing and gigantism in island biota of the present and the past, while critically considering the varied selective factors that might be involved

FIGURE A33 Adult *Micromalthus* beetle (Coleoptera: Micromalthidae) in Dominican amber (Poinar amber collection accession no. C-7-191); length of beetle = 1.8 mm

There are many cases today of morphologically indistinguishable insect populations that represent reproductively isolated sibling species.

37. MORPHOLOGICAL STASIS AND SPECIES LONGEVITY

Because of their detailed state of preservation, amber fossils can be used to study morphological stasis and, from these studies, inferences on species longevity can be drawn. Although insect species in general are thought to survive for only 2 to 3 million years (the late Frank Carpenter, pers. comm.), *Micromalthus* beetles in 20- to 30-million year old Dominican amber so closely resemble the only known extant species, *Micromalthus debilis*, that they have been considered conspecific (Figure A33). This is just one of a number of Holocene and Tertiary insects that have been considered conspecific with extant species (Hörnschemeyer et al., 2009, Table 3). It suggests that once a successful survival pattern is established between a species and its surroundings, evolutionary stasis may persist for long periods and result in species continuity. Of course, it will never be known whether the morphological resemblance between fossil and extant species is a result of convergence or that the two similar appearing taxa separated by millions of years represent sibling species.

SUMMARY AND CONCLUSIONS

1. COMMUNITY EVOLUTION BEHAVIORAL EVIDENCE

Mandic et al. (2008, Figure 5) provided excellent evidence from the Austrian North Alpine fore-deep basin of Middle Miocene behavioral fixity shown by razor clams (Solenacea) preserved in their vertical burrows, part of a shallow-water environment, with population densities and spacing like those of living razor clams such as *Ensis*.

5. CLASSES OF PALEONTOLOGIC DATA AND RATES OF CHANGE

Kaye et al. (2008) made it clear that the alleged discovery of soft, pliable tissues from dinosaur bone is mistaken, as study of this type of material from both dinosaur and mammalian fossil bones shows that it is actually bacterial biofilms adhering to the internal structures of bone and of Recent origin, as shown by carbon-14 dating.

REFERENCES

Abele, L.G. (1972). A note on the Brazilian bromeliad crabs (Crustacea, Grapsidae). *Arquivos de Ciências do Mar*, 12, 123–126.

Adam, K.D. (1994). Anomalien des Zahnwechsels bei *Elephas primigenius* aus dem Quartär des Oberrheins. *Stuttgarter Beiträge zur Naturkunde, Serie B (Geologie und Paläontologie)*, 211, 1–36.

Baoshun, F., Aharon, P., Byerly, G.R., and Roberts, H.H. (1994). Barite chimneys in the Gulf of Mexico slope: initial report on their petrography and geochemistry. *Geo-Marine Letters*, 14, 81–87.

Benton, M.J., Csiki, Z., Grigorescu, D., Redelstorff, R., Sander, P.M., Stein, K., and Weishampel, D.B. (in press). Dinosaurs and the island rule: the dwarfed dinosaurs from Haţeg Island. *Palaeogeography, Palaeoclimatology, Palaeoecology*.

Bishop, G.A. (1983). Two new species of crabs, *Notopocorystes* (*Eucorystes*) *eichhorni* and *Zygastrocarcinus griesi* (Decapoda: Brachyura) from the Bearpaw (Campanian) of north-central Montana. *Journal of Paleontology*, 57, 900–910.

Bogoch, R. and Shirav, M. (1978). Petrogenesis of a Senonian barite deposit, Judean Desert, Israel. *Mineralium Deposita*, 13, 383–390.

Bosetti, E.V., Grahn, Y., Horodyski, R.S., Mauller, P.M., Breuer, P., and Zabini, C. (submitted). An earliest Givetian "Lilliput Effect" in the Paraná Basin, and the collapse of the Malvinokaffric fauna. *International Journal of Earth Sciences*.

Boucot, A.J., Ed. (1990). *Evolutionary Paleobiology of Behavior and Coevolution*. Elsevier, Amsterdam, 735 pp.

Boucot, A.J. and Chen, X. (2009). Fossil plankton depth zones. *Palaeoworld*, 18, 213–234.

Bovien, P. (1937). Some types of association between nematodes and insects. *Videnskabellige Meddelelser fra Dansk Naturhistorisk Forening*, 101, 1–114.

Brasier, M., Cotton, L., and Yenney, I. (2009). First report of amber with spider webs and microbial inclusions from the earliest Cretaceous (c. 140 Ma) of Hastings, Sussex. *Journal of the Geological Society, London*, 166, 989–997.

Brunton, C.H.C. and Mundy, D.J.C. (1993). Productellid and plicatiferid (productoid) brachiopods from the Lower Carboniferous of the Craven Reef Belt, North Yorkshire. *Bulletin of the Natural History Museum London (Geology)*, 49, 99–119.

Buffetaut, E. (1985). Ein *Steneosaurus* (Crocodilia, Mesosuchia) mit regeneriertem Schwanzende aus dem Lias Epsilon (Toarcium) von Schwaben. *Stuttgarter Beiträge zur Naturkunde, Serie B (Geologie und Paläontologie)*, 113, 1–7.

Carter, R.M. (1968). On the biology and palaeontology of some predators of bivalved mollusca. *Palaeogeography, Palaeoclimatology, Palaeoecology*, 4, 29–65.

Chatterton, B.D.E. and Fortey, R.A. (2008). Linear clusters of articulated trilobites from Lower Ordovician (Arenig) strata of Bini Tinzoulin, north of Zagora, southern Morocco. In *Advances in Trilobite Research* (pp. 73–78), edited by I. Rábano, R. Gozalo, and D. García-Bellido. Instituto Geológico y Minero de España, Madrid.

Chattopadhyay, D. (2009). History of Cenozoic drilling predation: a multitaxon approach. In *Proceedings of the 9th North American Paleontological Convention (NAPC 2009), June 21–26, Cincinnati, OH*, p. 17.

Chin, K. and Bishop, J.R. (2007). *Exploited Twice: Bored Bone in a Theropod Coprolite from the Jurassic Morrison Formation of Utah, U.S.A.*, SEPM Special Publication No. 88. Society for Sedimentary Geology, Tulsa, OK, pp. 379–387.

Chin, K., Hartman, J.H., and Roth, B. (2009). Opportunistic exploitation of dinosaur dung: fossil snails in coprolites from the Upper Cretaceous Two Medicine Formation of Montana. *Lethaia*, 42, 185–198.

Clarke, M.R. and Fitch, J.E. (1979). Statoliths of Cenozoic teuthoid cephalopods from North America. *Palaeontology*, 22, 479–511.

Clarke, M.R. and Maddock, L. (1988). Statoliths of fossil Coleoid Cephalopods. In *The Mollusca*. Vol. 12. *Paleontology and Neontology of Cephalopods* (pp. 153–168), edited by M.R. Clarke and E.R. Truman. Academic Press, San Diego, CA.

Clarke, M.R., Maddock, L., and Steurbaut, E. (1980). The first fossil cephalopod statoliths to be described from Europe. *Nature*, 287, 628–630.

Conard, N.J. (2009). A female figurine from the basal Aurignacian of Hohle Fels Cave in southwestern Germany. *Nature*, 459, 248–252.

Davis, P.G. and Briggs, D.E.G. (1995). Fossilization of feathers. *Geology*, 23, 783–786.

De Schutter, P. (2009). The presence of *Megachasma* (Chondrichthyes: Lamniformes) in the Neogene of Belgium, first occurrence in Europe. *Geologica Belgica*, 12, 179–203.

Ebbestad, J.O.R., Lindström, A., and Peel, J.S. (2009). Predation on bellerophontiform mollusks in the Palaeozoic. *Lethaia*, 42, 469–485.

Ellingson, D.B. (2009). Preliminary report on a pathological *Bison antiquus* skeleton from Woodburn, Oregon, excavated as part of a high school curriculum. *Geological Society of America Abstracts with Programs*, 41(7), 455.

Engeser, T. (1990). Phylogeny of the fossil coleoid Cephalopods (Mollusca). *Berliner Geowissenschaftliche Abhandlungen (A)*, 124, 123–191.

Etches S., Clarke, J., and Callomon, J. (2009). Ammonite eggs and ammonitellae from the Kimmeridge Clay Formation (Upper Jurassic) of Dorset, England. *Lethaia*, 42, 204–217.

Förster, R. (1969). Epökie, Parasitimus und Regeneration bei fossilen Dekapoden. *Mitteilungen der Bayerische Staatssammlung für Paläontologie und Historische Geologie*, 9, 45–59.

Getty, P.K. and Hagadorn, J.W. (2009). Palaeobiology of the *Climatichnites* tracemaker. *Palaeontology*, 52, 753–778.

Gingerich, P.D., ul-Haq, M., von Koenigswald, W., Sanders, W.J., Smith, H., and Zaimout, I.S. (2009). New protocetid whale from the Middle Eocene of Pakistan: birth on land, precocial development, and sexual dimorphism. *PLoS ONE*, 4(2), e4366.

Gong, E., Martin, L.D., Burnham, D.A., and Falk, A.R. (2009). The birdlike raptor Sinornithosaurus was venomous. *Proceedings of the National Academy of Sciences U.S.A.*, pnas.org/cgi/doi/10.1073'pnas.0912360107.

Grant, R.E. (1963). Unusual attachment of a Permian linoproductid brachiopod. *Journal of Paleontology*, 37, 134–140.

Harrington, C.R. (2007). Late Pleistocene mummified mammals. In *Encyclopedia of Quaternary Science*, Vol. 4 (pp. 3197–3202), edited by S.A. Elias. Elsevier, Amsterdam.

Harzhauser, M. (2009). Aquitanian gastropods of coastal Tanzania and their biogeographic implications for the early western Indo–Pacific. *Palaontographica Abteilung A*, 289, 123–156.

Harzhauser, M., Reuter, M., Piller, W.E., Berry, B., Kroh, A., and Mandle, O. (2009). Oligocene and Early Miocene gastropods from Kutch (NW India) document an early biogeographic switch from Western Tethys to Indo–Pacific. *Paläontologische Zeitschrift*, 83, 333–372.

Heie, O.E. (1967). Studies on fossil aphids (Homoptera: Aphidoidea). *Spolia Zoologica Musei Hauniensis*, 26, 1–273.

Holloway, D.J. and de Carvalho, M. da G.P. (2009). The extraordinary trilobite *Fenestraspis* (Dalmanitidae, Synphoriinae) from the Lower Devonian of Bolivia. *Palaeontology*, 52, 933–950.

Hörnschemeyer, T., Wedmann, S., and Poinar, G. (2009). How long can insect species exist? Evidence from extant and fossil *Micromalthus* beetles (Insecta: Coleoptera). *Zoological Journal of the Linnean Society*. DOI: 10.111/j.1096-3642.2009.00549.x

Hou, X.-G., Siveter, D.J., Aldridge, R.J., and Siveter, D.J., (2009). A new arthropod in chain-like associations from the Chengjiang Lagerstätte (Lower Cambrian), Yunnan, China. *Palaeontology*, 52, 951–961.

Imms, A.D. (1948). *A General Textbook of Entomology*. E.P. Dutton, New York, 727 pp.

Jacobs, G.H. and Rowe, M.P. (2004). Evolution of vertebrate color vision. *Clinical and Experimental Optometry*, 87, 206–216.

Kaye, T.G., Gaugler, G., and Sawlowicz, Z. (2008). Dinosaurian soft tissues interpreted as bacterial biofilms. *PLoS ONE*, 3(7), 1–7.

Kelley, P.H. and Hansen, T.A. (2006). Comparison of class- and lower taxon-level patterns in naticid gastropod predation, Cretaceous to Pleistocene of the U.S. Coastal Plain. *Palaeogeography, Palaeoclimatology, Palaeoecology*, 236, 302–320.

Kelley, P.H. and Hansen, T.A. (2007). *Latitudinal Patterns in Naticid Gastropod Predation Along the East Coast of the United States: A Modern Baseline for Interpreting Temporal Patterns in the Fossil Record*, SEPM Special Publication No. 88. Society for Sedimentary Geology, Tulsa, OK, pp. 287–299.

Kiel, S., Amano, K., and Guenkins, R. (2008). Bivalves from Cretaceous cold-seep deposits on Hokkaido, Japan. *Acta Palaeontologica Polonica*, 53, 525–537.

Koteja, J. and Poinar, Jr., G.O. (2001). A new family, genus and species of scale insect (Hemiptera: Coccinea: Kukaspididae, new family) from Cretaceous Alaskan amber. *Proceedings of the Entomological Society of Washington*, 103, 356–363.

Krassilov, V.A. (2008). Mine and gall predation as top down regulation in the plant–insect systems from the Cretaceous of Negev, Israel. *Palaeogeography, Palaeoclimatology, Palaeoecology*, 261, 261–269.

Krassilov, V.A. and Rasnitsyn, A.P., Eds. (2008). *Plant–Arthropod Interactions in the Early Angiosperm History: Evidence from the Cretaceous of Israel*. Pensoft Publishers/Brill, Boston, MA.

Krassilov, V.A. and Shuklina, S. (2008). Arthropod trace diversity on fossil leaves from the mid-Cretaceous of Negev, Israel. *Alavesia*, 2, 241–247.

Labandeira, C.C. (2007). *Assessing the Fossil Record Of Plant–Insect Associations: Ichnodata Versus Body-Fossil Data*, SEPM Special Publication No. 88. Society for Sedimentary Geology, Tulsa, OK, pp. 9–26.

Lane, N.G. (1979). Paleontology and paleoecology of the Crawfordsville Fossil Site (Upper Osagian, Indiana). *University of California Publications in Geological Sciences*, 99, 1–141.

Lehmann, U. and Weitschat, W. (1973). Zur Anatomie und Ökologie von Ammoniten: Funde von Kropf und Kiemen. *Paläontologische Zeitschrift*, 47, 69–76.

Małkowski, K. (1976). Regeneration of some brachiopod shells. *Acta Geologica Polonica*, 26, 439–442.

Mandic, O., Harzhauser, M., and Roetzel, R. (2008). Benthic mass-mortality events on a Middle Miocene incised-valley tidal-flat (North Alpine Foredeep Basin). *Facies*, 54, 343–359.

Mapes, R.H. and Nützel, A. (2009). Late Palaeozoic mollusc reproduction: cephalopod egg-laying behavior and gastropod larval palaeobiology. *Lethaia*, 42, 341–356.

Martill, D.M. and Frey, E. (1995). Colour patterning preserved in Lower Cretaceous birds and insects: the Crato Formation of N.E. Brazil. *Neues Jahrbuch für Geologie und Paläontologie–Monatsheft*, 118–128.

Martinell, U. and Domènech, R. (2009). Commensalism in the fossil record: Eunicid polychaete bioerosion on Pliocene solitary corals. *Acta Palaeontologica Polonica*, 54, 143–154.

Maximov, V.V. (2000). Environmental factors which may have led to the appearance of color vision. *Philosophical Transactions of the Royal Society of London B*, 355, 1239–1242.

Mazin, J.-M., Billon-Bruyat, J.-P., and Padian, K. (2009). First record of a pterosaur landing trackway. *Proceedings of the Royal Society, Biological Sciences*, 276, 3881–3886.

Mottequin, B. and Sevastopulo, G. (2009). Predatory boreholes in Tournaisian (Lower Carboniferous) spiriferid brachiopods. *Lethaia*, 42, 274–282.

Müller, P., Krobricki, M., and Wehner, G. (2000). Jurassic and Cretaceous primitive crabs of the family Prosopidae (Decapoda: Brachyura): their taxonomy, ecology and biogeography. *Annales Societatis Geologorum Poloniae*, 70, 49–79.

Mundy, D.J.C. (1982). A note on the predation of brachiopods from the Dinantian reef limestones of Cracoe, north Yorkshire. *Transactions of the Leeds Geological Association*, 9, 73–83.

Noda, H. and Lee, Y.-G. (1989). Wood-boring bivalve *Martesia striata* from the Middle Miocene Sinhyeon Formation in the Ulsan Basin, Korea. *Annual Report of the Institute of Geosciences, University of Tsukuba*, 15, 61–67.

Nomura, S.-I. and Maeda, H. (2008). Significance of autochthonous fossil barnacles from the Miocene Natori Group at the Moniwa–Goishi area, northeast Japan. *Palaeontological Research*, 12, 63–79.

Parker, A.R. (1998). Colour in Burgess shale animals and the effect of light on evolution in the Cambrian. *Proceedings of the Royal Society, Biological Sciences*, 265, 967–972.

Paterson, J.R., Hughes, N.C., and Chattergton, B.D.E. (2008). Trilobite clusters: what do they tell us? A preliminary investigation. In *Advances in Trilobite Research* (pp. 313–318), edited by I. Rábano, R., Gozalo, and D. García-Bellido. Cuadernos del Museo Geominero, No. 9. Instituto Geológico y Minero de España.

Pavlovec, R. (1964). Interesting fossils from the island of Krk (Croatia). *Bulletin Scientifique, Conseil des Academies de la RSF de Yougoslavie*, 9(1–2), 5–6.

Podenas, S. and Poinar, Jr., G.O. (2009). New crane flies (Diptera: Limoniidae) from Burmese amber. *Proceedings of the Entomological Society of Washington*, 111, 470–492.

Poinar, Jr., G.O. (1979). *Nematodes and Biological Control of Insects*. CRC Press, Boca Raton, FL, 277 pp.

Poinar, Jr., G.O. (1996). A fossil stalk-winged damselfly, *Diceratobasis worki* spec. nov. from Dominican amber, with possible ovipositional behavior in tank bromeliads (Zygoptera: Coenagrionidae). *Odonatologica*, 25, 381–385.

Poinar, Jr., G.O. (2006). *Feroseta priscus* (Neuroptera: Mantispidae), a new genus and species of mantidflies in Dominican amber. *Proceedings of the Entomological Society of Washington*, 108, 411–417.

Poinar, Jr., G.O. (2009a). *Dominibrentus leptus*, n. gen., n. sp. (Curculionoidea, Brentidae, Cyphagoginae, Dominibrentini, n. tribe), a straight-snouted weevil in Dominican amber. *Historical Biology*, 21, 51–55.

Poinar, Jr., G.O. (2009b). *Cascoplecia insolitis* (Diptera: Cascopleciidae), a new family, genus and species of flower-visiting, unicorn fly (Bibionomorpha) in Early Cretaceous Burmese amber. *Cretaceous Research*, 31, 71–76.

Poinar, Jr., G.O. (2009c). Early Cretaceous protest flagellates (Parabasalia: Hypermastigia: Oxymonada) of cockroaches (Insecta: Blattaria). *Cretaceous Research*, 30, 1066–1072.

Poinar, Jr., G.O. and Brown, A.E. (2005). New Aphidoidea (Hemiptera: Sternorrhyncha) in Burmese amber. *Proceedings of the Entomological Society of Washington*, 107, 835–845.

Poinar, Jr., G.O. and Brown, A.E. (2006). Remarks on *Parvaverrucosa annulatus* (=*Verrucosa annulata* Poinar and Brown 2005) (Hemiptera: Aphidoidea). *Proceedings of the Entomological Society of Washington*, 108, 734–735.

Poinar, Jr., G.O., Kerp, H., and Hass, H. (2008). *Palaeonema phyticum* gen. n., sp. n. (Nematoda: Palaeonematidae fam. n.), a Devonian nematode associated with early land plants. *Nematology*, 10, 9–14.

Poinar, Jr., G.O. and Poinar, R. (1999). *The Amber Forest*. Princeton University Press, Princeton, NJ, 270 pp.

Prokop, R. (1965). *Argodiscus hornyi* gen. n. et sp. n. (Edrioasteroidea) z českého středního ordoviku. *Časopis národního muzea, oddil přírodovědeckého*, 134, 30–34.

Regan, B.C., Julliot, C., Simmen, B., Vienot, F., Charles-Dominique, P., and Mollon, J.D. (2001). Fruits, foliage and the evolution of primate color vision. *Philosophical Transactions of the Royal Society of London B*, 356, 229–283.

Ren, D., Labandeira, C. C., Santiago-Blay, J. A., Rasnitsyn, A., Shih, C., Bashkuev, A., Logan, M. A. V., Hotton, C. L., and Dilcher, D. (2009). A probable pollination mode before angiosperms: Eurasian, long-proboscid scorpionflies. *Science*, 326, 840–847.

Richards, B.C. (1975). *Longusorbis cuniculosus*: a new genus and species of Upper Cretaceous crab, with comments on Spray Formation at Shelter Point, Vancouver Island, British Columbia. *Canadian Journal of Earth Sciences*, 12, 1850–1863.

Rodriguez, G. (1981). Decapoda. In *Aquatic Biota of Tropical South America. Part 1. Arthropoda* (pp. 41–51), edited by S.H. Hurlbert, G. Rodriguez, and N. Dias Dos Santos. San Diego State University, San Diego, CA.

Sabath, K. (1991). Upper Cretaceous amniotic eggs from the Gobi Desert. *Acta Palaeontologica Polonica*, 36, 151–192.

Sánchez, M.V. and Genise, J.F. (2009). Cleptoparasitism and detritivory in dung beetle brood balls from Patagonia, Argentina. *Palaeontology*, 52, 837–848.

Sartori, A.F. and Harper, E.M. (2009). Sticky bivalves from the Mesozoic: clues to the origin of the anomalodesmatan arenophilic system. *Lethaia*, 42, 486–494.

Schiffbauer, J.D., Leighton, L.R., and Kowalewski, M. (2009). Testing the microstructural record of predation: an experimental approach to examine the reliability of predatory microtraces. In *Proceedings of the 9th North American Paleontological Convention (NAPC 2009)*, June 21–26, Cincinnati, OH, p. 19.

Schoenemann, B., Liu, J.-N., Shu, D.-G., Han, J., and Zhang, Z.-F. (2009). A miniscule optimized visual system in the Lower Cambrian. *Lethaia*, 42, 265–273.

Schöne, B.R. (1999). Scleroecology: implications for ecotypical dwarfism in oxygen-restricted environments (Middle Devonian, Rheinisches Schiefergebirge). *Senckenbergiana Lethaea*, 79, 35–41.

Schweitzer, C.E. and Feldmann, R.M. (2009). Revision of the genus *Cyclothyreus* Remeš, 1895 (Decapoda: Brachyura: Dromiloidea). *Neues Jahrbuch für Geologie und Paläontologie–Abhandlungen*, 253, 357–372.

Selden, P.A. and Nudds, J. (2004). *Evolution of Fossil Ecosystems*. University of Chicago Press, Chicago, IL, 160 pp.

Shen, S., Fan, B., Zhang, C., and Zhang, X. (1994). A new species of permianellids (Brachiopoda): taxonomic and palaeoecologic significance. *Geobios*, 27, 477–485.

Smith, A.B. and Stockley, B. (2005). The geological history of deep-sea colonization by echinoids: roles of surface productivity and deep-water ventilation. *Proceedings of the Royal Society, Biological Sciences*, 272, 865–869.

Sprinkle, J. and Sumrall, C.D. (2008). New parablastoids from the western United States. *The University of Kansas Paleontological Contributions*, 16, 1–14.

Stewart, J.D. and Carpenter, K. (1990). Examples of vertebrate predation on cephalopods in the Late Cretaceous of the Western interior. In *Evolutionary Paleobiology of Behavior and Coevolution* (pp. 203–208), A.J. Boucot, Ed., Elsevier, New York.

Sumrall, C.D., Brett, C.E., and McKinney, M.L. (2009). A new agelacrinitid edrioasteroid attached to a large hardground clast from the McKenzie Member of the Mifflintown Member (Silurian) of Pennsylvania. *Journal of Paleontology*, 83, 794–803.

Sun, Yuanlin, and Baliński, A. (2008). Silicified Mississippian brachiopods from Muhua, southern China: lingulids, craniids, strophomenids, productids, orthotetids, and orthids. *Acta Palaeontographica Polonica*, 53, 485–524.

Taylor, T.N., Taylor, E.L., and Krings, M. (2009). *Paleobotany: The Biology and Evolution of Fossil Plants*. Elsevier, Amsterdam, 1230 pp.

Twitchett, R.J. (2001). Incompleteness of the Permian–Triassic fossil record: a consequence of productivity decline? *Geological Journal*, 36, 341–363.

Twitchett, R.J. (2007). The Lilliput effect in the aftermath of the end-Permian extinction event. *Palaeogeography, Palaeoclimatology, Palaeoecology*, 252, 132–144.

Underwood, C.J. (1993). The position of graptolites within Lower Palaeozoic planktic ecosystems. *Lethaia*, 26, 189–202.

Urbanek, A. (1993). Biotic crises in the history of the Upper Silurian graptolites: a palaeobiologic model. *Historical Biology*, 7, 29–50.

Urlichs, M., Wild, R., and Ziegler, B. (1994). Der Posidonien-Schiefer des unteren Juras und seine Fossilien. *Stuttgarter Beiträge zur Naturkunde Serie C*, 36, 1–95.

Vega, F.J., Zúñiga, L., and Pimentel, F. (2009). First formal report of a crab in amber from the Miocene of Chiapas and other uncommon Crustacea. *Geological Society of America Abstracts with Programs*, 47(7), 631.

Vijaya, Prasad, G.V.R., and Singh, K. (2009). Late Triassic palynoflora from the Pranhita-Godavari Valley, India: evidence from vertebrate coprolites. *Alcheringa*, 33, 91–111.

Vinther, J., Briggs, D.E.G., Prum, R.O., and Saranathan, V. (2008). The colour of fossil feathers. *Biology Letters*, 4, 522–525.

Vinther, J., Briggs, D.E.G., Clarke, J., Mayr, G., and Prum, R.O. (2009). Structural coloration in a fossil feather. *Biology Letters*, DOI:10.1098/rabl.2009.0524.

Vizcaíno, S.F., Zárate, M., Bargo, M.S., and Dondas, A. (2001). Pleistocene burrows in the Mar del Plata area (Argentina) and their probable builders. *Acta Palaeontologica Polonica*, 46, 289–301.

Vullo, R., Cavin, L., and Clochard, V. (2009). An ammonite–fish association from the Kimmeridgian (Upper Jurassic) of La Rochelle, western France. *Lethaia*, 42, 462–468.

Weiler, W. (1968). Otolithi piscium. In *Fossilium Catalogus*. I. *Animalia*, Pars 117, edited by F. Westphal. W. Junk, The Hague, The Netherlands.

Whittington, H.B. (1985). *The Burgess Shale*. Yale University Press, New Haven, CT, 151 pp.

Wigglesworth, V.B. (1965). *The Principles of Insect Physiology*, 6th ed. Methuen & Co., London, 741 pp.

Wilson, J.A., Mohabey, D.M., Peters, S.E., and Head, J.J. (2010). Predation upon hatchling dinosaurs by a new snake from the Late Cretaceous of India. *PLoS Biology*, 8(3), 1–10.

Wolff, E.D.S., Salisbury, S.W., Horner, J.R., and Varricchio, D.J. (2009). Common avian infection plagued the tyrant dinosaurs. *PLoS ONE*, 4(9), c7288.

Yokoyama, S. (2002). Molecular evolution of color vision in vertebrates. *Gene: An International Journal on Genes and Genomes*, 300, 69–78.

Yokoyama, S. and Yokoyama, R. (1996). Adaptive evolution of photoreceptors and visual pigments in vertebrates. *Annual Review of Ecology and Systematics*, 27, 543–567.

Žalohar, J., Hitij, T., and Križnar, M. (2009). Two new species of seahorses (Syngnathidae, *Hippocampus*) from the Middle Miocene (Sarmatian) Coprolitic Horizon in the Tunjice Hilla, Slovenia: the oldest fossil record of seahorses. *Annales de Paléontologie*, 95, 71–96.

Zuparko, R.L. and Poinar, Jr., G.O. (1997). *Aivalykus dominicanus* (Hymenoptera: Braconidae), a new species from Dominican amber. *Proceedings of the Entomological Society of Washington*, 99, 744–747.

Author Index

Taxonomic Index

Subject Index